Water in Karst

Water in Karst

Management, Vulnerability, and Restoration

Neven Kresic, Ph.D.

New York Chicago San Francisco
Lisbon London Madrid Mexico City
Milan New Delhi San Juan
Seoul Singapore Sydney Toronto

The McGraw-Hill Companies

Cataloging-in-Publication Data is on file with the Library of Congress

Water in Karst: Management, Vulnerability, and Restoration

1 2 3 4 5 6 7 8 9 0 QFR/QFR 1 8 7 6 5 4 3 2

ISBN 978-0-07-175333-3
MHID 0-07-175333-8

This book is printed on acid-free paper.

Sponsoring Editor
Larry S. Hager

Editing Supervisor
Stephen M. Smith

Production Supervisor
Pamela A. Pelton

Acquisitions Coordinator
Bridget L. Thoreson

Project Manager
Ruchira Gupta, Aptara, Inc.

Copy Editor
Bashishth Narayan Verma, Aptara, Inc.

Proofreader
Dev Dutt Sharma, Aptara, Inc.

Indexer
Sara Lynn Eastler

Art Director, Cover
Jeff Weeks

Composition
Aptara, Inc.

To Joanne and Miles

About the Author

Neven Kresic, Ph.D., is a hydrogeology practice leader at AMEC Environment and Infrastructure, Inc., an international engineering and consulting firm. A professional geologist and hydrogeologist, he works with a variety of national and international clients including industry, mining companies, water utilities, government agencies, and environmental law firms. He holds a bachelor's degree in hydrogeological engineering, a master's degree in hydrogeology, and a Ph.D. degree in geology, all from the University of Belgrade. Before coming to the United States as a Senior Fulbright Scholar at the U.S. Geological Survey in Reston, Virginia, and The George Washington University in Washington, D.C., Dr. Kresic was a professor of groundwater dynamics and hydrogeology at the University of Belgrade. He is Co-Chair of the Karst Commission of the International Association of Hydrogeologists, serves on the Management Committee of the Groundwater Management and Restoration Specialty Group of the International Water Association, and is formerly Vice President for International Affairs at the American Institute of Hydrology.

Contents

Introduction

Nous montons tous sur les épaules les uns des autres.
All of us climb on each others' shoulders.

JOVAN CVIJIĆ

Karst is a scientific term named after the geographic district between Slovenia and Trieste, Italy, which has a very distinct landscape. It is a Germanized word for "carso" (in Italian) and "kras" (in Slovene); all the three words are derived from the Indo-European word "kar" or "karra," which means rock. The following definitions of karst and karst topography are based on the 1960 edition of the *American Geological Institute Dictionary of Geology*.

In Karst, the limestone rocks are honeycombed by caverns and openings dissolved by groundwater. Most of the drainage is underground and the land is left dry and relatively barren. The landscape is dotted with sinkholes, interspersed with abrupt ridges, and irregular protuberant rocks. Dry valleys are common, and valleys containing streams often end abruptly where the latter plunge into underground channels and caves, sometimes to reappear as large springs miles away. These features are so characteristic of the Karst district of the eastern Adriatic coast that the name karst has been applied to them elsewhere.

Some common words from the Slavic languages of the Karst region have become international scientific terms describing karst features, mainly thanks to the Serbian geomorphologist Jovan Cvijić (Fig. I.1) who was the first to defend a doctoral thesis and publish a scientific monograph exclusively devoted to karst (Cvijić, 1893). These include *doline* (sinkhole), *ponor* (swallow hole, sink), *jama* (pithole), and *polje* (large closed depression with flat bottom characteristic of the Dinaric karst). During the following two decades, Cvijić laid the theoretical foundation of many past and current developments in karst geomorphology and hydrogeology by authoring over 40 publications devoted entirely or mostly to karst phenomena. Much of his original work and numerous drawings illustrating the origin of various karst forms and the nature of groundwater flow in karst terrains have since been republished, copied, or otherwise reaffirmed by many. Examples of typical hand drawings made by Cvijić in support of his concepts and illustrating karst features analyzed on his extensive journeys throughout the Balkans are provided in

Figure I.1 Cvijić on one of his numerous travels through the Balkans. (Courtesy of Jovan Cvijić Museum, Belgrade.)

Figs. I.2 through I.4. His insightful analysis of the karst relief and hydrography developed in the Dinaric orogenic belt led to the wide international acceptance of the term "classic Dinaric karst" as a synonym for fully developed mature karst. During his life, Cvijić received unanimous confirmation as the father of karstology from many scientific institutions and peers worldwide (Fig. I.5). Most importantly, he firmly established the famous Belgrade school of karst studies that ultimately produced three founding members of the Karst Commission of the International Association of Hydrogeologists (IAH) more than 40 years ago: Dr. Borivoje Mijatovic, Dr. Miomir Komatina, and Dr. Branko Stepanovic. They were soon joined by Dr. Petar Milanovic, the author of *Karst Hydrogeology* (1981), the first book on the topic published in English.

Groundwater in karst environments has intrigued scientists and ordinary people alike for millennia due to its many fascinating facets. It feeds the world's largest springs, many of which enabled the establishment of the first urban centers in human history and continue to serve as reliable sources of water supply to the present day; it creates a

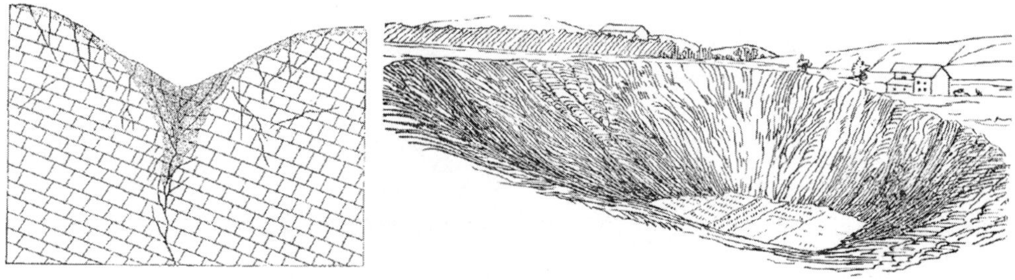

FIGURE I.2 *Left*: Doline (sinkhole) created by chemical dissolution of rock along fissures. *Right*: Deep doline in Kozina, classic karst of Slovenia. (From Cvijić, 1893, 1926; also available in Stevanovic and Mijatovic, 2005.)

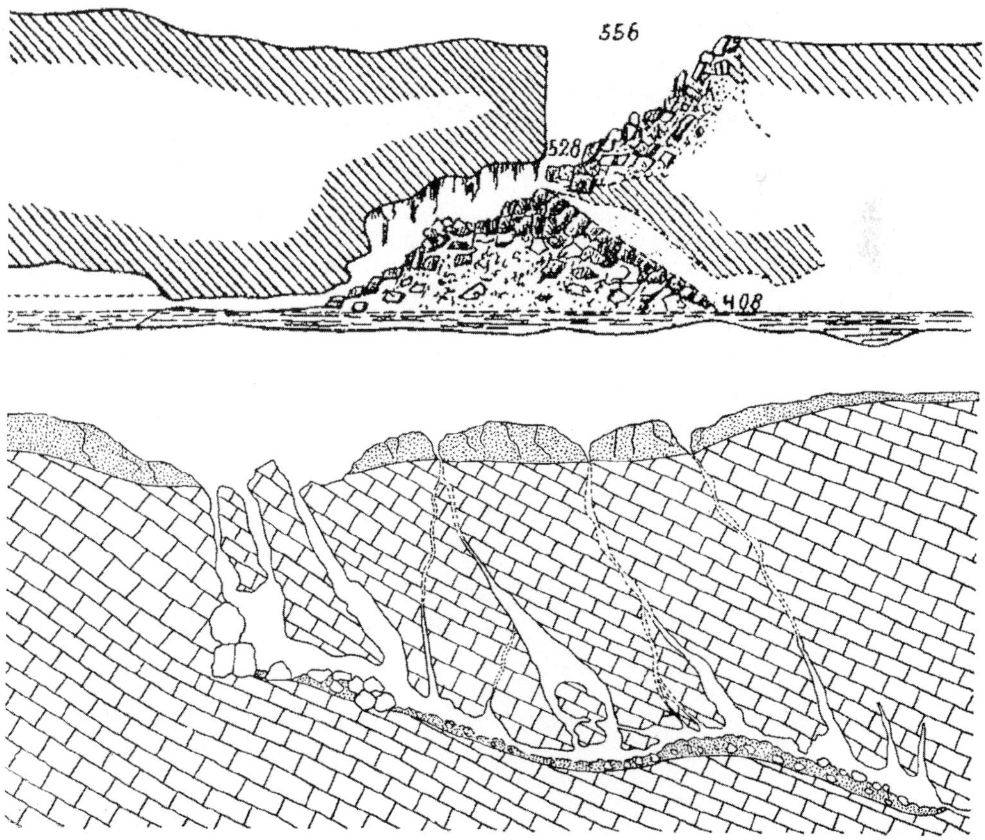

FIGURE I.3 Cvijić's drawings accompanying his explanations of the morphogenesis of different karst landforms. *Top*: Collapse sinkhole. *Bottom*: Alluvial sinkholes. (From Cvijić, 1893, 1926.)

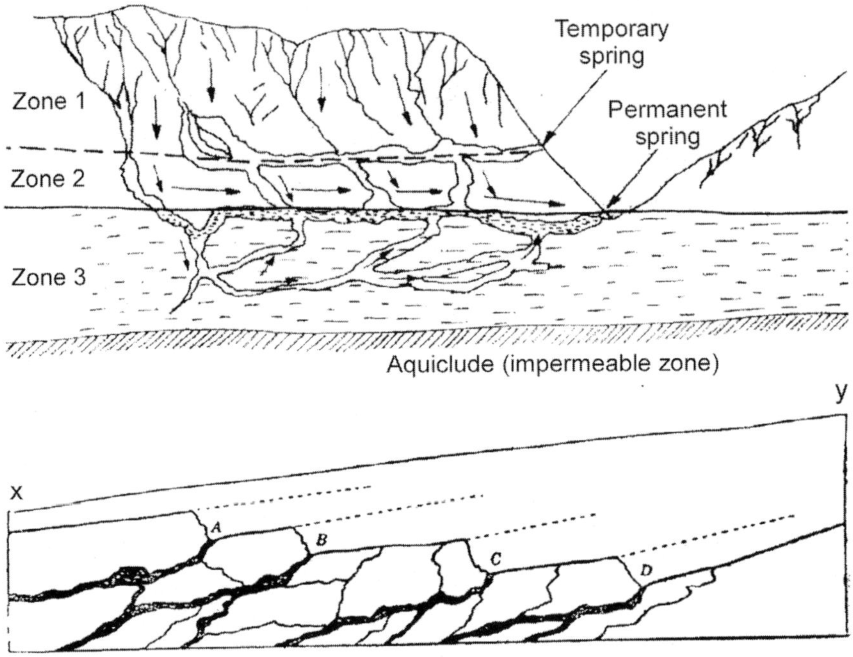

FIGURE I.4 Cvijić's sketches illustrating concepts of groundwater flow in karst (English annotation added). *Top*: Schematic section of a karst aquifer with three zones: (1) dry zone with percolation only; (2) transition zone of varying level with both percolation and saturated groundwater flow; (3) permanently saturated zone, including conduits with ascending flow. *Bottom*: Successive lowering of ponors (sinks) of streams in karst creating blind valleys; x-y denotes pre-karstic valley. (From Cvijić, 1918, and Stevanovic and Mijatovic, 2005.)

FIGURE I.5 The Cullum Geographical Medal of the American Geographical Society presented to Jovan Cvijić in 1924 "for the scholarship he displayed in his published works on the geography of the Balkan countries and for the originality of his field studies on the complicated physiography of the Karst." (Courtesy of Jovan Cvijić Museum, Belgrade.)

FIGURE I.6 Arrival of Charon on a boat from a cave believed to be an entrance to Hades, the Greek underworld. Illustration by Gustave Doré in the Pollard and Moss 1885 edition of Dante Alighieri's Inferno, Plate 9, Canto III—Charon.

mysterious and sometimes frightening underground world of caves (Fig. I.6); it supports living creatures often unique to specific locations (Fig. I.7); it behaves unpredictably as it can sometimes rise hundreds of feet after heavy rains in a matter of hours giving life to numerous temporary springs and increasing the flow of permanent springs 1000-fold; and it is extremely vulnerable to both natural and anthropogenic contamination, thus seriously limiting its unrestricted use in many parts of the world.

As discussed by LaMoreaux and LaMoreaux (2005), the earliest concepts of the hydrologic cycle, water source, occurrence, and quality were based on karst settings. During ancient times, underground rivers and springs were a part of Greek mythology and the subject of much discussion by Greek and Roman philosophers. Some notable references to karst, karst springs, and geomorphic features include

Figure I.7 The Texas blind salamander is a sightless, cave-dwelling creature that reaches a mature length of about 5 inches. It spends its life in complete darkness and is very sensitive to changes in water quality. Biologists know of only one population living in Edwards Aquifer caves around San Marcos, Texas. It is one of seven federally endangered species that depend on the Edwards Aquifer. (Photo copyright Gregg Eckhardt, printed with permission.)

- Cuneiform tablets that provide the first record of karst hydrological research as known and described during an expedition in 852 BC by the Assyrian King Salmanassar III to the headwaters of the Tigris. The source of the Tigris is a karst spring issuing from a cave that near the entrance has inscriptions stating that this source of water to the Tigris immortalizes Salmanassar III.
- Greek philosophers Empedocles (490 to 430 BC) and Aristotle (384 to 322 BC) theorized on the movement of groundwater from karst springs as related to the water cycle.
- As reported by Strabo in Book 8, Eratosthenes (276 to 194 BC) describes the hydrologic connection between Katavothra (Ponors) in Pheneos polje and Ladon Spring in the Peloponnesus of Greece.
- Poseidonius (135 to 50 BC) reports on the springs of Timavo near present-day Trieste, Italy, and describes the River Timivaius where it disappears in a cave.
- Strabo (60 BC to 28 AD) devoted the eighth book of his 17-volume encyclopedia of geographic knowledge "Geographica" to Greece; in it he describes poljes, underground streams, and other karst phenomena.
- Lucius Annaeus Seneca (3 BC to 65 AD), better known for his tragedies, was perhaps the most important Roman writer on karst. In the third book of his *Naturales Quaestiones* devoted to water, Seneca describes solution processes, development of large caves, springs, and disappearance and reappearance of streams (LaMoreaux and LaMoreaux, 2005).

Figure I.8 Entrances to natural and man-made passageways in limestone at the Pool of Siloam, Jerusalem. (Photograph courtesy of Abraham, September 2005.)

Water in karst has also inspired the first masterpieces in water supply engineering. The intermittent siphon-type Gihon Spring issuing from a cave was the main source of water supply for ancient Jerusalem and the irrigation of King's Garden in the Kidron Valley. Its intermittent nature and the location outside city walls resulted in construction of several waterworks in order to secure reliable water supply. The first one was excavation of a storage reservoir in the southern, low part of the Tyropoeon Valley within city walls, named the Pool of Siloam (Fig. I.8). The reservoir and the spring were connected by an aqueduct, the Siloam Channel, cut in places 20 ft into the ground and partially covered with stone slabs. The entrance to the Warren's Shaft System, located within the ancient city's walls, was discovered by C. Warren in 1867 and named after him. The system consisted of a 41-m-long irregularly shaped underground passage with a shaft at its end leading to the Gihon Spring (Fig. I.9). Although most scholars were in agreement that the Warren's Shaft System was man-made, a hydrogeological study conducted at the beginning of the 1980s suggested that the shaft and most of the tunnel were natural karst features in the rock adapted and combined by the builders of the system to make subterranean passage from the city to the spring possible (IMFA, 2003).

During the reign of Hezekiah (739 to 687 BC), the Siloam Channel was supplemented by the Hezekiah's Tunnel (also known as the Siloam Tunnel) dug underneath the City of David in Jerusalem apparently in preparation for an impending siege by the Assyrians. Hydraulic plaster was applied throughout the tunnel in order to seal voids of dissolution and tectonic origin. The 553-m-long curving tunnel is one of the few intact, eighth century BC biblical structures that the public can still visit today. According to the Siloam Inscription, it was excavated by two teams, one starting at each end of the tunnel and then meeting in the middle. It has been proposed that the tunnel, at least partially, followed

Figure I.9 Waterworks at Gihon Spring in Jerusalem. *Top*: Spring cave; in the center of the image is a carved passageway that leads to the area of Warren's Shaft and Hezekiah's Tunnel. The water from the spring flows through this passageway—away from the viewer. (Photograph courtesy of www.HolyLandPhotos.org.) *Bottom left*: Part of the Warren's Shaft System. (Photograph courtesy of IMFA, 2003.) *Bottom right*: View of the interior of Hezekiah's Tunnel; water from the spring is still flowing through the tunnel that can be visited. (http://www.bibleistrue.com/qna/pqna21.htm.)

natural underground features through which some water already flowed between the Gihon Spring and the Siloam Pool; the Siloam Inscription mentions that "there was a 'zdh' in the rock," which could be interpreted as a fracture or a karst feature, or both (IMFA, 2003). Frumkin and Shimron (2006) suggest that the digging of the tunnel was accomplished by directing the two teams from above using sounds generated by hammering on the solid rock through which the hewers were digging (arguably the first application of engineering geophysics in karst).

Using karst aquifers and springs for water supply was mastered by the Romans who preferred their water to any other sources. In virtually all Mediterranean countries

FIGURE I.10 *Top*: Part of the aqueduct built in the third century AC by Roman Emperor Diocletian for his summer residence on the Adriatic coast. The aqueduct brought water from a large karst spring still being used for the water supply of the Croatian port city of Split. (Photograph courtesy of Ivo Eterović.) *Bottom*: Reconstruction of Diocletian's palace in Split by architect Ernest Hébrard. (From E. Hébrard and J. Zeiller, *Spalato, le Palais de Dioclétien*, Paris, 1912.)

that were part of the Roman Empire, there are remains of numerous Roman aqueducts originating at karst springs, some of which are preserved in almost full glory (Fig. I.10). The Roman custom of building handsome fountains at the end point of aqueducts that brought karst water to Rome was revived in the fifteenth century, in the Renaissance. The Trevi Fountain shown in Fig. I.11 is the largest Baroque fountain in the city and one of the most famous fountains in the world, fabled for its water clarity. The city of Rome still relies on large karst springs for its water supply.

Karst aquifers are present on all continents and numerous oceanic islands and crop out on over 20% of the land surface, with more being covered by insoluble rocks. As the importance of water resources and their sustainability is becoming critical worldwide,

Figure I.11 The Baroque Trevi Fountain in Rome designed by Nicola Salvi and finished in 1762. Water for the fountain is brought via Aqueduct Vergine, the revived ancient underground aqueduct Aqua Virgo, from the 13-km-distant karst spring Salone. According to a legend, thirsty Roman soldiers asked a young girl for water. She directed them to the springs that later supplied the aqueduct. The source was named the Aqua Virgo after her. (Photograph courtesy of Dr. Zoran Stevanovic.)

karst aquifers are receiving rapidly increasing attention due to the many challenges related to their characterization, management, and protection. This book is a contribution to the ongoing efforts by the scientific, engineering, and regulatory communities to meet these challenges. I wish to thank numerous colleagues and karst enthusiasts for generously providing materials of their research. I am especially grateful to Phil Lucas, Bill Jones, and John Ackerman for letting me use many of their fascinating photographs throughout the book and illustrate what even the best drawings could not. I am also thankful to many other colleagues who provided some of their best photographs and made the book and the color plates so much more illustrative. Special thanks go to my colleagues at AMEC Environment and Infrastructure, Inc., for their support and permission to use project materials and photographs from the George Sowers Collection. During preparation of the book, I have greatly benefitted from discussions with the associates of the Center for Karst Hydrogeology, University of Belgrade, which continues the strong tradition of karst research and engineering founded by Jovan Cvijić and his contemporaries. I am also grateful to Dr. Neno Kukuric and Dr. Holger Treidel of UNESCO for the opportunity to participate in the project entitled Protection and Sustainable Use of the Dinaric Karst Transboundary Aquifer System (DIKTAS) and again interact with and learn from karst specialists and colleagues from Croatia, Bosnia and Herzegovina, Montenegro, Albania, and Serbia. Finally, I am indebted to my colleagues Dr. Petar Milanovic and Dr. Zoran Stevanovic, and my mentors from the IAH Karst Commission, the late Dr. William Back and the late Dr. Philip LaMoreaux, for the friendship and generosity in sharing their tremendous experience.

References

Cvijić, J., 1893. Das Karstphänomen. Versuch einer morphologischen Monographie. *Geographische Abhandlungen Herausgegeben von Prof.* Dr A. Penck, Wien, Bd. V. Heft. 3, pp. 1–114.

Cvijić, J., 1918. Hydrographie souterraine et évolution morphologique du karst. Recueil des Travaux de l'Institut de Géographie alpine, Grenoble, t. VI, fasc. 4, pp. 1–56.

Cvijić, J., 1926. Geomorfologija (Morphologie Terrestre). Knjiga druga (Tome Second). Državna štamparija Kraljevine Srba, Hrvata i Slovenaca, Beograd, 506p.

Frumkin, A., and Shimron, A., 2006. Tunnel engineering in the Iron Age: Geoarchaeology of the Siloam Tunnel, Jerusalem. *J Archaeol Sci* 33(2): 227–237.

IMFA (Israel Ministry of Foreign Affairs), 2003. Jerusalem–Water Systems of Biblical Times. Available at: http://www.mfa.gov.il/mfa/early%20history%20-%20archaeology/

LaMoreaux, P. E., and LaMoreaux, J., 2005. Karst: Foundation for concepts in hydrogeology. In: Stevanovic, Z., and Milanovic, P. (eds.), *Water Resources and Environmental Problems in Karst, Proceedings of the International Conference and Field Seminars,* Belgrade & Kotor, Serbia & Montenegro, 13–19 September 2005, Institute of Hydrogeology. University of Belgrade, Belgrade, pp. 3–8.

Milanovic, P. T., 1981. *Karst Hydrogeology*. Water Resources Publications, Littleton, CO, 434p.

Stevanovic, Z., and Mijatovic, B. (eds.), 2005. *Cvijić and Karst*. Serbian Academy of Science and Arts, Board on Karst and Speleology, Special Edition, Belgrade, 405p.

Water in Karst

PART 1

Karst Hydrogeology and Hydrology

CHAPTER 1

Karst Aquifers

1.1 Introduction

What sets karst areas apart from all other hydrogeologic environments is the nature and dynamics of surface water–groundwater interactions. The most obvious examples are the disappearance of an entire river into one large sink or cave (Fig. 1.1) and the formation of a major surface stream from a single spring (Fig. 1.2). In fact, when asked to define karst, in addition to caves and sinkholes, most professionals and nonprofessionals alike mention sinking streams and underground rivers. This is not surprising since, in many cases, the same surface stream can disappear and appear multiple times along its course (Figs. 1.3 to 1.5). It is sometimes possible to directly observe this journey of water from the surface to the subsurface and back. One can do it by standing at the right location (Fig. 1.6), or by walking, crawling, boating, diving, repelling, or otherwise following the flow of water (Figs. 1.7 and 1.8). However, on a karst watershed scale, even the most dedicated cavers–divers would not be able to follow all sinking streams on their entire journeys. Moreover, surface streams in some karst watersheds do not have definitive locations where all their water disappears into the subsurface; rather, they interact with the underlying aquifer similarly to streams in nonkarstic terrains, via diffuse exchange of water through the carbonate rock matrix and fractures/fissures. In other words, the flow of water between surface and subsurface does not have to be restricted to the locations such as sinks and springs, and their absence cannot be used to classify a watershed or parts of it as nonkarstic. Just as in terrains with noncarbonate, nonsoluble rocks and sediments, rivers in karst can form alluvial plains with unconsolidated fluvial deposits in which groundwater flows through interstitial pore space (slow, "diffuse" flow). In such cases, the exchange of water between the surface stream, the alluvial deposits aquifer, and the underlying karst aquifer can be quite complex with each of the three compartments possibly losing and/or gaining water to the other two during different seasons or recharge conditions.

Unfortunately, it is common to use the absence of sinking streams as an argument against classifying a terrain as karstic. The absence of sinkholes is another reason used to misclassify a terrain as nonkarstic by inexperienced geologists and hydrogeologists. As explained in the following chapters, this misunderstanding of the complex nature of karst may have major implications on management, protection, and restoration of its water resources.

Depending on the scale of observation (distances traveled), a surface stream and its "underground river" continuation in karst can exhibit a wide range of behavior in terms of both water quantity and quality. For example, the stream may be losing water in one of

Figure 1.1 The Rak River disappearance in the Tkalca Jama cave, Slovenia.

its segments and gaining water in another during a particular season and this situation may be completely reversed the following season. While flowing in the subsurface, a sinking stream may sometimes bifurcate and its water may appear in two or more surface (topographic) watersheds at multiple locations including springs. Much more common, however, and similar to surface drainage networks is convergence of subsurface flows in preferential flow paths into progressively higher-order karst conduits and channels (Figs. 1.9 and 1.10). This subsurface convergence of karst networks can happen over several topographic watersheds, thus presenting the main challenge for accurate determination of groundwater–surface water interactions. A typical example is spring contaminated

Figure 1.2 After sinking in the Tkalca Jama cave, the Rak River appears again as a large spring issuing from the Planinska Jama cave and forming the Unica River, Slovenia (see also Fig. 1.10). Views are from inside (*right*) and outside the cave (*left*).

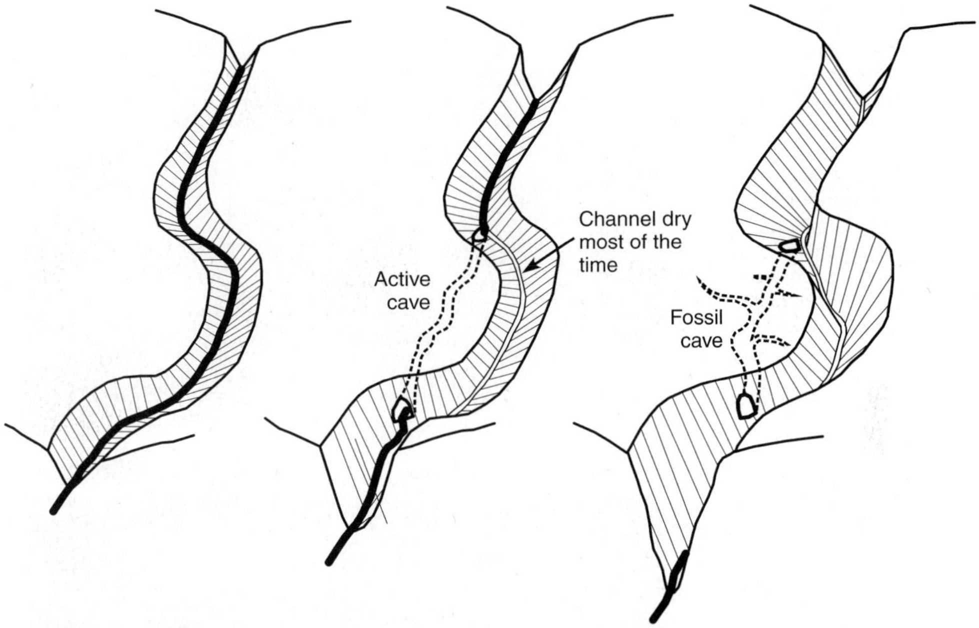

Figure 1.3 Schematic of cave formation in a river canyon. *Left*: Initial stage of canyon development. *Middle*: The river starts flowing underground through a cave, first most of the time and then only occasionally during high water levels in the aquifer. *Right*: As the canyon deepens, the cave is "fossilized" (becomes dry and inactive), while the same process continues at a deeper level, leaving longer segments of the river dry most of the time.

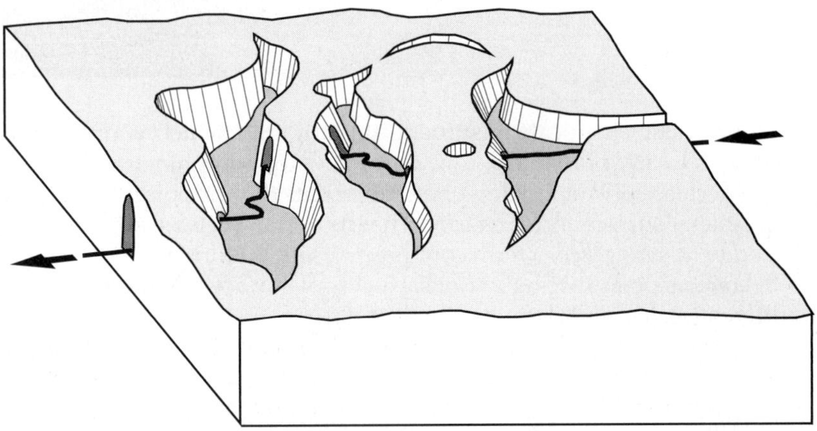

Figure 1.4 Multiple sinks and resurgences of the Notranjska River in the St. Skocjan village, Slovenia (see also Fig. 1.5). The river ultimately disappears into the Skocjanske Jame cave, protected by UNESCO as a world natural heritage site, and flows through inaccessible passages toward the Timavo Springs near Trieste, Italy.

FIGURE 1.5 Lezeca Ponor (*top*), and Naravni most (Natural Bridge) during high flow of the Notranjska River, Skocjanske Jame National Park, Slovenia. (Courtesy of Borut Petric, Skocjan Caves National Park.)

with a synthetic chemical constituent, the origin of which is unknown and cannot be traced back to any potential source within the topographic drainage area of the spring. A large permanent karst spring often represents the endpoint where an integrated karst water system (surface and subsurface) has its outlet. Such a system is therefore sometimes referred to as *springshed*, a term commonly used in Florida (Fig. 1.11).

In conclusion, resolving various aspects of surface water–groundwater interactions at different scales, and defining the overall karst water system (both at the surface and in the subsurface), is the key for successful management and restoration of its water resources. The remaining sections of this chapter discuss key characteristics of karst aquifers, methods for their delineation, and methods for quantifying their water budgets.

1.2 General Aquifer Characteristics

Karst aquifer is the main part of a karst water system. On a volume basis, it receives, stores, and transmits more water than all surface streams within the system combined.

Figure 1.6 The Nostrils Cave in Sohodol Gorge, Carpathian karst of Romania. (Courtesy of Dr. Zoran Stevanovic, Center for Karst Hydrogeology, University of Belgrade.)

Defining its extent, as well as mechanisms and areas of recharge and discharge, can be more or less complicated depending on the type of the hydrogeologic structure involved (Fig. 1.12):

1. *Open hydrogeologic structure*. Recharge and discharge zones are fully defined (known). Recharge from precipitation takes place over the entire areal extent of the aquifer, which is directly exposed to the land surface. This type of recharge is often called *autogenic*. Surface streams coming from nonkarstic areas lose water to the underlying karst aquifer via concentrated sinks, or over some distance through fractures and openings in the streambed, or both. This is called *allogenic* recharge. Discharge of the aquifer (via springs or diffuse) is either at the contact with the impermeable base (case 1a) or along main erosional basis such as a large permanent river or seacoast (case 1b).
2. *Semi-open hydrogeologic structure*. The aquifer is not entirely exposed to the surface, but the discharge zone is fully defined. The aquifer is partially covered by unconsolidated or consolidated rocks that may have varying degrees of permeability, thus allowing or preventing its indirect vertical recharge from precipitation. Recharge zones are mostly or partially known (cases 2a and 2b, respectively).
3. *Semi-closed hydrogeologic structure*. Recharge and discharge zones are only partially known (case 3a) or mostly unknown (case 3b).

FIGURE 1.7 Entrance to a partially submerged cave spring receiving water from a sinking stream on the Pester Plateau, western Serbia. The cave is impounded by the Uvac Reservoir. Note small boat for scale.

4. *Closed hydrogeologic structure*. The aquifer is completely isolated by impermeable geologic units and does not receive recharge. In practice, such aquifer can only be discovered by drilling. The absence of any significant recharge (from the land surface or from adjacent aquifers) is manifested by large, continuingly increasing drawdowns during pumping.

In some cases, an aquifer may indeed be completely isolated from the "rest of the world." The presence of freshwater in it is testimony to a very different hydrogeologic past when the aquifer was receiving natural recharge from one or more sources such as precipitation, surface water, or adjacent aquifers. Various subsequent geologic processes, including faulting and folding, may have resulted in its complete isolation. Such aquifers are called fossil aquifers or nonrenewable aquifers. In general, any aquifer that is part of a groundwater system that does not receive natural recharge, regardless of the hydrogeologic structure in which it is formed, is called nonrenewable. Typical examples are aquifers and groundwater systems in arid regions with little or no modern precipitation and without flowing surface streams.

FIGURE 1.8 Stream in Organ Cave, West Virginia. (Photograph courtesy of William Jones, Karst Waters Institute.)

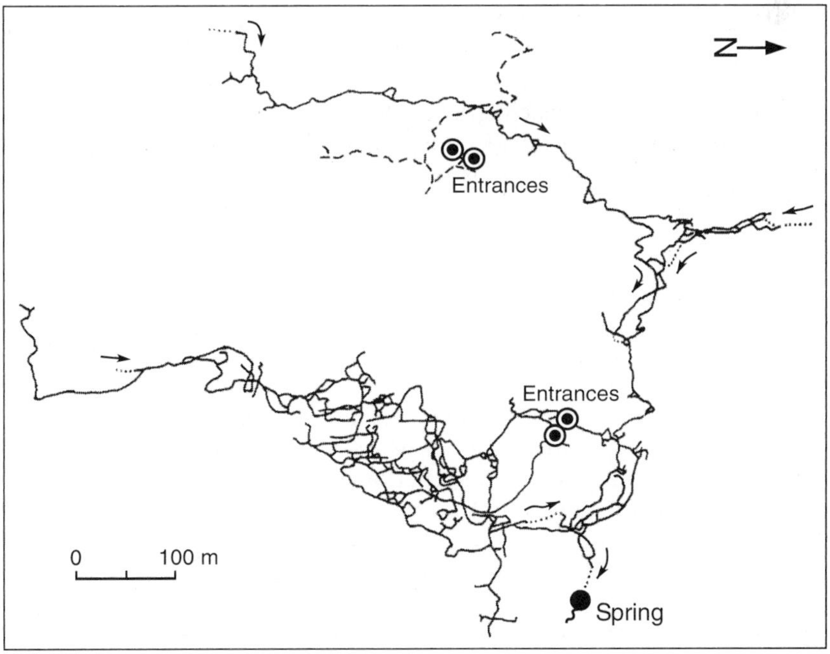

FIGURE 1.9 Cave system of Jopica and Gvozdenica Caves in Kordun, Croatia. (Survey and drawing by M. Čepelak; from Božičević, 1984.)

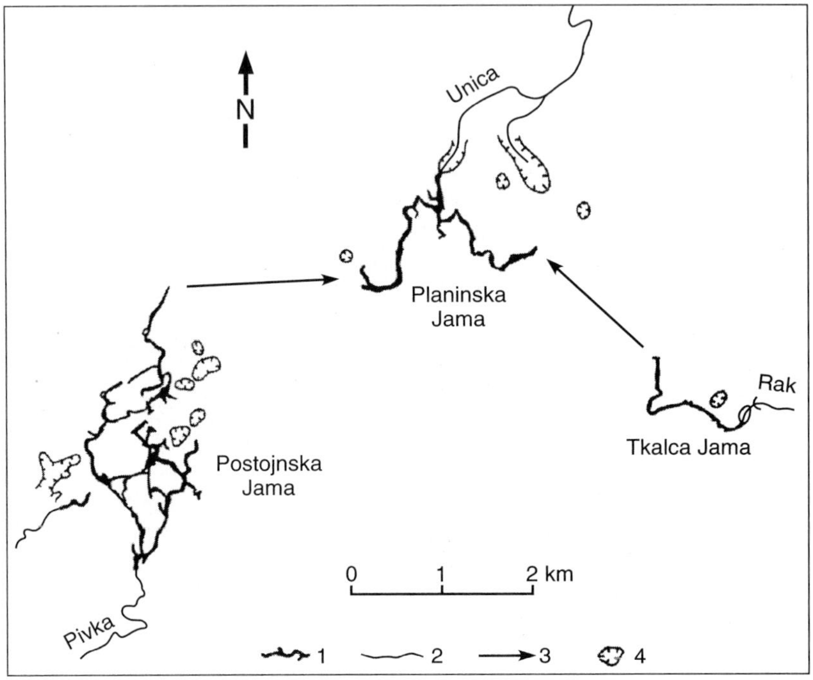

FIGURE 1.10 Schematic of the longest explored cave system in classic karst of Slovenia. The Unica River issuing from the Planinska Jama cave (see Fig. 1.2) is formed by the subsurface confluence of two major sinking streams, the Pivka and the Rak Rivers. Postojnska Jama cave is one of the oldest and most visited show caves in Europe. (Modified from IZRK, 1982.)

The only relevant characteristics of an aquifer to be qualified as karstic is that a portion of groundwater flow in it is taking place through a network of interconnected channels, conduits, and various other voids created primarily by dissolution of rock matrix. These voids can be the result of a number of physical and chemical processes involving descending (infiltrating) water, as well as ascending groundwater that may also carry acidic fluids such as dissolved H_2S, sulfuric acid, and CO_2 from greater depths. Regardless of the actual mechanism involved, the relative size and the number of interconnected karst voids are irrelevant; as long as they are present, at any scale, the aquifer is karstic. As explained further, what sets karst aquifers apart is their unique porosity, often referred to as triple porosity, which consists of the rock matrix porosity, the porosity of fractures, and the porosity of solution openings (karst channels, conduits, and cavities, in general). As a consequence, groundwater flow in karst does not conform to relatively straightforward principles governing flow in intergranular porous media and described by Darcy's law. Probably the only common characteristics karst aquifers share with other aquifer types on a regional scale is that the groundwater has to move from the areas of aquifer recharge to the areas (points) of aquifer discharge due to the hydraulic gradient in between, aquifer transmissivity and the effective porosity. In karst, this flow can often take quite unexpected local turns defying the expectations of professionals not used to working in such a complex groundwater environment. In many karst aquifers, it is not uncommon to have a couple of wells only few meters apart and screened at the same

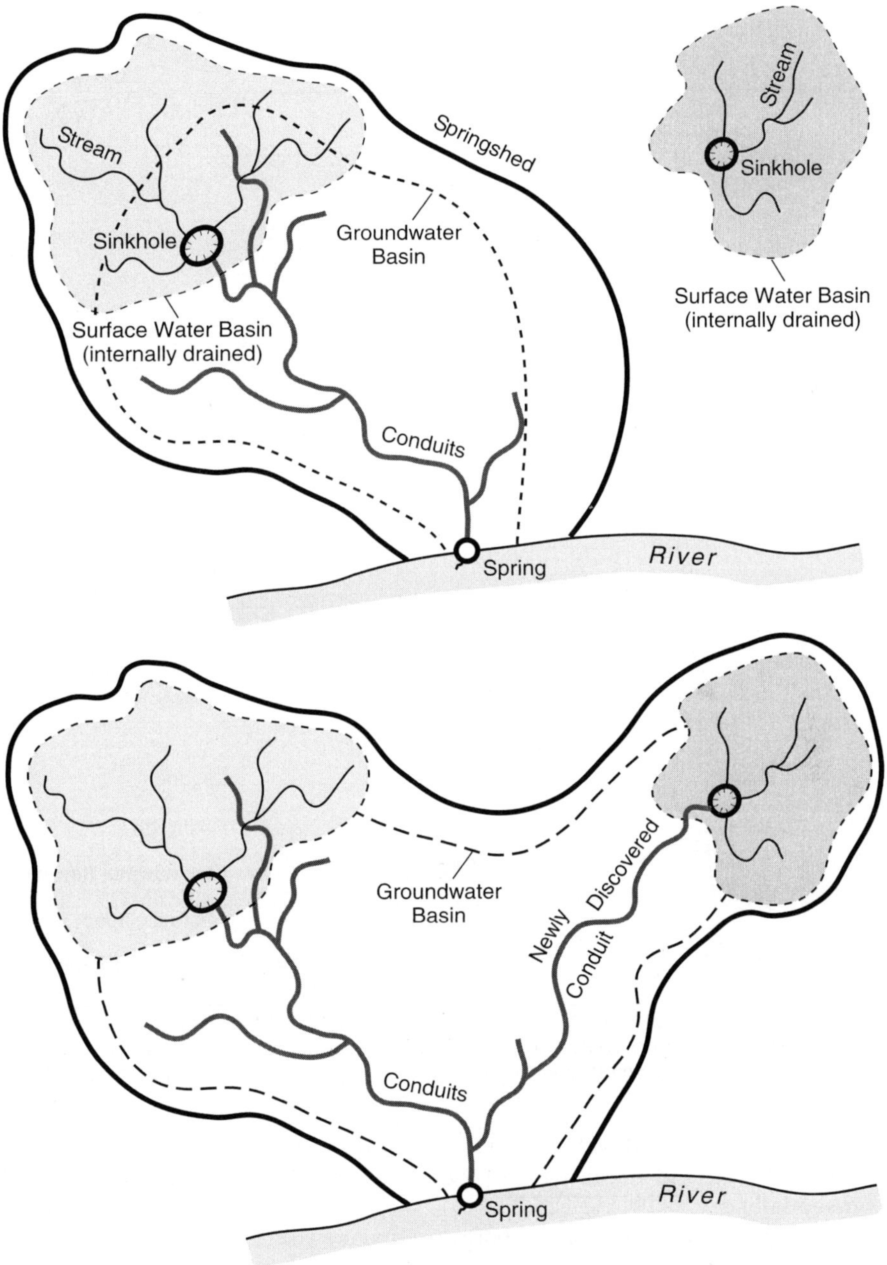

FIGURE 1.11 *Top*: Springshed boundary depicting surface water and groundwater basins. *Bottom*: Revised springshed boundary based on results of cave diving exploration and tracer tests that confirm presence of conduits connecting a nearby surface water basin to the springshed. (From Arthur, D.A., Upchurch, S., Skiles, W., Kincaid, T., Davis, H., and McGurk, B., 2004. Panel II. Delineation of springshed boundaries. Workshop to develop blue prints for the management and protection of Florida's springs, May 8–9, 2002, Ocala, Florida. Florida Geological Survey Special Publication No. 51.)

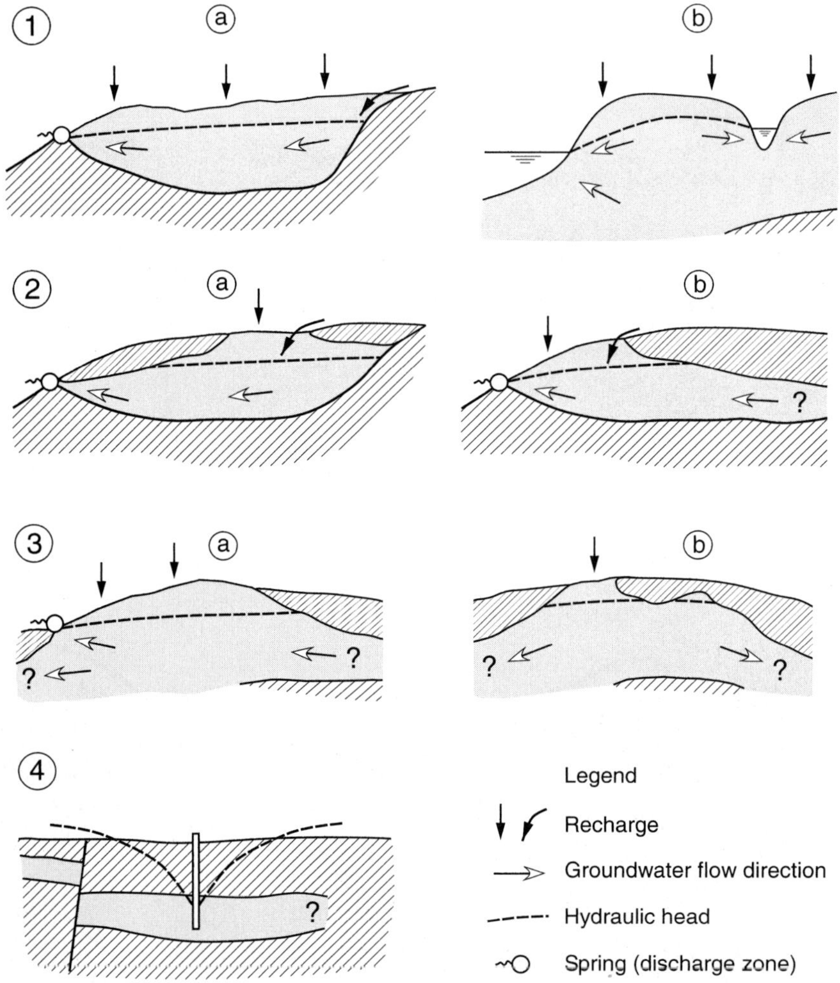

Figure 1.12 Types of hydrogeologic structures: (1) open; (2) semiopen; (3) semiclosed; (4) closed. Explanation in the text. (Modified from Kresic, N., 2009. *Groundwater Resources. Sustainability, Management, and Restoration*. McGraw Hill, New York, p. 852.)

depth below ground surface, with one well influenced by a major conduit and being quite prolific, and the other one completed in solid rock yielding very little water or essentially dry.

A good starting point for characterizing a karst aquifer is to determine its regional carbonate lithostratigraphy and geologic setting. Large epicontinental carbonate shelf platforms, at the scale of hundreds to thousands of kilometers, and thousands of meters thick, together with isolated carbonate platforms in open-ocean basins, have developed through the Paleozoic and Mesozoic eras, with the trend continuing into the Cenozoic era. At the same time, smaller (tens to hundreds of kilometers wide) carbonate platforms and buildups associated with intracratonic basins have also developed (James and Mountjoy, 1983). Both types have been redistributed and reshaped during various phases of plate tectonics and can presently be found throughout the world, both adjacent to oceans and seas or deep inside the continents where they may also be buried by younger nonsoluble rocks. The region of classic Dinaric karst in the Balkans, Europe, is an example of a tectonically disturbed large thick Mesozoic carbonate platform where the Adriatic Sea is the regional erosional basis for groundwater discharge. The sequence of carbonate sediments is thousands of meters thick (Fig. 1.13) and acts as an open hydrogeologic

FIGURE 1.13 Examples of karstified limestones in the famous Dinaric karst of former Yugoslavia developed on thick Mesozoic carbonate platform. *Top*: The Dinara Mountain in Dalmatia, Croatia, that gave name to the Dinaric orogenic belt. *Bottom*: The Biokovo Mountain, Makarska Riviera, Croatia.

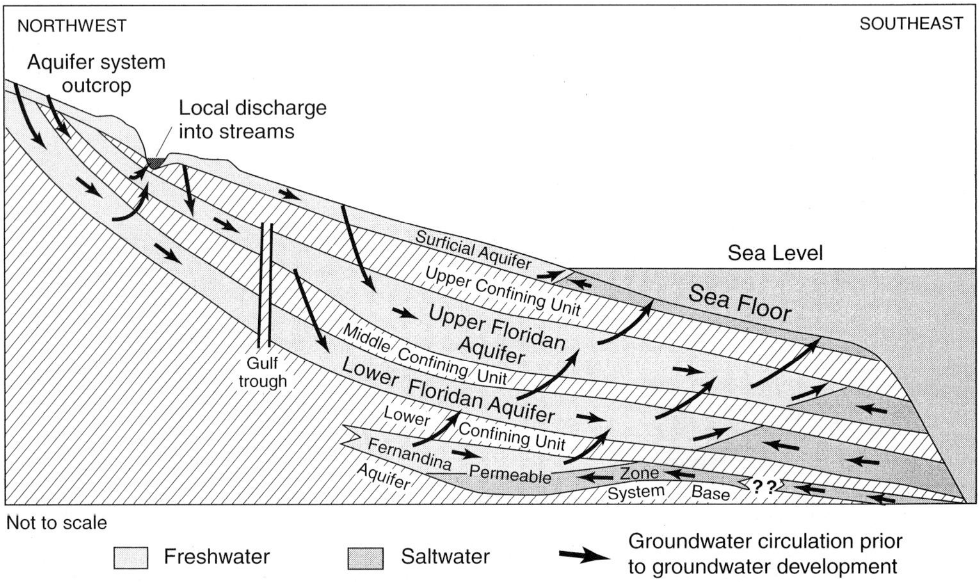

FIGURE 1.14 Schematic cross section of Floridan aquifer system from northwest to southeast Georgia. The Lower Floridan aquifer in this area includes a highly permeable unit called the Fernandina permeable zone. This zone is the source of a considerable volume of fresh to brackish water that leaks upward through the middle confining unit and ultimately reaches the Upper Floridan aquifer. (Barlow, P.M., 2003. Ground water in freshwater-saltwater environments of the Atlantic coast. U.S. Geological Survey Circular 1262, Reston, Virginia, 113p, modified from Krause and Randolph, 1989.)

structure. Deep oil exploration wells have encountered large open cavities more than 3 km below ground surface.

Like the Euro-Asian Mesozoic platforms, the Floridan aquifer in the southeast United States (parts of North and South Carolinas, Georgia, and Florida) is also developed on a thick epicontinental platform but with gently sloping undisturbed carbonate layers mostly covered with less permeable clastic sediments (Fig. 1.14). The aquifer consists primarily of Tertiary limestone and dolomite of Paleocene to Miocene age. Regional flow directions are from the inland outcrops toward the Atlantic Ocean and the Gulf of Mexico, with submerged discharge zones along the continental shelf. Florida has more than 20 well-documented large offshore springs and a number of undocumented ones. These springs provided resources to prehistoric people and wildlife when the sea level was lower. Evidence for occupation of offshore sites has been discovered by researchers from the Florida State University Department of Anthropology, who have conducted surveys at and near some offshore springs and have recovered an abundance of chert tools (Scott et al., 2004). Although some of the offshore springs may be discharging brackish to saline water today, they almost certainly discharged freshwater during times of lower sea levels when prehistoric human occupation occurred at these sites. Where limestone is exposed at or very near the land surface, as in north-central Florida, karst features such as sinkholes, large springs, and caves are fully developed. Many caves are completely and permanently filled with groundwater. Quite a few such caves providing water to large karst springs have been explored by cave divers.

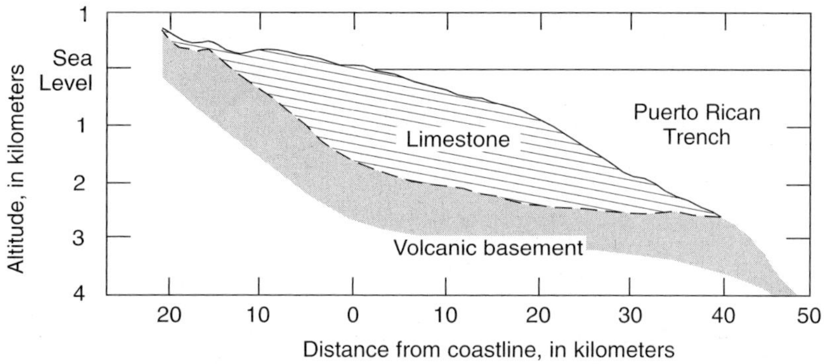

FIGURE 1.15 Section of the north coast limestone belt, Puerto Rico. (From Giusti, 1978, modified from Shubert and Ewing, 1956.)

Karst of large Caribbean islands, such as Cuba, Puerto Rico, Hispaniola, and Jamaica, is an example of relatively small carbonate platforms lying on a low permeable, usually magmatic, base (Fig. 1.15). Subsurface and surface karst features are fully developed including caves, sinkholes, and cone-shaped hills called "mogotes" in Cuba and "pepinos" in Puerto Rico. Figure 1.16 illustrates typical karst landscape on the island of Hispaniola (in Jamaica, such landscape is called "cockpit karst"). Regional groundwater

FIGURE 1.16 Convoluted pattern of sinkholes and residual limestone hills in the aerial photograph taken by George Sowers in the Dominican Republic. *Arrows* point to couple of houses for scale. (From the George Sowers collection, AMEC Environment and Infrastructure, Inc., printed with permission.)

FIGURE 1.17 Divers exploring submerged passages in Ponderosa Cave, Yucatán, Mexico. Note high vuggy (dissolution) porosity along the cave walls. (Photograph courtesy of David Rhea, Global Underwater Explorers.)

flow direction is from the upland recharge areas toward the coastline, with the submerged discharge zones along the freshwater–saltwater interface.

When of considerable thickness, young carbonate sediments may constitute important aquifers for both local and centralized water supply worldwide. In addition to the Bahamas and numerous other islands in the Caribbean, examples include the Yucatán peninsula of Mexico (Fig. 1.17), much of British Honduras and Guatemala, Jamaica, Barbados, Bermuda, the Cebu limestone of the Philippines, the Jaffna limestone in Sri Lanka, low-lying coral islands of the Indian Ocean such as the Maldives, and portions of Cambodia and Vietnam. The high infiltration capacity of young carbonate sediments in coastal areas and islands means that there are few streams or rivers, and groundwater may be the only available source of water supply. This source is often very vulnerable to salt (sea) water encroachment due to overpumping of fresh groundwater.

The majority of karst areas in the United States constitute portions of carbonate platforms now located away from the coastlines. Examples include Edwards Group limestone in Texas and karst of Kentucky, Tennessee, Missouri, and Indiana, all of which have very prolific aquifers. Figure 1.18 shows the Catfish Farm well near San Antonio, Texas, which discharged more than 115,000 L/s under pressure to create a vertical column of water rising 10 m above the surface when first drilled. Regional groundwater flow in these aquifers is directed toward large karst springs located at the lowest contact between karst and nonkarst, or toward the lowest permanent surface stream intersecting the carbonates. Groundwater discharge along such streams is commonly through large springs, often submerged due to present-day river damming. Probably, the best-known

Figure 1.18 *Top*: The highest-capacity drilled well in the world, capable of producing 40,000 gallons per minute (2.5 m^3/s), located on the Ron Pucek's Catfish Farm near San Antonio, Texas. The well is completed in karstic Edwards Aquifer. *Bottom*: Water from the well being conveyed toward the catfish basins. The well is now decommissioned through agreement with the Edwards Aquifer Authority to preserve groundwater for other uses. (Photographs copyright by Gregg Eckhardt, printed with kind permission).

karst terrain in the United States is in the Mammoth Cave area of central Kentucky (Fig. 1.19). Recharge water enters the aquifer through sinkholes, swallow holes, and sinking streams. Surface streams are scarce because most of the water is quickly routed underground through solution openings. In the subsurface, water moves through caverns and other types of large solution openings, which riddle the Mississippian limestones that underlie the Mammoth Cave Plateau and the Pennyroyal Plain to the south and

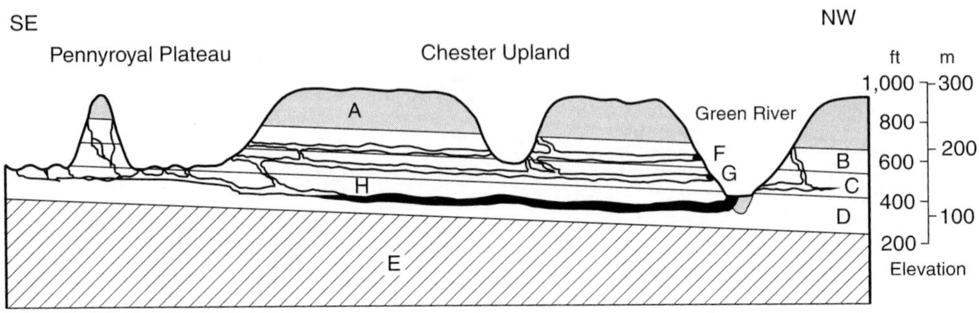

Figure 1.19 Generalized profile through Mammoth Cave (from Palmer, 1985). A = caprock of interbedded sandstone, shale, and limestone (Chesterian Series); B = Girkin Formation; C = Ste. Genevieve Limestone; D = St. Louis Limestone; E = impure limestone and detrital rocks; F = upper cave levels of late Tertiary age; G, H = typical levels of Quaternary age (partial flooding of the lowest level is due to late Pleistocene alleviation of the Green River).

southwest of it. Many of these cavities form the large, extensive passages of Mammoth Cave, the world's longest and best-studied cave system.

In Nevada and Utah in the United States, carbonates and other sedimentary rocks are deposited on the continental shelf of the ancestral coast of North America between 570 and 280 million years ago. The layers of ancient marine sediments have a cumulative thickness of as much as 40,000 ft (more than 12,000 m). All the sedimentary rocks in this sequence, where deformed and fractured, may transmit some groundwater flow, but the carbonate-rock layers, because of their brittleness and dissolution by flowing water, are the principal aquifers (Dettinger, 1989). Due to tectonic forces (faulting and folding), these carbonate aquifers have complex shapes and are connected to aquifers of other rock types. Some of the blocks may be highly productive aquifers, whereas others

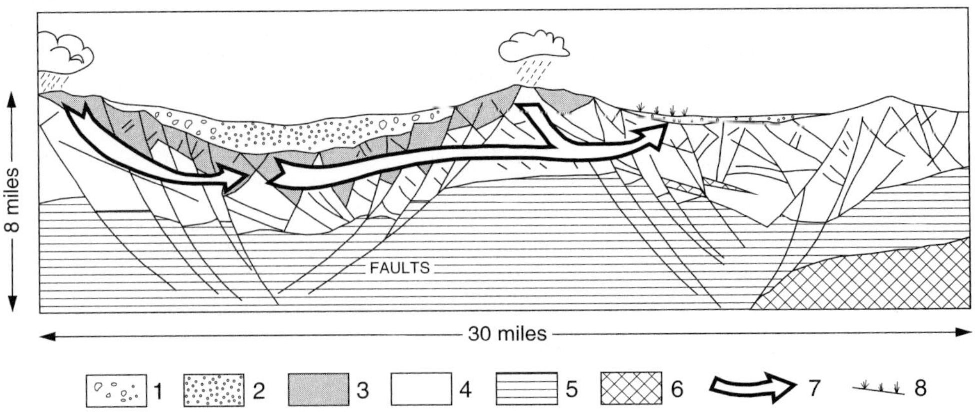

Figure 1.20 Schematic hydrogeologic section across mountain ranges and intervening basins, showing configuration at depth of aquifers and rocks that impede flow (from Dettinger, 1989). (1) Sand and gravel forming basin-fill aquifers; (2) Clay and silt: basin-fill deposits that impede groundwater flow; (3) Volcanic rocks: aquifers as well as layers that impede flow; (4) Carbonate rock aquifers; (5) Older, noncarbonate sedimentary rocks impeding flow; (6) Crystalline basement impeding flow; (7) Groundwater flow path; (8) Playas, springs, and native vegetation.

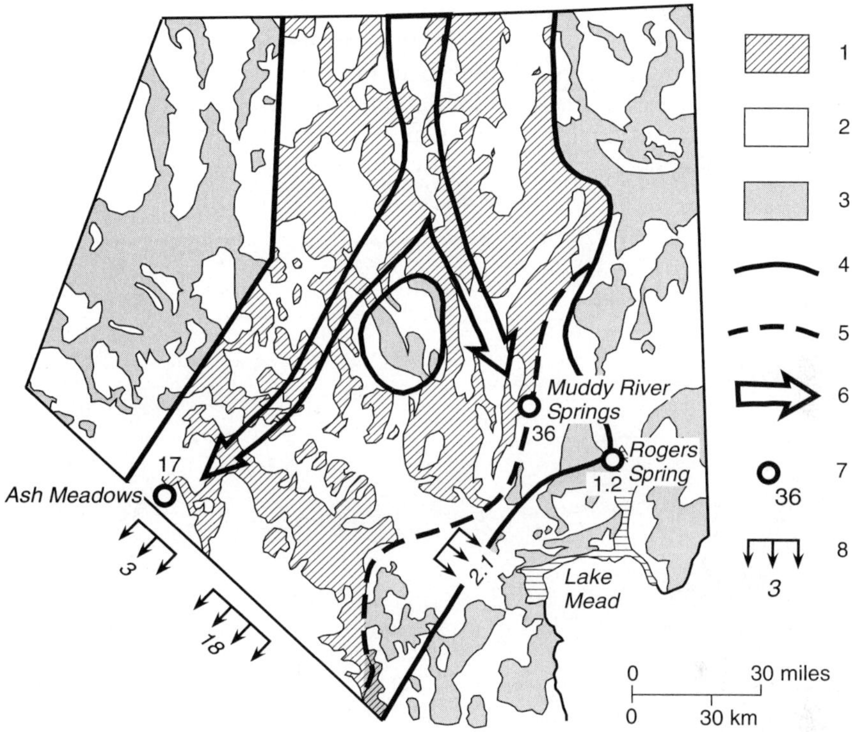

FIGURE 1.21 Components of regional groundwater budget within Central Corridor in southern Nevada (from Dettinger, 1989). (1) Carbonate rock outcrops within Central Corridor; (2) Basin fill; (3) Areas underlain by thin or isolated carbonate rocks, or carbonate rocks that may contain poor-quality water; (4) Boundary of Central Corridor; (5) Western limit of rock sequences containing salt-bearing minerals and overlying carbonate rocks. Groundwater east of this line is generally poor quality; (6) General direction of regional groundwater flow; (7) Spring with the annual discharge in acre-feet per year; (8) Regional groundwater outflow to adjacent aquifers/states, in acre-feet per year.

may transmit only moderate quantities of water or impede flow altogether. Regional groundwater flow, through both the carbonate-rock aquifers and basin-fill aquifers, is from the areas of recharge (rain and snow on the mountain ranges) toward the areas of discharge (springflow, playas).

Figures 1.20 and 1.21 show the results of a regional groundwater flow study in southern Nevada. The total rate of flow in carbonate-rock aquifers is estimated to be about 77,000 acre-ft per year [3 cubic meters per second (m^3/s); note that 1 acre-ft equals 1,233 cubic meters (m^3)]. The main discharge is at several warm-water springs or through the basin-fill sedimentary aquifers that partly fill valleys, while the rest flows out of Nevada into adjacent states, mostly California (Dettinger, 1989). It is interesting that, due to very low recharge (Nevada and Utah are high-desert terrains), the entire flow in this vast area is less than the average flow of some individual first-magnitude springs in Florida.

In the United States, sandstone and carbonate-rock deposits are often interbeded over large areas and form aquifers of a mixed type, called sandstone and carbonate-rock aquifers. This aquifer type is present mostly in the eastern half of the Nation, but also occurs in Texas and in Oklahoma, Arkansas, Montana, Wyoming, and South Dakota. The carbonate rocks generally yield more groundwater than do the sandstone rocks, owing to dissolution and larger open-pore space (Maupin and Barber, 2005).

In addition to the classic Dinaric karst of the Balkans (former Yugoslavia), the karst of southwest China is best known for its spectacular surface and subsurface karst topography (Fig. 1.22). As discussed by Zhaoxin and Chuanmao (2004), carbonate rocks

Figure 1.22 Tower karst as seen from the Li River, Guangxi Province, China.

	Flow Rate Range (L/s)				
Province	50–500	500–1,000	1,000–2,000	>2,000	Total
Yunnan	648	45	35	3	731
Guizhou	231	20	11	1	26
Guangxi	284	13	2	0	229
Total	1,163	78	48	4	1,293

From Zhaoxin and Chuanmao, 2004.

TABLE 1.1 Large Karst Springs with Different Flow Rates in Three Provinces in Southwest China

are widely spread and occupy about one-third of the total area of 500,000 km^2 in which mountains and hills are dominant morphologic features. Altitude decreases from 2,500 m in the northwest to less than 200 m in the southeast. Annual precipitation is more than 1,000 mm/year. Karst features are extensively developed and the rainfall infiltration rate is generally about 30% to 70%. It is estimated that karst aquifers store 40% to 70% of total groundwater resources in southwest China (Table 1.1). Large portion of discharge of karst aquifers is via springs; there are 1,293 registered significant springs with yields greater than 50 L/s (Table 1.2). Because of a humid climate and abundant surface water, groundwater has not been as extensively utilized for irrigation as it has been in northern China, but it is still important for urban and domestic water supply. Examples of significant use of karst aquifers and springs are in the provincial capitals Kunming and Guiyang, and the city of Tianjin (Zhaoxin and Chuanmao, 2004).

As indicated earlier, karst aquifers can be present at varying depths below ground surface and covered by nonkarstified sediments and rocks without any visible signs of typical karst topography such as sinkholes and sinking streams. The Panonian Basin of Central Europe (in Hungary and northern Serbia) is an example of extensive Mesozoic basement karst aquifers buried deep under younger clastic sedimentary rocks. These aquifers constitute complex systems with typical paleokarst features together with large bauxite deposits in Hungary. Karstified Mesozoic carbonates were subsequently covered by younger sediments including new carbonates that were also subject to several karstification episodes in Tertiary when exposed at the land surface. As

Province	Karst Water Resources	Total Groundwater Resources	Ratio (%)
Yunnan	3,250	7,420	43.7
Guizhou	1,680	2,290	73.2
Guangxi	4,840	7,760	62.3
Sichuan	2,940	6,300	46.6

From Zhaoxin and Chuanmao, 2004.

TABLE 1.2 Karst Water Resources in Four Provinces of Southwest China, in km^3/Year

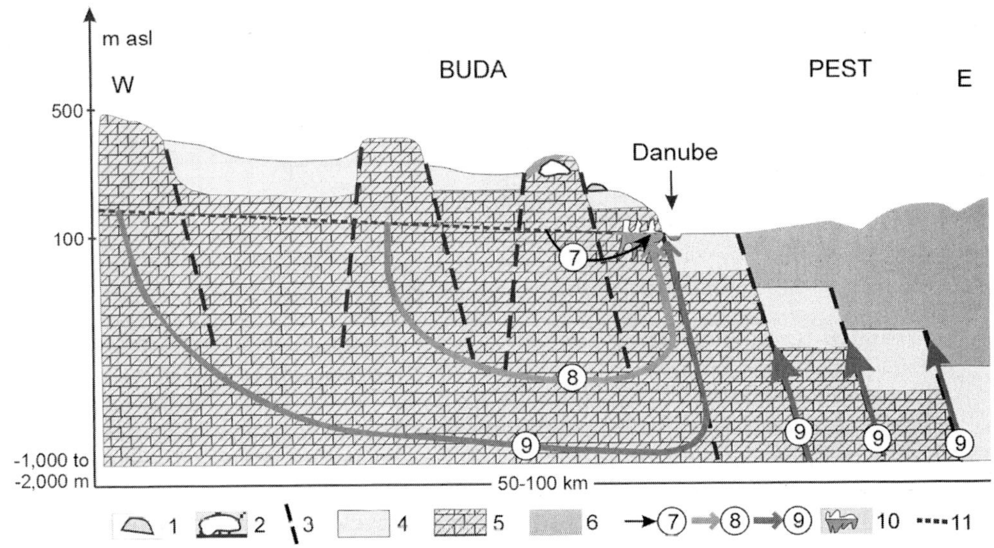

Figure 1.23 Generalized model for the flow systems of the Buda Thermal Karst (the western part of Budapest, Hungary. (1) travertine; (2) inactive, dry cave; (3) fault; (4) marls and clays; (5) carbonate rocks; (6) Neogene sediments; (7) local flow system; (8) intermediate flow system; (9) upward regional flow from the basement rocks; (10) active underwater cave; (11) karst water table. (Modified from Eröss et al., 2008.)

illustrated in Fig. 1.23, the already complicated hydraulics of groundwater flow in paleokarst (old karst) is further complicated by the ongoing karstification from both the infiltrating present-day cold water and the convective geothermal flow carrying deep hydrothermal acidic fluids such as sulfuric acid and hydrogen sulfide (Eröss et al., 2008; Jocha-Edelényi, 2005). The ascending thermal acidic groundwater creates the so-called hypogenic karst (e.g., see Klimchouk, 2011; Klimchouk and Ford, 2009) as opposed to exogenic (classic) karst created by descending groundwater from present-day infiltration. One of the reasons the Romans first colonized the area of present-day Budapest, capital of Hungary (called *Aquincum* in Roman times), is utilization of numerous thermal karst springs. There are still ruins visible today of the enormous baths that were built during that period. Many new baths were constructed during the Turkish period (1541 to 1686) and later so that today Budapest is indisputably the spa capital of Europe. The city is underlain by extensive cave systems developed at multiple levels including long spectacular passages submerged in thermal water (see color plates).

Karst aquifers developed in carbonate rocks are present on all continents and roughly 25% of the global population depends largely or entirely on karst for water supply. An overview of various types of carbonate rocks and their distribution in different parts of the world is provided by Ford and Williams (2007).

1.3 Porosity and Karstification

The nature of the porosity of all geologic porous media (sediments and all rocks, in general) is the single most important factor in determining the storage and movement of groundwater in any aquifer type. Many quantitative parameters describing "life cycle" of water and contaminants (when present) within a groundwater system directly or indirectly depend on porosity. Here are just a few: infiltration of precipitation into the subsurface, rock (sediment) permeability, groundwater velocity, volume of water that can be extracted from the groundwater system, and diffusion of contaminants into the rock matrix.

Porosity (n) is defined as the percentage of voids (empty space occupied by water or air) in the total volume of rock, which includes both solids and voids:

$$n = \frac{V_v}{V} \times 100\% \tag{1.1}$$

where V_v is the volume of all rock voids and V is the total volume of rock (in geologic terms, rock refers to all of the following: soils, unconsolidated and consolidated sediments, and any type of rock, in general). Assuming the specific gravity of water equals unity, total porosity, as a percentage, can be expressed in four different ways (Lohman, 1972):

$$n = \frac{V_i}{V} = \frac{V_w}{V} = \frac{V - V_m}{V} = 1 - \frac{V_m}{V}[\times 100\%] \tag{1.2}$$

where n is the porosity, in percent per volume; V is the total volume; V_i is the volume of all interstices (voids); V_m is aggregate volume of mineral (solid) particles; and V_w is volume of water in a saturated sample. Porosity can also be expressed as

$$n = \frac{\rho_m - \rho_d}{\rho_m} = 1 - \frac{\rho_d}{\rho_m}[\times 100\%] \tag{1.3}$$

where ρ_m is the average density of mineral particles (grain density) and ρ_d is the density of dry sample (bulk density).

The shape, amount, distribution, and interconnectivity of voids influence the permeability of rocks. Voids, however, depend on the depositional mechanisms of unconsolidated and consolidated sedimentary rocks, and on various other geologic processes that affect all rocks during and after their formation. Primary porosity is formed during the formation of rock. It includes voids between the grains of gravel and sand (Fig. 1.24), discontinuities between minerals in consolidated rocks (Fig. 1.25), or bedding planes of sedimentary rocks (Fig. 1.26).

The porosity of unconsolidated sediments (gravel, sand, silt, and clay) is often called intergranular porosity because the solids are loose detritic grains. When such rocks become consolidated, the former intergranular porosity is called matrix porosity. In general, the term matrix porosity is applied to primary porosity of all consolidated (hard) rocks, such as porosity between mineral grains (minerals) in granite, gneiss, slate, or basalt. Some unconsolidated or loosely consolidated (semiconsolidated) rocks may also contain fissures and fractures, in which case the nonfracture portion of the overall porosity is also called matrix porosity. Good examples are fractured clays and glacial till sediments, or residuum deposits that have preserved the fabric of the original bedrock in the form of fractures and bedding planes.

Sometimes, microscopic fissures in rocks are also considered part of the matrix porosity as opposed to larger fissures and fractures called macroporosity. In general, rocks that

Figure 1.24 Glacial outwash gravels north of Otis Lake, Oconomowoc quadrangle, Waukesha County, Wisconsin. August 23, 1904. (Photograph courtesy of USGS, 2011.)

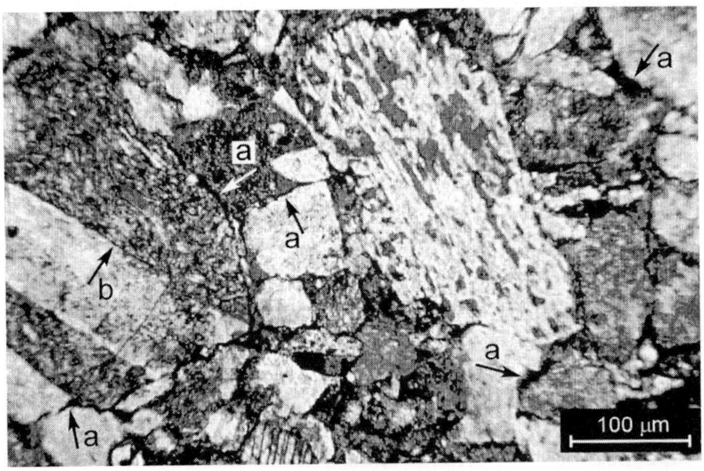

Figure 1.25 Photomicrograph of feldspar rich sandstone (arkose) collected from the Nutzotin Mountains of central Alaska. (a) *Arrows* point to pore spaces between individual sandstone grains. (b) *Arrow* points to an intragranular discontinuity between a feldspar crystal and surrounding ground mass. Note the scale of 100 microns. (Courtesy of Jeff Manuszak.)

FIGURE 1.26 *Top*: Limestone dissolution along bedding planes, Shalipayco, Peru. *Bottom*: Limestone bedding sculptured by the Li River, Guangxi Province, China.

have both the matrix and the fracture porosity are referred to as dual-porosity media. This distinction is important in terms of groundwater flow, which has very different characteristics in fractures compared with bulk of the rock. It is also important in contaminant fate and transport analysis, especially when contaminant concentrations are high causing its diffusion into the rock matrix where it can remain for long periods of time.

Secondary porosity is created after the rock formation, mainly due to tectonic forces (faulting and folding), which create micro- and macrofissures, fractures, faults, and fault zones in solid rocks (Fig. 1.27). In soluble carbonates, the secondary porosity is also

Figure 1.27 *Top*: Thrust fault and drag fold of thin-bedded limestone of the Nantoco formation about 50 m below the Gamma Member, Quebrada de Las Penas. Height of the cliff immediately underlying the fault plane is 3 to 8 m. Atacama Province, Chile. (Fig. 70.5 in U.S. Geological Survey Professional paper 450-C, 1962.) *Bottom*: Fractured Mesozoic limestone, the Beljanica mountain, eastern Serbia.

created by a variety of geochemical reactions between circulating fluids and mineral phases with dolomitization of limestone being one of the most important. Both the primary and secondary porosities can be successively altered multiple times, thus completely changing the original nature of the rock porosity. In general, total porosity decreases with depth in the upper portion of earth's crust. Intergranular and matrix porosity decrease mainly due to compaction, whereas decrease in the secondary fracture porosity is a combination of two main factors: (1) high fracture density near land surface and at shallow depths as the result of overrelaxation due to removal of the overlying solid matter by erosion (Fig. 1.28) and (2) lower fracture density and smaller fracture aperture at greater depths due to high pressures exerted by the overlying rock mass. In addition to its

FIGURE 1.28 *Top*: Densely fractured Mesozoic limestone, Durmitor National Park, Montenegro. *Bottom*: Houses built on fractured Mesozoic limestone in Split, Croatia.

general decrease with depth, the proportion of fracture porosity within the overall (bulk) volume of the aquifer is much lower than the volume of the aquifer occupied with matrix porosity.

Plots of average total porosity and porosity ranges for various rocks are shown in Figs. 1.29 and 1.30. As can be seen from the graphs, limestone, the principal rock of karst

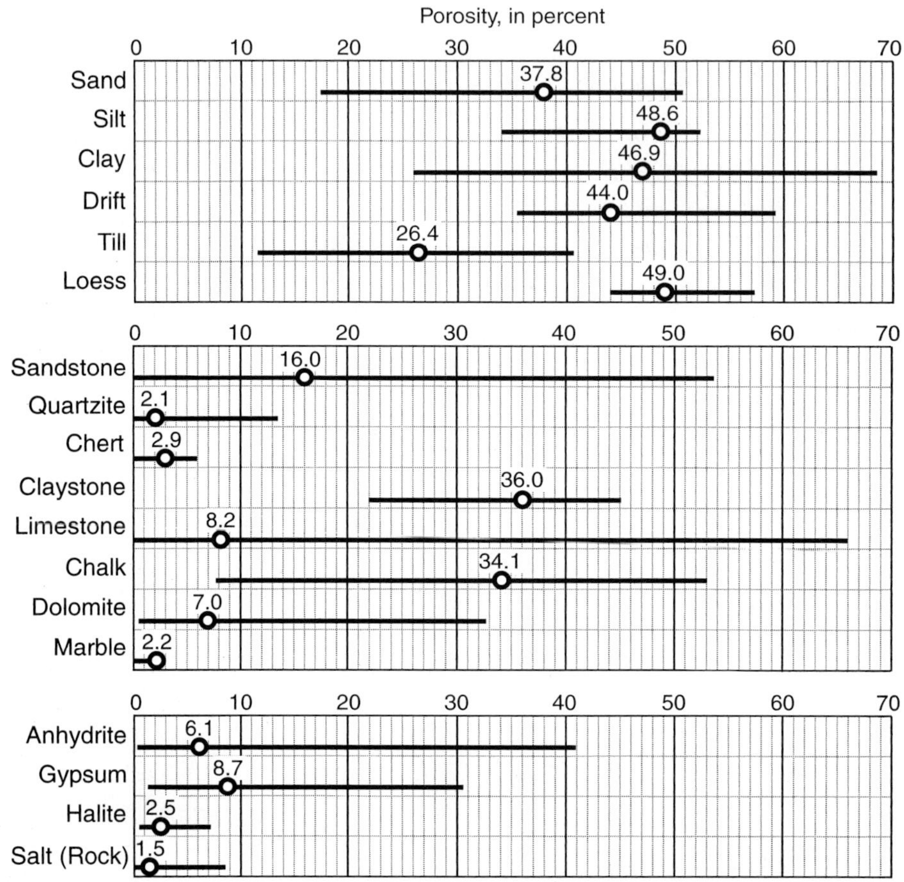

FIGURE 1.29 Porosity range (horizontal bars) and average porosities (circles) of unconsolidated and consolidated sedimentary rocks. (Kresic, 2007a; copyright Taylor & Francis Group, printed with permission.)

aquifers worldwide, has the widest porosity range of all rocks, thanks to many different depositional environments and processes of its diagenesis (conversion of carbonate sediment into consolidated rock). Since major oil fields around the world are hosted in limestone and dolomite reservoirs, there is abundant literature on the porosity, diagenesis, and classification of carbonate sedimentary rocks with classic works by Robert L. Folk (1959, 1980) and Robin G. C. Bathurst (1975) still holding on their own. Other excellent discussions on porosity of carbonate sediments and rocks are given by Scoffin (1987) and Moore (1989). In any case, hydrogeologists working in a particular karst terrain are strongly advised to thoroughly familiarize themselves with various aspects of aquifer porosity at their own site because any generalization in karst can potentially be erroneous.

Most importantly, all types of limestone as well as other soluble carbonate rocks such as dolomite, chalk, anhydrite, gypsum, and carbonate conglomerates are subject to

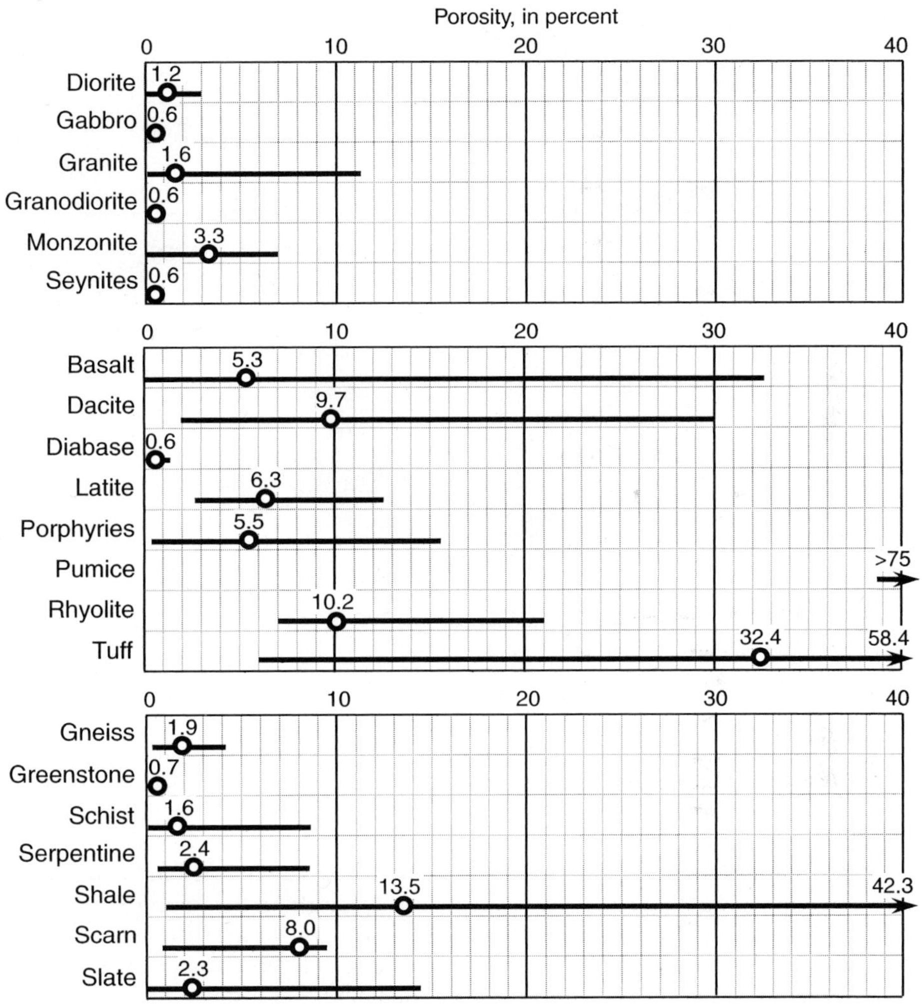

FIGURE 1.30 Porosity range (*horizontal bars*) and average porosities (*circles*) of magmatic and metamorphic rocks. (Kresic, 2007a; Copyright Taylor & Francis Group, printed with permission.)

karstification (dissolution of rock matrix), which typically increases aquifer porosity and storage capacity. The dissolution of rock by water infiltrating through secondary porosity features (Fig. 1.31) and by groundwater flowing in the saturated zone takes place in all soluble carbonate rocks. This is true regardless of their age and textural and mineral properties. All other factors being equal, different rocks will have different degree of karstification depending on the mineral composition and the presence of impurities (nonsoluble rock fraction). It is unfortunate, however, that some better known books on karst hydrogeology and geomorphology differentiate between limestones of different ages and state, without much elaboration, that "soluble rocks with extremely high primary porosity" usually have poorly developed karst. However, nothing can be farther

FIGURE 1.31 Example of dual-porosity media in karst. Fractures are enlarged by dissolution. Limestone landscape in northwest Ireland. (Photograph by George Sowers, printed with kind permission of Francis Sowers.)

from the truth as illustrated in Figs. 1.32 through 1.34. Enormous water-filled cave passages feed some of the world's largest springs draining young Eocene limestones of the Floridan aquifer (Fig. 1.32), which do have very high primary porosity, both total and effective (Fig. 1.33). Similarly, the longest submerged cave system in the world, with more than 170-mile-long explored passages so far, is developed in Yucatán, Mexico, with many more long underwater caves scattered throughout the young Miocene–Pleistocene limestones of the peninsula (see Fig. 1.17). Another example is karst development in the oölitic facies of the Pleistocene Miami Limestone of Florida (oöids are lime grains formed by chemical deposition around a detrital nucleus). Karstification is accompanied by calcite cementation, dissolution of the aragonite of the oöids, and the development not only of an oömouldic porosity but also of vugs with diameters from a few millimeters to several centimeters (Fig. 1.34; see also Bathurst, 1975; Stanley, 1966; and Robinson, 1967). Because of the karstification (dissolution) of already high primary oömouldic porosity, the Miami Limestone aquifer, also called the Biscayne aquifer, stores enormous volumes of groundwater and is extraordinarily prolific: Water supply wells often yield thousands of gallons per minute (hundreds of liters per second), sometimes with less than a foot (30 cm) of drawdown.

When analyzing porosity from the groundwater management perspective, it is very important to make a very clear distinction between the total porosity and the *effective porosity* of the rock. Effective porosity is defined as the volume of interconnected pore space that allows free gravity flow of groundwater. As opposed to younger Tertiary carbonates discussed earlier, older consolidated and compacted limestones of Mesozoic and Paleozoic ages typically have low primary porosity of 5% or less (Fig. 1.35), with the effective porosity of rock matrix still lower than that. However, significant volumes of

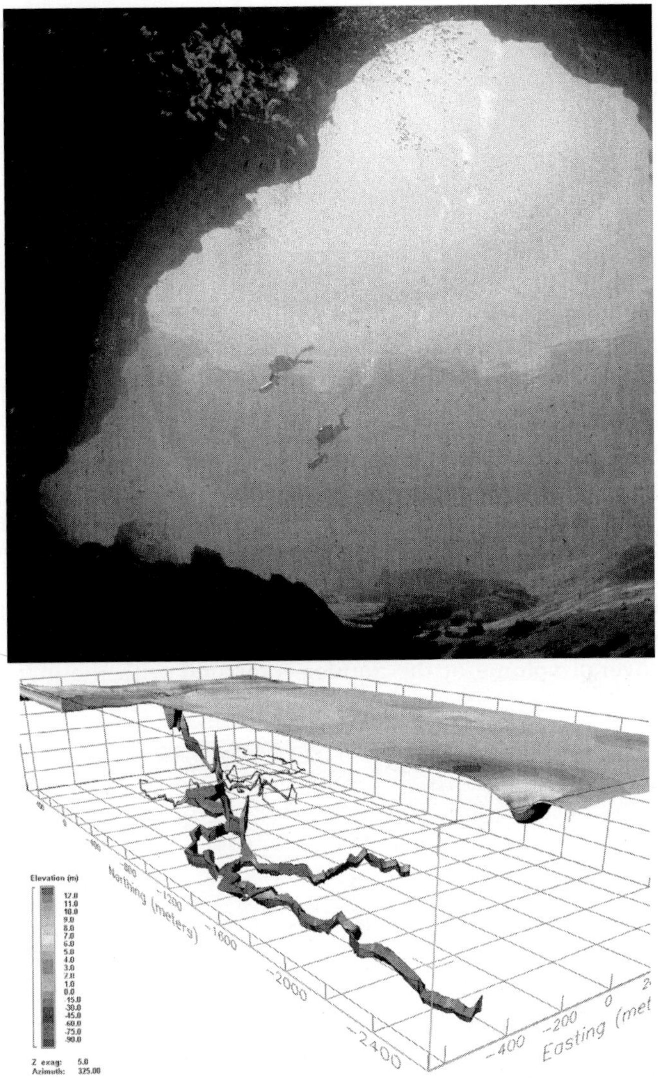

FIGURE 1.32 *Top*: View of the Wakulla Spring from cave diver's perspective. Tourist boats are visible 100+ ft above the divers. This first magnitude spring in Wakulla County, Florida, is one of the largest and deepest freshwater springs in the world with miles of explored submerged cave passages. Several of the early *Tarzan* movies starring Johnny Weissmuller were filmed in the Wakulla Springs state park. (Copyright David Rhea & Global Underwater Explorers 2006, printed with permission.) *Bottom*: Computer-generated view of the submerged Wakulla Cave passages based on the cave diving surveys. The passages all converge toward the Wakulla Spring. (Courtesy of Dr. Todd Kincaid.)

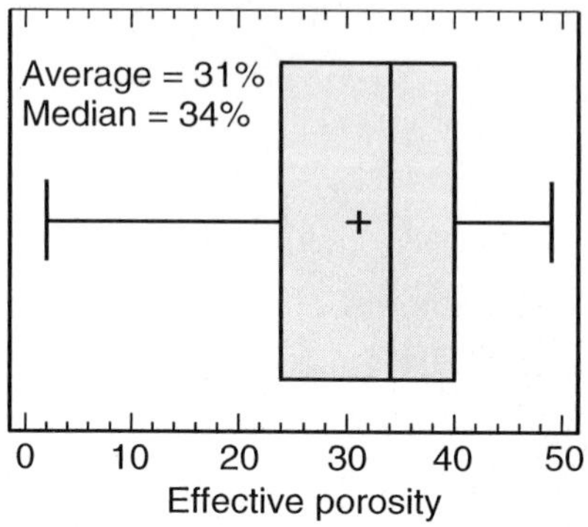

FIGURE 1.33 Box-and-Whisker plot of effective matrix porosity of the Upper Floridan carbonates in west-central Florida. The plot is based on analysis of 46 core samples from 10 different locations. (From Kresic, 2007a; raw data from Knochenmus and Robinson, 1996.)

water in such rocks can be stored in and flow through interconnected secondary porosity including fractures and various solution cavities formed along bedding planes and in faulted, jointed, and fractured zones (Figs. 1.36 through 1.39).

Although the interconnected (effective) secondary porosity enhanced by dissolution provides for most of the flow in older carbonate rocks, it still occupies a significantly smaller overall volume of the aquifer compared with the matrix (primary) porosity.

FIGURE 1.34 Large-diameter core from the Miami oolitic limestone (Biscayne aquifer), Miami, Florida. Hydraulic conductivity >1,000 ft/day. (From the George Sowers Collection, courtesy of AMEC Environment and Infrastructure, Inc.)

Figure 1.35 Cavities at a deep construction site in Paleozoic limestone, Hartsville, Tennessee. The matrix porosity is measured at 2%. (Photographs by George Sowers, printed with kind permission of Francis Sowers.)

FIGURE 1.36 Caves formed along fractures, Plitvice Lakes National Park, Croatia.

However, this general principle must be applied with great care at every karst site because the ratio between the total and effective porosities of any type (matrix, fracture, and solution) is scale dependent as illustrated in Fig. 1.40. Unfortunately, the concept of effective porosity is sometimes misinterpreted or misused in attempts to quantify and numerically model groundwater flow in karst with inadequate tools. For example, numeric groundwater models based on Darcy's law are still often used to model karst aquifers (see Sec. 6.5). This is justified by stating that the karst aquifer behaves as an equivalent porous medium (EPM) because all three porosity types are relatively uniformly distributed and interconnected at a particular scale (the Site) and therefore of equal importance for groundwater flow. It follows that the nature of groundwater flow through the overall karst aquifer volume is the same as, for example, through gravel deposits (see Fig. 1.24) and can therefore be described with the simple Darcy's equation (see Sec. 1.6.1).

At the same time, practitioners applying this approach may in many cases be confronted with the fact that groundwater levels in some monitoring wells are rising and falling very quickly in response to recharge episodes, whereas this is not the case in other wells nearby. They also may acknowledge that the linear groundwater velocity (v_L) is very fast in certain portions of the aquifer as indicated by a dye-tracing test. Note that the linear groundwater velocity in the EPM models is calculated as

$$v_L = \frac{K \times i}{n_{\text{eff}}} \tag{1.4}$$

Figure 1.37 Former entrance into a small natural cave formed in a fault zone, enlarged into a mining exploration tunnel (adit), Shalipayco, Peru. (Courtesy of Marc Etienne, AMEC.)

where K is the hydraulic conductivity of EPM, i is the hydraulic gradient, and n_{eff} is the effective porosity.

The problem arises when some karst modeling practitioners simulate high groundwater velocities in the preferential flow paths using a very small effective porosity in the model cells representing these paths, smaller than in the surrounding cells that are supposed to simulate aquifer matrix without cavities. This, of course, does not make any hydrogeologic sense as illustrated in Fig. 1.40: Cell B cannot have effective (or total) porosity smaller than cell A regardless of how convincing any related "expert" explanation or modeling justification may be. (Needless to say, both the total and the effective porosity of an empty karst cavity is 100%.)

Another question is the validity of publications that provide very exact numeric proportions of different porosity types for different karst aquifers. In other words, in

Figure 1.38 Two caves formed along major fractures and faults. *Left*: Dry overflow cave of the upper Lison Spring, France, active during peak flows. (Photograph courtesy of Dr. Zoran Stevanovic, Center for Karst Hydrogeology, University of Belgrade). *Right*: Petnica Cave near Valjevo, western Serbia.

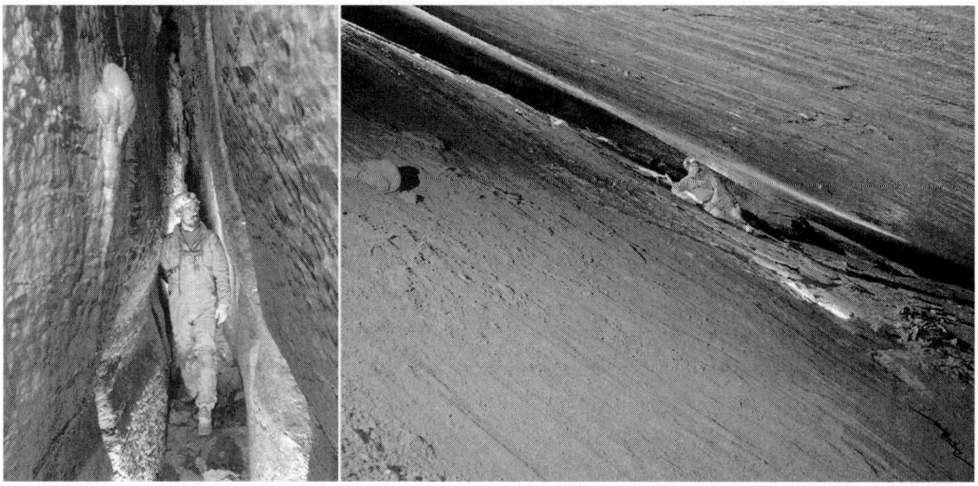

Figure 1.39 Cave passages formed along fractures and faults, Bath County, Virginia. *Left*: Canyon along fracture in Water Sinks Cave. *Right*: Slickensides, polished and grooved fault surfaces in Helictite Cave. (Photographs courtesy of Phillip Lucas.)

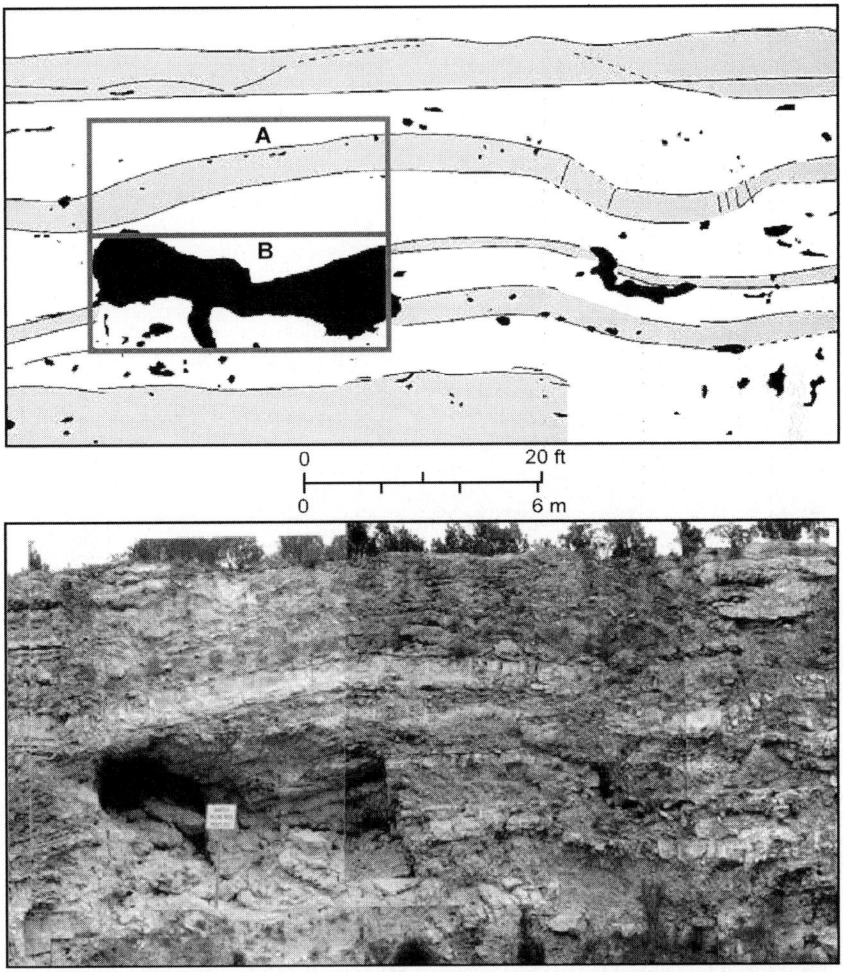

Figure 1.40 Interpretation of conduit porosity on a road cut in New Braunfels, Texas. Each outlined box is 25 × 10 ft, which is not an uncommon dimension for cells of local numeric models. Matrix porosity is reported at 1% to 4%. (Modified from Hovorka, 2009.)

many such publications, it is unclear how the exact volume of karst cavities and the exact volume of the karst aquifer in question were determined. The latter can sometimes be reasonably assumed based on three-dimensional contacts between carbonates and nonsoluble rocks defined after spending large funds on complex field investigations including extensive drilling and geophysics. However, it would be very difficult to determine the exact volume of karst cavities within the aquifer even after spending a lot more money. Klimchouk (2011; see Table 1.3) provides an illustrative quantitative comparison between estimated channel porosity of two different genetic types of cave systems based on actually surveyed cave passages. He also lists and discusses various assumptions used in the analysis. Based on these calculations, both genetic types of caves have significantly higher channel porosity, expressed as percentage of the bulk aquifer volume,

Cave	Length (km)	Area of Cave $m^2 \times 10^6$	Volume of Cave $m^3 \times 10^6$	Area of Cave Field (km^2)	Volume of Rock ($m^3 \times 10^6$)	Specific Volume (m^3/m)	Passage Density (km/km^2)	Cave Porosity (%)	Areal Coverage (%)
"Common" caves—speleogenesis in unconfined settings									
Blue Spring Cave, Indiana, USA, Carboniferous limestones	32.0	0.146	0.5	2.65	119.34	15.6	12.07	0.42 (0.08)	5.5 (1.1)
Mammoth Cave, Kentucky, USA, Carboniferous limestones	550.0	1.386	8.0	36.78	3310.2	14.5	14.95	0.24 (0.09)	3.77 (1.4)
Friars Hole System, WV, USA, Carboniferous limestones	70.0	0.3	2.7	4.37	349.92	38.6	16.00	0.77 (0.28)	6.86 (2.5)
Krasnaya Cave, Crimea, Ukraine, Jurassic limestones	17.3	0.064	0.27	0.74	37.0	15.5	23.23	0.15	8.55
Maze caves—speleogenesis in confined settings									
Jewel Cave, South Dakota, USA, Carboniferous limestones	148.0	0.67	1.49	3.01	135.63	10.00.0	49.11	1.10	22.20
Wind Cave, South Dakota, USA, Carboniferous limestones	143.2	0.43	1.13	1.36	61.0	7.9	105.68	1.86	31.73
Knock Fell Caverns, N. Pennines, UK, Carboniferous limestones	4.0	0.006	0.012	0.02	0.12	3.0	170.94	10.26	25.64
Fuchslabyrinth Cave, Germany, Triassic limestones (Muschelkalk)	6.4	0.0058	0.007	0.03	0.15	1.1	217.61	4.80	19.55
Moestroff Cave, Luxembourg, Triassic limestones (Muschelkalk)	4.0	0.004	0.0035	0.01	0.05	0.9	406.09	7.14	40.61

Botovskaya Cave, Siberia, Russia, Lower Ordovician limestones	23.0	0.067	0.104	0.11	1.37	4.5	201.75	7.62	58.51
Estremera Cave, Madrid, Spain, Neogene gypsum	3.5	0.008	0.064	0.06	0.71	18.3	59.32	9.04	13.56
Optimistychna Cave, W. Ukraine, Neogene gypsum	188.0	0.26	0.52	1.48	26.03	2.8	127.03	2.00	17.57
Ozerna Cave, W. Ukraine, Neogene gypsum	111.0	0.33	0.665	0.74	13.2	6.0	150.00	5.04	44.59
Mlynki Cave, W. Ukraine, Neogene gypsum	24.0	0.047	0.08	0.17	2.38	3.3	141.18	3.36	27.65
Kristalna Cave, W. Ukraine, Neogene gypsum	22.0	0.038	0.11	0.13	1.82	5.0	169.23	6.04	29.23
Slavka Cave, W. Ukraine, Neogene gypsum	9.0	0.019	0.034	0.07	0.98	3.7	139.14	3.47	29.05
Verteba Cave, W. Ukraine, Neogene gypsum	7.8	0.023	0.047	0.07	0.66	6.0	117.82	12.00	34.74
Atlantida Cave, W. Ukraine, Neogene gypsum	2.5	0.0045	0.0114	0.02	0.29	4.5	168.00	4.00	30.00
Ugryn Cave, W. Ukraine, Neogene gypsum	2.1	0.004	0.008	0.01	0.14	3.8	176.67	5.71	33.33
Jubilejna Cave, W. Ukraine, Neogene gypsum	1.5	0.002	0.0035	0.01	0.08	2.3	277.78	4.00	37.04

From Klimchouk, 2011, printed with permission.

TABLE 1.3 Characterization of Cave Patterns and Porosity in Unconfined Versus Confined Aquifers. Values in Brackets for the First Three Caves are Those Obtained by Worthington (1999) Using a "Rectangular" Method for Delineation of Cave Fields

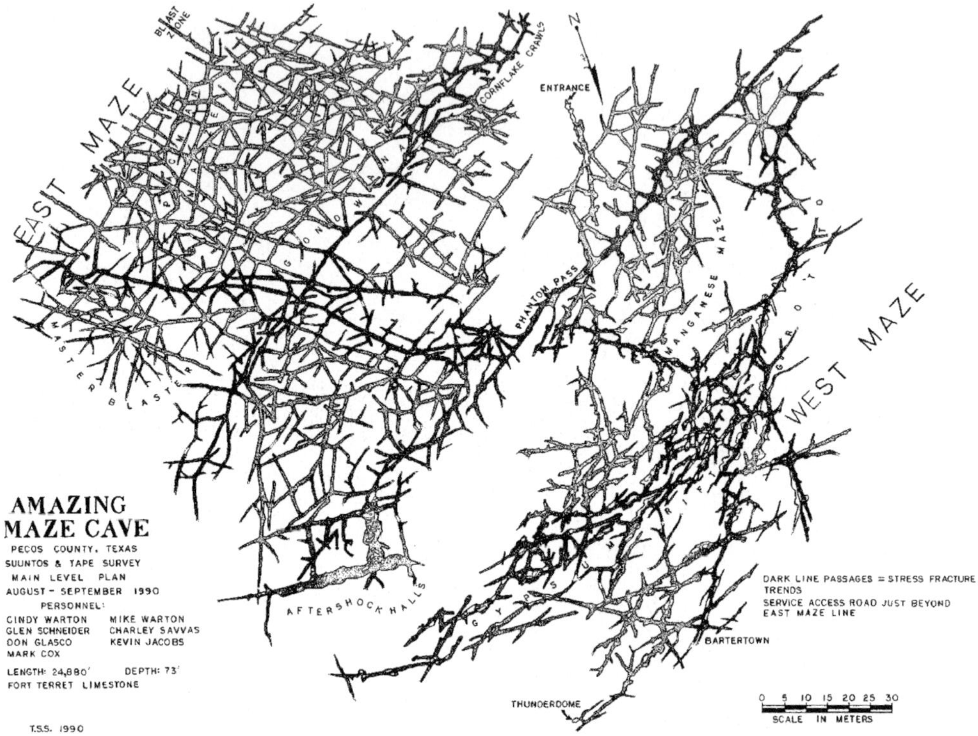

FIGURE 1.41 Amazing Maze Cave near Bakersfield in Pecos County, 30 miles east of Fort Stockton is the third longest in Texas with 4.7 miles of surveyed passages. The cave is developed in Fort Terret limestone most likely by hypogene speleogenesis similar to Guadalupe Mountains caves such as Carlsbad Caverns and Lechuguilla in New Mexico. (Courtesy of Texas Speleological Society.)

compared with some other estimates found in literature (e.g., see Ford and Williams, 2007; Worthington and Ford, 2009).

Without submerging in a rather academic discussion about validity of inevitable assumptions needed for quantifying proportions of the three porosity types in karst, the author wishes to emphasize the experience of many dedicated spelunkers and cavers worldwide. Some of them have witnessed, in their lifetime, a virtual "growth" of an accidentally discovered small cave few tens of meters long, to a major cave system with hundreds of channels at varying depths and of widely varying width and height, with the combined length of several miles to hundreds of miles (e.g., see Fig. 1.41). The estimates of the channel porosity proportion within the overall porosity of "their aquifer" would have been much different when they were young and inexperienced.

It is certain that in all similar cases, there are many more channels and cavities of varying sizes, including those permanently submerged, that will never be discovered even by the most dedicated cavers and cave divers. What is more important, however, is to realize that every karst site, from a 10-acre industrial facility to a watershed covering tens of square kilometers, is unique in terms of spatial distribution and relative importance of the three porosity types. Again, every generalization in karst based on some published

Figure 1.42 Erica Sughrue and Josh Tate in "M Survey" Blue Spring Cave, the longest in Tennessee. This dry passage extends in similar dimensions for over 1 mile in the 36-mile-long cave system. Note thick sediment deposit at the bottom of the dry channel. A book describing this complex system titled Blue Spring Cave has recently been published by Larry E. Matthews. (Photograph copyright Innermost Imagery, printed with permission; many more photos are available at http://innermostimagery.com.)

books or papers is potentially misleading and may result in project failure, regardless of the project goals.

One aspect of porosity in karst aquifers often receiving less attention is infill of cavities with unconsolidated sediments (see Figs. 1.42 and 1.43). This transient type of porosity may change depending on the hydraulics of groundwater flow and recharge mechanisms such that the loose sediments may be flushed out of cavities, new sediments deposited, and everything in between. Filling of karst cavities may also be more consolidated and therefore more permanent. However, under certain circumstances such as due to constantly changing and high hydraulic pressures, even these deposits may break down and be washed out. This is of particular concern when designing and constructing dams and reservoirs in karst (Fig. 1.44). Considering the presence and nature of intergranular porosity is also important for other aspects of karst aquifer management including quality of water supply and contaminant fate and transport.

1.3.1 Predictability and Depth of Karstification

In the past, the author has attended meetings of stakeholders evaluating the feasibility of groundwater remediation at several large federal facilities underlain by carbonate rocks

Figure 1.43 *Left*: Sheet-like deposit of solid calcium carbonate was formed over unconsolidated clayey sediments that were subsequently washed away leaving this nice suspended speleothem. (Photograph courtesy of John Ackerman, Minnesota Cave Preserve.) *Right*: An enthusiastic spelunker searching for a new cave passage that may be blocked by the sticky mud. (Photograph courtesy of Phil Lukas.)

Figure 1.44 Left spillway face, Green River Dam. Note filled convoluted karst features dissecting horizontal limestone beds. (Photograph by George Sowers, printed with kind permission of Francis Sowers.)

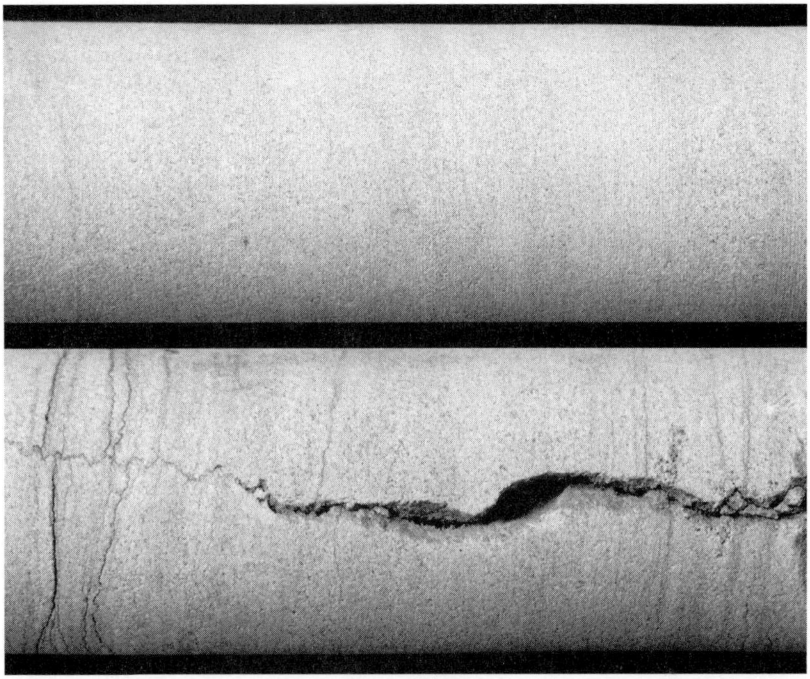

Figure 1.45 Two core samples from the same boring completed in Paleozoic limestone near Lexington, Kentucky. Core diameter is 4.5 cm (approximately 2 in.).

in one of the southern United States. At one such meeting, a consultant working for a federal client brought and passed around a boring core sample similar to the one shown in Fig. 1.45, top. On the basis of that core sample, the consultant qualified the aquifer underlying the Site as nonkarstic and therefore relatively easy to conceptualize. The groundwater at the Site was and still is contaminated with dissolved chlorinated solvents together with dense nonaqueous phase liquids. To date, the regulatory agency overseeing the Site characterization and remediation efforts refuses to use the word karst for the underlying aquifer. Instead, the aquifer is referred to as consisting of "partially weathered rock" and (deeper) "competent bedrock." More than 100 million U.S. dollars have been spent at the Site so far on various investigation and remediation activities spanning more than 20 years. Incidentally, the results of quite a few field investigation efforts including drilling and borehole geophysics revealed the presence of highly transmissive zones deep in the "competent bedrock" where the tested hydraulic conductivity of discrete borehole intervals exceeded 1,000 ft/day. In addition, great difficulties during drilling of some borings were reported including drop of drilling tools, loss of drilling fluid, and unstable/collapsing borehole walls. Finally, everyone involved in the project accepted the fact that the aquifer beneath the site is drained by a single large spring, which, on average, discharges more than 30 million gallons of water every day. Yet, the word karst is all but banned at the Site.

In contrast, two drilling core samples shown in Fig. 1.45 are from one boring completed in carbonate sediments near Lexington, Kentucky, the state in which citizens and environmental regulators alike understand the meaning of the word karst. Dozen or so

FIGURE 1.46 Road cut in karstified Paleozoic limestone near Lexington, Kentucky. The inset photograph, 2 ft across, shows a solution feature on the cut.

borings completed in limestone at this particular site were continuously cored up to 150-ft depths, but only a handful of solution features similar to the one shown in Fig. 1.45 (bottom) were found. Nevertheless, no one involved in the project had to be persuaded that the underlying aquifer is karstic because, among other things, visiting the site involved driving through road cuts such as the one shown in Fig. 1.46.

The preceding two discussions serve to emphasize the following main point about predicting locations of karst features in the subsurface: *The only predictable thing about karst is that it is unpredictable.* This is true regardless of the project scale and investigation funds available as expertly discussed by Milanovic in several major books on karst hydrogeology and water resources engineering in karst (e.g., Milanovic, 1979, 1981, 2004, 2006). For example, Fig. 1.47 shows that surprises can happen even after extensive and expensive investigations are performed by professionals well versed in the nature of karst. Borings no. 1 and no. 4 at a future dam site encountered "single small voids," whereas borings no. 2 and no. 3 were completed in solid limestone. A grout curtain designed to deal with "limited" karstification failed to achieve results even after injecting disproportionally more grout than what was designed. A 20-m-long exploration gallery (adit) was then excavated into the dam abutment and discovered a cave with a large chamber up to 20-m high. The cave was surveyed in detail resulting in redesign of the grout curtain. As can be seen, parts of the cave almost completely surround boring no. 3, which was completed entirely in solid limestone without any signs of dissolution voids of any kind.

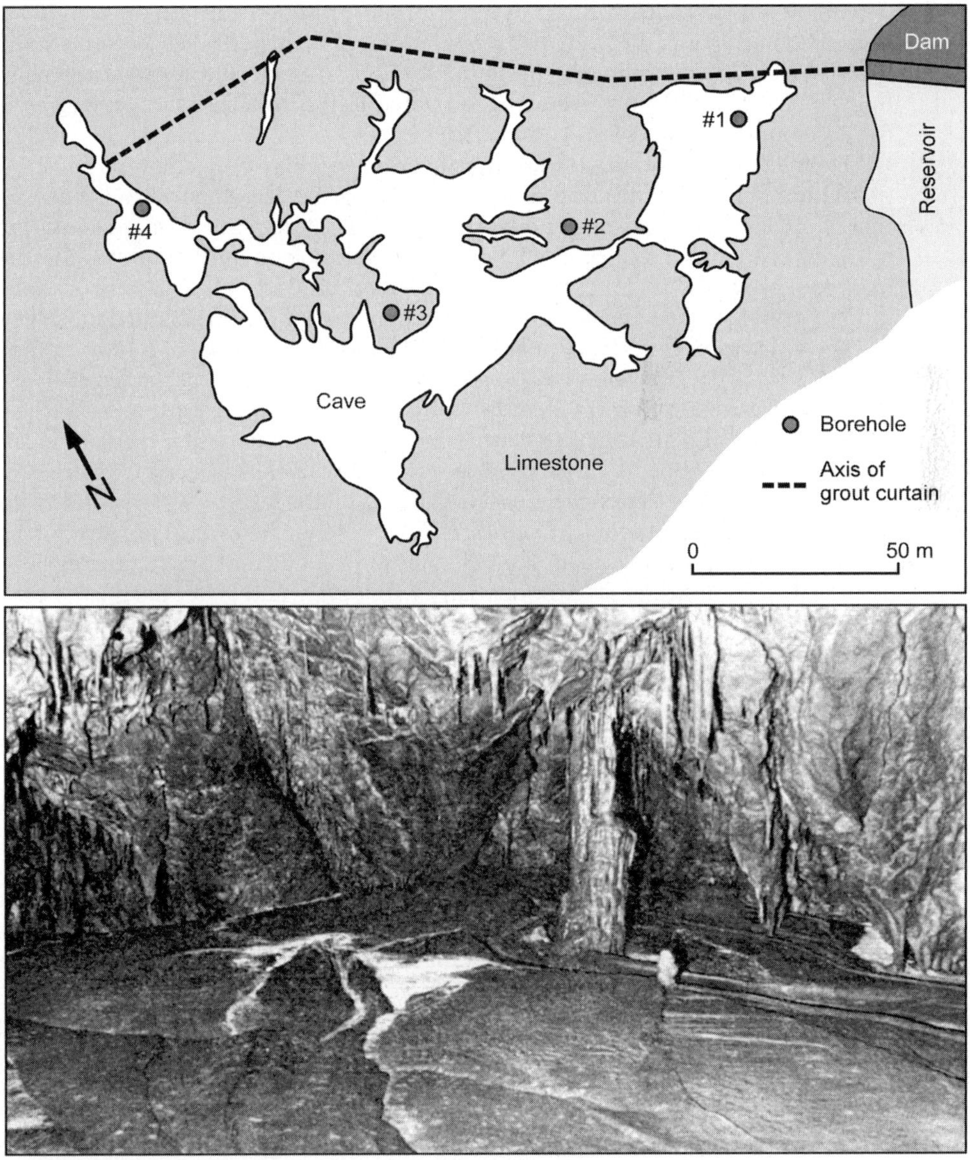

FIGURE 1.47 Cave in the left abutment of Sklope Dam, Croatia, discovered with an exploration gallery excavated after extensive grouting failed. Bottom photo shows the grout mass on the cave floor. (Modified from Božičević, 1971; photograph courtesy of Srećko Božičević.)

Unfortunately, some environmental regulators and agencies in the United States are the main drivers behind endless data collection at contaminated karst sites in a false hope that all risks of a perceived incomplete hydrogeological site characterization can be eliminated only if enough resources and funds were expended. The main contribution from karst hydrogeologists in such circumstances is to design and execute scientifically defensible site investigation and, as appropriate, present the case against endless drilling to find contaminants that may be hiding in one of the mysterious karst features at their site. This, by default, includes discussions about what constitutes unacceptable risks to human health and the environment as related to the issue at hand, as well as what are the acceptable project costs.

Defining depth of karstification at every particular project site is as uncertain as predicting spatial distribution of karst cavities or finding them. It has been proposed that, similarly to the primary porosity, the degree of karstification generally decreases with depth and reaches the point below which solution openings (karst voids) are absent (e.g., see Ford and Williams, 2007, p. 115; Milanovic, 2004, p. 10). This base of karstification is an irregular surface due to various heterogeneities of the carbonate deposits including uneven fracturing of the rock mass. Karstification also deepens in time as the level of aquifer discharge is lowered. The levels (locations) of aquifer discharge, such as a major surface stream or a spring at the contact with nonsoluble rocks, are lowered due to normal erosion of the land surface. In relatively thin carbonate deposits, the karstification will eventually reach nonsoluble rocks underlying the carbonates (see Fig. 1.19).

As suggested by Milanovic (2004) and based on the results of permeability investigations from 146-deep boreholes in eastern Herzegovina, the zone of greatest total porosity including karstification is from the land surface to depths of approximately 10 to 20 m. The karstification decreases exponentially with depth, and at depths greater than 300 m, the rock mass can be considered nonkarstified at regional scale. Locally, in tectonic zones, the karstification can extend deeper than 300 m, but these zones do not represent a significant rock volume and are expressed in two dimensions. The largest volume of karst voids (storage) is developed within the zone of water table fluctuation. Karstification decreases below the lowest groundwater levels and the base of karstification and the minimum water table level coincide. In general, the slope of the base of karstification leads to the zone of aquifer discharge. The most active karst channels are directly above the base level. The existence of karst drains below this level is not excluded, but they are rare and of limited transport capacities (Milanovic, 2004).

Although the above-proposed mechanism of karstification may be applicable to certain situations, it does not explain numerous documented hydrogeological systems developed in thick carbonate sequences that extend well below locations of aquifer discharge and the minimum water table. In fact, the largest springs in the world are of ascending type with deep confined conduits having extraordinary transport capacities often exceeding 10 m^3/s. Their drainage areas are measured in hundreds of square kilometers or more. Quite a few of these springs, also called *vauclusian* springs after the Fontaine de Vaucluse, the source of river Sorgue in Provance, France, have been explored by cave divers in recent years worldwide. The ascending character and the existence of a deep siphon at the Fontaine de Vaucluse were acknowledged by French explorers in the nineteenth century (Fig. 1.48). As of 2011, the vertical submerged conduit of the spring was explored to a depth of 315 m (950 ft) using a small submarine robot called MODEXA 350. The camera of the robot showed a sandy floor at this depth. The spring is capable of producing more than 200 m^3/s after heavy rainfall (Showcaves, 2011). General schematic of deep-sited karst conduits below minimum water table forming ascending springs is shown in Fig. 1.49. Jovan Cvijić understood this nature of groundwater flow in karst very well when developing his concepts (see Fig. I.4 in the Introduction).

Understanding and predicting vertical development of karstification is greatly complicated by the fact that sea level has fluctuated significantly in geologic time due to

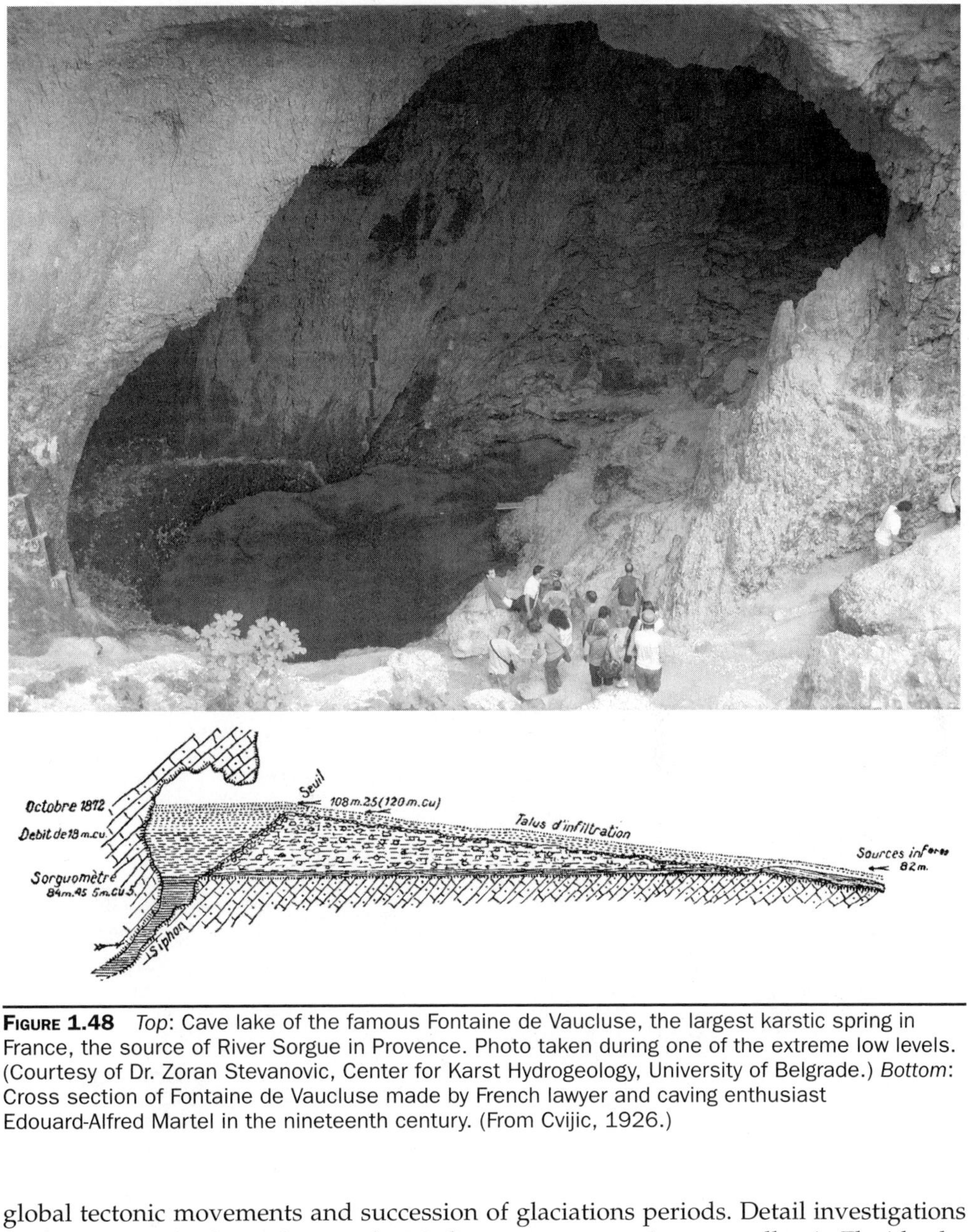

Figure 1.48 *Top*: Cave lake of the famous Fontaine de Vaucluse, the largest karstic spring in France, the source of River Sorgue in Provence. Photo taken during one of the extreme low levels. (Courtesy of Dr. Zoran Stevanovic, Center for Karst Hydrogeology, University of Belgrade.) *Bottom*: Cross section of Fontaine de Vaucluse made by French lawyer and caving enthusiast Edouard-Alfred Martel in the nineteenth century. (From Cvijic, 1926.)

global tectonic movements and succession of glaciations periods. Detail investigations of coastal karst hydrogeology in the Mediterranean countries—as well as in Florida, the United States, and Yucatán, Mexico—show vertical extent of karstification tens of meters below present-day sea level and an abundance of submerged, off-coast springs. Cave diving in coastal areas also revealed permanently submerged and extensive cave passages with speleothems that could have been formed only when exposed to air (Fig. 1.50).

Paleokarst is a term usually used to describe karst features (voids, cavities) at any depth explained by the burial of carbonate sediments below nonsoluble rocks. The

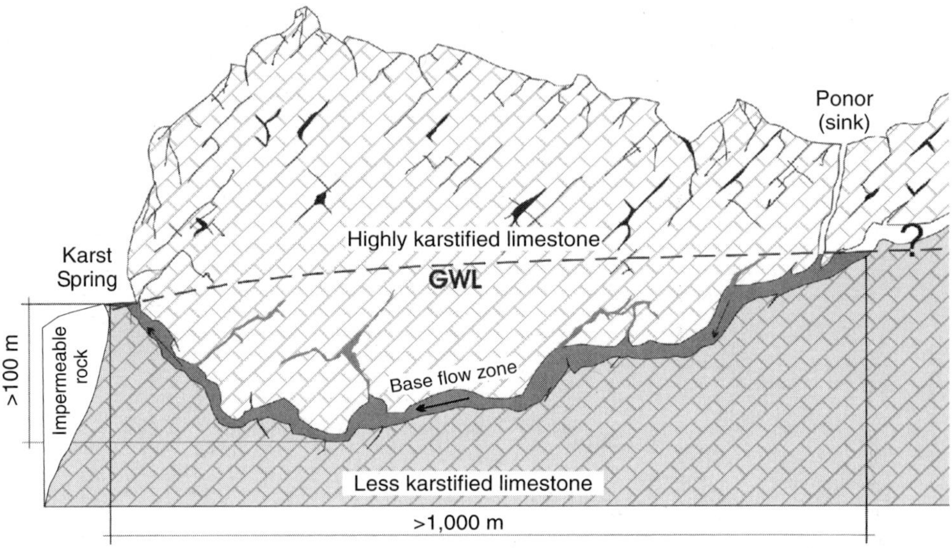

FIGURE 1.49 Common type of siphon karst springs developed in thick open hydrogeologic structures. GWL, groundwater level. (From S. Milanovic, 2005, printed with permission.)

FIGURE 1.50 Cave divers in submerged passages of the Nohoch Nah Chich in the Yucatán Peninsula. The abundant speleothems (stalactites, stalagmites, flowstone, and columns) were formed prior to the cave's submergence. (Photograph courtesy of David Rhea, Global Underwater Explorers.)

carbonates were karstified at earlier times when exposed at the surface or when in near-surface settings. These paleocavities may subsequently be filled with clastic deposists or may remain open. In some cases, the existence of karstification at great depths (several kilometers or so) can be explained by the fact that the same general area with carbonate rocks may have been subject to multiple periods of karstification depending on the geologic and tectonic history. In thick karst aquifers developed in active geologic environments, it is often possible to find very permeable zones, together with large karst conduits, at greater and varying aquifer depths. These zones may indicate position of a paleo water table where karstification intensity was the highest.

Open karst cavities at greater depths may be preserved even where pressure increases to lithostatic values, due to the high rigidity of limestone and dolomite. For example, Andreychouk et al. (2009) report that karst cavities more than 2 m in size in Devonian and Carboniferous limestone and dolomite were encountered at depths of 0.5 to 1.5 km on the western slope of the Kizelovski basin and the eastern slope of the North Urals oil and gas basin. Drops of drilling tools by 3 to 4 m and intense losses of drilling mud were encountered at depths of several kilometers when drilling carbonate strata underlying rock salt in the Peri-Caspian depression. Abundant cavities and karstified zones were identified at a depth of ca. 3 km in limestone and dolomite of the pre-Jurassic basement in Western Siberia. Smaller karst forms (vugs, caverns, solution-enlarged fractures) were detected in drill-hole cores retrieved from great depths in many petroliferous basins of the world. In southern Mangyshlak, the Triassic limestone and dolomite of the middle volcanogenic-carbonate unit are intensely karstified at a depth of 4 km and below. Their effective cavernosity ranges between 7.4% and 21.5%. Enhanced porosity and cavernosity was reported for Middle Carboniferous and Early Permian limestones at depths of 4 to 5 km with effective porosity up to 14% in many such zones. A general review of the data from North America reveals that Mesozoic and Paleozoic carbonate rocks may exhibit enhanced secondary porosity, up to 15% to 20%, at depths up to 10 km (Andreychouk et al., 2009).

Thanks to advances in caving and a tireless race to discover the deepest cave in the world without any help from drilling, it has now been documented that intensive karstification passable by humans can advance to great depths (see Fig. 1.51). At the time of this writing, the deepest mapped cave in the world (Krubera or Voronya Cave) is currently in the western Caucasus, Abkhazia, Georgia; its depth is 2,191 m (7,188 ft), which means that the unsaturated zone and the depth of karstification are at least that deep.

As concluded by Andreychouk et al. (2009), available data unequivocally indicate that karst can develop throughout the entire section of the earth's crust that is accessible by drilling. The distribution of karst voids at different depths is irregular. Relatively large cavities, with dimensions exceeding meters, occur in the top few kilometers. Smaller pores, vugs, and enlarged fractures prevail deeper. The authors note that this spatial heterogeneity can be exaggerated, at least partly, by the technical difficulties of identifying large cavities at great depths. Nevertheless, the available data from many regions of the world strongly suggest that the cavernosity of carbonate rocks within the 4- to 5-km-depth interval is increased relative to that of the overlying units.

Andreychouk et al. (2009) also present an appealing explanation for the presence of very deep karstification commonly regarded to be the result of dissolution by descending (infiltrating) water and CO_2 in it. This traditional concept of karst formed by groundwater flowing due to hydrostatic pressure or gravity in general is often referred to as *exogenic*

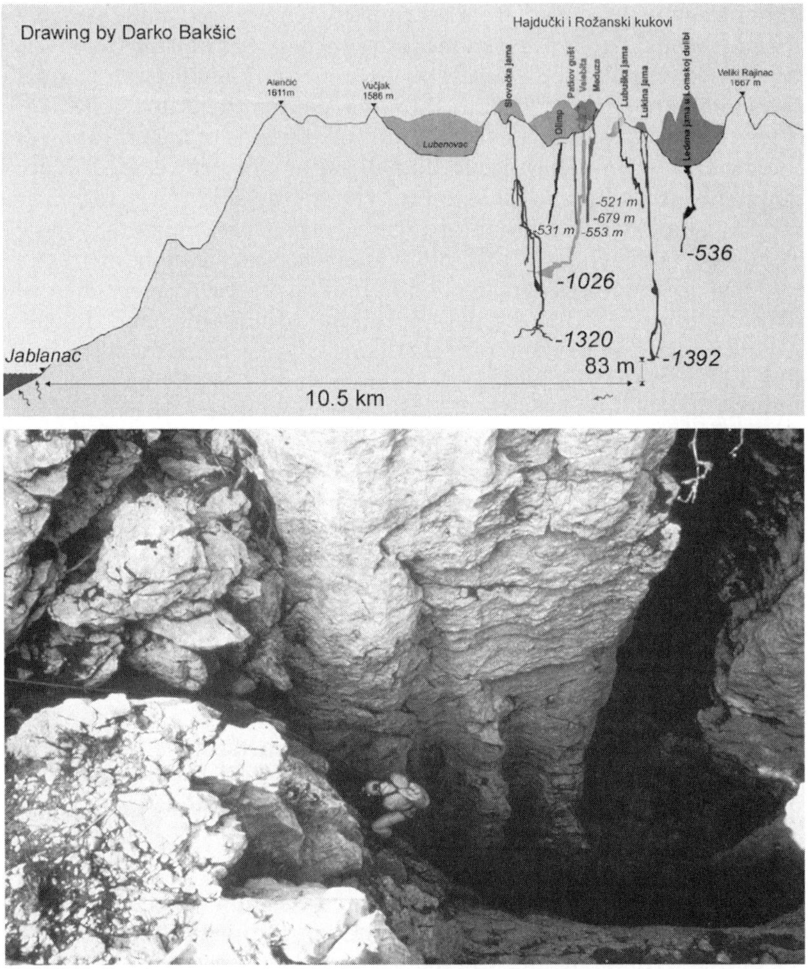

FIGURE 1.51 *Top*: Deep jamas (potholes) on the Velebit Mountain, Croatia. Posted depths are from the 2009 survey. As of August 2010, the depth of Lukina Jama, the deepest Croatian pothole, is extended to 1,421 m including 40 m of the surveyed submerged passage at the bottom. This part of the Dinaric karst is drained by numerous coastal and submarine springs in the Adriatic Sea such as The Jablanac (shown schematically on the left). Drawing by Darko Bakšić, courtesy of Croatian Speleological Server (www.speleologija.hr). *Bottom*: Spelunker entering the 1,421-m-deep Lukina Jama in the Velebit National Park, Croatia. (Photograph courtesy of Vlado Božić.)

karst. The authors argue that deep karstification can also be the result of dissolution by ascending fluxes of thermal liquid–vapor fluids, of endogenic and metamorphic origin. Listed below are key facts in support of this *endogenic* karst concept as discussed by Andreychouk et al. (2009):

- Lithostatic zone (LSZ) comprises the lowest parts of the earth's crust that have been studied (to a depth >12 to 15 km). The pressure in this zone is controlled by lithostatic load and by endogenic sources. A characteristic feature of the LSZ is the decreased density of rocks due to permanent natural hydro-fracturing. The data from very deep drilling suggest that hydro-fractured rocks are concentrated in certain horizons and zones along the migration pathways of the deep-seated fluids. The degree of porosity and fracturing in such zones is significantly higher than in the overlying mesozone, and may reach the values characteristic of the shallow karst (hydrostatic zone). Because of hydro-fracturing, "traditional" aquitard and aquiclude rocks lose their confining properties in the LSZ.

- Waters in the LSZ belong to sodium bicarbonate type and have relatively low mineralization. They are enriched in specific deep-seated components, such as CO_2, He, F, and As, some of which have characteristic isotopic properties. Compared with their counterparts from near-surface settings, the fluids in the LSZ have anomalously high aggressiveness with respect to many sedimentary, metamorphic, and igneous rocks. The predominantly ascending character of the fluxes of this water ensures removal of the dissolved matter from the LSZ.
- The development of karst in the upper, exokarst zone is primarily governed by external factors, such as the infiltration and inflow of meteoric water, the expulsion of formation waters under the pressure of the overlying rocks, by topographic relief, and climate. In the lower, endokarst zone not only the carbonates but also the silicate and aluminosilicate rocks, traditionally considered only weakly or very weakly soluble, can be subject to intense dissolution by fluxes of ascending deep-seated fluids. For example, one of the most abundant and stable resistant forms of silica under crustal conditions, quartz, at $T = 300°C$ to 350°C and $P = 200$ to 250 MPa, acquires an aqueous solubility similar to that of gypsum and anhydrite in typical near-surface conditions.
- The depth of penetration of infiltration waters into the earth's crust is finite, so that below some given depth the infiltration concept is no longer applicable. In any case, this concept cannot be valid at depths where pressure is much higher than hydrostatic ("suprahydrostatic"), because these zones represent impermeable barriers for gravity-driven infiltration waters.

In a series of landmark publications, Palmer (1991, 2007), Klimchouk (1992, 1994, 2011), and Klimchouk et al. (2000) firmly established the concept of hypogenic karst, previously largely ignored by karst hydrogeologists trained in traditional karst settings. As mentioned earlier, the well-established concept, applicable to many open hydrogeologic structures, explains the karstification to be the result of carbonate rock dissolution by infiltrating (descending) waters enriched with soil and atmospheric CO_2, the so-called exogenic or epigenic karst. In contrast, hypogenic karst is created by ascending waters that, in addition to CO_2, may carry upward other acidic fluids such as sulfuric acid and dissolved H_2S, the origin of which is deep (endogenic) such as due to metamorphism, magmatic, and tectonic processes. Dissolution of carbonates by ascending groundwater is enhanced in semiconfined and confined hydrogeologic settings where cross-formational diffuse flow of the acidic fluids can take place over large areas and create karst cavities by slowly "eating away" the rock mass (Fig. 1.52). This process can result in formation of very convoluted and extensive karst cavities along previously weakened portions of the rock (such as due to fracturing and faulting) as well as in relatively competent rock. Hypogenic karst does not have to be deep and it is not formed only by endogenic

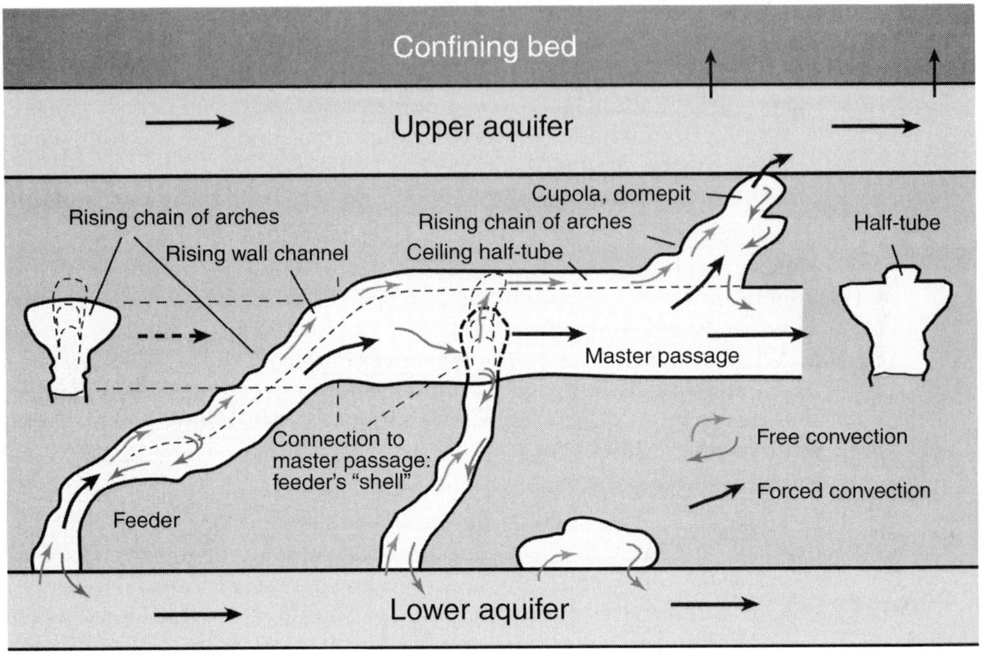

FIGURE 1.52 Morphologic suite of rising flow, diagnostic of hypogenic transverse origin of caves. (Modified from Klimchouk, 2011, printed with permission.)

fluids such as sulfuric acid. The sufficient triggers for its development are an ascending groundwater flow and some degree of the carbonate aquifer confinement with less permeable rocks. In many cases, there will be an overlap between the ascending and descending flows, especially at the locations of regional groundwater discharge.

A perfect example of a very complex mixture of endogenic, exogenic, and paleokarst, all in one spectacular setting, is deep buried karst aquifer in the Yellowstone National Park, the United States. Karstified limestones outcrop in a small area at the base of the Terrace Mountain, known as the Hoodoos, couple of miles south of the Mammoth Springs. Their extent below thousands of feet thick, young volcanic ash deposits and lava flows is unknown. Although the park is doted by countless hot springs and geysers, some of which discharge several thousand gallons per minute, the only visible discharge from limestone is at the Boiling River cave shown in Fig. 1.53. It is virtually certain that, in addition to Mammoth Springs (Fig. 1.54), the origin of at least some of the other spectacular hot springs is related to the deep-sited karst aquifer. Hot water rich in H_2S discharges under pressure from this primary groundwater reservoir via fractures and faults in the overlying volcanic rocks and feeds the thermal springs, some of which have very deep fascinating pools.

1.3.2 Surface and Subsurface Karst Landforms

All landforms specific to karst terrains, visible at the land surface and in the subsurface (Fig. 1.55), are the result of carbonate rock dissolution (karstification) to varying degrees.

FIGURE 1.53 Boiling River discharging from a limestone cave at the bottom of the Yellowstone River valley near Mammoth Springs, Yellowstone National Park. This thermal spring is one of the largest in the United States; the flow rate is estimated between 0.5 and 1 m^3/s (exact flow measurements are not available). The water temperature is 140°F (60°C). The small patch of limestone is uncovered by down cutting of the Yellowstone River through thousands of feet thick volcanic ash and multiple lava flows that buried the carbonate sediments. Bottom photo shows part of the collapsed cave ceiling behind the spring.

Other processes, such as mechanical erosion by flowing water, may play an important role in shaping a specific karst landscape, but the chemical dissolution of the carbonate rock mass is always the primary factor. Solution features on pure carbonate rock outcrops directly exposed at the land surface look very similar in different geographic and climatic conditions. Their scale and the actual morphology may vary, but the imprint of dissolution on the rock surface is unmistakable (Figs. 1.56 through 1.59). Secondary porosity of the rock, such as fissures and fractures, enhances dissolution and may result in a very

Figure 1.54 One of the Mammoth Springs in Yellowstone National Park, Wyoming. These thermal springs with high content of H_2S gases are of ascending type and drain a small portion of the deep buried karst aquifer, via faults through overlying magmatic rocks. The main discharge of the karst aquifer in this general area of Yellowstone is at the Boling River Spring shown in Fig. 1.53.

chaotic, dissected, and almost impassable *barren rocky terrain* (the origin of word *karst*; see Fig. 1.57). When taken to extremes, the dissolution of pure carbonates can create meters to decameters deep grooves and mini-canyons as illustrated in Fig. 1.60. Limestone and dolomite containing considerable fraction of sand and clay will commonly not display landscape shown in the preceding photographs. Instead, chemical weathering of impure carbonates will produce a layer of residuum soil consisting of insoluble minerals. This residuum layer is of highly variable thickness and transitions into the underlying karstified carbonates more or less abruptly (Fig. 1.61). Narrow zones of concentrated rock dissolution, called slots or chimneys, are usually filled with the residue and may not be visible at the surface (Fig. 1.62). The residuum layer and the upper, highly weathered (karstified) portion of the underlying rock as shown in Figs. 1.61 and 1.62 are sometimes together called *epikarst*.

Sinkholes and Other Depressions

Sinkholes are the signature of karst terrains. They usually form at mechanically weakened fracture intersections where water can easily percolate and flow deeper into the underlying rock carrying away dissolved carbonates. In time, this initial weak spot enlarges due to the continuing carbonate dissolution and a rudimentary solution sinkhole, usually of a circular shape, begins to form (Fig. 1.63). Together with the progression of

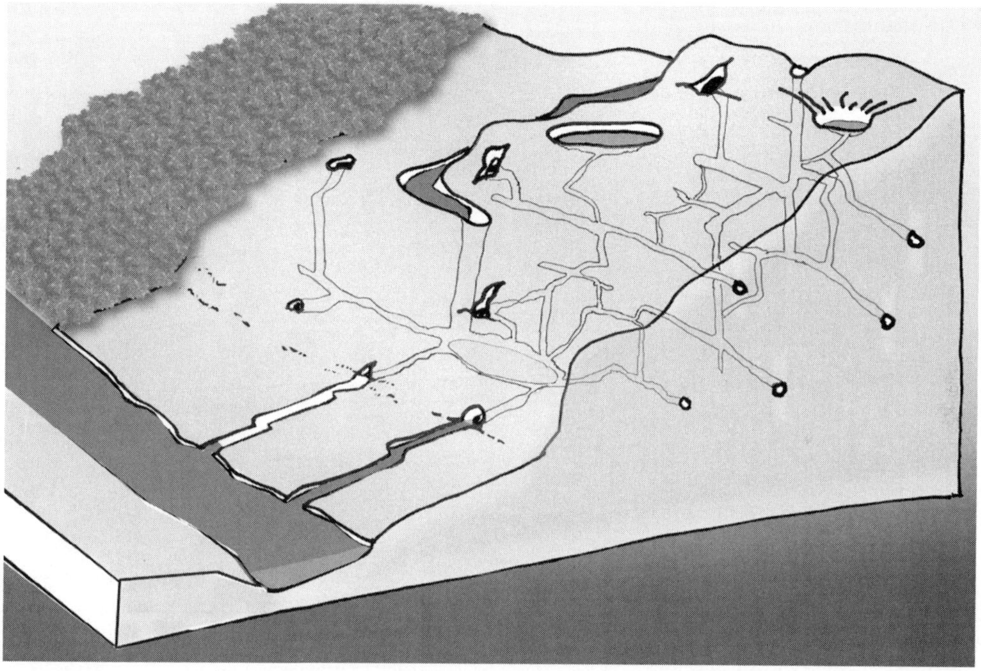

Figure 1.55 Sketch of most common karst landforms. Closed depressions at the land surface such as sinkholes and uvalas direct water into the subsurface and a maze of potholes and cave passages. Sinking stream brings more water to the underground drainage that feeds springs emerging from caves and discharging to a major stream. Two former springs at higher elevation are left dry by incision of the river and another one is active only after heavy rainfall. An area like this can be relatively barren or forested, depending on local climate and land use. (Courtesy of Marin Krešić.)

carbonate dissolution and development of a closed depression that collects an increasing amount of surface runoff, an insoluble residue may accumulate at the sinkhole bottom and its sides sometimes completely masking solid rock. Depending on the mineral composition of the parent rock (more or less impurities), the residue can form a continuous soil layer in between the sinkholes as well. In such case, the only evidence that there is consolidated carbonate rock below is the very presence of sinkholes. Taken to extremes, in pure, thick limestone deposits, sinkholes can grow to be hundreds of meters in diameter and more than hundred meters deep, sometimes encroaching onto each other (Fig. 1.64). However, a number of additional factors may be involved in the formation of sinkholes (Figs. 1.65 through 1.67), which therefore can have many different shapes, dimensions, and hydrologic functions (Figs. 1.68 through 1.70).

Cover-subsidence sinkholes are formed in areas where limestone is covered by relatively incohesive and permeable materials such as sand. Under these conditions, individual grains of sand move downward in sequence, replacing grains that have themselves moved downward to occupy space formerly held by the dissolved limestone. Spalling of incohesive sand into solution cavities developed along joints in the limestone may cause subsidence due to upward migration of the narrow cavities, the so-called piping

FIGURE 1.56 Karrens formed on inclined limestone surface by dissolution from rainfall. Note a hat in the lower left for scale. (From Cvijić, 1926.)

FIGURE 1.57 Intensely karstified pure limestone in Makarska, Croatia. Landscape like this is commonly called "angry karst" in Croatia, Herzegovina, and Montenegro, the lands of classic Dinaric karst.

FIGURE 1.58 Karstified dolomite on Sierra del Torcal, Spain. (Photograph by George Sowers, printed with kind permission of Francis Sowers.)

(Fig. 1.67, left). Areas where sand cover is as much as 50- to 100-ft thick may develop cover-subsidence sinkholes that are only a few feet in diameter and depth (Back and Sinclair, 1986).

Collapse sinkholes above cavities in both the unconsolidated cover and the karstified carbonate rocks present serious natural hazard in karst terrains. Clay component in the

FIGURE 1.59 *Left*: Corroded surface of dolomite near Shalipayco, Peru; photo is 1 m across. *Right*: Karren field on limestone near Ponoarele village in the Mehedinti karst area, Romania; photo is 1 m across. (Right photo courtesy of Dr. Zoran Stevanovic, Center for Karst Hydrogeology, University of Belgrade.)

FIGURE 1.60 *Left*: Detail from karst plateau above Karuc spring in Cetinje, Montenegro. Note human figure for scale. (Courtesy of Dr. Zoran Stevanovic, Center for Karst Hydrogeology, University of Belgrade.) *Right*: Karst Stone Forest in Shilin County, 90 km from Kunming, China. (Photograph by George Sowers, printed with kind permission of Francis Sowers.)

limestone cover provides a degree of cohesiveness that allows it to bridge a developing cavity in the limestone (Figs. 1.66, left, and 1.67, right). The result of failure of the bridge is a cover-collapse sinkhole whose dimensions are related to the size of the cavity and the bearing strength of the clay. A thick layer of dense clay may have sufficient strength to bridge a large-diameter cavity whose roof gradually crumbles and falls into a solutionally

FIGURE 1.61 Slots in limestone, road cut near Knoxville, Tennessee. (From the George Sowers collection, courtesy of AMEC Environment and Infrastructure, Inc.)

Figure 1.62 Slots in limestone, road cut east of Nashville, Tennessee. (From the George Sowers collection, courtesy AMEC Environment and Infrastructure, Inc., printed with permission.)

Figure 1.63 Rudimentary (initial) solution sinkhole, 4 ft across, forming in pure limestone near Makarska, Croatia. Compare with satellite image of the same general area shown in Fig. 1.64.

FIGURE 1.64 Astronaut photograph of the Biokovo mountain range on the Adriatic coast of Croatia, approximately 20-km across. Mega sinkholes dot this part of the classic Dinaric karst. The coastal town of Makarska is denoted with M. Bright white color at the mountain tops in the left part of the photograph is from snow. The photograph was acquired on December 29, 2008, with a Nikon D2Xs digital camera fitted with a 180-mm lens. It is provided courtesy of the ISS Crew Earth Observations experiment and the Image Science & Analysis Laboratory, NASA Johnson Space Center.

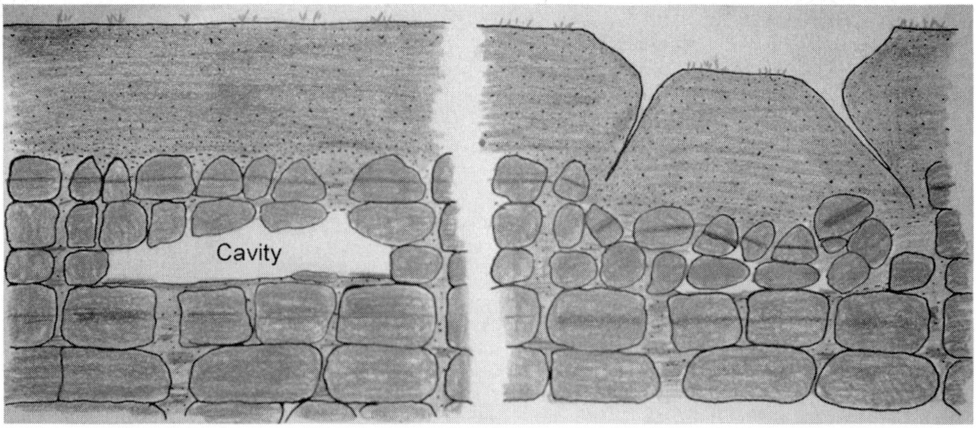

FIGURE 1.65 Schematic of sinkhole formation due to collapse of cavity ceiling. (From the George Sowers Collection, courtesy of AMEC Environment and Infrastructure, Inc.)

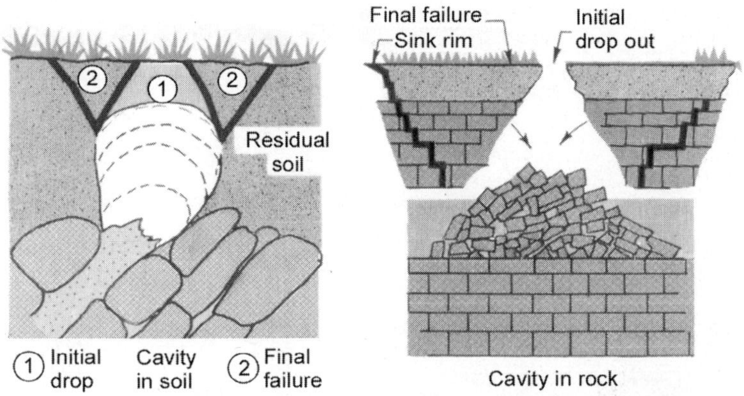

FIGURE 1.66 Examples of sinkholes formed by dome collapse in residual soil (*left*) and cavity collapse in consolidated carbonates (*right*). The initial drop in either case leaves an opening narrower than the cavity below. The usual sinkhole shape is attained after the final failure of the overhang. (From the George Sowers Collection, courtesy of AMEC Environment and Infrastructure, Inc.)

enlarged joint in limestone. When the clay roof finally collapses, the resulting sinkhole will likely form abruptly and will be relatively large (Beck and Sinclair, 1986). Subsequent collapse of the remaining roof and the side erosion of the unconsolidated materials will give the sinkhole its final common shape. Occasionally, very large sinkholes may form in the unconsolidated, insoluble limestone cover that is several hundred feet thick (Fig. 1.70).

Over time, the concentrated infiltration of rainfall in the sinkhole can form the so-called sinkhole throat, an enlarged solution feature that connects the sinkhole with the

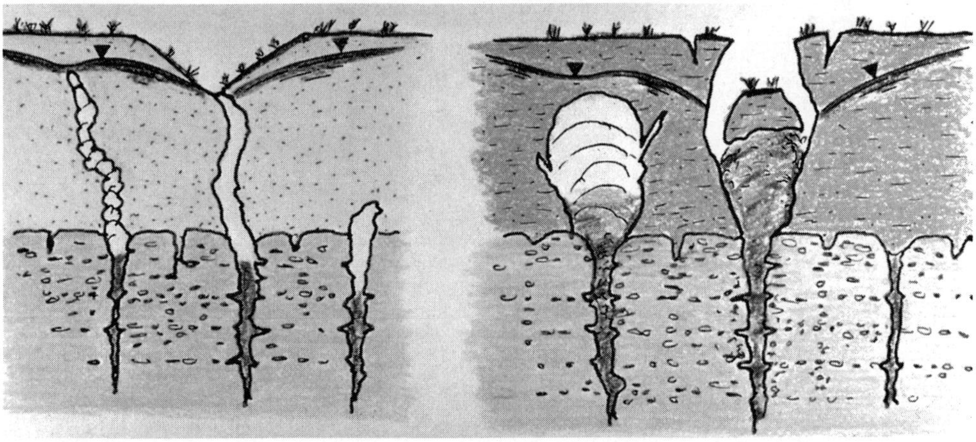

FIGURE 1.67 Sinkhole formation due to seepage erosion and raveling above karstified limestone. *Left*: Hourglass piping in clean sand. *Right*: Doming and dome collapse in clayey sand and sandy clay. (From the George Sowers Collection, courtesy of AMEC Environment and Infrastructure, Inc.)

FIGURE 1.68 Sinkholes of varying size and hydrologic function in different parts of the world all have one thing in common: They are the main signatures of the underlying karst aquifers. (A) Solution sinkhole with impermeable residue sediments at the bottom, serving as a cattle watering hole, Montenegro. (B) Solution sinkhole used for farming, western Serbia. (C) Deep sinkhole in classic Dinaric karst near Imotski, Croatia, approximately 130 m across, used as pasture (from Cvijić, 1926). (D) Shallow undrained (plugged) solution sinkhole characteristic of St. Louis limestone, Scottsville-Bowling Green Pike, 11 miles west of Scottsville, Allen County, Kentucky, 1917 (courtesy of USGS, 2011). (E) Late Jim Quinlan pointing at a collapse sinkhole, Mammoth Cave National Park, Kentucky. (F) Filled solution sinkhole near Barceloneta, Puerto Rico. (G) One of the largest sinkholes in the United States, formed by limestone cover collapse in Shelby County, Alabama, in December 1972. (H) Aerial view of a cover collapse sinkhole, south Hillsborough County, Florida. (Photographs G and H from the George Sowers Collection, courtesy of AMEC Environment and Infrastructure, Inc.)

FIGURE 1.69 Blue Lake in Imotski, Croatia, is one of the most impressive sinkholes in the classic Dinaric karst. As seen by the watermarks, the level of the karst aquifer (lake surface) fluctuates substantially. The lake occasionally dries up at the end of dry summers. It spilled over the sinkhole rim several times in the recorded history. Depth of the sinkhole is 190 m. See color plates for the nearby Red Lake, the deepest sinkhole in the world.

underlying cavities in the karstified rock. As illustrated in Figs. 1.71 through 1.73, the throat, which is often masked by the residuum sediments and rock debris, may be passable leading into a pit (pothole) that is sometimes part of a larger, more complex karst system of passable cavities including horizontal caves.

Polje is a large, closed depression with flat bottom and an internal drainage developed in thick mature karst. By convention, the area of a polje has to be larger than 1 km^2. When smaller, a closed flat-floored depression in karst is sometimes called *uvala* or just a large sinkhole. Two or more sinkholes can also combine to create an uvala. Poljes are especially well developed in the classic Dinaric karst, Greece and Turkey, but are also present in other carbonate platform settings around the world. Some of the world's

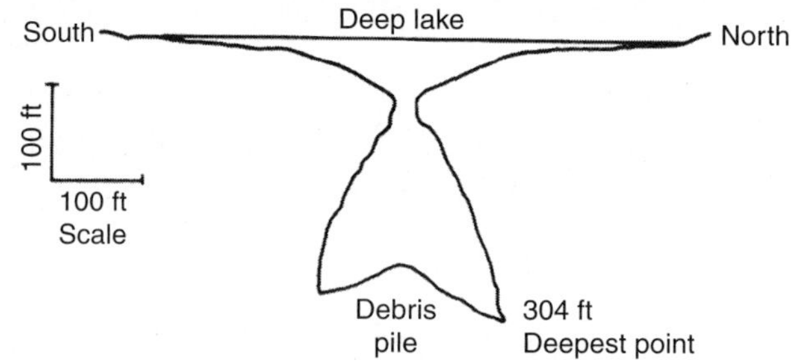

FIGURE 1.70 *Top*: Perfectly round sinkhole lake near Orlando, Florida (from the George Sowers Collection, courtesy of AMEC Environment and Infrastructure, Inc.). *Bottom*: Cross section of a sinkhole caused by collapse beneath Deep Lake near Arcadia, Florida. Survey by members of the National Speleological Society, Cave Diving Section. (From Back and Sinclair, 1986; based on I. S. Exley, written communication, 1982.)

highest karst poljes known to the author are in the Andes of Peru at elevations exceeding 4,600 m. Major faulting is usually the main reason for polje development as illustrated in Fig. 1.74. The tectonically formed depression is continuously filled with sediments brought in by colluvial and fluvial processes. In addition, as in the case of the Dinaric karst, some of the larger poljes were submerged in Neogene so that the total thickness of the combined lake and other unconsolidated polje deposits may exceed 1,000 m in some cases.

As the polje develops, karstification of the surrounding and underlying carbonate rocks continues resulting in its unique hydrologic role within the overall karst system (Fig. 1.75). During seasonally high water levels in the karst aquifer, or after heavy rainfall, the aquifer may drain into the polje (it literally spills over) via intermittent springs creating surface streams. Depending on the springs' discharge rate, saturation of the

Figure 1.71 *Left*: A sinkhole after the soil fill and loose rocks have been partially removed. The uncovered sinkhole opening ("throat") allowed John Ackerman to explore over 2 miles of cave passages. The cave was subsequently purchased by the Minnesota Cave Preserve. *Right*: This sinkhole opened suddenly after a torrential rain. John Ackerman (on right) reached a cave floor 65 ft from the surface and went on to explore over 4 miles of large passages. Exploration continues as this cave system owned by the Minnesota Cave Preserve reveals its secrets. (Photographs courtesy of John Ackerman, http://www.karstpreserve.com.)

Figure 1.72 Perfectly vertical pitholes can form below sinkholes and be masked by the residual fill. Excavation of the fill in this sinkhole led into a deep pit connecting with miles of horizontal passages. (Photograph courtesy of John Ackerman, http://www.karstpreserve.com.)

Figure 1.73 Neversink in Alabama is frequently visited by brave cavers and spelunkers. This grand pit is probably partially formed by a sinkhole collapse as evidenced from a large pile of debris at the bottom. (Photograph courtesy of Geary Schindel.)

polje deposits, and the swallowing capacity of the perimeter ponors (sinks), the inflow of groundwater may eventually flood the polje creating a temporary lake (Fig. 1.76). As the aquifer level drops below the elevation of the polje surface, the lake is drained via ponors and estavelles (former springs now acting as sinks) along the polje perimeter as well as via alluvial sinks on the polje floor and in streambeds. This somewhat simplistic succession of events can in reality be complicated by a variety of geologic and hydrogeologic factors including contacts between karst and nonkarst. A polje can have permanent spring(s) on the upper side, forming a permanent stream that sinks into a permanent sink (ponor) on the polje's down-gradient perimeter. The word permanent in the preceding sentence can be substituted with the word temporary (intermittent) to describe another polje, and a combination of these words can be used to describe yet another polje.

Some poljes are flooded only during exceptionally wet seasons, whereas others are always flooded regardless of the seasonal amount of snowmelt and rainfall recharging

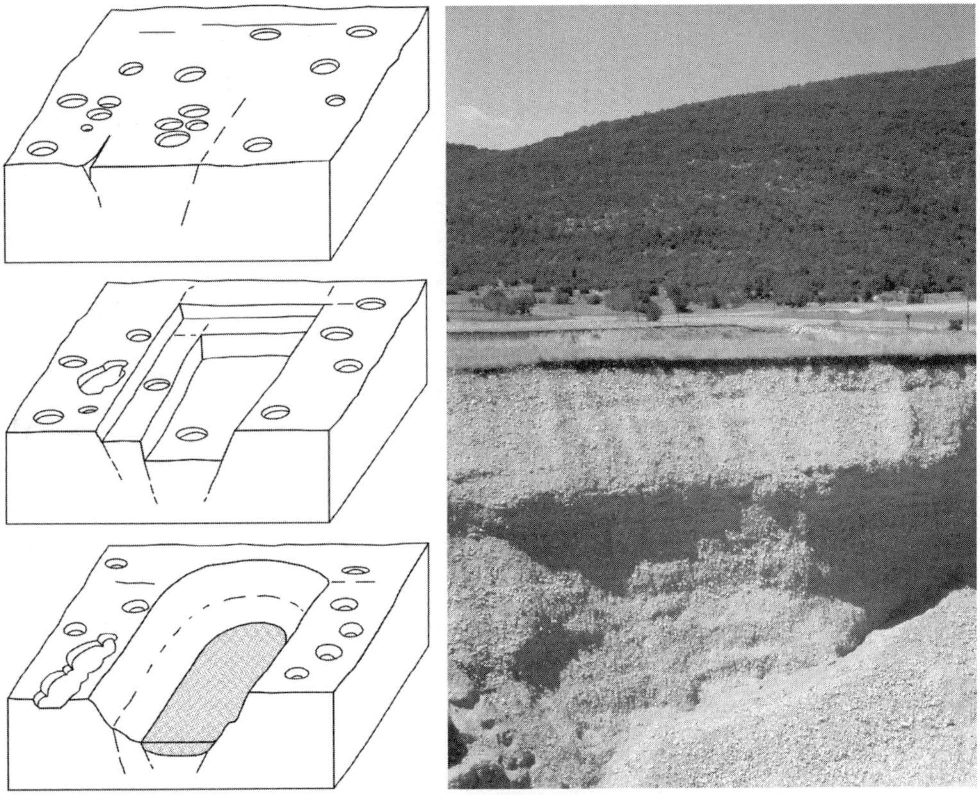

Figure 1.74 *Left*: Schematic of typical karst polje formation. *Right*: Thick gravel deposits in the Dabar Polje, Herzegovina.

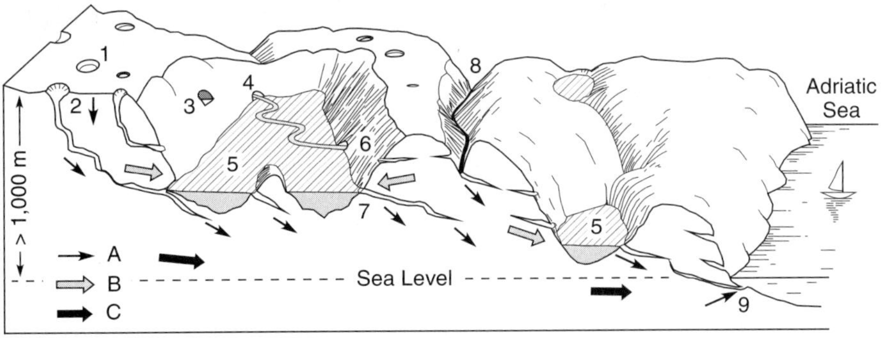

Figure 1.75 Karst and groundwater flow features in the thick Mesozoic carbonate platform of the Dinarides (modified from Kresic, 1988). (1) sinkhole (doline); (2) pithole (jama); (3) dry cave—former spring; (4) active spring; (5) karst polje with unconsolidated fill sediments, (6) active sink; (7) estavelle (interchangeable spring and sink); (8) deep canyon with the loosing stream; (9) submarine spring in the Adriatic Sea. A: predominant local groundwater flow direction and aquifer recharge; B: local groundwater flow direction during wet season and after heavy rains; C: regional groundwater flow.

Figure 1.76 *Top*: Popovo polje near Trebinje, Herzegovina, during rare recent flood of winter 2011. (Photograph courtesy of Željko Zubac.) *Bottom*: Uvala Loški potok in Slovenia during flood. (Photograph courtesy of Dr. Andrej Mihevc.)

the aquifer. Although the actual duration of the flood may vary, hydrologically active poljes are flooded every year in the wet season and eventually drain during dry months. A number of inactive caves and jamas, representing former springs, sinks, and estavelles, can be found at various elevations above the current polje level marking the progression of karstification. During all seasons, regardless of the local directions of groundwater flow along the polje perimeter, there is an ongoing regional groundwater flow underneath the polje toward the main, lower areas of discharge such as major permanent surface streams, springs, or, in case of the Dinaric karst, the Adriatic Sea (Fig. 1.75.)

Figure 1.77 *Left*: The Komarnica River canyon near Brezane, Montenegro. This part of the canyon is under retention formed by the Piva Dam on the Piva River. (Photograph courtesy of Dobrisav Bajović-Bajone.) *Right*: Rio Grande near the lower end of the Santa Elena Canyon, Big Bend National Park, Texas.

Canyons and Natural Dams

River canyons are not unique to karst terrains but represent perhaps their most fascinating landform. Rivers in limestone, which is very resistant to mechanical erosion, often create narrow, deep canyons with perfectly vertical walls in places (Fig. 1.77). Deepening of the canyon is accompanied by the karstification of the surrounding aquifer that continues to discharge into the river, often via springs (Fig. 1.78). Numerous dry caves on the canyon walls are the testimony of the aquifer discharge when the canyon was shallower (Fig. 1.79). Sometimes, after heavy rainfall, former cave springs are given a breadth of life creating spectacular but short-lived waterfalls on the canyon walls.

Natural cascades, dams, and lakes formed behind travertine deposits are surface landforms unique to karst terrains. Travertine is deposited from surface streams fed by groundwater saturated with carbonates. It is also commonly deposited directly at karst springs. A minute initial waterfall in the streambed, perhaps caused by faulting, selective erosion, or tree debris, is sometimes enough to start the deposition of calcium carbonate as the water aerates and loses CO_2. Travertine deposits consist of pure calcite and can sometimes grow into dams tens of meters high forming deep, crystal-clear lakes. Although travertine has high total porosity (see inset in Fig. 1.80), its effective porosity is virtually zero enabling the formation of impermeable natural dams.

The Plitvice Lakes National Park in Croatia, protected by UNESCO as the world's natural heritage site, is the prime example of natural dams in karst. There are 16 larger lakes formed in succession above each other and numerous smaller lakes and travertine cascades in the Korana River canyon (Fig. 1.81). The travertine deposits are currently growing at the rate of approximately 14 mm/year. The classic Dinaric karst is dotted with natural travertine dams and waterfalls. One such dam in the Krka Falls National

Figure 1.78 The Tara River canyon in Montenegro, deepest in Europe, is mostly inaccessible except by raft. Several large and numerous smaller karst springs discharge into the river, some as nice waterfalls over travertine deposits.

Figure 1.79 *Left*: Cave above Li River, China. *Right*: One of the numerous caves along limestone cliffs above the Colorado River, American Southwest. (Right photograph courtesy of Graham Schindel.)

Figure 1.80 Low-flow waterfall of Big Spring over massive travertine deposits near Šetonje, eastern Serbia. The inset, approximately 1 m across, shows a close-up of travertine.

Figure 1.81 Travertine cascades between larger natural lakes in the Korana River canyon, Plitvice Lakes National Park, Croatia.

FIGURE 1.82 The main waterfall over travertine deposits in the Krka Falls National Park, Croatia.

Park in Croatia (Fig. 1.82) was utilized for construction of the world's second alternating current (AC) hydropower plant which was put in service only 2 days after the first one on the Niagara Falls in the United States (on August 28, 1895). This plant was used to provide electricity and city lights for the coastal town of Šibenik. Interestingly, Nikola Tesla, the inventor of the AC electricity and the designer of the first plant, who also helped with the Krka Plant, was born in the Dinaric karst and, as a child, used to play building small dams and water wheels on a spring-fed stream by his house.

Cavities

Karst potholes and pits (international term for both is *jama*) are predominantly vertical karst cavities developed by water percolating through the vadose (unsaturated) zone (Fig. 1.83). Pit is defined as a deep hole, usually circular in outline, having vertical or nearly vertical walls, whereas pothole is a more general term for both a single shaft and an entire system of passages that is dominantly vertical (Field, 2002).

Caves, the predominantly horizontal karst cavities, are formed by the solutional and mechanical action of flowing groundwater. Numerous books have been written on the origin of different types of caves including classifications based on various geomorphologic, geologic, and other criteria. For example, Bögli (1980), Palmer (1991, 2007), and Gillieson (1996) provide detailed discussions on the effects the hydraulic gradients, chemical kinetics, geologic fabric, and different modes of groundwater recharge have on the development of particular cave patterns. Other informative references include Klimchouk et al. (2000), Culver and White (2004), and Gunn (2004). As a general rule, there is seldom one easy explanation how a certain cave was formed. Even in case of an

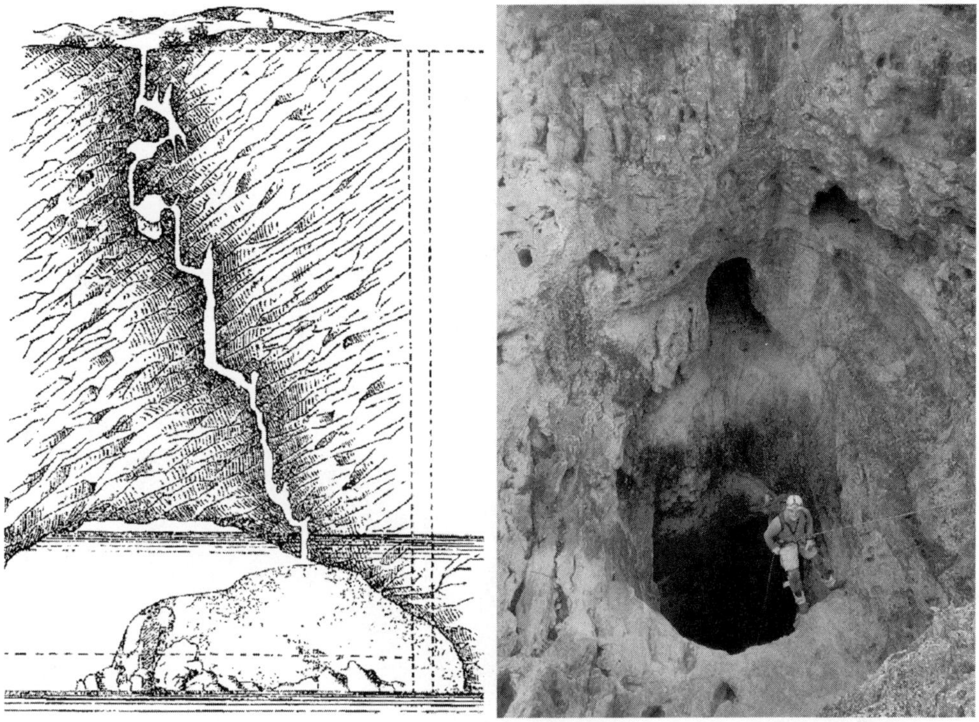

FIGURE 1.83 *Left*: Drawing of Trebička Jama (pithole) in Slovenia made by Jovan Cvijić for his doctoral dissertation on karst (from Cvijić, 1895). *Right*: Entrance to the Valdina Farm Sinkhole, intermittent sink of the Seco Creek, Texas. (Photograph courtesy of Dave Bunnell.)

apparent rudimentary flow-through cave, there may be surprises waiting to be discovered as shown in Fig. 1.84. The vertical shaft leading to a previously unknown, higher-elevation cave and the lateral connection between the two caves were discovered by accident. Both caves (now comprising one cave system) were obviously formed in a very similar manner, following the same general fracture/fault pattern. The lower cave has a blind dry valley leading to its upper entrance—the former sink of a surface stream. Yet, a sinking cave entrance to the higher horizon is absent, and there are no any indirect signs of its possible location at the land surface.

As the karstification progresses vertically downward, and provided there is sufficient thickness of soluble carbonate sediments, the passages of exogenic caves continue to be formed at successively lower levels (Fig. 1.85), sometimes creating a true three-dimensional maze of conduits, channels, and vertical shafts connecting different cave horizons. This is regardless of the origin of water involved in the process of cave development that can come primarily from one of the following:

1. Sinking streams (allogenic recharge).
2. Scattered vertical recharge from the land surface including via sinkholes and shafts (autogenic recharge).
3. Some combination of both.

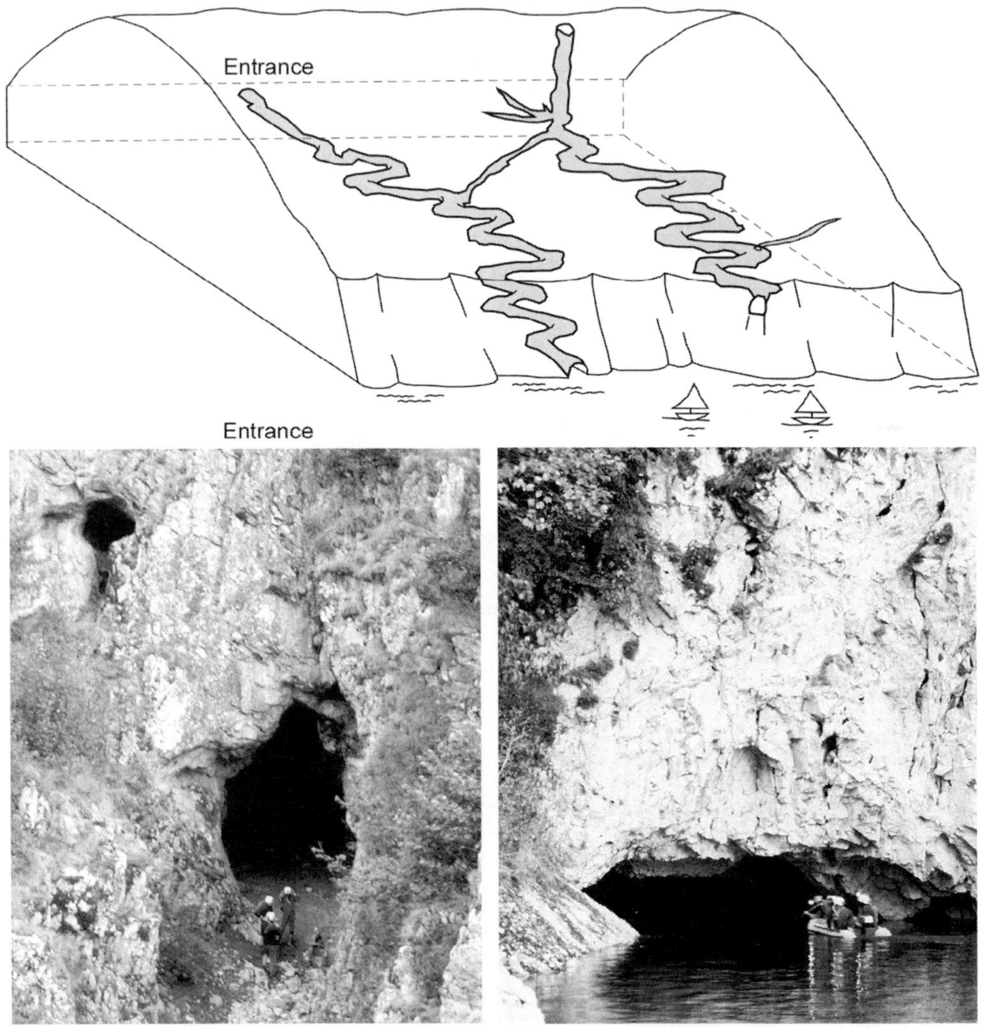

Figure 1.84 Schematic of the Usac cave system near Sjenica, Serbia. Lower cave entrance (former spring, now submerged) in the Uvac River canyon is accessible by boat only.

In any case, the dissolution of carbonate matrix progresses mostly along secondary porosity features such as bedding planes, fractures, and faults. This process of cave development creates passages that are organized into a network, often with progressively larger passages in the down-gradient direction toward the locations of aquifer discharge such as karst springs. In relatively humid climates, speleothems are created by the percolating water in the upper dry cave horizons, thus confirming the absence of any significant channel flow that may break them or otherwise prevent their formation. Well-developed and tall, thin stalagmites as well as thin fragile stalactites are the best

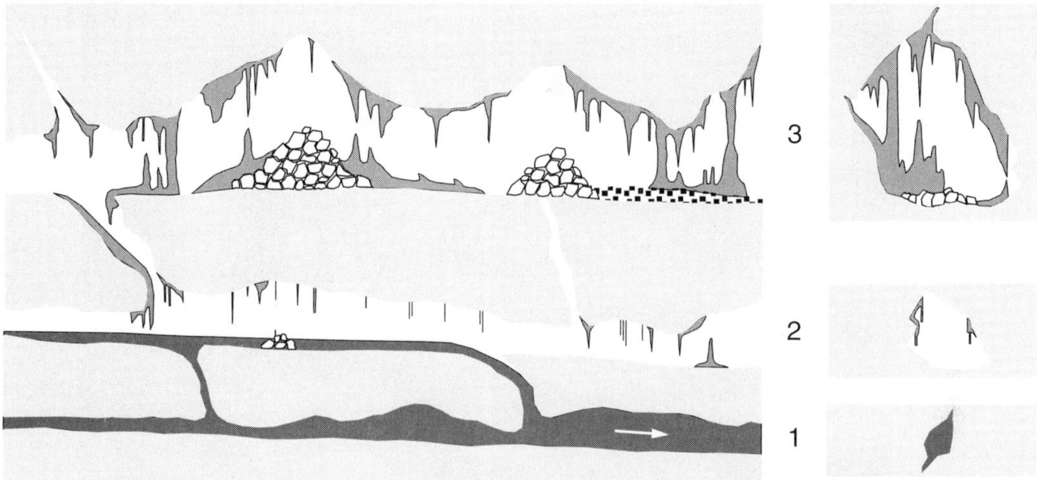

FIGURE 1.85 Main phases in cave development. (1) Phreatic conduit being developed below water table by dissolution. (2) Cave passage with permanent to intermittent flow on the floor; speleothems starting to form, occasional rock fall. (3) Last phase of cave development—permanently dry passages with a variety of abundant speleothems.

such evidence (Fig. 1.86). Massive speleothems can sometimes break off due to their own weight or tectonic disturbance only to confuse a novice caver (Fig. 1.87). In the lower horizons, where the groundwater flows more frequently or permanently, speleothems are rare or absent (Fig. 1.88).

The mechanical action of groundwater flowing through karst conduits and channels, including under high pressure, adds significantly to the enlargement of cave passages. The force of flowing water can dislodge and further break down pieces of rocks. When combined with the cave ceiling collapse due to gravity, this action of water can result in the creation of enormous caves of sometimes frightening dimensions (Figs. 1.89 through 1.91).

The exception to the above general description of cave development is formation of hypogenic caves. As mentioned earlier, such caves grow due to ascending flow of groundwater that may carry strong acidic fluids of deep (endogenic) origin such as sulfuric acid (see Fig. 1.52). In addition to structural control by fractures and bedding planes, hypogenic caves often exhibit the so-called ramiform and spongework patterns, or a seemingly random three-dimensional pattern resembling pores in a sponge (Palmer, 1991, 2007). Some of the longest and most convoluted caves in the world are of hypogenic origin (Fig. 1.92).

Hypogenic caves formed at greater depths may be subsequently uplifted by tectonic activity, placed in a near-surface setting, and partially or completely drained of initial water. At some point, a hypogenic cave may start receiving the flux of atmospheric descending water resulting in creation of classic speleothems (Fig. 1.93). Possible succession of morphogenic processes (endogenic/hypogenic followed by exogenic) may create very complex caves with confusing features, which defy simple explanation, thus making the lives of cavers and speleologists so much more interesting. The regular erosion of land

Figure 1.86 *Upper left*: Giant stalagmites are trademark of Vernjikica Cave near Bor in eastern Serbia. *Upper right*: Countless stalagmites of varying diameter and color are crowding this protected cave in southwestern Serbia. *Bottom*: Snow-white stalactites in the Ceremošnja Cave near Kučevo, eastern Serbia.

surface may create natural entrances to previously hidden hypogenic caves that are not related to any action of surface water or sinking streams. Figure 1.94 shows the main entrance to Carlsbad Caverns, one of the largest hypogenic caves in the United States, formed this way. The Carlsbad Caverns National Park has approximately 130 caves all of which are hypogenic, created by the ascending acid fluids originating from oil and gas fields.

FIGURE 1.87 A huge fallen speleothem formation in Lechuguilla cave, New Mexico. The formation fell perhaps thousands of years ago judging by the accumulation of new speleothem growth on it. (Photograph courtesy of Dave Bunnell; many more photographs are available at http://www.goodearthgraphics.com/under_earth/.)

1.4 Aquifer Parameters and Hydraulic Head

Virtually all concepts and principles of groundwater flow through porous media in classic hydrogeology were developed from studies of sand and gravel aquifers. These concepts and the related aquifer parameters such as hydraulic conductivity, transmissivity, and storage are also applicable to other unconsolidated sediments with intergranular porosity such as silts and clays, as well as their various mixtures with sand and gravel. The majority of laboratory tests, field measurements, and field tests conducted to determine the aquifer parameters are based on the assumption that the groundwater flow in porous media is accurately described with the Darcy's equation. Darcy was a French civil engineer who was first to quantitatively analyze the flow of water through sands as part of his design of water filters for the city of Dijon. Darcy's findings, published in 1856, are the foundation of all modern studies of fluid flow through porous media. On the basis of numerous laboratory tests with varying hydraulic conditions and sand sizes, Darcy established the following empirical relationship, now known as Darcy's law:

$$Q = KA\frac{\Delta h}{l}[m^3/s] \tag{1.5}$$

Figure 1.88 *Top*: Hydrologically semiactive passages in a cave owned and explored by Minnesota Cave Preserve. The presence of speleothems indicates that the passages are seldom completely submerged. *Bottom*: This horizon of narrow conduits is always active and the flow is frequently under pressure after heavy rainfall. In one such instance, the spelunker on the left photo managed to narrowly escape the quickly rising muddy water and the complete flooding of the conduit by clinging to and then climbing the fixed ladder placed in a drilled shaft. (Photographs courtesy of John Ackerman, Minnesota Cave Preserve; many more photographs are available at http://www.karstpreserve.com.)

Figure 1.89 *Left*: "Really Big Rock." *Right*: "Holy Moley." (Highland County caves, Virginia; photographs courtesy of Phil Lucas.)

Figure 1.90 Fossil river passage in the Chiquibul cave system, Belize. Down cutting of the cave river has left this passage high and dry. (Photograph courtesy of Dave Bunnell; many more photographs available at http://www.goodearthgraphics.com/under_earth/.)

This linear law states that the rate of fluid flow (Q) through porous medium is directly proportional to the cross-sectional area of flow (A) and the loss of the hydraulic head between two points of measurements (Δh), and it is inversely proportional to the distance (l) between these two points of measurement. K is the proportionality constant of the law called *hydraulic conductivity* and has units of velocity. This constant is arguably the most important quantitative parameter characterizing the flow of water and other fluids through porous media. Other common forms of Darcy's equation are

$$v = K\frac{\Delta h}{l}[\text{m/s}] \tag{1.6}$$

$$v = Ki[\text{m/s}] \tag{1.7}$$

where v is the so-called Darcy's velocity, and i is the hydraulic gradient. Note that Darcy's velocity is not the real groundwater velocity with which water particles move through pore space between sediment grains. It is an imaginary velocity describing flow through the entire cross-sectional area [A in Eq. (1.5)] of the porous medium including solid grains. In Russian literature, it is adequately called "smeared velocity." In order to somehow account for this fact, a concept of "linear velocity" (v_L) was later introduced (not by Darcy) in an attempt to better represent the true groundwater velocity that cannot be feasibly measured:

$$v_L = \frac{v}{n_{\text{eff}}}[\text{m/s}] \tag{1.8}$$

FIGURE 1.91 Collapse from a surface sinkhole into a river passage has made a new entrance to this cave near Dagu Dong, Guangxi Province, China. Note small figure of a spelunker standing on a rock in the lower right for scale. (Photograph courtesy of Dave Bunnell; many more photographs available at http://www.goodearthgraphics.com/under_earth/.)

where n_{eff} is the effective porosity (porosity of interconnected pore space between solid grains).

The principle of the hydraulic head and the hydraulic gradient is illustrated in Fig. 1.95. At the bottom of monitoring well no. 1, where the well screen is open to the

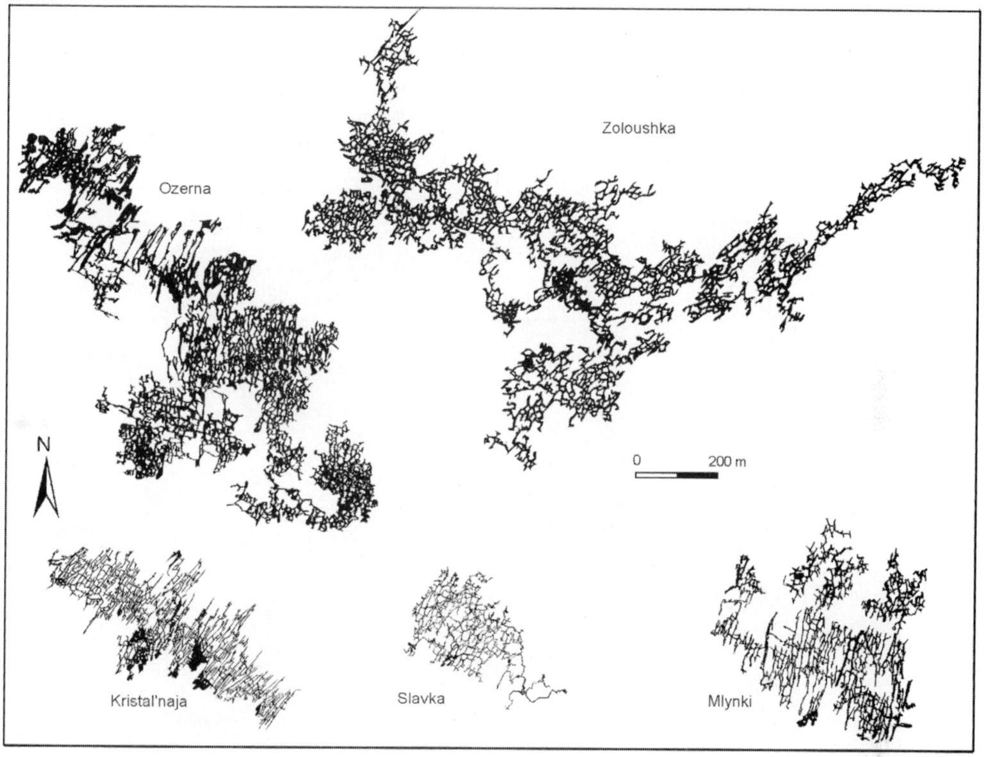

Figure 1.92 Patterns of hypogenic maze caves in Miocene gypsum in the western Ukraine. The region contains the five longest gypsum caves in the world. The second and third longest gypsum caves are Ozerna (117 km) and Zoloushka (90.2 km) shown here. The longest is Optymistychna, with 214 km of mapped passages (not shown). Maps are courtesy of the speleological clubs of Ternopil (Ozerna, Kristalnaja, and Mlynki), Chernivtsy (Zoloushka), and Kiev (Slavka). (From Klimchouk, 2011, printed with permission.)

saturated zone, the total energy (E) or the driving force for water flow at that point in the aquifer is

$$E = z + h_p + \frac{v^2}{2g} \tag{1.9}$$

where z is elevation above datum (datum is usually mean sea level, but it could be any reference level), h_p is the pressure head due to the pressure of fluid (groundwater) above that point, v is groundwater velocity, and g is acceleration of gravity.

Since the groundwater velocity through intergranular porous media in most cases is quite low, the third factor on the right-hand side may be ignored for practical purposes and the Eq. (1.9) becomes:

$$E = h = z + h_p \tag{1.10}$$

FIGURE 1.93 Carlsbad Caverns in Carlsbad, New Mexico, is packed with fascinating speleothems created after this hypogenic cave was uplifted and exposed to percolating atmospheric water during more humid times. (Photographs courtesy of Peter Jones, National Park Service; available at http://www.nps.gov/cave/photosmultimedia/index.htm.)

where h is the *hydraulic head*, also called *piezometric level*. The pressure head represents the pressure of fluid (p) of constant density (ρ) at that point in the aquifer:

$$h_{\mathrm{p}} = \frac{p}{\rho g} \tag{1.11}$$

In practice, the hydraulic head is determined in monitoring wells or piezometers by subtracting measured depth to the water level from the surveyed elevation of the top of the casing:

$$\mathrm{h} = \text{elevation of top of casing} - \text{depth to water in the well} \tag{1.12}$$

As the groundwater flows from well no. 1 to well no. 2 (Fig. 1.95), it loses energy due to friction between groundwater particles and the porous media grains. This loss equates to a decrease in the hydraulic head measured at the two wells:

$$\Delta h = h_1 - h_2 \tag{1.13}$$

Figure 1.94 Tourist path leading down to the main entrance to Carlsbad Caverns, The Carlsbad Caverns National Park, New Mexico. (Photograph courtesy of Peter Jones, National Park Service; available at http://www.nps.gov/cave/photosmultimedia/index.htm.)

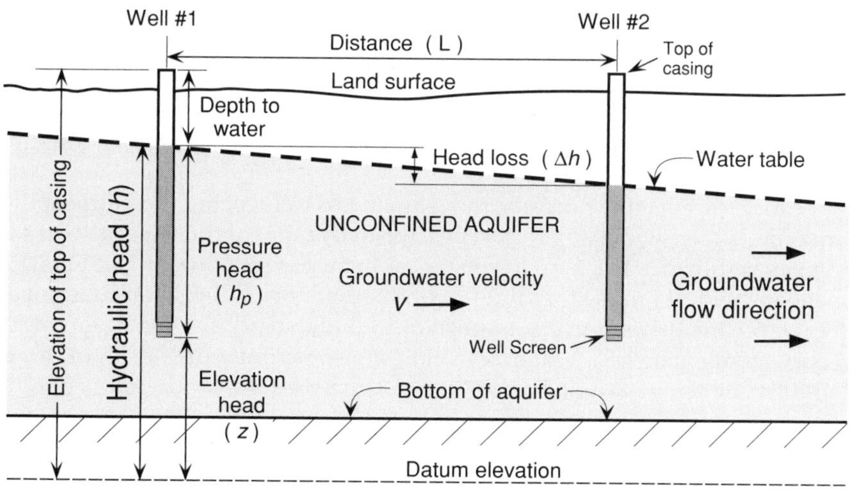

Figure 1.95 Schematic presentation of key elements for determining the hydraulic head and the hydraulic gradient in an unconfined aquifer. (Kresic, 2007a; copyright Taylor & Francis Group, printed with permission.)

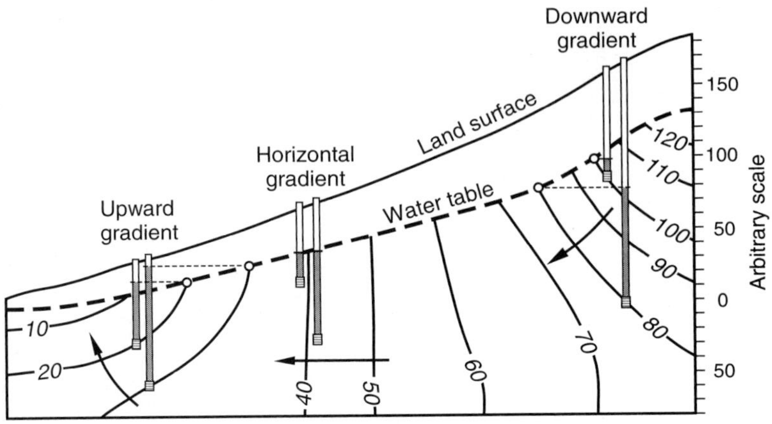

FIGURE 1.96 Movement of groundwater in an unconfined aquifer showing the importance of both vertical and horizontal hydraulic gradients (modified from Winter et al., 1998).

The *hydraulic gradient* (i) between the two wells is obtained when this decrease in the hydraulic head is divided by the distance (L) between the wells:

$$i = \frac{\Delta h}{L} \text{[dimensionless]} \tag{1.14}$$

Groundwater flow in the intergranular porous media always takes place from the higher hydraulic head toward the lower hydraulic head (just as in the case of surface water: "water cannot flow uphill"). However, as explained later, this is not always true in karst conduits where the flow velocity can be rather high and cannot be ignored as in Eq. (1.10). It is also important to understand that, except in case of a very limited portion of an aquifer, there is no such thing as strictly horizontal groundwater flow. In an area where aquifer recharge is dominant, the flow is vertically downward and laterally toward the discharge area; in a discharge area, such as surface stream, this flow has an upward component (Fig. 1.96).

Since the effective porosity, the main factor influencing the permeability of a porous medium, varies widely by rock types, the hydraulic conductivity also has wide reported ranges as shown in Fig. 1.97. In general, as in the case of porosity (see Fig. 1.29), limestones have the widest range of hydraulic conductivity of all rocks, which is not apparent from Fig. 1.97. It is therefore very important to understand that the reported high hydraulic conductivity values for "cavernous limestone and dolomite" in Fig. 1.97 are for the bulk aquifer volume including karst cavities. The hydraulic conductivity of old consolidated limestone with low matrix porosity (e.g., 2%) and without cavities approaches that of massive igneous and metamorphic rocks.

The following examples illustrate the variability of hydraulic conductivity of carbonate rocks. Chalk and some limestones may have high porosity, but since the pores are small (usually less than 10 micrometers, or μm), primary permeability is low and specific retention is high (Cook, 2003). For example, the mean interconnected porosity of the Lincolnshire Limestone in England is 15%, whereas the mean matrix hydraulic conductivity is only 10^{-9} m/s (Cook, 2003, after Greswell et al., 1998). The groundwater flow is

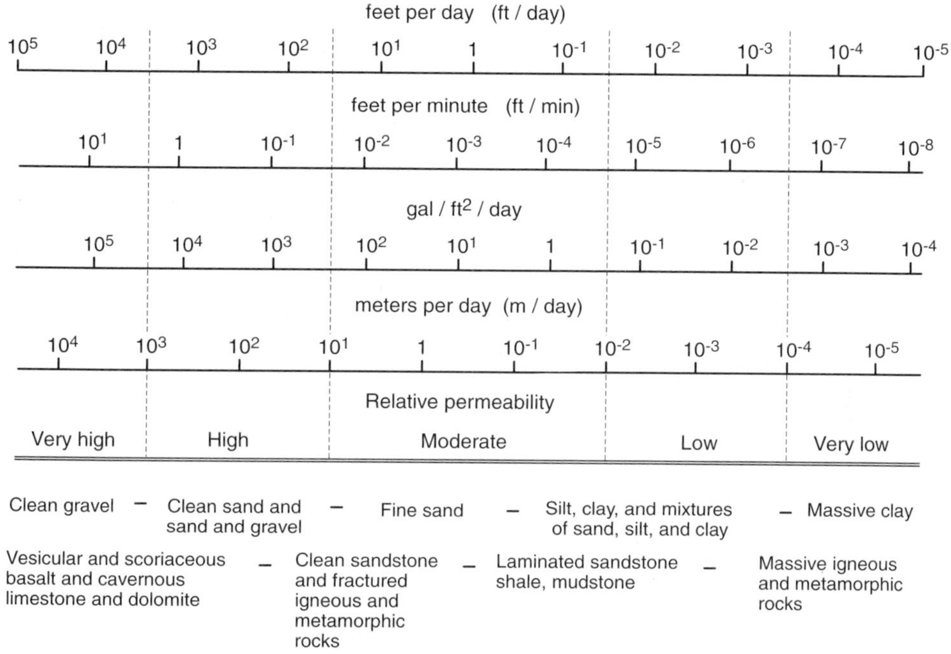

FIGURE 1.97 Range of hydraulic conductivity for different rock types (USBR, 1977).

largely restricted to the fractures. The aquifer hydraulic conductivity determined from pumping tests ranges between approximately 20 and 100 m/day, which is more than five orders of magnitude greater than the matrix hydraulic conductivity (Cook, 2003, after Bishop and Lloyd, 1990). The San Antonio segment of the Edwards Aquifer in Texas, the United States, consists of Cretaceous limestones and dolomites that have undergone multiple periods of karstification. The average aquifer hydraulic conductivity, based on more than 900 well pumping tests, is approximately 7 m/day, whereas the mean matrix hydraulic conductivity is approximately 10^{-3} m/day (Cook, 2003, after Halihan et al., 2000).

Based on the results of 191 aquifer pumping tests, the median horizontal hydraulic conductivity of the Upper Floridan aquifer in Georgia, the United States, is 140.3 ft/day (Fig. 1.98). Such high median value is, in part, the result of generally high effective matrix porosity of the Upper Floridan aquifer Tertiary carbonates, as illustrated with the data from west-central Florida (see Fig. 1.33). In contrast, the median vertical hydraulic conductivity of 17 core samples collected from the same aquifer is three orders of magnitude lower and shows scatter of data over seven orders of magnitude (Fig. 1.99). However, great caution should be exercised when interpreting results of laboratory tests such as these. The main reason for caution is simple fact that the laboratory tests are performed on small samples (few centimeters/inches in diameter/length), which by default cannot include any significant fracture or solution feature. Another issue is the utilization of any ratio of horizontal-to-vertical hydraulic conductivity, the values of which are obtained with two different methods. The laboratory tests performed on small core samples cannot be related to large-scale aquifer pumping tests that integrate response of all types

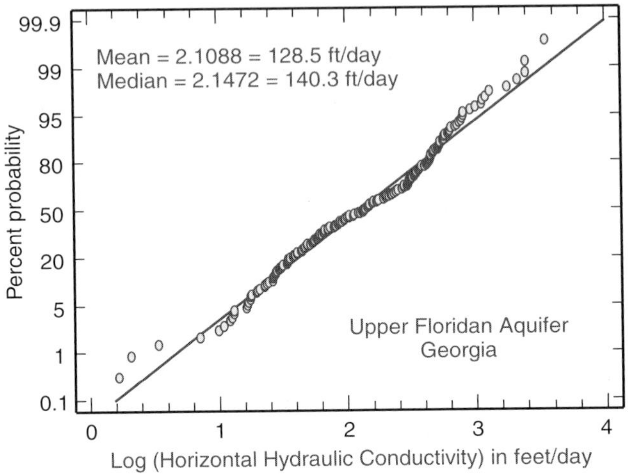

FIGURE 1.98 Distribution of horizontal hydraulic conductivity of the Upper Floridan aquifer in Georgia, the United States, based on the results of 191 aquifer pumping tests. (Raw data from Clarke et al., 2004.)

of porosity present in the aquifer (karst cavities of all kinds included), in essentially all directions including the vertical. In case of Fig. 1.99, the hydraulic conductivity values are only representative of the carbonate rock matrix samples oriented vertically. Only if there were equivalent laboratory test data for matrix samples oriented horizontally (horizontal core plugs taken from the larger vertical core), one would be able to estimate possible vertical anisotropy of the carbonate rock matrix. The anisotropy ratio obtained

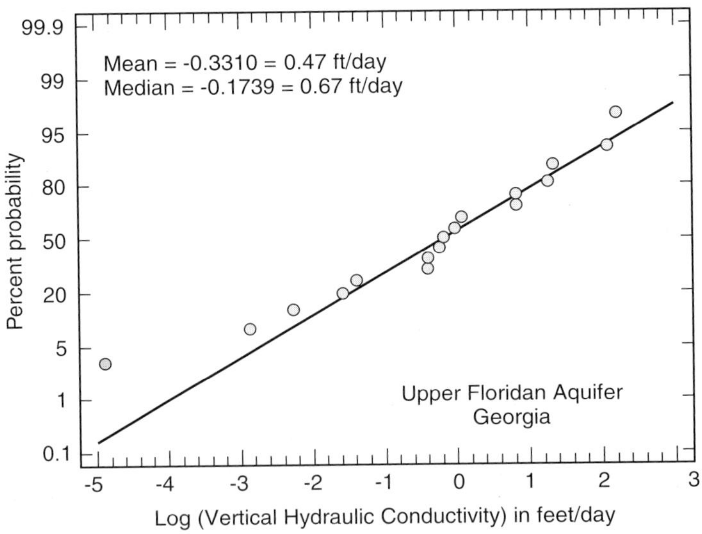

FIGURE 1.99 Distribution of vertical hydraulic conductivity of the Upper Floridan aquifer in Georgia, the United States, based on the results of 17 laboratory tests on cores. (Raw data from Clarke et al., 2004.)

this way would still not be representative of the possible overall vertical anisotropy of the karst aquifer, which includes karst voids and cavities that may be oriented in any number of three-dimensional directions.

Budd and Vacher (2004) provide comparison of reported matrix, fracture, and channel permeability for 12 different karst aquifers in northern America, Australia, England, and Germany and discuss the importance of high matrix permeability for groundwater flow in the Floridan aquifer, the United States.

Vesicular basalts can have very high hydraulic conductivity, but they are on average less permeable than medium to coarse sand and gravel, which are rock types with the highest average hydraulic conductivity. Pure clays and fresh igneous rocks generally have the lowest permeability, although some field-scale bedded salt bodies were determined to have permeability of zero (Wolff, 1982). This is one of the reasons why salt domes are considered as potential depositories of high radioactivity nuclear wastes in some countries.

Except in rare cases of uniform and nonstratified, homogeneous unconsolidated sediments, hydraulic conductivity and permeability vary in space and in different directions within the same rock mass due to its heterogeneity and anisotropy. Most practitioners tend to simplify these inherent characteristics of porous media by dividing the complex three-dimensional hydraulic conductivity tensor into just two main components: horizontal and vertical hydraulic conductivity. Unfortunately, it seems common practice to apply some "rules of thumb" indiscriminately, such as vertical conductivity is 10 times lower than the horizontal conductivity, without trying to better characterize the underlying hydrogeology. This difference in the two hydraulic conductivities can vary many orders of magnitude in anisotropic rocks and, in many cases, it may be completely inappropriate to apply the concept altogether: A highly transmissive fracture or a karst conduit may have any shape and spatial extent (direction), at any depth.

One parameter often used for various calculations of groundwater flow rates is *transmissivity* (T) of porous media. For practical purposes, it is defined as the product of the aquifer thickness (b) and the hydraulic conductivity (K):

$$T = b \times K \tag{1.15}$$

It follows that an aquifer is more transmissive (more water can flow through it) when it has higher hydraulic conductivity and when it is thicker.

The capacity of a karst aquifer to transmit groundwater flow ranges from very low to very high, depending on the location and aquifer heterogeneity, that is, the relative importance of different porosity types. Good example is the analysis of carbonate-rock aquifers in southern Nevada by Dettinger (1989). Coyote Spring Valley aquifer transmissivity at the MX wells is extremely high (about 200,000 sq ft/day) and productivity is also high (3,400 gallons per minute with only 12 ft of drawdown). However, transmissivities elsewhere in the Central Corridor region, based on tests at 33 other water wells, are between 5,000 and 11,000 sq ft/day, while average well capacity is similar to those at Army Well 1 near Mercury, where 455 gallons per minute can be pumped for long periods of time with 85 ft of drawdown at the well. The same study shows that within 10 miles of regional springs, aquifers are an average of 25 times more transmissive than they are farther away. These are areas where flow is converging, flow rates are locally high, and the conduit type of flow likely plays a significant, if not predominant, role.

The *storativity* of an aquifer is the volume of water it releases from, or takes into, storage due to change in the hydraulic head. The physical mechanism that releases or stores water in the storage is not the same for unconfined and confined aquifers. When an unconfined aquifer is pumped, which results in the change of the hydraulic head (lowering of the water table), the water is released due to gravity drainage of the affected portion of the aquifer. The ratio between the volume of water the unconfined aquifer will yield due to gravity drainage and the total affected volume of the aquifer is called *specific yield* (S_y). The volume of water that remains in the aquifer after this drainage has taken place is called *specific retention*. Consequently, the specific yield is equal to total porosity minus specific retention. This definition is very similar to *effective porosity* and it is not uncommon to use the two terms interchangeably. However, specific yield is a dynamic quantity, which also depends on the rate of change of the water table, and can be relatively reliably determined only by aquifer testing in the field. In other words, lowering of the water table (release of water) and aquifer recharge (storage of water) occur over short periods of time, so that gravity drainage is rarely or never complete (Lohman, 1972). Specific yield of aquifer porous media obtained from field testing is therefore usually somewhat lower than the equivalent effective porosity. It should be noted that field determination of specific yield using aquifer pumping tests often has an associated uncertainty due to nonuniqueness of data analysis. In laboratory, the effective porosity and specific yield are usually determined using American Society for Testing and Materials (ASTM) standard test D425 from measurements of porosity (n) and irreducible water saturation, that is, specific retention (S_r) using the following equation:

$$S_y = n - S_r \tag{1.16}$$

In two classic publications (1923a, 1923b), Meinzer describes in much detail the concept of specific yield and early experimental determinations of it by Hazen (1892) and King (1899). As always with Meinzer, his observations and conclusions are timeless and it is worthwhile to present them here:

- The specific yield has frequently been called the "effective porosity" or "practical porosity," because it represents the pore space that will surrender water to wells and is therefore effective in furnishing water supplies.
- The distinction between gravity water and that which is retained by the rock or soil is not entirely definite, because the amount of water that will drain out depends on the length of time it is allowed to drain; on the temperature and mineral composition of the water, which affect its surface tension, viscosity, and specific gravity; and on various physical relations of the body of rock and soil under consideration. For example, a smaller proportion of water will drain out of a small sample than out of a large body of the same material. As the methods of determining specific yield have not been standardized, and as it may continue to be desirable to use different methods for different purposes and under different conditions, data as to specific yield should always be accompanied by a statement of the methods used in determining it.
- The importance of water that a saturated rock will furnish, and hence its value as a source of water supply, depends on its specific yield—not on its porosity. Clayey or silty formations may contain vast amounts of water and yet be unproductive

and worthless for water supply, whereas a compact but fractured rock may contain much less water and yet yield abundantly.

Values of specific yield for unconfined intergranular aquifers generally range between 0.05 and 0.3, although lower or higher values are possible, especially in cases of finer grained and less uniform material (lower values), and uniform coarse sand and gravel. When interpreted as EPM, pumping tests in karst aquifers with high matrix porosity show high values of specific yield as well. In general, the coarser sediments drain more quickly so that S_y shows less time dependency and variability than for the finer sediments (Meinzer, 1923a, 1923b; Healy and Cook, 2002).

One additional mechanism contributes to changes in storage of unconfined aquifers—compressibility of the water and aquifer material in the saturated zone. In most cases, however, the changes in water volume due to unconfined aquifer compressibility are minor and can be ignored for practical purposes. On the contrary, storage of confined aquifers is entirely dependent on compression and expansion of both aquifer water and solids. Namely, it is dependent on aquifer *elastic properties*. Because of a decline in hydraulic head, the water derived from storage in a confined aquifer comes from expansion of the water and compression of the aquifer solids. Water added to storage because of a rise in hydraulic head is accommodated partly by compression of the water and partly by expansion of the aquifer solids (Lohman, 1972). Storage properties (storativity) of confined aquifers are defined by the *coefficient of storage*. Although rigid limits cannot be established, the storage coefficients of confined aquifers may range from about 0.00001 to 0.001. In general, denser aquifer materials have smaller coefficient of storage. The value of coefficient of storage in confined aquifers may not be directly dependent on void content (porosity) of the aquifer material (USBR, 1977). The concept of elasticity does not apply to the portion of water storage in karst cavities, which behave similarly to a pressurized hydraulic system of pipes and reservoirs.

Specific storage (S_s) of confined aquifers is the volume of water released (or stored) by the unit volume of porous medium, per unit surface of the aquifer, due to unit change in the component of hydraulic head normal to that surface. The unit of specific storage is length^{-1} (e.g., m^{-1} or ft^{-1}) so that, when the specific storage is multiplied by aquifer thickness (b), it gives the coefficient of storage (S), which is a dimensionless number: $S = S_s b$. The specific storage is given as

$$S_s = \rho_w g(\alpha + n\beta) \tag{1.17}$$

where ρ_w is the density of water, g is the acceleration of gravity, α is the compressibility of the aquifer skeleton, n is the total porosity, and β is the compressibility of water.

Note that, when using transmissivity and storage coefficients in quantitative calculations and groundwater modeling, horizontal flow is assumed by default, since the aquifer thickness is included within their values. It is more common, especially when studying fate and transport of contaminants (contaminant particles), that the analysis is performed using values of horizontal and vertical hydraulic conductivities, specific storage, and actual aquifer thickness. This enables calculations of three-dimensional velocity vectors and aquifer water balance, which, in turn, are used to calculate changes in the contaminant concentrations as it moves in three-dimensional space.

It cannot be emphasized enough that the preceding general discussion on aquifer parameters is in reality applicable only to intergranular, unconsolidated aquifers and porous media (sand, gravel, silt, clay, and their mixtures). Nevertheless, many working professionals and regulatory agencies continue to insist on the application of Darcy's equation in karst justifying this approach by stating that "their" aquifer at their site (whatever the scale of that site may be) behaves as an EPM. In almost all such cases, there is not a rigorous quantitative or sensitivity analysis of any kind that would provide a defensible basis for this position. The following routine approach of characterizing an aquifer is applied in karst indiscriminately because it has also been routinely applied in many other hydrogeologic settings and because it may be endorsed (enforced) by the regulations and the regulators in some countries such as the United States:

- Install a number of monitoring wells.
- Measure water elevation in the wells and draw a contour map of the potentiometric surface (hydraulic head).
- Determine groundwater flow direction and hydraulic gradients from the potentiometric surface map.
- Perform few slug tests in the monitoring wells, and maybe a 24-hour aquifer pumping test to determine the hydraulic conductivity and storage parameters.
- Calculate the groundwater velocity using Darcy's equation and the values of hydraulic conductivity obtained from the slug (aquifer) tests, the calculated hydraulic gradient, and an assumed effective porosity.
- Calculate an optimum pumping rate and the radius of influence (drawdown) of a pumping well or wells (if that is part of the project scope).
- Perform quarterly monitoring of water levels (and contaminant concentrations if applicable) to determine if there are any significant seasonal changes.

This approach is destined to fail in karst, sooner or later, for the following main reasons:

- The potentiometric surface in karst is not continuous and smooth as in intergranular aquifers so that the concept of water table (hydraulic head) contour lines does not apply. This is illustrated in Fig. 1.100. Karst aquifer A and intergranular aquifer B both have the same general flow direction from the north to the south as shown on the map view. A triplet of close wells, installed at same depths anywhere in the intergranular aquifer, would reasonably accurately determine the general north–south flow direction based on the measured hydraulic head. The same cannot be stated for the karst aquifer case where the three-well principle may give very different directions depending on the position of individual wells relative to the preferential flow paths (karst conduits and/or large fracture zones).
- A group of closely spaced wells (meter to dekameter scale) may show a completely "random" distribution of the measured hydraulic head as illustrated in Fig. 1.101. One well may be completed in a homogeneous rock block, without any significant fractures and with low matrix porosity, and may even exhibit the so-called glass effect (no fluctuation of water level regardless of the precipitation–infiltration dynamics). On the other hand, a well 10 or so meters away and maybe close to or in a preferential flow path (conduit, channel, intensely fractured rock mass) may show the hydraulic head fluctuation of several meters or more in response to a major rainfall.

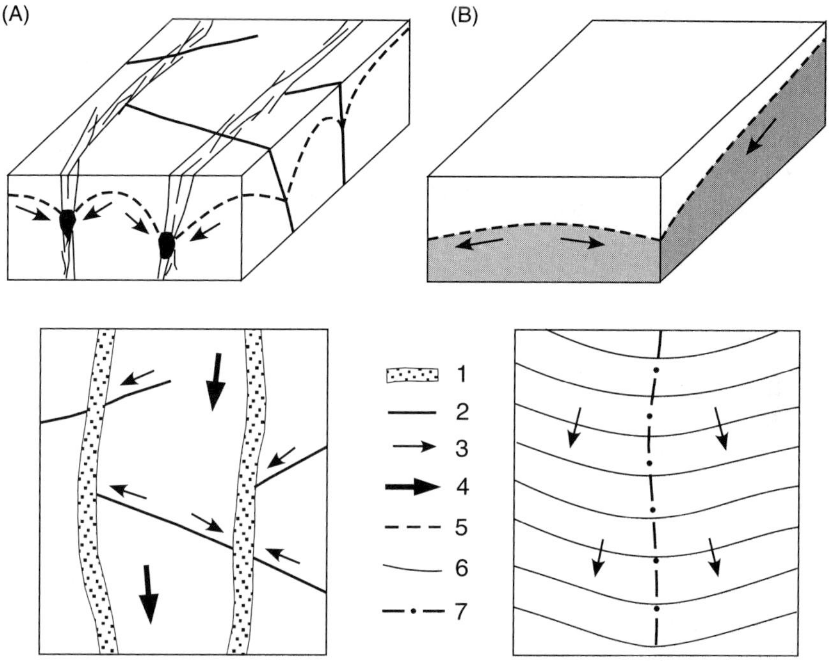

Figure 1.100 Schematic block diagram of groundwater flow in three dimensions (*top*) and its map presentation (*bottom*) for a karst (A) and an intergranular (B) aquifer. (1) Preferential flow path (e.g., fracture or fault zone or karst conduit/channel); (2) fracture/fault; (3) local flow direction; (4) general flow direction; (5) position of the hydraulic head (water table in the intergranular aquifer); (6) hydraulic head contour line; (7) groundwater divide. (From Kresic, 1991.)

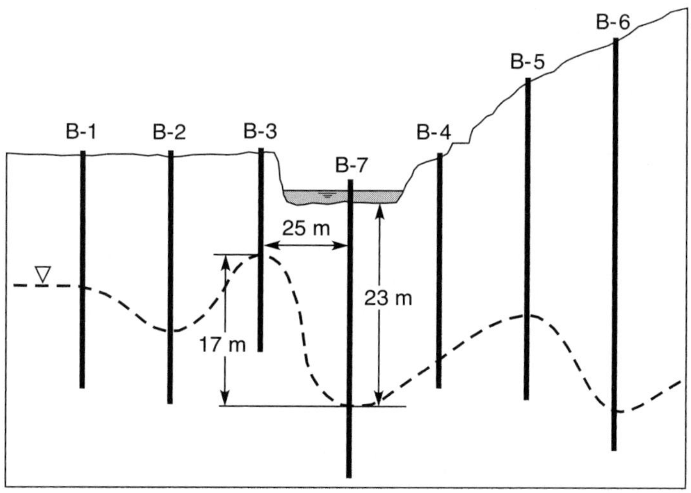

Figure 1.101 Hydraulic head measured at a transect of piezometers perpendicular to the Trebišnjica River in eastern Herzegovina, the largest sinking stream in Europe. (From Milanovic, 1979.)

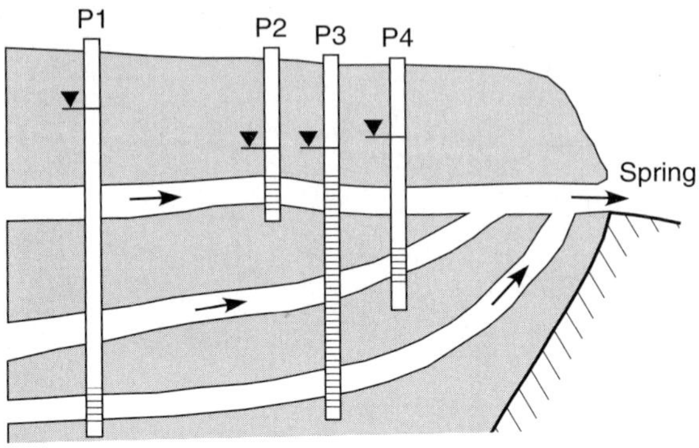

Figure 1.102 Example of how closely spaced monitoring wells in clusters may register very different hydraulic heads depending on the depth and length of well screens (open borehole). (Modified from Kupusović, 1989).

- Hydraulic heads measured at closely spaced wells intersecting different karst conduits (preferential flow paths) at different depths may be completely misinterpreted as illustrated in Fig. 1.102. The situation shown represents actual field conditions near a large permanent karst spring in the classic Dinaric karst of Herzegovina. Well P3 with the longest screen connects all three conduits and shows the same water level as the shallowest well P2. Considering the measured water level at P4 and either at P2 or P3, one could erroneously conclude that the groundwater is flowing away from the spring, which, in fact, is fed by all three channels.
- As every hydrogeologist working in karst can attest, fluctuations of water levels at monitoring wells can vary widely following rainfall or snowmelt episodes; water levels can sometimes rise tens of feet or meters in the matter of hours or days, only to drop back within days or weeks. Measuring water levels once every 3 months would provide no relevant information about the aquifer hydrodynamics. Figure 1.103 is a nice example of fluctuating water levels in the aquifer (cave) and bursts of groundwater flow force that may never be registered with quarterly monitoring.
- Darcy's law and the hydraulic conductivity and effective porosity concepts for calculating groundwater velocity and flow rate are not applicable to channel or conduit flow. Figures 1.104 and 1.105 illustrate these simple facts: Free water is flowing over solid surface (channel flow) and through "empty spaces" (conduit flow), which all have effective porosity of 100% regardless of their geometry and size.
- Slug tests and their quantitative interpretation in karst aquifers are meaningless and should be avoided. The only potential application is strictly qualitative, to assess relative differences between monitoring wells.

In conclusion, using only the hydraulic head information and some "equivalent hydraulic conductivity" for the purposes of determining groundwater flow directions and velocity in a karst aquifer would be insufficient. The hydraulic head measurements in

Figure 1.103 Cavers looking up at wood logs brought in and left at a ledge near cave ceiling by the underground stream flood waters. Note wood debris at the cave floor. Culverson Creek Cave System, West Virginia. (Photograph courtesy of Phil Lukas.)

monitoring wells should be continuous and accompanied with a continuous recording of precipitation. They should also be combined with dye tracing whenever possible, including in different seasons. In addition, any hydraulic testing of the aquifer in the field should be designed to gain an understanding of various karst-specific hydraulic factors including possible flow through preferential flow paths and under pressure.

1.5 Aquifer Tests

In a karst aquifer, the inflow of water to a pumped well would most likely occur from discrete intervals where the borehole intersects preferential flow paths (fractures and solution openings) within the surrounding rock. The presence of such intervals may be

FIGURE 1.104 Underground stream channel flow through passages of the Culverson Creek Cave System, West Virginia. (Photographs courtesy of Phil Lukas.)

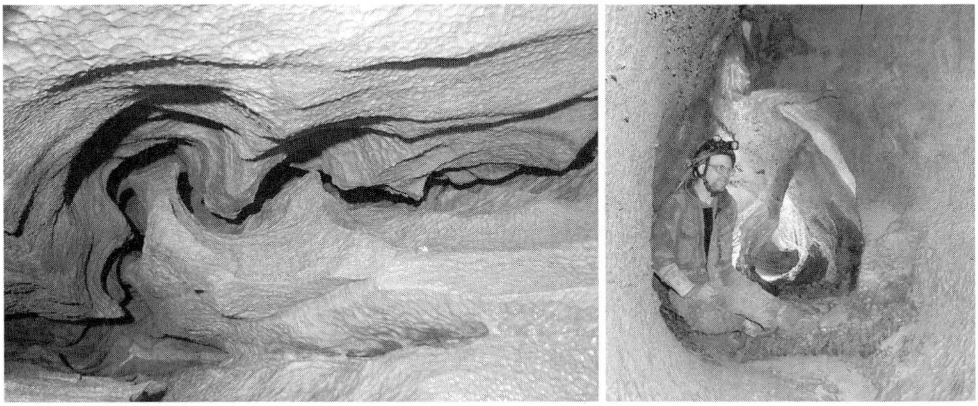

FIGURE 1.105 *Left*: View down at tortuous downcutting by flowing water through former cave floor. Scallops, irregular oval-shaped hollows on the rock surface, indicate the flow direction down and to the left (see explanation in Sec. 1.6.7). *Right*: Young phreatic tube formed below water table by rock dissolution, now intermittently active, in a Bath County, Virginia cave. (Photographs courtesy of Phil Lukas.)

indicated by various methods of geophysical logging, and their actual flow contribution may be measured and calculated using borehole flow meters (flowmeters). The flowmeters can be utilized in various ways, with or without pumping of the well. During well pumping, packers can be used to isolate portions of the open borehole for a more precise characterization. Classic packer tests, during which water is injected under pressure into discrete intervals, can be performed to determine permeability of the tested intervals. Finally, long-term pumping tests are utilized to assess the amount and quality of water available for extraction from the aquifer.

Simultaneous use of geophysical logging tools and flowmeters is the best available method for in situ characterization of fractured rock and karst aquifers (Paillet, 1994; Paillet and Reese, 2000). Integration of new geophysical methods with conventional logging techniques can be used to define flow zones, lithology, structure, and their relations (Fig. 1.106). Borehole flowmeters, whether vertical or horizontal, are used to identify and quantify water-producing zones in a well. Some horizontal flowmeters can measure the direction of flow through the borehole and thus identify intervals that either contribute water to or take water out of the borehole. Flow logging tests between boreholes ("cross-hole" tests) can indicate the degree of connectivity of preferential flow intervals (zones) beyond individual borings (Fig. 1.107), while transient tests can be used to estimate hydraulic properties such as transmissivity and storage coefficient.

FIGURE 1.106 Diagram showing orientation of fractures intersecting boreholes URL14 and URL15 determined from acoustic televiewer logs and projections of fractures between the two boreholes. (From Paillet, 1989.)

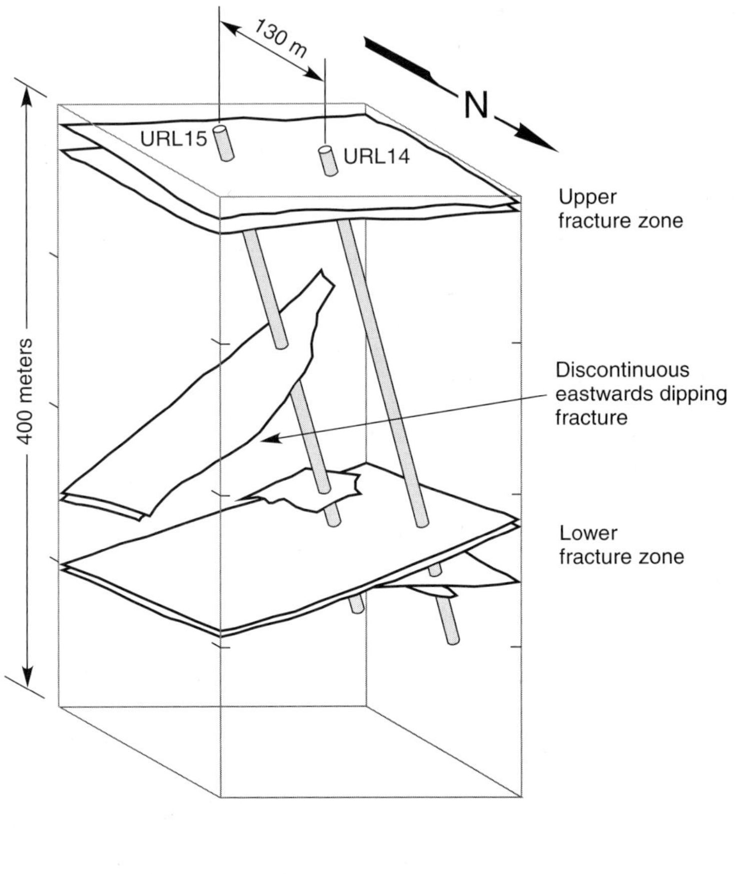

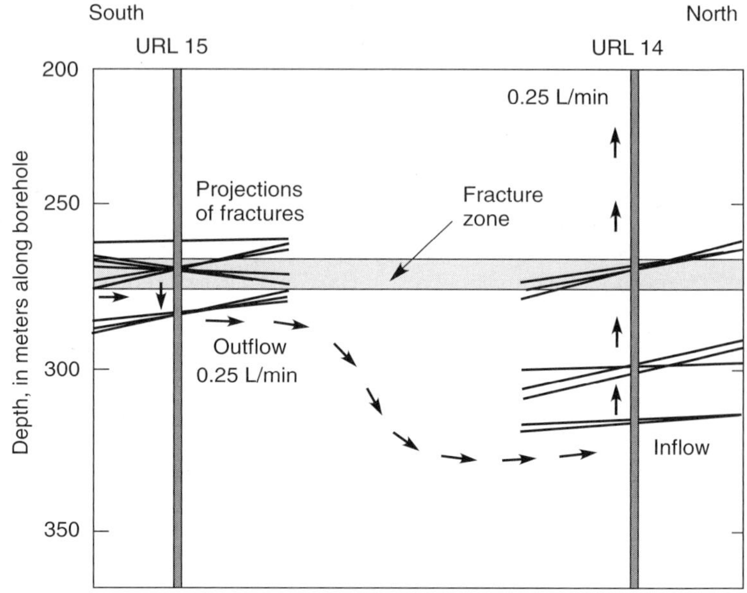

Figure 1.107 Diagram showing inferred hydraulic connection between boreholes URL14 and URL15 during cross-hole pumping tests. (From Paillet, 1989.)

1.5.1 Borehole Geophysical Methods

Borehole radar provides a method to detect fracture zones at distances as far as 30 m or more from the borehole in electrically resistive rocks. Radar measurements can be made in a single borehole (transmitter and receiver in same borehole) or by cross-hole tomography (transmitter and receiver in separate boreholes). Single-hole, directional radar can be used to identify the location and orientation of fracture zones, and cross-hole tomography can be used to delineate fracture zones between boreholes. The movement of a saline tracer through fracture zones can be monitored by borehole radar (Williams and Lane, 1998).

Acoustic televiewer (ATV) and *optical televiewer* (OTV) imaging results in continuous and oriented 360-degree views of the borehole wall from which the character, relation, and orientation of lithologic and structural planar features can be defined for studies of fractured-rock aquifers. Fractures are more clearly defined under a wider range of conditions on acoustic images than on optical images including dark-colored rocks, cloudy borehole water, and coated borehole walls. However, optical images allow for the direct viewing of the character of and relation between lithology, fractures, foliation, and bedding. The most powerful approach is the combined application of acoustic and optical imaging with integrated interpretation. Imaging of the borehole wall provides information useful for the collection and interpretation of flowmeter and other geophysical logs, core samples, and hydraulic and water quality data from packer testing and monitoring (Williams and Johnson, 2004).

Lithology and structures such as voids, fractures, fracture infillings, foliation, and bedding planes are viewed directly on the OTV images. OTV images can be collected in air- or clear-water-filled intervals of boreholes. Unflushed drilling mud, chemical precipitation, bacterial growth, and other conditions that affect the clarity of the borehole water or produce coatings on the borehole wall impact the quality of OTV images.

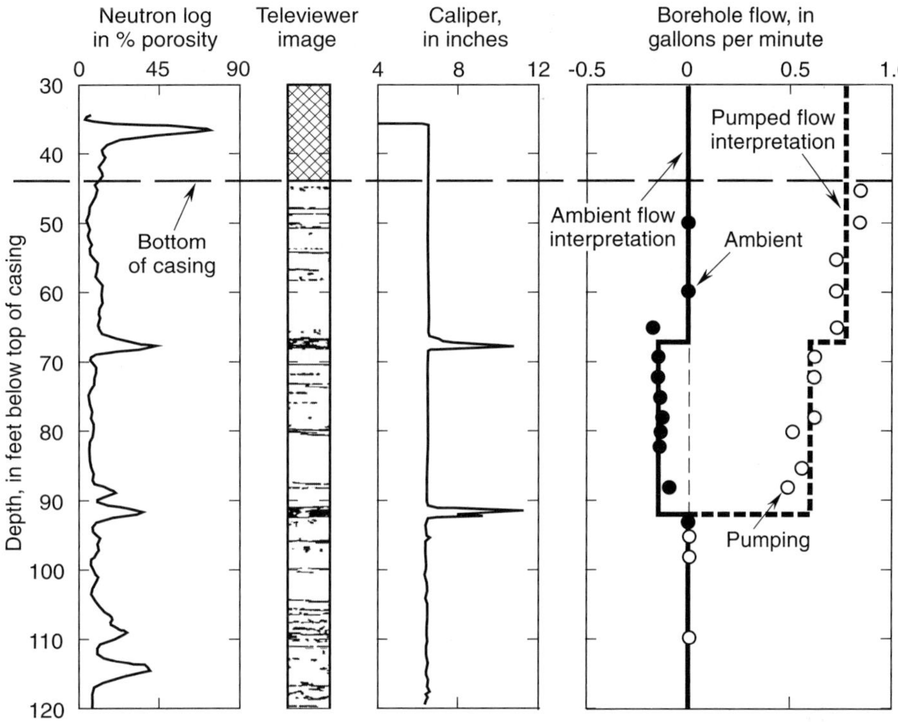

FIGURE 1.108 Background geophysical and vertical-flow logs with multiple geophysics for well FC-29 at Fort Campbell, Kentucky, the United States. (Simplified from Wilson et al., 2001.)

ATV images can be collected in water or light mud-filled intervals of boreholes. Borehole enlargements related to structures such as voids, fractures, foliation, and bedding planes scatter energy from the acoustic beam, reduce the signal amplitude, and produce recognizable features on the images. Acoustic impedance contrast between the borehole fluid and the wall indicates the relative hardness of the borehole wall. Lithologic changes, foliation, bedding, and sealed fractures may be detected even when there is no change in borehole diameter if there is sufficient acoustic contrast (Fig. 1.108). Multiecho systems record the full wave train of the reflected acoustic signal and are capable of imaging behind plastic casing. Such systems are useful for imaging poorly competent intervals that will not stay open without being cased, and for inspecting annular grout seals. The newer digital televiewer systems allow interactive determination of fracture orientation (Williams and Lane, 1998; Williams and Johnson, 2004).

Television cameras commonly are used in groundwater studies to inspect the condition of well casing and screens; they can also be used to directly view (1) lithologic texture, grain size, and color; (2) water levels and cascading water; and (3) bedrock fractures and cavities. Television logs can be obtained in clear water and above the water level. The most sophisticated television systems are magnetically oriented and provide a 360-degree digital image of the borehole wall (Williams and Lane, 1998). With an additional light source, television cameras can sometimes provide useful information on the extent of larger karst cavities crossed by the boring.

Flowmeters can be used to measure borehole flow under ambient and pumped conditions. Measurements made under ambient conditions can help to delineate transmissive fractures and other permeable zones and to indicate the direction of vertical hydraulic gradients; they are also useful in interpreting fluid-conductivity logs and borehole water quality data. Measurements made under pumped conditions can be used to develop hydraulic conductivity profiles of aquifers (Molz et al., 1990; Paillet et al., 1987; Williams and Lane, 1998; Young et al., 1998).

Differences in hydraulic head between two transmissive geologic units or preferential flow paths may produce vertical fluid flow in a borehole. Water enters the borehole at the unit with the higher head and flows toward and out of the unit with the lower head (Fig. 1.109). The vertical flow rate is limited by the geologic unit or zone with the lower transmissivity. If the heads of the different transmissive zones are the same, no vertical flow will occur in the well or borehole. However, it is recognized that the borehole

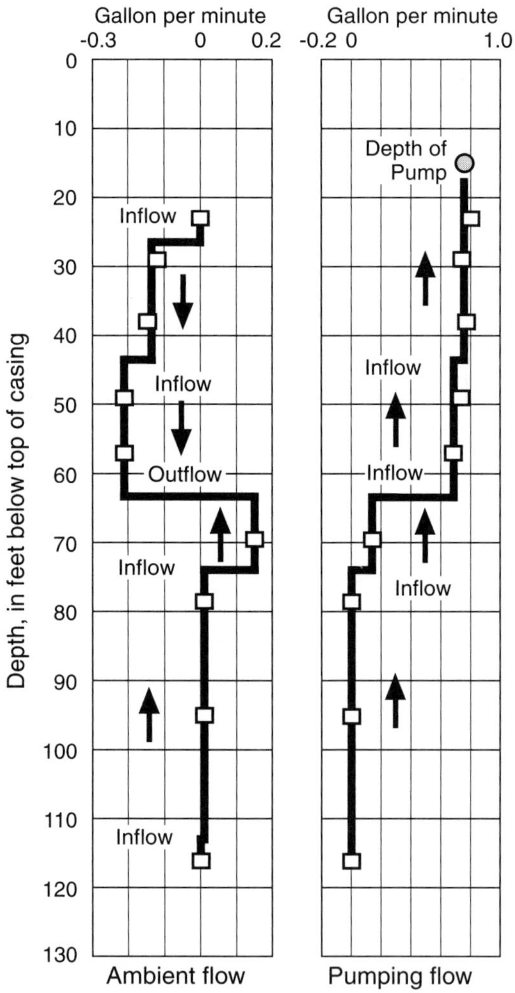

FIGURE 1.109 Sample heat-pulse flowmeter data from borehole in fractured rock aquifer. Flow was measured under ambient (*left*) and pumping (*right*) conditions. *Arrows* indicate interpreted direction of flow, as fluid moves from zones of higher head to lower head (USGS, 2004).

may facilitate vertical flow between aquifers and fractures that would not normally be present, so any interpretation of vertical flows must be made with caution (United States Geological Survey, USGS, 2004).

Single borehole flowmeter logging can be used to

- Measure the rate of vertical flow.
- Identify the direction of vertical flow.
- Establish relative hydraulic gradients.
- Identify transmissive zones or geologic units open to borehole that are the source(s) of water or that could act as conduits for cross-flow in an open hole.

In cross-hole flowmeter profiling, a flowmeter is placed in one borehole while fluid is pumped out of or into another borehole at a constant rate. Cross-hole flowmeter profiling is used to identify cross-hole connections and provide data that can be used to solve for

- Transmissivity.
- Hydraulic head.
- Storage coefficient of the geologic unit or more permeable zones between the boreholes.

The nature of the aquifer heterogeneities can affect the flowmeter measurements, which are particularly sensitive to flowmeter positioning relative to the preferential flow zone (Molz and others, 1989; Kearl and others, 1994, from Wilson et al., 2001). Steeply inclined fractures may produce results that are difficult to interpret because of nonhorizontal flow across the borehole. The acoustic Doppler velocimeter is the only tool of the three point-measurement methods capable of measuring three-dimensional flow (Wilson et al., 2001).

Vertical flowmeter measurements can be analyzed to provide qualitative and quantitative aquifer characteristics. Qualitative measurements attribute inflow and outflow to individual geologic units or transmissive zones. In conjunction with the drawdown data, they can be used to estimate interval-specific hydraulic conductivity and transmissivity. Methods commonly used for quantitative hydraulic analysis of flow zones from flowmeter log data include proportion, analytical solution, and numerical modeling techniques. Proportion and analytical solution methods provide estimates of transmissivity for the flow zones, while numerical model methods provide estimates of transmissivity and hydraulic head for the flow zones (Molz and Melville, 1996; Paillet, 1989, 1998, 2000, 2001; USGS, 2004; Wilson et. al., 2001).

Borehole flowmeter measurements are often combined with *hydrophysical logging*, which involves replacing the borehole fluid with deionized water, followed by a series of temperature and fluid-electrical-conductivity (FEC) logs. The logs profile the entire borehole to determine where formation water is entering and leaving (Tsang and others, 1990; Pedler and others, 1992, after Wilson et al., 2001). A time series of FEC logs can identify the locations and rates of inflow and outflow. Hydrophysical logging can identify both vertical and horizontal flows, since it surveys a length of the borehole, rather than providing point measurements. This attribute makes hydrophysical logging a valuable

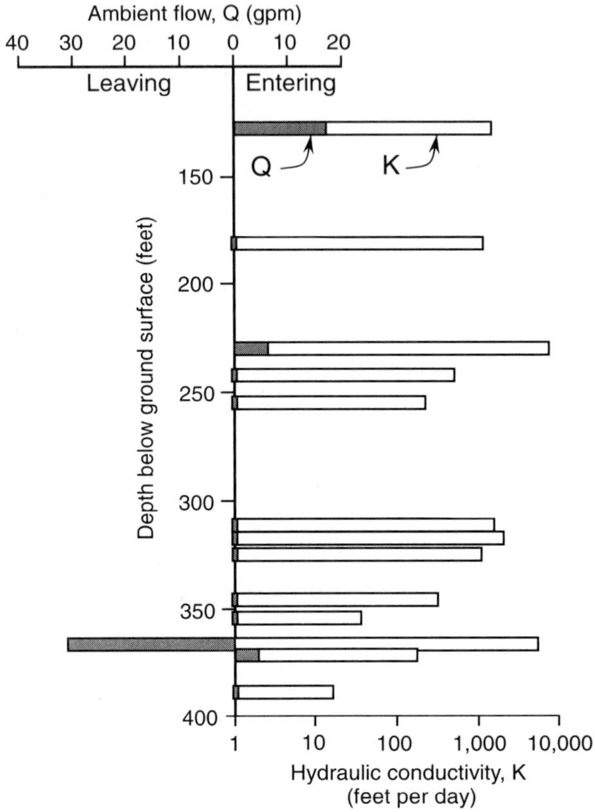

FIGURE 1.110 Results of hydrophysical logging in a deep boring in a carbonate aquifer. K, hydraulic conductivity; Q, ambient flow. (From Kresic, 2007a; data courtesy of Army Environmental Center.)

method for obtaining profiles of flow characteristics in uncased wells or in wells with long screens (Fig. 1.110).

The theory of hydrophysical logging is based on the law of mass balance and the linear relation between FEC and dissolved mass. By recording the changes in the electrical conductivity in the fluid column with depth, the locations of the water-producing zones can be determined and the volumetric rate of inflow can be calculated (Wilson et al., 2001).

1.5.2 Borehole Hydraulic Methods

Packer tests of hydraulic and chemical data. Borehole packers are pneumatic or mechanical devices that isolate sections of a borehole by sealing against the borehole wall. Hydraulic tests or collecting groundwater samples for chemical analyses then can be conducted on the isolated section of the borehole. Packer testing should test both the more permeable portions of the borehole (preferential flow paths) and less permeable portions without fractures and cavities, which may be representative of the matrix porosity.

Figure 1.111 shows three most common methods of packer testing. First, widely applied method of pressure permeability testing was developed by Lugeon, primarily for design of geotechnical grouting during construction of dams (Lugeon, 1933).

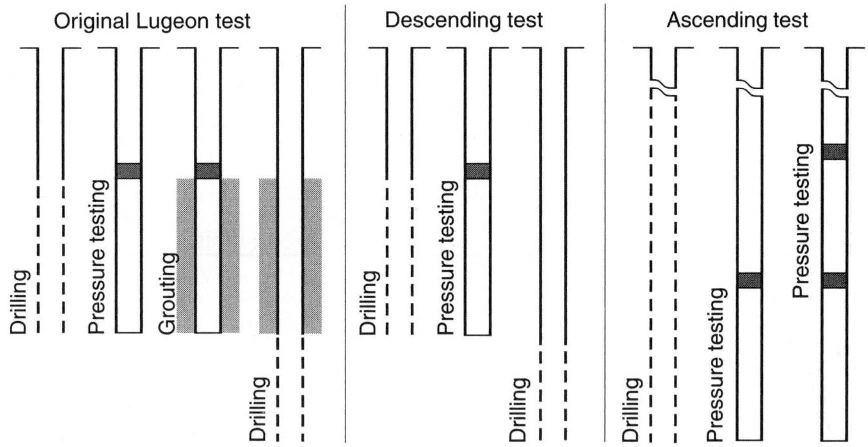

FIGURE 1.111 Possible configurations for packer testing with water injection under pressure. (From Cvetković-Mrkić, S., 1995. Metode geotehničkih melioracija, prva knjiga (in Serbian; Methods of geotechnical meliorations, volume 1). Rudarsko-geološki fakultet Univerziteta u Beogradu, Beograd, 177p.)

Original Lugeon method included grouting of the tested interval before drilling to deepen the borehole for the next test interval. Because of the high cost and time requirements, this method of pressure testing is now seldom used. However, the term Lugeon test is still widely used for describing multiple pressure tests regardless of the drilling method.

Descending test is used mostly in unstable rock. The hole is drilled to the bottom of each test interval and an inflatable packer is set at the top of the interval to be tested. After the test, the boring is cased and then drilled to the bottom of the next test interval. Ascending test is suited for stable rock where the boring can be drilled to the total depth without casing. Two inflatable packers, usually 5 to 10 ft (1.5 to 3 m) apart, are installed on the drill rod or pipe used for performing the test. The rod section between the packers is perforated. Tests start at the bottom of the hole. After each test, the packers are raised to the length of the test section, and another test is performed.

The length of the test section is governed by the character of the rock, but generally a length of 10 ft (3 m) is acceptable. Occasionally, a good packer seal cannot be obtained at the planned depth because of bridging, raveling, fractures, and cavities, or a rough hole. If a good seal cannot be obtained, the test section length should be increased or decreased, or test sections overlapped to ensure that the test is made with well-seated packers. In case of highly permeable fractures and cavities, the 10-ft (3-m) section may take more water than the pump can deliver, and no stabilization of pressure (back pressure) can be developed. If this occurs, the length of the test section should be shortened until back pressure can be developed (USBR, 1977).

Figure 1.112 shows test parameters used to determine the hydraulic conductivity of the tested borehole interval. The hydraulic conductivity is given as (USBR, 1977)

$$K = \frac{Q}{C_s r H} \tag{1.18a}$$

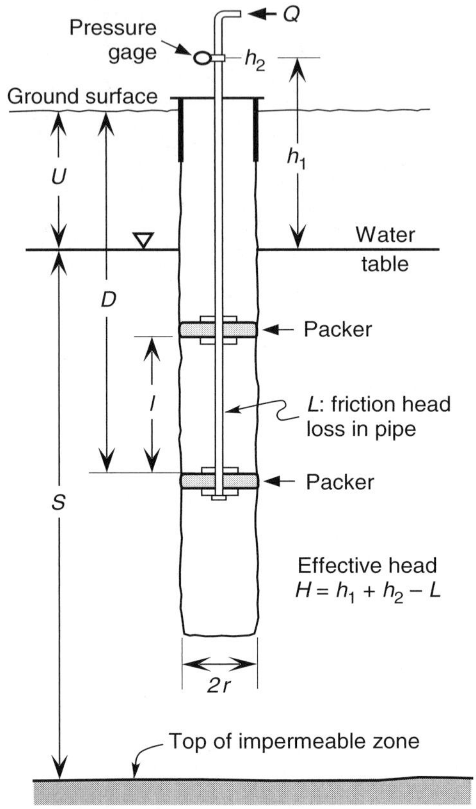

FIGURE 1.112 Pressure permeability test parameters for determining the hydraulic conductivity of consolidated rocks in saturated zone. l: length of test section; D: distance from ground surface to bottom of test section; U: thickness of unsaturated material; S: thickness of saturated material; h_1: distance between gage and water table; h_2: applied pressure at gage; $H = h_1 + h_2 - L =$ effective head; L: head loss in pipe due to friction; ignore head loss for $Q < 4$ gal/min in $1^{1/4}$ in. pipe; use length of pipe between gage and top of test section for computations; r: radius of test hole; Q: steady flow into well. (Modified from USBR, 1977.)

where C_s is the conductivity coefficient for semispherical flow in saturated material through partially penetrating cylindrical test wells, H is the effective head, and r is the borehole radius. Values of C_s are found from the graph shown in Fig. 1.113 for different values of l/r, where l is the length of the tested interval.

Multiple pressure permeability tests (Lugeon test) apply the pressure in three or more approximately equal steps. Each pressure is usually maintained for 20 minutes or so under the requirement of steady water intake. Flow readings are made at 5-minute intervals. The pressure is then raised to the next step. After the highest step, the process is reversed and the pressure maintained for approximately the same time as during the first cycle. A plot of water taken against pressure for all test steps is then used to evaluate hydraulic conditions (Figs. 1.114 and 1.115).

Pressure permeability is often expressed in Lugeon units (Lu), which is equal to the water intake of 1 liter (L) per 1 meter (m′) of the tested interval, during time interval of

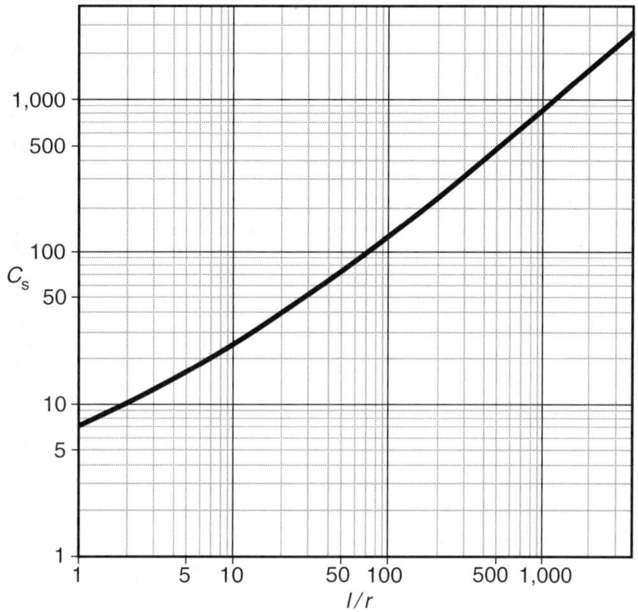

FIGURE 1.113 Conductivity coefficients for semispherical flow in saturated materials through partially penetrating cylindrical test well. (From USBR, 1977.)

1 minute (min), under an injection pressure of 1 mega Pascal (MPa) or approximately 10 atmospheres: 1 Lu = 1 L/min/m′ at 1 MPa.

Specific permeability (q) is another measure of rock permeability obtained from the pressure and water intake measurements. It is defined as volume of water injected into the borehole per 1-m length, during 1 minute, under a pressure of 1-m of water

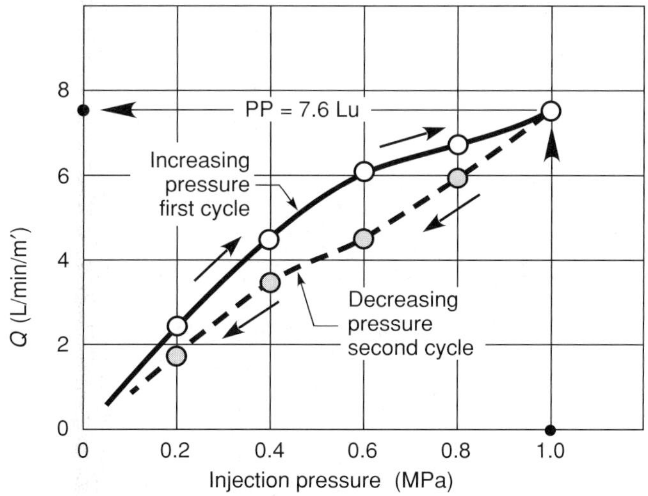

FIGURE 1.114 Typical graph of a Lugeon pressure test results. PP, pressure permeability; in the case shown, it is 7.6 Lugeons for the default injection pressure of 1 MPa. (From Cvetković-Mrkić, 1995.)

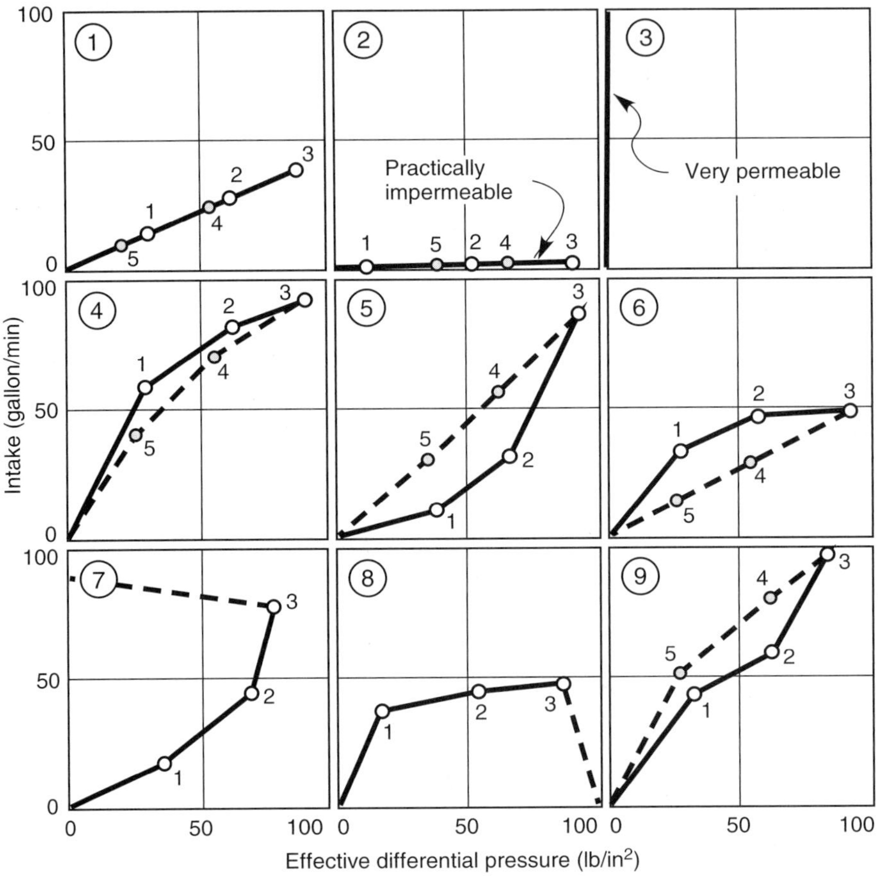

FIGURE 1.115 Plots of simulated, multiple pressure, permeability tests. (From USBR, 1977.)

(0.1 atmospheres):

$$q = \frac{Q}{H \times l} \qquad [\text{L/min/m}' \text{ at } 0.1 \text{ at}] \tag{1.18b}$$

where q is the water injection rate, H is the effective injection pressure, and l is the length of test interval (see Fig. 1.112).

Hypothetical test results of multiple pressure tests are plotted in Fig. 1.115. The curves are typical of those often encountered. The test results should be analyzed using confined flow hydraulic principles combined with data obtained from the core or boring logs. Probable conditions represented by plots in Fig. 1.115 are

1. Very narrow, clean fractures. Flow is laminar, permeability is low, and discharge is directly proportional to head.
2. Practically impermeable material with tight fractures. Little or no intake regardless of pressure.

3. Highly permeable, relatively large, open fractures indicated by high rates of water intake and no back pressure. Pressure shown on gauge caused entirely by pipe resistance.
4. Permeability high, with fractures that are relatively open and permeable but contain filling material that tends to expand on wetting or dislodges and tends to collect in traps that retard flow. Flow is turbulent.
5. Permeability high, with fracture filling material that washes out, increasing permeability with time. Fractures probably are relatively large. Flow is turbulent.
6. Similar to point 4, but fractures are tighter and flow is laminar.
7. Packer failed or fractures are large, and flow is turbulent. Fractures have been washed clean; highly permeable. Test takes capacity of pump with little or no back pressure.
8. Fractures are fairly wide but filled with clay gouge material that tends to pack and seal when under pressure. Takes full pressure with no water intake near end of test.
9. Open fractures with filling that tends to first block and then break under increased pressure. Probably permeable. Flow is turbulent.

The USGS has developed a Multifunction Bedrock Aquifer Transportable Testing Tool (BAT^3), which can be used to test discrete borehole intervals by either injecting or withdrawing water. The equipment is designed to perform the following operations by isolating a fluid-filled interval of a borehole using two inflatable packers (Shapiro, 2001):

- Collect water samples for chemical analysis.
- Identify hydraulic head.
- Conduct a single-hole hydraulic test by withdrawing water.
- Conduct a single-hole hydraulic test by injecting water.
- Conduct a single-hole tracer test by injecting and later withdrawing a tracer solution.

The equipment can be configured to conduct these operations with only one of the borehole packers inflated, dividing the borehole into two intervals (above and below the inflated packer). The Multifunction BAT^3 is designed with two inflatable packers and three pressure transducers that monitor fluid pressure in the test interval (between the packers), as well as above and below the test interval; pressure transducers above and below the test interval are used to ensure that the borehole packers seal against the borehole wall during applications.

Sections of a borehole containing highly transmissive fractures are most easily tested by withdrawing water, whereas fractures with low transmissivity are tested by injecting small volumes of fluid. The Multifunction BAT^3 is configured with both a submersible pump and a fluid-injection apparatus in the test interval to accommodate hydraulic tests that either withdraw or inject water. With this capability, the Multifunction BAT^3 can estimate transmissivity ranging over approximately eight orders of magnitude (Shapiro, 2001).

1.5.3 Aquifer Pumping Tests

Aquifer pumping tests are among the most costly and labor-intensive field tests in hydrogeology. It is therefore crucial to approach their planning and execution by following well-established and accepted practices and guidelines. There is an abundant published literature on the topic and numerous recommendations and guidance documents by the USGS (Ferris et al., 1962; Stallman, 1971), the United States Environmental Protection Agency (USEPA; Osborne, 1993), the United States Army Corps of Engineers (USACE, 1999), the United States Bureau of Reclamation (USBR, 1977), and the ASTM (1999). Other excellent reference books include *Groundwater and Wells* by Driscoll (1989), which covers all aspects of well design, drilling, and testing; *Analysis and Evaluation of Pumping Test Data* by Kruseman et al. (1991); and *Aquifer Testing: Design and Analysis of Pumping and Slug Tests* by Dawson and Istok (1992).

In general, an aquifer pumping test consists of the following main phases:

- Definition of test objectives.
- Development of a focused conceptual site model (CSM).
- Test design.
- Data collection (recording).
- Data analysis.

The question of defining the test objectives could simply be rephrased to "what exactly do you want to learn about the aquifer?" Is the test being conducted as part of a water budget study where the concern is defining transmissivity and storativity? Or, is the test part of a water supply study where the concern is specific capacity and safe well yield? Is the test part of a groundwater contaminant transport study where the ultimate question is the velocity and direction of the groundwater (contaminant) flow? Is there any concern between the possible interconnection of two or more separated aquifers, such as a near-surface water table aquifer and a deeper artesian aquifer separated by an aquitard? A careful definition of the test objectives is therefore essential to ensure a successful test (USACE, 1999).

Although aquifer pumping tests are, among other things, performed to provide data for development of a more detailed CSM, at the same time it is very important that already available hydrogeologic information be fully utilized in the test design. Ideally, this focused CSM includes locations (or their estimates) of any hydraulic boundaries that may impact drawdown measured in the pumped well and the monitoring wells. Hydraulic boundaries are contacts with less permeable formations (*aquitards*) in both the vertical and the horizontal directions, and all natural and artificial surface water drainage features near the test site including more exotic ones such as leaky sewers. In karst aquifers, it is equally important not to oversee possible existence of subsurface equipotential boundaries such as flowing karst conduits and channels. Possible well interference due to other wells pumping in the vicinity should also be examined.

Preliminary estimates of the aquifer thickness and hydraulic conductivity, including its anisotropy, as well as of the underlying and/or overlying aquitards (where applicable), are useful in predicting possible pumping rates for the test, radius of influence, and drawdown at the test well and various distances from it (i.e., at the existing or possible future monitoring wells). These estimates and other available information on the site hydrostratigraphy integrated within the CSM are then used in the next phase—the test design. In contaminant hydrogeology, every attempt should be made to delineate horizontal and vertical extent of groundwater contamination prior to any aquifer field test. This delineation should be followed by an analysis of possible impact of the proposed test on contaminated groundwater flow directions, both horizontal and vertical.

General design of an aquifer pumping test includes the following key elements:

- Design of test well (radius, depth of screened interval).
- Design of monitoring wells (locations, depth of screened interval).
- Pumping design (pump size, energy source, provision for maintaining constant rate).
- Duration of the test.
- Water discharge control, measurements, and disposal.

When performing aquifer tests in karst aquifers, it is particularly important to dispose of the extracted water far enough from the pumping well so to exclude possibility of its percolation back to the aquifer where it may impact the drawdown. As explained further, it is also important to plan for a longer test in order to analyze possible effects of the dual porosity nature of the aquifer.

Pumping tests are much less frequent in contaminant hydrogeology primarily because of the two main concerns: (1) disposal issues related to large quantities of contaminated water extracted during the test and (2) possible alteration of contaminated groundwater (plume) flow directions. Consequently, various forms of slug tests are still considered default for contaminated aquifers because the contaminated groundwater is not extracted to the surface. When a detail evaluation or implementation of groundwater remediation alternatives includes a pump and treat system, it would often be required that an aquifer pumping test is performed to collect information necessary for the remedial design. This is because slug tests are very short (minutes or less) and characterize a very limited volume of the aquifer porous media in the immediate vicinity of the test well, as opposed to the pumping tests that last much longer and evaluate response of a significantly larger aquifer volume. As indicated earlier, slug tests in karst are meaningless because, in addition to their very short duration, the underlying assumption for all methods of their analyses is that the porous media is intergranular.

Most methods for characterizing aquifer parameters with pumping tests, including in karst, are based on quantitative analysis of the observed drawdown versus the time of pumping at the well. The drawdown is measured in one or more observation wells (piezometers) and/or the pumping well itself. Because the time of pumping is explicitly included in mathematical formulas describing relationship between the rate of pumping, the drawdown, and the aquifer parameters, these methods are called "transient" or time dependent. Virtually all such methods routinely applied today are in one way or another based on the pioneering work of Theis (1935) and involve some type of grapho-analytical solution during which the observed time–drawdown data are matched against theoretical model curves ("type curves"). Unfortunately, practicing hydrogeologists often forget the following key assumptions of the Theis equation and apply the method indiscriminately including in karst: The aquifer is homogeneous, isotropic, confined, of unlimited extent, with an impermeable horizontal base, horizontal flow lines, and receives no leakage (recharge) of any kind. It is obvious that hydrogeologic conditions in the field often prevent direct application of the Theis method not only in karst but also in intergranular porous media (sand and gravel) and fractured rock as well.

As discussed by Kresic (2007a, 2007b), various analytical methods have been continuously developed to account for situations such as the presence of leaky aquitards with or without storage and above or below the pumped aquifer, delayed gravity drainage in unconfined aquifers, aquifer anisotropy, and the presence of fractures. Regardless of the aquifer type and the nature of aquifer porosity, it is important to understand that often more than just one type of curve or analytical method may be fitted to the observed time–drawdown curve. Selection of the correct one therefore depends on the thorough overall hydrogeologic knowledge of the aquifer in question (Fig. 1.116). This is particularly true

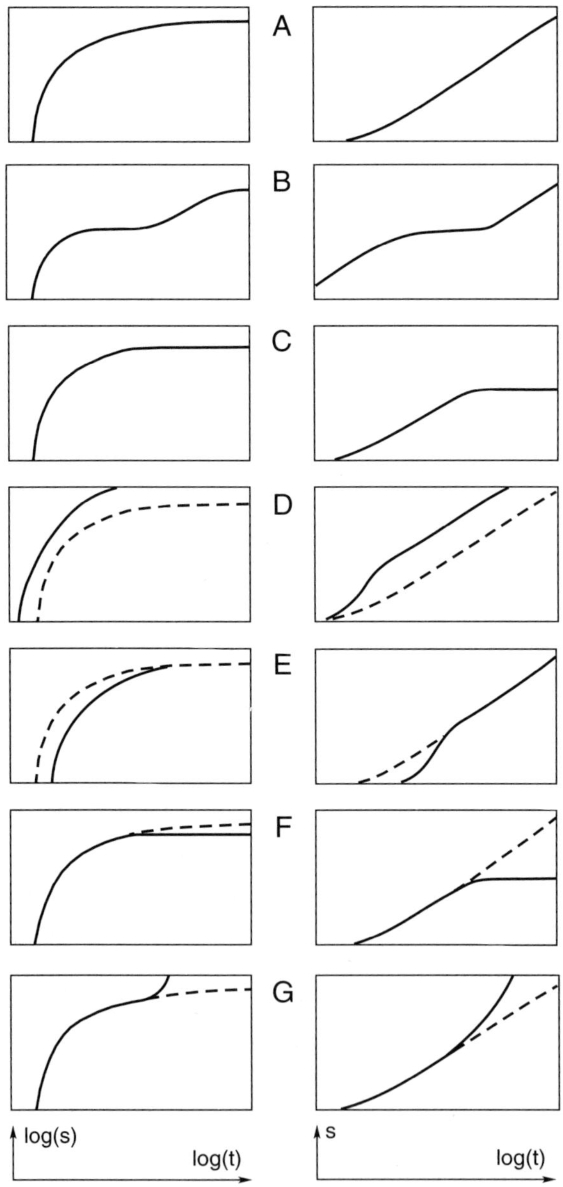

FIGURE 1.116 Log–log (*left*) and semilog (*right*) curves of drawdown versus time. A: confined aquifer; B: unconfined aquifer; C: leaky (or semiconfined) aquifer; D: effect of partial penetration; E: effect of well-bore storage (large diameter well); F: effect of recharge boundary; G: effects of an impervious boundary. (From Griffioen and Kruseman, 2004.)

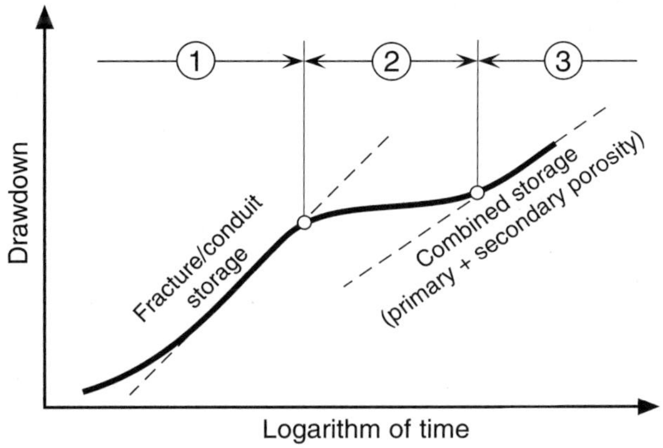

FIGURE 1.117 Theoretical response of the time–drawdown curve caused by effects of specific porosity of karst aquifers. (1) Drawdown due to the initial drainage of secondary porosity (fractures and solution cavities/conduits). (2) Transitional drawdown. (3) Drawdown due to stabilized drainage of all porosity types, including matrix porosity. (From Kresic, 2007b.)

in the case of karst aquifers because of the absence of any mathematically rigorous and widely agreed-upon analytical method of aquifer pumping test analysis. As is the case with numeric groundwater modeling of karst aquifers using EPM approach, analyses of aquifer pumping tests in karst are mostly performed by applying and combining methods developed for intergranular and simple fractured rock aquifers.

A typical time–drawdown curve in response to groundwater withdrawal from a well completed in a karst aquifer is shown in Fig. 1.117. Given enough pumping time and the presence of all porosity types, the time–drawdown curve would show three distinct segments. The first portion of the curve, with a uniform slope, indicates a quick response from a well-connected network of secondary porosity, which may include large dissolutional openings and/or fractures. Drainage of this type of porosity, in the early stages of the test, is characterized by storage properties generally similar to that of confined aquifers. Unconfined intergranular aquifers often exhibit similar early response to pumping, due to sudden change in hydraulic pressure, with the storage coefficient significantly less than 1% (1×10^{-3} or less). The flattening of the curve (curve portion 2) indicates that the initial source of water in large solutional openings (channels, conduits, cavities, in general) and/or fractures is being supplemented by water coming from another set of porosity. This additional inflow of water starts when the fluid pressure in the conduits and/or large fractures decreases enough, resulting in the hydraulic gradients from the smaller fractures and fissures toward the larger fractures and/or karst conduits. This is a transitional portion of the curve. Again, a similar response often happens in unconfined intergranular aquifers when additional water, due to gravity drainage, starts reaching the lowered water table (this is called delayed gravity response to pumping).

As the sources of water from different sets of secondary porosity features, possibly including water from the primary (matrix) porosity, attain similar level of influence, the drawdown curve exhibits another relatively uniform slope (curve portion 3). Such rock formation is often referred to as a dual-porosity formation because of the distinct hydraulic characteristics of different types of porosity present (Barenblatt et al., 1960). An exact determination of individual storage properties of different porosity types in a karst aquifer is beyond capability of common aquifer pumping tests. Aquifer storage parameters related to the effective matrix porosity may be approximately determined by laboratory tests, which would ideally include core samples from multiple locations and depths within the aquifer.

Depending on the duration of the test and the characteristics of the karst aquifer system itself, the time–drawdown curve may exhibit additional changes due to some factors external to the main portion of the aquifer being pumped. These factors may

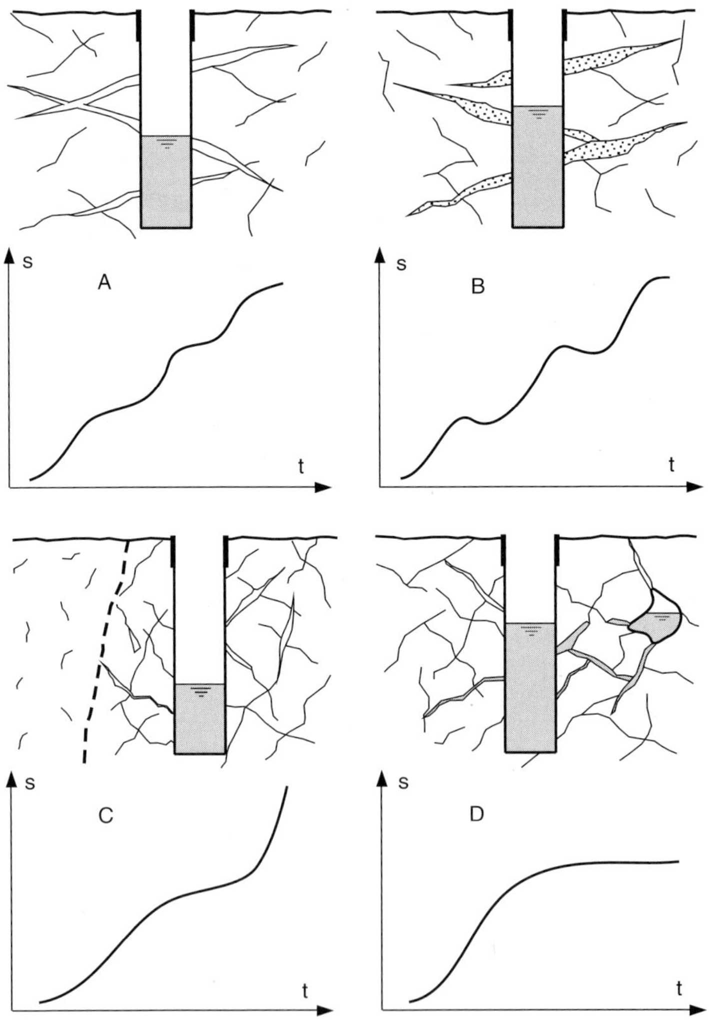

FIGURE 1.118 Characteristic time–drawdown curves for some examples of well pumping tests in karst aquifers. (A) Pumping from a limited number of large fracture and conduit sets that are being consecutively dewatered; (B) Pumping from large fractures/conduits filled with clastic sediments that are being washed out; (C) Cone of depression reaches a less permeable (less fractured) portion of the aquifer; (D) Cone of depression reaches an equipotential boundary such as a large karst conduit or a surface stream. (Modified from Larsson, 1982.)

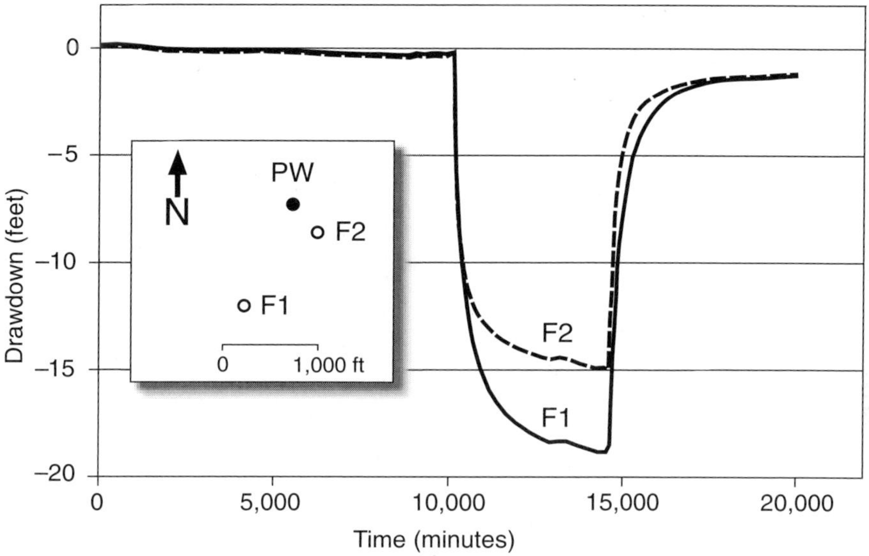

FIGURE 1.119 Time versus drawdown observed at the test well (PW) and two observation wells (F1 and F2) during the Upper Floridan aquifer pumping test at Pateville, Georgia. (From Kresic, 2007b; data courtesy of Geologic Survey Branch of the Georgia Environmental Protection Division.)

include influence of impermeable boundaries, recharge, or constant head boundaries. Given the nature of karst aquifer storage properties, and almost inevitable external influences, time–drawdown curves of long-term pumping tests in karst aquifers exhibit a wide variety of shapes, often nontypical. It is therefore critical that such tests be analyzed using overall geologic and hydrogeologic knowledge about the tested aquifer, rather than formally applying a "type curve formula" typically developed for nonkarstic porous media.

Figure 1.118 illustrates just some of the possible responses of karst aquifers to groundwater withdrawal. In addition to the discussion presented within the figure caption, it is important to emphasize the general importance of step-drawdown tests in karst aquifers. Such tests, routinely performed to determine well efficiency and optimum pumping rate for the main aquifer test (e.g., Rorabaugh, 1953; Roscoe Moss Company, 1990; Kresic, 2007a), may also reveal the nature of aquifer porosity surrounding the well.

Figures 1.119 and 1.120 show comparative analysis of the aquifer pumping test conducted in the Upper Floridan aquifer in southern Georgia at Test Site 14 near Pateville, the United States. The test was performed at 290-ft (88-m)-deep irrigation well with the pumping phase lasting 72 hours and the pumping rate of 1,425 gallons per minute (90 L/s). The thickness of the Floridan aquifer in the test area is approximately 340 ft (105 m). A residuum layer, approximately 60-ft (20-m) thick, overlies the Floridan aquifer limestone at the test site. Drawdown was recorded at two partially penetrating observation wells, F1 and F2, which are located 380 and 90 m from the test well, respectively. The drawdown at the closer observation well F2 is smaller than at the farther well F1 indicating that the aquifer is anisotropic. Comparison of various grapho-analytical methods applied to the time–drawdown data at F1 is given for illustrative purposes only. The results of all the analyses reflect adjustments for partial penetration of the well and

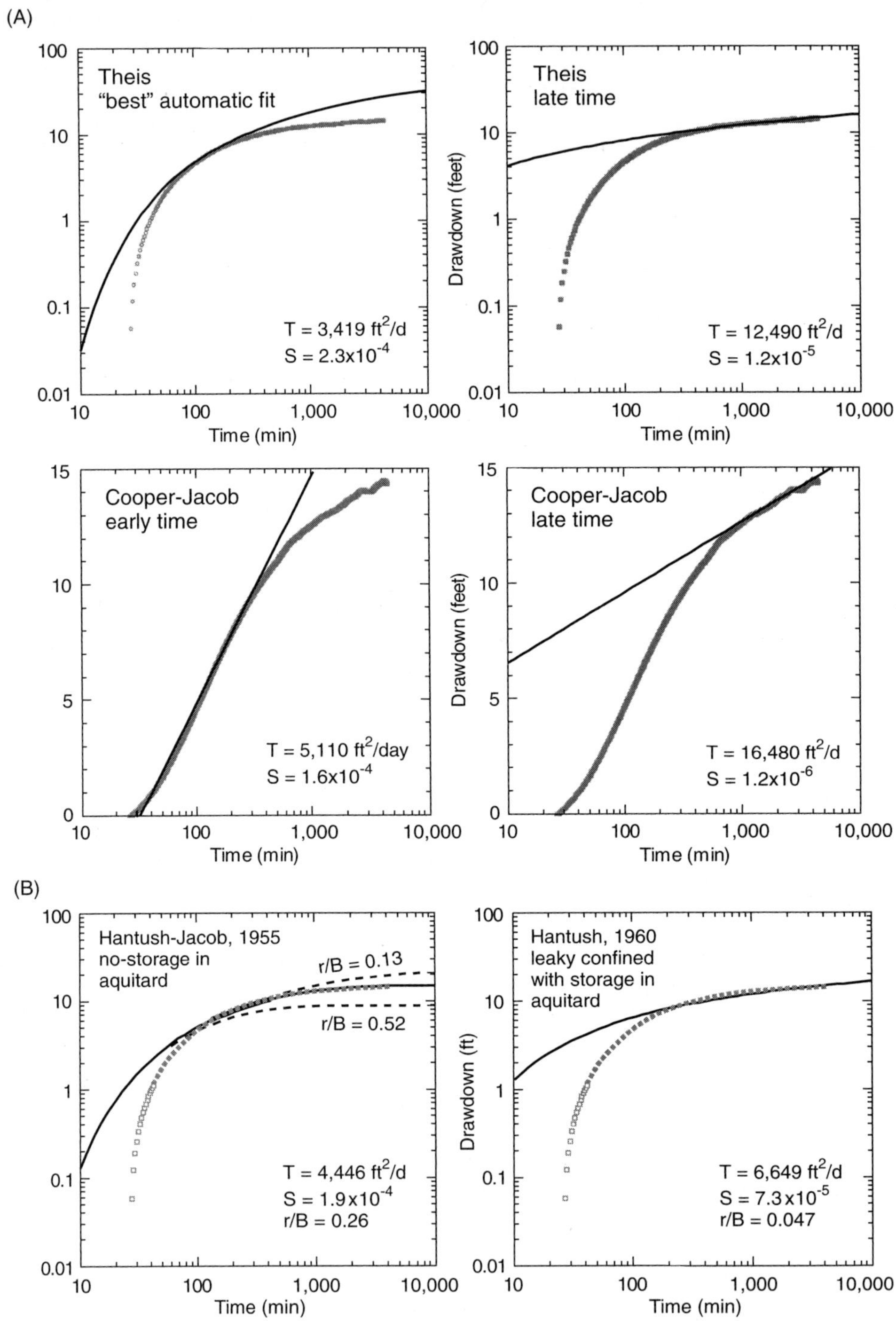

FIGURE 1.120 (A) Theis and Cooper–Jacob solutions for the time versus drawdown data at the Upper Floridan aquifer observation well F1, Test Site 14, Pateville, Georgia. See text for discussion. (B) Leaky aquifer solutions for the same data. (From Kresic, 2007b; data courtesy of Geologic Survey Branch of the Georgia Environmental Protection Division.)

the hydraulic head trend observed prior to pumping. Figure 1.120 clearly shows that the Floridan aquifer does not behave as an ideal "Theisian" confined aquifer and that one simple Theis type curve (or a Cooper–Jacob straight line) cannot provide satisfactory estimates of the aquifer parameters: transmissivity—T and storativity—S.

Deviation from the Theis curve indicates that some less than an ideal (e.g., not purely confining) mechanism is influencing the drawdown. At least several possible explanations immediately come to mind: (1) the cone of depression is reaching a more permeable portion of the aquifer, which is reflected in the slower rate of the drawdown increase; (2) the overlying residuum is starting to provide water due to the hydraulic head decrease in the pumped aquifer (leaky aquitard solution); (3) the effects of dual-porosity are just starting to show up; and (4) an equipotential boundary, such as a karst conduit, is starting to exert its influence on the drawdown. Given significant differences in the results between the methods shown, it is obvious that obtaining "true" (i.e., more representative) aquifer parameters would require additional investigations, including a longer pumping test and possibly a different analytical method in order to assess the dual porosity nature of the aquifer. In any case, the drawdown data later in the test is more representative of the overall aquifer characteristics and its future response to a long-term pumping such as for water supply, irrigation, or aquifer restoration purposes. In the absence of a longer observation period, it is advisable to focus on curve fitting and analytical explanation of the late drawdown data, rather than trying to "approximately" fit the entire data set.

A schematic of a formation having a network of solution openings intersecting a network of "diffuse" fractures (in which the groundwater flow is slow), and being pumped by a fully penetrating well, is shown in Fig. 1.121. The equations and boundary conditions governing the pumping of water from a well in a dual-porosity karst formation are given by Greene et al., (1999) and are described below.

Flow in dissolutional openings:

$$S\frac{\partial h}{\partial t} - T\frac{1}{r}\frac{\partial}{\partial r}\left(r\frac{\partial h}{\partial r}\right) = \beta\left(h_{\mathrm{f}} - h\right) \tag{1.19}$$

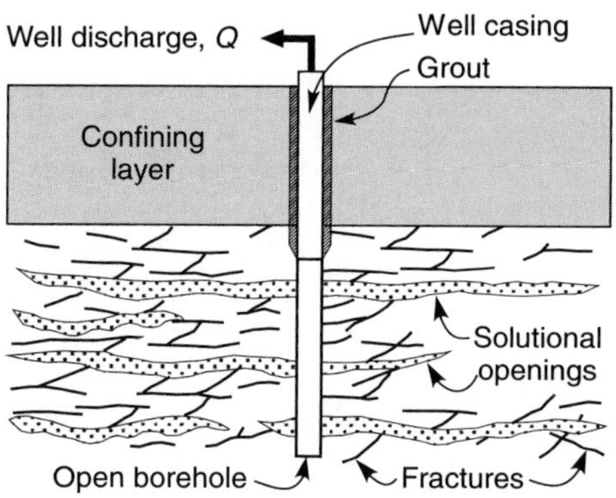

FIGURE 1.121 Conceptual model of groundwater flow to a well pumping in a karst formation consisting of solutional openings and a network of diffuse fractures. (From Greene et al., 1999.)

Initial condition in dissolutional openings:

$$h(r, t = 0) = H \tag{1.20}$$

Boundary conditions in dissolutional openings:

$$2\pi r T \left.\frac{\partial h}{\partial r}\right|_{r \to 0} = Q \tag{1.21a}$$

$$h(r \to \infty, t) = H \tag{1.21b}$$

Flow in fractures:

$$S_\mathrm{f} \frac{\partial h_\mathrm{f}}{\partial t} = -\beta\,(h_\mathrm{f} - h) \tag{1.22}$$

Initial condition in fractures:

$$h_\mathrm{f}(r, t = 0) = H \tag{1.23}$$

where S is the storativity of dissolutional openings, S_f is the storativity of the network of diffuse fractures, t is the time, T is the transmissivity of dissolutional openings, h is the hydraulic head in the dissolutional openings, h_f is the hydraulic head in the network of diffuse fractures, H is the initial head, r is the radial distance from the pumping well, and β is the rate of fluid exchange between the network of fractures and the dissolutional openings.

Assuming that the aquifer is homogeneous and isotropic, its properties are estimated by solving the above equations for various choices of T, S, S_f, and β and the calculated type curves are compared with the observed data until an acceptable fit is obtained. Figure 1.122 shows the results of such curve fitting to the data observed during testing of the Dickey well in the confined Madison limestone karst aquifer near Spearfish, South Dakota, the United States (Greene et al., 1999). The observed data are for the Kyte well, located 1,800 ft from the Dickey well, which was pumped for 6 days at 680 gallons per minute. Note that, if the test had lasted only 1 day, for example, the effects of dual porosity would have not been apparent. As mentioned earlier, unconfined aquifers often exhibit similar time–drawdown curves due to delayed gravity response.

For illustration purposes, Fig. 1.123 shows the results of fitting the Kyte well data to a type curve obtained using Neuman solution for unconfined aquifers with delayed gravity response. Arguably, the fitted type curve looks even better than the one on Fig. 1.122. Boulton (1973) showed analytically that the same set of equations can be applied to (1) the delayed flow to a well in an unconfined aquifer and (2) leakance from the upper aquifer through the aquitard overlying the pumped confined aquifer. Kresic (2007a) argues that possible interpretation of the aquifer test results shown in Figs. 1.122 and 1.123 thus become more ambiguous knowing that the Madison limestone aquifer at the test location is overlain by the Minnelusa aquifer which in its lower part acts as a confining layer consisting of interbedded sandstones and dolomitic limestone.

Whatever the reason for the apparent dual porosity-like response to the pumping test is (e.g., true dual porosity nature of the karst aquifer or leakance from the overlying aquitard, or both), the selection of aquifer parameters for some future quantitative

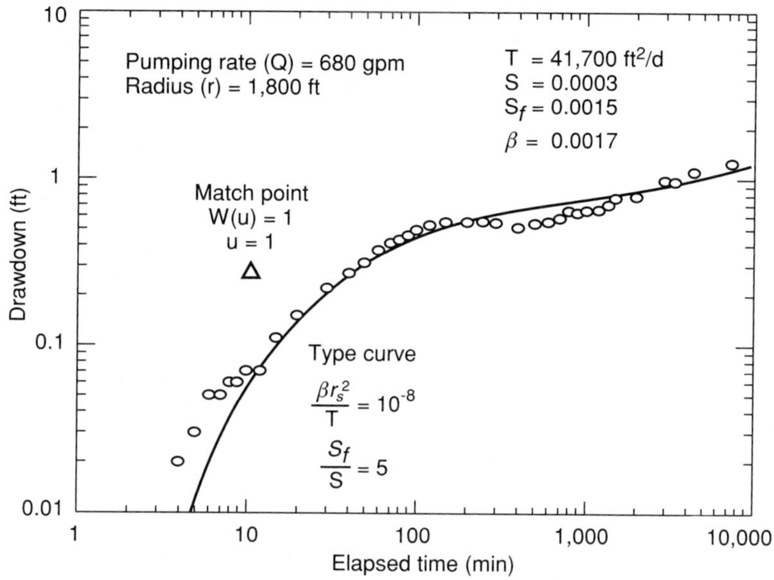

FIGURE 1.122 Best fit type curve match to the drawdown data observed at the Kyte well during the 6-day aquifer pumping test at the Dickey well, Madison limestone aquifer. (From Greene et al., 1999.)

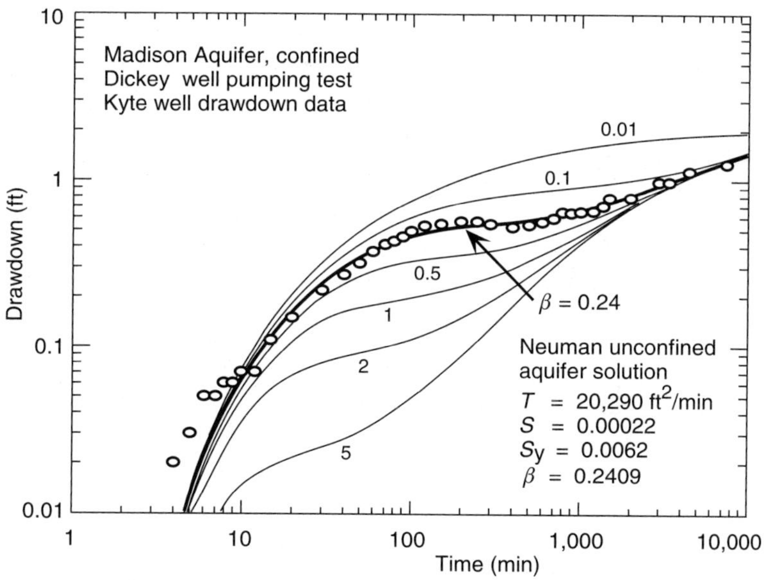

FIGURE 1.123 Neuman solution type curve match to the drawdown data observed at the Kyte well, Madison limestone aquifer, during the 6-day aquifer pumping test at the Dickey well. (From Kresic, 2007a; time–drawdown data from Greene et al., 1999.)

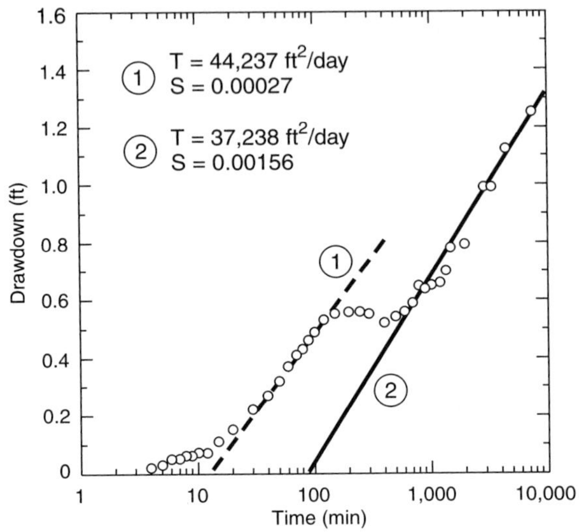

FIGURE 1.124 Cooper–Jacob solution to the early and late drawdown data observed at the Kyte well, Madison limestone aquifer, during the 6-day aquifer pumping test at the Dickey well. (From Kresic, 2007a; time–drawdown data from Greene et al., 1999.)

analysis should always consider the late drawdown information. Interestingly enough, as illustrated by Fig. 1.124, a simple Cooper–Jacob solution for both the early and the late drawdown data gives values of T and S very similar to the dual-porosity solution of Greene et al. (1999). The Cooper–Jacob early storativity is almost identical with the fracture/conduit storativity of Greene et al., while the Cooper–Jacob late storativity is very similar to the "diffuse fracture" storativity of Greene et al. (see Fig. 1.122).

Aquifer Anisotropy

Karst aquifers are by definition heterogeneous and anisotropic when analyzed at a realistic field scale. Determining quantitative parameters of these two very important characteristics of the karst porous media is virtually impossible using aquifer pumping test results alone. At least some information on tectonic (fracture) fabric, geometry and nature of aquifer top and bottom, and carbonate depositional environment, should be available in order to successfully interpret results of an aquifer test in karst. As mentioned earlier, Fig. 1.119 shows that the drawdown observed at the monitoring well F2 is smaller than the one at the monitoring well F1, even though F2 is much closer to the pumping well (PW) and both wells exhibit identical shapes of the time–drawdown curve. This anomaly could not be explained if it were observed in an isotropic aquifer. Knowing the karstic nature of the Upper Floridan aquifer in the test area, one immediate suspect is aquifer anisotropy. However, because of the lack of at least three monitoring wells at various distances and directional angles (azimuths) from the pumping well, determining aquifer anisotropy analytically using the test results would not be possible.

Various methods of determining anisotropy of aquifer transmissivity, developed for homogeneous intergranular aquifers, have been more or less successfully applied in karst using the EPM approach. Warner (1997) applies methodology developed by Papadopulos

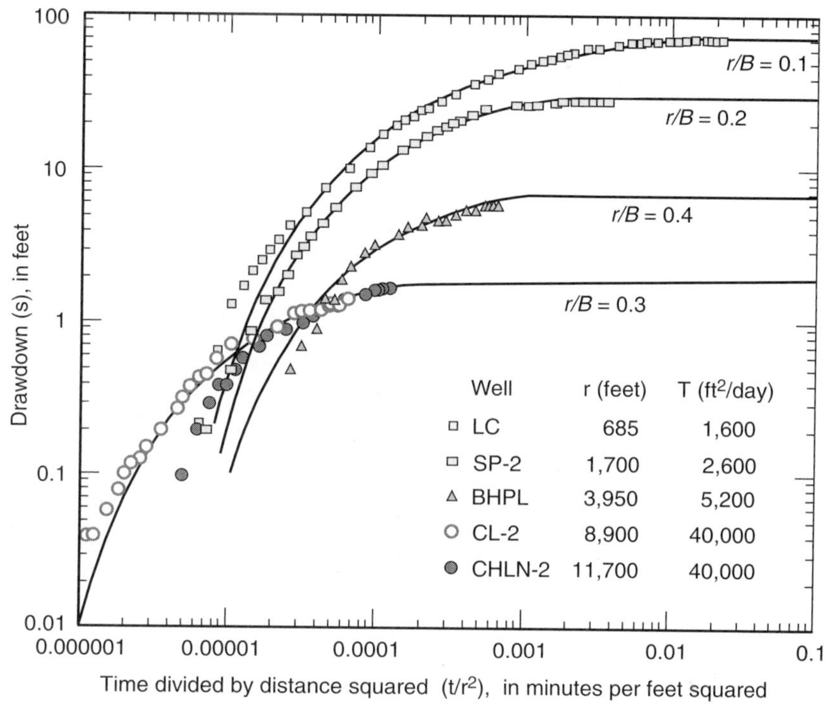

FIGURE 1.125 Drawdown data for five observation wells superimposed on the best fit leaky confined aquifer-type curves for RC-5 aquifer test, Madison limestone aquifer, South Dakota, the United States. (From Greene, 1993.)

(1965) and a related computer program by Maslia and Randolph (1986) to estimate anisotropy of the Upper Floridan aquifer transmissivity using pumping test data from the southwestern Albany area, Georgia, the United States. A degree of professional judgment is exercised by excluding from the analyses, as an outlier, a well with the highest transmissivity. Although debatable, such practice of a selective interpretation of the data is often present in karst hydrogeology given its complexity and nonconformity with many methods developed for the intergranular porous media.

Greene (1993) explains in detail a method developed by Hantush (1966) using an example from a pumping test conducted in the Madison limestone aquifer in the western Rapid City area, South Dakota, the United States (see Figs. 1.125 and 1.126). The method assumes homogeneous, leaky confined aquifer conditions where anisotropy is in the horizontal plane. Figure 1.125 shows field-data plots for five observation wells located at different distances (r) and angles from the production well. The five plots would theoretically fall on the same curve if the transmissivity were not anisotropic or leakage did not occur. The displacement of the plots from one another is the result of directional transmissivities in the Madison aquifer and because leakage through the confining bed does not equally affect drawdown in the five wells (Greene, 1993). The results of the anisotropy analysis are presented in Fig. 1.126. The major axis of transmissivity is 56,000 sq ft/day (5,200 m^2/day) at an angle of 42 degrees clockwise from north. The minor axis of

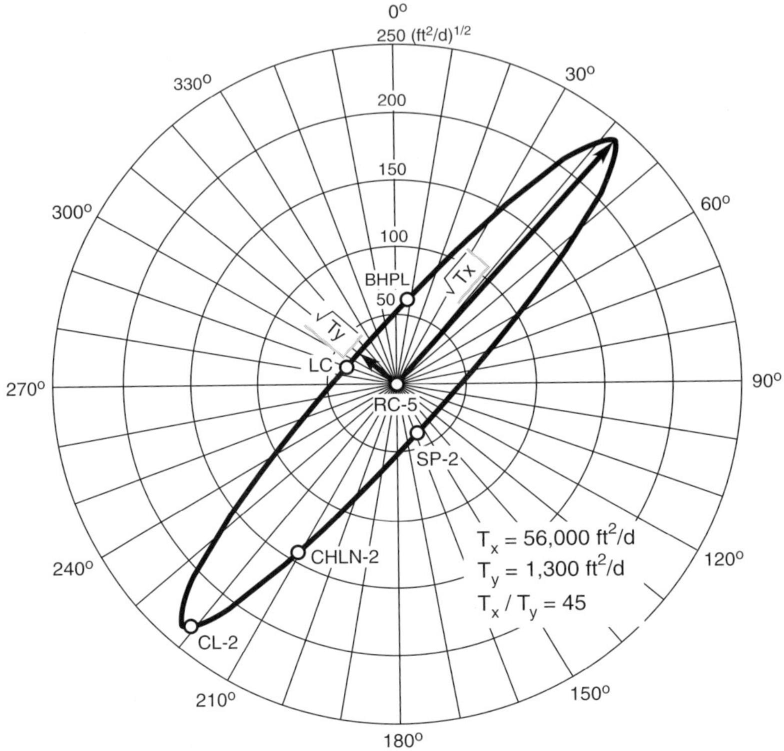

FIGURE 1.126 Theoretical ellipse showing the angle and magnitude of the major and minor axes of transmissivity from the anisotropic analysis of the RC-5 aquifer test, Madison limestone aquifer, South Dakota, the United States. (From Greene, 1993.)

transmissivity is 1,300 sq ft/day (120 m^2/day) at an angle of 48 degrees counterclockwise from north. It should be noted that any such analysis, based on the EPM approach, is just an indicator of a possible anisotropy due to, for example, some preferential flow paths (conduits) in the karst aquifer.

Transmissivity of Discrete Aquifer Zones

In a karst aquifer with heterogeneous secondary porosity (fractures and dissolutional cavities), the inflow of water to a pumped well would most likely occur from discrete intervals where the borehole intersects preferential flow paths within the surrounding rock. One such example is illustrated in Fig. 1.127 that shows results of point flow calculations based on the flowmeter data collected in the pumping well 12K147 during three different overall pumping rates (Warner, 1997). The majority of the flow entering the well is from two discrete zones at about 118 and 122 ft (35 and 37 m) below ground surface. The graph also illustrates that the lower tested portion of the well does not contribute any flow to the well at the lowest pumping rate of 1,080 gallons per minute (4,087 L/min). Based on just this data, one could erroneously conclude that the permeability of the 123- to 136-ft section is very low: When the pumping rate is doubled and then tripled, the section in question does contribute significantly to the overall pumping rate. Although

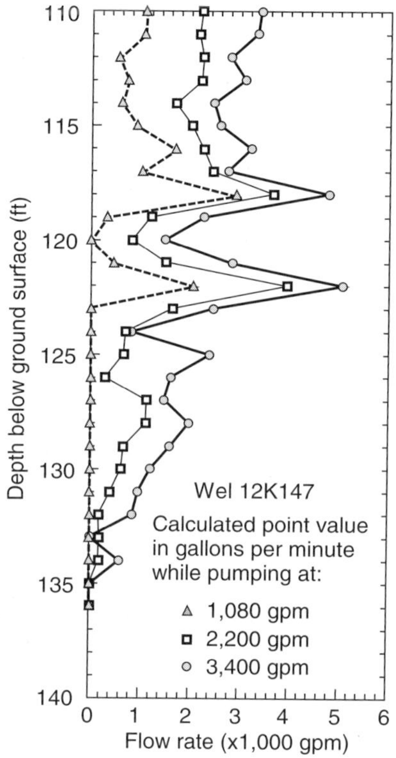

FIGURE 1.127 Calculated point flow rates entering a test well in the Upper Floridan aquifer near Albany, Georgia, the United States, during three different overall pumping rates. (Modified from Warner, 1997.)

water entering the well along this section may still be coming indirectly from the two preferential zones located above, it is clear that the rock mass is capable of transmitting this flow and can certainly not be qualified as "low permeable."

Figure 1.128 illustrates a pumping test showing influence of one, and possibly two, discrete and highly transmissive zones in the Gray Limestone aquifer in southern Florida (time–drawdown data courtesy of Ronald R. Reese, USGS). The aquifer at the test location is unconfined, 25-ft thick, and underlain by a low permeable confining unit. The major flow zone identified by geophysical logging and flowmeter testing in the well is 5-ft thick, from 23 to 28 ft below ground surface. Another possible thin transmissive zone was identified at 17 ft (Ronald R. Reese, personal communication). The test well was pumped for 24 hours at 297 gallons per minute and the drawdown was observed at six monitoring wells screened at various depths. Drawdown observed at well C-1145, located 152 ft from the pumping well and screened in the Gray Limestone aquifer, is shown in Fig. 1.128. Core recovery at the nearby well C-1135 for the interval 20 to 30 ft was only 7% indicating the presence of large solution openings (Reese and Cunningham, 2000). Although the absolute value of the highest measured drawdown is very small (less than 0.07 ft), the graph clearly shows a typical curve characteristic of delayed gravity response and/or dual porosity. This test also illustrates the importance of collecting continuous (frequent) drawdown data using precise pressure transducers.

Two methods of the aquifer test analysis, Cooper–Jacob (modified for unconfined conditions) and Neuman (unconfined aquifer with delayed gravity response) give very

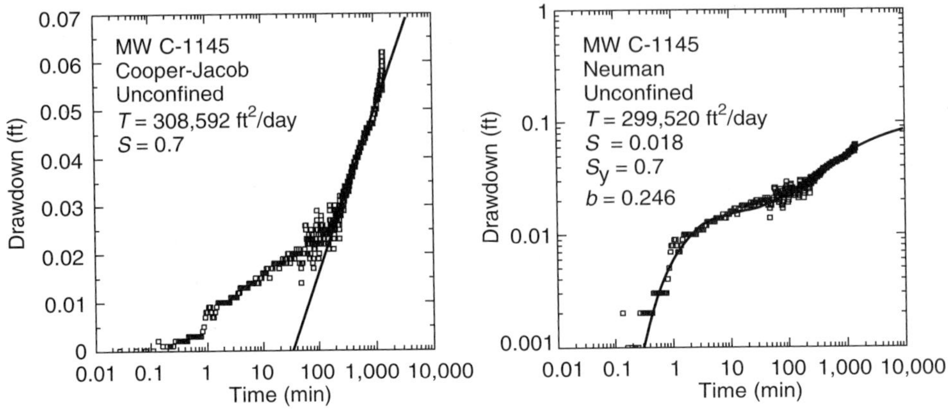

FIGURE 1.128 Time–drawdown analysis for the data collected at monitoring well C-1145 during the Gray Limestone aquifer test at the FAA Radar site, southern Florida. (Data courtesy of Ronald R. Reese, USGS.)

similar, high values of transmissivity: 308,592 and 299,520 sq ft/day. Specific yield of 0.7 in both cases is also very high indicating the presence of interconnected macroscopic voids such as solutional openings (note that aquifer tests in intergranular aquifers cannot produce specific yields this high). Very interesting disturbances in the recorded drawdown data between approximately 50 and 300 minutes into the test could be explained by pressure redistribution within the system, as the upper transmissive zone at 17 ft below ground surface was being dewatered near the pumping well. Namely, the drawdown at the pumping well was lowered close to the 17-ft zone and then eventually below it, which first resulted in partial and then complete dewatering of the conduit(s) near the extraction well. This dewatering is marked by the steep change in the slope of the curve.

Location of Equipotential Boundary

Karst conduits, filled with the flowing water partially or fully (under pressure), may act as equipotential boundaries of varying strength, just like surface streams do when hydraulically connected with the underlying aquifer. The areal extent and exact locations of such invisible boundaries are difficult (if not impossible) to determine without extensive and costly field investigations. When their presence is suspected based on geologic and geomorphologic indicators (e.g., fracture fabric, linear orientation of sinkholes), or a quick stabilization of drawdown during aquifer pumping tests, it is sometimes possible to approximately locate such boundary using time–drawdown data from at least two monitoring wells.

Figures 1.129 and 1.130 show the general setup and time–drawdown data for a karst aquifer test performed in the Sabana Seca/Vega Baja area, Puerto Rico (Torres and Diaz, 1984). There is little doubt that the drawdown recorded at the pumping Ceiba well was almost immediately influenced by a strong equipotential boundary (after less than 10 minutes) and then remained unchanged for the remainder of the test. Just a few hours after the test was initiated, shrimp and fish started flowing out of the test well clearly indicating a cavernous hydraulic connection to Rio Cibuco located about 500 ft southwest of the test well (Arturo Torres, personal communication). An approximate location of the

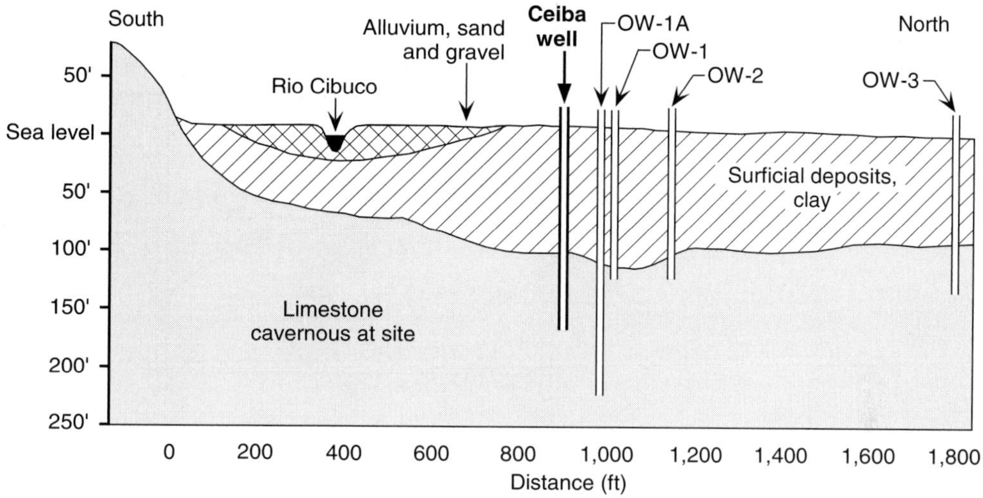

FIGURE 1.129 Vertical cross section of observation wells and pumping well of the Cibuco aquifer test near Vega Baja, Puerto Rico. (From Torres and Diaz, 1984.)

cavernous system influencing the Cibuco aquifer test could be determined using simple methodology based on the theory of image wells and the general Theis equation of flow toward a fully penetrating well in a confined aquifer (see Figs. 1.131 through 1.133).

The stabilization of the drawdown S_r at the pumping well in Fig. 1.131 is balanced out by the negative drawdown S_i at the image well, which is injecting water into the aquifer.

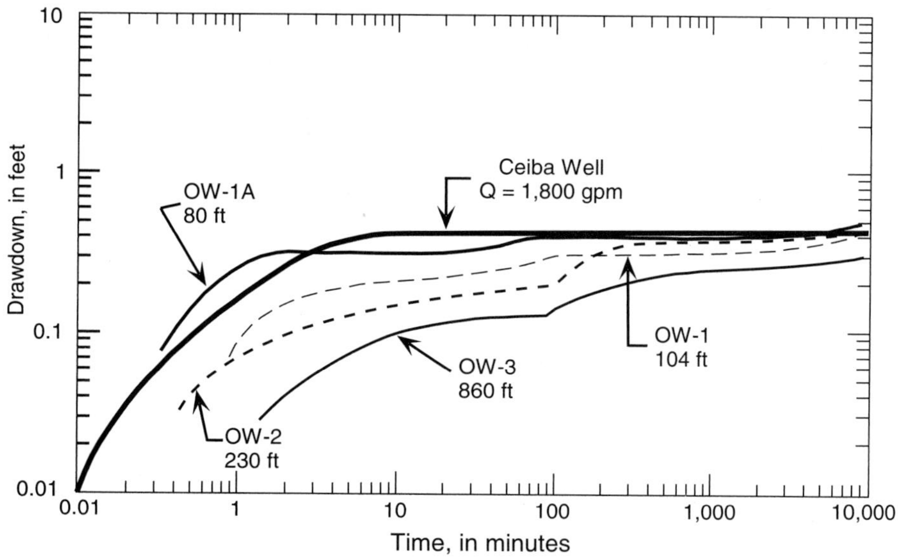

FIGURE 1.130 Time–drawdown curves at Ceiba well and four observation wells recorded during the Cibuco aquifer test near Vega Baja, Puerto Rico. (From Torres and Diaz, 1984.)

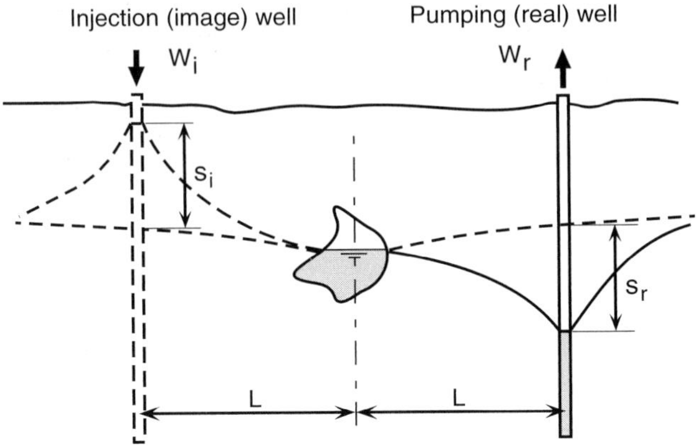

FIGURE 1.131 Superposition of flow toward a well located near a karst conduit with constant potential, using theory of image wells. (From Kresic, 1991.)

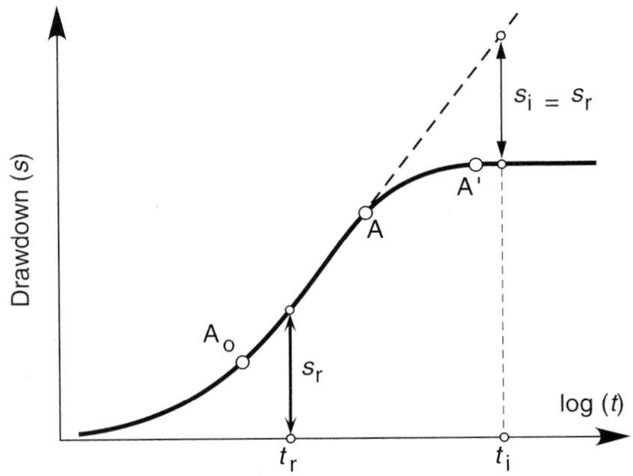

FIGURE 1.132 Determination of times t_r and t_i corresponding to equal drawdowns s_r and s_i in the real and imaginary wells, respectively; explanation in text. (From Kresic, 1991.)

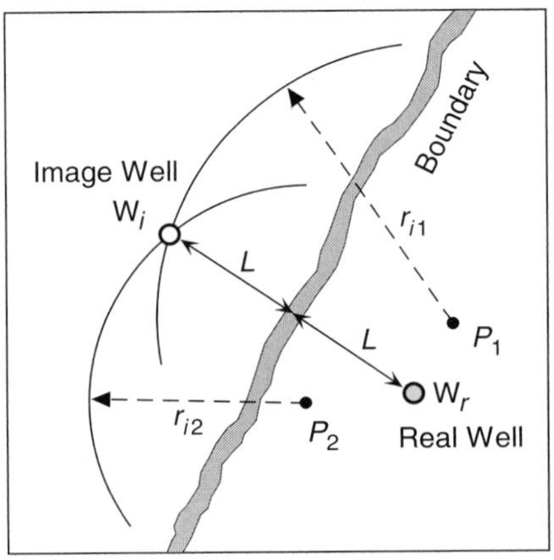

FIGURE 1.133 Location of the equipotential boundary and the image well using time–drawdown information for the two observation wells P1 and P2; explanation in text. (From Kresic, 1991.)

The injection of water at the imaginary well represents the equipotential boundary and happens at the same flow rate as the actual pumping of water from the aquifer at the real well. The drawdowns at the two wells are equal and balanced out:

$$s_r + s_i = \frac{Q}{4\pi T} W(u_r) + \frac{(-Q)}{4\pi T} W(u_i) \tag{1.24}$$

where $W(u_r)$ and $W(u_i)$ are the Theis well functions for the real and imaginary wells, respectively, and

$$u_r = \frac{r_w^2 S}{4Tt_r} \quad \text{and} \quad u_i = \frac{(2L)^2 S}{4Tt_i} \tag{1.25}$$

where r_w is the drawdown in the real well assuming no well losses, L is the distance from the real well to the boundary (which is the same as the distance from the boundary to the image well), T is the aquifer transmissivity, S is the storage coefficient, and t_r and t_i are times as shown in Fig. 1.132.

The relationship in Eq. (1.25) is reflected in the drawdown stabilization shown in Fig. 1.132. The same shape of time–drawdown graphs applies to both the pumping and the observation wells. Drawdown S_r at an observation well graph should be selected anywhere on the straight line portion of the graph, between points A_0 and A. Drawdown S_i, which is equal to S_r, should be selected after the influence of the boundary is complete (after point A′). The corresponding times for the two drawdowns are noted and then used in further calculations.

Since the real and the image wells are completely equivalent (except for the sign of the pumping rate Q), the following relation is true:

$$L = \frac{r_w}{2}\sqrt{\frac{t_i}{t_r}} \tag{1.26}$$

The next step is to determine the location of the image well knowing that the drawdown measured at the observation well is the sum of two drawdowns—one from the influence of the real well (S_r) and the other from the influence of the image well (S_i):

$$s_p = s_r - s_i = \frac{Q}{4\pi T}[W(u_r) - W(u_i)] \tag{1.27}$$

The distance from the observation well to the image well is found using times t_r and t_i, the corresponding drawdowns S_r and S_i, and knowing that $W(u_r) = W(u_i)$:

$$r_i = r_r \times \sqrt{\frac{t_i}{t_r}} \tag{1.28}$$

The same calculation is performed for at least one more observation well, which then enables location of the image well as shown in Fig. 1.133. The boundary is drawn at the half distance between the image and the real wells, perpendicular to the line connecting the two wells.

1.6 Groundwater Flow

1.6.1 Darcy's Flow (Diffuse Flow)

As illustrated in previous sections, interpretation of the hydraulic head measurements in karst aquifers is often nonunique and should not be used to estimate rates of groundwater flow based solely on the principles established by Darcy. The Darcy's flow equation is applicable to intergranular porous media and laminar flow conditions only. It is given as a product of the hydraulic conductivity (K), the hydraulic gradient (i), and the entire cross-sectional area of flow (A) inclusive of solids and all voids:

$$Q = K \times i \times A \tag{1.29}$$

For a karst aquifer, in which all three types of porosity are present (see Fig. 1.134), it is very difficult, if not impossible, to determine reasonably representative values of the three main quantities in the above equation. The hydraulic gradient determined based on the hydraulic head measurements in just several wells may be misleading as illustrated earlier (see Figs. 1.100 through 1.102). The same is true when selecting the most appropriate wells and times of the hydraulic head measurement (see well hydrographs in Fig. 1.134). Best professional judgment would therefore have to be made when deciding if and how to average the hydraulic gradients, in both space and time, over an area covered by monitoring wells.

The importance of field-measured hydraulic heads for any type of groundwater flow calculations cannot be emphasized enough. They are part of all flow equations presented in the following sections. They are also the main calibration target in numeric groundwater modeling, describe the initial conditions for the model, and are part of the majority of model boundary conditions. Without sufficient information on the time-dependent hydraulic head distribution within the area of interest, any calculation of the groundwater flow rates is highly speculative.

The hydraulic heads measured in monitoring wells (piezometers), surface streams and lakes (ponds), and at springs are routinely presented with groundwater contour maps. Common practice is to collect one datasets per different hydrologic season, draw the respective groundwater contour maps for the area of interest, and make conclusions as to possible varying seasonal groundwater flow directions and gradients. Unfortunately, this approach is completely inadequate for karst aquifers as illustrated in Fig. 1.134. Without continuous recording of water levels and precipitation, one would not be able to make any conclusions as to the real dynamics of groundwater flow in the aquifer. Nevertheless, seasonal contours maps of karst aquifers are routinely made by many professionals who often exclude certain anomalous data points while trying to develop a normal-looking map. Figure 1.135 shows that, in karst, all questionable hydraulic head measurement should be further examined to find possible causes.

The most incorrect and misleading case, for any porous media including karst, is when data from monitoring wells screened at different depths in a groundwater system with vertical gradients are lumped together and contoured as one dataset. A perfect example would be a karst aquifer with thick residuum (regolith) deposits or a layer of unconsolidated sediments overlying it. If data from wells screened at different depths in the unconsolidated sediments and the bedrock were lumped and contoured together, it would be impossible to interpret where the groundwater is actually flowing for the

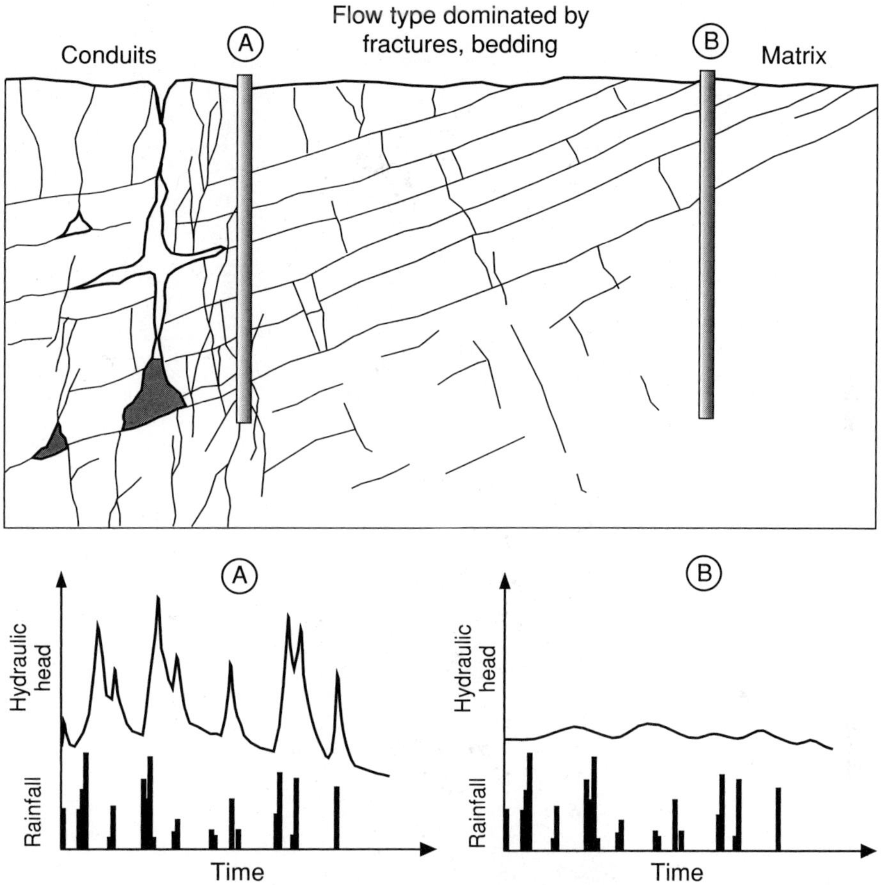

Figure 1.134 Response of the hydraulic head measured in monitoring wells to different types of flow in karst aquifers. A: Rapid conduit flow after major recharge events and no significant storage in rock matrix. B: Delayed and dampened response of the matrix. Flow additionally influenced by fractures may produce any combination of the two hydrographs. (From Kresic, 2007a.)

following reasons: (1) The shallow sediments act primarily as an intergranular porous medium in unconfined conditions, and horizontal flow directions in them may be influenced by local (small) surface drainage features. (2) The karst bedrock has discontinuous flow through fractures and conduits at different depths, which are often under pressure (confined conditions) and may be influenced by regional features such as major rivers or springs. The flow in these two distinct porous media (the residuum/unconsolidated sediments and the karst bedrock) may therefore be in two different general directions at a particular site, including strong vertical gradients from the residuum toward the underlying bedrock. Creating one "average" contour map for such system would not make any hydrogeologic sense.

However, once the judgment on representative hydraulic gradients is made, an equally difficult task of determining the actual (effective) or "representative" cross-sectional area of groundwater flow comes next. In other words, the flow takes place

FIGURE 1.135 Erica Sughrue observes water issuing from this ceiling dome and continuing to flow into the Rusty's Cave canyon. (Photograph copyright Innermost Imagery, printed with permission; many more photos are available at http://innermostimagery.com.)

through all of the following: interconnected pores within the matrix porosity (often referred to as "diffuse flow"), fractures, and solutional cavities that may be fully or partially submerged and with almost any imaginable shape (Fig. 1.136).

The simplest quantitative method for estimating groundwater flow rate in a karst aquifer is to assume that the porous media all behave similarly at some "representative" scale and then apply the Darcy's equation as shown in Fig. 1.137. Steady state flow in a homogeneous, isotropic, confined aquifer of constant thickness, which is resting on a horizontal base, can be calculated for the unit width of the aquifer, or for some other width of interest, by directly applying Darcy's equation in its simplest form [same as Eq. (1.29)]:

$$Q = KA\frac{h_1 - h_2}{L}[m^3/s] \tag{1.30}$$

This is illustrated in Fig. 1.137, left, where the cross-sectional area of the flow segment of interest is $A = a \times b : a$ is the width of the flow, b is the aquifer thickness that stays

FIGURE 1.136 *Left*: Jim Goodbar in a circular gypsum cave passage near Carlsbad, New Mexico. *Right*: Bedding planes and fractures in limestone enlarged by dissolution; outer rim of the Montezuma Lake, Arizona.

constant between the two known hydraulic heads h_1 and h_2 located along the same flow path, distance L apart, and K is the hydraulic conductivity.

The unconfined steady-state flow shown in Fig. 1.137, right, is nonlinear because the saturated thickness (position of the water table) does change between h_1 and h_2. The

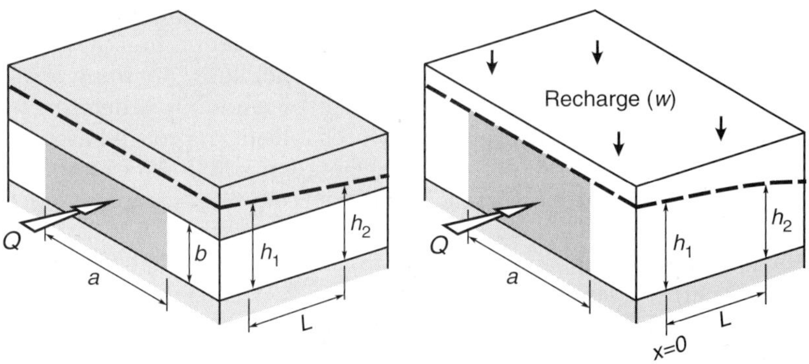

FIGURE 1.137 Confined (*left*) and unconfined (*right*) steady state planar flow in intergranular porous media. The groundwater flow rate is calculated for a selected width (*a*) of the aquifer in both cases.

following equations describe the unconfined conditions including possible recharge rate (w):

$$Q = a \times K\frac{h_1^2 - h_2^2}{2L} + w\left(x - \frac{L}{2}\right) \quad \text{for } x > 0 \tag{1.31}$$

$$Q = a \times K\frac{h_1^2 - h_2^2}{2L} + w\frac{L}{2} \quad \text{for } x = 0 \tag{1.32}$$

In reality, flow conditions are almost always more complicated, including changing aquifer thickness, possible transition between confined and unconfined flow, nonhorizontal base, heterogeneous porous media, and time-dependent (changing in time) recharge from precipitation and adjacent aquifers or aquitards. Various analytical equations have been developed for different flow conditions, as described in hydrogeology textbooks (e.g., see Freeze and Cherry, 1979; Domenico and Schwartz, 1990; Kresic, 2007a).

Although the EPM approach still seems to be the predominant in karst hydrogeologic practice, it is hardly ever justified. Its inadequacy is emphasized when dealing with contaminant fate and transport analyses where all field scales are equally important, starting with contaminant diffusion into the rock matrix and ending with predictions of most likely contaminant pathways and travel velocities in the subsurface.

In the analytically most complicated, but at the same time the most realistic case, the groundwater flow rate in a karst aquifer is calculated by integrating the Darcy's equation of flow through the rock matrix with the hydraulic equations of flow through various sets of fractures, conduits (pipes), and free-flowing subsurface channels (i.e., unconfined pipes). This integration, or connectivity between the four different flow components, can be deterministic, stochastic, or some combination of the two. Fracture connectivity is established by a direct translation of actual field measurements of the geometric fracture parameters such as dip and strike (orientation), aperture, and spacing between individual fractures in the same fracture set, and then doing the same for any other fracture set. Cavities (conduits and channels) are connected in the same way, by measuring the geometry of each individual cavity. Finally, all of the secondary porosity features (fractures and cavities) are connected based on the three-dimensional field measurements and mapping. As can be easily concluded, this deterministic approach must include many uncertainties and assumptions by default. For example, there is hardly any answer to the following question: You have walked and measured this cave, but what if there is another one somewhere in the vicinity you do not know anything about?

Stochastic interconnectivity is established by randomly generating fractures or pipes, possibly using some statistical and/or probabilistic approach based on field measures of the geometric fracture (pipe) parameters (Fig. 1.138). An example of combining deterministic and stochastic approaches is when computer-generated fracture and/or pipe sets are intersected by a mapped preferential flow path such as cave.

Once the overall geometric connectivity between the four flow types is established, the most difficult task of developing physically defensible mathematical integration of the following different sets of flow equations still remains:

- Darcy's flow through the rock matrix ("diffuse flow").
- Laminar and turbulent flow through fractures.

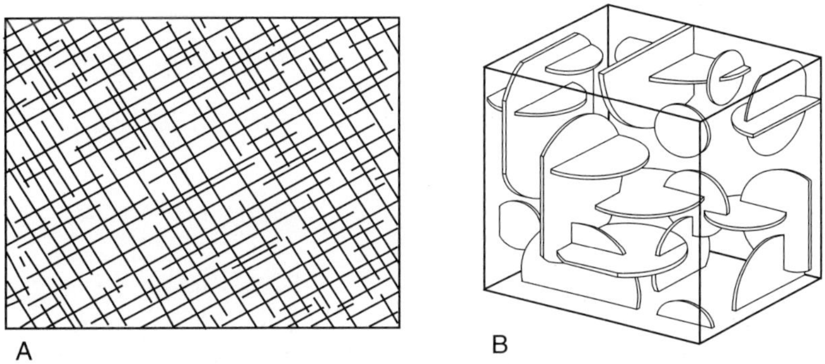

Figure 1.138 Two approaches of representing fracture networks. (A) Two-dimensional, nearly orthogonal random fractures; (B) three-dimensional orthogonal disks. (From Long et al., 1985; copyright American Geophysical Institute.)

- Laminar and turbulent flow under pressure through fully submerged conduits (pipes).
- Laminar and turbulent flow with free surface through karst channels (partially submerged conduits/pipes).

From the above discussion, it is obvious that, except for relatively simple analytical equations used to test basic concepts, most other quantitative methods of karst groundwater flow calculations include some type of modeling. Unfortunately, even if one were to satisfactorily characterize the three-dimensional geometry of all karst flow domains, there are no available modeling programs today capable of fully integrating the related flow equations. This general lack of user-friendly modeling tools, as well as the complexity of several recently developed numeric models that include certain elements applicable to karst, is the main reason why some working professionals select to apply simple EPM calculations, and some of them continue to defend the use of inadequate EPM numeric models in karst at any price.

1.6.2 Flow in Fractures and Bedding Planes

Fractures and bedding planes in brittle carbonate sediments play identical roles in terms of groundwater flow. Aperture and thickness of these discontinuities are two parameters used most often in various single-fracture flow equations, while spacing between the fractures (bedding planes) and fracture orientation is used when calculating flow through a set of fractures. However, these actual physical characteristics are not easily defined for the following reasons:

1. Fracture aperture is not constant and there are voids and very narrow or contact areas called asperities as illustrated in Fig. 1.139. Various experimental studies have shown that the actual flow in a fracture is channeled through narrow, conduit-like tortuous paths (Fig. 1.140) and cannot be simply represented by the continuous, uniformly spread flow between two parallel plates (Cacas, 1989; Chilès and de Marsily, 1993). Deep road cuts in carbonate sediments often enable direct observation of this mechanism as illustrated in Fig. 1.141.

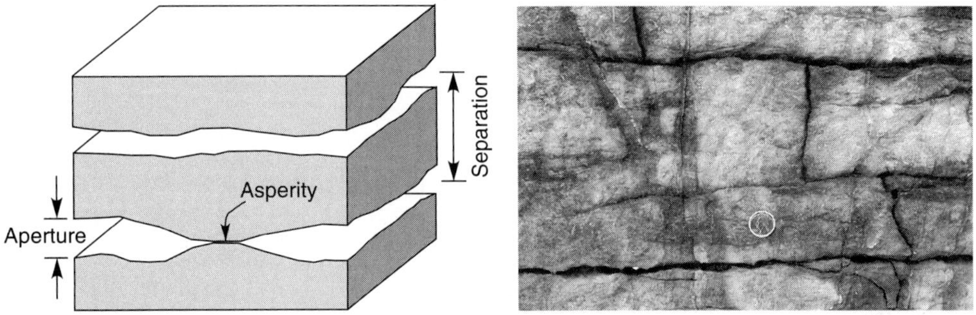

FIGURE 1.139 *Left*: Fracture aperture, asperity, and separation between two parallel fractures in the same fracture set. *Right*: complex sets of discontinuities (bedding planes and fractures) are one of the main characteristics of carbonate sediments (note the outlined 25-cent coin for scale in the middle bottom).

2. Due to stress release, the aperture measure at outcrops or in accessible cave passages is not the same as an in situ aperture. Aperture measured on drill cores and in borings is also not a true one—The drilling process commonly causes bedrock adjacent to fractures to break out, thereby increasing the apparent widths of fracture openings as viewed on borehole-wall images (Williams et al., 2001).
3. Fractures have limited length and width, which can also vary between individual fractures in the same fracture set. Spacing between individual fractures in the same set can also vary. Since all these variations take place in the three-dimensional space, they cannot be directly observed, except through continuous coring or logging of multiple closely spaced boreholes, which is the main cost-limiting factor.

Evolution of analytical equations and various approaches in quantifying fracture flow is given by Whiterspoon (2000) and Faybishenko et al. (2000). In the simplest form, the hydraulic conductivity of a fracture with an aperture B ($B = 2b$ in Whiterspoon's notation) and represented with two parallel plates is

$$K = B^2 \frac{\rho g}{12\mu} \tag{1.33}$$

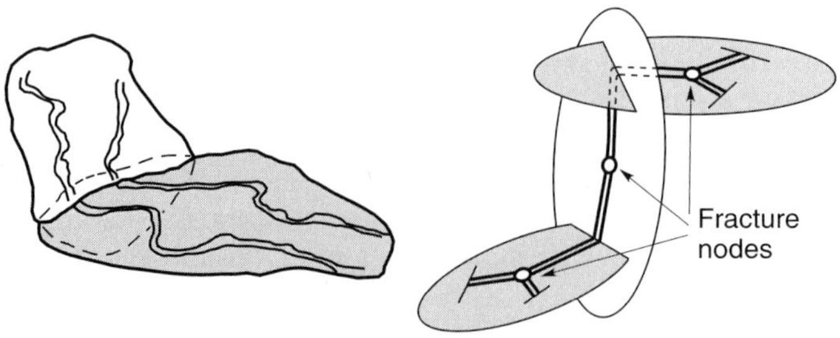

FIGURE 1.140 Channeling in a fracture plane and equivalent channel model. (After Cacas, 1989, from Chilès and Marsily, 1993).

FIGURE 1.141 Discrete seeps from a thinly bedded zone in limestone; road cut near the Kentucky River, US Interstate Highway 75 southeast of Lexington, KY.

where ρ is the fluid density, g is the acceleration of gravity, and μ is the fluid viscosity. For the steady laminar flow, the flux (Q) per unit drop in head (Δh) is given by the so-called cube (or cubic) law:

$$\frac{Q}{\Delta h} = CB^3 \tag{1.34}$$

where C is a constant that depends on the geometry of the flow field. The cubic law applies to ideal smooth parallel fractures without asperities.

The concept of roughness (f) is introduced to account for deviations from ideal conditions, which assume smooth fracture walls and laminar flow. The roughness f is related to the Reynolds number (Re; indicator if the flow is turbulent or laminar), and the friction factor (Ψ) through the following equation:

$$f = \frac{\Psi \mathrm{Re}}{96} \tag{1.35}$$

The friction factor is given by

$$\Psi = \frac{D}{v^2/2g}\Delta h \tag{1.36}$$

where v is the flow velocity.

The Reynolds number is given as

$$Re = \frac{Dv\rho}{\mu} \tag{1.37}$$

where D is the fracture hydraulic diameter assumed to be equal to four times the hydraulic radius, v is the mean velocity of the fluid, ρ is the density of the fluid, g is the acceleration of gravity, and μ is the dynamic viscosity of the fluid. In fluid mechanics, the hydraulic radius is generally described as the cross-sectional area of flow divided by the wetted perimeter of flow.

For a relatively smooth fracture, where the ratio between the absolute height of the asperities (ε) and aperture width (B) is less than 0.1, the transition to turbulent flow is at Reynolds number of about Re = 2,400. As the roughness factor ($f = \varepsilon/B$) increases, in the range of 0.5 to 0.8, the Reynolds number for transition to turbulent is significantly less than 2,400.

In the case of roughness, $f>1$, the cubic law in Eq. (1.34) becomes (Witherspoon, 2000):

$$\frac{Q}{\Delta h} = \frac{CB^3}{f} \tag{1.38}$$

The hydraulic conductivity of a set of parallel fractures (K_f), with N fractures per unit distance, average fracture aperture B, and the fracture set porosity n_f (where $n_f = NB$), is (Snow, 1968):

$$K_f = \left(\frac{\rho g}{\mu}\right)\frac{NB^3}{12} \tag{1.39a}$$

This approximation assumes that the representative aquifer volume acts as an EPM (Darcian continuum). Snow concludes that fracture porosity of a representative aquifer volume with an isotropic three-dimensional system of similar fractures is $n_f = 3NB$, while the hydraulic conductivity (K_{3f}) of such three-dimensional system is double that of any one of its individual fracture sets:

$$K_{3f} = \left(\frac{\rho g}{\mu}\right)\frac{NB^3}{6} \tag{1.39b}$$

Equations of fluid flow in nonideal fractures with influences of various geometric irregularities and fracture network modeling approaches are discussed in detail by Bear et al. (1993), Zimmerman and Yeo (2000), and Faybishenko et al. (2000).

1.6.3 Flow in Conduits

In many karst aquifers, the most important portion of groundwater flow takes place in interconnected solution cavities that are commonly referred to as conduits or channels. By definition, a conduit (or a channel) has one dimension, length, disproportionally larger than the other two. The most complicating and challenging factor when attempting to apply principles of pipe and channel hydraulics to real karst conduits is that their orientation and interconnectivity within the three-dimensional aquifer space is for the

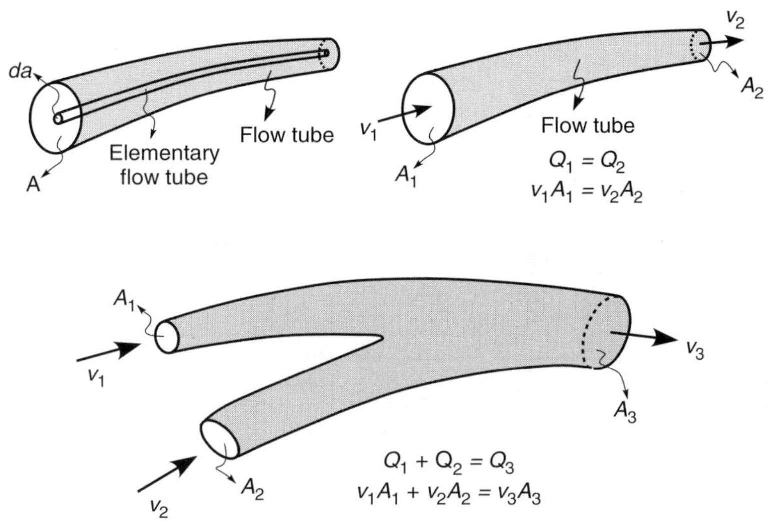

FIGURE 1.142 *Top*: Flow tube in porous media; *da* = cross section of elementary flow tube; A, A_1, A_2, = cross-sectional area of flow; Q_1 and Q_2 = flow rate; v_1 and v_2 = average flow velocity. *Bottom*: Principle of flow continuity for three flow tubes.

most part unknown and unpredictable. Notably, a network of accessible cave passages that has been mapped in the field often does not represent the conduit network where the actual flow is taking place most of the time. The accessible cave passages may have some visible flow along their floors, and may become more active or even fully submerged during major recharge episodes, but the flow in the permanently saturated zone is taking place at lower horizons, usually inaccessible even to well-equipped, skilful cave divers.

Most importantly, the traditional hydraulics of pipes (channels) is based on the principle of flow continuity, which assumes that there is no inflow or outflow of water through pipe walls. This is illustrated in Fig. 1.142. The elementary flow rate (dQ) through an elementary flow tube is directly proportional to the elementary cross-sectional area of the tube (da) and the flow velocity (v):

$$dQ = vda \tag{1.40}$$

"Many" such elementary flow tubes can be combined into a realistic three-dimensional portion of an aquifer through which the groundwater flow is taking place; as long as there is no loss or gain of water inside this (combined) realistic flow tube, however, the principle is the same:

$$Q = \int_A dQ = \int_A vda \tag{1.41}$$

$$Q = v_{\text{av}} A \tag{1.42}$$

where A is the sum of all cross sections of the elementary flow tubes and v_{av} is the average flow velocity within the realistic flow tube.

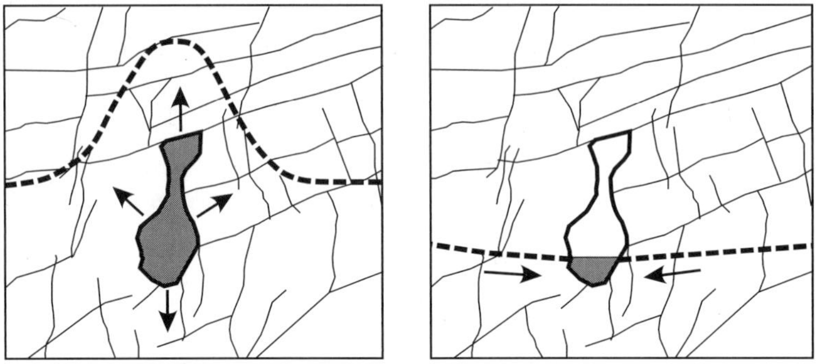

FIGURE 1.143 *Left*: Rapid buildup of pressure in a karst conduit after major recharge episode results in groundwater flow from the conduit into the surrounding rock matrix where the hydraulic head is lower. *Right*: Flow from the surrounding matrix to the conduit during recession periods (the conduit may or may not be fully submerged). Thick *dashed lines* are hydraulic heads in the matrix, *arrows* indicate flow direction between the conduit and the matrix.

Two or more flow tubes can merge into one and the resulting flow rate is additive (Fig. 1.142, bottom). Consequently, one flow tube can split into several where the sum of the new flows is equal to the initial flow. It should be noted that the term "tube" in aquifer hydraulics does not imply that the portion of the aquifer under consideration must look like (has shape of) a tube; what it means is that the flow lines (streamlines) of groundwater particles inside the tube remain inside, and there are no new particles entering the tube.

In a karst aquifer, however, there is an ongoing exchange of water between the conduits (tubes) and the surrounding rock matrix including small fissures and fractures where the pipe or channel flow hydraulics is not applicable. As illustrated in Fig. 1.143, when the hydraulic head (pressure) in a fully submerged conduit is higher than in the surrounding matrix, which can happen after rapid episodes of recharge, the flow of water is from the conduit toward the matrix. During recession periods, the flow is from the matrix toward the conduit, whereas any combinations of these two cases including their duration will depend on various site-specific factors such as the dynamics and nature of aquifer recharge and the relative significance of different porosity types.

Where most, if not all, common applications of the hydraulic equations of flow through pipes (channels) in karst fail is during episodes of rapid recharge and hydraulic head changes. In a pressurized karst conduit system, the propagation of hydrostatic pressure changes from one portion of the system to another can be practically instantaneous (on the order of hours) at distances measured in kilometers. This is schematically illustrated in Fig. 1.144 and with real-world photographs in Fig. 1.145. Karst aquifers with well-developed conduit networks and low matrix porosity typically exhibit high amplitudes of the hydraulic head fluctuations, with rises sometimes exceeding tens of feet (meters) in the matter of hours (Fig. 1.146). The simple mechanism of rapid pressure propagation in karst aquifers is often erroneously referred to as "fast" or "quick" flow even though the flow of the newly recharged water through conduits is not directly responsible for the initial increase of the spring discharge rate. In carbonate aquifers with

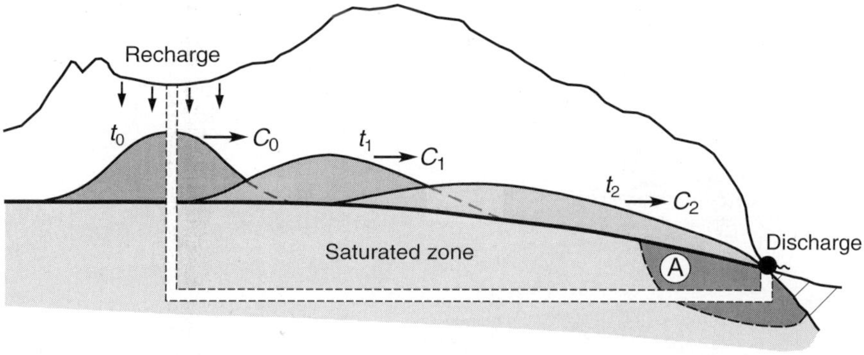

FIGURE 1.144 Formation and movement of a "groundwater wave" caused by localized recharge event. Velocity of the wave is C_0 at time t_0, C_1 at time t_1, and C_2 at time t_2, where $C_0 > C_1 > C_2$ due to decreasing hydraulic gradients. A is the volume of old water stored in the aquifer and discharged under pressure at the spring due to the recharge event. The newly infiltrated water will start discharging at the spring with delay. Hypothetical conduit system representing the flow under pressure and instantaneous pressure propagation is depicted with *dashed lines*. (Modified from Yevjevich, 1981.)

significant matrix porosity, the effects of rapid pressure redistribution in conduits are dampened by the high permeability of conduit walls (see Fig. 1.17).

1.6.4 Bernoulli Equation

Flow through a pipe of varying cross-sectional area is described by the Bernoulli equation for real viscous fluids as illustrated in Fig. 1.147. Since there is no gain or loss of water in the pipe, the flow rate remains the same while other hydraulic elements change from one cross section to another. The total energy line (E) of the flow decreases from the upgradient cross section toward the downgradient cross section of the same flow tube (pipe or conduit) due to energy losses. These losses or resistance to flow are the result of internal friction within the flowing fluid, friction between the fluid and the conduit walls, and the general flow turbulence including local losses caused by various conduit irregularities and wall roughness. Bernoulli equation is based on the principle of energy conservation where the energy losses represent an irreversible conversion of mechanical (flow) energy into heat energy. The following general equation is for the three flow cross sections, denoted as 1–1, 2–2, and 3–3 on Fig. 1.147 (note that at the starting cross section 0–0, the flow velocity is zero):

$$E = z_1 + \frac{p_1}{\rho g} + \frac{\alpha v_1^2}{2g} + h_{w(0-1)} = z_2 + \frac{p_2}{\rho g} + \frac{\alpha v_2^2}{2g} + h_{w(0-2)}$$

$$= z_3 + \frac{p_3}{\rho g} + \frac{\alpha v_3^2}{2g} + h_{w(0-3)} \tag{1.43}$$

where z is the elevation head above the reference level, p is the fluid pressure, ρ is the fluid density, g is the acceleration of gravity, v is the average flow velocity, α is the Coriolis

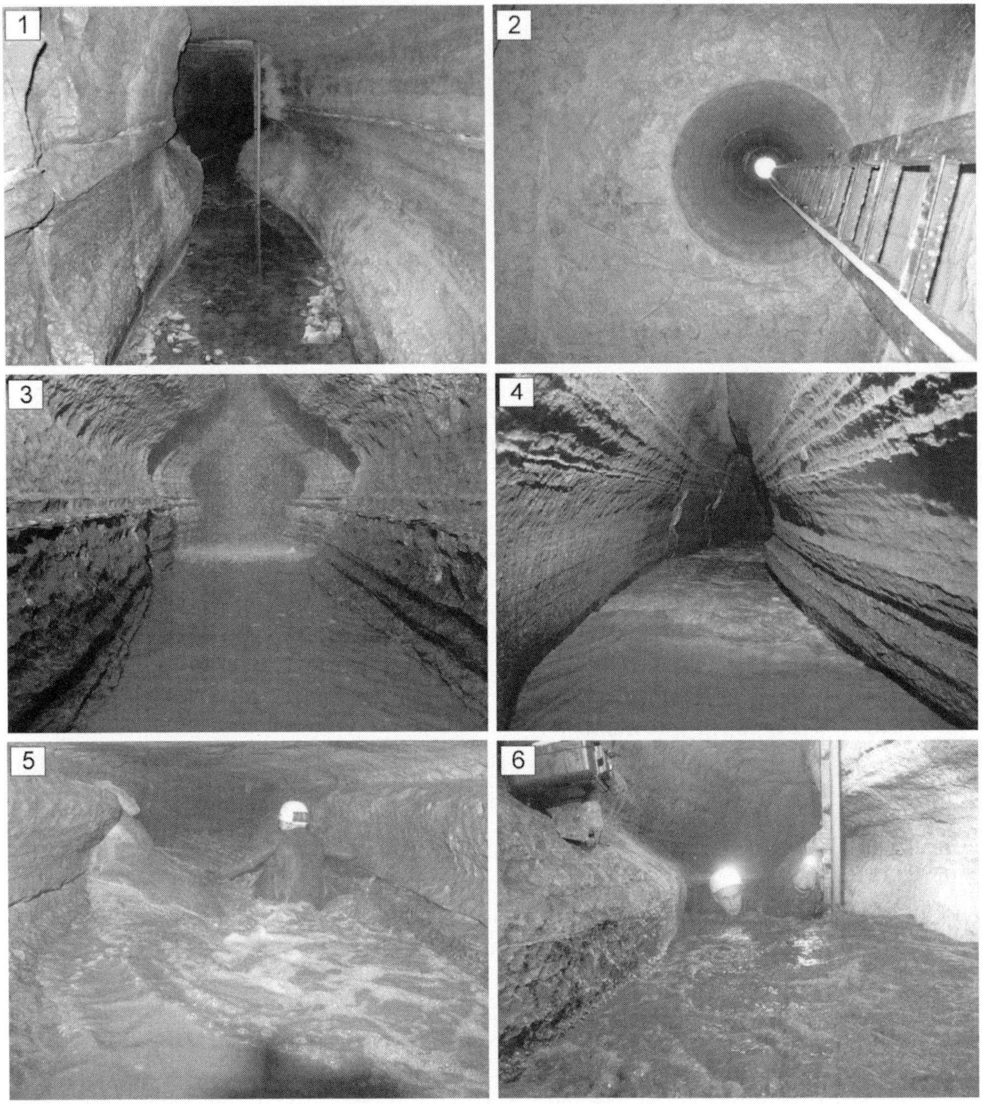

FIGURE 1.145 Encounter with a summer storm in Bat River Cave, Minnesota Cave Preserve. (1) Ladder extending from the new drilled vertical entrance to the cave; visible is normal, typically low, and calm water flow; (2) looking up the ladder in the drilled shaft; (3) after sudden intense storm on August 18, 2007, a waterfall occurred where dry ceiling existed before (4:03 PM); (4) the water begins to rise and becomes turbulent (4:17 PM); (5) cavers round the bend and struggle against the current to retreat (4:22 PM); (6) seconds after this photo was taken, it was almost impossible to stand in the passage without getting swept away (4:59 PM). (Photographs courtesy of John Ackerman; available at http://www.karstpreserve.com/index.html.)

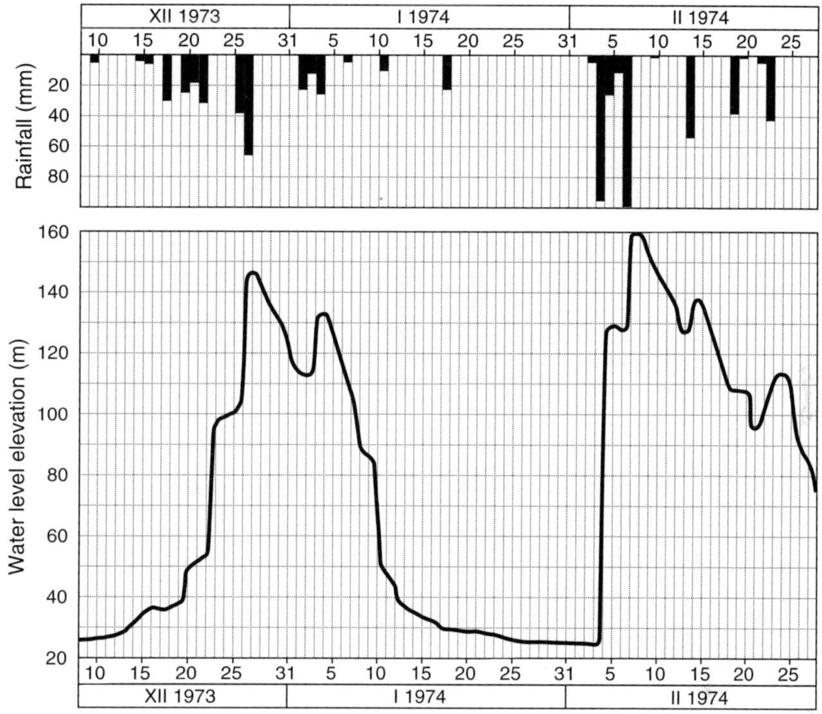

FIGURE 1.146 Detail of a piezometer hydrograph in the drainage area of the Ombla Spring near Dubrovnik, Croatia. The piezometer is located in Herzegovina, several kilometers from the Adriatic Sea and the spring. After rainfall of 96 mm on February 4, 1974, the water level rose 104 m in less than 24 hours. The largest recorded amplitude of water levels at piezometers in the area is 198 m or 650 ft. (Modified from Milanovic, 1979.)

coefficient of velocity nonuniformity at the flow cross section ($1.06 < \alpha < 1.13$), and h_w is the total energy loss between two flow cross sections.

Unlike the energy line (E), the hydraulic head line (H) may go "up" and "down" along the same flow tube as the cross-sectional area increases or decreases, respectively. In other words, when the flow velocity decreases due to widening of the conduit, the hydraulic head increases because of the increased fluid pressure. The total energy surface, which includes the velocity component ($\alpha v^2/2g$), can be directly measured only by the Pitot device whose installation is not feasible in most field conditions. Monitoring wells and piezometers, on the other hand, only record the hydraulic head, which does not include the flow velocity component. It is therefore conceivable that two piezometers in or near the same karst conduit may not provide useful information for calculation of the real flow velocity and flow rate between them and may even falsely indicate the opposite flow direction. In fact, as discussed by Bögli (1980), water rising through a tube in an enlargement passage can flow backward over the main flow conduit and into another tube that begins at a narrow passage in the same main conduit.

There are various complicating factors when attempting to calculate flow through natural karst conduits using the pipe approach and Bernoulli equation:

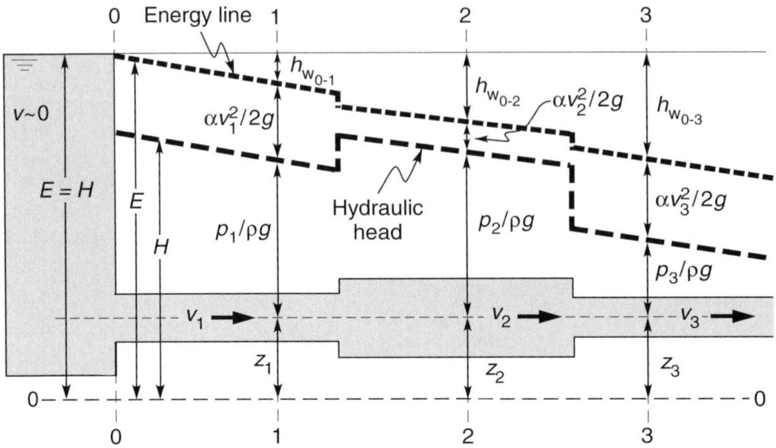

FIGURE 1.147 Illustration of the Bernoulli equation for flow of real viscous fluids through a pipe with the varying cross-sectional area. Energy line (E) at any given cross section is the sum of the elevation head (z), the fluid pressure head ($p/\rho g$), and the velocity head ($\alpha v^2/2g$). The fluid pressure head and the elevation head give the hydraulic head (H).

1. Flow through the same conduit may be both under pressure and with free surface, and the conduit or part of it may eventually be completely drained of water as the permanently saturated zone is lowered.
2. Since conduit walls are more or less irregular ("rough"), the related coefficient of roughness that contributes to energy losses has to be estimated and inserted into the general flow equation.
3. Conduit cross section may vary significantly causing various local energy losses, particularly where it abruptly constricts or widens.
4. The flow may be both laminar and turbulent in the same conduit, depending on flow velocity, cross-sectional area, and wall irregularities. The energy losses and velocity distribution are quite different for these two flow regimes (see next section).
5. Portions of the conduit system, such as large chambers and vertical shafts, may serve as storage reservoirs during and following periods of high recharge rather than simple flow-through conduits. Possible impacts of unknown reservoirs are unpredictable but can be assumed and tested with simple hydraulic equations as illustrated further.

Despite the limitations of the hydraulic calculations listed above, pipe and reservoir approaches to karst aquifer systems have been practiced considerably and with varying success. Figure 1.148 illustrates schematic of a system approximated with three subsystems or reservoirs (R) interconnected with pipes all of which ultimately drain at a single outlet such as a karst spring (modified from Avdagic, 1990). The reservoirs may represent actual surface impoundments such as flooded karst poljes, or subsurface aquifer compartments represented by the governing hydraulic heads (H) and receiving recharge (P).

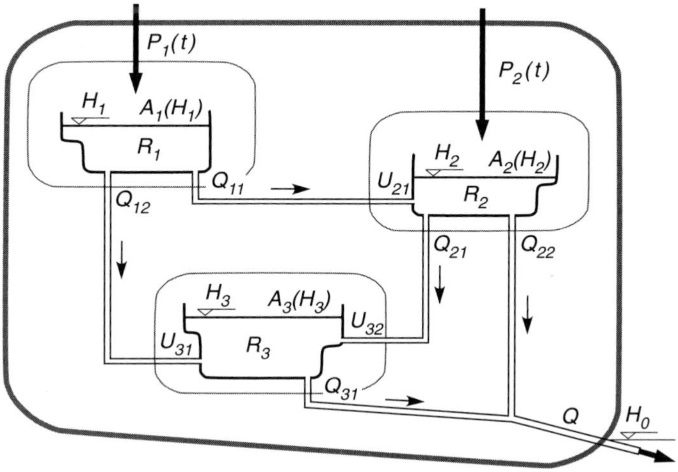

FIGURE 1.148 Karst aquifer system with three subsystems approximated with reservoirs connected by pipes; explanation in text. (Modified from Avdagic, 1990.)

The hydraulic head and the recharge can vary as functions of time (t), whereas the areas of the reservoirs (A) can vary with the hydraulic head: $A_i(H_i)$, where index "i" denotes reservoir number. The output (Qi) from and the input (U_i) to the reservoirs can also vary in time as a function of the governing hydraulic heads.

Equations describing the first subsystem are

$$A_1(H_1) \times \frac{dH_1}{dt} = P_1(t) - \sum_{i-1}^{2} Q_{1i} \tag{1.44}$$

$$Q_{11} = f_{11}(H_1, H_1 - H_2) \tag{1.45}$$

$$Q_{12} = f_{12}(H_1, H_1 - H_3) \tag{1.46}$$

Equations describing the second subsystem are

$$A_2(H_2) \times \frac{dH_2}{dt} = P_2(t) + U_{21} - \sum_{i-1}^{2} Q_{2i} \tag{1.47}$$

$$Q_{21} = f_{21}(H_2, H_2 - H_3) \tag{1.48}$$

$$Q_{22} = f_{22}(H_2, H_2 - H_0) \tag{1.49}$$

Equations describing the third subsystem are

$$A_3(H_3) \times \frac{dH_3}{dt} = \sum_{i-1}^{2} U_{3i} - Q_{31} \tag{1.50}$$

$$Q_{31} = f_{31}(H_3, H_3 - H_0) \tag{1.51}$$

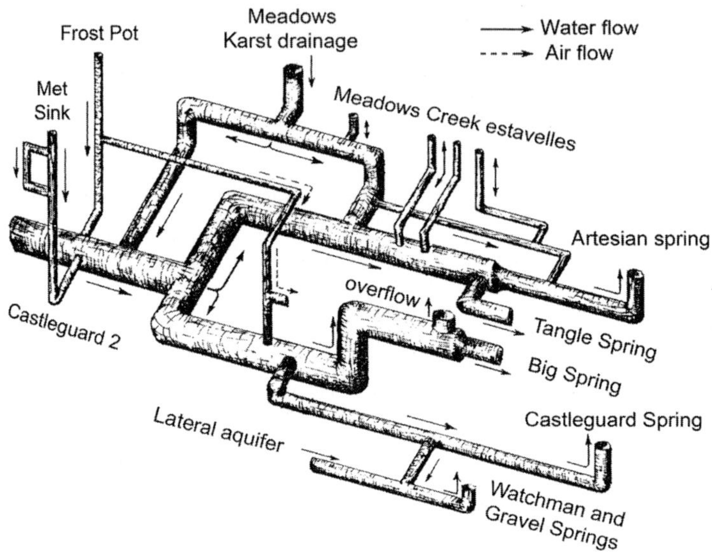

FIGURE 1.149 Pipe diagram of the Castleguard Valley karst aquifer, Alberta, based on dye tracing and cave observations. (Smart, 1983; from Palmer, 2007.)

Connectivity equations are

$$Q_{11} = U_{21} \tag{1.52a}$$

$$Q_{12} = U_{31} \tag{1.52b}$$

$$Q_{21} = U_{32} \tag{1.52c}$$

The final system output equation is

$$Q = Q_{31} + Q_{22} \tag{1.53}$$

Wide availability of software packages for hydraulic calculations of pipes and reservoirs can make this and similar approaches quite appealing. However, similarly to all other flow calculations in karst, the number of assumptions, and the required input parameters, most of which cannot be measured, quickly multiply with each additional pipe and reservoir such that the final solution is by default nonunique and the model is overparameterized. Nevertheless, the appeal of a seemingly effortless play with hypothetical configurations of pipes (caves) offered by computer programs, and the possibility of eventually reaching the just-right solution, cannot be ignored as illustrated in Fig. 1.149.

1.6.5 Laminar and Turbulent Flow

Flow in conduits is laminar when it takes place in parallel concentric layers (lamina) that do not mix. The maximum velocity is along the centerline of the conduit and decreases toward the walls where it becomes infinitesimally small, theoretically equal to zero as illustrated in Fig. 1.150. This is because the flow is along individual tubular fluid lamina

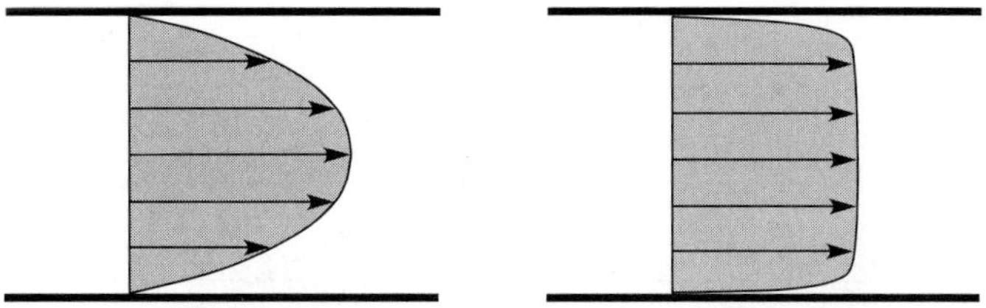

Figure 1.150 Comparison of laminar (*left*) and turbulent (*right*) velocity profiles in a pipe.

between which there is viscous resistance to flow, with the highest resistance between the fluid and the conduit walls. The velocity profile for turbulent flow is also shown in Fig. 1.150 for comparison. At some critical velocity, the laminar flow will become turbulent with the formation of eddies and chaotic motion (mixing) of fluid particles in all three dimensions. In turbulent flow regime, however, there still is a thin layer of laminar flow next to the conduit walls, called boundary layer. Importantly, the turbulence increases the resistance to flow dramatically so that large increases in hydraulic head will be required to further increase the flow rate in the conduit.

Viscosity (μ) is the property of fluid that resists shear stress acting on it. Terms viscosity, absolute viscosity, and dynamic viscosity are used interchangeably, whereas kinematic viscosity (ν) is related to the dynamic viscosity as follows:

$$\nu = \frac{\mu}{\rho} \tag{1.54}$$

where ρ is the fluid density.

Units of dynamic viscosity are Newton second per meter squared (Ns/m^2), or kilogram per meter second (kg/ms), or Pascal second (Pa s) with all three units being equal. The unit for kinematic viscosity (m^2/s) is impractically large so that another unit, centistoke (cSt), is more commonly used (1 cSt = 10^{-6} m^2/s). The unit for fluid density is kg/m^3. As illustrated in Fig. 1.151, viscosity and density of liquids decrease with increasing temperature.

In comparison with water, for example, honey is much more viscous and it moves much slower under the same hydraulic gradient because the resistance to flow (friction) between individual honey lamina is much higher than between the lamina of water. Water and gasoline are relatively low viscosity fluids that flow very easily. Motor oil and molasses are other examples of more viscous fluids that offer great resistance to flow. In the presence of suspended material, the viscosity of a liquid increases because its density is higher.

The Hagen–Poiseuille equation is commonly used in hydraulics to calculate the pressure loss (Δp) between two cross sections of laminar flow of incompressible fluids in a circular tube:

$$\Delta p = \frac{128\mu L Q}{\pi d^4} \tag{1.55a}$$

or

$$\Delta p = \frac{8\mu L Q}{\pi R^4} \tag{1.55b}$$

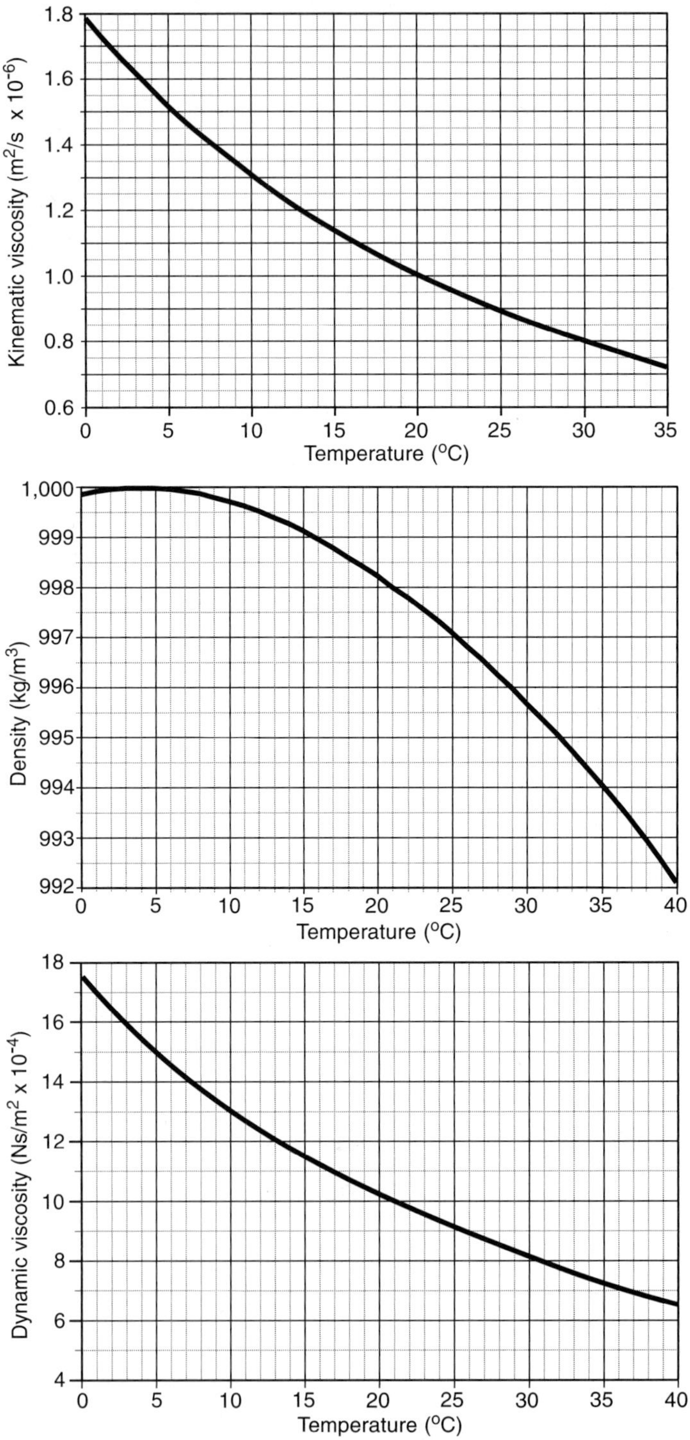

FIGURE 1.151 Kinematic viscosity, density, and dynamic viscosity of water for different temperatures. Density and viscosity data from Maidment, 1993; dynamic viscosity data from Giles, Evett, and Chiu, 1994.

where Q is the flow rate, μ is the dynamic (absolute) viscosity, L is the distance between the two cross sections, d is the tube diameter, R is the tube radius, and π is the mathematical constant "pi."

The pressure loss (Δp) and the hydraulic head loss (Δh) are related as follows:

$$\Delta p = \rho g \Delta h \tag{1.56}$$

where ρ is the fluid density and g is the acceleration of gravity.

The Hagen–Poiseuille equation can also be expressed with the hydraulic head loss (Δh) and the average flow velocity (denoted with capital V to distinguish from kinematic viscosity, ν):

$$\Delta h = \frac{32 \mu L V}{\rho g d^2} \tag{1.57}$$

$$V = \frac{\pi R^4}{8 L \nu} \Delta h \tag{1.58}$$

In terms of the flow rate (Q), Eq. (1.55) can be written as

$$Q = \frac{\Delta p}{L} \frac{\pi d^4}{128 \mu} = \frac{\Delta h}{L} \frac{\pi d^4 \rho g}{128 \mu} \tag{1.59}$$

The flow velocity at any radial distance r from the center of the tube is

$$V_r = \frac{1}{4\mu} \frac{\Delta p}{L} (R^2 - r^2) \tag{1.60}$$

whereas the maximum flow velocity in the center of the tube is

$$V_{\max} = \frac{1}{4\mu} \frac{\Delta p}{L} R^2 \tag{1.61}$$

where R is the internal tube radius.

The maximum velocity of laminar flow in a circular tube is two times higher than the average flow velocity of the entire cross-sectional area of the tube (A) so that the flow rate can also be calculated as

$$Q = A \times \frac{V_{\max}}{2} \tag{1.62}$$

The Darcy–Weisbach equation can be used to calculate hydraulic head losses for both laminar and turbulent flows in a circular tube:

$$\Delta h = f \frac{L}{d} \times \frac{V^2}{2g} \tag{1.63}$$

where f is the dimensionless experimental friction coefficient also known as the Darcy friction factor. As can be seen, the losses are proportional to the squared average velocity, whereas for laminar flow, this relationship is linear.

From Eq. (1.63), the flow rate (Q) through a conduit can be calculated as

$$Q = A \times \sqrt{\frac{\Delta h d 2g}{fL}} \tag{1.64}$$

where A is the cross-sectional area perpendicular to flow.

The Darcy friction factor for common commercial pipes can be estimated from the experimental graph of Moody (1944), which relates the relative roughness of pipe walls and the Reynolds number (Fig. 1.152). The relative roughness (ε/d) is the ratio of the absolute length of irregularities (ε) protruding from the pipe walls to the pipe internal diameter (d). For example, for a natural conduit 3 m (300 cm) in diameter that has wall irregularities 1.8-cm long (which is not uncommon for scalloped walls), the relative roughness is 0.006. For the Reynolds number of 20,000 (bottom axis of the Moody diagram) and the relative roughness line of 0.006, the friction factor is 0.036. In the complete turbulence zone, the friction factor is a function of the relative roughness only; it remains constant for the same relative roughness regardless of the Reynolds number. For purely laminar flow, where the Reynolds number (Re) is less than 2,000, the friction factor is independent of the relative roughness:

$$f = \frac{64}{\text{Re}} \tag{1.65}$$

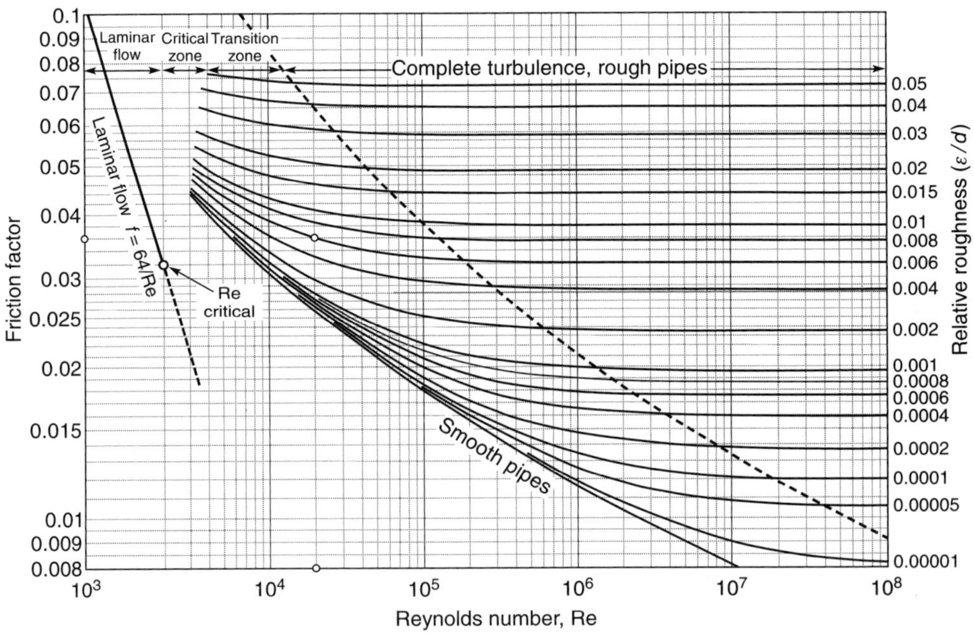

FIGURE 1.152 Moody diagram for determining Darcy friction factor in pipes based on relative roughness of pipe walls and the Reynolds number. (Modified from Moody, 1944.)

The Darcy friction factor for turbulent flow (Reynolds number greater than approximately 4,000) in rough conduits can also be calculated by numerically solving the Colebrook equation (Colebrook, 1939) where it is given implicitly (i.e., it is on both sides of the equation):

$$\frac{1}{\sqrt{f}} = -2\log_{10}\left(\frac{\varepsilon/d}{3.7} + \frac{2.51}{Re\sqrt{f}}\right) \tag{1.66}$$

Another form of the Colebrook equation can be used to calculate the friction factor for flow with free surface such as in a pipe that is flowing partially full of fluid:

$$\frac{1}{\sqrt{f}} = -2\log_{10}\left(\frac{\varepsilon}{12R_h} + \frac{2.51}{Re\sqrt{f}}\right) \tag{1.67}$$

where R_h is the hydraulic radius. For completely filled circular conduits, $R_h = d/4 =$ (inside diameter)/4. For partially filled and conduits of other shapes, the hydraulic radius is given as the ratio of the cross-sectional area of flow (A) to its wetted perimeter (P_w): $R_h = A/P_w$.

The Haaland equation (Haaland, 1983) can be used to directly calculate the Darcy friction factor for a fully flowing circular pipe. It is an approximation of the implicit Colebrook equation with acceptable accuracy for practical purposes:

$$\frac{1}{\sqrt{f}} = -1.8\log_{10}\left[\left(\frac{\varepsilon/d}{3.7}\right)^{1.11} + \frac{6.9}{\mathrm{Re}}\right] \tag{1.68}$$

The Chézy equation is another widely used formula for calculating velocity in pipes (conduits) as well as in open channels (for which it was originally derived) using the hydraulic radius (R_h):

$$V = C \times \sqrt{R_h J} \tag{1.69}$$

where C is the Chézy coefficient and J is the loss of hydraulic head due to friction.

The Chézy coefficient, which is a function of wall roughness, and the Darcy friction factor (f) are related as follows:

$$C = \sqrt{\frac{8g}{f}} \tag{1.70}$$

The Darcy friction factor and the empirical Manning's roughness coefficient (n) are related as

$$n = R_h^{1/6}\sqrt{\frac{f}{8g}} \tag{1.71}$$

where the Manning's roughness coefficient has dimension of $\mathrm{m}^{-1/3}\mathrm{s}$. Note that all the three friction factors, Darcy, Manning, and Chézy, have been determined empirically by various researchers and their selection for any particular flow setting is subjective.

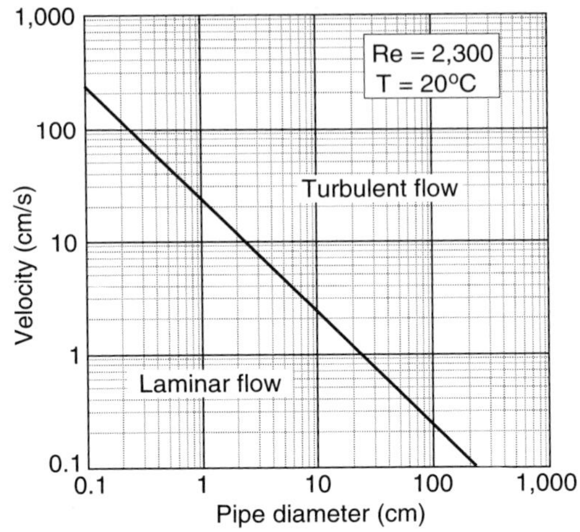

FIGURE 1.153 Graph of critical velocity separating laminar and turbulent flow regime for different pipe diameters and critical Reynolds number of 2,300. The kinematic viscosity of water (ν) at 20°C is approximately 1.01×10^{-6} m^2/s. The critical velocity is calculated as $V_{cr} = \text{Re}_{cr} \times \nu \times 1/d$. (Modified from Vuković and Soro, 1985.)

In the traditional hydraulics of pipes, it has been demonstrated with experiments and widely accepted that the critical Reynolds number at which the flow is not laminar any more is somewhere between 2,000 (experiments of Moody) and 2,300 (experiments performed by Osborn Reynolds). For smaller Reynolds numbers, the influence of wall roughness will be dampened and the flow will remain laminar; at higher Re numbers, the viscous force becomes smaller compared with the inertia force so that even small disturbances at the walls will be allowed to grow into turbulence. As can be seen from the Moody diagram (Fig. 1.152), the transition from laminar to turbulent flow is not abrupt and there is a relatively wide transition zone where various factors influence formation of turbulence. The conduit diameter (d), fluid kinematic viscosity (ν), dynamic viscosity (μ), density (ρ), and flow velocity (V) all determine the Reynolds number:

$$\text{Re} = \frac{\rho V d}{\mu} = \frac{V d}{\nu} \tag{1.73}$$

In general, laminar flow occurs for low velocities, small diameters, low densities, and high viscosities, while turbulent flows occur for the opposite conditions: high velocities, large diameters, high densities, and low viscosities (e.g., see Fig. 1.153). Extensive conduit wall irregularities also result in faster transition to turbulent flow (Fig. 1.154). However, in the hydraulics of karst conduit, there are many unknown factors related to the geometry of submerged conduits as well as irregularities of conduit walls, so that any calculations described earlier are highly speculative and should be applied with care, advisably only at a hypothesis-testing level. For example, for the same flow rate in a karst conduit, the velocity may increase and decrease based on the conduit cross-sectional area and the flow may transition from laminar to turbulent and back. This includes a possibility that laminar flow in the center of the conduit has higher velocity than along some segments of the same conduit where the flow is completely turbulent. Turbulence also increases

FIGURE 1.154 Ceiling pendants illustrating excessive roughness in a passage of the Culverson Creek Cave System, West Virginia. (Photograph courtesy of Phil Lukas.)

resistance to flow as a large part of the mechanical energy in the flow is spent on the formation of eddies that eventually dissipate their energy as heat. Consequently, much larger increases in the conduit entry pressure will be required to further increase the turbulent flow rate as compared with laminar flow.

Unfortunately, in various literatures on caves and karst, there is a wide range of Reynolds numbers for which authors state that the flow is turbulent. This includes Reynolds numbers as low as 10. None of such publications presents a basis for the reported Reynolds numbers including experimental measurements of laminar and turbulent flow in actual karst conduits and their possible deviations from the well-established principles of classic hydraulics. Notwithstanding the fact that nothing is "certain" in karst, the author's advice is to avoid using unsubstantiated numbers and input parameters for hydraulic calculations in karst regardless of the authority that presented them.

1.6.6 Open Channel Flow

Flow rate in karst channels, which are not submerged (where the flow is with free surface—see Fig. 1.104), can be calculated using the Chézy–Manning equation:

$$Q = A\frac{1}{n}R^{2/3}i_0^{1/2} \tag{1.74}$$

where A is the cross-sectional area of flow, n is the Manning's roughness coefficient, R is the hydraulic radius, and i_0 is the slope of the channel bottom assumed to be equal to

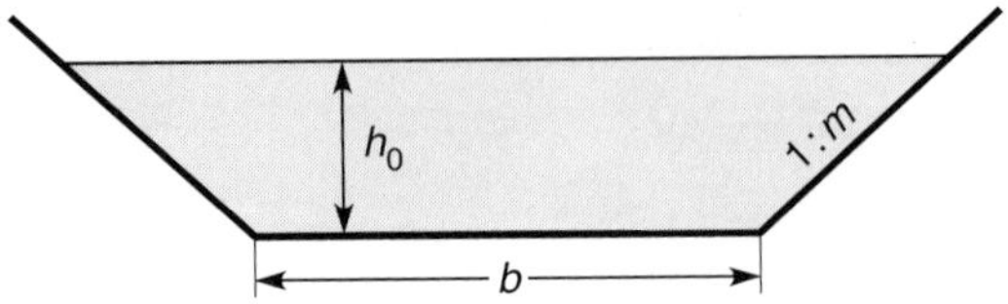

FIGURE 1.155 General dimensions of a trapezoidal channel.

the hydraulic gradient (slope) of the free water surface (i.e., the flow is assumed to be uniform).

As indicated earlier, the hydraulic radius is calculated by dividing the cross-sectional area of flow with the wetted perimeter. For a trapezoidal channel shown in Fig. 1.155, the cross-sectional area of flow (A) and the wetted perimeter (P_{w}) are

$$A = (b + mh_0)h_0 \tag{1.75}$$

$$P_{\mathrm{w}} = b + 2h_0\sqrt{1 + m^2} \tag{1.76}$$

which gives the following flow rate:

$$Q = (b + mh_0)h_0 \frac{1}{n}\left[\frac{(b + mh_0)h_0}{b + 2h_0\sqrt{1 + m^2}}\right]^{2/3} \times i_0^{1/2} \tag{1.77}$$

For a rectangular channel, where $m = 0$, the flow rate is

$$Q = bh_0 \frac{1}{n}\left[\frac{bh_0}{b + 2h_0}\right]^{2/3} \times i_0^{1/2} \tag{1.78}$$

For wide channels, where the normal depth (h_0) is much smaller than the channel width, the wetted perimeter can be approximated by the channel width ($P_{\mathrm{w}} \sim b$), and the hydraulic radius by the normal depth ($R \sim h_0$). In that case, the flow is

$$Q = \frac{1}{n} bh_0^{5/3} \times i_0^{1/2} \tag{1.79}$$

For channels with smooth surfaces, such as solid cement, the Manning's roughness coefficient ranges from 0.010 to 0.015 as finish gets rougher. For cemented rock, it is between 0.017 and 0.030; for channel bottom lined with unconsolidated sediments, it ranges between 0.022 and 0.030 depending on the sediment grain size; for rough channels with broken rocks or incised irregularly into rock, it ranges from 0.035 to higher than 0.045.

1.6.7 Flow Velocity

Because of widely varying hydraulic conductivity and effective porosity of karstified carbonates, even within the same aquifer system, the groundwater velocity in karst can vary over many orders of magnitude. One should therefore be very careful when making a (surprisingly common) statement such as "groundwater velocity in karst is generally

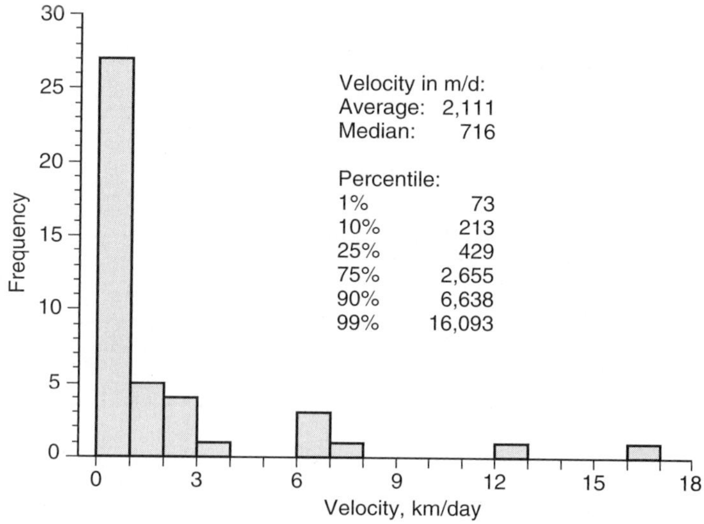

FIGURE 1.156 Apparent groundwater flow velocities in the West Virginia karst based on 43 dye-tracing tests. (Data from Jones, 1977.)

very high." Although this may be true for flow taking place in karst conduits, a disproportionally larger volume of any karst aquifer has relatively low groundwater velocities through small fissures and rock matrix. One common method for determining groundwater flow directions and apparent flow velocities in karst is dye tracing. However, most dye-tracing tests in karst are designed to analyze possible connections between known (or suspect) locations of surface water sinking or focused recharge, and the locations of groundwater discharge such as springs. Since such connections commonly involve some type of preferential flow paths, the apparent velocities calculated from the dye-tracing tests are usually biased toward the high end. Figure 1.156 shows the results of 43 tracing tests in karst regions of West Virginia, the United States. The median groundwater velocity is 716 m/day, while 50% of the tests show values between 429 and 2,655 m/day (25th and 75th percentile of the experimental distribution, respectively). It is interesting that, based on 281 dye-tracing tests, the most frequent velocity (14% of all cases) in the classic Dinaric karst of Herzegovina as reported by Milanovic (1979) is quite similar: between 864 and 1,728 m/day (Fig. 1.157). Twenty-five percent of the results show groundwater velocity greater than 2,655 m/day in West Virginia and greater than 5,184 m/day in Herzegovina. Neither the Herzegovina nor the West Virginia data show any obvious relationship between the apparent groundwater velocity and the hydraulic gradient (Fig. 1.158).

Various approximate calculations of flow velocity have been made based on the geometry of scallops (Fig. 1.159) sculptured on the conduit walls by turbulent flow eddies (e.g., see Curl, 1974; Bögli, 1980; Palmer, 2007; Ford and Williams, 2007). Scallops are common in limestone and gypsum caves but much less so in dolomite as they require homogeneous host rock with uniform mineral size. According to Palmer (2007), scallops are not formed if the water velocity is less than about 1 cm/s, because the turbulent eddies are not stable or uniform enough in such slow-moving water. At velocities greater than about 3 m/s, the bedrock surfaces are usually abraded by sediment, which prevents scallops from forming. The same author provides a graph of flow velocity versus scallop length and advises its use for general estimates only. A quick estimate can also be made

FIGURE 1.157 Apparent groundwater flow velocities in the eastern Herzegovina karst based on 281 dye-tracing tests. (From Milanovic, 1979.)

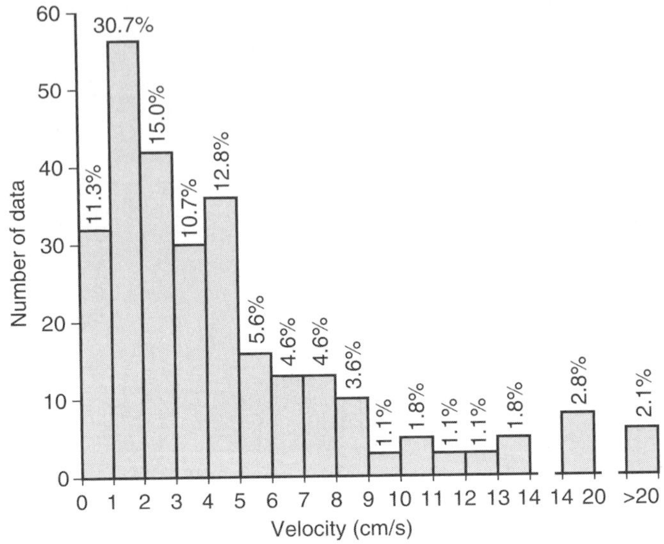

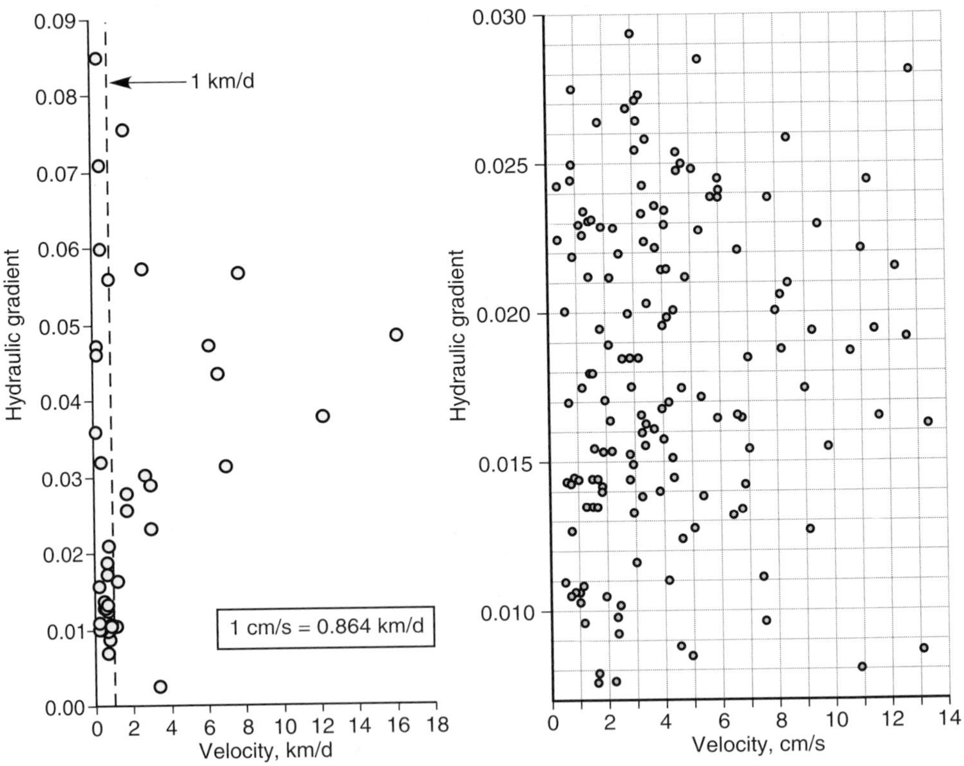

FIGURE 1.158 *Left*: Apparent groundwater flow velocity versus hydraulic gradient for the 43 dye-tracing tests in the West Virginia karst. (Data from Jones, 1977.) *Right*: Apparent groundwater flow velocity versus hydraulic gradient for the 281 dye-tracing tests in the Herzegovina karst. (From Milanovic, 1979.)

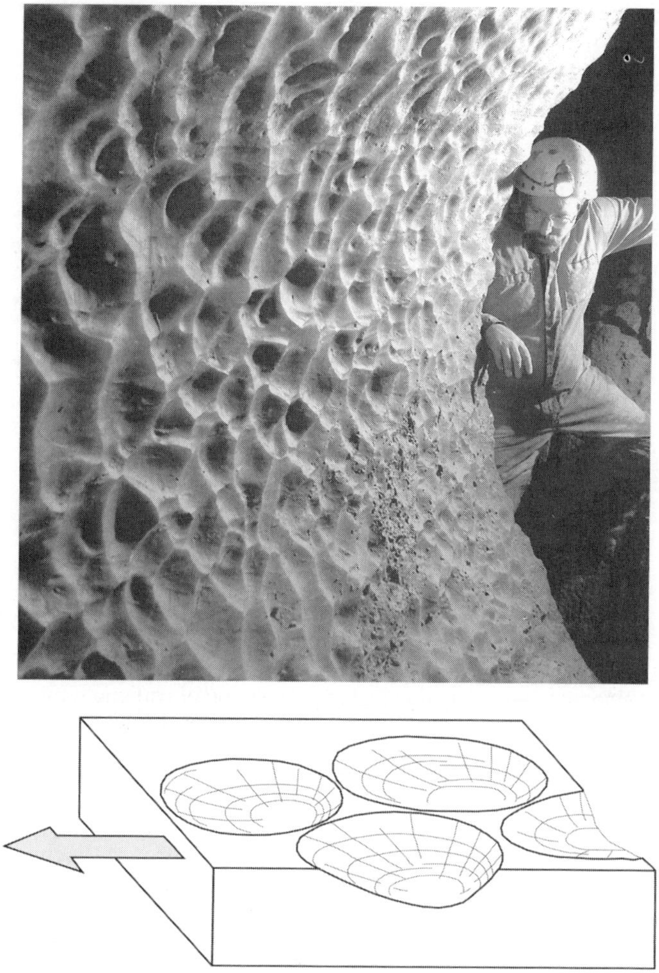

Figure 1.159 *Top*: Scalloped wall in a Bath County, Virginia cave. (Photograph courtesy of Phil Lukas). *Bottom*: Schematic of individual scallops and determination of flow direction based on their shape: The steeper slopes are on the upgradient side of flow. Note that individual scallops always overlap and the entire wall surface is covered as seen in the photograph.

using the following relationship (Palmer, 2007):

$$V(cm/s) = X/L \tag{1.80}$$

where L is the mean scallop length in centimeters and X is a number that depends on the temperature of the water that formed the scallops. $X = 375$ at 0°C, 275 at 10°C, 210 at 20°C, and 170 at 30°C. For comparison, the calculated flow velocity for a canyon passage in White Lady Cave, Little Neath Valley, United Kingdom, is 1.21 m/s and the flow rate is 9.14 m^3/s, for the cross-sectional area of flow of 7.6 m^2 and scallop length of 4.1 cm (White, 1988).

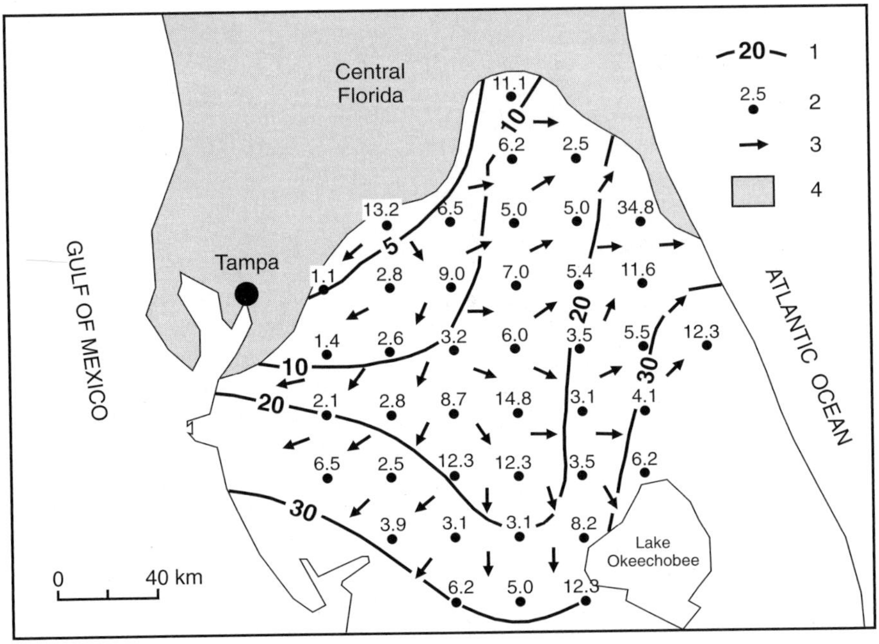

FIGURE 1.160 Velocity of groundwater in the Floridan aquifer determined from ^{14}C content of dissolved carbonate species and adjusted for solution of carbonate minerals. Arrows show direction of flow determined from potentiometric surface. 1: ^{14}C hydroisochrons, in 1,000 years; 2: Velocity of groundwater flow, in meters per year; 3: Approximate direction of groundwater flow; 4: Approximate area of significant mixing. (Hanshaw and Back, 1974.)

Confined karst aquifers in which major concentrated discharge points such as large springs are distant or absent generally have lower groundwater flow velocities. This is regardless of the predominant porosity type because the whole system is under pressure and the actual displacement of "old" aquifer water with the newly recharged one coming through the overlying nonkarstic rocks (sediments) is rather slow. Figure 1.160 shows the results of groundwater flow velocity estimates using carbon-14 isotope dating for the confined portion of the Floridan aquifer in central Florida (Hanshaw and Back, 1974). The average groundwater velocity based on 40 values is 6.9 m/year or 0.019 m/day.

1.7 Aquifer Recharge

Together with natural aquifer discharge and artificial groundwater extraction, aquifer recharge is the most important water budget component of a karst system. Understanding and quantifying recharge processes is the prerequisite for any analysis of the resource sustainability. It also helps policy makers come to better-informed decisions regarding land use and water management, since protection of natural recharge areas is paramount to the sustainability of the karst water resource. The first important step in recharge analysis is to consider the scale of study, since the approach and methods of quantification are directly influenced by it. For example, it may be necessary to characterize local recharge within several acres or tens of acres at a contaminated site where contaminants are introduced into the subsurface. This scale of investigation would obviously not be feasible or needed for water resources assessment at a karst drainage basin scale. However, as illustrated further, karst aquifer recharge is variable at all scales, in both space and time. It is therefore by default that any estimate of recharge involves averaging a number of quantitative parameters and their extrapolation–interpolation in time and space. This also means that there will always be a degree of uncertainty associated with quantitative estimates of recharge, and that this uncertainty would also have to be analyzed and quantified. In other words, aquifer recharge is both a probabilistic and a deterministic process: if and when it rains (laws of probability), the infiltration of water into the subsurface and the eventual recharge of the saturated zone (water table) will follow physical (deterministic) laws. At the same time, both sets of laws (equations) use parameters that are either measured directly or estimated in some (preferably quantitative) way, but would still have to be extrapolated–interpolated in space and time. For this reason, the quantification of aquifer recharge is one of the most difficult tasks in hydrogeology. Unfortunately, this task is too often reduced to simply estimating a percentage of total annual precipitation that becomes groundwater recharge and then using that percentage as the recharge rate for various calculations or groundwater modeling at all spatial and time scales. Following are some of the examples illustrating the importance of not reducing the determination of karst (or any other) aquifer recharge to a "simple" nonconsequential task of estimating "average annual recharge."

- Recharge is usually calibrated as part of karst aquifer modeling studies based on an EPM approach that in itself is not valid. In any case, the model developer should clearly discuss uncertainties related to the "calibrated" recharge rates and the sensitivity of all other key model parameters. When matching "average annual" field measurements of the hydraulic head, a difference of 10% or 15% in aquifer recharge rate may not be all that sensitive compared with some assumed equivalent aquifer transmissivity (hydraulic conductivity); however, this difference is very significant in terms of aquifer water budget and any analysis of aquifer sustainability.
- Rate and locations of aquifer recharge have implications on the shape and transport characteristics of groundwater contaminant plumes. More or less direct recharge from land surface may result in a more or less diving plume, respectively. For example, transport of contaminants from a sinking surface stream is quite different than transport of contaminants from nonpoint sources at the land surface. Different recharge rates also result in different overall contaminant concentrations—higher recharge usually results in lower concentrations due to dilution (assuming that the recharging water is not contaminated).
- Depending on the local hydrogeologic setting, rainwater that successfully infiltrates into the subsurface (i.e., without evaporating back to the atmosphere or running off to a surface stream) and percolates past the root zone may take anywhere from hours to years to traverse the vadose zone and reach the water table. The effects of any recharge estimates may thus be rather abstract if not clearly tied to specific goals of the study.
- Natural and anthropogenic climate cycles and changes alter long-term groundwater recharge patterns, the effects of which may be faced by future generations.

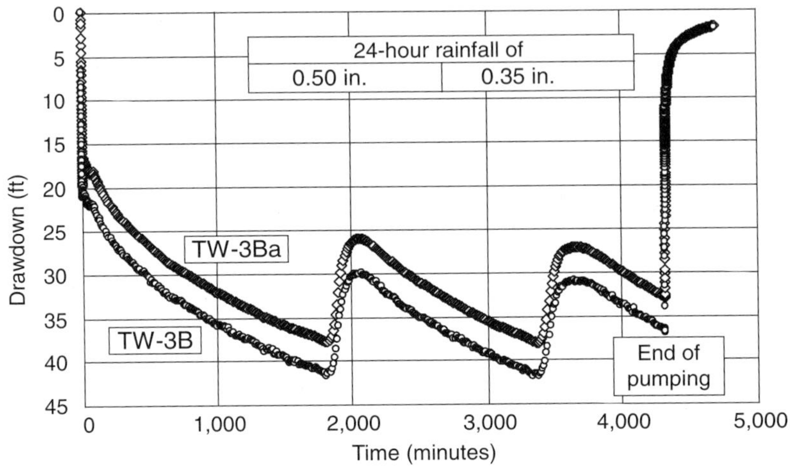

FIGURE 1.161 Hydrograph of a pumping test in a karst aquifer, Frederick, Maryland. The two temporary drawdown recoveries during the test were caused by summer storms, not cessation of pumping that remained constant during the test.

Various water budget and groundwater recharge terms are used interchangeably, sometimes causing confusion. In general, infiltration refers to any water movement from the land surface into the subsurface. This water is called potential recharge indicating that only a portion of it may eventually reach the water table (saturated zone). The term "actual recharge" should be used to avoid any possible confusion: It is the portion of the infiltrated water that reaches the aquifer and it is confirmed based on groundwater studies. The most obvious confirmation that the actual groundwater recharge is taking place is a rise in the hydraulic head (water table) elevation (Fig. 1.161). However, the hydraulic head can also rise due to cessation of groundwater extraction (pumping) in the vicinity and this possibility should always be examined. Effective infiltration and deep percolation refer to water movement below the root zone and are often used to describe the actual recharge. Since evapotranspiration (loss of water to the atmosphere) refers to water at the surface and in the subsurface, this should be clearly indicated.

1.7.1 Autogenic Recharge

Direct recharge by precipitation of a karst aquifer exposed (outcropping) at the land surface is usually called autogenic. In mature karst where residuum is thin or absent, this type of recharge is arguably the easiest to assess. It commonly exceeds 80%, regardless of climatic conditions, as there are very little, if any, factors that could prevent instantaneous infiltration of precipitation (Fig. 1.162). It is in these settings with rapid aquifer recharge over extensive bare karst areas that sudden intermittent springs occur including some of the world's largest (Fig. 1.163), while permanent springs exhibit the highest amplitudes of discharge fluctuations. The exceptions are karst areas where sinkholes have bottoms covered by sediments that, when clay fraction dominates, may retain water for longer periods including formation of semipermanent impoundments (see sinkhole photographs in Fig. 1.68). When the residuum sediments are thicker and when the karstified

FIGURE 1.162 Extensively folded and faulted limestone beds in the Durmitor National Park, Montenegro, an example of karst terrain with maximum recharge exceeding 80% of precipitation.

FIGURE 1.163 Sopot Spring in the Boka Kotorska Bay, the Adriatic coast of Montenegro, is one of the largest intermittent karst springs in the world. Photograph on the left shows discharge rate of $>200\ m^3/s$ less than 24 hours after a major summer storm. There are no sinking surface streams in the drainage area of the spring and the spring is dry most of the year. Inset shows the entrance to the spring cave when dry. (Left photograph courtesy of Dr. Sasa Milanovic, right photograph courtesy of Dr. Igor Jemcov.)

rocks are covered with other unconsolidated sediments, the complexity of estimating autogenic aquifer recharge increases. In such cases, the following major factors, also applicable to nonkarstic terrains, will have varying degrees of influence depending on local settings (more detail discussion is provided in Kresic and Mikszewski, 2009).

Precipitation

As emphasized repeatedly, an analysis of precipitation at all temporal scales is highly recommended for any project in karst. If at all possible, this should be done in conjunction with an analysis of stream and/or spring flow hydrographs and water level fluctuations at monitoring wells. At a minimum, and regardless of the project scale and scope, long-term precipitation data for the closest available climate gauging station should be analyzed. It is also generally recommended to include a rain gauge at the site, as precipitation can vary greatly over small distances. Barometric pressure is another parameter typically monitored at the site level, as many pressure transducers placed in monitoring or production wells are not vented to the atmosphere, meaning that the barometric pressure must be subtracted from all measurements in order to obtain the water level measurement.

One common limiting factor for many groundwater projects is a relatively short period of site-specific data on precipitation, flow rate, and hydraulic head fluctuations. Even when this record is several years long, it still will not provide for a more detailed analysis of natural cycles of aquifer recharge. In order to determine where the site-specific data belongs in terms of natural hydrologic and climatic cycles, it is highly recommended to analyze long-term data from precipitation gauging stations, wells, and springs in the same or similar hydrogeologic and climate settings that may be available from government agencies.

A simple visual analysis of precipitation versus flow rate such as the one shown in Figs. 1.164 and 1.165 will provide useful information as to the likely effects of recharge from precipitation and will help select specific quantitative analyses. For example, even in the absence of daily precipitation and flow record, it may be concluded that the main factor for assessing relevant characteristic of a particular karst aquifer is not its rapid response to rainfall. This is illustrated in Fig. 1.164. The Niangziguan Spring has a quite uniform discharge rate for a karst aquifer, varying less than 50% on a monthly basis for any given year. At the same time, despite the semiarid climate in north China, the average spring flow is very high, thanks to the large drainage area, estimated at 7,217 km^2 including bare karst land over approximately 2,250 km^2 (Liang et al., 2008). Notable is the months-long delay of the overall aquifer response to recharge from precipitation as emphasized in Fig. 1.164, bottom. In comparison, the response of the Maramec Spring in Missouri, the United States, is much more rapid as illustrated in Fig. 1.165. In this case, analyzing only the average monthly data on precipitation and flow rate may be inadequate for a particular project scope.

Temporal Distribution and Probability of Precipitation

Autocorrelation, frequency, and probability analyses of time series such as precipitation and stream or spring discharge are standard in all hydrologic studies and also constitute the key steps in developing various predictive quantitative models as explained in Chap. 6. The interdependence of the successive members of a time series, or memory of

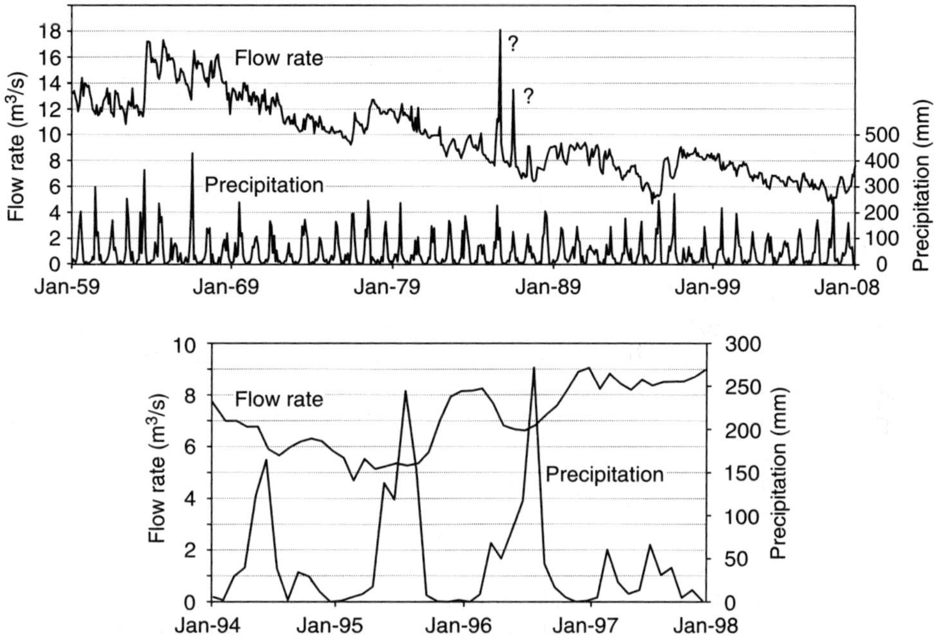

Figure 1.164 Average monthly flow rate of the Niangziguan Spring in Shanxi province, north China, versus average monthly precipitation in its drainage area. Part of the same graph for a shorter period of 4 years is shown on the bottom. (Data courtesy of Liang Yongping, Institute of Karst Geology, Guilin, China.)

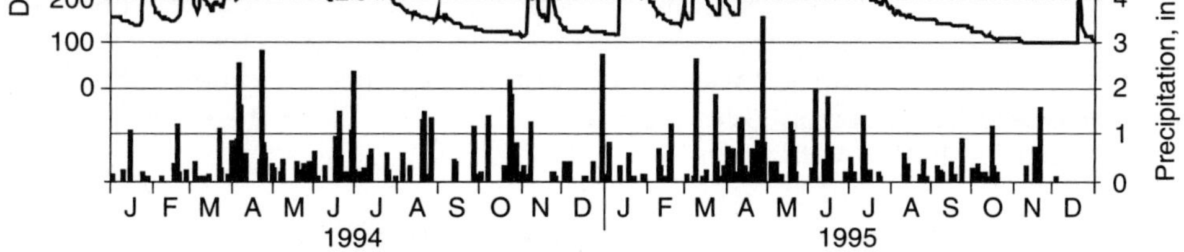

Figure 1.165 Average daily flow rate of the Maramec Spring and daily precipitation at Rolla-UMR gauging station, 1994 to 1995. (Modified from Vandike, 1996.)

the time series, is evaluated by the *autocorrelation coefficient* (also called *serial correlation coefficient*), whose estimate for any time-lag k is

$$r_k = \frac{\frac{1}{n-k}\sum_{i=1}^{n-k}(x_i - x_{\text{av}})(x_{i+k} - x_{av})}{\frac{1}{n}\sum_{i=1}^{n}(x_i - x_{\text{av}})^2} = \frac{COV(x_i, x_{i+k})}{VARx_i} \tag{1.81}$$

where n is the total number of data in the sample; x_i is the value of the variable (e.g., monthly rainfall) at time $t = i$; x_{i+k} is the value of the variable at time $t = i + k$; x_{av} is the average value of the data in the sample. The numerator in Eq. (1.81) is called the *autocovariance* (or just covariance, COV) and the denominator is called the *variance* (VAR) of the time series (note that the square root of the variance is called *standard deviation*).

Equation (1.81) means, for example, that if precipitation (a hydrologic variable) is measured on a monthly basis, for lag $k = 1$, we pair values recorded at months 1 and 2, months 2 and 3, months 3 and 4, and so on. The number of pairs in the autocorrelation is therefore $n - 1$, where n is the number of data. For lag $k = 2$, we pair months 1 and 3, months 2 and 4, and so on. Consequently, the number of pairs in correlation decreases and it is now $n - 2$. Autocorrelation coefficients are calculated for various lags and then plotted on a graph called *autocorrelogram*. If there is some predictability based on past values of the series to its present value, the series is *autocorrelated*. Terms that are also often used to describe an autocorrelated series are *persistence* and *memory*. If a series is not autocorrelated, it is called random or independent (i.e., persistence is absent; the series is without memory). The hypothesis that a time series is dependent (autocorrelated) is tested by various statistical tests. One of the simpler tests is proposed by Bartlett (from Gottman, 1981); to be significantly different from zero at the level of confidence 0.05 (i.e., with 95% probability), the autocorrelation coefficient must be

$$r_{\text{k}} > \frac{2}{\sqrt{n}} \tag{1.82}$$

where n is the total number of data in the sample. This test is in hydrologic practice often performed just for the first, or the first two lags. It is advisable, however, to perform a test for the entire correlogram introducing limits of confidence. This may uncover possible periodic components in the time series that would otherwise be considered as independent if, say, lag 2 was found not to be significantly different from zero. A test proposed by Anderson gives limits of confidence (*LC*) for the entire autocorrelogram (Prohaska, 1981):

$$LC(r_{\text{k}}) = \frac{1 \pm Z_\alpha \cdot \sqrt{n - k - 2}}{n - k - 1} \tag{1.83}$$

where n is the sample size, k is the lag, Z_α is the value of the normally distributed standardized variable at the α level of confidence; the values of Z_α are provided in Table 1.4. Note that the one-sided test in statistics is used to test a hypothesis that one process

Confidence level	0.10	0.05	0.01	0.005	0.002
Value of Z for on-sided test; or	−1.28 1.28	−1.645 1.645	−2.33 2.33	−2.58 2.58	−2.88 2.88
Value of Z for two-sided test; and	−1.645 1.645	−1.96 1.96	−2.58 2.58	−2.81 2.81	−3.08 3.08

From Spiegel and Meddis, 1980.

TABLE 1.4 Critical Values of Z Used for Determining Statistical Confidence Levels

is more probable than the other when the interest in the extreme values is only on one side of the probability distribution.

Figure 1.166 shows autocorrelogram for monthly precipitation in the Niangziguan Spring drainage area created using *Statgraphics* commercial software program (StatPoint Technologies, Inc., 2012; Manugistics, Inc., 2000). A perfect periodicity is clearly visible as indicated by the statistically significant peaks of the same sign separated by 12 months. In *Statgraphics*, the statistical significance of the autocorrelation coefficients is calculated using a critical value of the standard normal distribution for 100(1-α)% probability limits around zero:

$$0 \pm z_{\alpha/2} se[r_k] \tag{1.84}$$

where *se* is the standard error for autocorrelation coefficient (r_k) calculated as

$$se[r_k] = \sqrt{\frac{1}{n}\left\{1 + 2\sum_{i=1}^{k-1} r_k^2\right\}} \tag{1.85}$$

If α is selected to be 0.05, any sample autocorrelations that fall outside the limits in Eq. (1.84) are statistically significantly different from zero at the 95% significance level.

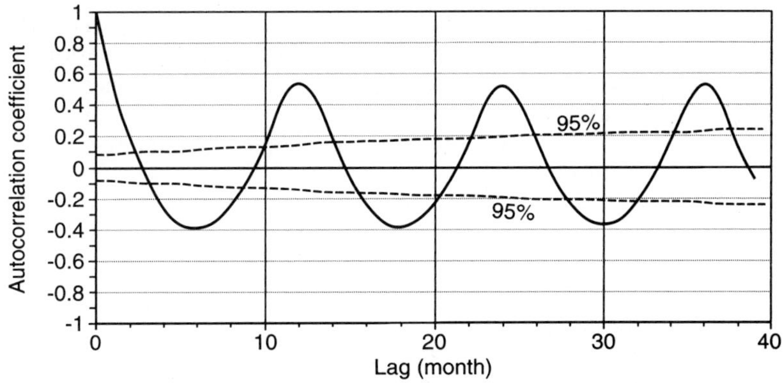

FIGURE 1.166 Autocorrelogram of monthly precipitation in drainage area of the Niangziguan Spring.

Constructing a periodogram is another common tool that can be helpful in identifying randomness and seasonality in time series data. It is a plot of frequency and ordinate pairs based on a Fourier decomposition of the data into a sum of waves described by harmonically related trigonometric functions (sines and cosines). This is so-called *frequency domain analysis* as opposed to *time domain analysis* of autocorrelation. Periodogram shows the contribution to the overall variance of the time series data attributable to oscillations at different frequencies for a specific time period used to construct a frequency spectrum (Polhemus, 2005). Discrete Fourier decomposition for estimating periodic component of a time series $Y(t)$ is

$$Y(t) = y_{\text{av}} + \sum_{i=1}^{n} (a_i \cos(2\pi f_i t) + b_i \sin(2\pi f_i t)) \tag{1.86}$$

where y_{av} is the mean (average) of the series, t is the time, and f_i is the *i-th* Fourier frequency given as

$$f_i = \frac{1}{n} \tag{1.87}$$

for $i = 0, 1, \ldots, n/2$ if n is even, and $i = 0, 1, \ldots, (n\text{-}1)/2$ if n is odd (n is the number of data).

The coefficients a_i and b_i are

$$a_i = 2f_i \sum_{i=1}^{n} y_i \cos(2\pi f_i t) \tag{1.88}$$

$$b_i = 2f_i \sum_{i=1}^{n} y_i \sin(2\pi f_i t) \tag{1.89}$$

The periodogram calculates the power in the data at each Fourier frequency:

$$I(f_i) = \frac{n}{2}\left(a_i^2 + b_i^2\right) \tag{1.90}$$

which is scaled so that the sum of the periodogram ordinates across all of the Fourier frequencies, except for $i = 0$, yields the sum of squared deviations of the time series about its mean. In effect, the periodogram generates an analysis of variance by frequency (StatPoint Technologies, Inc., 2012).

If the periodogram contains a prominent spike, the data may not be random and the period T (seasonality) is then given as

$$T = \frac{1}{f} \tag{1.91}$$

where f is the frequency of the spike.

The fastest frequency is called the Nyquist frequency. It is 0.5 ($T = 1/f = 1/0.5 = 2$). The slowest frequency we can detect is zero. This is a frequency that never repeats. Trend lines have zero frequencies and infinite period. For these reasons, the frequencies

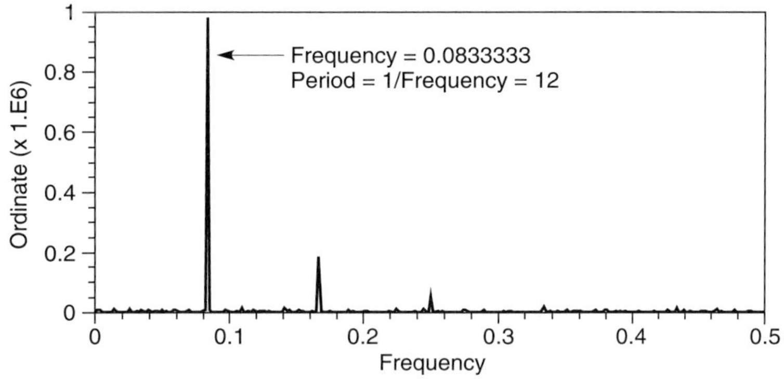

FIGURE 1.167 Periodogram of monthly precipitation in drainage area of the Niangziguan Spring.

on graphs always range between 0 and 0.5. A pure *cycle* (i.e., time series transition from peak to valley and back to peak) can be hidden by noise so that visual inspection of plotted raw data alone is inadequate. The spectral density can often pick out the pure cycle.

A frequency domain analysis of monthly rainfall in the drainage area of Niangziguan Spring performed using *Statgraphics* is shown in Fig. 1.167. As expected, based on the autocorrelogram in Fig. 1.164, the highest frequency is 0.08333 (period of 12 months), followed by two lower peaks at 0.166667 (6 months) and 0.25 (4 months). The result only confirms a well-known fact that many hydrologic variables have regular annual and semiannual cycles of wet and dry seasons.

In contrast to the monthly precipitation data, the analysis of annual rainfall in the Niangziguan Spring drainage area does not show any statistically significant autocorrelation or a dominant frequency, with scattered peaks indicating a random time series (Fig. 1.168). This is not surprising as many climatic stations around the world show similar results for the annual precipitation, in part due to relatively short periods of data recording. It is possible that a long cycle exists for the annual precipitation in the Niangziguan Spring drainage area but could not be determined from the limited available information. In any case, as can be seen from Figs. 1.164 and 1.168, both the monthly spring flow and the annual precipitation series exhibit a long-term decreasing trend.

An example of a very long period of observation of both precipitation and spring discharge is the famous Fontaine de Vaucluse, the largest karst spring in France (Fig. 1.169). The drainage area of the spring is approximately 1,130 km^2 (Cognard-Plancq et al. 2006a, 2006b) and the average annual rainfall in the 1878 to 2004 period is 1,096 mm, with the minimum and maximum observed values of 641 mm (1953) and 1,740 mm (1977), respectively. The historical rainfall database was created from six gauging stations and adjusted for the mean altitude of the drainage area (Cognard-Plancq et al., 2006b). Interestingly, while both series exhibit very similar trends until 1950s, the annual flow rate shows a decreasing and the precipitation an increasing trend since. To the best of the author's knowledge, there still is no detail, definitive explanation of the two opposite trends. Some possible reasons for the decreasing trend in the spring flow include reduced recharge due to land cover/land use changes, increased groundwater withdrawal in the drainage area, and redistribution of the subsurface karst drainage areas. The increasing precipitation trend in this part of Europe may be the result of an ongoing climate change.

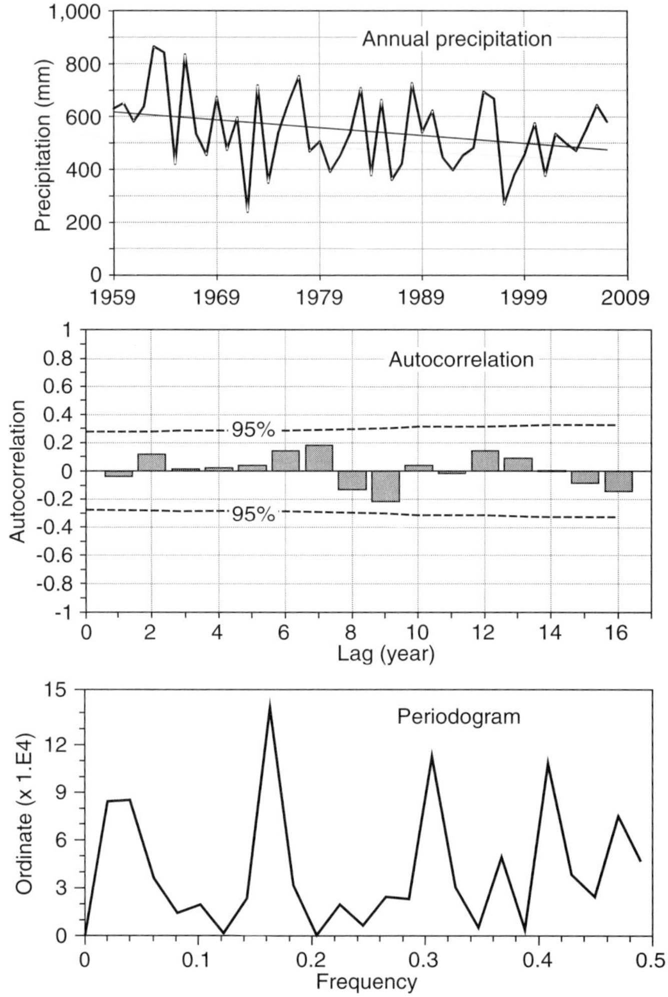

FIGURE 1.168 *Top*: Annual precipitation in drainage area of the Niangziguan Spring with a linear trend. Autocorrelogram and periodogram are shown in the middle and the bottom, respectively.

In general, when a time series is sufficiently long as in this case, the periodogram is usually calculated for the autocorrelation coefficients rather than the raw data. When the number of years with record is short (e.g., 49 as in the case of Niangziguan Spring), the number of unbiased autocorrelation coefficients would be approximately one-third (16), such that the calculation of the periodogram would not make sense. Note that many statisticians believe that the number of calculated autocorrelation coefficients should not exceed 30% of short data sets; in any case, there should be at least 30 pairs of raw data left for the calculation of the last autocorrelation coefficient (one with the largest lag). The spectral density function, $S(f)$, for the autocorrelation coefficients is (Jevdjević, 1974)

$$S(f) = 2\left(1 + 2\sum_{k=1}^{m} D_k r_k \cos(2\pi k f)\right) \tag{1.92}$$

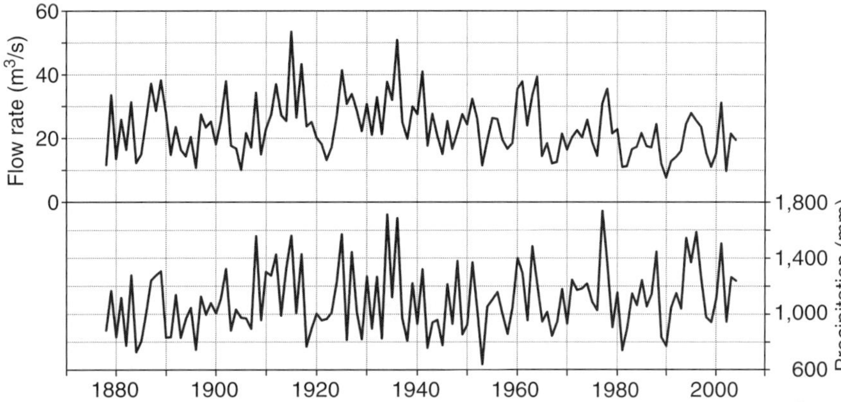

FIGURE 1.169 Annual discharge rate and precipitation in drainage area of Fontaine de Vaucluse. (Data courtesy of Anne-Laure Cognard-Plancq and Christophe Emblanch from Laboratoire d'Hydrogéologie, Faculté des Sciences, Université d'Avignon et Pays de Vaucluse, Avignon, France.)

where k is the number of the autocorrelation coefficient ($k = 1, 2, \ldots, m$), D_k is the filter function, r_k is the coefficient of autocorrelation for lag k, and f is the frequency.

Filter proposed by Bartlett relates the length of the autocorrelogram (m) and the lag (k) as follows (Jevdjević, 1974):

$$D_k = 1 - \frac{k}{m} \tag{1.93}$$

Another commonly used filter proposed by Hanning is (Jevdjević, 1974)

$$D_k = \frac{1 + \cos(2\pi k f)}{2} \tag{1.94}$$

As can be seen in Fig. 1.170, the periodogram of the raw annual precipitation data for Fontaine de Vaucluse does not show a single predominant frequency. Multiple scattered peaks suggest that the time series is random, similarly to the results of the autocorrelation analysis. In contrast, the periodogram of the autocorrelation coefficients shows one significant frequency of 0.08247, which corresponds to a period of 12.125 years. This may indicate existence of a long-term cycle of annual precipitation.

In addition to somewhat more demanding analyses described earlier, moving averages of varying lengths are routinely used to smooth time series and emphasize less obvious longer variations. Many public-domain programs are available to perform this task of simple linear filtering. Less utilized but much more powerful are various weighted moving average filters such as the Henderson's family of trend filters, which smooth the data and reveal possible seasonality (Ladiray and Quenneville, 2001). Henderson filters are used in preference to simpler moving averages because they can reproduce polynomials up to third degree. This means that they are able to adequately capture trend turning points and points of inflection (Australian Bureau of Statistics, 2005). An example of the 13-point moving average Henderson's filter applied to the annual precipitation in the Fontaine de Vaucluse drainage area is shown in Fig. 1.171. Interestingly, the average

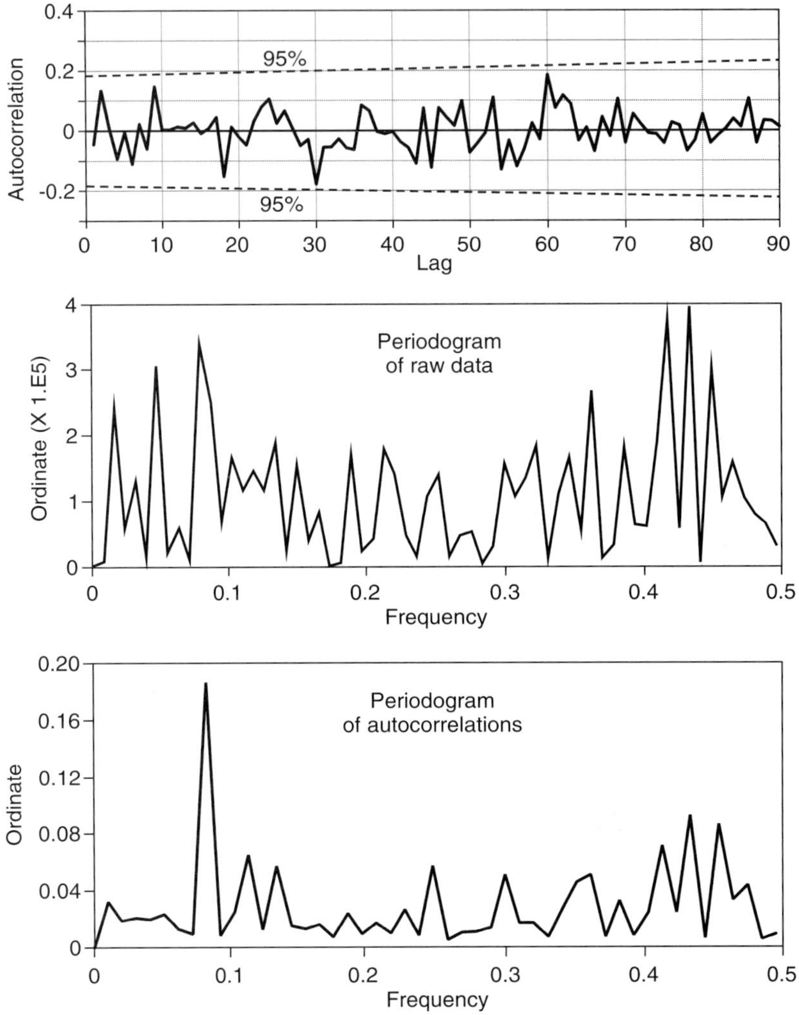

FIGURE 1.170 Autocorrelogram (*top*) and periodograms of the annual precipitation in the Fontaine de Vaucluse drainage area calculated for the raw data (*middle*) and the autocorrelation coefficients (*bottom*) using *R* and *Statgraphics* computer programs.

distance between the same-sign peaks is very close to the period of 12 years indicated on the bottom periodogram in Fig. 1.170.

Determining probability distribution of various time series is a default procedure in hydrology. In case of precipitation, this analysis is performed for series with the same discrete time interval of data collection—for example, hourly, daily, weekly, monthly, or annual. When the precipitation is recorded continuously, such as with data recorders, the time series would have to be transformed into a discrete series with a desired interval for the purposes of analysis. High-intensity precipitation events, such as summer storms that cause excessive spring turbidity or flash flooding, have to be analyzed for time intervals

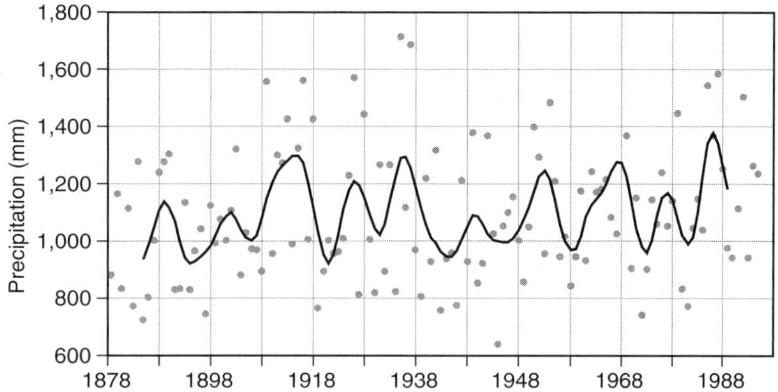

FIGURE 1.171 Henderson's 13-point weighted moving average filter applied to annual precipitation in the Fontaine de Vaucluse drainage area.

measured in tens of minutes (e.g, 10, 20, 30, 45, 60, and 90 minutes), whereas weekly or monthly intervals will suffice for a general analysis of source reliability for water supply. The main limiting factor of probability analyses is the number of available data, since it has been widely accepted that at least 30 data are needed for statistically defensible conclusions. For example, determining the probability of maximum monthly precipitation based on 5 years of data does not make sense even though one may think that 60 data are enough. In other words, as illustrated in Figs. 1.166 and 1.167, monthly precipitation is a well-defined cyclical time series and one would have to separately analyze at least 30 Januaries, 30 Februaries, and so on. At the same time, each monthly series is random (see Fig. 1.172) and therefore well suited for a probability analysis, since it cannot be stated with the absolute certainty what the July precipitation will be next year, for example.

When familiarizing with or describing a probability analysis of any time series, one should make a clear distinction between various terms to avoid any possible confusion. For example, term *frequency analysis* is sometimes used as a synonym for probability analysis; in this context, it is completely different than the *frequency domain analysis* (also called *spectral analysis*) described with Eqs. (1.86) through (1.94) and illustrated in Fig. 1.167. In probability analysis, the term frequency connotes a count (number) of events of a certain magnitude, such as the July precipitation equal to or less than 150 mm. In hydrologic studies, the probability of some magnitude being exceeded (or not exceeded) is usually the primary interest. A very informative public-domain publication entitled *Hydrologic Frequency Analysis* (Engineering Manual 1110-2-1415) is available on the USACE Web site at http://140.194.76.129/publications/eng-manuals/. It covers all major aspects of the subject and provides guidance in fitting probability distributions and construction of confidence limits. Techniques are presented, which can possibly reduce the errors caused by small sample sizes.

Probability analysis can be graphical (empirical) and analytical (theoretical). As emphasized by USACE (1993), every set of frequency data should be plotted graphically, even though the final frequency (probability) curves are obtained analytically. It is important to visually compare the observed data with the derived curve. The graphical method of probability-curve determination can be used for any type of study, but analytical methods have certain advantages when they are applicable. The principal

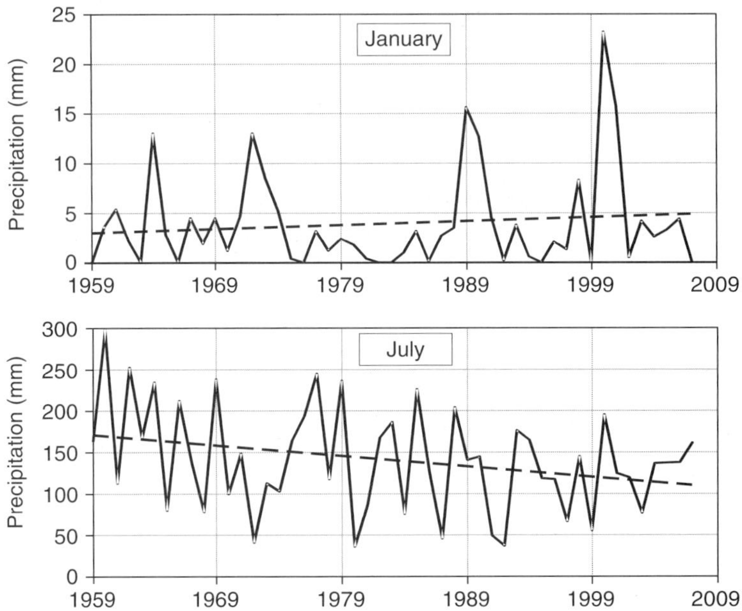

FIGURE 1.172 January and July precipitation in the Niangziguan Spring drainage area. Note opposite linear trends for the 2 months indicating that climate change includes possible redistribution of monthly precipitation within the annual cycle.

advantages of graphical methods are that they are generally applicable, that the derived curve can be easily visualized, and that the observed data can be readily compared with the computed results. However, graphical methods of probability analysis are generally less consistent than analytical methods, as different individuals would draw different curves. Also, graphical procedures do not provide means for evaluating the reliability of the estimates, the so-called goodness of fit.

The fitting of data analytically consists of selecting a theoretical probability distribution, estimating the parameters of the distribution from the data by some fitting technique, and then evaluating the distribution function at various points of interest. Most major commercially available statistical software programs offer a wide variety of probability distributions, often more than 20, which can be quickly tested for a specific data set, compared visually, and the most appropriate one selected based on quantitative goodness of fit. It is therefore recommended that more than just one or two be tested in any given case. This is particularly true when fitting probability distributions to precipitation data because different months will, by default, have different characteristics of probability. This includes a possibility that none of the tested theoretical functions will fit the measured data at the 95% confidence limit.

Figure 1.173, top, illustrates several theoretical probability functions fitted to the January precipitation in the Niangziguan Spring drainage area. There is a relatively good agreement between the recorded data (circles) and several theoretical curves, whereas some of the curves do not show any fit. However, statistical tests indicate that none of the reasonably looking probability distributions fits the data at the 95% confidence limit. Nevertheless, one can still use the curve with the best visual or statistical fit to assess the

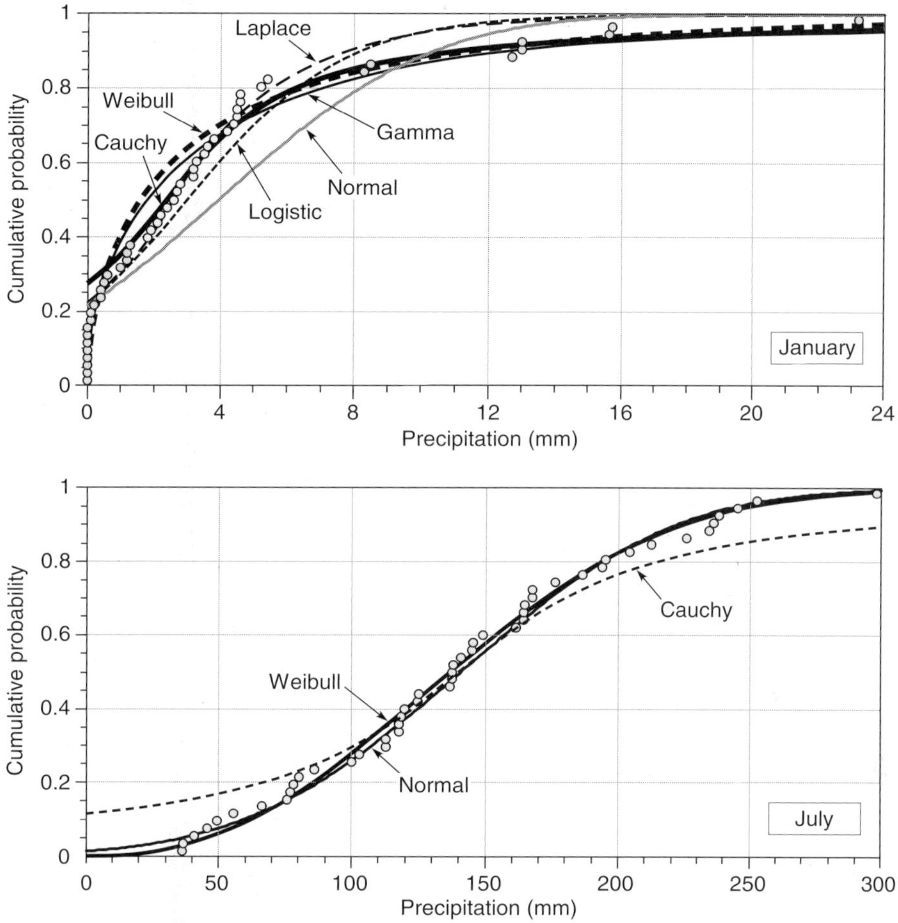

FIGURE 1.173 Probability distribution functions fitted to January and July precipitation in the Niangziguan Spring drainage area.

probability of January precipitation for different levels. For example, the probability of precipitation higher than 20 mm is only about 5% based on the curve with the best fit (Cauchy distribution), whereas the probability of zero precipitation is about 20% based on several curves. One tip for fitting various probability curves to experimental data that include zeros (note that there were 8 years out of 49 with zero precipitation in January in this case) is to replace zeros with a very small number such as 0.001 mm. This will enable testing probability distributions that require data to be >0, such as Gamma and Weibull, for example, as they are quite versatile in describing various data sets.

Figure 1.173, bottom, illustrates fitting of the July precipitation to three probability distributions of which both the Normal and the Weibull distributions describe the data with the 95% accuracy. Together with August, July is the month with the highest precipitation, comprising about 30% of the total annual precipitation in the Niangziguan Spring drainage area. It is therefore not surprising that the Normal and the Weibull distributions

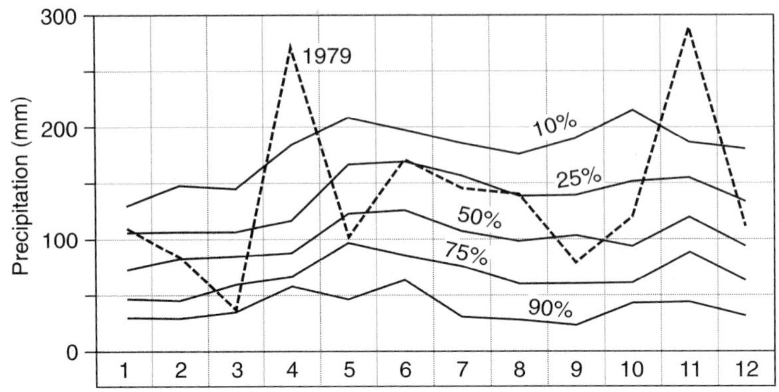

FIGURE 1.174 Probabilities of monthly precipitation at the Krnjaca gauging station based on 31 years of data, together with the 1979 data. (Modified from Kresic, 1991.)

best describe the July data as it has been demonstrated that annual precipitation data in most cases are best fitted to these two distributions as well.

In addition to percent probability, two other terms are often used: the *return period* (sometimes called *recurrence interval*) and the *percentiles* (or *quantiles*) of a probability distribution. The theoretical return period is the inverse of the probability that the event will be exceeded in any 1 year. For example, a 10-year maximum July precipitation has a 1 in 10 = 0.1 or 10% chance of being exceeded in any 1 year, and a 50-year maximum precipitation has a 0.02 or 2% chance of being exceeded in any 1 year. Conversely, a 10-year minimum has 90% chance of being exceeded in any 1 year.

Percentiles are values below which specific percentages of the data are found. For example, for the 49-year July data for the Niangziguan Spring, the 25th percentile is 96 mm based on the Weibull distribution. This means that 25% of the time the precipitation is lower, and 75% of the time it is higher. Percentiles are often used to plot monthly or weekly average precipitation and other hydrologic variables of interest (e.g., spring flow) for the period of record and then compare them with the short-term data. This practice can be very useful in water resource management including assessment if short-term observations are representative of the long-term behavior of the system. For example, Fig. 1.174 illustrates that relying only on 1 year of a spring flow recordings may be misleading when evaluating its reliability as a source of water supply. In 1979, during two main months of the Bucje Spring aquifer recharge, April and May, the recorded precipitation was well above or at the average, followed by the above-average summer months. It is therefore very likely that a long-term, average discharge during summer months, when the demand is the highest, is lower than the one recorded in 1979. Conversely, when examining the Niangziguan Spring graph in Fig. 1.164, it is easily concluded that a below-average monthly precipitation in July–August would most likely result in a below-average spring discharge the following winter. Having a continuously updated graph similar to the one shown in Fig. 1.174 would help prepare water management plans for the upcoming shortage.

As mentioned earlier, a probability analysis of high-intensity rainfall storms is very important in karst environments with fast response to precipitation. This response may be on the order of hours in well-developed, bare karst and may sometimes result in flash floods that pose various hazards to water supply and other infrastructure. Return

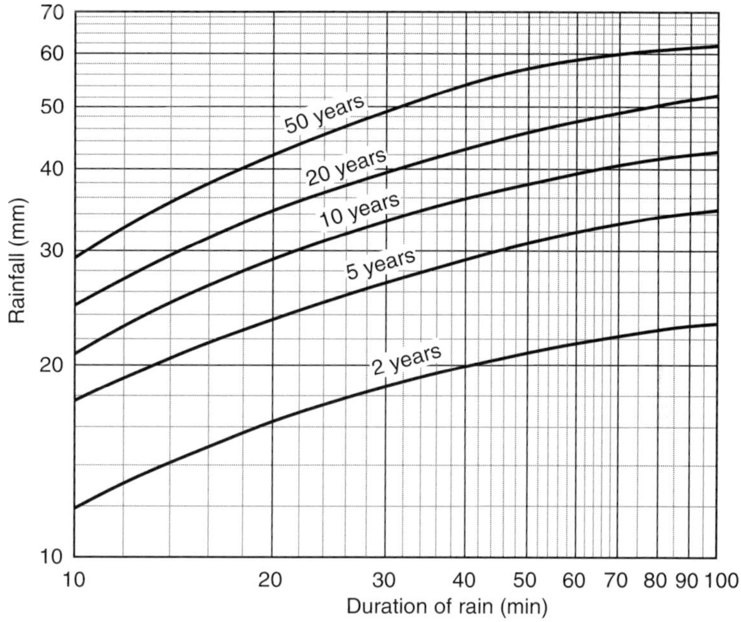

FIGURE 1.175 Example graph of return periods of high-intensity rainfall storms. (Modified from Vukmirović and Despotović, 1983.)

periods for varying intensity and duration storms can be determined using the same probability analysis methods described previously. Some of the probability distributions with proven record of fitting to short-duration rainfall storms include the Gamma family: 2-parameter, Pearson III, log-Pearson III, and Weibull. The results of one such analysis are shown in Fig. 1.175. Further discussion on application of probability analysis to spring and surface streamflows is provided in Sec. 2.6.4.

Spatial Distribution of Precipitation

Determining spatial distribution of precipitation is very important for assessing aquifer recharge in case of large drainage areas. It is also important for relatively small areas of interest where topography and elevation change significantly. The simplest method for determining average precipitation (P_{av}) over an area is the Thiessen graphical method illustrated in Fig. 1.176, left. The general equation for any number of polygons (note that there are six in Fig. 1.176) is

$$P_{av} = \frac{\sum_{i=1}^{n} (P_i a_i)}{\sum_{i=1}^{n} a_i} = \frac{\sum_{i=1}^{n} (P_i a_i)}{A} \tag{1.95}$$

where P_i is the recorded precipitation at station i, a_i is the area of the polygon that belongs to station i, A is the total area, and n is the number of gauging stations. Equation (1.95) is also applicable to the method of isohyets illustrated in Fig. 1.176, right. Here, P_i corresponds to the average value between two successive contour lines, and a_i is the area between the two successive contours within the drainage basin boundary. For example, the average precipitation in the subarea a_2 is 70 mm, and in the subarea a_4, it is 67.5 mm.

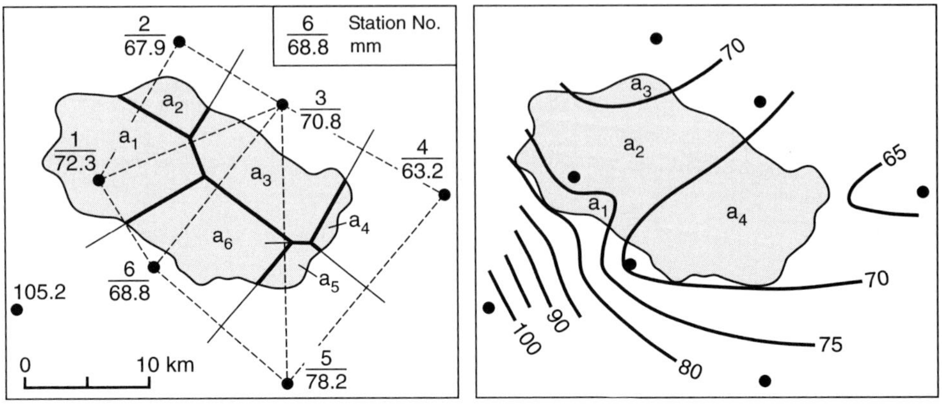

FIGURE 1.176 Thiessen polygons method (*left*) and isohyetal method (*right*) for determining average precipitation in a drainage area (*shaded*). Monthly precipitation at individual gauging stations is given in mm.

In most regions of the world, precipitation increases with elevation due to uplifting of the horizontally moving air masses when they encounter a topographic barrier. As the air masses move up and flow over the barrier, they cool down and precipitate an increasing amount of rain and snow. The downward movement of air on the lee side causes adiabatic warming that dissipates the clouds and creates a rain shadow (Fig. 1.177). Although this orographic effect exhibits many variations based on local characteristics (e.g., wind speeds, steepness and height of the barrier), it commonly accounts for much of the spatial variation of precipitation and must be taken into account when estimating average precipitation over an area. If there is no sufficient number of gauging stations at different altitudes in the drainage area with a significant relief, the methods illustrated in Fig. 1.176 would produce inaccurate estimates unless appropriate corrections were made. As discussed by Dingman (1994), the orographic effect can carry over into many hydrologic relations. For example, average streamflows, dry-season streamflows, and floods were all significantly related to elevation in central New England, the United States.

FIGURE 1.177 Influence of topography on precipitation, the so-called orographic effect. (From Jevdjević, 1956.)

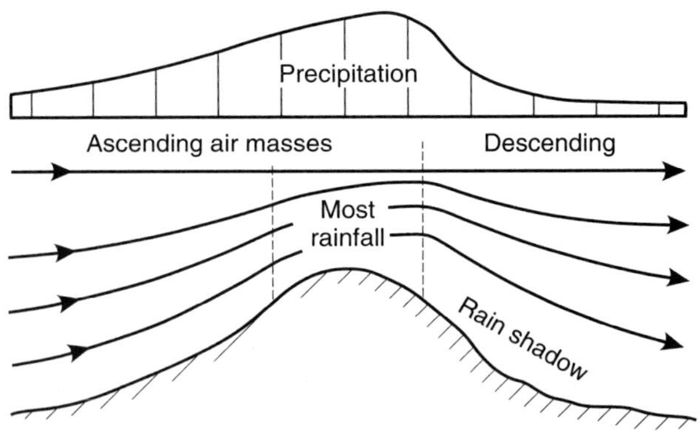

Precipitation deposited as snow is another important factor in evaluating both autogenic and allogenic aquifer recharge. First, snow acts as an insulating blanket that helps retain heat in the soil, which is important hydrologically as well as biologically: If soil is prevented from freezing, its ability to accept infiltrating water is generally enhanced. The principal effect of snow, however, is to delay the input of precipitated water into the land phase of the hydrologic cycle and thus to affect the seasonal distribution of surface runoff and aquifer recharge (Dingman, 1994).

Data Augmentation and Extension

For different reasons, there is often an incomplete, discontinuous, or short precipitation record for a gauging station. This prohibits application of various statistical and probability methods described earlier so that it may be desirable to augment or extend the record by using nearby gauging stations with which there is high correlation. Often, a simple arithmetic average of four gauging stations relatively evenly surrounding the one with the missing record would be sufficient. In case such ideal situation does not exist, the next simplest estimate is made based on regression with the closest station (note that data from close stations are usually well correlated so that multiple regression with several intercorrelated stations would not be statistically correct). Various more sophisticated methods have been introduced including variograms (geostatistics) in time domain. Interested reader should consult Little and Rubin (2002) and Schafer (1997) for more detail discussions.

Evapotranspiration

Just like precipitation, evapotranspiration (*ET*) depends on general climatic conditions in the study area. Globally, approximately 65% of all precipitation falling on landmass returns to the atmosphere through *ET*, which can be defined as the rate of liquid water transformation to vapor from open water, bare soil, or vegetation with soil beneath (Shuttelwort, 1993). Transpiration is defined as the fraction of total *ET*, which enters the atmosphere from the soil through the plants. The rate of *ET*, expressed in inches per day or millimeters per day, has traditionally been estimated using meteorological data from climate stations located at particular points within a region, and parameters describing transpiration by certain types of vegetation. By far, most methods for estimating *ET* are developed for agricultural cropland and the purposes of irrigation with an abundant related literature available. The majority of such methods are based on a rather ambiguous concept of potential evapotranspiration (*PET*) introduced by Thornthwaite (1946, 1948), which is of little practical use for determining actual ET in most real field settings. *PET* is defined as the rate at which *ET* would occur from a large area completely and uniformly covered with growing vegetation, which has access to an unlimited supply of soil water and without advection or heat-storage effects.

There are many proposed and often rather complex empirical equations for estimating *PET* using various parameters such as air temperature, solar radiation, radiation exchange for free water surface, hours of sunshine, wind speed, vapor pressure deficit (VPD), relative humidity, and aerodynamic roughness (e.g., see Singh, 1993; Shuttleworth, 1993; and Dingman, 1994). The main problem in applying empirical equations to a very complex physical process such as evapotranspiration is that in most cases, such equations produce very different results for the same set of input parameters. As pointed out by Brown (2000), even in cases of the most widely used group of equations referred to as "modified Penman equations," the results may vary significantly (Penman proposed his equation in 1948 and it has been modified by various authors ever since).

The *PET* from an area can be estimated from the free water evaporation assuming that the supply of water to the plant is not limited. Actual evapotranspiration, ET_{act}, equals the potential value, *PET*, as limited by the available moisture (Thornthwaite, 1946). On a natural watershed with many vegetal species, it is reasonable to assume that ET_{act} rates do vary with soil moisture, since shallow-rooted species will cease to transpire before deeper-rooted species (Linsley and Franzini, 1979). A moisture-accounting procedure can be established using the continuity equation:

$$P - R - G_0 - ET_{act} = \Delta M \tag{1.96}$$

where P is the precipitation, R is the surface runoff, G_0 is the subsurface outflow, and ΔM is the change in moisture storage. ET_{act} is estimated as

$$ET_{act} = PET\frac{M_{act}}{M_{max}} \tag{1.97}$$

where M_{act} is the computed soil moisture stage on any date and M_{max} is an assumed maximum soil moisture content (Kohler, 1958, from Linsley and Franzini, 1979).

In addition to available soil moisture, plant type is another important component of actual evapotranspiration, as certain species require more water than others. A Colorado State University study in eastern Colorado examined the water requirements of different crops in 12 agricultural areas (Broner and Schneekloth, 2007). On average, corn required 24.6 in./season, sorghum required 20.5 in./season, and winter wheat required 17.5 in./season. Allen et al. (1998) present detailed guidelines for computing crop water requirements together with representative values for various crops. Native vegetation to semiarid and arid environments is more adept at surviving in low-moisture soils and generates much lower ET_{act} values than introduced species. ET_{act} increases with increasing plant size and canopy densities. With regard to stage of development, actively growing plants will transpire at a much greater rate than dormant plants (Brown, 2000). Wind augments *PET* by actively transporting heat from the air to vegetation, and by facilitating the transfer of water vapor from vegetation to the atmosphere. Humidity and temperature together determine the VPD, a measure of the "drying power" of the atmosphere. The VPD quantifies the gradient in water vapor concentration between vegetation and the atmosphere and increases with increasing temperature and decreasing humidity (Brown, 2000).

The main problem with various *PET* and ET_{act} calculations is illustrated in Fig. 1.178: namely, the results often imply that during summer months, in many climatic settings including temperate humid climates, there should be no aquifer recharge from rain since *ET* exceeds the amount of available rainfall. This, of course, is not the case as illustrated further in Sec. "Hydraulic Head and Infiltration."

On the basis of statistical analysis of 254 drainage areas from around the world, Turc proposed the following simple method of estimating ET_{act} from precipitation and air temperature data (Gray, 1973):

$$ET_{act} = \frac{P}{\left[0.90 + \left(\frac{P}{I_T}\right)^2\right]^{1/2}} \tag{1.98}$$

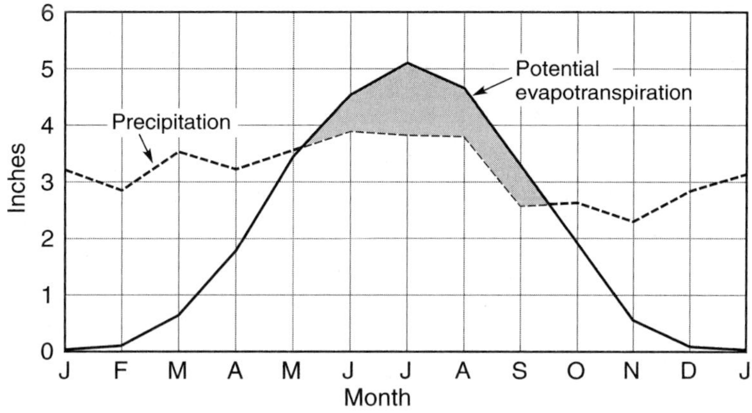

FIGURE 1.178 Monthly precipitation and adjusted potential evapotranspiration for Lewisburg, West Virginia, calculated by the Thornthwaite (1948) method. (From Jones, 1973.)

where P is the annual precipitation (mm), I_T is the evaporative capacity of air = 300 + $25T + 0.05T^3$, and T is the average annual air temperature (°C).

For a shorter period of 10 days, ET_{act} is determined using a more complex equation that takes into account soil moisture and crop type:

$$ET_{\text{act}} = \frac{P + E_{10} + K}{\left[1 + \left(\frac{P+E}{I_T} + \frac{K}{2I_T}\right)^2\right]^{1/2}} \tag{1.99}$$

where E_{10} is the estimated evaporation from bare soil, not higher than 10 mm, during 10 days without precipitation; P is the actual precipitation during the 10-day period; and K is the crop factor given as

$$K = 25\left(\frac{Mc}{G}\right)^{1/2} \tag{1.100}$$

where M is the final growth of dry matter (kg/ha), G is the length of vegetative period (days), c is the crop constant (1.33 for short grass), and I_T is the evaporative capacity of air given as

$$I_T = \frac{(T+2)\sqrt{R_{\text{si}}}}{16} \tag{1.101}$$

where T is the average air temperature in the 10-day period (°C) and R_{si} is the incoming solar energy (cal/cm^2/day).

Hydraulic Head and Infiltration

When a saturated zone developed in unconsolidated sediments overlies a karst aquifer, the recharge of the karst aquifer will depend on the hydraulic gradient between these two saturated zones and the hydraulic conductivity of the overlying sediments as illustrated

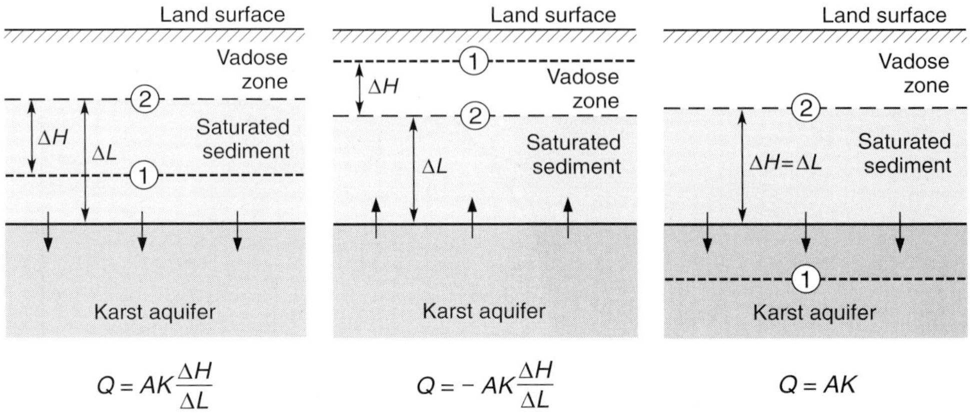

FIGURE 1.179 Karst aquifer recharge expressed as the flow rate (Q) from the overlying saturated unconsolidated sediments. 1: Hydraulic head in the karst aquifer; 2: Hydraulic head (water table) in the unconsolidated sediments. A is the area perpendicular to flow direction, K is the hydraulic conductivity of the unconsolidated sediments.

in Fig. 1.179. Notably, when the hydraulic head in the karst aquifer is higher than the hydraulic head in the overlying sediments, groundwater will flow from the karst aquifer upward so that its recharge will not take place. When the hydraulic head in the karst aquifer is below its contact with the saturated overlying unconsolidated sediments, the downward groundwater flow through these unconsolidated sediments will depend only on their hydraulic conductivity.

As emphasized repeatedly, the only accurate evidence that the actual recharge of an aquifer is taking place is a rise in the saturated hydraulic head. Although, in karst, the response of the hydraulic head to recharge can vary at short distances in the same aquifer depending on the predominant porosity type (see Fig. 1.134), the rise in the hydraulic head can still be used to estimate the recharge based on the precipitation record. This is illustrated in Fig. 1.180 that shows the impact of two summer storms at a monitoring well screened in karstified limestone. The limestone in the area is generally covered by a thin residuum of variable thickness, generally about 5 to 10 ft. The first peak on the well hydrograph was caused by a 0.5-in. (12.7-mm) summer rainfall followed by another 0.35-in. (8.9-mm) rainfall the following day. Based on this information, recharge (R) can be estimated using simple equation:

$$R = S_y \Delta h \tag{1.102}$$

where S_y is the specific yield (dimensionless) and Δh is the water table rise (L).

The specific yield of the limestone in this part of Frederick, Maryland, is estimated at 0.037 based on an aquifer pumping test in the vicinity. This gives the recharge of 0.19 in. for the first peak and 0.28 in. for the second peak, or 38% and 80% of the rainfall, respectively. The result shows that recharge from rainfall can be significant even when potential evapotranspiration is the highest such as during summer months with high air temperatures (96°F or 36°C in this particular case). Moreover, for 6 days prior to these two rain events, the water table was exhibiting a decreasing trend in the absence of precipitation, while

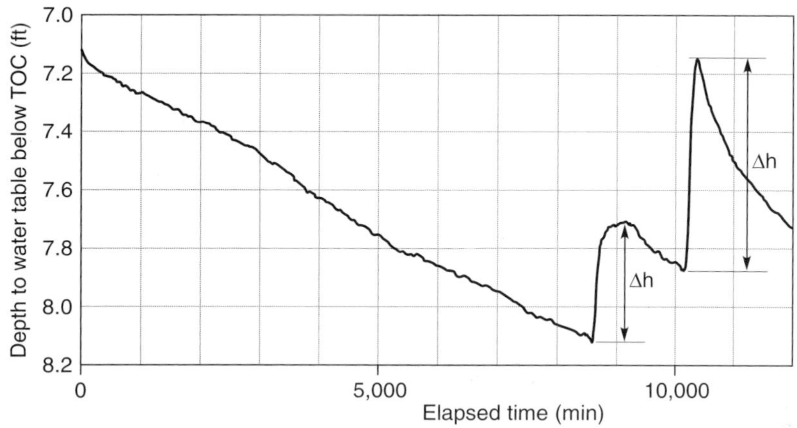

FIGURE 1.180 Hydrograph of water level at monitoring well TW-7 in Frederick, Maryland, showing the impact of two summer storms. Hydrographs such as this one can be used to estimate aquifer recharge rates provided there is information on the amount and duration of rainfall.

air temperatures were similarly high. This illustrates inadequacy of the general concept of potential evapotranspiration for estimating either actual evapotranspiration or karst aquifer recharge in many cases (see graph in Fig. 1.178). Notably, the second rainfall event, even though of a lower magnitude, resulted in a much higher recharge rate as the soil moisture deficit was largely replenished by the first rain.

Because of its simplicity and general availability of water level measurements in groundwater projects, the water table fluctuation method may be the most widely used method for estimating recharge rates in humid regions. The main uncertainty in applying this approach is the value of specific yield in karst, which in many cases would have to be assumed. Another key factor to consider is water-level measurement frequency. From Fig. 1.180, it is obvious that one could not rely on weekly or even daily measurements for recording representative peaks in the karst aquifer water table. In addition, if the water table in the unconsolidated sediments is close to the land surface, it may be necessary to account for any *ET* losses from the saturated zone to the vadose zone and the atmosphere.

When the vadose zone in the sediments overlying a karst aquifer is relatively thick, and in the absence of accurate, continuous hydraulic head measurements, the recharge rate through soil (sediment) column can be estimated using equations of Darcy, Richards, Brooks and Corey, and van Genuchten, as well as analyzing concentration of environmental isotopes with depth (see Kresic and Mikszewski, 2009). However, the processes of infiltration and water percolation through unsaturated zone are highly dependent on various soil properties (of which the hydraulic conductivity is only one) such that even in relatively homogeneous sediments, the infiltration rates vary greatly over short distances. For example, Fig. 1.181 illustrates the range of cumulative percolation rates at seven lysimeters all installed on a 100-sq ft plot located within a 2.8-mi^2 watershed developed in fractured rock in Pennsylvania, the United States. This watershed was subject to a comparative study of several different methods for estimating aquifer recharge (Risser et al., 2005). Even at the 100-sq ft scale, the coefficient of variation between monthly rates at individual lysimeters is greater than 20% for 6 months, with the June, July, and August values of about 50%, 100%, and 60%, respectively. Figure 1.182 shows mean

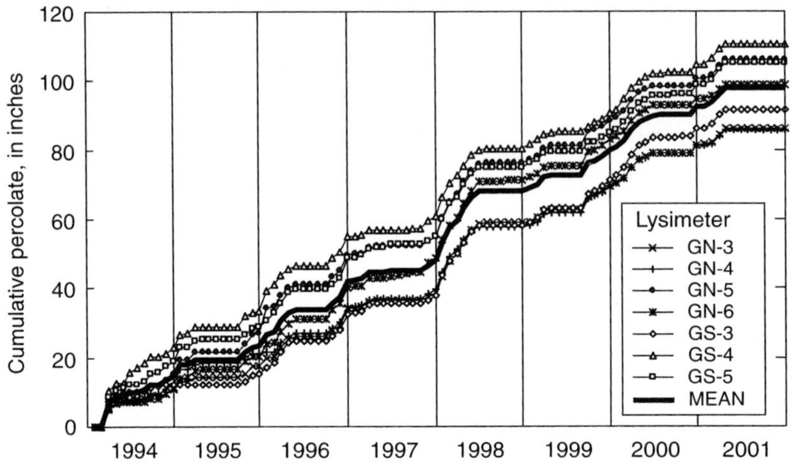

FIGURE 1.181 Range of cumulative percolation rates at seven lysimeters installed on a 100 sq ft lot located within a 2.8-m^2 watershed in Pennsylvania, the United States. (From Risser et al., 2005.)

monthly recharge rates for seven lysimeters during the 1994 to 2001 period of record. The mean annual recharge rate was 12.2 in. or 29% of precipitation. In comparison, this rate was 24% as calculated with the water table fluctuations method. Based on this example, it is reasonable to expect that the infiltration rates in native residuum developed on karstified rocks would vary similarly as such residuum is anything but homogeneous.

Three physical characteristics of the soil [permeability, available water capacity (AWC), and slope] affect the availability of water for plant consumptive use by regulating both infiltration and the ability of the soil profile to store water. Infiltration is largely a function of permeability and slope, while the water storage capacity is determined by the product of the AWC and the root-zone depth. Finer soils (silts and clays) on steeper slopes generally have less deep percolation than those with coarser textures and lesser slopes. They also have lower permeabilities and higher AWC, which slows down infiltration but increases the overall soil–moisture storage capacity. Steeper slopes, with

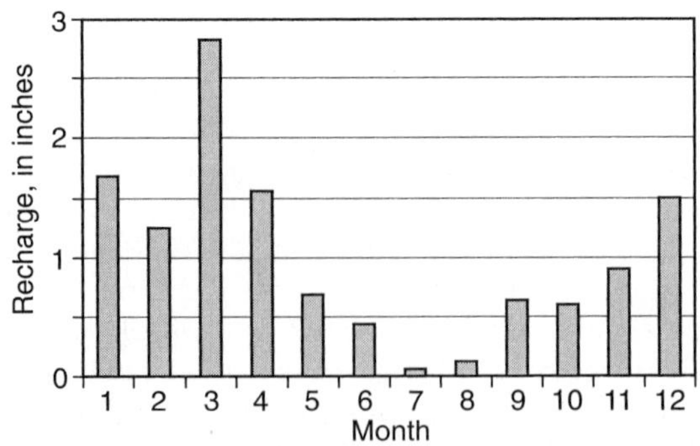

FIGURE 1.182 Mean monthly recharge rates for seven closely spaced lysimeters during 1994 to 2001 period of record. (Data from Risser et al., 2005.)

the same vegetative cover and soil permeability, generally have more runoff and less infiltration. Sandy soils limit runoff, thus increasing infiltration, which results in greater deep percolation and potential recharge. The higher actual ET associated with finer-textured soils indicates greater AWC; therefore, more water is available for consumptive water use by vegetative cover (Dugan and Peckenpaugh, 1985). Rainfall intensity also plays an important role in recharge generation on nonkarstified sediments. In general, higher rainfall intensities are indicative of less infiltration and more runoff; the converse occurs with lower rainfall intensities.

Infiltration can lead directly to recharge when the wetting front reaches the regional water table while infiltration is still occurring. In the arid and semiarid regions, however, deep water tables in nonkarstified sediments and short infiltration times lead to infiltration ending long before the wetting front crosses the water table. In these instances, water that has infiltrated into the unsaturated zone can be removed by ET or, if the water has moved past the root zone, remain in storage until it is displaced by water from a subsequent infiltration event. Unsaturated zone water in storage below the root zone can continue to redistribute over time. This long-term movement of water, or percolation flux, occurs both vertically, by drainage, and horizontally in response to heterogeneity or anisotropy of the sediments. Long-term percolation can eventually result in recharge of the saturated zone (Coes and Pool, 2005).

Accessible cave passages in karst in some cases offer a unique opportunity to directly observe percolation of water from the land surface. However, a nonkarst hydrogeologist may be puzzled by the fact that, in a particular case, there is no visible water dripping from the ceilings in sometimes hundreds of meters long passages during and following high and long precipitation events. Such situations are not uncommon in aquifers that are either overlain by nonkarstified (or residuum) sediments or directly exposed at the surface, since both the vertical and the horizontal movement of water in karst are quite unpredictable. For example, it is possible to observe, in deep vadose zone shafts, an inflow of water from some of the lateral feeder conduits but not from the others, at varying vertical depths including tens or hundreds of feet below ground surface (see Fig. 1.183). This situation may also change both vertically and in time depending on local conditions and redistribution of flows through the three-dimensional underground maze of vadose zone conduits (see Fig. 1.184).

A very illustrative discussion describing the influence of the type of infiltrating water on the character of abandoned cave passages is provided by Palmer (2007) and briefly presented here (see Fig. 1.185). Passages overlain by insoluble, low-permeability beds are likely to receive only sparse capillary seepage. Speleothems are limited to evaporative minerals, and gypsum crusts are the most abundant deposits. Toward the eroded edge of the insoluble cap, the chemistry of infiltrating water is zoned as follows:

Zone 1: Evaporative minerals are tightly bonded to the bedrock surfaces on which they grow. Water enters the cave at such a low rate that there is no observable water film. Evaporation consumes nearly all the moisture.

Zone 2: Capillary seepage makes the walls moist enough that gypsum and other evaporative mineral deposits cannot form. As this zone encroaches on zone 1, the evaporative minerals drop of the ceiling and walls.

Zone 3: Capillary water fed by infiltration through the sandstone cap-rock reaches calcite saturation in a closed system, which nearly depletes the CO_2 content. Where the water enters the cave, it absorbs CO_2 from the cave atmosphere and

Figure 1.183 Erica Sughrue hangs above the Sinkhole, a 165-ft drop into a short passage. The water is issued from a small crack in the limestone, and after hitting the bottom, it disappears in areas too tight for people to continue. The Sinkhole is now closed and should not be visited. (Photograph copyright Innermost Imagery, printed with permission; many more photos are available at http://innermostimagery.com.)

becomes aggressive once more. It forms only a microscopic film of moisture on the walls and is only able to etch and recrystallize the bedrock surfaces. The crystal size becomes smaller and a white weathering rind forms. Conditions are too moist for gypsum and other evaporative minerals to form. Quartz and opal precipitate in places because of the sudden drop in pH.

Zone 4: Water similar to that of zone 3 enters the cave as diffuse seepage but has enough flow to form etched surfaces and rills in the cave walls below the input points.

Zone 5: Water infiltrates through the sandstone cap-rock and concentrates active drips and flows that stay aggressive over their entire paths through the cave. The

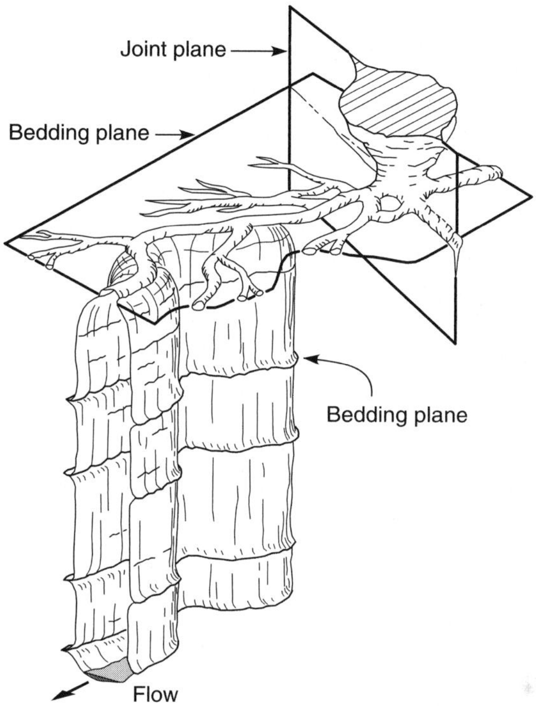

Figure 1.184 Common structural control of cave passages and vertical shafts in Mammoth cave, Kentucky, the United States. (From Southeastern Friends of the Pleistocene, 1990.)

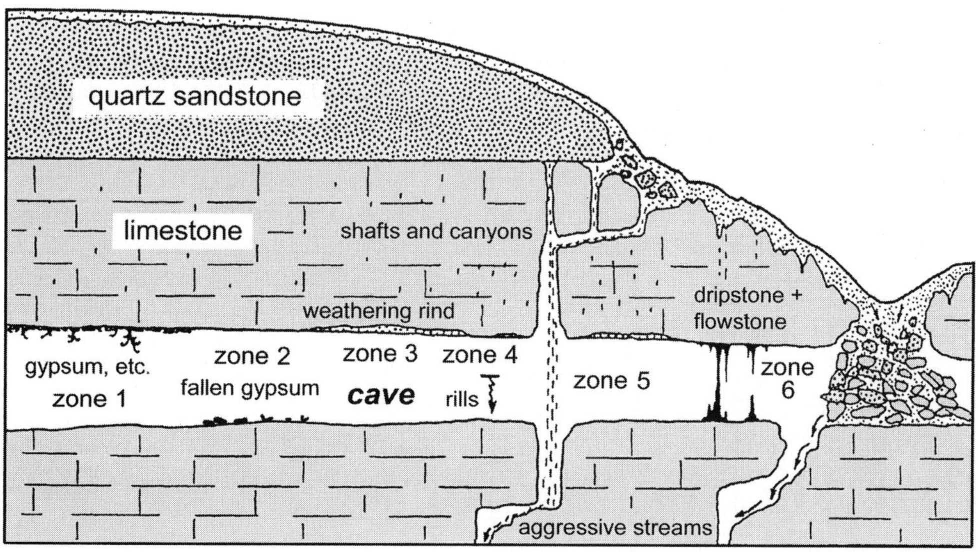

Figure 1.185 Variation in the nature of infiltration into a sandstone-capped cave, such as those in the karst plateaus of east-central United States, and the resulting effects on the cave. See text for zone descriptions. (From Palmer and Palmer, 1995, and Palmer, 2007, printed with permission.)

sandy soil has a PCO_2 of about half as great as in zone 6, where the sandstone cap has been removed by erosion, but in zone 5, the water reaches the cave with little or no dissolved carbonate. No solutional capacity is lost at the soil–bedrock contact, and so dissolution is expended mainly on cave development. Shafts and canyons are typical in the upstream portions.

Zone 6: Water infiltrates through soil-covered limestone and reaches equilibrium with dissolved calcite at the same high CO_2 values as those of the soil. Water loses much CO_2 when it enters the cave, and calcite is forced to precipitate as speleothems. This is typical of small flows and drips. Larger flows fed by sinkholes generally remain aggressive and produce solution openings (Palmer, 2007).

Land Cover and Land Use

Bare soil surface can become encrusted with, or sealed by, the accumulation of fines or other arrangements of particles that prevent or retard the entry of water into the soil. As rainfall starts, the fines accumulated on bare soil may coagulate and strengthen the crust and/or enter the soil pores and seal them off. A soil may have excellent subsurface drainage characteristics but still has a low infiltration rate because of the retardant effect of surface crusting or sealing (Tackett and Pearson, 1965; King, 1992; Rawls et al., 1995). This low-permeable crust is the main reason why bare soil has much lower infiltration capacity than soil with vegetative cover (Fig. 1.186). For example, experiments conducted by Nassif and Wilson (1975) on bare and grassed soil plots showed that bare soil had more than 100% lower infiltration rate than the grassed soil regardless of the water-input rate. Consequently, bare soil is much more easily eroded by surface runoff. The presence of vegetation protects the soil surface from the impact of rainfall. Root systems of vegetation tend to enhance soil porosity and permeability. Organic matter of the soil lowers the bulk density, greatly increases pore sizes and pore-size distribution, and hence increases the infiltration (Rawls, 1983; King, 1992).

The quantity and quality of groundwater are profoundly affected by the historic and current land use and land cover at both regional and local scales. Understanding historic land use/land cover and comparing it with the current and projected land use/land cover

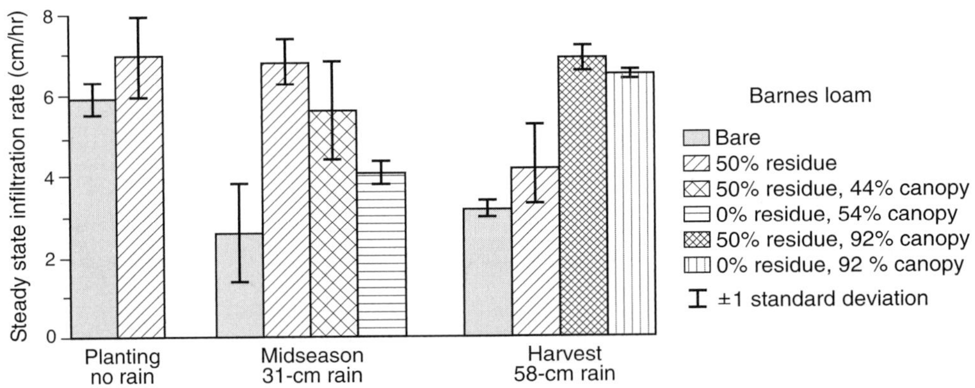

FIGURE 1.186 Temporal effects of soybean canopy and residue on the steady-state infiltration rate on Barnes loam. The bare soil has the lowest infiltration capacity for all infiltration rates and seasons. (Modified from Rawls et al., 1993.)

is therefore of critical importance to the project at hand. Three main ongoing trends in land use and the associated human-induced changes in land cover worldwide that are disrupting natural hydrologic cycles include the following (Kresic and Mikszewski, 2009):

1. Conversion of forests into agricultural land, mostly a practice in undeveloped countries.
2. Rapid urbanization in undeveloped and developing countries converting all other land uses into urban land.
3. Rapid decentralization of cities (i.e., suburbanization) and reforestation of former agricultural land, particularly in the United States and other developed countries.

Urban development and the creation of impervious surfaces inevitably result in increase of runoff and soil erosion. In turn, this generally reduces direct infiltration potential and groundwater recharge. Increased sediment load carried by surface streams often results in the formation of fine-sediment deposits along stream channels, which reduces hydraulic connectivity and exchange between surface water and groundwater. Clear-cutting of forests also alters the hydrologic cycle and results in increased erosion and sediment loading to surface streams. Conversion of low-lying forests into agricultural land may increase groundwater recharge, especially if it is followed by irrigation. However, it is important to remember that often such recharge is irrigation return, and the origin of water may be the underlying aquifer in question (irrigation return may also be a small fraction of the groundwater originally pumped). In contrast, cutting of forests in areas with steeper slopes will generally decrease groundwater recharge because of the increased runoff, except in cases of karstified carbonates outcropping at the land surface.

In general, urban development results in decreased infiltration rates and increased surface runoff because of the increasing area of various impervious surfaces (rooftops, asphalt, and concrete). However, the infiltration rate varies significantly within an urban area based on actual land use. Influence of different land cover and land use on aquifer recharge is illustrated in Tables 1.5 and 1.6. Notably, the infiltration rate varies

Vegetation Type	Infiltration	Surface Runoff	CWR	STD	AET	CIR	DP	DP (%)
Row crop	36.45	8.25	32.14	7.36	24.78	8.98	11.67	26.11
Alfalfa	39.51	5.19	40.59	6.82	33.77	9.13	5.74	12.84
Small grain	39.51	5.19	32.35	2.55	29.80	3.26	9.71	21.72
Grassland	39.51	5.19	33.29	6.02	27.27	7.61	12.24	27.38
Woodland	39.51	5.19	34.76	4.41	30.35	7.92	9.16	20.49
Fallow	36.45	8.25	22.26	1.19	21.07	2.10	15.38	34.41

Note: Mean annual precipitation is 44.70 in., mean annual potential evapotranspiration is 51.53 in. CWR, consumptive water requirements; STD, soil-moisture deficit under dryland conditions; AET, actual evapotranspiration (consumptive water use under dryland conditions); DP, deep percolation (aquifer recharge); CIR, consumptive irrigation requirements.
From Dugan and Peckenpaugh, 1985.

TABLE 1.5 Mean Annual Values, in Inches, of Selected Output from the Soil-Moisture Program in the CMRASA Study Area for Soil Group 1 at Paris, Arkansas, the United States

Land Use and Land Cover	Area (mi^2)	Precipitation (in./year)	Recharge (in./year)	Recharge (%)
Undeveloped and nonbuiltup	641	44.2	24.1	54.5
Residential	13	43.3	12.7	29.3
Builtup	35	45.0	13.3	29.6
Urban	99	43.7	8.1	18.5
All categories	788	44.2	21.4	48.4

Modified from Lee and Risley, 2002.

TABLE 1.6 Estimates of Mean Annual Recharge on the Basis of Mean Annual Precipitation, Generalized Surficial Geology, and Land-Use and Land-Cover Categories from the Willamette Lowland Regional Aquifer System Analysis

significantly within an urban area based on the actual land use. This is particularly important when evaluating fate and transport of contaminant plumes, including development of groundwater models for such diverse areas. For example, a contaminant plume may originate at an industrial facility with high percentage of impervious surfaces resulting in hardly any actual infiltration, and then migrate toward a residential area where infiltration rates are considerable because of the open space (yards) and various vegetative covers that may also be irrigated. In karst with thin residuum, water running off impermeable surfaces may quickly enter the subsurface, thus locally increasing both the recharge and loading of the contaminants where present.

In undeveloped countries, where the booming of megacities and associated slums cause many societal and environmental problems, groundwater recharge rates may actually increase due to leaky water lines and unregulated sewage disposal, contaminating groundwater resources and adding yet another problem to the list. However, this situation is not unique to undeveloped countries. An interesting recent study of the karstic Edwards Aquifer within limits of Austin, Texas, suggests that leaky utility lines, urban irrigation, and/or other unidentified sources of water may be causing a noticeable long-term rise in water levels of the aquifer beneath the city (Garcia-Fresca and Sharp, 2005).

A figure showing land cover/land use change in the area of Rainbow Springs near Dunnellon, Florida, between 1995 and 2007 is included in the color plates. Discharge of this first magnitude spring exhibits a decreasing linear trend as illustrated in Fig. 1.187. Likely explanations for this trend include reduced aquifer recharge due to increased residential, industrial, commercial, and transportation developments (i.e., significantly more red color on the 2007 map north and northeast of the spring—see color plates). Another likely reason is larger groundwater withdrawals via wells for water supply and irrigation. In any case, as can be seen in Fig. 1.187, the precipitation pattern did not change as notably during the same period indicating reason(s) other than reduced natural discharge.

Geology and Topography

When soil cover (residuum) is thin or absent, the lithology, tectonic fabric, and the degree of surficial bedrock karstification play dominant role in aquifer recharge. Highly

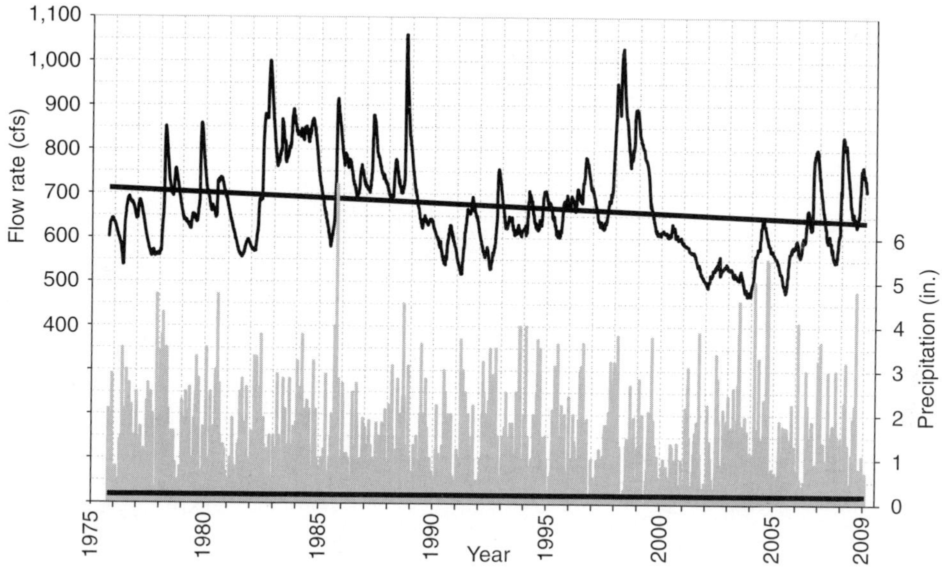

FIGURE 1.187 Hydrograph of daily flow of Rainbow Springs near Dunnellon, Florida, versus daily precipitation at Blitchton Tower (October 1975 to July 1996) and Rainbow Springs (August 1996 to January 2009) gauging stations. Straight thick lines represent linear trend. (From Kresic and Mikszewski, 2012.)

fractured and folded bedrock, steep bedding, and mature karst result in the highest possible rate of infiltration that, as mentioned earlier, exceeds 80%. Examples of such terrain from different parts of the world are shown in Figs. 1.188 and 1.189 (many other photographs throughout the book illustrate this point). Importantly, brittle carbonate sediments such as well-consolidated limestone and dolomite are always fractured regardless of the orientation and thickness of sedimentary bedding planes. This includes massive carbonate sediments. It is therefore not coincidence that the mapping symbols for limestone and dolomite resemble regular and skewed brick walls, respectively, while massive carbonate rocks are presented with scattered intersecting lines depicting fractures (see Fig. 1.190). For one reason or another, however, few practicing hydrogeologists forget these simple facts from time to time and argue that massive limestone, horizontal limestone layers, or beds sloping at the same angle as the topographic slopes impede infiltration. Although this characterization may be true to some extent at a local scale of, say, several square meters, it hardly has any practical meaning for karst aquifer management, protection, and restoration. Water running over a limestone surface will quickly find a fracture and disappear into the subsurface. While doing so, it will also keep dissolving the fracture walls making this disappearance a bit quicker the next time.

When karstified rock is covered by thick residuum and other sediments, topographic slopes will be a factor in generating surface runoff, thus reducing the direct infiltration. In karst depressions filled with sediments, such as sinkholes and uvalas, direct precipitation and any runoff coming off the depression slopes may pond resulting in both an increased infiltration and more evaporation.

An ill-defined term epikarst has been gaining popularity, bringing more confusion into various aspects of developing hydrogeological CSMs in karst. For many karst

FIGURE 1.188 Typical karst landscape in the Dinarides of Montrenegro. An example of "super karst" with the highest infiltration rate. (Photograph courtesy of Dobrisav Bajović-Bajone.)

hydrogeologists involved in solving practical engineering problems throughout the world, this term simply refers to the shallow, more weathered (more karstified) portion of the subsurface. Higher karstification is due to higher density of fractures at and near the land surface (which is typical for any solidified rock) and an abundant supply of carbon dioxide from the atmosphere and the soil layer where present (as a reminder, dissolution of CO_2 in water creates carbonic acid, the main agent of carbonate rock dissolution and karstification). Figures 1.61 and 1.62 in Sec. 1.3.2 illustrate highly weathered zones, equivalent to epikarst, with the abundance of unconsolidated residuum and large in-filled dissolutional features.

Detailed description of the epikarst concept is provided by Ford and Williams (2007) among others. Common to this and other descriptions is that epikarst is given important hydrogeologic and hydraulic roles. It is viewed as a perched aquifer sitting well above the main permanently saturated zone of the karst aquifer. From this perched aquifer (epikarst), water percolates to the main, deeper aquifer primarily via isolated vertical drains (shafts) that usually form beneath sinkholes. Because of the contrast in permeability between the epikarst (permeability is higher in the shallower, more weathered rock) and the underlying competent rock, as well as the limited transfer capacity of the isolated drains (which are also narrowing down with depth), the epikarst is underdrained. The result of this is formation of a permanent or semipermanent perched saturated zone (perched aquifer) in which most groundwater moves horizontally toward the vertical drains. Epikarst characteristics vary based on recharge pattern (climate of the site), the degree of weathering, and the presence of soil/residuum.

FIGURE 1.189 Steep limestone beds and a sinkhole in the Andes, Peru. Another example of "super karst" with the highest infiltration rate. (Photograph courtesy of Marc Etienne, AMEC Environment and Infrastructure, Inc.)

FIGURE 1.190 Bedded (layered) and massive limestone and other sedimentary carbonates are always intensely fractured due to their brittle character. This is the characteristics selected for their graphical presentation on geologic maps and cross sections. *Top left*: Limestone and marly limestone beds in the Djerdap National Park, Serbia, with short vertical fractures connecting bedding planes, and fractures intersecting multiple bedding planes. *Top right*: Massive Mesozoic limestone with various fracture sets near Resava, Serbia. *Bottom*: Geologic map symbols for various carbonate sedimentary rocks. (From USGS, 2006.)

Unfortunately, the above-described concept of epikarst is being applied indiscriminately including at sites where there is clear evidence of its complete absence. In fact, many karst areas worldwide show absence rather than presence of perched aquifers sitting above some main, deeper, permanently saturated karst aquifer zone. Many photographs in this book do not need additional explanation why this concept of epikarst does not apply in a particular case (e.g., see Figs. 1.162, 1.188, and 1.189). It would be erroneous, at any karst site, to assume that transfer of infiltrating (percolating) water from the land surface to deep permanently saturated karst aquifer zone is happening via vertical shafts only, whatever their number and density may be. Countless worldwide examples of dry shallow monitoring wells, tunnels, galleries, and surveyed cave passages at various depths, as well as groundwater tracing tests, show that the concept of epikarst as explained earlier does not hold water in many site-specific settings. People in many parts of the world cannot drill or dig any shallow wells to obtain drinking water from karst terrains for any reasonable period because the subsurface is completely dry from the land surface down to depths of hundreds of meters. All precipitation that falls over the karst terrain quickly finds its way deep into the subsurface through fractures and dissolution features without creation of any shallow perched aquifer. Trickles of water or waterfalls that appear in dry cave passages at various depths or in accessible, observable vertical shafts may last for some time, but many will disappear after cessation of rainfall as quickly as they appeared (see Fig. 1.83). For example, if the hydraulic explanation of epikarst were applicable to Fig. 1.51, then the epikarst zone would have to be more than 1,000-m (3,300-ft) thick. Red Lake near Imotski, Croatia, the deepest sinkhole in the world, does not show any perched water seeping out of "epikarst" along the entire depth of its vertical walls (Fig. 1.191).

The author of this book has been engaged in discussions about the hydrogeologic significance of epikarst during which its advocates were, in support of their theory, citing examples from various carbonate terrains in the United States; at the same time, they were qualifying some other karst terrains, such as the classic Dinaric karst, as an exception to the rule. The main argument explaining the unique hydraulic role of epikarst in many such discussions appears to be flat-layered and supposedly sparsely fractured carbonates. This rather unusual geologic condition, in otherwise brittle limestone and dolomite that are easily and extensively fractured and faulted, would presumably result in the horizontal groundwater flow toward sparse vertical drains. Several practical questions arise from this theoretical, academic concept of groundwater flow in epikarst:

- Can the presumed perched saturated zone in epikarst be utilized for water supply, even at a very local scale?
- Is this zone permanent, semipermanent, or intermittent (and for how long)?
- If there is dissolved contamination detected in the perched saturated epikarst zone, can it be hydraulically contained by extraction wells or linear drains for example?
- Is it possible, as part of groundwater remediation efforts, to locate occasional vertical drains in epikarst and focus the aquifer remediation efforts there? (Because contaminated groundwater in the areas between these occasional drains moves laterally toward the drains and not downward at any appreciable rate.)

An illustrative example provided by Palmer (2007) may help examine possible answers to these and other practical questions. Figure 1.192 shows the distribution of

FIGURE 1.191 Vertical wall of the Red Lake near Imotski, Croatia, the deepest sinkhole in the world. Note spruce forest at the rim for scale (total depth of the sinkhole is 513 m). The water level in the lake is also the level of the permanently saturated zone in the surrounding karst aquifer. Depth of water in the sinkhole during average low conditions is about 250 m.

sinkholes over Blue Spring Cave, Indiana, the United States. The map includes only those sinkholes that appear on the 1:24,000 topographic quadrangle. The actual number of sinkholes is *several times greater* (emphasis added) than is shown on the topographic map. The main passable (nonsubmerged) passages in the cave are about 8 to 12 m in diameter. Limestone is exposed at the surface, beneath 2 to 5 m of soil, over the entire cave. The surface rock unit is mainly the prominently bedded St. Louis limestone. Most of the cave is located in the underlying massive Salem Limestone. Hundreds of small sinkholes seem to have a random distribution in relation to the cave, but the deepest are located over the main stream passage. The stream in the main passage is an important agent in removing sediment and collapse debris from deep sinkholes. As suggested by Palmer,

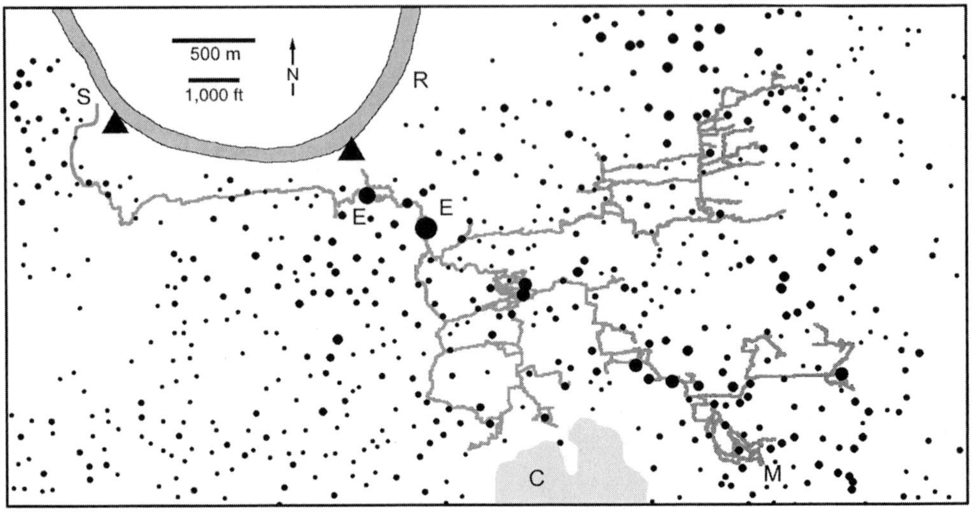

Figure 1.192 Sinkhole distribution over Blue Spring Cave, Indiana. Nearly all sinkholes are in the bedded St. Louis Limestone. The largest penetrate into the underlying prominently jointed Salem Limestone. Cave passages are shown in gray. E, main cave entrances; M, inflow of main cave stream; S, Blue Spring; R, east fork of White River; C, sandstone cap-rock over limestone. *Black triangle* symbol denotes abandoned spring alcove, largest black dot is sinkhole at least 25-m deep, smallest dot is sinkhole about 3-m deep; intermediate sinkhole depths are shown in proportion to their dot size. (From Palmer, 2007, printed with permission.)

most of the infiltration in the region enters small sinkholes from where it is drained down by at least one sollutionally enlarged conduit. However, most small sinkholes have no apparent relation to the mapped cave passages, so that it is uncertain where all this water flows. As suggested by the author, it is possible that many such sinkholes drain into small inaccessible passages and, in general, the sinkhole drainage converges into relatively few streams before it joins the traversable parts of the cave (Palmer, 2007).

By reading this brief discussion and by looking at the map in Fig. 1.192, it appears that the answer to all previously posed practical questions regarding epikarst would have to be negative, including its very existence: namely, the density of sinkholes would effectively invalidate the concept and any possibility that there is some significant, continuous, shallow, perched aquifer zone in the area.

Without engaging in additional academic discussions about representative scales of the epikarst hydraulic concept (e.g., perched area of 100 sq ft, 1 acre, and 1 hectare; thickness of 10 and 100 ft; and horizontal spacing between possible vertical shafts 100 ft and 1,000 ft), and notwithstanding that these exceptional hydraulic characteristics may be applicable at some sites, the author cautions novice hydrogeologists to carefully design field investigations and continuously monitor groundwater at their site. This will enable them to make informed conclusions as to the presence of "perched water in epikarst" and its practical importance for their particular project.

1.7.2 Allogenic Recharge

Sinking of surface streams is one of the main characteristics of karst terrains. It provides direct aquifer recharge with large quantities of water on a continuous or, in case of

FIGURE 1.193 Sinks of Potts Creek in Alleghany County, Virginia. (Photograph courtesy of Phil Lucas.)

intermittent streams, semicontinuous basis. It also facilitates rapid development of the preferential flow paths, as this water is usually undersaturated with respect to calcite and can readily dissolve the rock mass, thus enlarging the existing conduits or creating new ones along fractures and bedding planes. Figure 1.193 illustrates allogenic recharge via large sink on a surface stream under relatively high flow conditions. Large capacity sinks are commonly formed in significant fault zones like the one visible on the photograph. The rock mass was initially disturbed sufficiently enough to allow concentrated subsurface flow. In many cases, large sinks also develop near contacts between soluble carbonates and insoluble rocks and may often have enough capacity to receive entire surface streams, thus giving them common names such as "Lost River" or "Lost Creek" (see Fig. 1.194). In time, sinks in river channels may lose their function as the karstification and surface erosion progress downward and upgradient. Dry cave entrances to former sinks and blind dry valleys are evidence of former allogenic recharge (e.g., see Fig. I.4, bottom, in "Introduction").

As it flows over karstified carbonates, a surface stream can also lose its water through fractures and solution opening in the streambed, which can be more or less obvious or even unnoticed if the stream channel is underlain by alluvium sediments (Fig. 1.195). In such cases, the only evidence that the stream is losing water to the underlying karst aquifer would be a confirmed downgradient decrease in the measured flow rate of the stream.

When a stream sinks within a well-defined area, such as a single sink or a group of closely spaced sinks that are not submerged, determining the allogenic recharge rate is

FIGURE 1.194 Sinking of the Lost Creek in a cave near Drvar, Bosnia and Herzegovina.

relatively straightforward and involves streamflow measurement immediately upgradient of the sink. When the sinking location is under retention (submerged), the task at hand is a lot more complicated, since it is often impossible to accurately measure the representative flow velocity in the sink itself or immediately upgradient of it. In favorable conditions, the representative average flow velocity in the sink can be determined with an array of submerged flowmeters or Pittot tubes, and the flow (Q) calculated using the standard equation:

$$Q = v_{av} A \tag{1.103}$$

where v_{av} is the average flow velocity and A is the cross-sectional area of the receiving subsurface channel (tube).

It is also possible to perform a hydraulic simulation of the sinking during dry periods by blowing air into the sink with large-capacity fans and measuring the air (fluid) velocity

FIGURE 1.195 Significant portion of the Texas' Edwards Aquifer recharge takes place along streambeds through cavities, faults, and fractures. *Top left*: Whirlpool over Crippled Crawfish Cave during a recharge event on the Onion Creek near Austin, Texas. (Photograph courtesy of David Johns.) *Top right*: Fractures in the Helotes Creek, which loses water to the underlying karst aquifer. (Photograph copyright Gregg Eckhardt, printed with permission). *Bottom*: In 2002, record-high spillway waters of the Canyon Dam on the Guadalupe River washed away loose sediments and exposed these fractures in limestone. Woodruff and Abbott (1986) suggest that infiltrating water of the Guadalupe River contribute to San Marcos spring discharge. (Photograph copyright Gregg Eckhardt, printed with permission; many other photographs of Edwards Aquifer features available at http://www.edwardsaquifer.net.)

with an array of instruments. Since the experiment conditions can be easily controlled, this method allows for an accurate determination of the cross-sectional distribution of the flow velocities for varying airflow rates. Once a functional relationship is established, it is possible to measure the actual sinking water flow velocity at only one favorable location, which can be securely equipped. The flow rate is then calculated as

$$K = \frac{v_{av}}{v_{rec}} \quad Q = K v_{rec} A \tag{1.104}$$

where v_{av} is the average velocity for the airflow simulation and v_{rec} is the recorded water flow velocity during actual sinking.

FIGURE 1.196 Partially (*left*) and completely (*right*) flooded sinkhole in front of several entrances into the Water Sinks Cave, Highland County, Virginia. Note muddy water full of debris and limestone cliff face above the flooded cave entrances. (Photographs courtesy of Phil Lukas.)

In many field situations, the above-explained methods are impracticable to implement due to high-pressure conditions during floods and loads of debris that can easily damage instruments (see Fig. 1.196). In such cases, it is advisable to calculate the sinking flow rates using one of the applicable hydraulic equations for pipe flow described earlier, and an established relationship between the flood levels and the representative hydraulic heads in the subsurface. Two examples illustrating this hydraulic relationship are shown in Figs. 1.197 and 1.198. In the first case, the relationship is simple: As soon as the sink (ponor) is flooded, the sinking flow becomes constant at the limit of the channel's capacity. For the capacity of the ponor shown in Fig. 1.198, the relevant hydraulic pressure is one measured in the channel when the flow is under pressure (case a); when

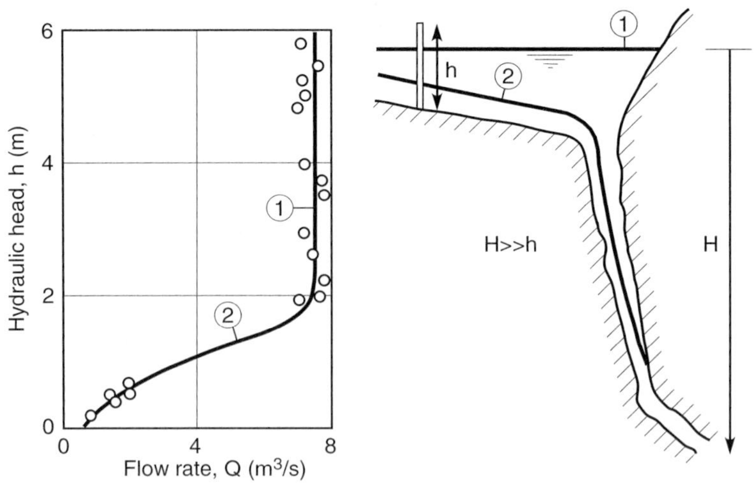

FIGURE 1.197 Sinking flow rate versus hydraulic head of surface water above the sink in case of large difference in elevation between the surface water levels and the subsurface sink channel, $H >> h$. (From Avdagić, 1985.)

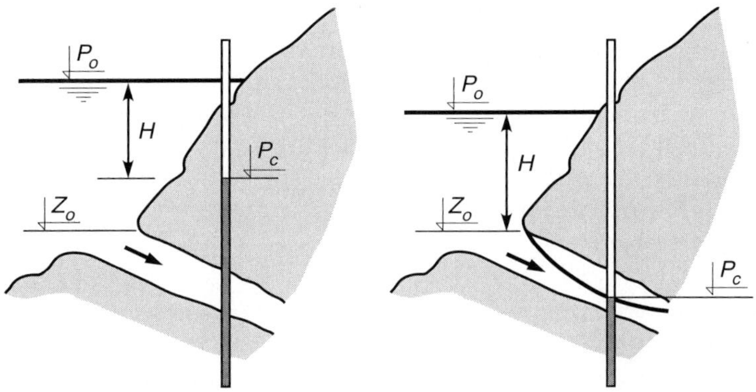

FIGURE 1.198 Analysis of the representative hydraulic head, H, for determining the sinking capacity of flooded ponors with respect to the flow regime in the drain channel. *Left*: When the system is under pressure ($P_C > Z_O$), the head is $H = P_O - P_C$. *Right*: When the system is not under pressure ($P_C < Z_O$), the head is $H = P_O - Z_O$. (From Avdagić, 1976.)

the flow in the channel is not under pressure (case b), the elevation of the sink itself is used in calculations.

Flow hydrograph shown in Fig. 1.199 is typical for most flooded karst poljes in the Dinaric karst, which have hydraulic connections with large springs acting as their erosion basis. The amount of sinking (loss of water from the polje to the underlying aquifer) and the spring discharge rate are inversely proportional. During high-intensity precipitation events and the related rapid increase of the hydraulic head in the aquifer, the rate of sinking decreases and may even become negative such that the ponor (sink) becomes a spring, the so-called estavelle, discharging water into the flooded polje (case a in

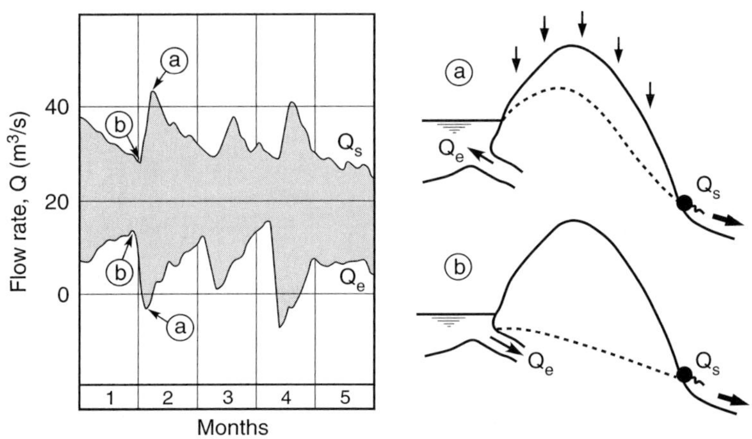

FIGURE 1.199 Relationship between flow rates at a spring (Q_S) and a ponor (Q_P) with the proven connection. The aquifer hydraulic head is depicted with *dotted line*. (a) Discharge rate at the ponor (sink) is negative, that is, the ponor acts as a spring or estavelle. (b) Sinking rate at the ponor increases and the flow rate at the spring decreases as the hydraulic head in the aquifer is lowered. (From Kresic, 1991.)

Fig. 1.199). At the same time, the flow rate at the base spring increases due to the steep hydraulic gradient in the aquifer. As the hydraulic head in the aquifer starts decreasing following the cessation of heavy rainfall, the sinking flow rate increases while the flow of the base spring decreases due to the lower hydraulic gradient in the aquifer (case b in Fig. 1.199). The difference between the measured input of water to the aquifer via the ponor and the measured output at the base spring (the shaded area on the graph in Fig. 1.199) is due to other recharge mechanisms such as direct infiltration of precipitation and inflow from other smaller sinks.

1.8 Karst Springs

Spring is a location at the land surface where groundwater discharges from the aquifer, creating a visible flow. This discharge is caused by difference in the elevation of the hydraulic head in the aquifer and the elevation of the land surface where the discharge is taking place. The opening through which groundwater discharges is called *spring orifice*. Springs issuing from consolidated rocks usually have a well-defined orifice, as opposed to springs in unconsolidated sediments, although the latter ones may also have a clearly visible orifice. *Fracture* (or *fissure*) *spring* refers to concentrated discharge of water from bedding planes, joints, cleavage, faults, and other breaks in the consolidated fractured rock. In karst, fracture springs are in many cases modified by dissolution and referred to as *tubular springs* or *cave springs*.

Secondary springs issue from locations located away from the primary spring discharge, which is covered by colluvium (rock fall fragments) or other natural debris and therefore not visible. When capturing such springs, every attempt should be made to remove all the debris and locate the primary spring orifice(s), since the secondary discharge locations tend to migrate in time.

Although karst is best known for its spectacular large cave springs (Fig. 1.200), the water seldom issues entirely from a single opening at any given location. Often near large cave springs, there are scattered smaller springs issuing from a variety of openings at different elevations, some of them active only after heavy rains. Also, what is called a single spring in one locality may be equivalent to what in another locality is regarded as a group of springs, each of which has an individual name. Some springs are designated in the singular and some in the plural—for example, Silver Spring and Comal Springs. The idea that underlies this usage is that if the water issues from a single opening or from several openings that are close together, it forms a spring, whereas if it issues from a number of openings that are farther apart, it forms springs (Fig. 1.201). However, local terminology is so variable in this respect that there is no consistent distinction between "a spring" and "springs" and often there is no uniformity in usage of terms even for the same group of openings (Meinzer, 1927).

All springs (except some associated with young volcanism and hydrothermal activity where discharge is driven by gases and temperature gradients) ultimately discharge at the land surface because of the force of gravity. However, regardless of the rock type, they are divided into two main groups based on the nature of the hydraulic head in the underlying aquifer at the point of discharge:

- *Gravity springs* emerge under unconfined conditions where water table intersects land surface. They are also called *descending springs*.

Figure 1.200 The third largest karst spring in France, Le Loue, issuing from Jurassic limestones, Franche-Comte, French Jura. (Photograph courtesy of Dr. Zoran Stevanovic.)

- *Artesian* springs discharge under pressure due to confined conditions in the underlying aquifer. They are also called *ascending* or *rising springs*.

In general, when the contact between the water-bearing rock and the impermeable underlying rock is sloping toward the spring, in the direction of groundwater flow, and the aquifer is above this impermeable contact, the spring is called *descending contact spring* (Fig. 1.202a). When the impermeable contact slopes away from the spring, in direction opposite of the groundwater flow, the spring is called *overflowing contact spring* (Fig. 1.202b).

In karst, descending springs are often intermittent and reflect periodically active conduits that are sloping down toward the spring. When the rate of vertical erosion of a surface stream incised in carbonates exceeds the rate of karstification in the surrounding

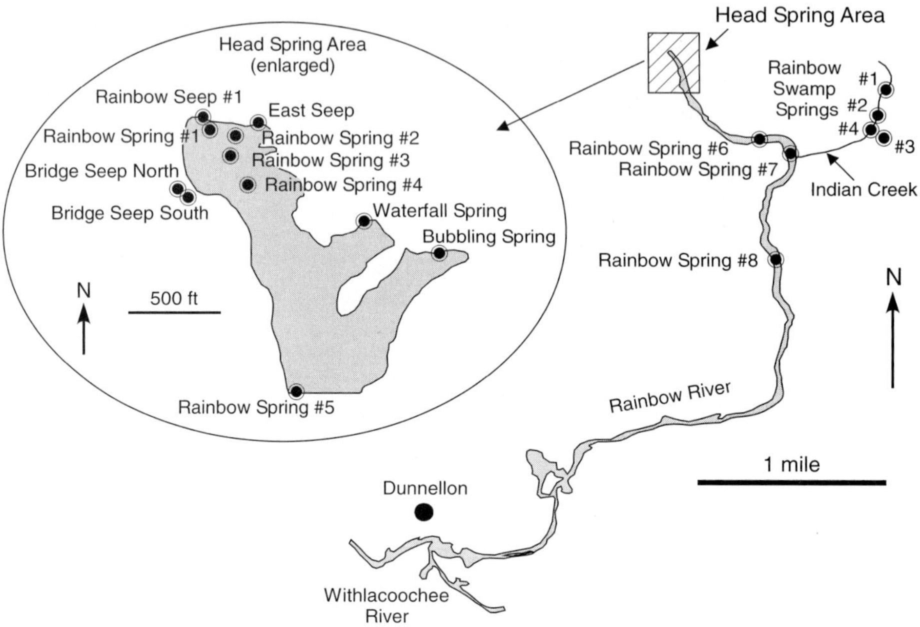

FIGURE 1.201 Locations of spring vents in the Rainbow Springs complex. (From Jones et al., 1996.)

aquifer, springs may form high on the canyon walls resulting in spectacular waterfalls. Large vertical displacements along active faults can have similar effect. Some examples of descending springs of this type are shown in Figs. 1.203 and 1.204.

Figure 1.205 shows some examples of *barrier springs*, the term generally referring to springs at any lateral contact between the aquifer and a low-permeable rock (sediment). Such contact can have many different shapes caused by depositional processes and tectonic movements that form a variety of faults and folds. As explained earlier, in addition to the lateral growth of groundwater drainage basins in karst across topographic

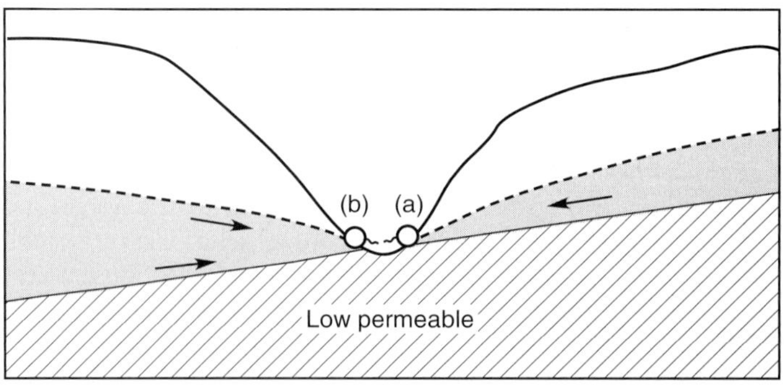

FIGURE 1.202 Contact gravity springs: (a) descending, and (b) overflowing. (Modified from Milojevic, 1966.)

Figure 1.203 *Left*: Typical descending gravity spring of the Sava River in Slovenia formed in a fault zone; the spring is discharging about 0.5 m^3/s (Photograph courtesy of Dr. Zoran Stevanovic.) *Right*: Spring issuing from Da Long Dong (Big Dragon Cave) near Jishou, western Hunan. (Photograph courtesy of Mark Graham; www.geospace.co.uk.)

Figure 1.204 A typical intermittent descending gravity spring, issuing from limestones of the Cretaceous Kometan Formation near Dokan Lake in northern Iraq. (Photograph courtesy of A. Holm.)

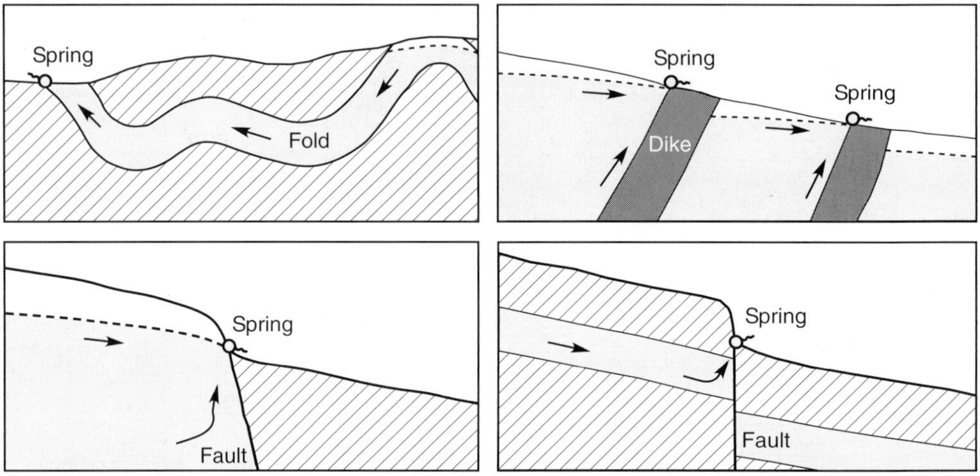

FIGURE 1.205 Several types of barrier springs issuing at lateral contacts between water-bearing carbonates and barriers to groundwater flow formed by low-permeable rocks. (From Kresic, 2010; copyright Elsevier, Inc., printed with permission.)

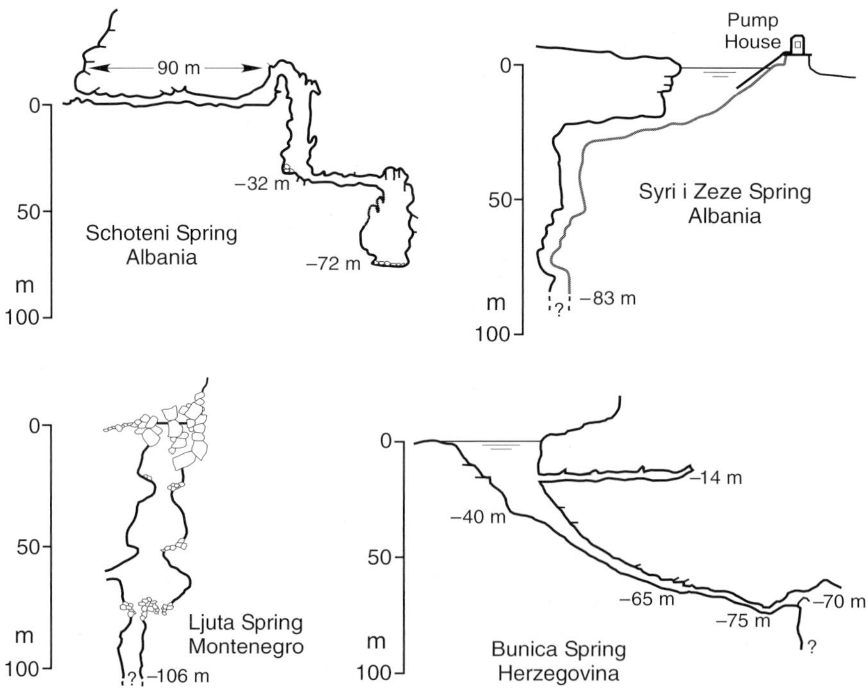

FIGURE 1.206 Examples of large ascending springs in karst, with conduits explored by divers. (From Touloumdjian, 2005.)

divides, there is another important aspect of the karstification process—it also continues with depth regardless of the erosion basis for surface water flow. In other words, provided there is enough thickness of the carbonate sediments below surface water bodies or lateral contacts with less permeable rocks, the dissolution process will result in development of deep confined karst conduits, which feed ascending karst springs called *vauclusian* springs after Fontaine de Vaucluse, the source of river Sorgue in Provance, France (see Figs. 1.48 and 1.49). In fact, many of the large permanent karst springs in the world, some of which are shown in Figs. 1.206 through 1.208, are vauclusian.

In some cases, regardless of the size, ascending springs in karst may issue from pools, which have bottoms covered by sand, gravel, or rock fragments, and without visible conduits. However, the conduits or fractures may be masked and bridged over by clastic sediment—a possibility that cannot be excluded. For example, Meinzer (1927) gives the following account of the Bennett Spring in Dallas County, Missouri (see Fig. 1.208):

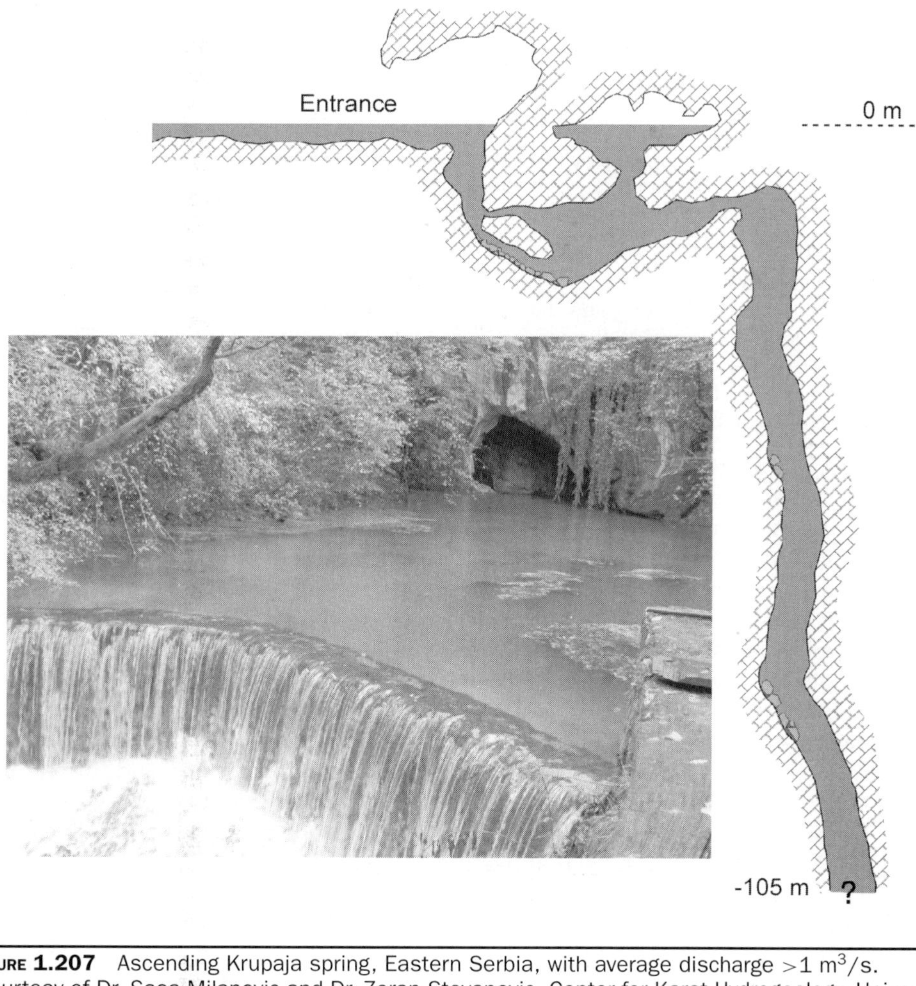

FIGURE 1.207 Ascending Krupaja spring, Eastern Serbia, with average discharge >1 m^3/s. (Courtesy of Dr. Sasa Milanovic and Dr. Zoran Stevanovic, Center for Karst Hydrogeology, University of Belgrade.)

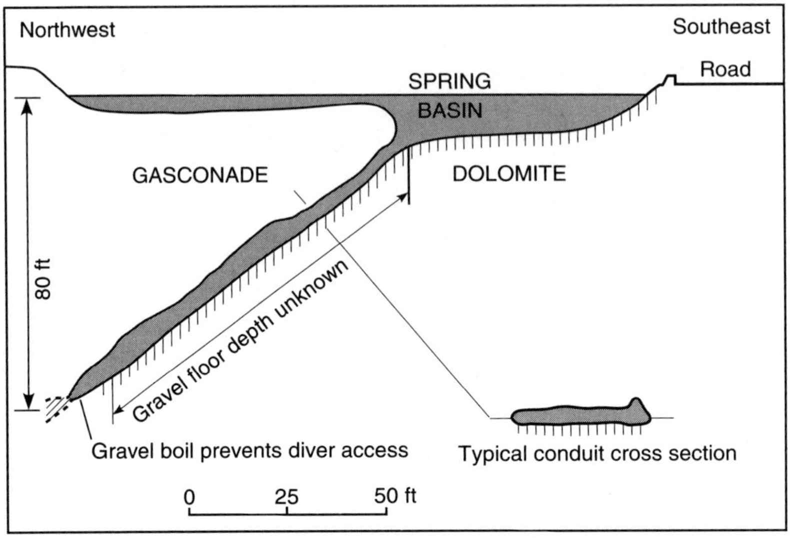

FIGURE 1.208 Northwest–southeast longitudinal profile through first magnitude Bennett Spring in Missouri from data supplied by Donald N. Rimbach, Michael R. Tatalovich, and M. Grussemeyer. (From Vineyard, J.D. and Feder, G.L., 1982. Springs of Missouri. Missouri Department of Natural Resources, Division of Geology and Land Survey, Water Resources Report No. 29, Rola, MO, 212p.)

> It issues from a circular basin in gravel about 30 feet in diameter and gives rise to a stream that is about 1½ miles long and that empties into Niangua River at a level about 22 feet below the spring. The spring furnishes power for a mill in the village of Brice. The temperature of the water is about 58°F. The spring was visited in 1903 by Shepard, who states that at that time it boiled up with great force from a vertical cavelike opening through the limestone into a large oval basin.

Quite a few springs in Florida, where limestone is covered by a thick layer of surficial sands, issue from pools without clearly visible conduits and with only sporadically visible small limestone outcrops.

One practical advice to novice karst hydrogeologists is that, if a spring is large, behaves like a karst springs, and appears to have a very significant deficit in its topographic (land surface) drainage area, then it is most likely a karst spring. In other words, even when such a spring discharges from a nonkarstic rock, but there are sedimentary carbonates somewhere in the vicinity, one should not rule out a possibility that the primary aquifer providing water to the spring is actually a karst (limestone) aquifer. A very illustrative example is Giant Spring in Great Falls, Montana. This first magnitude spring has a very uniform discharge of about 200 million gallons per day and forms probably the shortest river in the United States called the Roe River that flows into the Missouri River. The springwaters originate as precipitation that falls on the Madison limestone of Little Belt Mountains over 4,000 ft above the Missouri River. The limestone has all characteristics of karst, including losing streams and dry caves. At Giant Springs, the Madison limestone is less than 400 ft below the surface and groundwater from this karstic aquifer escapes upward through fractures in the overlying Kootenai sandstone (Fig. 1.209).

Outlets of karst springs can be buried below unconsolidated alluvial deposits due to a rise in base level of surface streams, or they can be buried by glacial or marine deposits (see Fig. 1.210). In any such case, it may not be immediately apparent that the spring is

FIGURE 1.209 Vertical fractures in Kootenai Sandstone at Giant Springs State Park near Great Falls, Montana. The fractures are extension fractures and are open, providing an easy conduit for groundwater in the Madison limestone aquifer to rise to the surface and be discharged in and next to the Missouri River. (Photograph courtesy of late Dr. David Baker.)

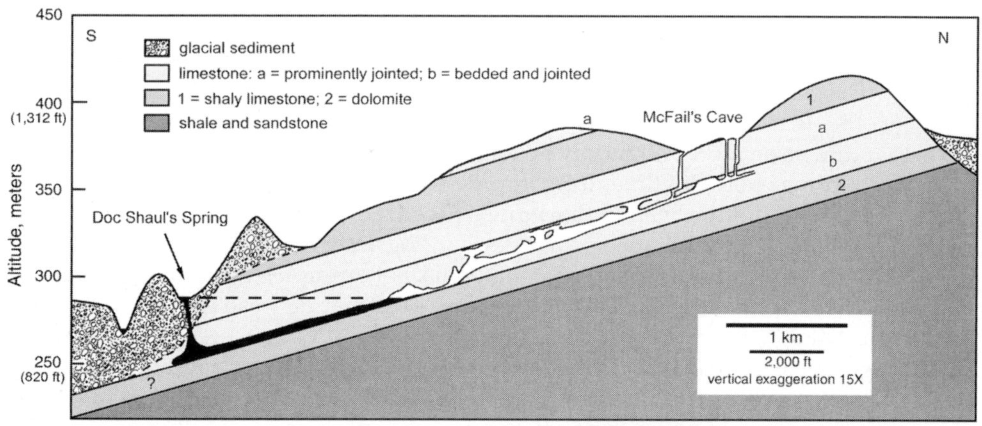

FIGURE 1.210 Profile through McFail's Cave, New York. The former spring outlet was buried by about 30 m of glacial and river deposits during the past 120,000 years, so that water is forced to rise through the sediment to an artesian spring. (From Palmer, 2007, printed with permission.)

Figure 1.211 Boiling Spring on Potts Creek, Alleghany County, Virginia. (Photo courtesy of Phil Lukas.)

karstic, but a combination of hydrogeological field mapping and spring sampling and monitoring should be sufficient to solve the mystery in most cases.

Discharge of groundwater into surface streams and other water bodies may occur below the surface water level and may not be immediately visible. This discharge can be either diffuse or, more commonly in karst, concentrated via *submerged (subaqueous) springs*. Such springs can sometimes form impressive boils in streams, especially after heavy rains when the hydraulic head in the aquifer rises well above the surface stream (Fig. 1.211).

Submerged freshwater springs discharging at sea floor (*submarine* springs) have been known to and intrigued people for millennia. For example, the Roman geographer Strabo, who lived from 63 BC to 21 AD, mentioned a submarine fresh groundwater spring 4 km from Latakia, Syria, near the Mediterranean island of Aradus. Water from this spring was collected from a boat, utilizing a lead funnel and leather tube, and transported to the city as a source of freshwater. Other historical accounts tell of water vendors in Bahrain collecting potable water from offshore submarine springs for shipboard and land use, Etruscan citizens using coastal springs for "hot baths" (Pausanius, ca. second century AD), and submarine "springs bubbling freshwater as if from pipes" along the Black Sea (Pliny the Elder, ca. first century AD; from UNESCO, 2004). Karstified rocks comprise 60% of the shoreline of the Mediterranean and are estimated to contribute 75% of its freshwater input, mostly via direct discharge to the sea (UNESCO, 2004). The Adriatic

Figure 1.212 *Top*: Submarine spring Vrulja in the Plominski Bay, east coast of Istria, Croatia. The clear springwater contrasts the muddy water brought into the bay by the Boljuncica River after heavy rains. (Photograph courtesy of Andrija Rubinić.) *Bottom*: One of submarine springs near Croatian town of Brela on Makaraska Riviera, the Adriatic Sea.

coast of Montenegro and Croatia is dotted with submarine springs often discharging tens of cubic meters per second (Fig. 1.212).

The groundwater flow toward the coast and its submarine discharge are caused by the hydraulic gradient between the inland recharge areas and the sea level (Fig. 1.213). If the aquifer is confined and well protected by a thick aquitard, the flow of fresh groundwater may continue well beyond the coastline with the ultimate discharge taking place along distant submarine aquifer outcrop as in case of the Floridan aquifer off the coast of Georgia and South Carolina, the United States (Fig. 1.214).

Estavelle is a spring with dual function: It acts as a spring during high hydraulic heads in the aquifer, and as surface water sink during periods when the hydraulic head

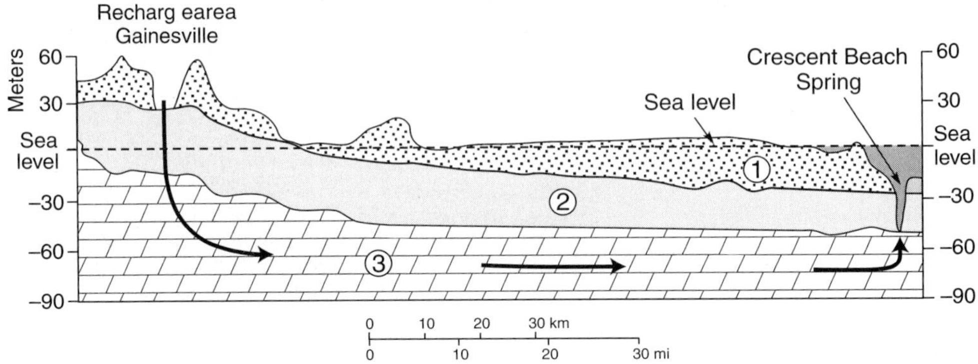

FIGURE 1.213 Idealized cross section of groundwater flow to Crescent Beach Spring, Florida. (1) Post-Miocene deposits (green clay, sand, and shell); (2) confining unit (Hawthorn Formation); (3) Upper Floridan aquifer (Eocene Ocala Limestone). (Modified from Barlow, 2003.)

in the aquifer is lower than in the body of surface water (Fig. 1.215). Estavelles are located within or adjacent to surface water features.

Discharge of *ebb-and-flow springs*, or periodic springs, occurs in relatively uniform time intervals (periods) and is explained by the existence of a siphon in the rock mass behind the spring (Fig. 1.216). The siphon fills up and empties with regularity, regardless of the recharge (rainfall) pattern. A better-known example from the United States is Periodic Spring in the Bridger-Teton National Forest near Jackson, Wyoming. Situated at the base of limestone cliffs, the spring discharges about 285 gallons per second. Springwater gushes from an opening for several minutes, stops abruptly, and then begins a new cycle a short time later. Intermittent water flows range anywhere from 4 to 25 minutes and the water is clear and cold (US Forest Service Intermountain Region, 2008).

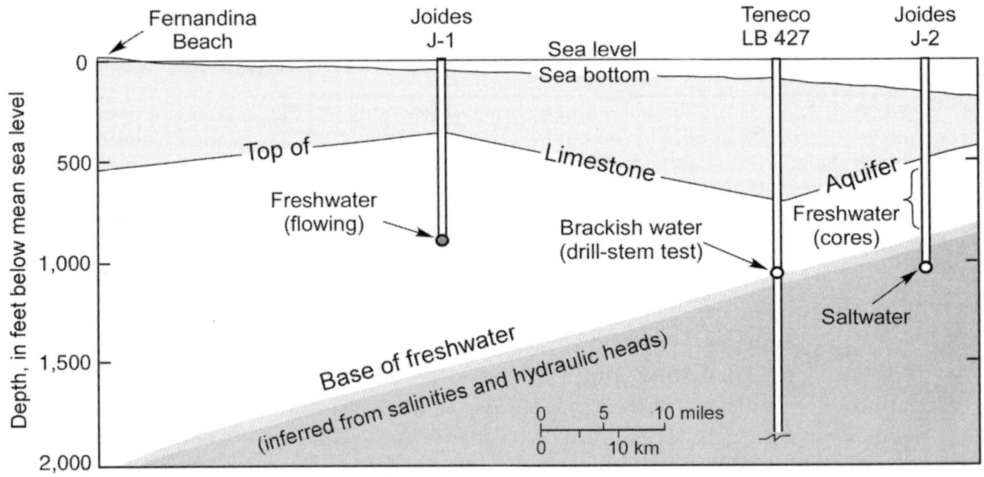

FIGURE 1.214 Inferred position of the freshwater–saltwater interface based on hydraulic testing and water analyses at offshore exploratory oil wells, Atlantic Ocean off the coast of Georgia and Florida. (from Johnston et al., 1982.)

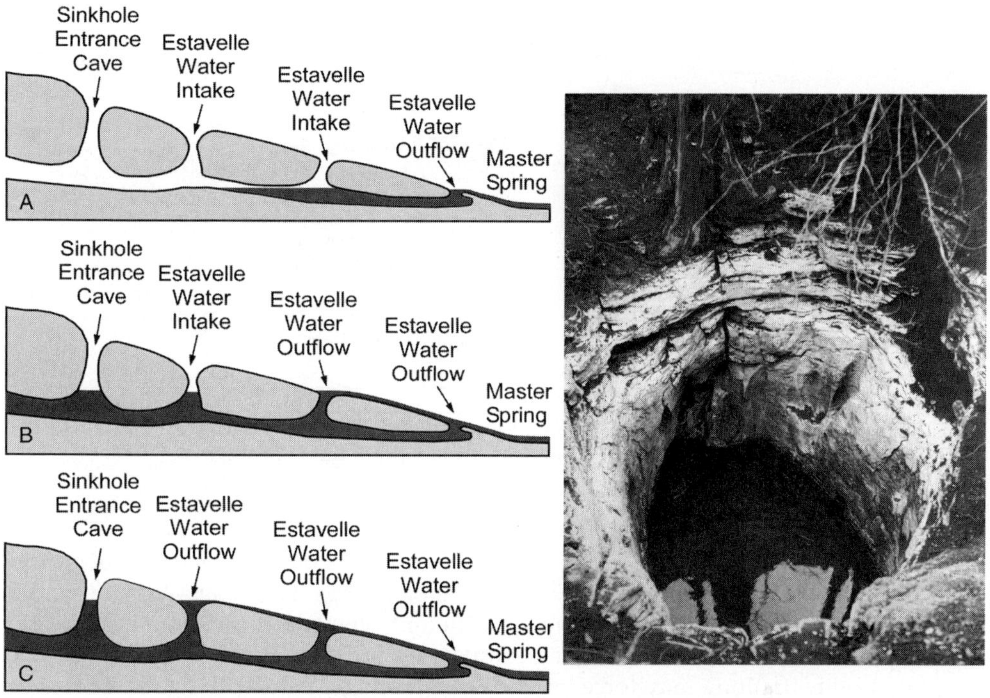

FIGURE 1.215 *Left*: Hypothetical stream profile showing sequential reversal of flow in estavelles. (A) Low flow: only lower sections of karst drainage system flooded; estavelles are sinkholes (swallow holes); (B) medium flow: headward flooding of karst drainage system; lowest estavelle becomes spring (rising); and (C) High flow: storage capacity of karst drainage system exceeded; estavelles become springs, water backs up in headward caves. (From Vineyard and Feder, 1982.) *Right*: The Oval Sink, a reversible sinkhole (estavelle) in Springfield, Missouri. (From Vineyard and Feder, 1982; photograph by Jerry Vineyard.)

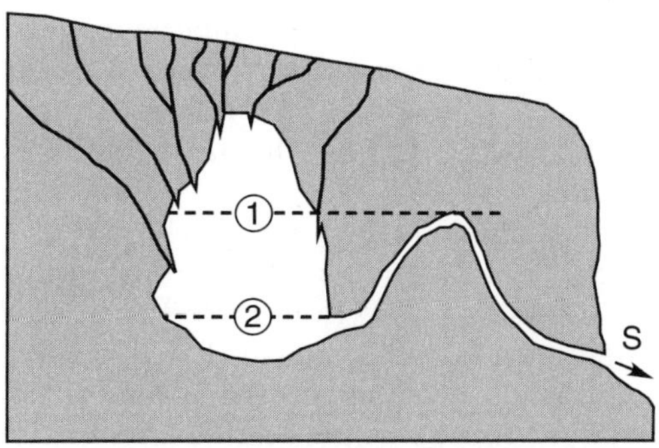

FIGURE 1.216 Schematic of an ebb-and-flow (periodic) spring. When water in the cavity reaches level 1, the siphon becomes active and the spring (S) starts flowing. When the level drops to position 2, the spring stops flowing. (From Radovanovic, 1897.)

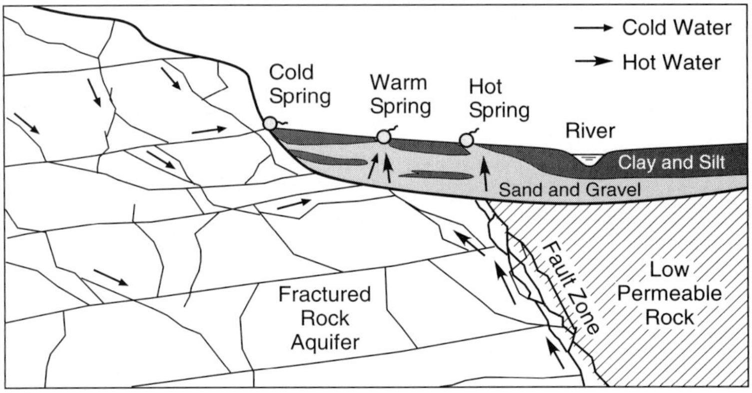

FIGURE 1.217 Three springs of different temperature in a barrier fault zone overlain by alluvium. (From Kresic, 2010; copyright Elsevier, Inc., printed with permission.)

Topography and geologic fabric (rock type and tectonic features such as folds and faults) play the key role in the emergence of springs. When site-specific conditions are rather complex, springs of formally different type based on some classifications may appear next to each other causing confusion. For example, a lateral impermeable barrier caused by faulting may force groundwater from greater depth to ascend and discharge at the surface through the overlying alluvial sediments in the stream valley (Fig. 1.217). This water may have elevated temperature due to normal geothermal gradient in the earth's crust and discharge at a thermal spring. At the same time, groundwater of normal temperature may issue at a gravity spring located very close to the thermal spring, at the contact with less permeable alluvial fine-grained sediments such as clays. Yet, a third spring may be present with its temperature varying depending on the precipitation pattern, seasonal influences, and the mixing mechanism of waters with different temperatures.

Thermal springs are divided into *warm springs* and *hot springs* depending on their temperature relative to the human body temperature of 98°F or 37°C: Hot springs have a higher and warm springs a lower temperature. Warm spring has temperature higher than the average annual air temperature at the location of discharge. Water temperature of both groups of thermal springs can fluctuate in time reflecting more or less surficial influence.

As discussed by Meinzer (1923a, 1927, 1940), many of the thermal springs in the western United States issue along faults, and most of them probably derive their heat from hot gases or liquids that rise from underlying bodies of intrusive rock. In many cases, the host aquifers for thermal water are karstified carbonates covered by nonkarstic rocks. Basin-fill sediment may act as a thermal blanket that traps heat in relatively shallow aquifers beneath large areas of some of the basins. Nice examples of springs produced by the rise of deep waters through fault openings are found along the edges of the mountain ranges of Nevada and western Utah. Many of these springs have large yields, some of them discharging several cubic feet a second. The abundance of these springs and the copious flow of some of them are even more impressive because of the aridity

FIGURE 1.218 Ascending thermal spring in the channel of the Rio Grande River, Big Bend National Park, Texas. The spring is located adjacent to a cliff composed of thinly bedded limestone. The higher hydraulic head in the little pool maintains clear water as opposed to often muddy water of the river. Note limestone outcrop in the pool. (From Kresic, 2007a; copyright Taylor & Francis Group, printed with permission.)

of the region in which they occur. The mountain ranges of this region consist largely of tilted fault blocks, and in many places, there are recent fault scarps in the alluvial slopes at the foot of the mountains. The springs are situated along the general courses of the fault scarps and the yield of many of them is larger than would be expected from their topographic drainage areas. Some of the largest springs occur along narrow dry ranges that supply little water and have relatively uniform flow throughout the year, whereas ordinary springs in the region fluctuate more with the season. Notably, quite a few thermal springs issue from pools that are believed to be associated with fractures and fault conduits. Similar settings are also found in other parts of the American West. An example in Fig. 1.218 shows a spring located in the Rio Grande River fault zone, at the contact between carbonate bedrock and alluvium.

While in the western United States thermal springs commonly occur along normal faults, in the eastern United States they occur in regions of folded and thrust-faulted rocks. Figure 1.219 shows three possible models of fluid circulation in such areas, according to Sorey et al. (1983); other models are presented by Breckenridge and Hinckley (1978) and Hobba et al. (1979). Early descriptions of thermal springs in several provinces of the Appalachian Mountains primarily dealt with their therapeutic and recreational values (Moorman, 1867; Crook, 1899; Fitch, 1927). The springs occur in areas of steeply dipping folded rocks that are transected by nearly vertical east–west-trending fracture zones.

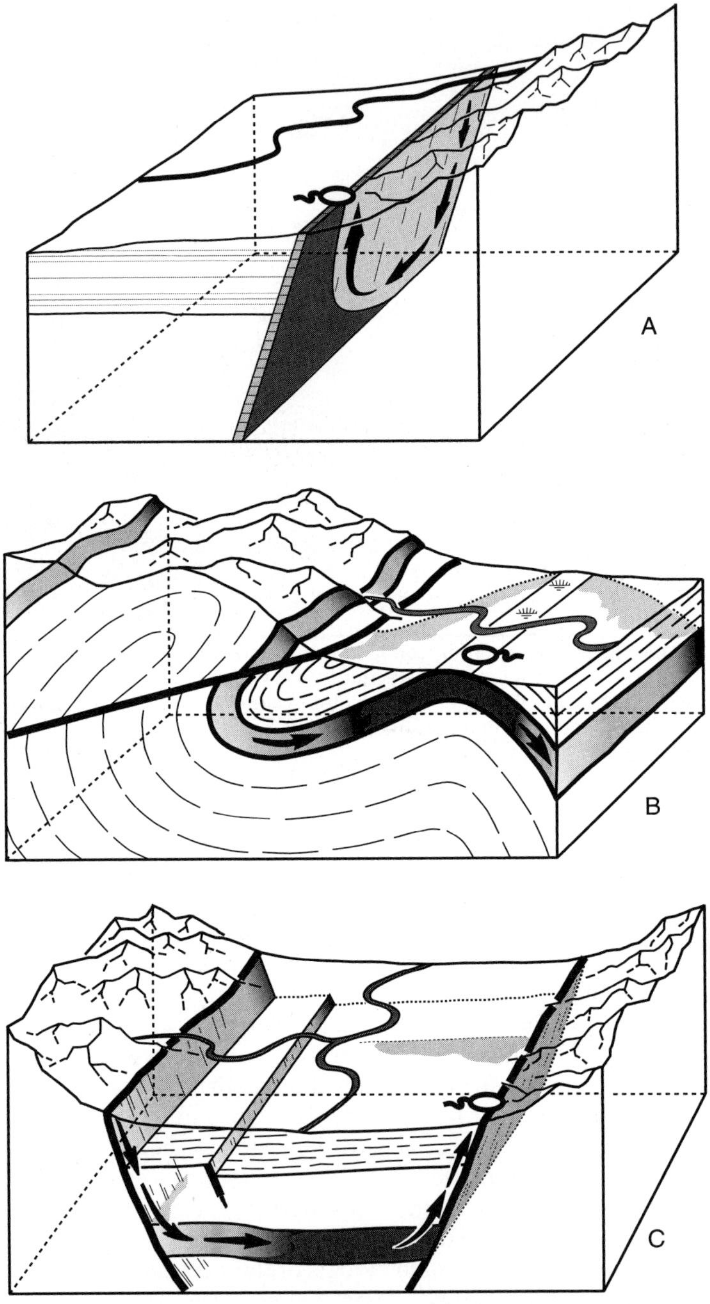

Figure 1.219 Conceptual models for types of hydrothermal-convection systems in which low-temperature geothermal-resource areas in category of isolated thermal springs and wells occur. (A) Fault plane. (B) Margin of anticline. (C) Deep reservoir. *Arrows* indicate direction of fluid circulation; *dark shading* shows location of reservoir containing low-temperature geothermal resources. (From Sorey et al., 1983.)

Figure 1.220 Spa pool fed by thermal water issuing from the historic Octagon Spring (inset) in the Homestead resort, Warm Springs, Virginia, established in 1766. Archeological evidence indicates use of the mineral and thermal springs in the area dating back to 7000 BC. (From Kresic and Mikszewski, 2012; copyright CRC/Taylor & Francis, printed with permission.)

Correlation of springs with topographic lows, or gaps, apparently results from the fact that easily eroded areas correspond to zones containing many fractures, which, in turn, provide the increased vertical permeability needed to establish a hydrothermal-convection system. Thermal springs always issue from either carbonate rocks or sandstones. Chemical analyses of the springwaters issuing from carbonate rocks exhibit consistently low concentrations of dissolved silica and high concentrations of magnesium and calcium, which indicate that the flow of warm water is restricted to the carbonate rocks. Analyses of waters from springs issuing from fractured sandstone show higher concentrations of dissolved silica and lower concentrations of magnesium and calcium, which indicates that flow is restricted to the sandstone beds (Sorey et al., 1983).

Geochemical considerations suggest that reservoir temperatures are not substantially higher than the measured surface temperatures at most eastern thermal springs; observed temperatures range from 18°C to 41°C. The occurrence of these springs in areas of average heat flow and relatively low temperature gradients (Costain et al., 1976; Perry et al., 1979) indicates that the depths of hydrothermal circulation are generally between 1 and 3 km. Figure 1.220 shows the Homestead's spa pool fed by the historic Octagon Spring

FIGURE 1.221 Men's bath house (also known as Jefferson Pools) in Hot Springs, Bath County, Virginia.

Figure 1.222 Aqua spring has the largest flow rate and the largest variation of flow of all springs in the Bullpasture George, Virginia. *Left*: Average flow conditions. *Right*: High flow rate of flood water carrying a heavy sediment load. The boils or mounds of water above the spring are several feet high. (Photographs courtesy of Phil Lukas; many more photos of Virginia springs are available at www.virginiacaves.org/karstsprings.html.)

in Warm Springs, Bath County, Virginia. The spring's temperature is steady 98°F (37°C). Several more hot springs ranging between 90°F and 108°F comprise this group of the warmest thermal springs in Virginia. There are also cold karst springs on the grounds of the resort, one of the best known and oldest spas in the United States. A hiking trail through the forest above the resort leads to one of the largest, albeit hidden, sinkholes in the Appalachians. Only several miles away are historic Hot Springs where the men's bath house (also known as Jefferson Pools) was built in 1761 and has been in use ever since (Fig. 1.221). This part of Virginia also has some of the best-kept secrets: swimming caves with thermal waters on private lands that can be visited only by select few.

The world's largest springs are karstic and they also have the highest amplitude of discharge rates compared with springs in all other aquifer types. In addition, water of many karst springs with large discharge fluctuation becomes loaded with suspended fine sediments following rainstorms (Fig. 1.222). These characteristics make karst springs the most attractive and, at the same time, the most challenging sources of water supply (detail discussion on karst spring development for water supply is provided in Chaps. 8 and 9). Meinzer's classification of springs based on average discharge expressed in the United States units is still widely used in the United States (Table 1.7). The same table also includes Meinzer's classification based on the metric system. However, the classification based solely on average spring discharge, without specifying other discharge parameters, is not very useful when evaluating the potential for spring utilization. For example, a spring may have a very high average discharge, but it may be dry or just trickling most of the year. This discharge variability is typically much higher in carbonate rocks with low matrix porosity, which cannot store significant volumes of groundwater that would provide relatively sustained flow between rainfall (recharge) episodes. Practice in most countries is that the springs are evaluated based on the minimum discharge recorded over a long period, typically longer than several hydrologic years (hydrologic year is defined as spanning all wet and dry seasons within a full annual cycle). The simplest

1. Classification Based on the Metric System		
Magnitude	**Discharge in Metric Units**	**Discharge in English Units (Approximate)**
First	10 cubic meters per second or more	353 second-feet (cubic feet per second)
Second	1–10 cubic meters per second	35–353 second-feet
Third	0.1–1 cubic meter per second	3.5–35 second-feet
Fourth	10–100 liters per second	158 gallons per minute to 3.5 second-feet
Fifth	1–10 liters per second	16–158 gallons per minute
Sixth	0.1–1 liter per second	1.6–16 gallons per minute
Seventh	10–100 cubic centimeters per second	1.25 pints to 1.6 gallons per minute
Eight	Less than 10 cubic centimeters per second	Less than 1.25 pints per minute
2. Classification Suggested for Practical Use in the United States		
Magnitude	Discharge	
First	100 second-feet (cubic feet per second) or more	
Second	10–100 second-feet	
Third	1–10 second-feet	
Fourth	100 gallons per minute to 1 second-foot	
Fifth	10–100 gallons per minute	
Sixth	1–10 gallons per minute	
Seventh	1 pint to 1 gallon per minute	
Eight	Less than 1 pint per minute	

From Meinzer, 1923a.

Table 1.7 Classification of Springs Based on Average Discharge Rate

Discharge (gpm)	No. of Springs
<100	653
100–450	155
450–4,500	147
4,500–45,000	5
>45,000	0

Table 1.8 Classification of 960 Springs in Tennessee According to the Magnitude of Flow Expressed in Gallons Per Minute

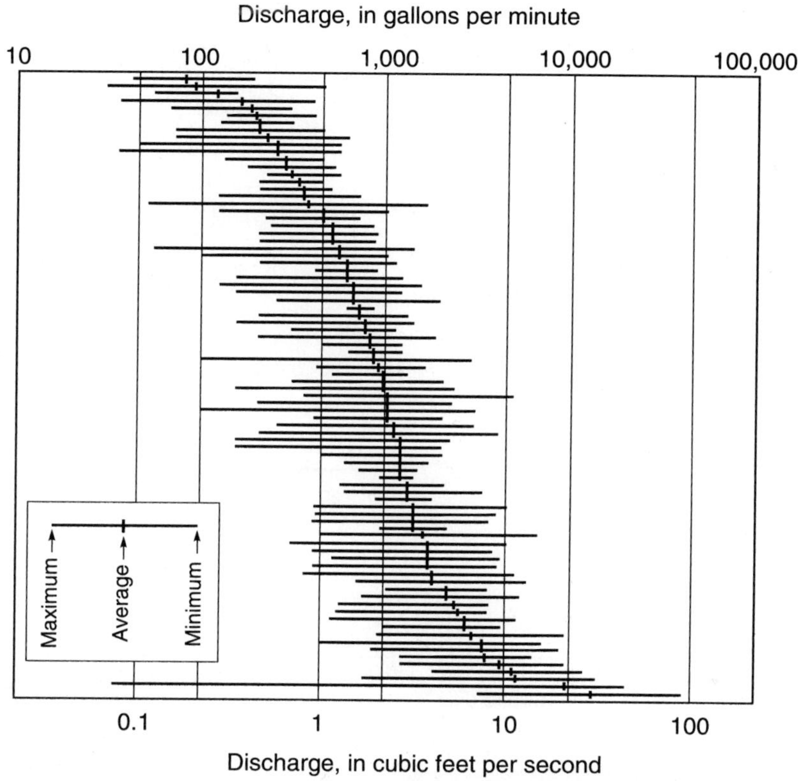

FIGURE 1.223 Minimum, average, and maximum discharges of 84 large springs in eastern Tennessee, the United States. (From Sun et al., 1963.)

measure of variability is the ratio of the maximum and minimum discharge:

$$I_v = \frac{Q_{max}}{Q_{min}} \tag{1.105}$$

Springs with the index of variability (I_v) greater than 10 are considered highly variable, and those with $I_v < 2$ are sometimes called *constant* or *steady springs*. Meinzer

Magnitude	No. of Springs
First	None
Second	4
Third	62
Fourth	16
Fifth	2

From Sun et al., 1963.

TABLE 1.9 Classification of 84 Large Springs in East Tennessee

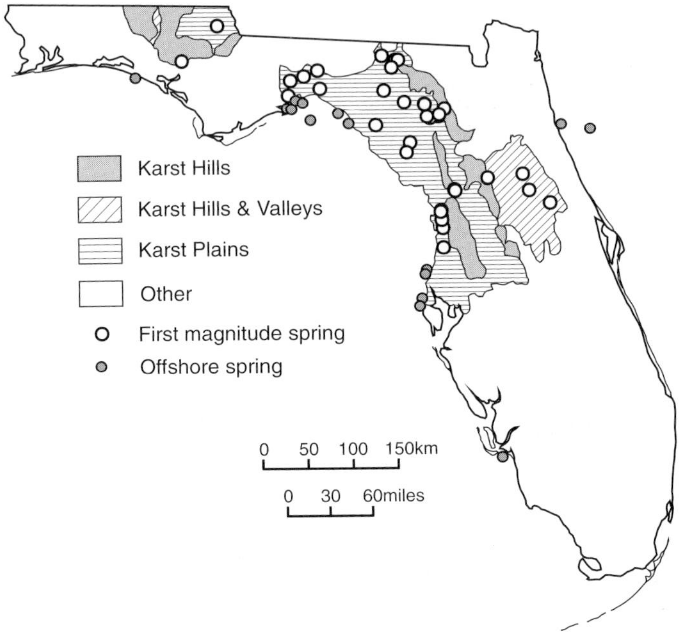

FIGURE 1.224 First magnitude springs and large offshore springs in Florida, the United States, and their relation to main geomorphologic karst types. (Modified from Scott et al., 2004.)

FIGURE 1.225 Nikšićka vrela, one of the springs which form the longest and largest sinking stream in Europe, the Trebišnjica River in eastern Herzegovina. The maximum flow rate of all the springs exceeds 800 m^3/s. The springs are now submerged by the Bileća Lake reservoir, which has the largest volume of many artificial reservoirs built in the classic Dinaric karst. (Photograph courtesy of Dr. Petar Milanovic.)

FIGURE 1.226 A: Spring in the Cirque de Consolation, French Jura. (Photograph courtesy of Dr. Nico Goldscheider.) B: Small karst spring in the Andes, Peru. C: First magnitude spring of the Bosna River near Sarajevo, Bosnia. D: Bekhal Spring in Rawanduz area, Zagros Mountains, northern Iraq, has maximum discharge >17 m^3/s. (Photograph courtesy of Dr. Zoran Stevanovic.) E: Grza Spring in Eastern Serbia discharging >1 m^3/s. (Photograph courtesy of Dr. Sasa Milanovic.) F: Historic San Pedro Springs in San Antonio, Texas, flow only during times of heavy rainfall due to heavy pumping demands on the Edwards Aquifer. (Copyright Gregg Eckhardt, printed with permission.)

proposed the following measure of variability expressed in percentage:

$$V = \frac{Q_{\text{max}} - Q_{\text{min}}}{Q_{\text{av}}} \times 100(\%) \tag{1.106}$$

where Q_{max}, Q_{min}, and Q_{av} are maximum, minimum, and average discharge, respectively. Based on this equation, a constant spring would have variability less than 25%, and a variable spring would have variability greater than 100%.

FIGURE 1.227 A: Looking out from the submerged feeder conduit of Sweetwater Spring, Ocala National Forest, Florida. (Photograph courtesy of Clayton Coates, U.S. Forest Service.) B: Cueva del Gato spring, Sierra de las Nieves, Southern Spain. (Photograph courtesy of Dr. Zoran Stevanovic.) C: Fontaine de Fontestorbes near village of Belesta, the French Pyrenees. The rate of this ebb-and-flow spring changes cyclically from about 50 to 1,800 L/s approximately every 30 minutes. (Photograph courtesy of William Jones, Karst Waters Institute.) D: Maiden Spring drains a large karst area in Tazewell County, Virginia. (Photograph courtesy of Phil Lukas.)

An extensive study of springs in east Tennessee, the United States, is illustrative of the natural variability of spring flows and the role of the underlying geology. More than 960 springs were observed and described by De Buchananne and Richardson (1956). During their reconnaissance, discharge measurements were made at many of the springs and flow estimates for the remainder. The total of these measurements was about 265 million gallons per day. All but a few of the measurements were made during the relatively dry period June through September; thus, this volume represents a near minimum, or at least a below average, total for the 960 springs considered. The 960 springs described by De Buchananne and Richardson are classified according to the magnitude of their flow as shown in Table 1.8.

The results of a follow-up study by Sun et al. (1963) are shown in Fig. 1.223 for 84 large undeveloped springs in east Tennessee for which records of minimum, average, and maximum flows are available. The horizontal line shows the range of discharge, and the vertical line indicates the average discharge for the spring's period of record. Table 1.9 shows the classification of these springs based on their magnitude. The least variable springs issue from shale of the Conasauga Group, and all these are of relatively small magnitude. The most variable springs issue from solution openings and drain karst aquifers developed in the formations of the Knox Group, Chickamauga Limestone, or limestones of the Conasauga Group that all have low matrix porosity. The karst cavities are of varying size and degree of interconnection so that, as the hydraulic head fluctuates from wet to dry seasons, spring discharge fluctuates in accordance with the ability of the saturated cavities to transmit the water to the spring orifices.

It is estimated that there are nearly 700 karst springs in Florida, the United States, of which 33 are the first magnitude springs with an average discharge greater than 100 cubic feet per second (ft^3/s) (64.6 million gallons per day or 2.83 m^3/s; see Fig. 1.224). With abundant precipitation, favorable recharge conditions, and exceptional storage capacity of young porous limestone, Florida has one of the largest concentrations of freshwater springs on Earth. In the United States, only karst of Missouri with nine first-magnitude springs approaches Florida. Worldwide, the classic Dinaric karst is unrivaled with most first-magnitude springs of which the largest one, the Trebišnjica Springs ($Q_{max} > 800\ m^3/s$), is now submerged by a reservoir (Fig. 1.225). The Buna Spring near Mostar, Herzegovina, is the largest spring issuing from a single cave, with the minimum and maximum flow rates exceeding 10 and 300 m^3/s, respectively (see color plates). Karst aquifers of Turkey and China also host tens of first-magnitude springs (Günay, 2010; Wu et al., 2010; Zhaoxin and Chuanmao, 2004), whereas a few such springs are found in each of the other well-known karst areas including France, the European Alps, the Middle East, Mexico, and Papua New Guinea (Milanovic, 1979, 2004; Ford and Williams, 2007; Kresic and Stevanovic, 2010). Examples of karst springs from around the world are included in Figs. 1.226 and 1.227.

1.9 References

Andreychouk, V., Dublyansky Y., Yezhov Y., and Lysenin G., 2009. *Karst in Earth's Crust: Distribution and the Main Types*. University of Silesia and Ukrainian Institute of Speleology and Karstology, Sosnowiec—Simferopol, 72p.

ASTM (American Society for Testing and Materials), 1999. *ASTM Standards on Determining Subsurface Hydraulic Properties and Ground Water Modeling*, 2nd ed. West Conshohocken, Pennsylvania, PA, 320p.

Australian Bureau of Statistics, 2005. *Information Paper: An Introductory Course on Time Series Analysis—Electronic Delivery*. 1346.0.55.001, January 2005, p. 124. Available at: www.abs.gov.au/websitedbs/D3310144.nsf/home/home?opendocument

Barenblatt, G. E., Zheltov, I. P, and Kochina, I. N., 1960. Basic concepts in the theory of seepage of homogeneous liquids in fissured rocks. *J Appl Math Mechs*, 24:1286–1303.

Bathurst, R. G. C., 1975. Carbonate sediments and their diagenesis. Second Enlarged Edition. Developments in Sedimentology, vol 12, Elsevier, Amsterdam, The Netherlands, 658p.

Bear, J., Tsang, C. F., and de Marsily, G. (eds.), 1993. *Flow and Contaminant Transport in Fractured Rock*. Academic Press, San Diego, CA, 548p.

Beck, B. F., and Sinclair, W. C., 1986. *Sinkholes in Florida: An Introduction*. Report 85–86-4, The Florida Sinkhole Research Institute, Orlando, FL, 16p.

Bishop, P. K., and Lloyd, J. W., 1990. Chemical and isotopic evidence for hydrogeological processes occurring in the Lincolnshire Limestone. *J Hydrol*, 121:293–320.

Boulton, N. S., 1973. The influence of delayed drainage on data from pumping tests in unconfined aquifers. *J Hydrol*, 19(2):157–169.

Božičević, S., 1984. Kroz naše špilje i jame. Mala znanstvena knjižica Hrvatskoga prirodoslovnog društva, II izdanje. Zagreb, Croatia, 72p.

Božičević, S., 1971. Primjena speleologije pri injektiranjima u krsu (Application of speleology in grouting of karst terranes; in Croatian). 1st Yugoslav Symposium on Hydrogeology and Engineering Geology, Herceg Novi.

Bögli, A., 1980. *Karst Hydrology and Physical Speleology*. Springer-Verlag, New York, 284p.

Breckenridge, R. M., and Hinckley, B. S., 1978. Thermal Springs of Wyoming: Geological Survey of Wyoming Bulletin 60, 104p.

Broner, I. and Schneekloth, J., 2007. *Seasonal Water Needs and Opportunities for Limited Irrigation for Colorado Crops*. Colorado State University Extension. Available at: http://www.ext.colostate.edu/Pubs/crops/04718.html; accessed 23rd August 2007.

Brown, P., 2000. Basis of evaporation and evapotranspiration. *Turf Irrigation Management Series: I*, The University of Arizona College of Agriculture, Tucson, Arizona, 4p.

Budd, D. A., and Vacher, H. L., 2004. *Matrix Permeability of the Confined Floridan Aquifer, Florida, USA*. *Hydrogeol J*, 12:531–549.

Cacas, M. C., 1989. Développment d'un modèle tridimensionel stochastique discret por la simulation de l'écoulement et des transports de masse et de chaleur en milieu fracturé. *Ph.D. Thesis*, Ecole des Mines de Paris, Fontainebleau, France.

Chilès, J. P., and de Marsily, G., 1993. Stochastic models of fracture systems and their use in flow and transport modeling. In: Bear, J, Tsang, C.F., and de Marsily, G. (eds.), *Flow and Contaminant Transport in Fractured Rock.*, Academic Press, San Diego, CA, pp. 169–236.

Clarke, J. S., Leeth, D. C., Taylor-Harris, D., Painter, J. A., and Labowski, J. L., 2004. Hydraulic properties of the Floridan Aquifer System and equivalent clastic units in coastal Georgia and adjacent parts of South Carolina and Florida. *Georgia Geologic Survey Information Circular 109*, Atlanta, Georgia, 50p.

Coes, A. L., and Pool, D. R., 2005. Ephemeral-stream channel and basin-floor infiltration and recharge in the Sierra Vista subwatershed of the upper San Pedro basin, Southeastern Arizona. *U.S. Geological Survey Open-File Report 2005–1023*, Reston, Virginia, 67p.

Cognard-Plancq, A. -L., Gévaudan, C., and Emblanch, C., 2006a: Apports conjoints de suivis climatologique et hydrochimique sur le rôle de filtre des aquifères karstiques dans l'étude de la problématique de changement climatique; Application au système de la Fontaine de Vaucluse. *Proceedings of the 8th Conference on Limestone Hydrogeology*. Neuchatel, September 21–23, 2006, pp. 67–70.

Cognard-Plancq, A. -L., Gévaudan, C., and Emblanch, C., 2006b. Historical monthly rainfall-runoff database on Fontaine de Vaucluse karst system: Review and lessons. IIIéme Symposium International Sur le Karst "Groundwater in the Mediterranean Countries", Malaga, Spain. In: Duran, J. J., Andreo, B., Carrasco, F. Y. (eds.), *Karst, Cambio Climatico y Aguas Subterraneas*. Publicaciones des Instituto Geological y Minero de Espana. Serie: Hidrogeologia y Aguas Subterrraneas, vol. 18, pp. 465–475.

Colebrook, C. F., 1939. Turbulent flow in pipes, with particular reference to the transition region between smooth and rough pipe laws. *J Inst Civil Eng*, London, 11:133–156.

Cook, P. G., 2003. A guide to regional groundwater flow in fractured aquifers. *CSIRO Land and Water*, Seaview Press, Henley Beach, South Australia, 108p.

Costain, J. K., Keller, G. V., and Crewdson, R. A., 1976. Geological and geophysical study of the origin of the warm springs in Bath County, Virginia. *Virginia Polytechnic Institute and State University Report for U.S. Department of Energy under Contract E-(40–1)-4920*, Blacksburg, VA.

Crook, J. K., 1899. The Mineral Waters of the United States and Their Therapeutic Uses. Lea Brothers & Co., New York, 588p.

Culver, D. C., and White, W. B. (eds.), 2004. *Encyclopedia of Caves*. Elsevier, Academic Press, Burlington, MA, 654p.

Cvijić, J., 1895. *Karst; Geografska Monografija (Karst; A Geographic Monograph; in Serbian)*. Beograd, Štamparija Kraljevine Srbije, 176p.

Cvijić, J., 1926. *Geomorfologija (Morphologie Terrestre)*. Knjiga druga (Tome Second). Beograd, 506p.

Dawson, K., and Istok, J., 1992. *Aquifer Testing; Design and Analysis*. Lewis Publishers, Boca Raton, FL, 280p.

De Buchananne, G. D., and Richardson, R. M., 1956. *Ground-Water Resources of East Tennessee*. Tennessee Division of Geology, Nashville, TN, Bulletin 58, pt. 1, 393p.

Dettinger, M. D., 1989. Distribution of carbonate-rock aquifers in southern Nevada and the potential for their development, summary of findings, 1985–88. *Program for the Study and Testing of Carbonate-Rock Aquifers in Eastern and Southern Nevada, Summary Report No. 1*, Carson City, NV, 37p.

Dingman, S. L., 1994. *Physical Hydrology*. Macmillan, New York, 575p.

Domenico, P. A., and Schwartz, F. W., 1990. *Physical and Chemical Hydrogeology*. John Wiley & Sons, New York, 824p.

Driscoll, F. G., 1989. *Groundwater and Wells* (Third Printing). Johnson Filtration Systems Inc, St. Paul, MN, 1089p.

Dugan, J. T., and Peckenpaugh, J. M., 1985. Effects of climate, vegetation, and soils on consumptive water use and ground-water recharge to the central Midwest regional aquifer system, mid-continent United States. *U.S. Geological Survey Water-Resources Investigations Report 85–4236*, Lincoln, NE, 78p.

Eröss, A., Madl-Szönyi, J., and Csoma, A., 2008. Characteristics of discharge at Rose and Gellért Hills, Budapest, Hungary. *Central Eur Geol*, 51(3):267–281.

Faybishenko, B., Witherspoon, P. A., and Benson, S. M. (eds.), 2000. *Dynamics of Fluids in Fractured Rock, Geophysical Monograph 122*, American Geophysical Union, Washington, D.C., 400p.

Ferris, J. G., Knowles, D. B., Brown, R. H., and Stallman, R. W., 1962. Theory of aquifer tests. *U.S. Geological Survey Water Supply Paper 1536-E*, Washington, D.C., 173p.

Field, M. S., 2002. *A Lexicon of Cave and Karst Terminology with Special Reference to Environmental Karst Hydrology*. U.S. Environmental Protection Agency, Office of Research and Development, EPA/600/R-02/003, Washington, D.C., 214p.

Fitch, W. E., 1927. *Mineral Waters of the United States and American Spas*. Lea & Febiger, New York, 799p.

Folk, R. L., 1959. Practical petrographic classification of limestones. *Am Assoc Petroleum Geologists Bull*, 43(1):1–38.

Folk, R. L., 1980. *Petrology of Sedimentary Rocks*. Hemphill Publishing Company, Austin, TX, 182p.

Ford, D., and Williams, P., 2007. *Karst Hydrogeology and Geomorphology*. John Wiley and Sons Ltd, The West Sussex, England, 562p.

Freeze, R. A., and Cherry, J. A., 1979. *Groundwater*. Prentice-Hall, Englewood Cliffs, NJ, 604p.

Garcia-Fresca, B., and Sharp, J. M., 2005. Hydrogeologic consideration of urban development: Urban induced recharge. In: Ehlen, J., Haneberg, W. C., and Larson, R. A. (eds.), *Humans as Geologic Agents*. GSA Review in Engineering, Boulder, CO, vol. XVI, pp. 123–136.

Giles, R. V., Evett, J. B., and Chiu, L., 1994. *Fluid Mechanics and Hydraulics*, 3rd ed. Schaum's Outline Series, McGraw Hill, New York, 378p.

Gillieson, D., 1996. *Caves; Process, Development, Management*. Blackwell Publishers, Oxford, 324p.

Giusti, E. V., 1978. Hydrogeology of the karst of Puerto Rico. *U.S. Geological Survey Professional Paper 1012*, Washington, D.C., 68p.

Gottman, J. M., 1981. *Time-Series Analysis. A Comprehensive Introduction for Social Scientists*. Cambridge University Press, Cambridge, 400p.

Greene, E. A., 1993. Hydraulic properties of the Madison aquifer system in the western Rapid City area, South Dakota. *U.S. Geological Survey Water-Resources Investigations Report 93–4008*, Rapid City, South Dakota, 56p.

Greene, E. A., Shapiro, A. M., and Carter, J. M., 1999. Hydrogeologic characterization of the Minnelusa and Madison aquifers near Spearfish, South Dakota. *U.S. Geological Survey Water-Resources Investigations Report 98–4156*, Rapid City, South Dakota, 64p.

Greswell R., Yoshida, K., Tellam, J. H., and Lloyd, J. W., 1998. The micro-scale hydrogeological properties of the Lincolnshire Limestone, UK. *Q J Eng Geol*, 31:181–197.

Griffioen, J., and Kruseman, G. P., 2004. Determining hydrodynamic and contaminant transfer parameters of groundwater flow. In: Kovalevsky, V. S., Kruseman, G. P., and Rushton, K. R. (eds.), 2004. *Groundwater Studies: An International Guide for Hydrogeological Investigations*. IHP-VI, Series on Groundwater No. 3, UNESCO, Paris, France, pp., 217–238.

Günay, G., 2010. Geological and hydrogeological properties of Turkish karst and major karstic springs. In: Kresic, N., and Stevanovic, Z. (eds.), *Groundwater Hydrology of Springs; Engineering, Theory, Management and Sustainability*. Elsevier, New York, pp. 479–497.

Gunn, J. (ed.), 2004. Encyclopedia of Caves and Karst Science. Fitzroy Dearborn, Taylor & Francis Group, New York, 902p.

Haaland, S. E., 1983. Simple and explicit formulas for the friction factor in turbulent flow. *J Fluids Eng (ASME)*, 103(5):89–90.

Halihan T., Mace, R. E., and Sharp, J. M. Jr., 2000. Flow in the San Antonio segment of the Edwards aquifer: Matrix, fractures, or conduits? In: Wicks, C. M., and Sasowsky, I. D. (ed.), *Groundwater Flow and Contaminant Transport in Carbonate Aquifers*. AA Balkema, Rotterdam, The Netherlands, pp. 129–146.

Hazen, A., 1892. Experiments upon the purification of sewage and water at the Lawrence Experiment Station. *Massachusetts Board of Health Twenty-third Annual Report*, November 1, 1889, to December 31, 1891, pp. 428–434.

Healy, R. W., and Cook, P. G., 2002. Using groundwater levels to estimate recharge. *Hydrogeol J*, 10(1):91–109.

Hobba, W. A. Jr., Fisher, D. W., Pearson. F. J. Jr., and Chemerys, J. C., 1979. Hydrology and geochemistry of thermal springs of the Appalachians. *U.S. Geological Survey Professional Paper 1044-E*, E1-E36.

Hovorka, S. D., 2009. Stop 3, 337 Loop Road Cut, New Braunfels. In: Schindel, G., Johnson, S., Hoyt, J., Green, R. T., Alexander, E. C., and Krietler, C., *Hydrology of the Edwards Group: A Karst Aquifer Under Stress*. A Field Trip Guide for the US EPA Groundwater Forum, November 19, 2009, San Antonio, Texas.

IZRK, 1982. Krasko podzemlje pri nas i v svetu. Vodnik 7, Postojna, Slovenia, various pagination.

James, N. P., and Mountjoy, E.W., 1983. Shelf-slope break in fossil carbonate platforms: An overview. In: Stanley D. J., and Moore, G. T. (eds.), *The Shelfbreak: Critical Interface on Continental Margins*. SEPM Special Publication, No. 33, pp. 189–206.

Jevdjević, V., 1956. Hidrologija, I deo (Hydrology, Part 1; in Serbian). Hidrotehnički Institut Jaroslav Černi, Beograd, 404p.

Jevdjević, V., 1974. Vjerovatnoća i statistika u hidrologiji (Probability and statistics in hydrology, in Serbo-Croatian). Zavod za hidrotehniku Gradjevinskog fakulteta, Sarajevo, br. 15, 309p.

Jocha-Edelényi, E., 2005. Karsthydrogeology of the Transdanubian Range, Hungary: Geological constrains and human impact on a unique karst reservoir. *Occasional Papers of the Geological Institute of Hungary*, vol. 204, Geological Institute of Hungary, Budapest, pp. 53–58.

Johnston, R. H., Bush, P. W., Krause, R. E., Miller, J. A., and Sprinkle, C. L., 1982. Summary of hydrologic testing in tertiary limestone aquifer, Tenneco offshore exploratory well–Atlantic OCS, lease-block 427 (Jacksonville NH 17–5). *U.S. Geological Survey Water-Supply Paper 2180*, Washington, D.C., 15p.

Jones, W. K., 1973. *Hydrology of Limestone Karst in Greenbrier Countu, West Virginia*. Bulletin 36, West Virginia Geological and Economic Survey, Charleston, WV, 49p.

Jones, W. K., 1977. *Karst Hydrology Atlas of West Virginia*. Special Publication 4, Karst Waters Institute, Charles Town, WV, 111p.

Jones, G. W., Upchurch, S. B., and Champion, K. M., 1996. *Origin of Nitrate in Ground Water Discharging from Rainbow Springs, Marion County, Florida*. Ambient Ground-Water Quality Monitoring Program, Southwest Florida Water Management District, Brooksville, FL.

King, R. B., 1992. Overview and bibliography of methods for evaluating the surface-water-infiltration component of the rainfall-runoff process. *U.S. Geological Survey Water-Resources Investigations Report 92–4095*, Urbana, IL, 169p.

King, F. H., 1899. Principles and conditions of the movements of ground water. *U. S. Geological Survey Nineteenth Annual Report*, Washington, D.C.

Klimchouk, A. B., 1992. Large gypsum caves in the Western Ukraine and their genesis. *Cave Science*, 19(1):3–11.

Klimchouk, A. B., 1994. Speleogenesis under confined conditions, with recharge from adjacent formations. Publ. Serv. Geol. Luxembourg, vol. XXVII: Comptes Rendus du Coll. Intern. de Karstologie a Luxembourg, pp. 85–95.

Klimchouk, A., 2011. *Hypogene Speleogenesis: Hydrogeological and Morphogenetic Perspective*, 2nd ed. Special Paper No. 1, National Cave and Karst Research Institute, Carlsbad, New Mexico, 106p.

Klimchouk, A. B., Ford, D., Palmer, A., and Dreybrodt, W. (eds.), 2000. *Speleogenesis: Evolution of Karst Aquifers*. National Speleological Society, Huntsville, AL, 527pp.

Klimchouk, A., and Ford, D. C. (eds.), 2009. *Hypogene Speleogenesis and Karst Hydrogeology of Artesian Basins*. Ukrainian Institute of Speleology and Karstology, Simferopol, Ukraine, 292p.

Knochenmus, L. A., and Robinson, J. L., 1996. Descriptions of anisotropy and heterogeneity and their effect on ground-water flow and areas of contribution to public supply wells in a karst carbonate aquifer system. *U.S. Geological Survey Water-Supply Paper 2475*, Washington, D.C., 47p.

Kohler, M. A., 1958. *Meteorological Aspects of Evaporation*. Int. Assn. Sci. Hydr. Trans., General Assembly, Toronto, vol. III, pp. 423–436.

Krause, R. E., and Randolph, R. B., 1989. Hydrology of the Floridan aquifer system in southeast Georgia and adjacent parts of Florida and South Carolina. *U.S. Geological Survey Professional Paper 1403–D*, 65p.

Kresic, N., 1988. Karst i pećine Jugoslavije (Karst and Caves of Yugoslavia; in Serbo-Croatian). Naučna knjiga, Belgrade, 149p.

Kresic, N., 1991. Kvantitativna hidrogeologija karsta sa elementima zaštite podzemnih voda (Quantitative karst hydrogeology with elements of groundwater protection, in Serbo-Croatian). Naučna knjiga, Beograd, 196p.

Kresic, N., 2007a. *Hydrogeology and Groundwater Modeling*, 2nd ed. CRC Press, Taylor & Francis Group, Boca Raton, FL, 807p.

Kresic, N., 2007b. Hydraulic methods. In: Goldscheider, N., and Drew, D. (eds.), *Methods in Karst Hydrogeology*. International Contributions to Hydrogeology 26, International Association of Hydrogeologists, Taylor & Francis, London, pp. 65–92.

Kresic, N., 2009. *Groundwater Resources. Sustainability, Management, and Restoration*. McGraw Hill, New York, 852p.

Kresic, N., 2010. Types and classification of springs. In: Kresic, N., and Stevanovic, Z. (eds.), *Groundwater Hydrology of Springs; Engineering, Theory, Management and Sustainability*. Elsevier, New York, pp. 31–85.

Kresic, N., and Mikszewski, A., 2009. Chapter 3, Groundwater Recharge. In: Kresic, N., *Groundwater Resources. Sustainability, Management, and Restoration*. McGraw Hill, New York, pp. 235–292.

Kresic, N., and Stevanovic, Z. (eds.), 2010. *Groundwater Hydrology of Springs; Engineering, Theory, Management and Sustainability*. Elsevier, New York, 573p.

Kresic, N., and Mikszewski, A., 2012. *Hydrogeological Conceptual Site Models: Data Analysis and Visualization*. CRC, Taylor & Francis Group, Boca Raton, FL, 552p.

Kruseman, G. P., de Ridder, N. A., and Verweij, J. M., 1991. *Analysis and Evaluation of Pumping Test Data*, completely revised 2nd ed. International Institute for Land Reclamation and Improvement (ILRI) Publication 47, Wageningen, The Netherlands, 377p.

Ladiray, D., and Quenneville, B., 2001. Seasonal adjustment with the X-11 method. *Lecture Notes in Statistics*, vol. 158, Springer-Verlag, New York.

Larsson, I., 1982. Ground water in hard rocks. *Project 8.6 of The International Hydrological Programme*, UNESCO, Paris, 228p.

Lee, K. K., and Risley, J. C., 2002. Estimates of ground-water recharge, base flow, and stream reach gains and losses in the Willamette River Basin, Oregon. U.S. *Geological Survey Water-Resources Investigations Report 01–4215*, Portland, Oregon, 52p.

Liang, Y., Han, X., Xue, F., et al., 2008. *Water Protection of Karst Springs in Shanxi*. WaterPower Press, Beijing, China.

Linsley, R. K., and Franzini, J. B., 1979. *Water-Resources Engineering*, 3rd ed. McGraw-Hill, New York, 716p.

Little, R. J. A., and Rubin, D. B., 2002. *Statistical Analysis with Missing Data*, 2nd ed. Wiley-Interscience, New York, 408p.

Lohman, S. W., 1972. Ground-water hydraulics. *U.S. Geological Survey Professional Paper 708*, 70p.

Lugeon, M., 1933. *Barrages et Géologie: Méthodes de Recherches, Terrassement et Imperméabilization*. Dunod, Paris, 138p.

Maidment, D. R. (ed.), 1993. *Handbook of Hydrology*. McGraw Hill, New York, various paging.

Manugistics, Inc., 2000. Statgraphics Plus 5. Rockville, MD, variable paging.

Maslia, M. L., and Randolph, R. B., 1986. Methods and computer program documentation for determining anisotropic transmissivity tensor components of two-dimensional ground-water flow. *U.S. Geological Survey Open-File Report 86–227*, 64p.

Maupin, M. A., and Barber, N. L., 2005. Estimated withdrawals from principal aquifers in the United States, 2000. *U.S. Geological Survey Circular 1279*, Reston, VA, 46p.

Meinzer, O. E., 1923a. The occurrence of ground water in the United States with a discussion of principles. *U.S. Geological Survey Water-Supply Paper 489*, Washington, D.C., 321p.

Meinzer, O. E., 1923b. Outline of ground-water hydrology with definitions. *U.S. Geological Survey Water-Supply Paper 494*, Washington, D.C., 71p.

Meinzer, O. E., 1927. Large springs in the United States. *U.S. Geological Survey Water-Supply Paper 557*, Washington, D.C., 94p.

Meinzer, O. E., 1940. Ground water in the United States; a summary of ground-water conditions and resources, utilization of water from wells and springs, methods of scientific investigations, and literature relating to the subject. *U.S. Geological Survey Water-Supply Paper 836-D*, Washington, D.C., pp. 157–232.

Milanovic, P., 1979. *Hidrogeologija Karsta i Metode Istraživanja (in Serbian; Karst Hydrogeology and Methods of Investigations)*. HE Trebišnjica, Institut za korištenje i zaštitu voda na kršu, Trebinje, 302p.

Milanovic, P. T., 1981. *Karst Hydrogeology*. Water Resources Publications, Littleton, CO, 434p.

Milanovic, P. T., 2004. *Water Resources Engineering in Karst*. CRC Press, Boca Raton, FL, 312p.

Milanovic, P. T., 2006. *Karst Istočne Hercegovine i Dubrovačkog Priobalja (Karst of Eastern Herzegovina and Dubrovnik Litoral; in Serbian)*. ZUHRA, Belgrade, 362p.

Milanovic, S., 2005. Hydrogeological characteristics of some deep siphonal springs in Serbia and Montenegro karst. In: Stevanovic, Z., and Milanovic, P. (eds.), *Water Resources and Environmental Problems in Karst*, Proceedings of the International Conference and Field Seminars, Belgrade & Kotor, Serbia & Montenegro, September 13–19, 2005, Institute of Hydrogeology, University of Belgrade, Belgrade, pp. 451–458.

Moody, L. F. 1944. Friction factors for pipe flow. *Tran ASME*, 66(8):671–684.

Moore, C. H., 1989. *Carbonate Diagenesis and Porosity. Developments in Sedimentology*, vol 46, Elsevier, Amsterdam, The Netherlands, 338p.

Moorman, J. J., 1867. The mineral waters of the United States and Canada. Kelly & Piet, Baltimore, MD, 507p.

Nassif, S. H., and Wilson, E. M., 1975. The influence of slope and rain intensity on runoff and infiltration. *Hydrol Sci Bull*, 20:539–553.

Osborne, P. S., 1993. Suggested operating procedures for aquifer pumping tests. *Ground Water Issue, United States Environmental Protection Agency, EPA/540/S-93/503*, 23p.

Paillet, F. L., 1989. Analysis of geophysical well logs and flowmeter measurements in boreholes penetrating subhorizontal fracture zones, Lac du Bonnet Batholith, Manitoba, Canada. *U.S. Geological Survey Water-Resources Investigations Report 89–4211*, Lakewood, CO, 30p.

Paillet, F. L., 1994. Application of borehole geophysics in the characterization of flow in fractured rocks. *U.S. Geological Survey Water-Resources Investigations Report 93–4214*, Denver, CO, 36p.

Paillet, F. L., 1998. Flow modeling and permeability estimation using borehole flow logs in heterogeneous fractured formations. *Water Resources Res*, 34(5):997–1010.

Paillet, F. L., 2000. A field technique for estimating aquifer parameters using flow log data. *Ground Water*, 38(4):510–521.

Paillet, F. L., 2001. Hydraulic head applications of flow logs in the study of heterogeneous aquifers. *Ground Water*, 39(5):667–675.

Paillet, F. L., Hess, A. E., Cheng, C. H., and Hardin, E. L., 1987. Characterization of fracture permeability with high resolution vertical flow measurements during borehole pumping. *Ground Water*, 25:28–40.

Paillet, F. L., and Reese, R. S., 2000. Integrating borehole logs and aquifer tests in aquifer characterization. *Ground Water*, 38(5):713–725.

Palmer, A. N., 1985. The Mammoth Cave region and Pennyroyal Plateau. In: Dougherty, P. H. (ed.), *Caves and Karst of Kentucky*, Kentucky Geological Survey Special Publication 12, Series XI, pp. 97–118.

Palmer, A. N., 1991. Origin and morphology of limestone caves. *Geol Soc Am Bul*, 103:1–21.

Palmer, A. N., 2007, *Cave Geology. Cave Books*, Trenton, NJ, 454p.

Palmer, A. N., and Palmer, M. V., 1995. Geochemistry of capillary seepage in Mammoth Cave. *Mammoth Cave National Park, 4th Science Conference*, pp. 119–133.

Papadopulos, I. S., 1965. Nonsteady flow to a well in an infinite anisotropic aquifer. *Proceedings of the Dubrovnik Symposium on the Hydrology of Fractured Rocks*. International Association of Scientific Hydrology, pp. 21–31.

Perry, W. C., Costain, J. K., and Geiser, P. A., 1979. Heat flow in western Virginia and a model for the origin of thermal springs in the folded Appalachians. *J Geophys Res*, 84(B12):6875–6883.

Polhemus, N. W., 2005. *How to: Forecast time series data using STATGRAPHICS Centurion*. StatPoint, Inc., Available at: www.statlets.com/howtoguides.htm

Prohaska, S., 1981. Stohastički model za dugoročno prognoziranje rečnog oticaja (Stochastic model for long-term prognosis of river flow; in Serbian). *Vode Vojvodine*, Special Edition 1981, Novi Sad, 106p.

Radovanovic, S., 1897. *Podzemne Vode; Izdani, Izvori, Bunari, Terme i Mineralne Vode (Ground Waters; Aquifers, Springs, Wells, Thermal and Mineral Waters; in Serbian)*. Srpska književna zadruga, Beograd, vol. 42, 152p.

Rawls, W. J., 1983. Estimating soil bulk density from particle size analysis and organic matter content. *J Soil Sci*, 135(2):123–125.

Rawls, W. J., Ahuja, L. R., Brakensiek, D. L., and Shirmohammadi, A. 1995. Infiltration and soil water movement. Chapter 5. In: Maidment, D. R. (ed.), *Handbook of hydrology*. McGraw Hill, New York, pp. 5.1–5.51.

Reese, R. S., and Cunningham, K. J., 2000. Hydrogeology of the Gray Limestone aquifer in Southern Florida. *U.S. Geological Survey Water-Resources Investigations Report 99–4213*, Tallahassee, FL, 244p.

Risser, D. W., Gburek, W. J., and Folmar, G. K., 2005. Comparison of methods for estimating ground-water recharge and base flow at a small watershed underlain by fractured bedrock in the eastern United States. *U.S. Geological Survey Scientific Investigations Report 2005–5038*, Reston, VA, various pages.

Robinson, R. B., 1967. Diagenesis and porosity development in Recent and Pleistocene oolites from southern Florida and Bahamas. *J Sediment Petrol*, 37:355–364.

Rorabaugh, M. I., 1953. Graphical and theoretical analysis of step-drawdown test of artesian well. *Proc Am Soc Civil Eng*, 79(separate no. 362):23.

Roscoe Moss Company, 1990. *Handbook of Ground Water Development*. John Wiley & Sons, New York, 493p.

Schafer, J. L., 1997. *Analysis of Incomplete Multivariate Data*. Chapman & Hall/CRC, New York, 444p.

Scoffin, T. P., 1987. An introduction to carbonate sediments and rocks. Chapman and Hall, New York, 274p.

Scott, T. M., Means, G. H., Meegan, R. P., et al., 2004. *Springs of Florida. Florida Geological Survey*, Bulletin No. 66, Tallahassee, FL., 658p.

Shapiro, A. M., 2001, Characterizing ground-water chemistry and hydraulic properties of fractured rock aquifers using the Multifunction Bedrock-Aquifer Transportable Testing Tool (BAT^3). *U.S. Geological Survey Fact Sheet FS-075–01*, 4p.

Showcaves, 2011. Available at: http://www.showcaves.com/english/fr/springs/Vaucluse.html. Accessed 29th November 2011.

Shubert, G. L., and Ewing, M, 1956. Gravity reconnaissance survey of Puerto Rico. Geol Soc Am Bull, 67(4):511–534.

Shuttleworth, W. J., 1993. Evaporation, Chapter 4. In: Maidment, D. R. (ed.), *Handbook of Hydrology*. McGraw Hill, New York, pp. 4.1–4.53.

Singh, V. P., 1993. *Elementary Hydrology*. Prentice Hall, Englewood Cliffs, NJ, 973p.

Smart, C. C., 1983. Hydrology of a glacierized alpine karst. *Ph.D. Thesis*, McMaster University, Hamilton, Ontario, 343p.

Snow, D. T., 1968. Rock fracture spacings, openings, and porosities. *J Soil Mech Found Div, Am Soc Civil Eng*, 94:73–91.

Sorey, M. L., Reed, M. J., Foley, D., and Renner, J. L., 1983. Low-temperature geothermal resources in the Central and Eastern United States. In: Reed, M. J. (ed.), *Assessment of Low-Temperature Geothermal Resources of the United States-1982*. Geological Survey Circular 892, United States Department of the Interior, pp. 51–65.

Southeastern Friends of the Pleistocene, 1990. Hydrogeology and geomorphology of the Mammoth Cave area, Kentucky. 1990 Field excursion led by Quinlan, J. F., Ewers, R. O., and Palmer, A. N., Nashville, TN, 102p.

Spiegel, M. R., and Meddis, R., 1980. *Probability and Statistics. Schaum's Outline Series*, McGraw-Hill, New York, 372p.

Stallman, R. W., 1971. Aquifer-test, design, observation and data-analysis. *U.S. Geological Survey Techniques of Water-Resources Investigations*, book 3, chapter B1, 26p.

Stanley, S. M., 1966. Paleoecology and diagenesis of Key Largo Limestone, Florida. *Bull Am Assoc Petrol Geologists*, 50:1927–1947.

StatPoint Technologies, Inc., 2012. Statgraphics—Rev. 1/7/2010. Warrenton, VA. Available at: www.statlets.com/statgraphics_centurion.htm

Sun, P. -C. P., Criner, J. H., and Poole, J. L., 1963. Large springs of East Tennessee. *Geological Survey Water-Supply Paper 1755*, Tennessee Geological Survey, Nashville, 52p.

Tackett, J. L., and Pearson, R. W., 1965. Some characteristics of soil crusts formed by simulated rainfall. *Soil Sci*, 99:407–413.

Theis, C. V., 1935. The lowering of the piezometric surface and the rate and discharge of a well using ground-water storage. *Trans, Am Geophys Union*, 16:519–524.

Thornthwaite, C. W., 1946. The moisture factor in climate. *Trans Am Geophys Union*, 27:41–48.

Thornthwaite, C. W., 1948. An approach toward a rational classification of climate. *Geologic Rev*, January:55–94.

Torres, A., and Diaz, J. R., 1984. Water resources of the Sabana Seca to Vega Baja area, Puerto Rico. *U.S. Geological Survey Water-Resources Investigations Report 82–4115*, San Juan, Puerto Rico, 53p.

Touloumdjian, C., 2005. The springs of Montenegro and Dinaric karst. In: Stevanovic, Z., and Milanovic, P. (eds.), *Water Resources and Environmental Problems in Karst—Cvijić 2005*, Proceedings of International Symposium, University of Belgrade, Institute of Hydrogeology, Belgrade, pp. 443–450.

UNESCO (United Nations Educational, Scientific and Cultural Organization), 2004. Submarine groundwater discharge. Management implications, measurements and effects. *IHP-VI, Series on Groundwater No. 5*, IOC Manuals and Guides No. 44, Paris, 35p.

US Forest Service Intermountain Region, 2008. *Geologic Points of Interest by Activity; Springs/Falls*. Available at: http://www.fs.fed.us/r4/resources/geology/geo_points_interest/activities/springs_falls.shtml; Accessed 30th November 2008.

USACE (U.S. Army Corps of Engineers), 1993. *Hydrologic Frequency Analysis*. Engineering manual 1110–2-1415, Washington, D.C., various paging. Available at: http://140.194.76.129/publications/eng-manuals/

USACE (United States Army Corps of Engineers), 1999. *Groundwater Hydrology*. Engineer Manual 1110–2-1421, Washington, D.C., various pages.

USBR (U.S. Bureau of Reclamation), 1977. *Ground Water Manual*. U.S. Department of the Interior, Bureau of Reclamation, Washington, D.C., 480p.

USGS (United States Geological Survey), 2004. *Vertical Flowmeter Logging. Office of Ground Water, Branch of Geophysics*. Available at: http://water.usgs.gov/ogw/bgas/flowmeter/

USGS (United States Geological Survey), 2006. *FGDC Digital Cartographic Standard for Geologic Map Symbolization. Appendix A, 37-Lithologic patterns, 37.1-Sedimentary-rock lithologic patterns. FGDC-STD-013–2006*. Available at: http://ngmdb.usgs.gov/fgdc_gds/geolsymstd/download.php

USGS (United States Geological Survey), 2011. *USGS Photographic Library*. Available at: http://libraryphoto.cr.usgs.gov.

Vandike, J. E., 1996. *The Hydrology of Maramec Spring*. Missouri Department of Natural Resources, Division of Geology and Land Survey, Water Resources Report Number 55, Rola, MO, 104p.

Vukmirović, V., and Despotović, J., 1983. Osnovne faze statističke obrade jakih kiša (Key steps in statistical processing of storm rainfall; in Serbo-Croatian). Jugoslovenski simpozijum o inženjerskoj hidrologiji, Okrugli sto, Split, Novembar 9–12, pp. 32–45.

Vuković, M., and Soro, A., 1985. *Osnovi Hidraulike. Drugo Izdanje (Fundamentals of Hydraulics. Second Edition; in Serbian)*. University of Belgrade, School of Mining and Geology, Department of Hydrogeology, Belgrade, 232p.

Warner, D., 1997. Hydrogeologic evaluation of the Upper Floridan aquifer in the southwestern Albany area, Georgia. *U.S. Geological Survey Water Resources Investigations Report 97–4129*, Atlanta, GA, 27p.

White, W. B., 1988. *Geomorphology and Hydrology of Karst Terrains*. Oxford University Press, New York, 464p.

Williams, J. H., and Johnson, C. D., 2004. Acoustic and optical borehole-wall imaging for fractured-rock aquifer studies. *J Appl Geophys*, 55(1–2): 151–159.

Williams, J. H., and Lane, J. W., 1998. *Advances in Borehole Geophysics for Ground-Water Investigations: U.S. Geological Survey Fact Sheet 002–98*, U.S. Geological Survey, 4p.

Williams, J. H., Lane, J. H. Jr., Kamini, S., and Haeni, F. P., 2001. Application of advanced geophysical logging methods in the characterization of a fractured-sedimentary bedrock aquifer, *U.S. Geological Survey Water-Resources Investigations Report 00–4083*, Ventura County, California, 28p.

Wilson, J. T., Mandell, W. A., Paillet, F. L., et al., 2001. An evaluation of borehole flowmeters used to measure horizontal ground-water flow in limestones of Indiana, Kentucky, and Tennessee, 1999. *U.S. Geological Survey Water-Resources Investigations Report 01–4139*, Indianapolis, Indiana, 129p.

Winter, T. C., Harvey, J. W., Franke, O. L., Alley, and W. M., 1998. Ground water and surface water: A single resource. *U.S. Geological Survey Circular 1139*. Denver, CO, 79p.

Witherspoon, P. A., 2000. Investigations at Berkeley on fracture flow in rocks: from the parallel plate model to chaotic systems. In: Faybishenko, B., Witherspoon, P. A., and Benson, S. M. (eds.), *Dynamics of Fluids in Fractured Rock*, Geophysical Monograph 122, American Geophysical Union, Washington, D.C., pp. 1–58.

Wolff, R. G., 1982. Physical properties of rocks—Porosity, permeability, distribution coefficients, and dispersivity. *U.S. Geological Survey Open-File Report 82–166*, 118p.

Woodruff, C. M., and Abbott, P. L., 1986. Stream piracy and evolution of the Edwards Aquifer along the Balcones Escarpment, Central Texas. In: Abbott, P. L, and Woodruff, C. M. (eds), *The Balcones Escarpment, Geological Society of America*, San Antonio, TX, pp. 51–54.

Worthington, S. R. H., and Ford, D. C., 2009. Self-organized permeability in carbonate aquifers. In: Kresic, N. (guest editor), *Theme Issue: Ground Water in Karst. Ground Water*, 47(3), pp. 326–336.

Wu, Q., Xing, L., and Zhou, W., 2010. Utilization and protection of large karst springs of China. In: Kresic, N., and Stevanovic, Z. (eds.), *Groundwater Hydrology of Springs; Engineering, Theory, Management and Sustainability*. Elsevier, New York, pp. 543–565.

Zhaoxin, W., and Chuanmao, J., 2004. Groundwater resources and their use in China. In: Zekster, I. S., and Everet, L. G. (eds.), *Groundwater Resources of the World and Their Use. IHP-VI, Series on Groundwater No. 6*, UNESCO, Paris, pp. 143–159.

Zimmerman, R. W., and Yeo, I. W., 2000. Fluid flow in rock fractures: from the Navier-Stokes equations to cubic law. In: Faybishenko, B., Witherspoon, P. A., and Benson, S. M. (eds.), *Dynamics of Fluids in Fractured Rock,* Geophysical Monograph 122, American Geophysical Union, Washington, D.C., pp. 213–224.

CHAPTER 2

Flow Measurements and Analysis

2.1 Introduction

Although the principles and techniques of flow measurements in karst are the same as in any other hydrologic setting, in many instances the challenges of applying them in the field will be significantly greater. This is particularly true for flow measurements in submerged conduits or for karst springs with very high flow amplitudes issuing from locations that are difficult to access. In any case, the following three general methods can be applied to flow measurements in surface streams and in subsurface karst channels: (1) weirs and flumes, (2) area velocity, and (3) dye tracers. These and other more specialized techniques such as Acoustic Doppler Current Profiler (ADCP) are described in great detail in a series of technical publications by the USGS available for free download at: http://pubs.usgs.gov/twri/. The National Streamflow Information Program by USGS offers general information on the topic (www.water.usgs.gov/nsip/). Particularly useful is a joint publication by several US government agencies entitled *Water Measurement Manual* (Bureau of Reclamation, 2001), also available online or for free download at http://www.usbr.gov/pmts/hydraulics_lab/pubs/wmm/.

As discussed by NNPSMP (2008), the flow variables and the frequency with which they are measured depend on the project objectives and data analysis plans. For example, single measures of instantaneous spring or stream discharge are highly unlikely to satisfy any of the common objectives of water management in karst because they represent only a snapshot in time. If conducted as part of a synoptic survey within a study watershed, however, such data might be useful in comparing the hydrologic behavior of subwatersheds, characterizing the relative magnitudes of sediment (or contaminant) loads or flows from subwatersheds, or calibrating a hydrologic model for the study watershed.

Systematic collection of low-flow and peak-flow data has wide application for flood management, dam design, stormwater projects, artificial aquifer recharge by surface water, nonpoint source projects, water supply design, and habitat restoration efforts. In urban watersheds where streams are shaped by peak discharges, management of water quantity is often the first objective of watershed projects, for both stream morphology and biologic concerns. Peak stage is relatively easy and inexpensive to monitor, and a comparison of peak stage before and after a program of stormwater best management practices (BMPs) could be quite useful. Knowledge of changes in peak water levels can also be critical in stream restoration projects for both physical channel work and restoration of biotic communities. It will be necessary in these cases to monitor precipitation and other important explanatory variables to interpret changes in peak flow values. At the same time, continuous recording of low discharges during baseflow conditions and prolonged droughts is as important as recording of peak flows. This is true for establishing both water quality and quantity criteria for a variety of management goals including design of reliable water supply and maintenance of minimum environmental flows in case of competing interests of different water users.

Streams and springs without permanent gages and sufficient data record present special challenge when determining characteristic flows needed for various calculations. Long-term minimum baseflow is of particular importance for various applications such as determining maximum allowable loading rates of a groundwater contaminant that are still protective of in-stream water quality standards. In the United States, this flow is typically referred to as 7Q10, or 7-day, consecutive low flow with a 10-year return frequency (i.e., the lowest streamflow for seven consecutive days that would be expected to occur once in 10 years).

Flow measurement at successive stream segments is a common method of determining if the stream is losing or gaining water between the segments. However, because of the variability of flow conditions in the same stream and the associated potential measurement errors, this method should be applied with great care in order to avoid false conclusions. For example, if only one set of measurements is made at a few locations, with several or more hours separating them, and the streamflow is under the influence of recent precipitation, the results would almost certainly be misleading since the flow wave is moving rapidly. It is therefore best if the flow measurements are performed after a long period without precipitation and the applied method is based on a continuous recording of the stream stage at successive stream segments. Flow hydrographs derived in this way provide information on the actual change of volume of water between the segments, which is the only real measure of gain or loss.

Continuous discharge data play an important role in the design of sampling programs for many objectives. Because concentrations of many contaminants in surface streams and the karst subsurface are strongly associated with discharge, many sampling programs are stratified by flow conditions—more samples are taken at higher discharges, for example. Discharge data can also be used to diagnose water quantity problems in watersheds and may itself be a variable expected to respond to implementation of BMPs. For example, Baker et al. (2004) proposed a *Flashiness Index*—a measure of the frequency and rapidity of short-term changes in streamflow, calculated from mean daily stream discharge data. Flashiness is an important component of a stream's hydrologic regime; land use and land management changes may lead to increased or decreased flashiness, often impairing aquatic life. The Index can be used to quantify the hydrologic impacts of watershed change and to evaluate programs aimed at restoration of natural streamflow regimes. Flow data may also be useful in evaluating agricultural programs where BMPs may promote infiltration over runoff or where drainage practices influence surface water flows. The relationship between discharge and pollutant concentration may also change in response to BMP implementation. Suspended sediment concentrations might be lower after implementation of conservation tillage, for example, at comparable flows; good discharge data would be important to document this change.

Presented further are several common methods of flow measurements applicable to karst field settings.

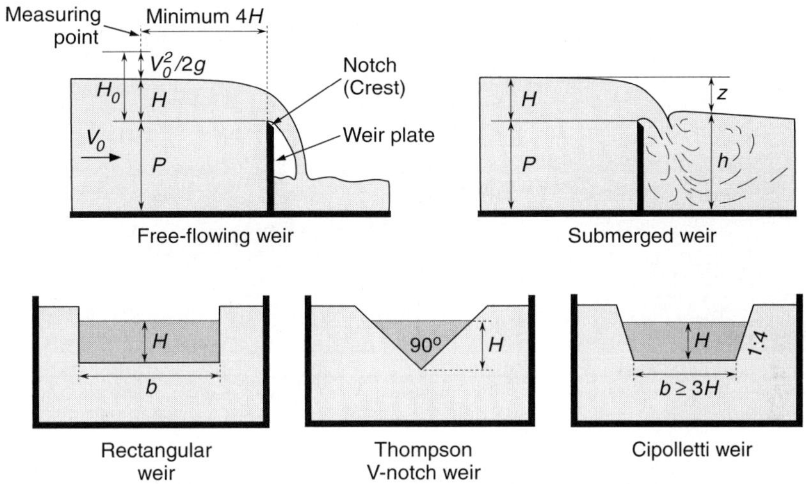

FIGURE 2.1 Sharp-crested weirs for stream and spring flow measurement (explanation in text).

2.2 Weirs

For long-term projects, discharge can be measured using a weir or flume, often referred to as a primary device. A primary device is a structure that water flows through or over that has a known relationship between stage and flow. If a device can be used, discharge measurement can be as simple as observing the stage of water just upstream of the device and consulting a table or using a simple equation to calculate discharge.

Weirs are commonly named by the shape of their blade overflow opening shape for sharp-crested weirs (Fig. 2.1) or the flow control section shape for broad-crested weirs. Thus, weirs are classified as rectangular, trapezoidal, or triangular. In the case of sharp weirs, the triangular weir is also called a V-notch weir, and one popular kind of trapezoidal weir is the Cipolletti weir (Fig. 2.1). In the case of rectangular or Cipolletti weirs, the bottom edge of the notch in the thin plate is the *crest*, and the side edges (which are vertical or flare up and outward) are the sides or ends. The point of the triangle is the crest of a V-notch weir. The lowest elevation of the overflow opening of the sharp-crested weirs or the control channel of broad-crested weirs is the head measurement *zero reference* elevation (Bureau of Reclamation, 2001).

The water surface drops noticeably as flow approaches the crest. This continuing drop of water surface or *drawdown* results from the acceleration of the water as it approaches the weir. The drop in water surface between the upgradient measuring station and the notch is equal to the change of velocity head, or $V^2/2g$, between the two (Fig. 2.1). Velocity of approach is equal to the discharge divided by the flow section area at the head measuring station. Velocity of approach is important because it can change weir calibrations by effectively reducing the crest length and/or measuring head. In addition, a variable weir discharge coefficient results as increasing velocity changes the curvature of flow springing from the weir plates.

When the distances from the sides of the weir notch to the sides of the weir pool are greater than two measurement heads, the water will flow relatively slowly along the bulkhead face toward the overflow opening. As the water from the sides of the channel

Figure 2.2 *Left*: Sharp-crested 120 degrees V-notch weir. *Right*: Prefabricated H flume. (Photographs courtesy of Don Meals, Tetra Tech, Inc.; From NNPSMP, 2008.)

nears the notch, it accelerates and has to turn to pass through the opening. This turning cannot occur instantaneously, so a curved flow path or *side contraction* results in which the water springs free to form a jet narrower than the overflow opening width.

Free flow occurs when a thin-plate weir allows free access of air under the falling jet sheet or *nappe*. With free flow, head measurements at an upstream location determine discharge with knowledge of weir size and shape. Downstream water rising above the weir crest elevation produces a *submerged weir* condition (Fig. 2.1). When the downstream water surface is near or above the crest elevation of a sharp-crested weir, accuracy of measurement should not be expected. The use of a submerged weir as a water measurement device is not good practice and should only be done as a temporary procedure. Because of the large loss of accuracy, designing thin-plate weirs for submergence should be deliberately avoided. However, submergence may happen unexpectedly or may be temporarily necessary as some karst springs may exhibit extremely high but rare flow rates.

In practice, weirs can range from small wood or metal plates temporarily mounted across small ditches or streams (Fig. 2.2) to more permanent installations involving concrete walls and other structures. Note that erecting any obstruction in a stream will create a pool upstream and care must be taken to avoid creating the potential for flooding during high flows. In some cases, a permit may be required.

A weir discharge measurement consists of measuring depth or *head*, H, relative to the crest at the proper upstream location in the weir pool (Fig. 2.1), and then using a table or equation for the specific kind and size of weir to determine discharge. Commonly, a *staff gage*, having a graduated scale with the zero placed at the same elevation as the weir crest, measures head. Putting staff gages in stilling wells dampens wave disturbances when reading head (Fig. 2.3). Using vernier hook point gages in stilling wells produces much greater accuracy than staff gages. These staff gages must be zero referenced to the weir crest elevation.

When the weir is not submerged, the general flow equation for a rectangular sharp-crested weir is

$$Q = mb\sqrt{2g}\, H_0^{3/2} \tag{2.1}$$

where m is the overflow coefficient, b is the width of the weir (overflow), g is the acceleration of gravity, and H_0 is the total energy of the overflow that includes the overflow height, H, and the velocity component. The overflow coefficient takes values

FIGURE 2.3 *Top*: Bastašica Spring near Drvar, Bosnia and Herzegovina. The overtopped rectangular weir is undersized for this flow rate. Stilling well with data logger is on the right. *Bottom*: Weir at Carter Hall Spring, Millwood, VA, the United States; note stilling well on the right. (Photograph courtesy of William Jones, Karst Waters Institute.)

$0.40 < m < 0.48$, with $m = 0.42$ used most often in practice. As the velocity of the overflow is usually not measured, this equation is replaced by a more practical one

$$Q = m_0 b\sqrt{2g}H^{3/2} \tag{2.2}$$

where the overflow coefficient, m_0, is calculated by the empirical formula of Bazen:

$$m_0 = \left(0.405 + \frac{0.0027}{H}\right) \times \left[1 + 0.55\left(\frac{H}{H+P}\right)^2\right] \tag{2.3}$$

Triangular V-notch sharp-crested weir with the 90-degree angle (Fig. 2.1), also known as Thompson weir, is often used for precise measurements of low flow rates such as in small streams and laboratories. The flow rate is calculated as

$$Q = 0.32\sqrt{2g}\,H^{5/2} \tag{2.4}$$

$$Q = 1.4\,H^{5/2} \tag{2.5}$$

In Eq. (2.5), the height of the overflow (H) is given in meters, and the flow rate (Q) is obtained in m^3/s.

For the widely used Cipolletti weir, the flow rate is calculated as

$$Q = 0.42\,b\sqrt{2g}H^{3/2} \tag{2.6}$$

$$Q = 1.86\,bH^{3/2} \tag{2.7}$$

In Eq. (2.7), the height of the overflow (H) is given in meters, and the flow rate (Q) is obtained in m^3/s.

When the downgradient elevation of the water surface is higher than the elevation of the rectangular weir crest, that is, for the condition that $h > P$ (see Fig. 2.1), the weir is submerged. The general conditions for a weir to be submerged is $z/P<0.7$, in which case the flow rate is a function of both the upgradient and the downgradient head: $Q = f(H,h)$. The flow rate can be calculated using

$$Q = \sigma_S m_0 b\sqrt{2g}\,H^{3/2} \tag{2.8}$$

where σ_S is the coefficient of submergence calculated using the empirical formula of Bazen:

$$\sigma_S = 1.05\left(1 + 0.2\frac{H-z}{P}\right) \times \sqrt[3]{\frac{z}{H}} \tag{2.9}$$

and the overflow coefficient, m_0, is calculated using Eq. (2.3).

Cipolletti and rectangular weirs are often used for measuring discharge rate of large karst springs that can have wide range of flows, from tens of L/s to many m^3/s.

Flumes are specially shaped open channel flow sections that restrict the channel area, resulting in increased velocity and a change in water level as water flows through a flume. The discharge through a flume is determined by measuring the stage in the flume at a specific point, depending on the type of flume. In general, flumes are used to measure discharge where weirs are not feasible. Flumes are well suited to measurement of flows containing sediment because the increased velocity through the flume tends to make it self-cleaning. Commonly used flumes include the Parshall, Palmer-Bowlus, and H flumes (see Fig. 2.2). Flumes come in a wide range of sizes denoting the maximum depth of flow they can accommodate and can be purchased as prefabricated units or built on-site. Most devices have exacting specifications for installation that may present construction challenges. More information on weirs and flumes is available at http://www.usbr.gov/pmts/hydraulics_lab/winflume/.

2.3 Velocity-Area Method

The velocity of moving water varies both across a stream channel and from the water surface to the bottom of the stream because of friction and irregularities in cross section and alignment. Friction caused by the rough channel surfaces slows the water near the bottom and sides of a channel, so that the fastest water is usually near the center of the channel and near the surface. On a river bend, the water on the outside of the bend moves faster than the water on the inside of the bend, as it has to cover more distance in the same time. It is therefore necessary to perform a number of velocity measurements across the stream channel and at different depths. Figure 2.4 is a schematic of the velocity-area method applicable to any surface or subsurface stream channel. The discharge in each subsection is computed by multiplying the subsection area by the measured average velocity that, in this case, is the average of two velocities measured at 0.2 and 0.8 subsection depths. The total discharge is then computed by summing the discharge of each subsection. To apply the method, a straight reach, reasonably free of large rocks or obstructions, with a relatively flat streambed, away from the influence of abrupt changes in channel width should be selected.

Determining the cross-sectional area of a flowing stream usually involves measuring water depths at a series of points across the stream and multiplying by the width of the stream within each segment represented by the depth measurement. The segment areas are then summed to determine the entire cross-sectional area. Commonly, the stream

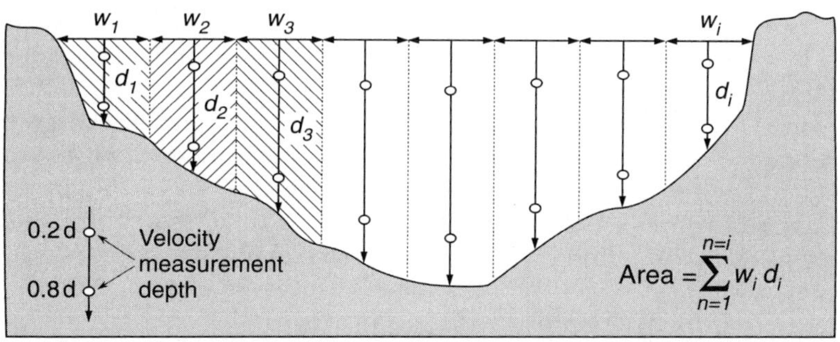

FIGURE 2.4 Schematic of the velocity-area method of streamflow measurements. Stream is divided into segments and discharge rate in each segment is calculated by multiplying the segment area with the flow velocity in that segment. The flow velocity is measured at two depths equaling 0.2 and 0.8 of the total depth of a segment. The total stream discharge is obtained by summing the segment discharges.

width is divided into 20 to 25 segments [streams less than 10 ft (3 m) wide may not allow as many segments] using tape or string to mark the center of each segment on a cable stretched across the stream perpendicular to the flow; typically, the stream is divided into enough segments so that each one has no more than 10% of the total streamflow. Segment width is generally measured using a cable, steel tape, or similar piece of equipment. Segment depth is measured using a wading rod, if conditions permit, or by suspending a sounding weight from a calibrated cable and reel system off a bridge, cableway, or boat or through a hole drilled in ice.

To deal with the variability in stream velocity within any cross-sectional area segment, studies by USGS support several general rules of thumb:

- Maximum velocity occurs at 5% to 25% of the depth, this percentage increases with increasing stream depth.
- Mean velocity in a vertical profile is approximated by the velocity at 0.6 depth.
- Mean velocity in a vertical profile is more accurately represented by the mean of the velocities at 0.2 and 0.8 depth.
- The mean velocity in a vertical profile is 80% to 95% of the surface velocity, the average of several hundred observations being 85%.

In some shallow streams and cave channels, it may be possible to measure flow velocity at only one depth.

The most common current meter used by the USGS for measuring streamflow velocity is the Price AA current meter that has a wheel of six metal cups that revolve around a vertical axis (Fig. 2.5). An electronic signal is transmitted by the meter on each revolution allowing the revolutions to be counted and timed. Because the rate at which the cups revolve is directly related to the velocity of the water, the timed revolutions are used to determine the water velocity. The Price AA meter is designed to be attached to a wading rod for measuring in shallow waters or to be mounted just above a weight suspended from a cable and reel system for measuring in fast or deepwater. In shallow water, the Price-Pigmy current meter can be used (Fig. 2.5). It is a two-fifths scale version of the Price AA meter and is designed to be attached to a wading rod. A third mechanical current meter, also a variation in the Price AA current meter, is used for measuring water velocity beneath ice. Its dimensions allow it to fit easily through a small hole in the ice, and it has a polymer rotor wheel that hinders the adherence of ice and slush.

There is a strong relationship between river stage and discharge and, as a result, a continuous record of stream discharge can be determined from the continuous record of stage. The development of an accurate stage–discharge relation requires numerous discharge measurements at all ranges of stage and streamflow (Fig. 2.6). In addition, these relations must be continually checked against ongoing discharge measurements because stream channels are constantly changing. Changes in stream channels are often caused by erosion or deposition of streambed materials, seasonal vegetation growth, debris, or ice. Reliable streamgaging therefore involves obtaining a continuous record of stage, making periodic discharge measurements, establishing and maintaining a relationship between the stage and discharge, and applying the stage–discharge relation to the stage record to obtain a continuous record of discharge that is critical for water resources management.

FIGURE 2.5 Measurement of underground flow with Price-Pigmy meter and wading rod. (Photograph courtesy of William Jones, Karst Waters Institute.) *Inset*: Price AA current meter on a round wading rod.

Most USGS streamgages measure stage and consist of a structure in which instruments used to measure, store, and transmit the stream–stage information are housed. Stage, sometimes called gage height, can be measured using a variety of methods. One common approach is with a stilling well in the riverbank (Fig. 2.7) or attached to a bridge pier. Water from the river enters and leaves the stilling well through underwater pipes allowing the water surface in the stilling well to be at the same elevation as the water surface in the river. The stage is then measured inside the stilling well using a float or a pressure, optic, or acoustic sensor. The measured stage value is stored in an electronic data recorder on a regular interval, usually every 15 minutes.

Most of the stage and streamflow information produced by the USGS is available in near real time through the National Water Information System (NWIS) Web site

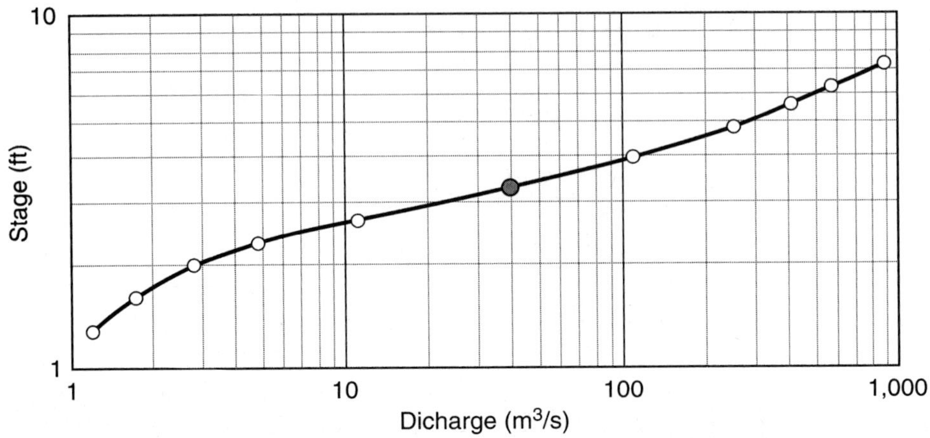

FIGURE 2.6 Example of a typical stage–discharge relation; here, for the large dark dot, the discharge of the stream is 40 cubic feet per second (ft^3/s) when the stage is 3.30 feet (ft). The smaller dots on the curve represent concurrent measurements of stage and discharge. (Modified from USGS, 2005.)

(http://waterdata.usgs.gov/nwis/). In addition to real-time streamgage data, the NWIS Web site also provides access to daily discharges and annual maximum discharges for the period of record for all active and discontinued streamgages operated by the USGS. This includes monitoring wells and springs in karst areas throughout the United States.

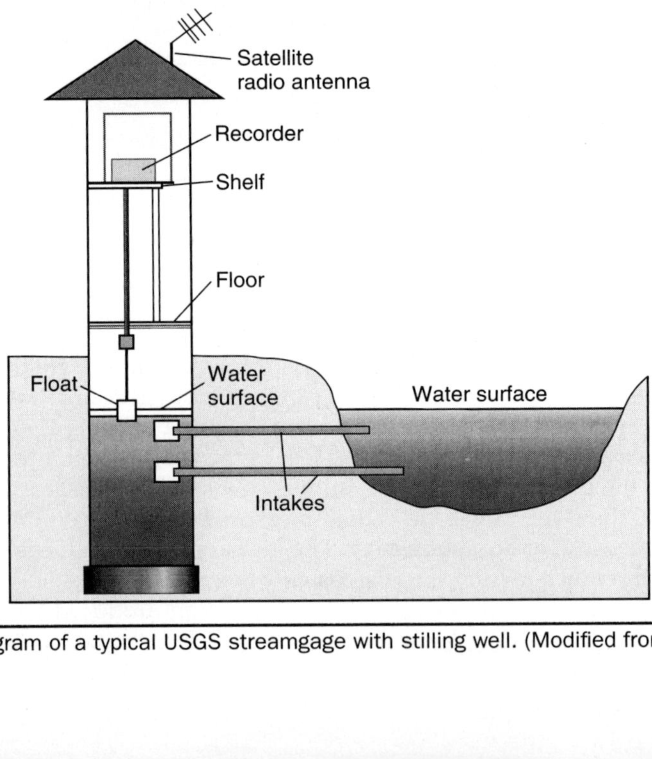

FIGURE 2.7 Diagram of a typical USGS streamgage with stilling well. (Modified from USGS, 2005.)

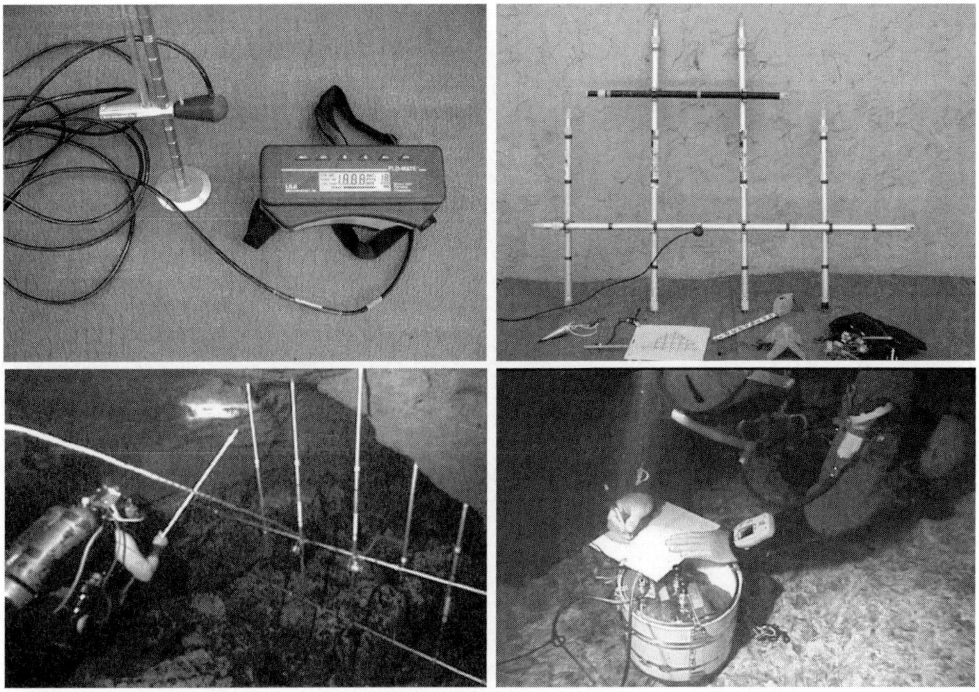

FIGURE 2.8 The area-velocity method applied in a submerged conduit feeding a spring in Florida, the United States. The velocity is measured by a standard flowmeter configuration shown in the top-right photo. The electronic flow sensors are attached to a grid made of telescopic rods (top right) spanning the conduit cross section. The grid is installed by cave divers and the velocity at different station points on the grid is recorded by the flowmeter in the underwater housing shown in bottom right. (Modified from Butt, 2003; photographs courtesy of Karst Environmental Services, Inc.)

When conditions are favorable, the velocity-area method can be successfully applied for flow measurements in submerged conduits as illustrated in Figs. 2.8 through 2.10. For more information, interested readers can visit www.karstenvironmental.com.

2.4 Slope-Area Method

The slope-area method consists of using the slope of the water surface in a uniform reach of channel and the average cross-sectional area of that reach to give a rate of discharge. The discharge may be computed from the Manning formula:

$$Q = \frac{1.486}{n} A R_{\mathrm{h}}^{2/3} S^{1/2} \tag{2.10}$$

where Q is discharge (ft^3/s), A is mean area of the channel cross section (ft^2), R_{h} is mean hydraulic radius of the channel (ft), S is energy slope of the flow, and n is a roughness factor depending on the character of the channel lining

A straight reach of the channel should be chosen at least 200 ft and preferably 1000 ft in length. If the reach is free of rapids, abrupt falls, or sudden contractions or

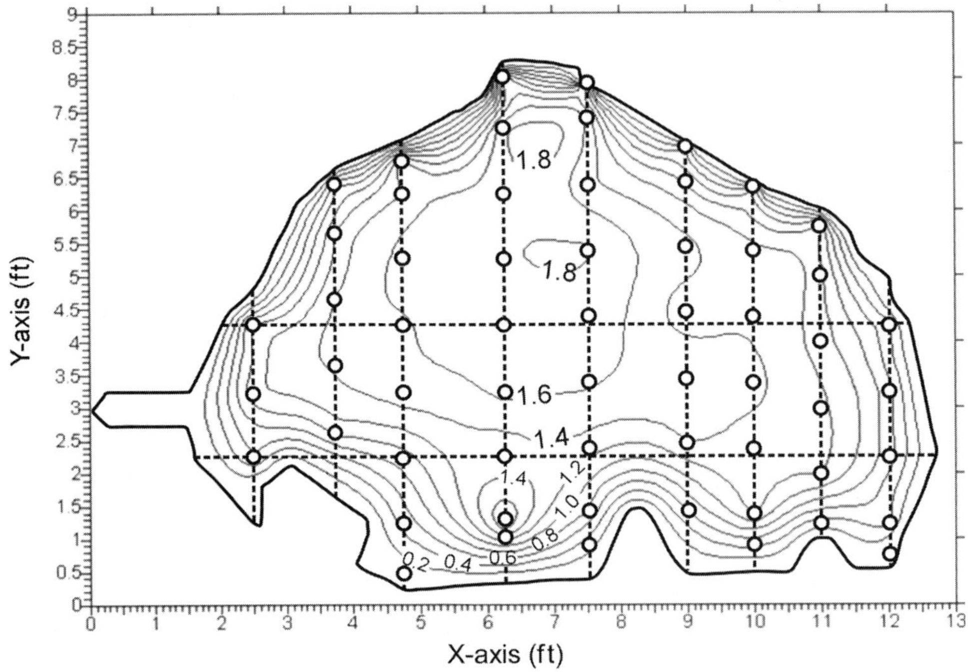

Figure 2.9 Contours of flow velocity at a submerged conduit cross section used to calculate flow rate at the Croaker Hole Spring, Putnam County, FL. Total cross-sectional area is 62.73 ft^2, average station point velocity is 1.248 ft/s, and total flow is 78.3 ft^3/s or 2.2 m^3/s. (From Butt, 2003; diagram courtesy of Karst Environmental Services, Inc.)

Figure 2.10 Underwater discharge measurements at Jackson Blue Spring, Merritt's Mill Pond, Jackson County, FL. (Photograph courtesy of Karst Environmental Services, Inc.)

expansions, then the water surface slope is the same as the energy slope. The slope, S, may be determined by dividing the difference in the water surface elevations at the two ends of the reach by the length of the reach. A gage point, carefully referenced to a common datum level, should be placed on each bank of the channel and in the center of the reach, in stilling wells if possible.

The hydraulic radius, R_h, is defined as the area of the cross section divided by its wetted perimeter. Where the channel is of regular cross section, and the depths at the ends of the course are equal, and the area and the wetted perimeter will be constant throughout the course. In irregular channels, the area and the wetted perimeter at several cross sections will be required, and a mean value will be used in computing the hydraulic radius. A static pressure tube can be used to measure depth of flow.

The roughness factor, n, depends on the character of the channel. It may vary from 0.010, where conditions approaching the ideal are maintained to 0.060, where the channel is strewn with stones and debris or is about one-third full of vegetation. Because the proper selection of the roughness for many streams is difficult and is, at best, an estimate, the discharge determined by the slope-area method is only approximate. Care must be taken to determine the slope and areas simultaneously if the water levels are changing.

Sections 1.6.3 through 1.6.6 provide other flow equations, including friction factors that can be used with this method; various other references with tables of friction factors can be found in the *Water Measurement Manual* by Bureau of Reclamation (2001).

2.5 Dye-Tracing Methods

Tracer methods can be used to determine discharge with accuracies that can vary considerably from about +/−1% to over 30%, depending on the equipment used and the care in applying the techniques. Accessible closed conduit measurements are typically more accurate than open channel measurements because of better area measurements and better tracer dispersion in conduits. In contrast, mixing can be a problem at low velocities in open channels. In general, a tracer is considered anything that mixes with or travels with the flow and is detectable. A detectable tracer can be timed as it passes through a reach, or tracer concentration profiles can be measured in a reach. Some tracers that have been used include dyes of various colors, other chemicals such as fertilizer, salt, and gases, and heat. Radioisotopes are now rarely used because of their safety and risks of contamination. In addition, isotope handlers must be licensed. In the United States, use of any chemical or anything that can affect ecological characteristics of the water or conveyance boundaries may require clearance from several federal and state authorities such as the U.S. Food and Drug Administration, the U.S. Environmental Protection Agency, and state fish and wildlife and natural resource departments. Government regulations and limits change with time and should be checked prior to a measurement program. However, even when operating within government regulations, public complaints related to taste and color and particles in the resulting water may occur (Bureau of Reclamation, 2001).

Salt tracers are sensed and quantified by measuring evaporated dry weight, chemical titration, or by measuring electrical conductivity. Dye concentrations are measured by fluorimetry or color comparison standards. Sometimes, visual observation of a dye cloud is used, but considerable loss of accuracy occurs. The most widely used dyes are fluorescein, rhodamine B, rhodamine WT, and Pontacyl Pink partially because they are easily visible in very dilute solutions. Rhodamine B and rhodamine WT have been cleared as nontoxic by the U.S. Food and Drug Administration. Rhodamine and Pontacyl Pink are also quite stable with respect to fading by sunlight and to changes caused by waterborne chemicals. They do not tend to deposit on flow surfaces, sediments, or weeds. These dyes are usually available in powder form, and solutions are easily prepared. Before conducting a discharge measurement program, selected dyes should be tested with water samples or earth canal embankment material samples and exposed to check for possible adsorption, chemical reaction, and fading effects on dye stability (Bureau of Reclamation, 2001).

Salt and dye tracers are used to determine discharge in two basic ways: (1) the velocity-area method, in which time of tracer travel through a known channel length and average cross-sectional area determine discharge and (2) the dilution method, in which discharge is determined by the downstream concentration of fully mixed tracer, which has been added upstream at a constant rate, and by accounting for the mass of tracer.

2.5.1 Velocity-Area and Tracer-Dilution-Discharge Methods

The discharge using velocity-area method is computed by the following general equation applicable to both open channel and closed conduit flow:

$$Q = \frac{AL}{T} \tag{2.11}$$

where Q is discharge in cubic feet or meters per second (ft^3/s; m^3/s), A is average cross-sectional area of reach length in square feet or square meters (ft^2, m^2), L is reach length between detection stations in feet or meters (ft, m), and T is recorded time required for the tracer solution to travel between the detection stations at each end of the measurement reach in seconds (s).

Either salt or dye may be conveniently used in tracer-velocity-area discharge measurements with equal potential accuracy. The only difference is that different detection equipment is needed. Dyes have an added advantage in that they can be detected visually, allowing simpler measurements of less accuracy that may be sufficient for some needs.

Commonly, sodium chloride (NaCl) is the selected salt used in the tracer injection solution. Finely ground salt should be purchased for ease in mixing the solution. Enough salt must be added to significantly increase the electrical conductivity of the water so that concentrations can be measured accurately. The required amount of salt can be estimated by analyzing the water for existing background quantity of salt in the measurement flow, estimating the amount of flow to be measured, and using chemical handbook data from conductivity to salinity tables. Trial runs may be needed to determine the optimum amounts, which may vary with discharge depending on the range to be measured.

For a measurement, a quantity of tracer solution is usually forced into the stream under pressure to provide better initial distribution and assure thorough mixing before arrival at the detection stations. The pop valve injector described by Thomas and Dexter (1955) will provide the faster and better mixing required to produce code accuracy. As discussed by Kilpatrick and Wilson (1989), dyes injected into a stream behave in the same manner as the water particles themselves. A measure of the movement of the tracer will in effect be a measure of the movement of an element of fluid in the stream and of its dispersion characteristics. The dispersion and mixing of the tracer in the receiving stream takes place in all three dimensions of the channel (Fig. 2.11). Vertical mixing is normally completed first, and lateral mixing later, depending on stream characteristics and velocity

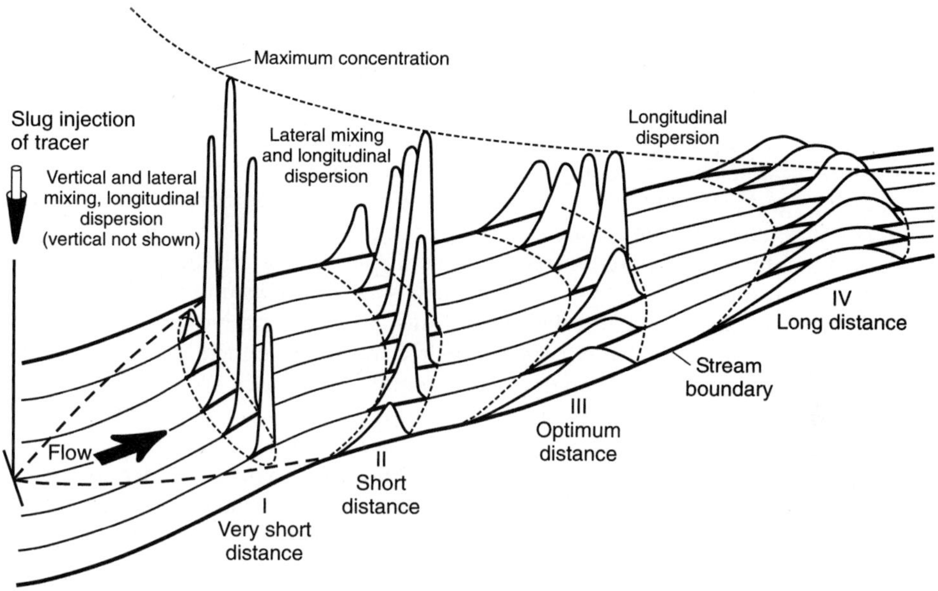

FIGURE 2.11 Lateral mixing and longitudinal dispersion patterns and changes in distribution of concentration downstream from a single, center slug injection of tracer. (From Kilpatrick and Wilson, 1989.)

variations. Longitudinal dispersion, having no boundaries, continues indefinitely and is the dispersion component of primary interest.

Most important, therefore, is a measurement of the response curve far enough downstream that mixing is almost complete in a cross section. At short distance downstream from a slug injection, the tracer is not fully mixed in the total flow of the stream, being more in the center than along the banks (see Fig. 2.12). Furthermore, the response curve measured in the center may be much shorter in duration, T_d, than for those along the

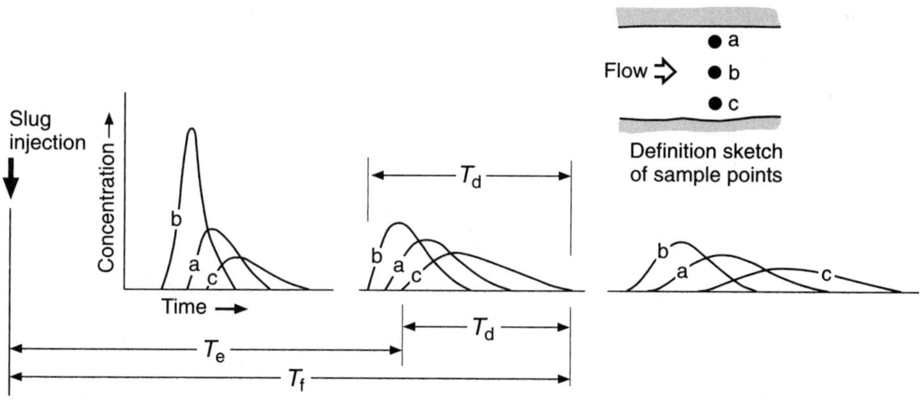

FIGURE 2.12 Typical response curves observed laterally and at different distances downstream from a slug injection of a tracer in the center of a stream. (From Kilpatrick and Cobb, 1985.)

banks. This is a common occurrence, because flow along the banks is usually slower and the channel banks tend to slow and elongate the tracer cloud. At such a short distance, an accurate measurement of discharge by dilution cannot be made by ordinary methods. When uniform mixing is reached, the areas under the time-concentration curves are essentially the same regardless of shape; at too short a distance, they are not. From a practical standpoint, complete mixing does not need to be attained. A good dilution-discharge measurement can be made at what is defined here as an optimum distance, L_O, downstream. The distance is optimum because T_n is not too long, and thus sampling of the complete response curves at several points laterally across a section is feasible. Note that the peaks of the response curves are not the same, and that their lengths, or durations, and arrival and departure times are different. Nevertheless, the areas under the individual response curves are nearly the same, which indicate good mixing and allow a good dilution-discharge measurement (Kilpatrick and Cobb, 1985).

Note that in Fig. 2.12 the duration or time of passage of a tracer response at a section, T_D, is the difference between the slowest trailing time along one bank and the fastest leading edge time, usually as observed in the center. The difference between the values of T_d and T_D can be significant. Unless otherwise indicated, common reference to the duration of the response curve is to T_D, as in most time-of-travel studies $T_D \approx T_d$ for the long stream lengths involved (Kilpatrick and Wilson, 1989).

The discharge as measured by the slug injection tracer-dilution technique is

$$Q = \frac{M}{A_C} \tag{2.12}$$

where Q is the volume rate of flow of the stream, M is the mass of tracer injected, and A_C is the area under the response curve obtained after adequate mixing of the tracer in the flow.

Equation (2.12) may be written as

$$Q = 5.89 \times 10^{-7} \frac{S_G V_I C}{A_C} \tag{2.13}$$

where Q is the discharge of the stream, in cubic feet per second (ft^3/s), V_I is the volume of concentrated dye solution injected into the stream, in milliliters (mL), C is the concentration of the dye solution injected into the stream, in micrograms per liter (μg/L), A_C is the area under the time-concentration curve, in units of minutes time micrograms per liter (min μg/L), and S_G is the specific gravity of the injected solution.

Successful application of Eq. (2.13) requires that the mass of tracer injected be fully recovered at the sampling point. This recovery is calculated by multiplying the area of the response curve or time-concentration curve, A_C, by the discharge, Q. Thus, it is vital that A_C accurately represents the dilution of the injected tracer. As shown in Fig. 2.11, a typical dye cloud may travel faster in the center of the stream than along the banks, where it may also be more elongated. To recover the injected tracer totally, therefore, requires complete measurement of the time-concentration curve at each of several locations laterally, such as at a, b, and c as depicted in Fig. 2.12. The necessity for extensive sampling is the reason this method has found less favor in the United States than the constant injection method described further (Kilpatrick and Cobb, 1985). Nevertheless, the method may be used where a constant-rate injection apparatus is not available or its use is not feasible.

The volume of tracer required for a slug injection discharge measurement is a function of the stream discharge, the measurement reach length, the stream velocity, and the peak concentration to be achieved at the sampling site. The following easy-to-use empirical equation for estimating the dosage requirements of rhodamine WT 20% dye for natural streams was developed by Kilpatrick (1970):

$$V_S = 3.79 \times 10^{-5} \frac{QL}{v} C_P \tag{2.14}$$

where C_P is the peak concentration at the sampling site, in micrograms per liter; L is the length of the measurement reach, in feet; Q is the stream discharge, in cubic feet per second; V_S is the volume of rhodamine WT 20% dye, in milliliters; and v is the mean-stream velocity, in feet per second.

A plot of Eq. 2.14 for a value of $C_P = 1.0\ \mu g/L$ is shown in Fig. 2.13. For most discharge measurements, a peak concentration of 10 to 20 micrograms per liter ($\mu g/L$) is recommended. Experience indicates that in lined channels and smooth pipes flowing

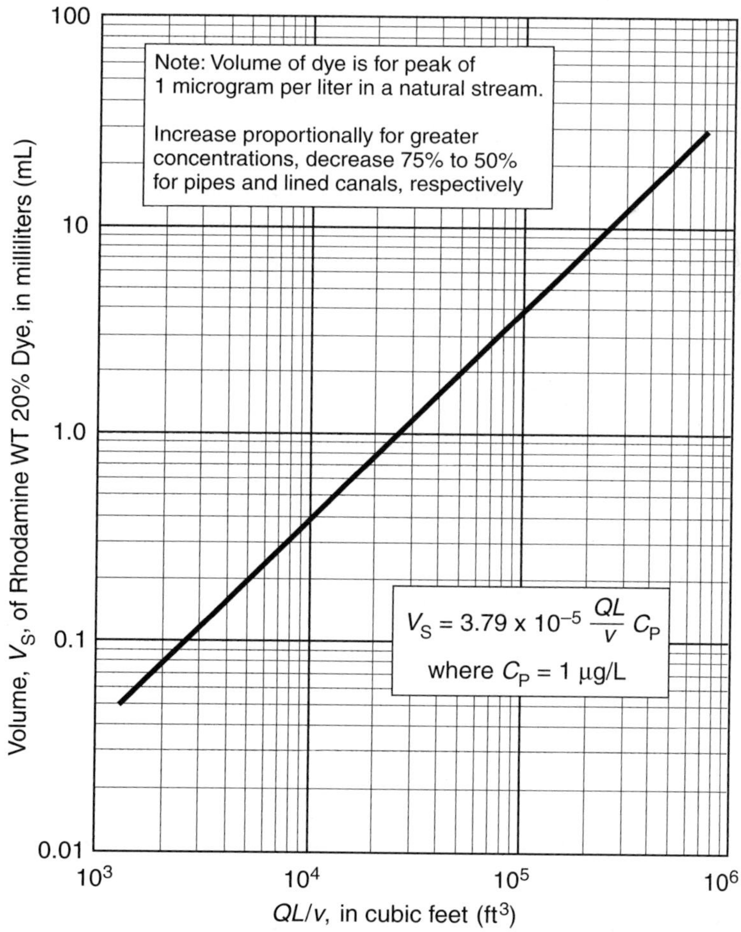

Figure 2.13 Quantity of rhodamine WT 20% dye required for slug injection to produce a peak concentration of 1 microgram per liter ($\mu g/L$) at a distance downstream, L, at a mean velocity, v, and with a discharge, Q, in the reach. (From Kilpatrick and Cobb, 1985.)

full, the dosages may be reduced to one half to one fourth, respectively, of the amount needed in a natural stream, as defined by Eq. (2.14). In such cases, longitudinal dispersion is less and, hence, less dye is required to produce a desired peak.

Pontacyl Pink dye comes as a 100% pure powder. It compares favorably with rhodamine WT as a water tracer and can be analyzed on most fluorometers using the same lamps and filters as for rhodamine WT. The weight in grams of Pontacyl Pink needed for a slug injection can be determined from Eq. (2.14) or Fig. 2.13 by multiplying the volume, in milliliters, for V_S by 0.24.

As illustrated earlier, the response curve at any point downstream from an instantaneous (slug) tracer injection is normally represented by plotting concentration against elapsed time. Figure 2.14 shows two such curves defined by the analysis of water samples taken at selected time intervals during the dye-cloud passage. They form the basis for determining time of travel (i.e., flow velocity) and dispersion characteristics of streams. To find velocity for Eq. (2.11) using salt tracers, a pair of conductivity or ion-specific electrodes is installed in the cross section at each end of a measured length of channel well downstream from the injection system. The distance between the pairs of electrodes should be sufficient to ensure accurate measurement of the time of travel between them. The electrodes are electrically energized and connected to a central data-logger instrument that continuously records the electrical conductivity (or ion concentration) at each pair of electrodes with respect to time. Measurements of dye concentration in flowing water are most efficiently performed with field fluorimeters including advanced fiber-optic sensors connected to data loggers. In either case, continuous tracer measurements in the field enable creating accurate plots of tracer concentration versus time that are needed to determine tracer mass recovery and flow velocity (Fig. 2.14).

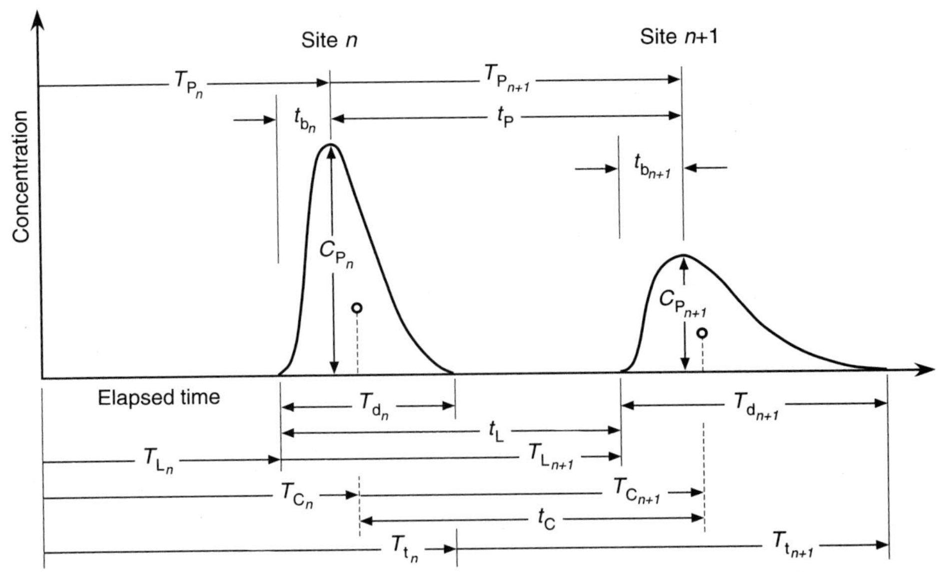

FIGURE 2.14 Definition sketch of time-concentration curves along a selected streamline resulting from an instantaneous dye injection (explanation of symbols in text). (From Kilpatrick and Wilson, 1989.)

The characteristics of the time-concentration curves along a streamline shown in Fig. 2.14 may be described in terms of elapsed time after an instantaneous dye injection. In addition to their peaks, the tracer cloud plots have a leading and a trailing edge of low conductivity (for salt) or low concentration (for dyes) approaching the baseline conductivity (concentration) of the flowing water. Definitions of different elapsed times shown in Fig. 2.14 are as follows:

T_{L} is elapsed time to the arrival of the leading edge of the response curve at a sampling point.
T_{P} is elapsed time to the peak concentration, C_{P}, of the response curve at a point.
T_{C} is elapsed time to the centroid of the response curve at a point.
T_{t} is elapsed time to the trailing edge of the response curve at a point.
t_{L} is travel time of the leading edge.
t_{P} is travel time of the peak concentration.

The mean travel time for the flow along a streamline is the difference in elapsed time of the centroids of the time-concentration curves defined upstream and downstream on the *same streamline*:

$$t_{\mathrm{c}} = T_{\mathrm{C}_{(n+1)}} - T_{\mathrm{C}n} \tag{2.15}$$

where n is the number of the sampling site.

The time, T_{d}, necessary for the response to pass a sampling point in a section is

$$T_{\mathrm{d}} = T_{\mathrm{t}_n} - T_{\mathrm{L}n} \tag{2.16}$$

The tracer-velocity-area method in open channels has limitations and drawbacks. The air entrained by surface velocities and spray above the surface may hinder visual detection of the position of the center of mass of the colored water in high-velocity flows (Hall, 1943). Also, slow flows are more likely to cause mixing problems.

2.5.2 Constant-Rate Tracer Injection Method

The constant-rate injection method generally is preferred to the slug-injection method in spite of the additional requirement for a device that will release the tracer at small, constant rates. Various such devices are available commercially or can be fabricated. Graphs of concentration versus time for three locations in a cross-sectional channel area are shown in Fig. 2.15. As can be seen, plateaus develop earlier in the main flow than they do close to the streambanks, where the flow is slower. Other factors to consider when selecting sample locations and times are as follows.

Short Distance: (1) Curve areas are not the same, lateral mixing is incomplete; (2) There is rapid buildup to plateaus of different levels; (3) Dilution-discharge measurement can be obtained by using discharge proportional sampling, or by discharge weighting the concentration data.

Optimum Distance: (1) Curve areas are about the same, mixing is nearly complete; (2) There is fast buildup to plateaus nearly the same in concentration, mixing is adequate; (3) Good dilution-discharge measurement can be made with moderate length injection, T_{D}.

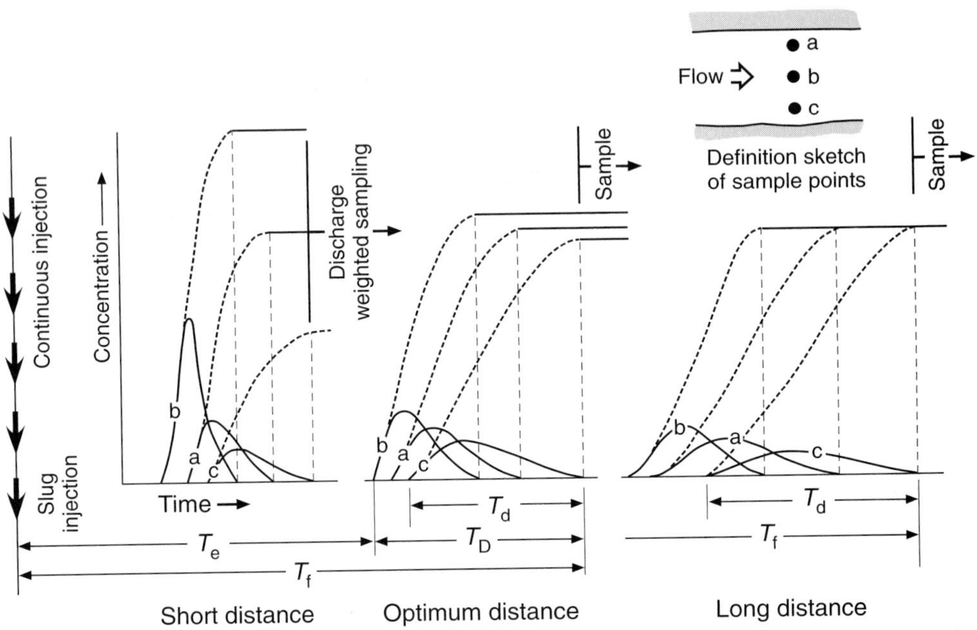

FIGURE 2.15 Time-concentration curves for slug and simulated constant injections observed at three points laterally across the channel at three different distances downstream from the injection point. (From Kilpatrick and Cobb, 1985.)

Long Distance: (1) Curve areas are identical, mixing is perfect; (2) There is slow buildup to plateau concentrations exactly the same but only after long period of injection; (3) Excellent dilution-discharge measurement can be made if injection is long enough to allow sampling after slowest point in section has reached an equilibrium plateau.

General: Minimum length of constant injection required, T_D, is equal to longest time slug dye cloud would be present at any location in the stream. Danger is in not injecting long enough and not sampling late enough, as samples taken prior to T_f would give impression mixing was poor and (or) yield data suggesting, incorrectly, a tracer loss.

A common fallacy is that shallow, turbulent whitewater streams produce rapid mixing. Turbulence is effective in yielding rapid vertical mixing, but unless there is substantial depth, lateral dispersion occurs gradually and the mixing length may be quite long. Mixing in pool-and-riffle streams sometimes occurs in relatively short distances, although the time to reach an equilibrium plateau should not be underestimated. Streams with deadwater areas, large backflow eddies, or shallow side flows should be avoided; such conditions produce the long, delayed, attenuated tracer clouds requiring long injection times.

An advantage of the constant-rate injection method is that the entire response curves need not be measured, only the plateau concentrations. Once equilibrium plateau conditions have been reached, the conservation of mass principle and the continuity equation are used to calculate the stream discharge rate:

$$QC_0 + qC_1 = (Q + q)C_2 \tag{2.17}$$

$$Q = q\frac{C_1 - C_2}{C_2 - C_0} \tag{2.18}$$

where C_0 is the natural or background concentration of the tracer, C_1 is the concentration of the injected strong tracer solution, C_2 is the equilibrium, or plateau concentration, averaged or weighted across the sampling section, including the background concentration of the stream, Q is the discharge being measured, and q is the discharge of the strong solution injected into the flow.

The required variables and Eq. (2.18) show that the constant-rate injection method does not need measurement of channel geometry or time measurement. Only the final concentration plateau value, C_2, at a downstream location must be recorded rather than a complete record of the passing cloud that is needed with the velocity-area method.

Eq. (2.18) can also be expressed in a more usable form assuming negligible natural background of the tracer in the stream:

$$Q = 5.89 \times 10^{-7} q \frac{C_1}{C_2} \tag{2.19}$$

where the concentrations are given in micrograms per liter, the rate of injection of the tracer, q, in milliliters per minute, and the discharge of the stream is calculated in cubic feet per second.

The lines in Fig. 2.16 are computed from Eq. (2.19) for a plateau concentration, C_2, of 10 μg/L. A dye concentration of 10 μg/L in the stream is well above background levels commonly experienced. In most cases, a good dye dilution-discharge measurement can be made with concentrations as low as 2 μg/L; however, higher concentrations are preferred, to lessen the significance of background levels and thus to improve accuracy. Figure 2.16 is used only to estimate the injection rate needed; the allowable latitude in

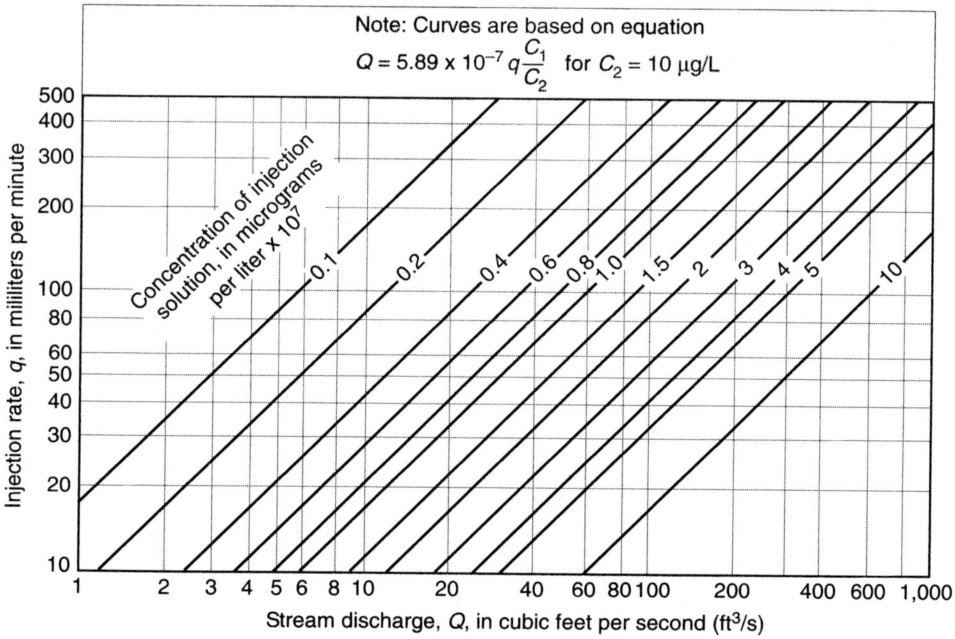

FIGURE 2.16 Graph used for estimating rates of dye injections for different stream discharges and concentrations of injection solutions. (From Kilpatrick and Cobb, 1985.)

C_2 virtually guarantees a successful measurement, no matter how poor the estimate may be. Furthermore, if 2 μg/L is viewed as a lower limit for C_2, discharges five times those in Fig. 2.16 may be measured, using the concentrations and injection rates shown.

As discussed by Kilpatrick and Cobb (1985), experience shows that measurement errors are often falsely attributed to dye losses. As can be seen from Eq. (2.19), a dye loss would produce a low value for C_2, and the computed discharge would be high. In many instances, failure to inject long enough or sampling too soon results in low values of C_2, particularly along the stream banks. Using a sufficiently long injection and waiting long enough to sample will permit the slowly developing plateaus along the banks to stabilize. Dye losses do occur, but the above cautions need to be emphasized.

Real dye losses occur where there is loss of surface water to the underlying aquifer; where excessively long stream reaches are involved; where there is finely suspended sediment, particularly clay or flocculent organic particles; and where chemicals like chlorine oxidize or quench the dye. To avoid quenching, measurements should not normally be made on streams having a pH less than five. Where a low pH does exist, the use of Pontacyl Pink dye is advisable. Dye losses have also been noted on very turbulent streams. The high oxygen levels may oxidize the dye in the same way that chlorine does.

If the stream to be measured may contain substances that can cause a dye loss, a gallon of the stream water should be obtained for laboratory use in preparing a standard to be compared with a duplicate one prepared from distilled water. The laboratory work must be done promptly, because the chemistry of the river water may change if it is stored too long.

2.5.3 Applications of Tracer-Dilution Method

Tracer dilution gauging is one of the most precise methods for measuring flow of smaller streams. In addition, when accompanied with contaminant chemical analyses, it is very

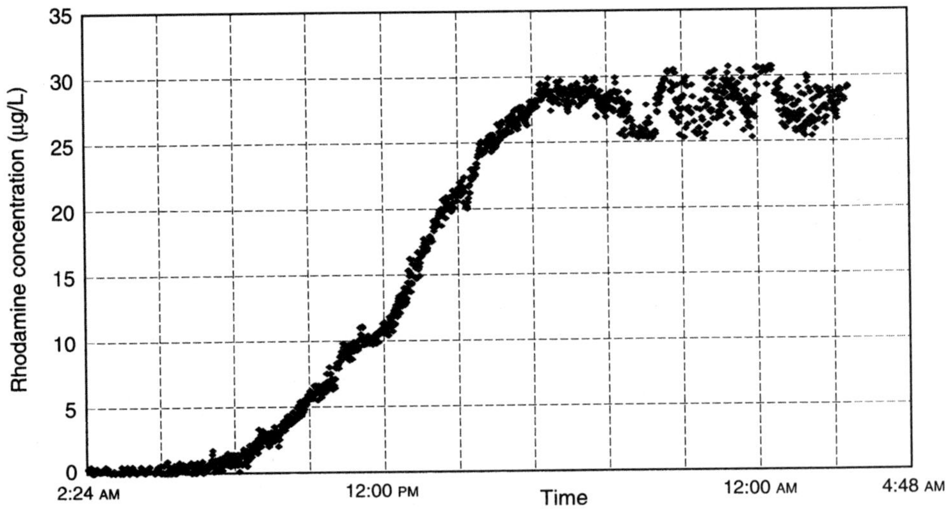

FIGURE 2.17 Rhodamine concentration at the downgradient location during an in-stream dye tracing study used to determine baseflow in a small stream. (Diagram courtesy of Lisa Pfau and Larry Neal, AMEC.)

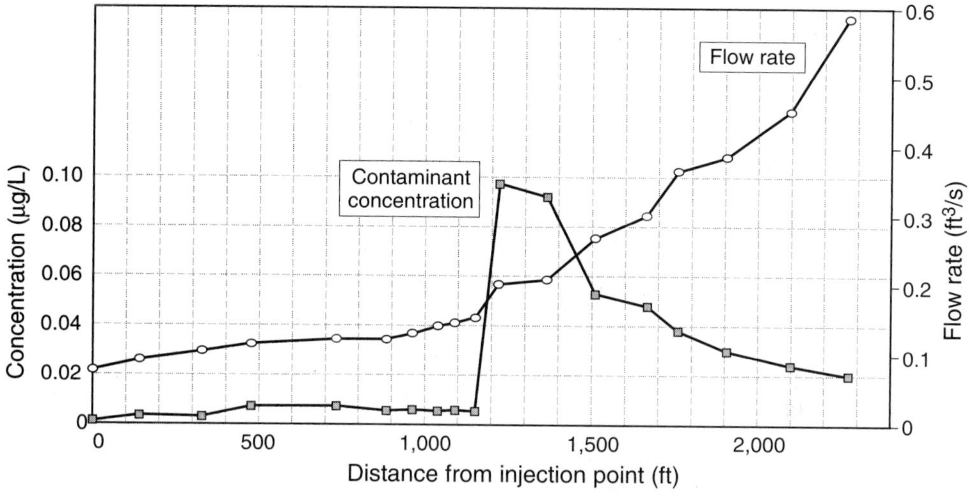

FIGURE 2.18 Contaminant concentration and streamflow rate along a stream segment determined from an in-stream dye tracer study. (Diagram courtesy of Lisa Pfau and Larry Neal, AMEC.)

useful when determining if and where there is an increased inflow of contaminated groundwater. Figures 2.17 and 2.18 show results of a dye tracer study used to assess the dry-weather baseflow of a stream and groundwater seepage inflows to the stream segment. The rhodamine dye was injected at the most upstream location of the study segment at a constant rate and constant known concentration. A water quality meter with a rhodamine sensor was placed in-stream at the most downstream sampling location and programmed to continuously record rhodamine concentrations. The dye was injected continuously into the stream until the most downstream dye concentrations reached a plateau. Note that if salt solution were used as the tracer, one would measure in-stream conductivity at the downstream location. Once dye concentrations along the stream reached a plateau, surface water samples were collected at each sampling location and analyzed for rhodamine concentrations, as well as concentrations of a constituent of concern (COC). Streamflow rate was determined for each stream sampling location using the rhodamine concentration based on the observed dilution of the tracer. As can be seen in Fig. 2.18, near the middle of the study reach the concentration of the COC increases notably and then decreases gradually indicating an influx of impacted groundwater to the stream is occurring along a preferential flow path through a short stream segment.

2.6 Hydrograph Analysis

The following discussion is based primarily on the information presented in Kresic (1997, 2007, 2009), and Kresic and Bonacci (2010). Hydrograph of a surface stream or spring discharge is the final result of various processes that govern the transformation of precipitation and other water inputs in the drainage area into the stream and spring flows. In many cases, discharge hydrograph of a karst spring closely resembles hydrographs of surface streams, particularly if the aquifer is unconfined and reacts relatively quickly

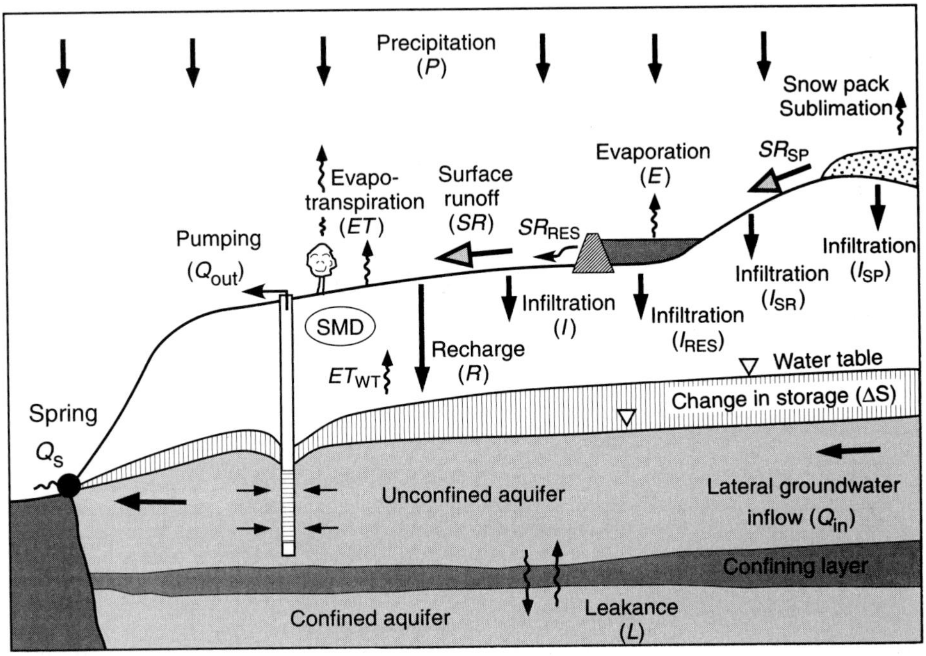

FIGURE 2.19 Elements of water budget in a spring drainage area ("springshed") (explanation in text). (Modified from Kresic, 2009.)

to water inputs. Just like in case of surface streams, discharge of many karst springs can increase several times or even orders of magnitude in the matter of hours after heavy rains. On the other hand, deep-ascending springs (often, thermal) that are isolated from a direct influence of surficial processes such as infiltration of precipitation may show only slight, delayed seasonal changes in discharge characteristics.

Figure 2.19 shows a number of possible springwater budget components that, to varying degree, may influence the spring discharge (Q_S). If the spring is replaced by a surface stream, Q_S would become a direct groundwater discharge into the stream or the so-called stream baseflow. Two additional components contributing to the streamflow would also have to be added: overland runoff and near-surface runoff or interflow. In general, the water budget of a surface stream watershed or a spring drainage area (springshed) states that the rate of change in water stored in the area is balanced by the rate at which water flows into and out of the area:

$$\text{Water Input} - \text{Water Output} = \text{Change in Storage} \tag{2.20}$$

The storage relates to three general hydrologic reservoirs: surface reservoirs including wetlands, the vadose zone, and the saturated zone. The water budget equations can be written in terms of volumes (for a fixed time interval), fluxes (volume per time, such as cubic meters per day), and flux densities (volume per unit area of land surface per time, such as millimeters per day).

Common to most components of the water budget is that they cannot be measured directly and have to be estimated from measurements of related quantities (parameters) and estimates of other components. Exceptions are direct measurements of total precipitation, streamflows, spring discharge rates, and well pumping rates. Other important quantities that can be measured directly and used in water budget calculations as part of various equations are the hydraulic head (water level) of both groundwater and surface water, and soil moisture content.

The complexity of the water budget determination depends on many natural and anthropogenic factors in the general area of interest, such as climate, hydrography and hydrology, geologic and geomorphologic characteristics, hydrogeologic characteristics of the surficial soils and subsurface porous media, land cover and land use, presence and operations of artificial surface water reservoirs, surface water and groundwater withdrawals for consumptive use and irrigation, and wastewater management. Following are some of the relationships between the components shown in Fig. 2.19 that can be utilized in quantitative water budget analyses of such a system:

$$\begin{aligned} I &= P - SR - ET \\ I &= I_{SR} + I_{RES} + I_{SP} \\ R &= I - SMD - ET_{WT} \\ Q_S &= R + Q_{IN} + L - \Delta S - Q_{OUT} \end{aligned} \tag{2.21}$$

where I is infiltration, in general, SR is surface water runoff, ET is evapotranspiration, I_{SR} is infiltration from surface runoff (including from sinking and loosing streams), I_{RES} is infiltration from surface water reservoirs, I_{SP} is infiltration from snow pack and glaciers, R is groundwater recharge, SMD is soil moisture deficit, ET_{WT} is evapotranspiration from water table, Q_S is spring discharge, Q_{IN} is lateral groundwater inflow to the aquifer feeding the spring, L is leakance from/to the underlying aquitard/aquifer, Q_{OUT} is well pumpage from the aquifer, and ΔS is change in storage of the aquifer. If the area is irrigated, yet one more component would be added to the list: infiltration of the irrigation water.

Ideally, most of the above relationship would have to be established to fully quantify the processes governing the water budget of a groundwater system feeding the spring or the surface stream including volumes of water stored in, and flowing between the three general reservoirs—surface water, vadose zone, and saturated zone. By default, a change in one of the many water budget components causes a chain reaction and thus influences all other components. These changes take place with more or less delay, depending on both the actual physical movement of water and the hydraulic characteristics of the three general reservoirs.

Transformation of precipitation into flow in channels has been the primary focus of applied surface water hydrologists for decades. As mentioned before, because hydrographs of many karst springs closely resemble hydrographs of surface streams, most if not all of the quantitative methods developed in surface water hydrology are applicable to karst hydrogeology as well. The following discussion that draws parallels between these two closely related systems (surface streams and karst springs) is based on the groundbreaking publication by James Dooge entitled *Linear Theory of Hydrologic Systems* and published by the United States Department of Agriculture (Dooge, 1973; see also Dooge and O'Kane, 2003). In subsequent years, the principles and mathematical models described by Dooge and other surface water hydrologist have been increasingly implemented by karst hydrogeologists.

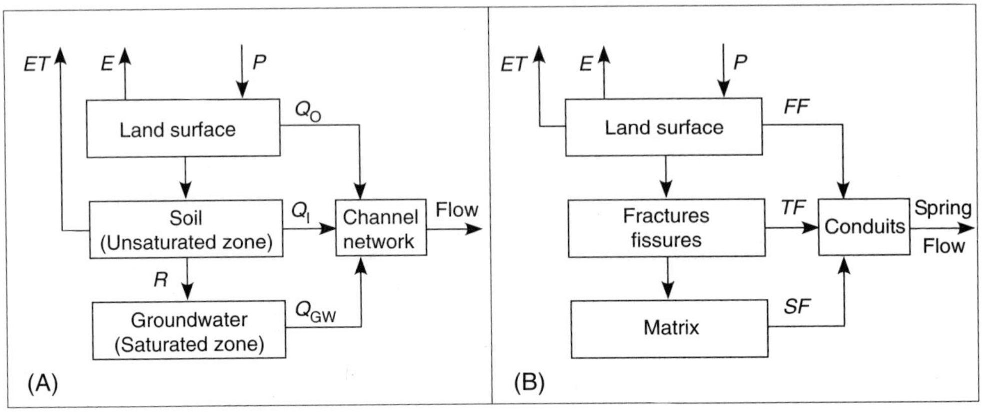

FIGURE 2.20 (A) Drainage area (catchment) of a surface stream as a hydrologic system. *P*, precipitation; *ET*, evapotranspiration; *R*, recharge; Q_O, overland flow; Q_I, interflow; Q_{GW}, baseflow from groundwater (modified from Dooge, 1973). (B) Aquifer drained by a spring as a hydrologic system. *FF*, fast flow (conduits); *TF*, transitional flow (fractures); *SF*, slow flow (matrix, small fissures).

The distinction shown in Fig. 2.20A between overland flow, interflow, and groundwater flow is not generally made in applied surface water hydrology. The analogy of this scheme in the hydrology of karst springs is shown in Fig. 2.20B. Overland flow, the fastest component of the streamflow generated after rainfall, is equivalent to the conduit type flow in karst aquifers. Interflow is equivalent to fracture/fissure systems, and groundwater (baseflow of surface streams) is equivalent to the matrix storage with diffuse flow or spring baseflow component of the overall spring discharge.

Practicing hydrologists distinguish between surface streamflow and baseflow and use a simplified model of the hydrologic cycle shown in Fig. 2.21A. The analogy with springs is shown in Fig. 2.21B. The precipitation is divided into (1) precipitation excess and (2) infiltration and other losses. The precipitation excess produces direct storm runoff at the land surface. In hydrology of karst springs, the precipitation excess is equivalent

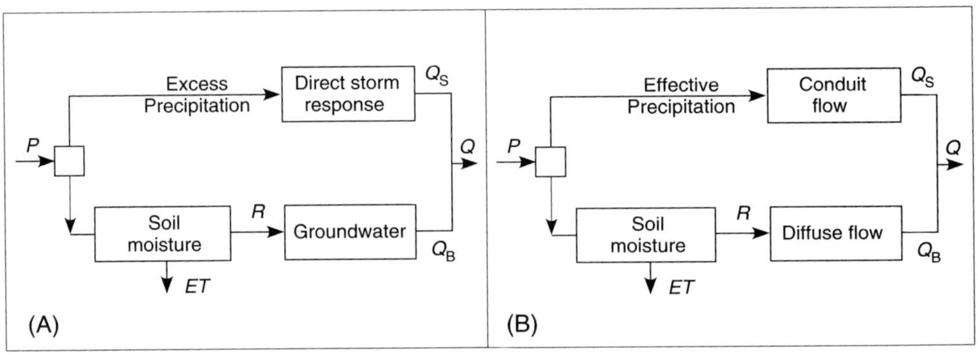

FIGURE 2.21 (A) The simplified surface stream catchment model. Q_S, storm response; Q_B, baseflow; *P*, precipitation; *ET*, evapotranspiration; *R*, recharge of groundwater; *Q*, surface streamflow. (Modified from Dooge, 1973.) (B) Equivalent aquifer-spring model. Q, spring flow.

to rapid infiltration to the saturated zone that produces the first response at the spring. The infiltration replenishes soil storage that is drawn down upon by transpiration. Any excess infiltration after soil moisture is satisfied forms recharge to groundwater, which eventually emerges as baseflow. In hydrology of springs, "groundwater" is storage in the aquifer matrix, and baseflow is flow of springs during recession periods.

As discussed by Healy et al. (2007) and applied to karst, an understanding of the water budget and underlying hydrologic processes provides a foundation for effective water management. Observed changes in the water budget over time can be used to assess the effects of climate variability and human activities on the available water resource. Comparison of surface stream or springwater budgets from different areas allows the effects of factors such as geology, soils, vegetation, and land use to be quantified. Human activities affect the natural hydrologic cycle in many ways. Modifications of the land to accommodate agriculture, such as installation of drainage and irrigation systems, alter infiltration, runoff, evaporation, and plant transpiration rates. Buildings, roads, and parking lots in urban areas tend to increase runoff and decrease infiltration.

Because stream and spring discharge rate is one of the few water budget elements that can be measured directly, the analysis of hydrographs provides a means for evaluating availability and sustainability of a water supply. In addition, a thorough analysis of discharge hydrographs in karst often provides useful information on the aquifer characteristics such as the nature of its storage and transmissivity.

Figure 2.22 shows the main elements of a discharge hydrograph. The beginning of discharge after a rainfall episode is marked with point *A*, and the time between the beginning of rainfall and the beginning of discharge, called *the starting time*, with t_s. The time in which the hydrograph raises to its maximum (point C) is called *the concentration time*—t_c. The time from maximum discharge until the end of the hydrograph, when the discharge theoretically equals zero (point E), is *the falling time*—t_f. Together, the concentration time and the falling time are called *the base time* of the hydrograph—t_b. The time between the centroid of the precipitation episode (C_P) and the centroid of the hydrograph (C_H) is called

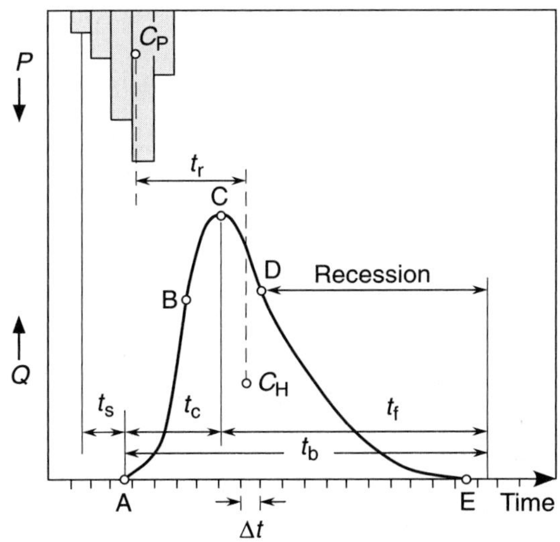

FIGURE 2.22 Components of a discharge hydrograph (explanation in text).

the retardation time (t_r). The time interval for recording the amount of precipitation and the flow rate at the spring is Δt.

The shape of the hydrograph is defined by its base (AE), the rising limb (AB), the crest (BCD), and the falling limb (DE). The falling limb corresponds to the *recession period*. B and C are *inflexion points* where the hydrograph curve changes its shape from convex to concave and vice versa. For surface streams, point D is the end of direct runoff after the rain. In general, the part of the hydrograph from point D is called the *recession curve*. The shape of a discharge hydrograph depends on the size and shape of the drainage area, as well as the precipitation intensity. When a rainfall episode lasts longer, and the intensity is lower, the hydrograph has a longer time base and vice versa: intensive short storm events cause sharp hydrographs with short time bases. The area under the hydrograph is the volume of discharged water for the recording period. In reality, unless a spring or a surface stream is intermittent, the recorded hydrograph has a more complex shape, which reflects the influence of antecedent precipitation and other possible water inputs. Such hydrographs are formed by the superposition of single hydrographs corresponding to separate precipitation events (Fig. 2.23) and other water inputs such as arrival of water from a sinking stream.

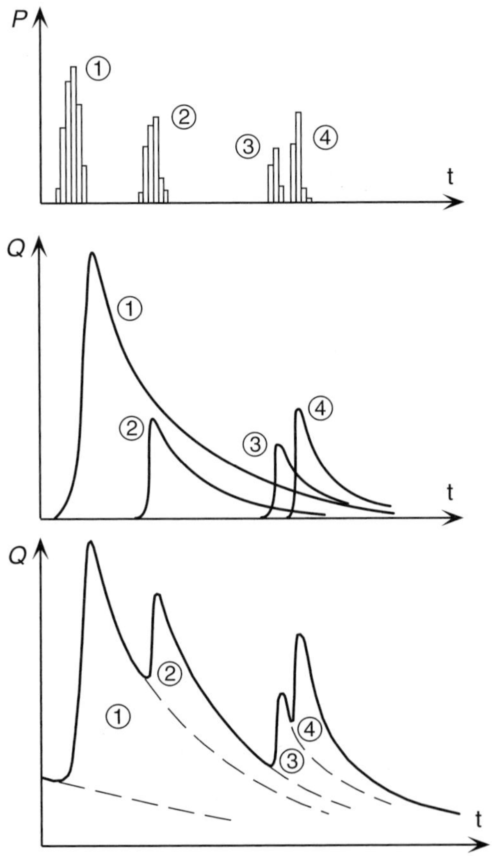

FIGURE 2.23 Complex discharge hydrograph (*bottom*) composed of single hydrographs (middle) as a result of several rainfall events (*top*). (Modified from Jevdjevic, 1956.)

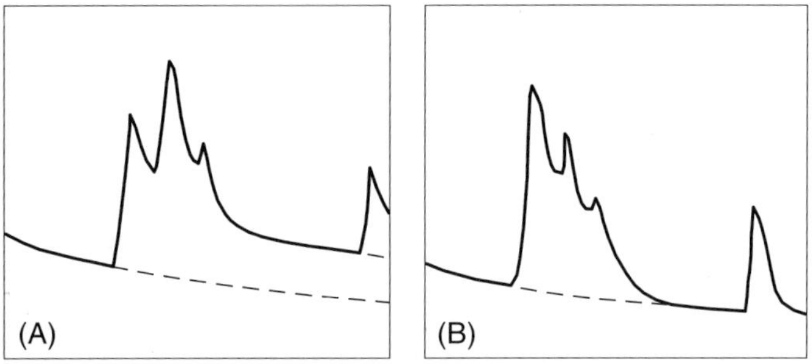

FIGURE 2.24 Possible reaction of karst aquifers to water inputs such as infiltration of precipitation as seen from spring discharge hydrographs. (A) Lifting of the baseflow to a higher hydrograph level because of the increased storage. (B) Discharge without increase in storage. (From Kresic, 1991.)

The impact of newly infiltrated water on spring discharge varies with respect to the predominant type of porosity and the stage of groundwater level. However, the first reaction to recharge of karst aquifers developed in open hydrogeologic structures is in many cases the consequence of pressure propagation through karst conduits and/or large fractures, and not the outflow of newly infiltrated water (see Fig. 1.144). The new water arrives at the spring with certain delay and its contribution is a fraction of the overall flow rate. After a recharge episode and the initial response of the system are over, two cases may occur as illustrated in Fig. 2.24: (a) an increase in volume of groundwater accumulated in storage, which is reflected in the shifting of the recession curve to a higher hydrograph level, and (b) new water is mostly transmitted through a well-developed network of fractures or karst channels/conduits, and is discharged without significant accumulation of groundwater in the surrounding matrix porosity. The recession curve continues along the same extrapolated line as before the rainfall event. The first case (a) is common during periods of main aquifer recharge, such as March through June in moderate climates in the Northern Hemisphere: both groundwater level and moisture content in the unsaturated zone are high, while evapotranspiration loss is relatively small compared to peak vegetative period of the year. Newly infiltrated water raises an already high hydraulic head in the aquifer and the groundwater is more easily injected into the aquifer matrix porosity and narrow fissures, including from the conduits. The second case (b) may reflect the summer/autumn period when the hydraulic heads and the hydraulic gradients are low and the newly infiltrated water from summer storms is transmitted quickly through well-connected large fractures and/or conduits. This case may also indicate lack of significant matrix/fissure porosity in the aquifer.

2.6.1 Recession Analysis

Analysis of the falling hydrograph limb shown in Fig. 2.25, which corresponds to a period without significant precipitation, is called the *recession analysis*. Knowing that the spring or surface water discharge is without disturbances caused by a rapid inflow of new water into the aquifer, the recession analysis provides good insight into the aquifer structure. By establishing an appropriate mathematical relationship between spring discharge and time, it is possible to predict the discharge rate after a given period without precipitation, and to calculate the volume of discharged water. For these reasons, recession analysis has been a popular quantitative method in spring discharge analysis for a long time.

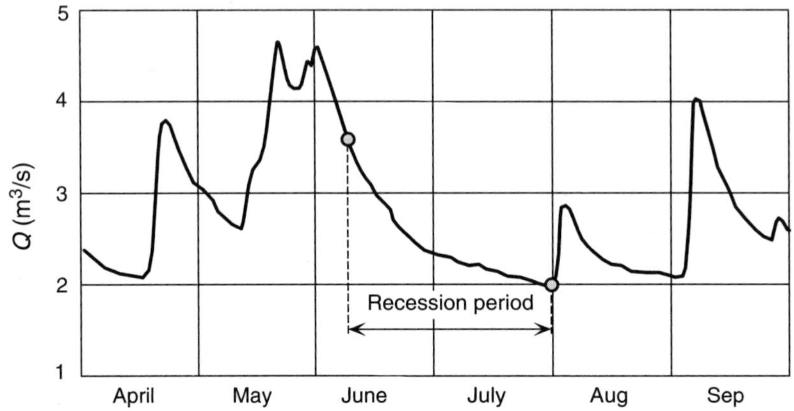

FIGURE 2.25 Spring discharge hydrograph with a recession period. (From Kresic, 1991.)

The ideal recession conditions—a long period of several months without precipitation—are rare in moderate/humid climates. Consequently, frequent precipitation can cause various disturbances in the recession curve that may not be removed unambiguously during analysis. It is therefore desirable to analyze as many recession curves from different years as possible. Larger sample allows for derivation of an average recession curve as well as the envelope of minima (Fig. 2.26) that enables a more accurate quantification of the expected long-term minimum discharge.

Two well-known mathematical formulas that describe the falling limb of hydrographs and the baseflow are proposed by Boussinesq (1904) and Maillet (1905). Both

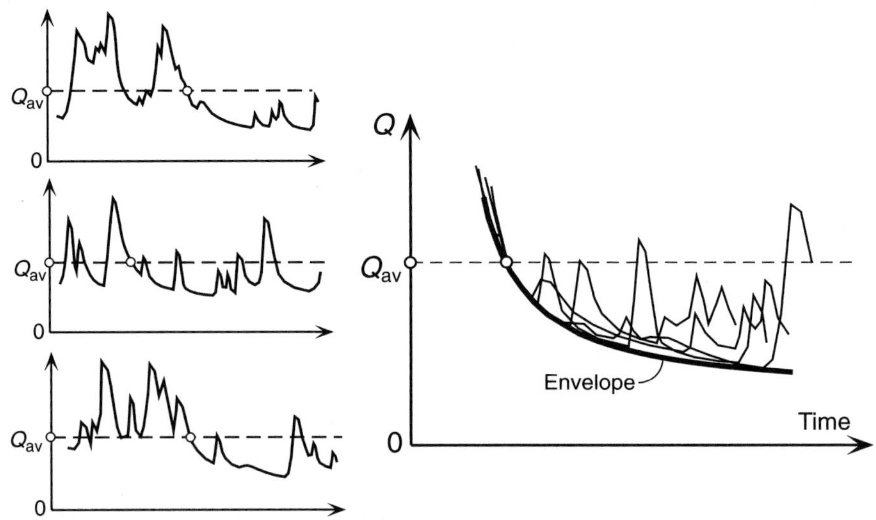

FIGURE 2.26 Three annual discharge hydrographs with main recession curves overlaid at the point of average discharge rate. The minima on the three overlaid curves are connected with the envelope that represents a long-term recession curve. (From Kresic, 2007.)

equations give dependence of the flow at specified time (Q_t) on the flow at the beginning of recession (Q_0). The Boussinesq equation is of hyperbolic form:

$$Q_t = \frac{Q_0}{[1 + \alpha(t - t_0)]^2} \tag{2.22}$$

where t is time since the beginning of recession for which the flow rate is calculated; t_0 is time at the beginning of recession usually (but not necessarily) set equal to zero.

The Maillet equation, which is more commonly used, is an exponential function:

$$Q_t = Q_0 \cdot e^{-\alpha(t-t_0)} \tag{2.23}$$

The dimensionless parameter α in both equations represents the *coefficient of discharge* (or *recession coefficient*) that depends on the aquifer transmissivity and specific yield. The Maillet equation, when plotted on a semilog diagram, is a straight line with the coefficient of discharge (α) being its slope:

$$\log Q_t = \log Q_0 - 0.4343 \cdot \alpha \cdot \Delta t$$
$$\Delta t = t - t_0 \tag{2.24}$$

$$\alpha = \frac{\log Q_0 - \log Q_t}{0.4343 \cdot (t - t_0)} \tag{2.25}$$

Introduction of the conversion factor (0.4343) is a convenience for expressing discharge in Eq. (2.25) in cubic meters per second, and time in days. Dimension of α is day^{-1}.

Figure 2.27 is a semilog plot of time versus discharge rate for the recession period shown in Fig. 2.25. The recorded daily discharges form three straight lines which mean

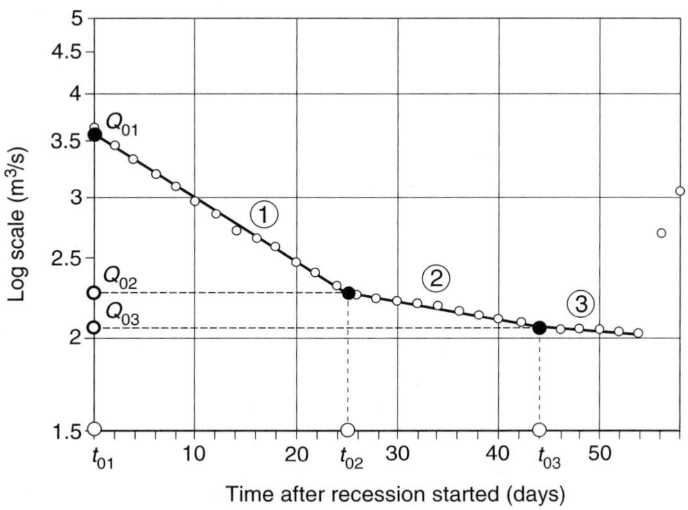

FIGURE 2.27 Semilog graph of time versus discharge for the recession period shown in Fig. 2.25. The duration of the recession period is 54 days. (Modified from Kresic, 1991.)

that the recession curve can be approximated by three corresponding exponential functions with three different coefficients of discharge (α). The three lines correspond to three *micro-regimes of discharge* during the recession. The coefficient of discharge for the first micro-regime, using Eq. (2.25), is (Kresic, 2007):

$$\alpha_1 = \frac{\log Q_{01} - \log Q_{02}}{0.4343 \cdot (t_{01} - t_{02})}$$

$$\alpha_1 = \frac{\log(3.55 \text{ m}^3/\text{s}) - \log(2.25 \text{ m}^3/\text{s})}{0.4343 \cdot 24.5 \text{ days}} = 0.019$$

The coefficient of discharge for the second micro-regime is:

$$\alpha_2 = \frac{\log(2.25 \text{ m}^3/\text{s}) - \log(2.06 \text{ m}^3/\text{s})}{0.4343 \cdot (44 \text{ days} - 24.5 \text{ days})} = 0.0045$$

The third coefficient of discharge, or slope of the third straight line, is found by choosing a discharge rate anywhere on the line, including its extension if the actual line is short. In our case, the discharge rate after 60 days (not shown on the graph) is 2.01 m^3/s, and α_3 is then

$$\alpha_3 = \frac{\log(2.06 \text{ m}^3/\text{s}) - \log(2.01 \text{ m}^3/\text{s})}{0.4343 \cdot (60 \text{ days} - 44 \text{ days})} = 0.0015$$

After determining the coefficients of discharge, the flow rate at any given time after the beginning of recession can be calculated using the appropriate Maillet equation. For example, discharge of the spring 35 days after the recession started, when the second micro-regime is active, is calculated as

$$Q_{35} = Q_{02} \cdot e^{-\alpha_2(35 \text{ days} - t_{02})}$$

$$Q_{35} = 2.25 \text{ m}^3/\text{s} \cdot e^{-0.0045 \cdot (35 \text{ days} - 24.5 \text{ days})} = 2.146 \text{ m}^3/\text{s}$$

Note that the initial discharge rate for the second micro-regime, Q_{02}, is 2.25 m^3/s and the corresponding time is 24.5 days.

Spring discharge 3 months (90 days) after the beginning of recession, assuming no precipitation for the entire period, may be predicted by using the characteristic values for the third micro-regime (see Fig. 2.27), where Q_{03} is the initial discharge for that regime:

$$Q_{90} = Q_{03} \cdot e^{-\alpha_3(90 \text{ days} - t_{03})}$$

$$Q_{90} = 2.06 \text{ m}^3/\text{s} \cdot e^{-0.0015(90 \text{ days} - 44 \text{ days})} = 1.923 \text{ m}^3/\text{s}$$

The coefficient of discharge (α) and the volume of free gravitational groundwater stored in the aquifer above spring level (i.e., groundwater that contributes to spring discharge) are inversely proportional:

$$\alpha = \frac{Q_t}{V_t} \tag{2.26}$$

where Q_t is the discharge rate at time t, and V_t is the volume of water stored in the aquifer above the level of discharge (spring level).

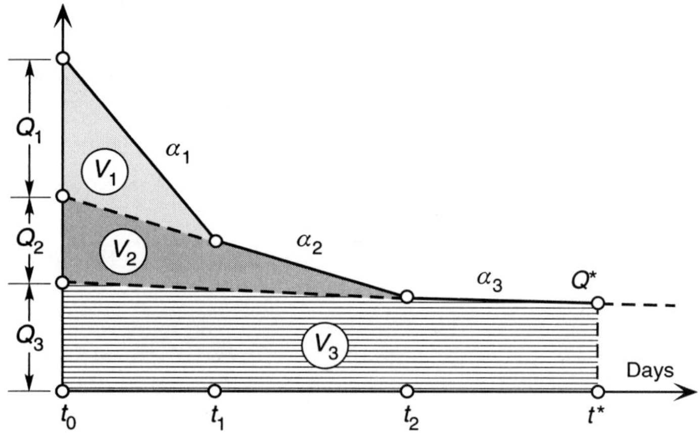

FIGURE 2.28 Schematic presentation of the recession with three micro-regimes of discharge, and the corresponding volumes of discharged water. (From Kresic, 2009.)

Equation (2.26) allows calculation of the volume of water accumulated in the aquifer at the beginning of recession, as well as the volume discharged during a given period of time. The calculated remaining volume of groundwater always refers to the reserves stored above the current level of discharge. The draining of an aquifer with three micro-regimes of discharge (as in our case) and the corresponding volumes of the discharged water are shown in Fig. 2.28. The total initial volume of groundwater stored in the aquifer (above the level of discharge) at the beginning of the recession period is the sum of the three volumes that correspond to three different types of storage (effective porosity):

$$V_0 = V_1 + V_2 + V_3 = \left[\frac{Q_1}{\alpha_1} + \frac{Q_2}{\alpha_2} + \frac{Q_3}{\alpha_3}\right] \cdot 86400\,\text{s}\,[m^3] \tag{2.27}$$

where discharge rates are given in m^3/s.

The volume of groundwater remaining in the aquifer at the end of the third micro-regime is the function of the discharge rate at time t^* and the coefficient of discharge α_3:

$$V^* = \frac{Q^*}{\alpha_3} \tag{2.28}$$

The difference between volumes V_0 and V^* is the volume of all groundwater discharged during the period $t^* - t_0$. In our case, the volume of groundwater stored in the aquifer at the beginning of recession is

$$V_0 = \left[\frac{(Q_{01} - Q_{02})}{\alpha_1} + \frac{Q_{02} - Q_{03}}{\alpha_2} + \frac{Q_{03}}{\alpha_3}\right] \cdot 86400\,\text{s}\,[\text{m}^3]$$

$$V_0 = \left[\frac{(3.55\ \text{m}^3/\text{s} - 2.55\ \text{m}^3/\text{s})}{0.019} + \frac{(2.55\ \text{m}^3/\text{s} - 2.20\ \text{m}^3/\text{s}}{0.0045} + \frac{2.20\ \text{m}^3/\text{s}}{0.0015}\right] \times 86400\,\text{s}$$

$$V_0 = 4.547 \times 10^6\ \text{m}^3 + 6.720 \times 10^6\ \text{m}^3 + 1.267 \times 10^8\ \text{m}^3 = 1.380 \times 10^8\ \text{m}^3$$

As can be seen, most water is stored in the storage that drains slowly in the third micro-regime.

The volume of water remaining in the aquifer above the spring level at the end of recession is

$$V^* = \frac{2.03\ \text{m}^3/\text{s}}{0.0015} \cdot 86400\ \text{s} = 1.169 \times 10^8\ \text{m}^3$$

which gives the following volume of water discharged at the spring for the duration of recession (54 days):

$$V = V_0 - V^* = 1.380 \times 10^{-8} - 1.169 \times 10^8\ \text{m}^3 = 21.1 \times 10^6\ \text{m}^3$$

Recession periods of large perennial karstic springs often have two or three micro-regimes of discharge as in this example. However, the recession curve can vary greatly from spring to spring and its shape can have various physical meanings.

It has been argued that, for karst springs, the initial steep portion of the curve represents turbulent drainage of large fractures and conduits (e.g., first micro-regime in Fig. 2.27), followed by a transitional portion of the curve where the flow is less turbulent and reflects contribution of smaller fractures and rock matrix (second micro-regime), and ending with the slowly decreasing curve, the so-called *master recession curve* where the drainage of rock matrix and small fissures is dominant. This last portion of the curve is also the most important for predicting future discharge in a complete absence of water inputs, that is, during prolonged droughts. However, as explained further, there may be other explanations for various breaks on the recession curve.

As discussed by Bonacci (1993), every break point on the recession line is caused by a change in a characteristic of the groundwater reservoir (aquifer). Two common reasons are illustrated in Fig. 2.29. The break points on the recession curve result from the decrease in the spring drainage area, and/or from the decrease in the effective porosity of the aquifer. Figure 2.30 presents some less common shapes of the recession curve and the changes in recession coefficients. The main difference between Figs. 2.29 and 2.30 is that $\alpha_2 > \alpha_1$ in the latter, that is, the recession coefficient α increases with time. The reasons for this can be numerous. For example, Fig. 2.30B gives a schematic presentation of a catchment that consists partially of limestone and schists. The outflow through limestone is much faster than through the schists, which causes a lag in the water inflow (in the period $t_i - t_j$) from the schists, and thus a change (increase) in the slope of the recession curve in the Mikro Vuono Spring in Greece (Soulios, 1991). Figure. 2.30C gives a schematic representation of a temporarily flooded cave in a karst aquifer. The bottom of the main outlet of the cave is at height H. As long as the groundwater level is below that height, the discharge occurs very slowly or not at all. A sudden lowering of the groundwater level leads to water outflow from the flooded cave. Understandably, this is not particular to caves; it can occur in a wider aquifer area with locally higher effective porosity (Bonacci, 1993). An identical situation is presented in Fig. 2.30D. In this case, there is a temporarily flooded polje in karst instead of an underground reservoir in the cave. The roles of the cave and the polje are identical.

The initial portion of the recession curve reflecting rapid discharge may not correspond to the simple exponential expression of the Maillet's type, and may be better explained by some other functions. Deviations from exponential dependence can be

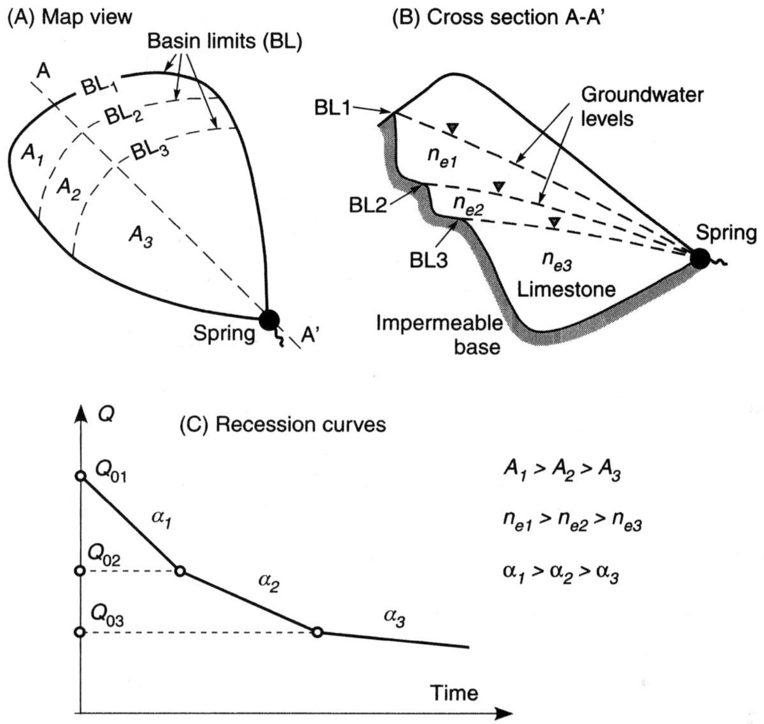

FIGURE 2.29 Map view (A) and cross-sectional view (B) of common reasons for changes in the value of recession coefficients (α) caused by changes in the size of catchment area (A) and effective porosity (n_e). (Modified from Bonacci, 1993.)

easily detected if the recorded data plotted on a semilog diagram do not form straight line(s).

Often, good approximation of the rapid drainage at the beginning of recession is the hyperbolic relation of the Boussinesq type. Its general form is

$$Q_t = \frac{Q_0}{(1+\alpha t)^n} \tag{2.29}$$

In many cases, this function correctly describes the entire recession curve. On the basis of 100 analyzed recession curves of karstic springs in France, Drogue (1972) concludes that among six exponents studied, the best first approximations of exponent n are 1/2, 3/2, and 2. The exact determination of exponent n and the discharge coefficient α for the function that best fits the measured data is performed graphically and by computation as follows (Kresic, 2007):

- The minimum recorded discharge at the end of recession is noted (Q_2= 0.057 m^3/s; see for example Fig. 2.31).
- Any discharge Q_1 on the recession curve, which is not the result of (possible) deviation due to recent rainfall, is chosen in the section between Q_2 and Q_0.

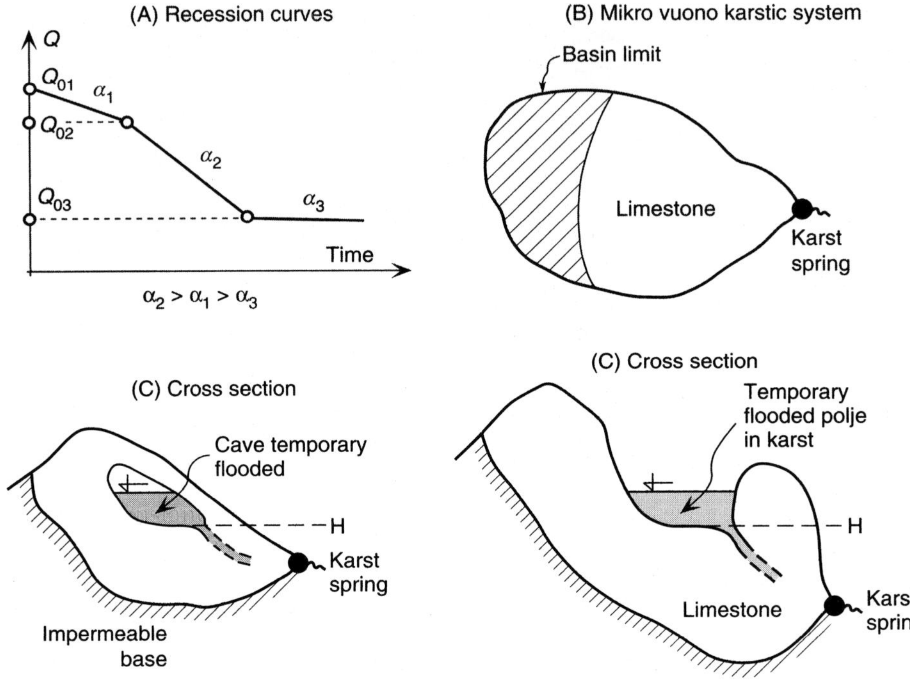

FIGURE 2.30 Explanation of other reasons for changes in the value of recession coefficients (α). (From Bonacci, 1993.)

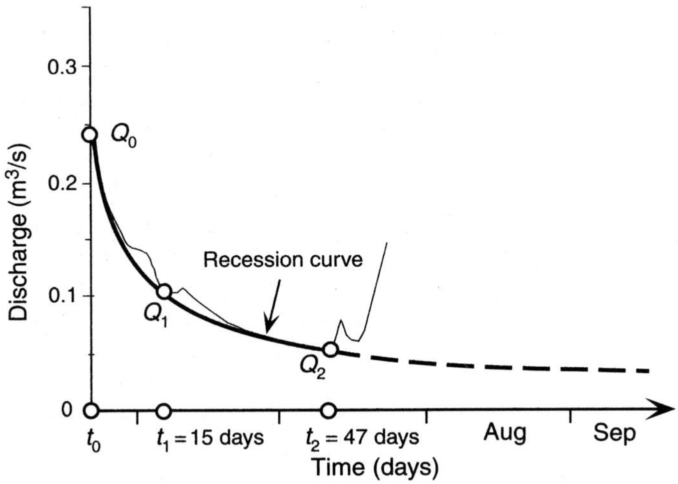

FIGURE 2.31 Recession curve of a spring with discharges used to determine parameters of a hyperbolic function. (Modified from Kresic, 2007.)

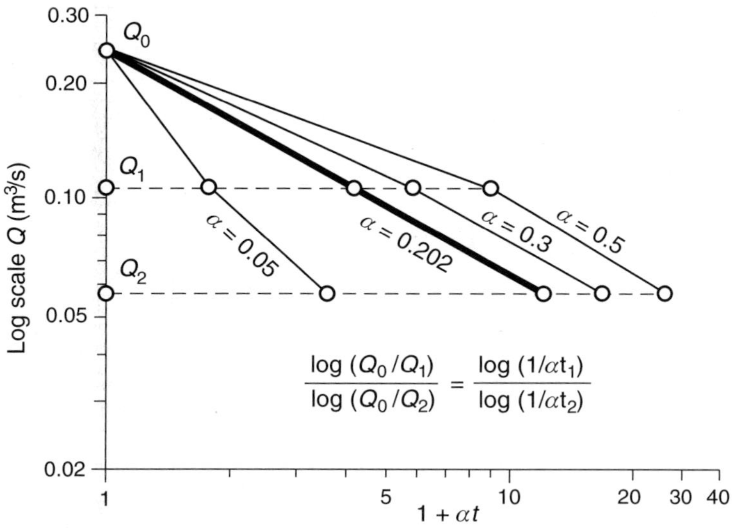

FIGURE 2.32 Graphical determination of discharge coefficient α (example shown in Fig. 2.31). (From Kresic, 2007.)

- The value of α that satisfies the following equation:

$$\frac{\log(Q_0/Q_1)}{\log(Q_0/Q_2)} = \frac{\log(1+\alpha t_1)}{\log(1+\alpha t_2)} \tag{2.30}$$

is determined by trial adopting an initial value for α (usually, 0.5). The result can be graphically checked as shown in Fig. 2.32: the correct coefficient of discharge gives the straight line through the points defined by Q_0, Q_1, Q_2, and the corresponding times t_0, t_1, t_2. The exact value of α for this example is 0.202.

$$\frac{\log(0.240\ \mathrm{m}^3/\mathrm{s}/0.105\ \mathrm{m}^3/\mathrm{s})}{\log(0.240\ \mathrm{m}^3/\mathrm{s}/0.060\ \mathrm{m}^3/\mathrm{s})} = \frac{\log(1+0.202\cdot 15\ \text{days})}{\log(1+0.202\cdot 47\ \text{days})}$$

$$0.5963 \approx 0.5929$$

The exponent n is calculated by substituting the determined value for α into either of the following two equations:

$$n = \frac{\log(Q_0/Q_1)}{\log(1+\alpha t_1)} \tag{2.31}$$

$$n = -\frac{\log(Q_1/Q_2)}{\log\left(\dfrac{1+\alpha t_1}{1+\alpha t_2}\right)} \tag{2.32}$$

For this example, Eq. (2.31) gives

$$n = \frac{\log\left(\dfrac{0.240\ \mathrm{m}^3/\mathrm{s}}{0.105\ \mathrm{m}^3/\mathrm{s}}\right)}{\log(1+0.202\times 15\ \text{days})} = 0.593$$

and the recession discharge equation is then

$$Q_t = \frac{0.24\ \text{m}^3/\text{s}}{(1 + 0.202 \cdot t)^{0.593}}$$

The coefficient of discharge α and the exponent n have the following general relationship:

$$\alpha = \frac{\sqrt[n]{Q_0} - \sqrt[n]{Q_t}}{t \cdot \sqrt[n]{Q_t}} \tag{2.33}$$

As in the case of the Maillet equation, the derived hyperbolic function can be used to calculate the volume of free gravitational water stored in the aquifer above the spring level. In general, this volume, in cubic meters, at any time t since the beginning of recession is

$$V_t = \frac{Q_0}{\alpha \cdot (n-1)} \left[1 - \frac{1}{(1+\alpha t)^{n-1}}\right] \times 86400\ \text{s} \tag{2.34}$$

2.6.2 Hydrograph Separation

One important aspect of separating baseflow of surface streams is its application in water budget and groundwater recharge studies (e.g., see Kresic and Mikszewski, 2009). In natural long-term conditions and the absence of artificial groundwater withdrawal, the rate of groundwater recharge in a drainage basin of a permanent gaining stream is equal to the rate of groundwater discharge into the stream. Assuming that all groundwater discharges into the surface stream, either directly or via springs, it follows that the stream baseflow equals the groundwater recharge in the basin. This simple concept is illustrated in Fig. 2.33. However, its application is not always straightforward, especially in karst terrains, and it should be based on a thorough understanding of the geologic and hydrogeologic characteristics of the drainage basin. The following general examples

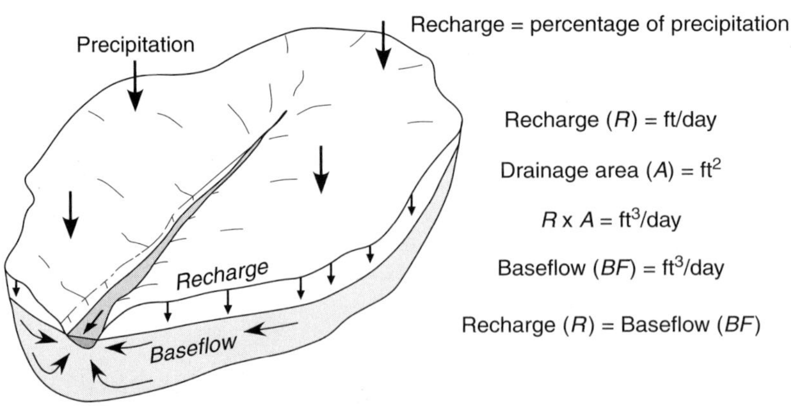

FIGURE 2.33 Estimation of aquifer recharge from surface stream baseflow. (Modified from Kresic, 2007.)

illustrate some situations where stream baseflow alone should not be used to estimate actual groundwater recharge (Kresic, 2007):

- Surface stream flows through a karst terrain where topographic and groundwater divides are not the same. The groundwater recharge based on baseflow may be grossly over- or underestimated depending on the circumstances.
- The stream is not permanent, or some river segments are losing water (either always or seasonally); locations and timing of the streamflow measurements are not adequate to assess such conditions.
- There is abundant riparian vegetation in the stream floodplain, which extracts a significant portion of groundwater via evapotranspiration.
- There is discharge from deeper aquifers, which have remote recharge areas in other drainage basins.
- A dam regulates the flow in the stream.

Most techniques for estimating baseflow are based on graphical separation of surface stream hydrographs into two major components: flow generated by surface and near-surface runoff, and flow generated by discharge of groundwater. The most common method of hydrograph component separation, shown in Fig. 2.34 with the line ABC, is applied to surface streams with significant groundwater inflow. Assuming that point C represents the end of all surface runoff, and the beginning of flow generated solely by groundwater discharge, the late near-straight line section of the hydrograph is extrapolated backward until it intersects the ordinate of the maximum discharge (point B). Point A, representing the beginning of surface runoff after rainfall, and point B are then connected with the straight line. The area under the line ABC is the baseflow, or the groundwater component of the surface streamflow. Obviously, for spring hydrographs, this area does not have an equivalent explanation since all discharged water is by definition groundwater. However, the portion of a spring hydrograph that corresponds to the baseflow component of a surface stream hydrograph could partially be attributed to a steady release of water from smaller voids (small fissures, matrix porosity) and partially to the release of water stored in the matrix due to additional pressure from the newly infiltrated water.

FIGURE 2.34 Two common methods of graphical hydrograph separation into flow components.

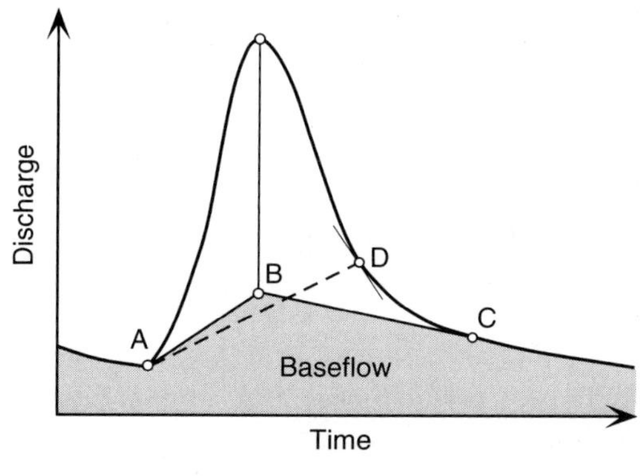

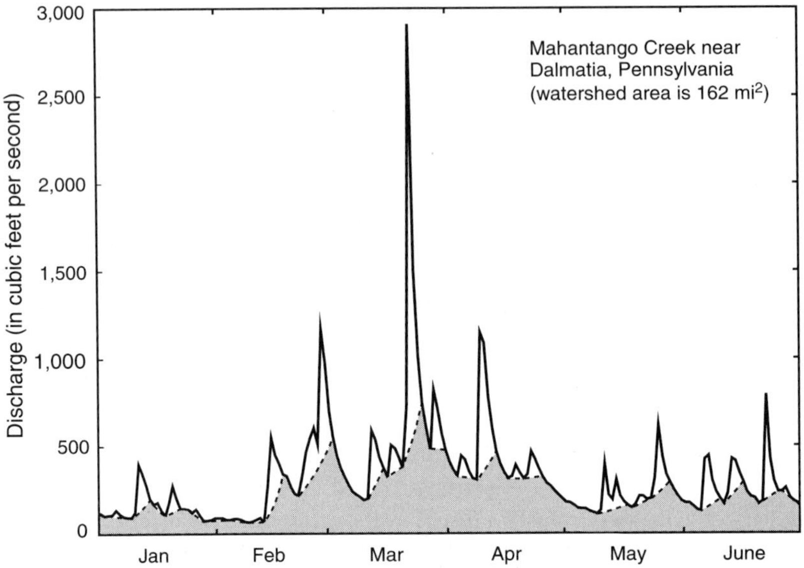

FIGURE 2.35 Streamflow hydrograph with separated baseflow component using PART program. (From Risser et al., 2005.)

Another method of visual (graphical) baseflow separation shown with the line ADC is somewhat more subjective as it involves selecting point D where the tangent to the falling hydrograph limb appears most distinct. Although some professionals view the graphical baseflow separation as a "convenient fiction" because of its subjectivity and lack of rigorous theoretical basis, it may sometimes provide useful information in the absence of detailed (and expensive) data on surface water runoff processes and drainage basin characteristics that contribute to streamflow generation. Risser et al. (2005) present a detailed application and comparison of two automated methods of hydrograph separation for estimating groundwater recharge based on data from 197 streamflow gaging stations in Pennsylvania. The two computer programs—PART and RORA (Rutledge, 1993, 1998, 2000) developed by the USGS—are in public domain and available for free download from the USGS Web site. The PART computer program uses a hydrograph separation technique to estimate baseflow from the streamflow record. As illustrated in Fig. 2.35, the baseflow constitutes all but the narrow peaks of the hydrograph. This, intuitively, seems applicable to karst terrains where the general flow of most water takes place within the aquifer before it becomes baseflow of surface streams, either via springs or via diffuse discharge.

The RORA computer program uses the recession-curve displacement technique of Rorabaugh (1964) to estimate groundwater recharge from each storm period. The RORA program is not a hydrograph-separation method; rather, recharge is determined from displacement of the streamflow-recession curve according to the theory of groundwater drainage.

Rorabaugh's method utilized by RORA is a one-dimensional analytical model of groundwater discharge to a fully penetrating stream in an idealized, homogenous aquifer with uniform spatial recharge. Because of the simplifying assumptions inherent to the

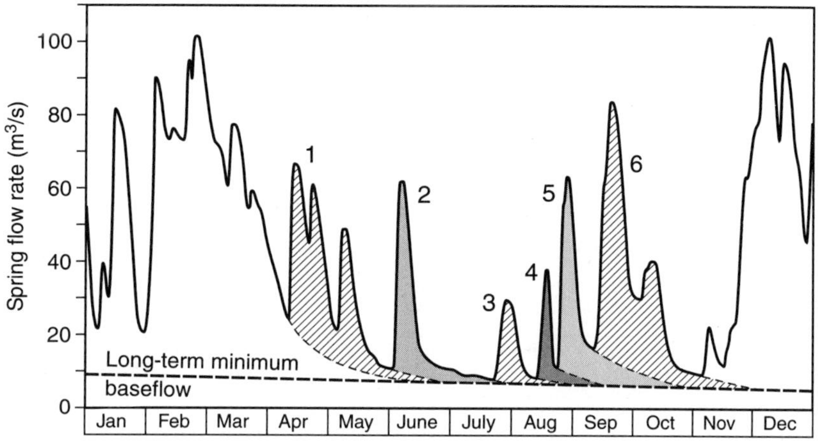

FIGURE 2.36 Determination of aquifer recharge using individual spring flow hydrographs (Ombla Spring near Dubrovnik, Croatia). See Table 2.1 for the results. (From Kresic, 1991.)

equations, Halford and Mayer (2000) caution that RORA may not provide reasonable estimates of recharge for some watersheds. In fact, in some cases, RORA may estimate recharge rates that are higher than the precipitation rates. Rutledge (2000) suggests that estimates of mean-monthly recharge from RORA are probably less reliable than estimates for longer periods and recommends that results from RORA not be used at timescales smaller than seasonal (3 months), because results differ most greatly from manual application of the recession-curve displacement method at small timescales.

Separation of spring hydrographs can be very useful for estimating aquifer recharge since an increase in spring flow after rainfall events is the main indicator of the overall aquifer recharge. Although it is often difficult to accurately determine a spring drainage area in karst, the spring hydrograph reflects the response to rainfall of all porosity types in the entire volume of the aquifer, as well as the response to all water inputs into the aquifer. In addition to direct recharge from precipitation, these inputs may include percolation of sheet surface runoff from less permeable areas, and direct allogenic recharge by losing (sinking) surface streams. For this reason, the spring hydrograph method of determining overall aquifer recharge, from all sources, is arguably the most objective one.

Figure 2.36 and Table 2.1 illustrate the principle of aquifer recharge determination from individual hydrographs and the corresponding rainfall amount. The volume of discharged water (shaded hydrograph areas) above the long-term baseflow is equal to aquifer recharge from individual rainfall events. The volume of rainfall is obtained by multiplying the spring drainage area (approximately 600 km^2) and the height (expressed in millimeters) of rainfall episodes that are clearly responsible for the individual hydrographs. The ratio of two volumes, multiplied by 100, gives the aquifer recharge expressed as a percentage of total rainfall (Table 2.1).

When spring hydrograph and precipitation data is available for multiple years, the above-described method can be used to estimate average seasonal or monthly recharge rates for the drainage area. These rates would vary from year to year depending on various seasonal factors such as antecedent saturation of the vadose zone, snow pack and evapotranspiration, and duration and intensity of individual precipitation events. In

Number of Hydrograph	Discharged Water (m^3)	Rainfall (mm)	Recharge (%)
1	8.704×10^7	201.5	72
2	3.451×10^7	69.3	83
3	1.614×10^7	35.4	76
4	1.067×10^7	27.8	64
5	5.046×10^7	164.9	51
6	1.011×10^8	271.8	62

From Kresic (1991).

TABLE 2.1 Determination of Recharge from Precipitation in the Ombla Spring Drainage Area Using Springflow Hydrograph Method Shown in Fig. 2.36. Drainage Area is Estimated at 600 km²; Data for Year 1969

general, however, such analysis provides a reasonable assessment of a long-term recharge pattern and characteristics in the spring drainage area. Figure 2.37 is an example of average monthly recharge rates in a limestone aquifer drained by a permanent springs with a well-defined drainage area. The horizontal lines represent base lengths of 29 individual hydrographs and the corresponding recharge rates determined as shown in Fig. 2.37 and Table 2.1. Average monthly rates (represented with dots) are time-weighted averages of the individual recharge rates within that month (note that bases of some individual hydrographs extend over parts of 2 months). As can be seen, the individual recharge rates show a wide scatter, particularly in May and June, 2 months with the largest number of hydrographs analyzed. However, the characteristic seasonal character of aquifer recharge in temperate humid climates is clearly visible. The recharge is the highest in the pre-and early-vegetation periods (March to June) when the air temperatures are moderate and the soil moisture deficit is minimal due to preceding snowmelt. In

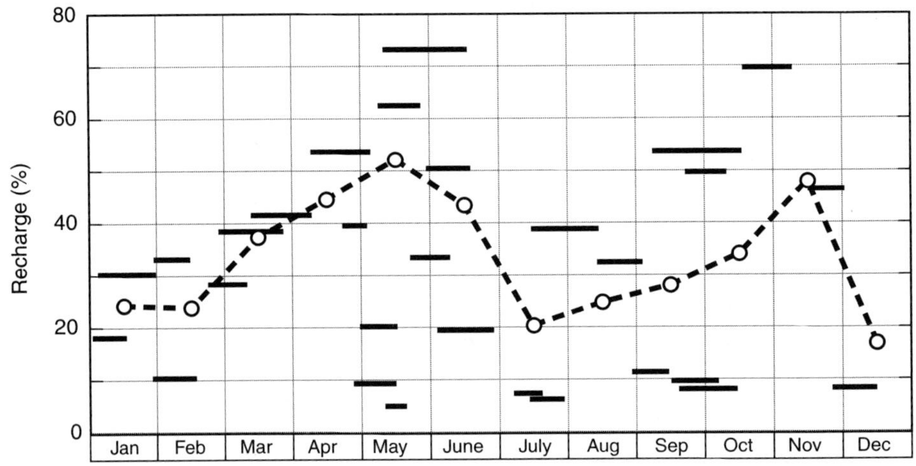

FIGURE 2.37 Average monthly recharge of the Grza Spring aquifer, eastern Serbia, determined using hydrograph method shown in Fig. 2.36 (explanation in text). (From Kresic, 1991.)

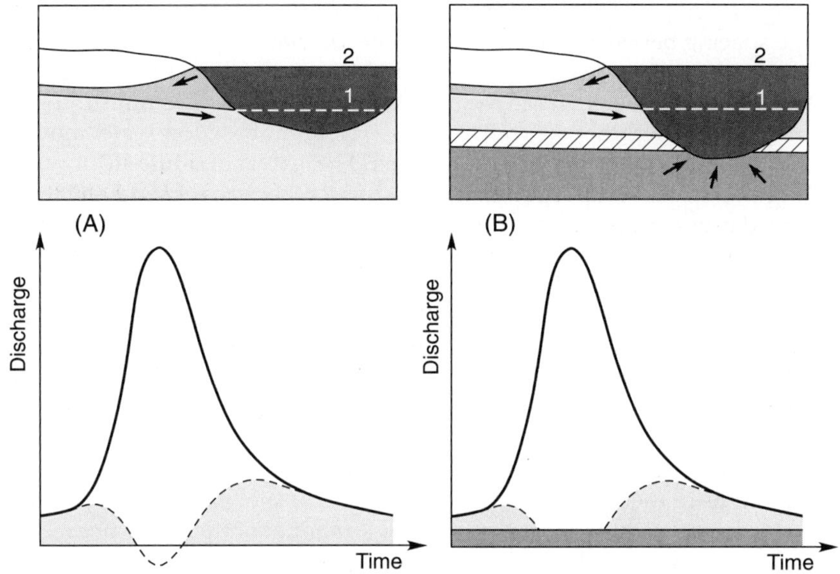

Figure 2.38 Stream hydrograph showing flow components after major rise due to rainfall when: (A) the stream stage is higher than the water table; (B) the stream stage is higher than the water table in the shallow aquifer, but lower than the hydraulic head in the deeper aquifer that is discharging into the stream. (1) Initial stream stage before rainfall and (2) stream stage during peak flow. (Modified from Kresic, 2009.)

addition, daily rainfall is the most frequent and monthly rainfall, the highest in May and June that all contributes to high aquifer recharge.

As illustrated in Fig. 2.38, graphical methods of baseflow separation may not be applicable at all in some cases. A stream with alluvial sediments having significant bank storage capacity may, during floods or high river stages, lose water to the subsurface so that no baseflow is occurring (Fig. 2.38A). The same would be true if the stream is losing water to the underlying karst aquifer. Or, a stream may continuously receive baseflow from a regional aquifer that has a different primary recharge area than the shallow aquifer and maintains a higher head than the stream stage (Fig. 2.38B). Although one may attempt to graphically separate either of the two hydrographs using some common method, it would not be possible to make any conclusions as to the groundwater component of the surface streamflow without additional field investigations.

Combining the hydrograph analysis with analyses of precipitation and the chemical and physical composition of discharged water can provide insights into the origin and residence time of water, aquifer recharge mechanisms, and the hydraulics of the spring flow generation. For example, various environmental and artificial isotopes, as well as groundwater chemical constituents such as major anions, cations, and conductance, can be used to plot chemographs and evaluate processes that lead to spring discharge. White (2010) explains why springs draining karstic carbonate aquifers are the best candidates for such approach as opposed to springs discharging from noncarbonate rocks that typically have little variation in concentrations of dissolved constituents. Karst springs fall into three broad categories: (1) those with response times much shorter with respect to the average spacing between storms, (2) those with response times much longer than the average spacing between storms, and (3) intermediate cases where the response times are comparable to the spacing between storms. Because of the rapid throughput of categories (1) and (3), stormwaters reach the spring before the water has time to come into chemical equilibrium with carbonate wall rock. As a result, the baseflow water moving to the spring is diluted with relatively fresh stormwater. The arrival of dilute stormwater at the spring is revealed by an abrupt drop in hardness and other chemical parameters (White, 2010).

As discussed throughout this book, karst aquifers have unique porosity resulting in two main types of groundwater flow: slow diffuse flow or baseflow and fast conduit flow (see also Atkinson, 1977). However, the percentages of these two components in karst spring hydrographs vary greatly as the result of many possible water inputs and different characteristics of a karst aquifer. Presence of sinking streams recharging the aquifer, extensive dendritic conduit systems, and high matrix porosity can produce complicated chemographs and hydrographs with multiple peaks; some of which show rapid response to rainfall, and some exhibiting long delays for the same spring. Separation of such hydrographs may therefore be completely ambiguous without thorough field investigations of the system. For example, Fig. 2.39 shows how the same spring may react quite differently in two successive years with similar periods of major recharge during the first half of the year, followed by long recession periods from June/July to October. Both recession periods show general increase in specific conductance as the longer-residing water (from the aquifer matrix and/or from deeper parts of the aquifer) starts supplying most of the spring flow. However, in 1995, this increase is significantly larger and with considerable noise that cannot be simply explained because the precipitation pattern did not change notably. Vandike (1996) provides a very detailed discussion about precipitation and recharge episodes recorded at various gauging stations in the spring drainage area and corresponding variations in spring hydrographs and chemographs.

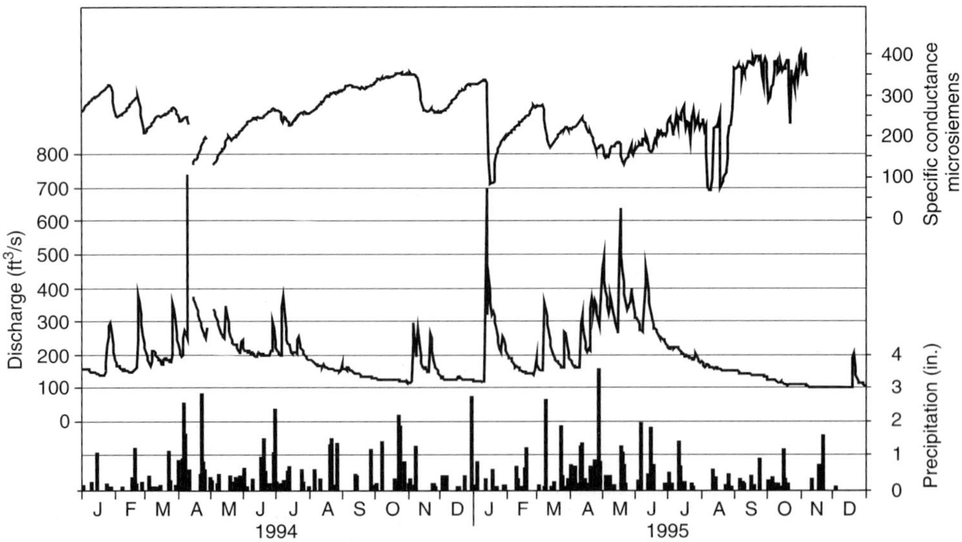

FIGURE 2.39 Average daily discharge and specific conductance at Maramec Spring, and daily precipitation at Rolla-UMR, 1994–1995. (Modified from Vandike, 1996.)

The outlet of Maramec Spring, Missouri's fifth largest spring, is a cave opening developed in the Lower Gasconade Dolomite. Cave diving has shown that the conduit which channels water to the spring reaches depths of at least 190 feet below pool elevation. The discharge of Maramec Spring varies from a low of about 56 ft^3/s to more than 1100 ft^3/s, and averages about 155 ft^3/s (Vandike, 1996).

Dye tracing shows that Maramec Spring is recharged from a 310 mi^2 area west and south of the spring in Dry Fork, Norman Creek, and Asher Hollow Watersheds. All of these watersheds are drained by losing streams that channel a significant part of their runoff into the subsurface. Hourly rainfall data collected at four locations in Dry Fork Watershed, combined with hourly discharge and specific conductance data collected at Maramec Spring, shows that discharge at the spring begins to increase as little as 4 to 6 hours after precipitation begins. The response time appears to be greater during relatively dry weather and less during wet weather when antecedent soil moisture is high. The rapid increase in Maramec Spring discharge is due to an increase in pressure head in the karst system as the water table elevation is increased in the recharge area. The actual water supplied by the recharge does not begin to arrive at the spring for several days, and the mass center of the recharge typically reaches the spring 12 to 15 days after heavy rainfall (Vandike, 1996).

Another example of combined use of hydrographs, chemographs, and turbidographs is shown in Fig. 2.40 from the work of Ryan and Meiman (1996) as interpreted by White (2010). Big Spring is a karstic limestone spring in Mammoth Cave National Park. It is in north of Green River and not related to Mammoth Cave and the associated large karstic drainage system, south of Green River. Big Spring has an elongate catchment that extends into agricultural lands north of the park boundary. The portion of the catchment within the park is forested and undeveloped. The hydrograph of Big Spring rose very quickly following a large storm, but the chemograph remained constant and only fell about a day later than the rise of the hydrograph. The turbidograph remained flat and low for another 18 hours and then a peak in turbidity occurred. The turbidity peak coincided with a rise in coliform bacteria in the springwater. The interpretation is that the discharge between the initial rise of the hydrograph and the drop in the chemograph represents

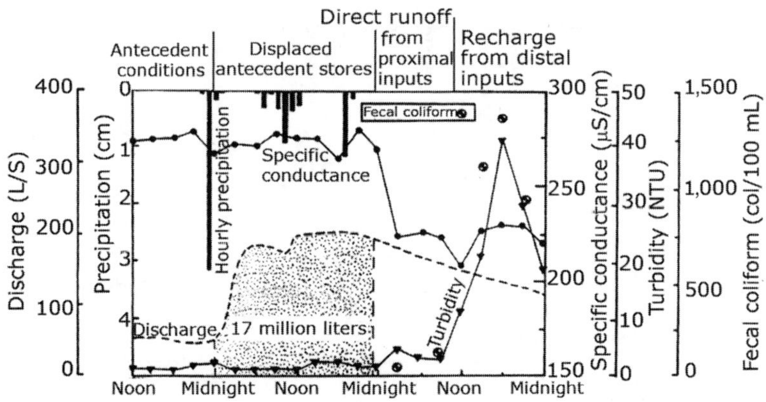

FIGURE 2.40 Storm response of Big Spring, Mammoth Cave National Park, Kentucky, the United States, showing arrival times for stormwater and for sediment load. (From Ryan and Meiman, 1996, as presented by White, 2010.)

water stored in flooded conduits. Increased discharge began immediately following the storm because of rising head in the upstream reaches of the system but only 24 hours later did the stormwater itself arrive at the spring. A further 18 hours later, stormwater from the upstream agricultural land, carrying both sediment and coliform bacteria, arrived at the spring. The latter interpretation was supported by a dye trace which showed that the travel time from the upstream end of the system to the spring was the same as the travel time indicated by the turbidograph (White, 2010).

Hydrochemical separation of the spring discharge hydrograph is based on the assumption that the constitution of the water entering the aquifer is considerably different than water already in it. When recharge by rain takes place, it is evident that the concentration of most cations characterizing groundwater, such as calcium and magnesium, is much lower in rainwater. Additional assumptions for the application of this method are that (after Dreiss, 1989):

- Concentrations of the chemical constituents in the rainwater chosen for monitoring are uniform in both area and time.
- Corresponding concentrations in the pre-storm water are also uniform over the active aquifer area and time.
- The effects of other processes in the hydrologic cycle during the rainfall event, including recharge by surface waters, are negligible.
- The concentrations and transport of constituents are not changed by chemical reactions in the aquifer.

This last condition assumes a minor dissolution of carbonate rocks during the fast flow of new water through secondary porosity. An example in Fig. 2.41 shows that the first two assumptions regarding the calcium ion are acceptable; its concentration in the springwater drops rapidly after the heavy rains that cause an increase in discharge rate.

Assuming a simple mixing of old aquifer water (Q_{old}) and newly infiltrated rainwater (Q_{new}), the total recorded discharge of the spring is the sum of the two (after Dreiss, 1989):

$$Q_{total} = Q_{old} + Q_{new} \tag{2.35}$$

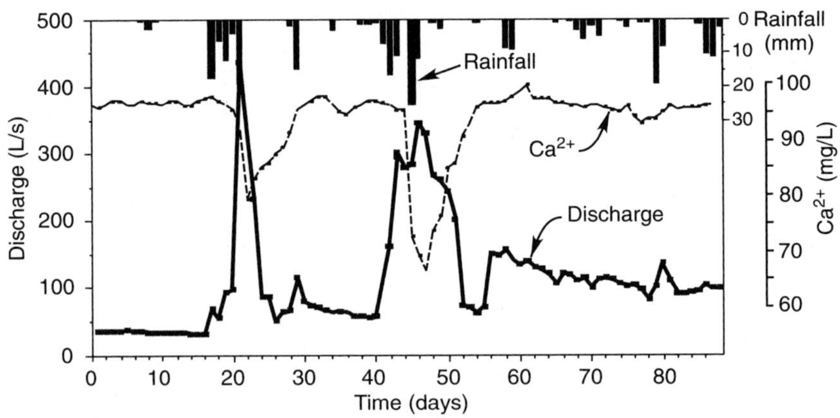

FIGURE 2.41 Spring discharge hydrograph, calcium chemograph, and rainfall in the spring drainage area. (From Kresic, 2007.)

If chemical reactions in the aquifer do not cause significant and rapid changes in the concentration of calcium ion in the infiltrating rainwater (which is often true for unconfined karst and intensely fissured aquifers where the flow velocity is high), the calcium ion balance in the springwater is

$$Q_{\text{total}} \cdot C_{\text{total}} = Q_{\text{old}} \cdot C_{\text{old}} + Q_{\text{new}} \cdot C_{\text{new}} \tag{2.36}$$

where Q_{total} is the recorded spring discharge, C_{rec} is the recorded concentration of calcium ion in the springwater, Q_{old} is the portion of the spring flow attributed to the "old" water (i.e., water already present in the aquifer before the rain event), C_{old} is the recorded concentration of calcium ion in the springwater before the rain event, Q_{new} is the portion of the spring flow attributed to the new water, and C_{new} is the concentration of calcium ion in the new water.

If C_{new} is much smaller than C_{old} (which is correct in this case since the concentration of the calcium ion in rainwater is usually less than 5 mg/L), the input mass of calcium ion is relatively small compared to its mass in the old aquifer water:

$$Q_{\text{new}} \cdot C_{\text{new}} << Q_{\text{old}} \cdot C_{\text{old}} \tag{2.37}$$

From Eq. (2.36) it follows, after excluding the (small) input mass, that

$$Q_{\text{old}} = \frac{Q_{\text{total}} \cdot C_{\text{total}}}{C_{\text{old}}} \tag{2.38}$$

Combining Eqs. (2.36) and (2.38) gives

$$Q_{\text{new}} = Q_{\text{total}} - \frac{Q_{\text{total}} \cdot C_{\text{total}}}{C_{\text{old}}} \tag{2.39}$$

By applying Eq. (2.39), it is possible to estimate the discharge component formed by the inflow of new rainwater if the spring discharge measurements and a continuous hydrochemical monitoring are performed before, during, and after the storm event.

The result of the hydrochemical hydrograph separation for the second major rainfall event, which started approximately 40 days after the beginning of monitoring (see Fig. 2.41), is shown in Fig. 2.42. One day after the major increase in the flow rate, only a minor amount of new water started discharging at the spring. This lag is 3 days for the first major increase in the discharge of new water. The lag between the maximum spring discharge and the maximum discharge of new water is again 1 day. New water stopped discharging at the spring 14 days after the rainfall started influencing the overall flow rate. The new water contributes with only 18.3% to the overall flow increase at the spring. The percentage of new water in the total spring discharge for the 14-day period is 14.6% (Kresic, 2007).

Imes et al. (2007) explain a very similar procedure to the one described above, where the initial specific conductance of the baseflow is accounted for. The mixing of the two flow components varies for different recharge events as illustrated in Fig. 2.43.

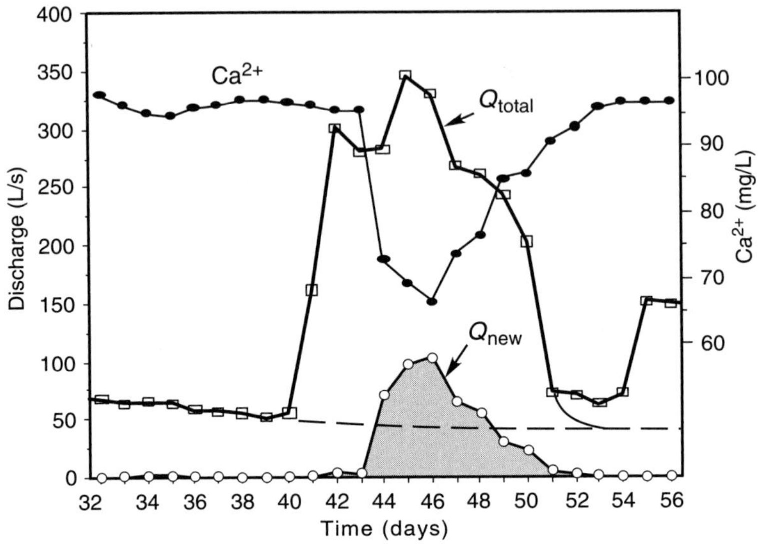

FIGURE 2.42 Spring hydrograph separation into new rainfall and pre-rainfall water. Note that pre-rainfall water is initially discharged rapidly due to increased pressure in the aquifer before any new water starts arriving at the spring. (Modified from Kresic, 2007.)

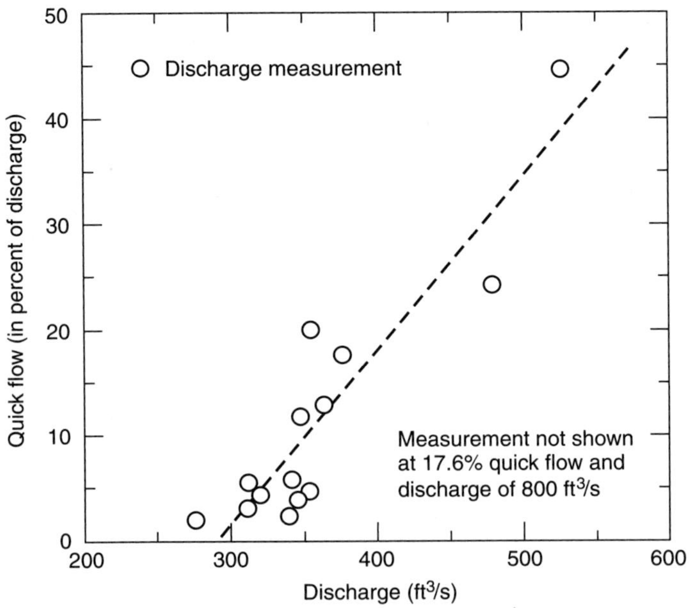

FIGURE 2.43 Discharge from Big Spring as a function of percentage quick flow, water years 2001 to 2004. (From Imes et al., 2007.)

One useful indicator of water residence in a carbonate aquifer is the calcite saturation index (SI_c) that can be calculated using the following formula:

$$SI_c = \log(IAP/K_T) \tag{2.40}$$

where IAP is the ion activity product of the mineral (calcite) and K_T is the thermodynamic equilibrium constant at a given temperature. PHREEQC, a public domain computer program for the simulation of various geochemical reactions developed at USGS (Parkhurst and Appelo, 1999), can be used to quickly perform this calculation. A value of SI_c equal to zero indicates that the water sample is saturated with calcite. A value for SI_c greater than 0 indicates that the sample is supersaturated with calcite, and value of SI_c less than 0 would indicate a water sample undersaturated with respect to calcite. The SI_c can be used to assess the hydrogeologic conditions contributing to the chemical characteristics of the springwater. For example, water flowing diffusely through carbonate rocks or water flowing through small fractures relatively quickly becomes saturated with respect to calcite. Conversely, water moving through large fractures and conduits requires longer flow paths and residence times to become saturated with respect to calcite (Adamski, 2000). If there is a notable, fast increase in spring discharge after major rainfall events, but the calcium concentration and SI_c do not change or even increase, this would be an indicator of expulsion of old water residing in the aquifer.

White (2010) cautions that karst springwaters exhibit a great variability in the state of saturation with respect to calcite and dolomite, and equilibrium calculations are of marginal value in the interpretation of karst water chemistry. The state of saturation is not, in itself, an indication of the presence of conduit drainage. Because of CO_2 degassing along conduits, springs discharging from conduits may be supersaturated. Some large springs of Florida that receive much of their recharge from overlying soils are fed by large conduits and discharge waters that range from near saturation to undersaturated (Katz et al., 1999).

The findings of Desmarais and Rojstaczer (2002) are typical of the fast conduit-flow dominated spring discharge without delay and without any significant influence of soil cover or matrix storage. Initially, the discharge from the Maynardville Limestone Spring in Bear Creek Valley, Tennessee, peaks approximately 1 to 2 hour from the midpoint of summer storms. This initial peak is likely due to surface loading, which pressurizes the aquifer and results in water moving out of storage. All of the storms monitored exhibited recessions that follow a master recession curve very closely, indicating that storm response is fairly consistent and repeatable, independent of the time between storms and the configuration of the rain event itself. Electrical conductivity initially increases for 0.5 to 2.9 days (longer for smaller storms), the result of moving older water out of storage. This is followed by a 2.1 to 2.5 day decrease in conductivity, resulting from an increasing portion of low conductivity recharge water entering the spring. Stable carbon isotope data and the calcite saturation index of the springwater also support this conceptual model. Spring flow following recharge episodes is to a greater extent controlled by displaced water from the aquifer rather than by direct recharge through the soil zone.

2.6.3 Environmental Isotopes

Environmental isotopes and other natural tracers are irreplaceable in general groundwater studies as they may provide answers on contemporary and historic recharge at timescales varying from days to thousands of years. They are useful in finding answers about possible mixing of groundwater of different age and origin within a groundwater system, or sources of groundwater recharge. In addition, as discussed by Sophocleous (2004), tracer studies are often surprising and cast doubt on the utility of physical methods of recharge assessment. A classic example is an attempt to quantify recharge at grassland and irrigated sites above the High Plains (Ogallala) aquifer. Using Darcian methodology based on physical vadose zone measurements, the calculated flux below the root zone ranged from 0.1 to 0.25 mm per year for a grassland site. Alternative analysis of chloride profiles led to estimates of 2.5 to 10 mm per year at the same exact location. This discrepancy of one to two orders of magnitude shows that preferential flow in some cases may be a dominant recharge mechanism, and illustrates the major need for more research in this area. This is even more emphasized in karst environments where preferential flow paths in both the residuum sediments and the underlying carbonates are the matter of fact. A very informative and detailed discussion on the application of environmental isotopes in hydrogeology, with special emphasis on karst is provided by Trček and Zojer (2010), and portions of it are briefly presented throughout this chapter.

The radioactive isotopes in the environment are of either natural origin or anthropogenic. The input source functions, which describe the time-varying global fluxes of isotopes deduced from atmospheric, cosmogenic, and anthropogenic production such as ^{3}H, ^{14}C, and ^{36}Cl, are well known. These isotopes have been measured monthly or annually in precipitation since 1950 and form peak-shaped curves (e.g., see Fig. 2.44). The input functions of radioactive isotopes that resulted from the nuclear power facilities or fuel reprocessing have either increased shapes steadily (e.g., ^{85}Kr) or remain elevated (e.g., ^{129}I). Similarly, air and water concentrations of some industrial chemicals such as chlorofluorocarbons (CFCs) exhibit well-defined shapes that can be incorporated as part of input functions used to date groundwater. Studying time-dependent concentrations

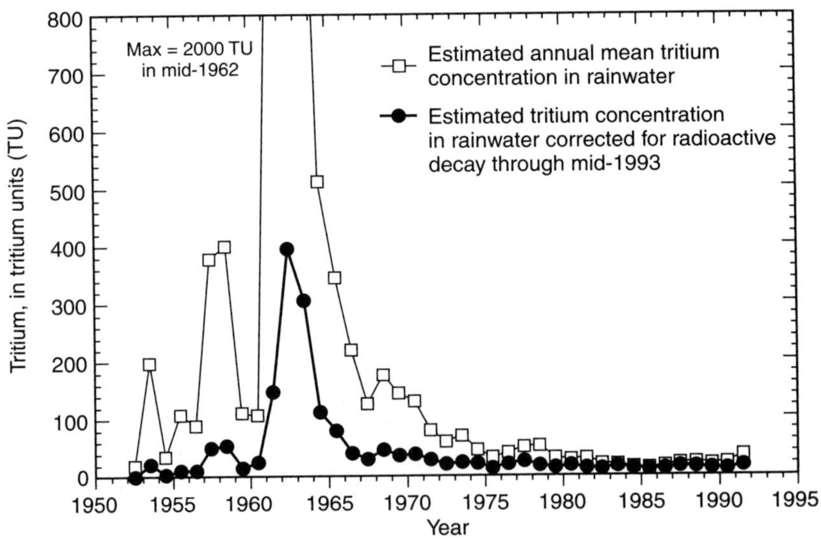

FIGURE 2.44 Estimated annual mean tritium concentration in rainwater, 1952 to 1991, southwestern Ohio. (From Rowe et al., 1999.)

of several environmental isotopes and anthropogenic chemicals such as CFCs in springwater often provides fairly informative age determinations over the past several decades.

Measurements of the stable isotopic composition of substances are conventionally reported in terms of a relative value δ:

$$\delta_x = \left(\frac{R_x}{R_{st}} - 1\right) \times 1{,}000 \tag{2.41}$$

where Rx is the isotope ratio (e.g., $^2H/^1H$, $^{18}O/^{16}O$, and $^{13}C/^{12}C$) in the substance x, R_{st} is the isotope ratio in the corresponding international standard substance, and δ is expressed in parts per thousand (‰).

A positive δ value means that the sample contains more of the heavy isotope than the standard. A negative δ value means that the sample contains less of the heavy isotope than the standard. A $\delta^{15}N$ value of +30‰ means that there are 30 parts per thousand or 3% more ^{15}N in the sample relative to the standard.

Various isotope standards are used for reporting isotopic compositions. Stable oxygen and hydrogen isotopic ratios are normally reported relative to the SMOW standard ("Standard Mean Ocean Water") or the virtually equivalent VSMOW (Vienna-SMOW) standard. Stable carbon isotope ratios are reported relative to the PDB (for Pee Dee Belemnite) or the equivalent VPDB (Vienna-PDB) standard. The stable oxygen isotope ratios of carbonates are also commonly reported relative to PDB or VPDB. Sulfur and nitrogen isotopes are reported relative to CDT (for Cañon Diablo Troilite) and AIR (for atmospheric air), respectively. VSMOW and VPDB are virtually identical to the now-unavailable SMOW and PDB standards; therefore, the use of VSMOW and VPDB is preferred.

Three temperature systems should be taken into account in the isotopic investigations of karst water: low-temperature (<90°C), medium-temperature (90 < T <150°C), and high-temperature (>150°C) systems. The groundwater isotopic composition is mainly unaffected in low-temperature systems due to relatively slow reactions. Therefore, the isotopic techniques have the widest range of application in hydrogeologic and hydrologic studies of these systems. For two other temperature systems, the knowledge of the magnitude and temperature dependence of the isotopic fractionation factors between the minerals and fluids (gases and water) is essential for the interpretation of changes in the spring stable isotopic composition.

Environmental tracers commonly used to estimate age of young groundwater (less than 50 to 70 years old) are the CFCs, and ratio of tritium and helium-3 ($^3H/^3He$). Because of various uncertainties and assumptions that are associated with sampling, analysis, and interpretation of the environmental tracer data, groundwater ages estimated by use of the CFC and $^3H/^3He$ methods are regarded as apparent ages and must be carefully reviewed to ensure that they are geochemically consistent and hydrologically realistic (Rowe et al., 1999). Isotopes typically used for determination of older groundwater ages are carbon-14, oxygen-18 and deuterium, and chlorine-36, while many other isotopes are increasingly studied for their applicability (Geyh, 2000). The best source for numerous studies on the application of various environmental isotopes in surface water and groundwater studies are the proceedings of the international conferences organized by the International Atomic Energy Agency (IAEA), as well as the related monographs published by the agency, many of which are available for free download at its Web site (www.iaea.org).

Although reference is often made to dating of groundwater, the age actually applies to the date of introduction of the tracer substance, and not the water. Unless one recognizes and accounts for all the physical and chemical processes that affect the concentrations of an environmental tracer in the aquifer, the tracer-based age is not necessarily equal to the transit time of the water (Plummer and Busenberg, 2007). The concentrations of all dissolved substances are affected, to some extent, by transport processes. For some tracers, the concentrations can also be affected by chemical processes, such as degradation and sorption during transit. For this reason, the term *age* is usually qualified with the word *model* or *apparent*, that is, *model age* or *apparent age*. The emphasis on model or apparent ages is needed because simplifying assumptions regarding the transport processes are often made and chemical processes that may affect tracer concentrations are usually not accounted for.

As discussed by Trček and Zojer (2010), environmental isotopes in karst studies are convenient tracers for the following reasons:

- Waters that are recharged at different times or at different locations have different isotopic composition, which is also reflected in the spring isotopic composition.
- Groundwaters that followed different flow paths preserve distinctive fingerprints, which are also reflected in the spring isotopic composition.
- An isotopically distinctive source along a groundwater flow path has a significant influence on the spring isotopic composition, which indicates a hydrodynamic connection.
- Meteoric waters retain their distinctive fingerprints until they mix with waters of different compositions or they react with minerals or other fluids.
- Groundwater solutes that originate from the atmosphere usually differ isotopically from those originating from geologic and biologic sources within the aquifer.
- Solute isotopic ratios change during the groundwater flow path due to the biological cycling of solutes and water–rock interactions; however, these changes are usually predictable and can be reconstructed from the spring isotopic compositions.

It follows that two contrasting properties of isotopes are used in isotopic studies of karst water. Generally, they refer to (1) changes in groundwater isotopic composition due to physical and chemical processes, such as mixing processes and water–rock interactions, and (2) the constancy of the isotopic composition along the groundwater flow path from the catchment area to the discharge location such as a spring. The isotopic investigations of water in karst most frequently include the following topics (Trček and Zojer, 2010):

- Identification of the aquifer recharge–discharge relationship.
- Assessment of possible origin of waters that contribute to the spring discharge.
- Determination of flow paths from the recharge to the discharge area of the spring.
- Groundwater residence time and related aquifer storage properties.
- Mixing of event waters (e.g., snowmelt and storms) and pre-event groundwater in the spring discharge.

- Determination of atmospheric sources that contribute to the solute isotopic composition of groundwater.
- Determination of weathering reactions along the flow paths that influence the spring isotopic composition.
- Assessment of influences of biologic cycling on the groundwater solute isotopic composition (e.g., nutrients within an ecosystem).
- Determination of groundwater geochemical evolution.
- Identification of sources and mechanisms of groundwater contamination.
- Biodegradation processes and transport phenomena.
- Testing hydrologic models using isotopic data.

Oxygen and Deuterium

The ^{16}O, ^{18}O, ^{1}H, and ^{2}H isotopes are the main isotopes that make up the water molecule. These isotopes of oxygen and hydrogen are stable and do not disintegrate by radioactive decay. In shallow groundwater systems with temperatures less than 50°C, the isotopic compositions of δ^2H and $\delta^{18}O$ in water are not affected by water–rock interactions (Perry et al., 1982). Since they are part of the water molecule, these isotopes are therefore ideal conservative natural tracers. Differences in the isotopic composition of groundwater and precipitation can be used to detect differences in the source water including recent precipitation. One important advantage of using δ^2H and $\delta^{18}O$ is that their composition in different components of the global water cycle can be predicted as illustrated in Fig. 2.45.

The difference in the mass of oxygen and hydrogen isotopes in water results in distinct partitioning of the isotopes (fractionation) as a result of evaporation, condensation, freezing, melting, or chemical and biological reactions. For example, δ^2H and $\delta^{18}O$ values in precipitation are isotopically lighter in areas with lower mean annual temperature. Strong seasonal variations are expected at any given location, whereas average annual values of δ^2H and $\delta^{18}O$ in precipitation show little variation at any one location

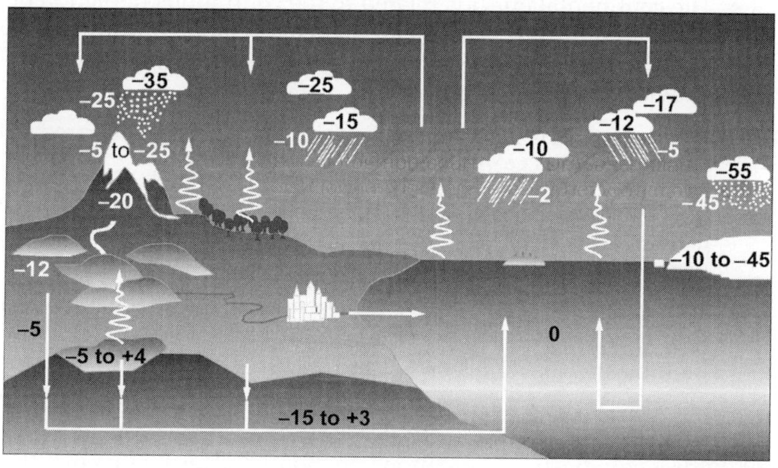

FIGURE 2.45 Typical values of ^{18}O composition (in ‰) in the global water cycle. (Modified from IAEA/WMO, 2006.)

(Dansgaard, 1964, from Key et al., 2002). In general, the average groundwater ^{18}O and ^{2}H composition equals to the weighted average of their composition in precipitation, whereas surface waters are enriched with heavier isotopes (more positive) in comparison to precipitation due to the evaporation process. The IAEA provides δ^2H and δ^{18}O precipitation data measured at various locations throughout the world (ftp://ftp.iaea.org).

The strong relationship between the δ^{18}O and δ^2H values of precipitation is reflected in the global meteoric water line (GMWL). The slope is approximately 8 and the so-called deuterium excess is approximately +10‰. The deuterium excess (d) is defined as

$$d_{\text{excess}} = \delta^2\text{H} - 8\delta^{18}\text{O} \tag{2.42}$$

The deuterium excess near the coast is smaller than +10‰ and approximately 0‰ only in Antarctica. In areas where, or during periods in which, the relative humidity immediately above the ocean is or was below the present mean value, d is greater than +10‰. An example is the deuterium excess of +22‰ in the eastern Mediterranean. The value of d is primarily a function of the mean relative humidity of the atmosphere above the ocean water. The coefficient d can therefore be regarded as a palaeoclimatic indicator (Geyh, 2000; Merlivat and Jouzel, 1979; Gat and Carmi, 1970). In a particular region, the local meteoric water line (LMWL) differs from GMWL in both the slope and the intercept. The GMWL/LMWL equations can be calculated from large data sets of precipitation stations from around the world that are included in the IAEA monitoring program and available at the agency's Web pages.

Following are some general rules for interpretation of the δ^2H and δ^{18}O content in sampled water (Trček and Zojer, 2010).

- Precipitation that occurs in polar regions is more depleted in isotopic composition compared to tropical regions. This latitude effect is due to successive rainouts from the clouds during moisture transport from the tropics to the poles. A latitude effect of −0.6 ‰ for δ^{18}O per degree is generally observed.
- The continental effect is related to the continuous depletion of the precipitation isotopic composition as the air mass moves inland from the coast. The precipitation isotopic composition changes from region to region as well as from season to season.
- Heavy rainfall is depleted in isotopic composition compared with light rain. This effect should be studied locally.
- δ^{18}O and δ^2H values decrease with an increase in altitude. This altitude effect generally varies from −0.15 to −0.5‰ for $\delta\ ^{18}$O and from −1 to −4‰ for δ^2H per 100 m rise in altitude.
- The seasonal effect reflects the dependence of the precipitation isotopic composition on local air temperatures; the lowest δ^{18}O and δ^2H values are in colder months and the highest δ^{18}O and δ^2H values are in warmer months.

Results of extensive isotopic studies performed in the framework of multidisciplinary karst hydrogeological investigations in Southwestern Slovenia that covered about 700 km^2 are illustrated in Figs. 2.46 and 2.47 (Stichler et al., 1997). During a period of 3 years, δ^{18}O and δ^2H contents in precipitation, six main karstic springs and two karst

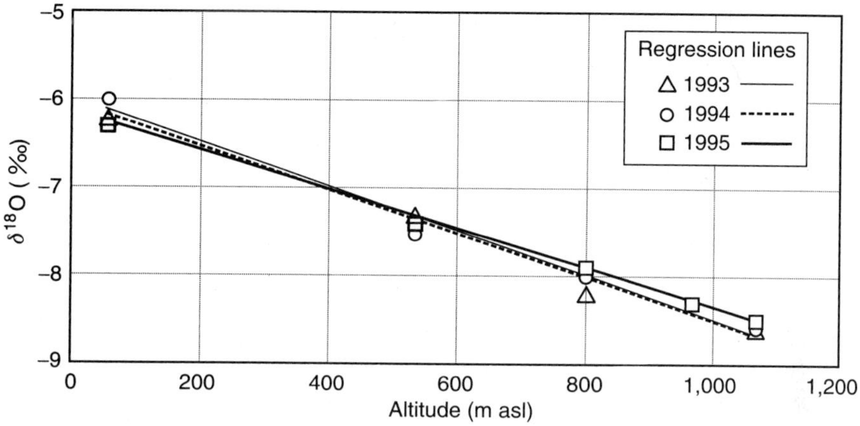

FIGURE 2.46 Correlation of $\delta^{18}O$ contents in precipitation from meteorological stations at Southwestern Slovenia and altitudes. (From Stichler et al., 1997.)

windows (sinkholes) were monitored. Precipitation was sampled at five stations that were spread over the project area at altitudes from 50 to 1070 m. Figure 2.46 illustrates the correlation between the weighted annual average of the precipitation $\delta^{18}O$ values and altitudes of the corresponding sampling stations. The springs and sinkholes were sampled at baseflow conditions, and their annual average $\delta^{18}O$ compositions were compared with Fig. 2.46 to estimate mean altitudes of the groundwater recharge areas.

Data of the isotopic investigations in Southwestern Slovenia were also used to determine mean groundwater residence times at six karst springs. This was based on differences between the heights of two amplitudes, namely (a) the amplitude of the precipitation^{18}O variation and (b) the amplitude of the groundwater ^{18}O variation

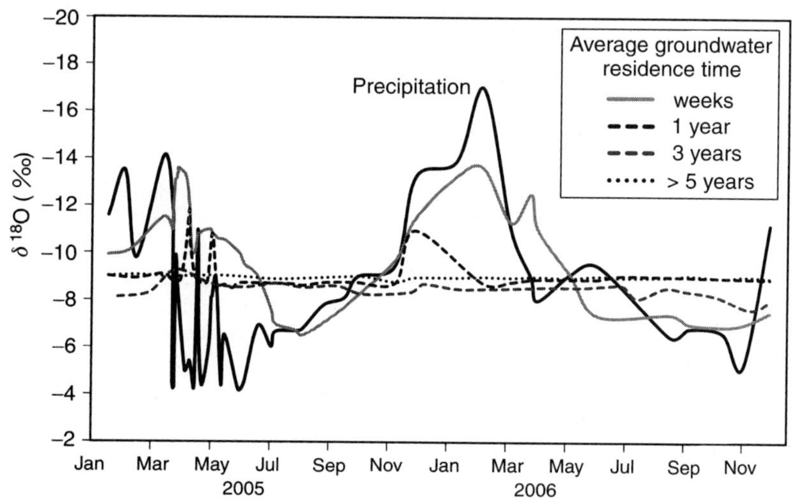

FIGURE 2.47 Comparison of the $\delta^{18}O$ variations in groundwater with different residence times. (From Trček and Zojer, 2010.)

(Stichler et al., 1997). The mean groundwater residence times were calculated by a dispersion model (Maloszewski and Zuber, 1996):

$$T = \frac{1}{2}\pi \left(\frac{-\ln f}{P_D} \right)^{1/2} \tag{2.43}$$

where T is average residence time, f is amplitude ratio: $f = B_0/A_0$, A_0 is amplitude of the precipitation ^{18}O variation, B_0 is amplitude of the groundwater ^{18}O variation, and P_D is dispersion parameter.

The residence times varied from 4.4 months in the Vipava Spring to 5.8 months in the Hubelj Spring. However, the analysis also indicated that two groundwater flow components contributed to the discharge of springs, with mean transit time of weeks in karst channels, and of years in the karst aquifer matrix. In case of the Hubelj Spring, a detail analysis showed that the amplitude of the groundwater ^{18}O variation decreases with the increasing residence time, and it becomes practically negligible after the groundwater average residence time exceeds 5 years (Fig. 2.47; see also Trček, 2003, 2007).

Geyh (2000) discusses various processes and factors that affect local isotopic compositions of precipitation and groundwater, as well as their deviations from the GMWLs and LMWLs. This includes groundwater mixing, reactions, evaporation, temperature, altitude, and continental effects. As illustrated in Fig. 2.48 for example, evaporation shifts both $\delta^{18}O$ and δ^2H values; the former are displaced as a result of isotopic exchange with volcanic CO_2 and limestone, the latter due to exchange with H_2S and silicate hydration. LMWL corresponds to the Mediterranean precipitation; the MWL of Pleistocene palaeowater may be apart from the MWL.

Tritium

Tritium (^3H), a naturally occurring radioactive isotope of hydrogen with the half-life of 12.43 years, has been used extensively as a hydrologic tracer and dating tool. It is

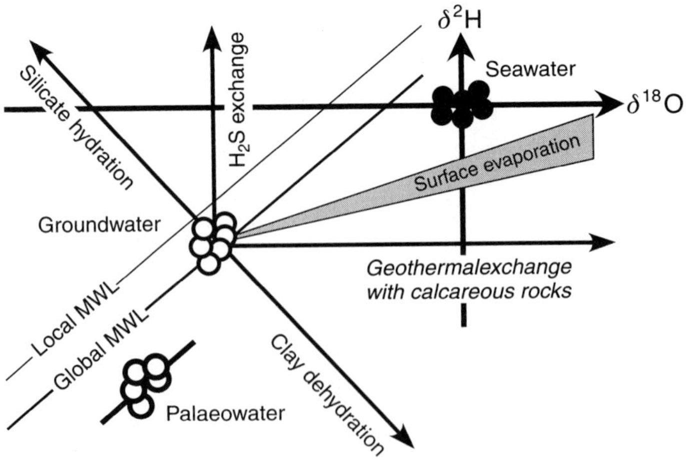

FIGURE 2.48 Various processes that shift the $\delta^{18}O$ and δ^2H values from the Meteoric Water Line, MWL. (From Geyh, 2000.)

produced naturally in the upper atmosphere by bombardment of nitrogen with cosmic radiation and, although few measurements are available, it is estimated that the natural concentration of tritium in precipitation is between 5 and 20 TU (Kauffman and Libby, 1954). Large quantities of tritium were released to the atmosphere during thermonuclear-weapons testing from 1952 until the late 1960s, with maximum releases occurring in the early 1960s. As a result, the amount of tritium in precipitation sharply increased during testing as tritium was introduced into the water cycle, and decreased after testing ended. The atmospheric-testing peak therefore provides an absolute time marker from which to estimate groundwater age. However, because radioactive decay and hydrodynamic dispersion have greatly reduced maximum tritium concentrations in groundwater, identification of the 1960s atmospheric-testing peak has become increasingly difficult. The interpretation of ages from tritium data alone is further complicated by the fact that monitoring and extraction wells are commonly screened over intervals that represent a wide range of groundwater ages; similarly, karst springs combine flow of groundwater in conduits and aquifer matrix that have very different residence times.

The amount of tritium in subsurface water at a given time is a function of the amount of tritium in the atmosphere when infiltration occurred and the radioactive decay rate of tritium. If flow in the unsaturated zone is assumed to be downward-vertical and piston type, the average infiltration flux (q_i) can be estimated by (Coes and Pool, 2005):

$$q_\mathrm{i} = \frac{\Delta z}{\Delta t} \cdot \theta_v \tag{2.44}$$

where Δz is depth to maximum tritium activity (L), Δt is elapsed time between sampling and maximum historic atmospheric tritium activity (T) and θ_v is volumetric soil–water content (L^3/L^3).

Changes in the amount of tritium in precipitation through time, radioactive decay, hydrodynamic dispersion, and mixing water of different ages in the subsurface typically preclude the use of this isotope for quantitatively estimating groundwater residence times. In many cases, qualitative observations may be the best use of tritium data (Clark and Fritz, 1997, from Key et al., 2002). The most accurate use of the data is to indicate pre- or post-1952 groundwater recharge. For a well-defined hydrologic system, analysis of the tritium-input function may provide sufficient information to make quantitative age estimates for groundwater in the system. More complex flow systems generally require more complex models to quantitatively use the tritium data (Plummer et al., 1993). Time-sequence data and multidepth samples also are required to accurately date groundwater horizons using tritium data.

Assuming that piston-flow conditions (no dispersion or mixing) are applicable, Clark and Fritz (1997, from Kay et al., 2002) provide the following guidelines for using the tritium data: (1) Groundwater that contains less than 0.8 TU and is underlying regions with continental climates has recharged the water table prior to 1952; (2) Water with 0.8 to 4 TU may represent a mixture of water that contains components of recharge from before and after 1952; (3) Tritium concentrations from about 5 to 15 TU may indicate recharge after about 1987; (4) Tritium concentrations between about 16 and 30 TU are indicative of recharge since 1953 but cannot be used to provide a more specific time of recharge; (5) Water with more than 30 TU probably is from recharge in the 1960s or 1970s; and (6) Water with more than 50 TU predominately is from recharge in the 1960s.

As explained in detail by Eastoe et al. (2012), the continuing depletion of the artificial tritium in the environment will likely reduce its future usefulness in the groundwater studies.

Tritium-Helium-3

The $^3H/^3He$ method was developed to remove the ambiguity associated with tritium-age estimation. Radioactive decay of tritium produces the noble gas helium-3 (^3He). Determination of the $^3H/^3He$ ratio, therefore, can be used to estimate the apparent date when a sample of water entered the aquifer as recharge. Because these substances virtually are inert in groundwater, unaffected by groundwater chemistry, and not derived from most of the typical sources of anthropogenic contamination, $^3H/^3He$ dating can be applied to a wide range of hydrologic investigations since the input function of ^3H does not have to be known (Geyh, 2000; Kay et al., 2002). By measuring both the mother and daughter activity, actual water ages can be calculated, provided the samples were unmixed and collected from an aquifer with a piston-flow groundwater movement. Incorporating the effects of dispersion and diffusion, which is a much more realistic scenario, requires application of modeling techniques.

The activity of ^3H in the sample ($^3H_{spl}$) is given as (Geyh, 2000):

$$^3H_{spl} = {}^3H_{init}e^{-\lambda \cdot t} \tag{2.45}$$

where $^3H_{init}$ is the activity of initial tritium, λ is the radioactive decay constant, and t is time since decaying started (absolute age). The growth of ^3He in a sample is given by

$$^3He_{spl} = {}^3H_{init}(1 - e^{-\lambda \cdot t}) \tag{2.46}$$

Combining Eqs. (2.45) and (2.46), the unknown and variable initial ^3H activity ($^3H_{init}$) is eliminated and the age of water (t) is obtained from

$$^3He_{spl} = {}^3H_{spl}(e^{-\lambda \cdot t} - 1) \tag{2.47}$$

$$t = -\frac{\ln\left(1 + \frac{^3He_{spl}}{^3H_{spl}}\right)}{\lambda} \tag{2.48}$$

The ^3He concentration in the sample has to be corrected for admixed ^3He from the earth's crust and from the atmosphere. Related procedures are described by Schlosser et al. (1988, 1989) and Shapiro et al. (1998).

The sum of tritium and its daughter product ($^3H + {}^3He_{trit}$) represents a conservative quantity equivalent to the amount of tritium in rainwater at the time of recharge, assuming that helium produced by tritium decay is not lost by upward diffusion to the unsaturated zone. This assumption is considered valid for aquifers where vertical flow velocities exceed approximately 1.5 ft per year (Schlosser et al., 1988, 1989; Poreda et al., 1988).

The concentration of tritiogenic ^3He will increase as tritium decays; thus, older waters will have higher $^3He_{trit}/^3H$ ratios. The age derived by use of this technique represents the time since the water became isolated from the atmosphere and traveled to the sampling point, and is independent of the tritium source function. One limiting factor for the application of the simple method described above in karst aquifers, as opposed to more complex modeling, is the assumption that diffusion, dispersion, or mixing at the sampling location did not affect the concentrations of ^3H or $^3He_{trit}$ in the sample (see Rowe et al., 1999).

Carbon-14

Radiocarbon (carbon-14 or ^{14}C) is the radioactive isotope of carbon with a half-life of 5,730 years. It occurs in atmospheric CO_2, living biosphere, and the hydrosphere after its production by cosmic radiation. Underground production is negligible. The ^{14}C activity is usually given as an activity ratio relative to a standard activity, about equal to the activity of recent or modern carbon. Therefore, the ^{14}C content of carbon-containing materials is given in percent modern carbon (pMC); 100 pMC (or 100% modern carbon) corresponds by definition to the ^{14}C activity of carbon originating from (grown in) 1950 AD (Geyh, 2000). In addition to radioactive isotope ^{14}C, two other stable carbon isotopes, ^{13}C and ^{12}C, are important for understanding the origin of CO_2 involved in the dissolved carbonate-CO_2 system in groundwater, and for correcting the age-dating results obtained from the ^{14}C isotope.

The ^{14}C composition of groundwater is the result of its radioactive decay, and of various chemical reactions between water and porous media in the unsaturated and saturated zones. These reactions include dissolution of carbon dioxide and dissolution of carbonate minerals. Recently infiltrating water and dissolved carbon dioxide gas in the unsaturated zone have a ^{14}C composition of about 100% modern because carbon dioxide gas diffuses from the atmosphere and plants respire carbon dioxide gas to the soil zone or unsaturated zone that is 100% modern. Water that has dissolved carbon dioxide gas and is infiltrating through the unsaturated zone or moving through the aquifer can also dissolve carbonate minerals, which increases dissolved inorganic carbon concentrations and can significantly reduce the ^{14}C composition of the water (Anderholm and Heywood, 2003). Groundwater with a long residence time in the aquifer was subject to similar processes at the time of its origin, and has undergone various transformations through reactions with the aquifer porous media since that time. Because of all these factors, ^{14}C is not a conservative tracer, and its use for groundwater dating studies is not straightforward.

Groundwater dating using ^{14}C is generally considered applicable for water up to 30,000 years old, although the original dating technique proposed for organic carbon samples is applicable to the 45,000 to 50,000-year range (Libby, 1946). Before determining how long groundwater has been isolated from the atmosphere or from the modern ^{14}C reservoir, the effect of chemical reactions on the ^{14}C composition of groundwater needs to be determined. Various models have been used to adjust or estimate the ^{14}C composition of water resulting from processes in the unsaturated zone and aquifer (Mook, 1980; Geyh, 2000; Anderholm and Heywood, 2003). These models range from simple ones that require little data to complex models that require much information about the carbon isotopic composition (ratio of stable carbon isotopes $^{13}C/^{12}C$ and ^{14}C) of gas in the unsaturated zone, carbon isotopic composition of carbonate minerals in the unsaturated zone and aquifer, and reactions that occur as water moves through the aquifer. The following equation can be used to estimate the apparent age of groundwater (Anderholm and Heywood, 2003):

$$t = \frac{5730}{\ln 2} \cdot \ln\left(\frac{A_0}{A_S}\right) \tag{2.49}$$

where t is apparent age, in years; A_0 is ^{14}C composition of water before radioactive decay and *after chemical reactions*, in percent modern; A_S is ^{14}C composition measured in the sample, in percent modern. An example of application of ^{14}C groundwater dating is shown in Fig. 1.160.

Notably, the experience with various correction models for the ^{14}C method shows that, using the same hydrochemical and isotope information, different models may produce corrections varying by many thousands of years (Geyh, 2000). An example of ambiguity associated with karst groundwater dating using ^{14}C is Giant Springs in Great Falls, Montana (see Fig. 1.209). The company bottling the springwater still advertises, as of February 2012, the extraordinary quality of its product partially based on the results of water age dating performed apparently by a graduate student using ^{14}C data. This analysis indicated that the springwater is 2,900 years old and thus protected from any modern influences (contamination). In contrast, a more recent study by the Montana Bureau of Mines and Geology shows that multiple samples from Giant Springs contain elevated tritium indicating a post-1950 recharge date. In addition, analytical results from CFC samples are consistent with the tritium results and suggest that water discharging from the springs is on the order of 20 to 30 years old (Patton, 2005). It is, of course, conceivable that both results are relatively correct since neither study addressed possible mixing of "old" and "new" groundwater discharged at the spring. Whatever the case may be, this example illustrates that studies of karst aquifers often produce unexpected results that need to be verified with multiple lines of evidence.

Chlorofluorocarbons and Sulfur Hexafluoride

CFCs, together with tritium and an emerging environmental tracer sulfur hexafluoride (SF6), can be used to trace the flow of young water (water recharged within the past 50 years) and to determine the time elapsed since recharge. CFCs also can be used to trace seepage from rivers into groundwater systems, provide diagnostic tools for detection and early warning of leakage from landfills and septic tanks, and to assess susceptibility of water-supply wells to contamination from near-surface sources.

CFCs are stable, synthetic organic compounds that were developed in the early 1930s as safe alternatives to ammonia and sulfur dioxide in refrigeration and have been used in a wide range of industrial and refrigerant applications. Probably better known to the public as Freon™, CFCs are nontoxic, nonflammable, and noncarcinogenic, but they contribute to ozone depletion. Therefore, in 1987, 37 nations signed an agreement to limit release of CFCs and to halve CFC emissions by 2000. Production of CFCs ceased in the United States as of January 1, 1996, under the Clean Air Act. Current estimates of the atmospheric lifetimes of CFC-11, CFC-12, and CFC-113 are about 45, 87, and 100 years, respectively.

Groundwater dating with CFC-11, CFC-12, and CFC-113 is possible because (1) their amounts in the atmosphere over the past 50 years have been reconstructed, (2) their solubilities in water are known, and (3) concentrations in air and young water are high enough that they can be measured. Age is determined from CFCs by relating their measured concentrations in groundwater back to known historical atmospheric concentrations and to calculated concentrations expected in water in equilibrium with air. For best results, the apparent age should be determined using multiple dating techniques because each dating technique has limitations. CFC dating is best suited for groundwater in relatively rural environments without localized, nonatmospheric CFC contamination from septic systems,

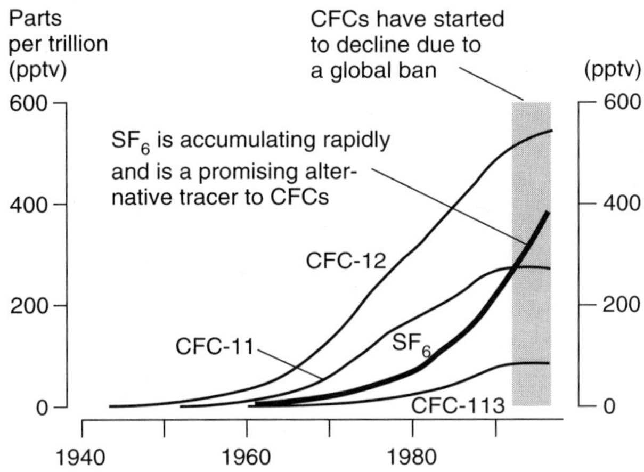

FIGURE 2.49 CFCs and sulfur hexafluoride (SF6) concentrations mixing ratios for air over North America during the last 50 years. (From Plummer and Friedman, 1999.)

sewage effluent, landfills, or urban runoff. The dating method appears to work well in shallow, aerobic, sand aquifers that are low in particulate organic matter, where results can be accurate within 2 to 3 years before the study date. Even where there are problems with CFC dating of groundwater, the presence of CFCs indicates that the water sample contains at least some post-1940s water, making CFCs useful as tracers of recent recharge. Where CFC and $^3H/^3He$ ages agree, or where all three CFCs indicate similar ages, considerable confidence can be placed in the apparent age (Plummer and Friedman, 1999).

The use of SF6 is an emerging alternative to using CFCs in dating groundwater as atmospheric CFC concentrations continue to fall. Industrial production of SF6 began in 1953 with the introduction of gas-filled high-voltage electrical switches. SF6 is extremely stable and is accumulating rapidly in the atmosphere. The historical atmospheric mixing ratio of SF6 is being reconstructed from production records, archived air samples, and atmospheric measurements. The mixing ratio is also being retrieved from concentrations measured in seawater and in previously dated groundwater. A preliminary reconstruction has been made of Northern Hemisphere SF6 mixing ratios (Fig. 2.49). As atmospheric CFC concentrations fall, an even more sensitive dating tool will be the ratio of SF6 to CFC-12. Although SF6 is almost entirely of human origin, there is likely a natural, igneous source of SF6 that will complicate dating in some environments. USGS scientists have successfully used SF6 to date shallow groundwater on the Delmarva Peninsula, Maryland, and water from springs in the Blue Ridge Mountains of Virginia, the United States, with SF6 (Plummer and Friedman, 1999).

2.6.4 Flow Probability Analysis

The principles of probability analysis of precipitation time series explained in Sec. Precipitation are also applicable to the analysis of flow probability. As with other hydrologic time series, an analysis of stream and spring flow hydrographs should, at the minimum, include determination of the general statistical parameters such as average, minimum, and maximum flows for the period of record, standard deviation (variance) of the flows, coefficient of variation, flow duration curves, and frequency of characteristic flows. Unfortunately, in many instances, a stream or spring under investigation will have relatively short periods of flow observation at irregular intervals. This simple fact may prevent any statistically rigorous probability analysis altogether. At the end, however, engineers and scientists engaged in a karst water-related project are still facing various decisions that must include some type of quantitative analysis. Here are just some of the examples: What is the likely minimum flow in summer? Will the spring ever go dry? How big should the storage reservoir be? What is the guaranteed quantity of water delivered to end-users? What should dimensions of the dam spillway be?

In any case, regardless of the available period of record and any specific question, most if not all answers will have to be given in terms of probability. The first and most important step in the flow probability analysis is to simultaneously evaluate the time series of major precipitation in the drainage area of interest, and display it together with the flow hydrograph on the same timescale. Any short-term or long-term anthropogenic influences such as seasonal groundwater withdrawal (pumpage) from the common aquifer or land-use changes (such as urbanization) that may impact the spring discharge should also be accounted for.

Depending on the goal(s) of the project, some characteristic flows may be more important than others. Common probability analyses include minimum and average monthly flows, minimum weekly flows for the period of highest demand (typically, during August–September in moderate humid climates), and absolute minimum daily flow. There is no universally accepted limit of the data points required for probability analysis, although quite a few textbooks on statistics give number 30 as a rule of thumb. As mentioned before, observation periods in practical hydrologic and groundwater studies are often short (several years or less), and this rule of thumb would eliminate many studies by default. However, in addition to the best professional judgment, probability analysis in such cases includes generation of synthetic flow time series and/or simulation of likely historic flows with stochastic time-series models (see Sec. 6.4).

Figure 2.50 illustrates important characteristics of any flow hydrograph that should be considered when performing probability analyses (note that this particular spring is not karstic but it exhibits characteristics very similar to discharge of streams and springs

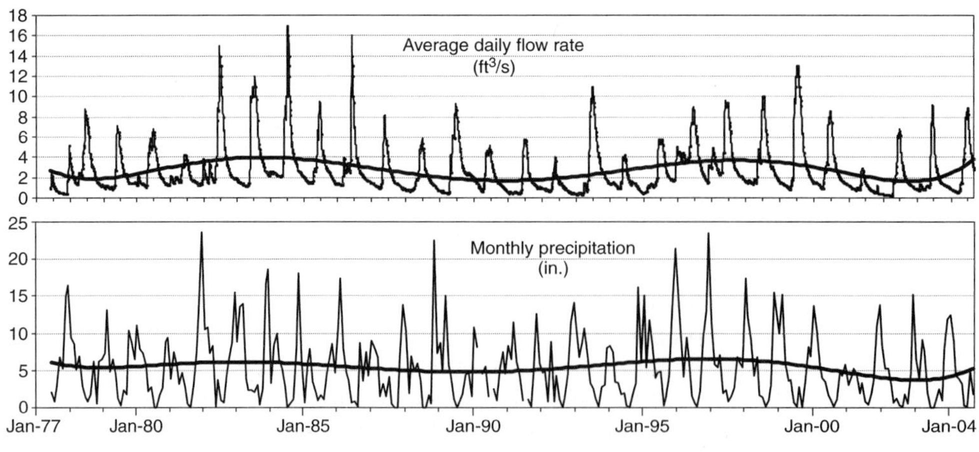

Figure 2.50 *Top*: Average daily discharge rate, in cubic feet per second (cfs), at the Crater Lake Spring, Oregon. *Bottom*: Monthly precipitation, in inches at Crater Lake, Oregon. The natural periodic cycle is presented with the sixth order polynomial line. (From Kresic and Bonacci, 2010.)

in karst). Some of the questions that can be easily answered from the long record of the Crater Lake Spring include

- Is there a regular pattern of discharge from one hydrologic year to another? (Yes)
- Is the discharge strongly influenced by precipitation? (Yes)
- Is there a general delay between major precipitation seasons and spring discharge? (Yes; the highest precipitation is in November to December, followed by the secondary peak in March to April; the highest flow rates are between mid-May and mid-July)
- Is there a long-term periodicity in spring discharge? (Yes; approximately 14 years between wet and dry periods)
- Is there a long-term linear trend unrelated to natural periodicity? (No; the spring behaves like a perfect reliable clock for almost 30 years now.)

It is obvious that, if the period of record included just one or two randomly selected years from the graph in Fig. 2.50, any answer to the above questions would be much less certain (less probable).

The term *frequency analysis*, which is used interchangeably with the term probability analysis, usually connotes a count (number) of events of a certain magnitude, such as the number of daily flows equal to or less than 2 m^3/s per year. In hydrologic studies, the probability of some magnitude being exceeded (or not exceeded) is usually the primary interest. Presentation of the data in this form is accomplished by accumulating the probability (area) under the probability density function. This curve is termed the *cumulative distribution function*. In most statistical texts, the area is accumulated from the smallest event to the largest. The accumulated area then represents nonexceedance probability or percentage. It is more common in hydrologic studies to accumulate the area from the largest event to the smallest. Area accumulated in this manner represents exceedance probability or percentage. Figure 2.51 shows the histogram and the cumulative probability distribution for the average daily flows at the Crater Lake Spring recorded in 1977 to 1985.

Two other terms are often used to describe flow probability: the *return period* (sometimes called *recurrence interval*) and the *percentiles* (or *quantiles*) of a probability distribution. The theoretical return period is the inverse of the probability that the event will be exceeded in any 1 year. For example, a 10-year maximum flow has a 1 in 10 = 0.1 or 10% chance of being exceeded in any 1 year and a 50-year maximum flow has a 0.02 or 2% chance of being exceeded in any 1 year. Conversely, a 10-year minimum flow has 90% chance of being exceeded in any 1 year.

Percentiles are values below which specific percentages of the data are found. For example, for the 5-year daily record of the Crater Lake, the 25th percentile for the empirical data (measured flows) is 1.3 cubic feet per second (cfs) which means that 25% of the time the flow rate is lower, and 75% of the time it is higher. Table 2.2 shows other characteristic percentiles. Percentiles are often used to plot flows of interest (e.g., monthly or daily flows) for the period of record and then compare them with the currently measured flow rate (Fig. 2.52). Combined with the similar plots of historic and current (recent) monthly precipitation, this practice can be very useful in water resources management by anticipating likely flows in the near future and implementing various management measures as needed. For example, if the currently measured flow is less than the 10th percentile

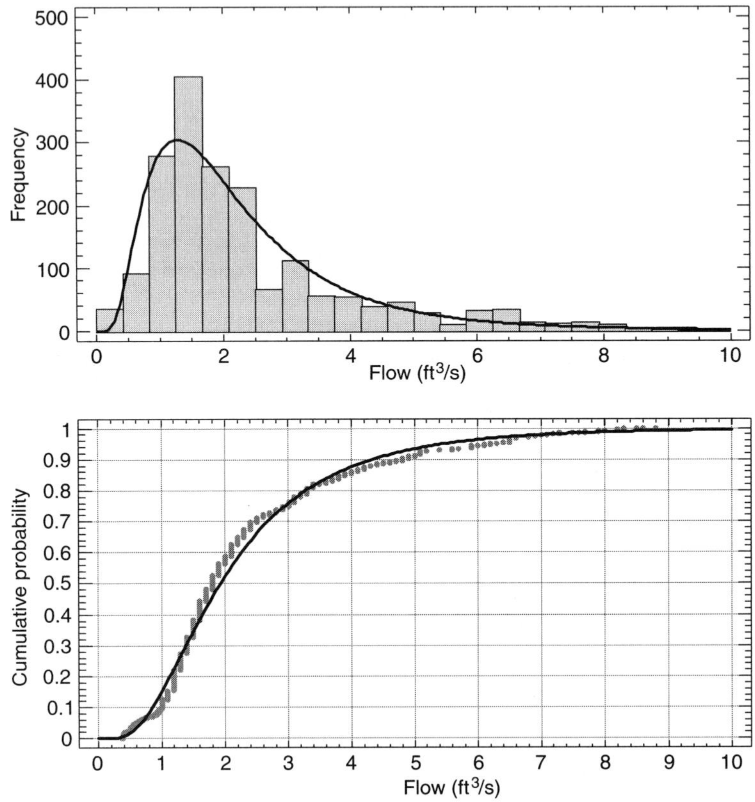

FIGURE 2.51 *Top*: Histogram of average daily flows at the Crater Lake Spring, Oregon, recorded in 1977 to 1985. The most frequent flow is about 1.5 cfs, occurring 400 days out of 2,920. *Bottom*: Cumulative empirical probability distribution (points) and fitted theoretical log-normal probability function. (From Kresic and Bonacci, 2010.)

Percentile (%)	Flow Rate (cfs)
1	0.39
5	0.62
10	1.0
25	1.3
50	1.8
75	3.0
90	4.8
95	6.2
99	7.9

From Kresic and Bonacci (2010).

TABLE 2.2 Percentiles of the Crater Lake Spring Daily Flow Rates for 1977 to 1985 Time Period

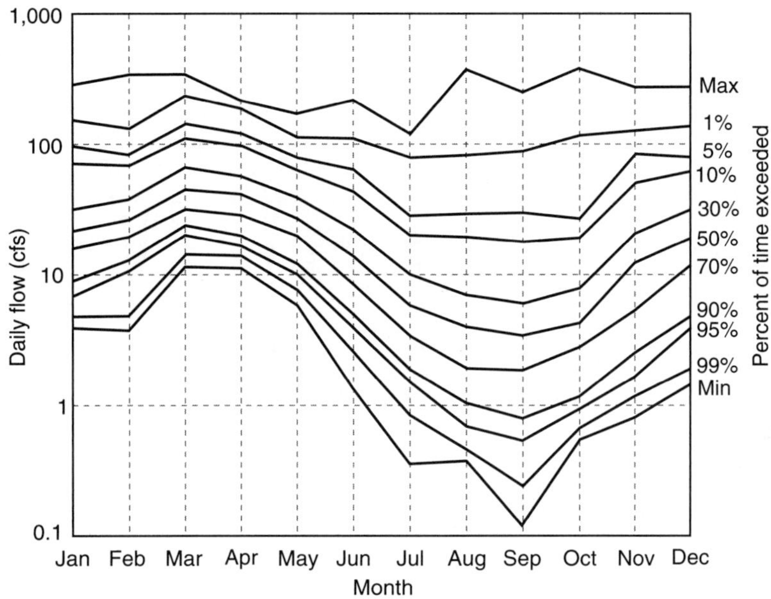

FIGURE 2.52 Example of daily flow duration curves for each month. (Modified from USACE, 1993.)

(i.e., the flow is higher 90% of the time) and the same is true for the recent precipitation, and if the spring discharge is significantly dependent on precipitation, it may be necessary to impose some type of use restriction before the hydrologic and meteorological conditions improve.

An example of a spring greatly influenced by groundwater withdrawal in its drainage area is shown in Fig. 2.53. Hydrographs of the average monthly discharge of Comal Springs in Texas for May and August, for the 73-year long period of record, show the impact of several droughts that were compounded by an increased pumpage from Edwards Aquifer. May typically has the highest recorded daily flows, and August the lowest. During the drought of the 1950s, the springs were dry from June to November of 1956. In cases like this one, it would not be possible to accurately estimate the natural recharge influences on the spring hydrograph and the nature of the aquifer, without subtracting the influences of pumpage. The same figure also illustrates how using average values, even when having an unusually long period of record, could lead to erroneous conclusions regarding secure discharge rates for any given time.

Probability graphs, such as the one shown in Fig. 2.54, are a much more appropriate tool for the assessment of long-term discharge records. For example, in this case, the theoretical probability that the average spring discharge in August would be less than 50 cfs is about 4%, and that it would be 0 cfs it is about 2% to 3% (note that we know from the record that the spring went dry in August 1956). However, it should be noted again that this probability analysis also reflects historic artificial groundwater withdrawals from the system and therefore should not be used alone for any planning purposes. In other words, such withdrawals may change in the future and their impact would have to be accounted for in some quantitative manner.

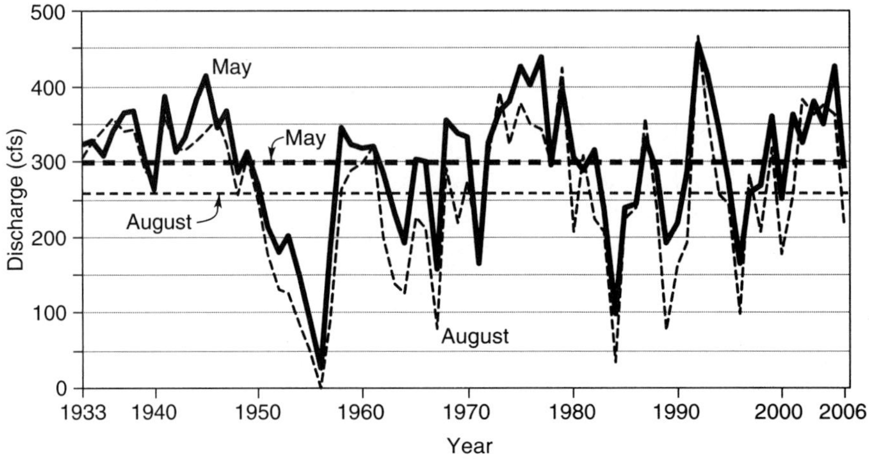

FIGURE 2.53 Average monthly discharges, in cubic ft per second, at Comal Springs, Texas, the United States, for May (*bold line*) and August (*dashed line*) for the 73-year-long period of record. (From Kresic, 2009.)

Some theoretical distributions widely used in hydrologic frequency analysis are the normal (Gaussian), log normal, exponential, two-parameter gamma, three-parameter gamma, Pearson type III, log-Pearson type III, extreme value (Gumbel), and log-Gumbel (Riggs, 1968; Stedinger et al., 1993).

Probability of High and Low Flows

Log normal, Pearson type III, and generalized extreme value distributions (Gumbel family of distributions; see Gumbel, 1941) are reasonable choices for describing flood flows using curve-fitting methods. However, it is advisable to use regional experience to select a distribution for a region and to reduce the number of parameters estimated for

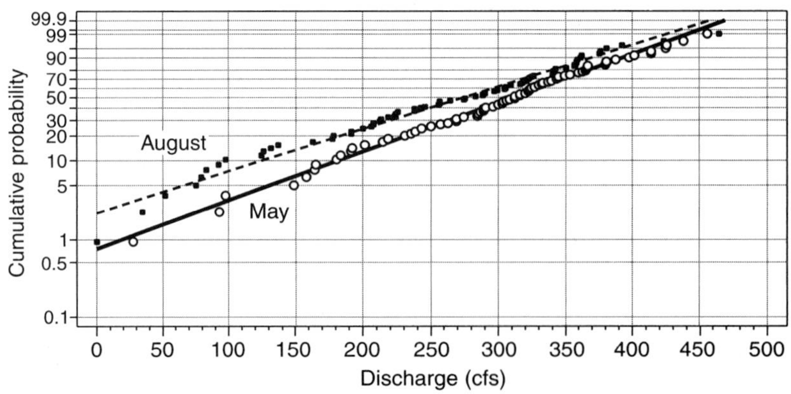

FIGURE 2.54 Extreme value probability distributions of average monthly flows in May and August for Comal Springs. (From Kresic, 2009.)

an individual site (Stedinger et al., 1993). In the United States, the log-Pearson type III distribution is recommended for high flow (flood) analyses by the Interagency Advisory Committee on Water Data (1982).

Many large karst springs have a limiting maximum discharge value independent of the catchment size, watershed conditions (e.g., antecedent soil moisture content and groundwater levels), and amount of precipitation (Bonacci, 2001; Panagopoulos and Lambrakis 2006; Herman et al. 2008). For example, observations of the Arch Spring karst drainage system in the United States during exceptional runoff from Hurricanes Frances and Ivan revealed quantitatively the magnitude of discharge necessary to override the karst drainage system and spill excess discharge into the surface channel (Herman et al., 2008).

Possible reasons for restricted maximum flow rates of some springs include limited dimensions of the karst conduits and spring orifice (Fig. 2.55), subsurface piracy between karst (conduit) drainage areas at high hydraulic heads in the aquifer, and overflow from a main spring to intermittent springs within the same or adjacent catchments. Similarly, in some cases the minimum discharge from a karst springs has a limit that cannot be exceeded regardless of duration and severity of droughts. For example, the Jadro Spring and the Ombla Spring in Croatia have minimum discharge of about 4 m^3/s and are therefore very reliable sources of water supply for the coastal cities of Split and Dubrovnik, respectively. Their high long-term minimum discharge can be explained by large and prolific karst aquifers that provide constant water supply during the long-lasting dry periods (Bonacci, 1995, 2001).

The minimum and maximum daily discharge rates of the Ombla Spring measured on annual basis from 1968 to 2007 at the Komolac gauging station are presented in Fig. 2.56. The minimum daily discharges never drop below 4 m^3/s, while the maximum

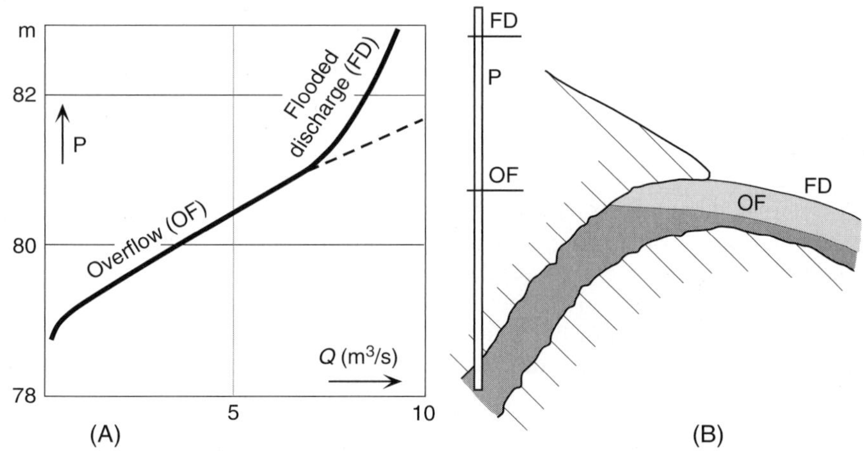

Figure 2.55 (A) Flow rate (Q) versus hydraulic head at piezometer P, Ljubovija intermittent spring near Mostar, Herzegovina. (B) Overflow (OF): An increasing discharge causes increase in the flow cross section at the outlet and thus slight increase in the hydraulic head at the piezometer P. Flooded discharge (FD): The outlet cross section cannot increase further as the entire orifice is flooded; further increase in the flow rate causes rapid increase in the hydraulic head at P. (From Hajdin, 1981.)

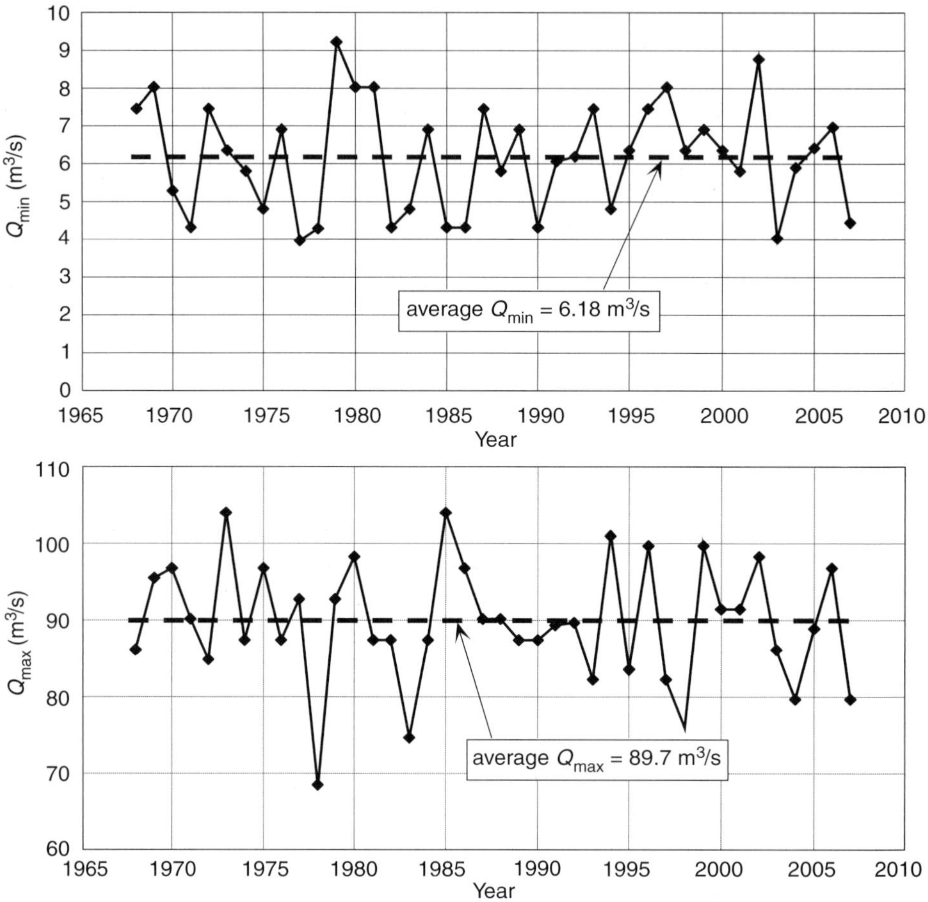

FIGURE 2.56 *Top*: Time series of minimum (*top*) and maximum (*bottom*) discharges of the Ombla Spring for the period 1968 to 2007. (From Kresic and Bonacci, 2010.)

discharges never exceed 104 m^3/s despite the large drainage area and very intensive rainfalls falling on it. The maximum spring discharge is limited by the capacity of the feeding conduits and a very low matrix porosity of the Mesozoic limestones. These limiting factors are clearly reflected in the hydraulic relationship between the spring discharge and the hydraulic head measured at two piezometers in the spring drainage area shown in Fig. 2.57. The portion of the karst aquifer in close proximity to the spring acts predominantly as a hydraulic system of pipes under pressure.

Low Flows

The demand for low-flow information is growing rapidly as a consequence of ever-increasing competition for water resources. In the United States, USGS districts are urged to analyze their low-flow data promptly and make it available to the public. The so-called 7Q10 flow is typically used to access low flows that may affect water management decisions for surface streams and springs. It is defined as the discharge at 10-year recurrence

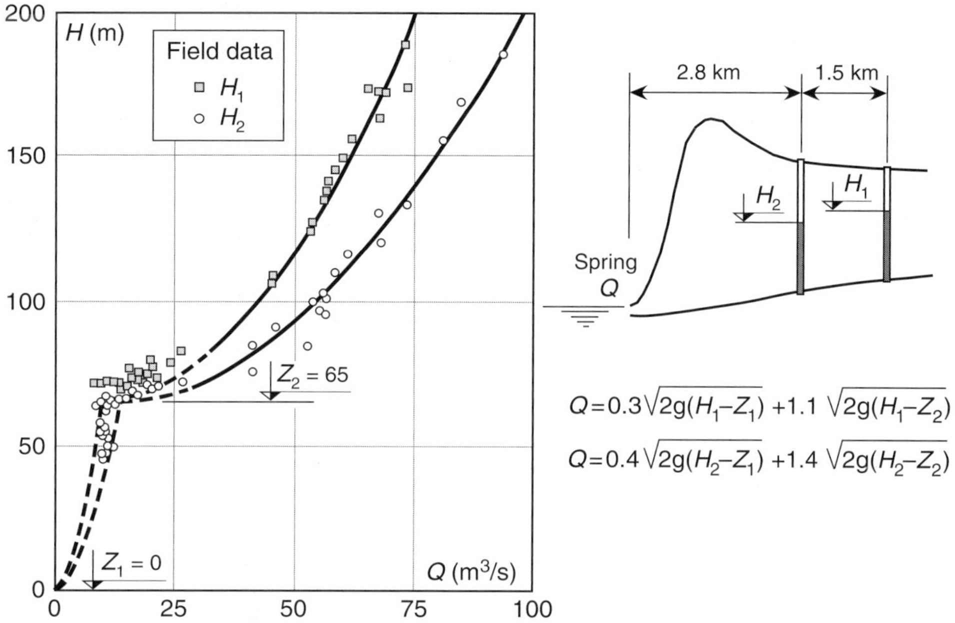

Figure 2.57 Quadratic relationships between the Ombla Spring discharge (Q) and the aquifer hydraulic head (H) measured at two piezometers. (Modified from Hajdin and Ivetic, 1976.)

interval taken from a frequency curve of annual values of the lowest mean discharge for 7 consecutive days (the 7-day low flow). This means that the 7-day low flow will be less than the 7-day 10-year low flow at intervals averaging 10 years in length; or the probability is 1/10 that the 7-day low flow in any 1 year will be less than the 7-day 10-year low. In general, the annual minimum n-day discharge is the minimum mean of any n (such as 2, 7, or 10) consecutive daily mean discharges in each water year. Each successive water year (starting on April 1) is analyzed to generate a time series of annual minimum n-day discharges for a gauging station using the entire period of record.

As recommended by the USGS (Raines and Asquith, 1997), frequency analysis of low flows is best performed using an alternative to traditional sample product moments (mean, variance, skewness, and kurtosis) that summarize the data and fit a probability distribution. This alternative approach uses L-moment statistics as introduced by Hosking (1990). The sample L-moment estimates are linear combinations of the ordered data and do not involve squaring or cubing the values as do the product moments. As a result, the L-moments are almost unbiased and exhibit small sampling variance, whereas the product moments are highly biased and can exhibit large sampling variance (Stedinger et al., 1993).

The Pearson type III distribution, which is in many cases the best one for fitting low-flow data, is a three-parameter probability distribution. The L-moment statistics of the data are used to estimate the parameters of the distribution. Either the data or logarithmic transformations of the data generally are used to estimate the moments (mean, standard deviation, and skewness). The log-Pearson type III distribution, which is recommended for flood (high-flow) frequency analyses (Interagency Advisory Committee on Water

Data, 1982), uses a logarithmic transformation of the measured flood peaks. Logarithmic transformations often are effective for reducing skewness of peaks that range over orders of magnitude. However, logarithmic transformations give greatly increased weight to peaks of small magnitude (low flows). When L-moment statistics are used, it is not necessary to use a logarithmic transformation. Also, a minimum 7-day discharge of zero is possible, which, if present, requires removal from the analysis because the logarithm of zero is undefined. A more complete description of frequency analysis using L-moments and the Pearson type III distribution can be found in Stedinger et al. (1993).

The low-flow discharge of a regulated stream might be affected by the operation of an upstream reservoir. Similarly, a low-flow discharge of a spring may be affected by groundwater withdrawals. Such effects might not be consistent from year to year depending on the timing and magnitude of reservoir releases or groundwater withdrawals. This inconsistency might affect the frequency-analysis assumption that the low-flow discharges are independent. For this reason, frequency analysis typically is used only on unregulated (natural) streams and unaffected karst aquifers. However, frequency analysis can be used meaningfully on regulated streams if no significant temporal trends exist in the regulated record and if the operational policies remain constant for the period. It is also desirable to test unregulated discharge data for temporal trends due to natural causes that could affect the assumption of independence among low-flow data.

The Mann-Kendall test (Helsel and Hirsch, 1992) is commonly used to determine if the annual minimum 7-day discharges for unregulated and regulated periods of record have significant trends. The test is a rank-based procedure (and therefore resistant to the effect of a small number of unusual values or outliers) used to determine whether a variable increases or decreases with time. The Mann-Kendall test in this application is a two-sided hypothesis test to indicate whether there is an increasing or decreasing trend in annual minimum 7-day discharges at the $\alpha = 0.05$ significance level. The null hypothesis (what is assumed to be true before the test) is that there is no trend; the alternative hypothesis (the situation anticipated to be true if the data show that the null hypothesis is unlikely) is that there is an increasing or decreasing trend. The null hypothesis is rejected in favor of the alternative hypothesis if the p-value from the test is less than or equal to the α-level. The p-value is the attained significance level (the significance level attained by the data), which is the probability of obtaining the computed test statistic, or one even less likely, when the null hypothesis is true (Helsel and Hirsch, 1992, p. 108).

Setting $\alpha = 0.05$ results in a 5% chance that the null hypothesis will be rejected when in fact it is true—that is, the conclusion from the test is that there is a trend when the true situation is that there is no trend. Thus, the confidence level of the test is 95%. The test yields a correlation coefficient, Kendall's tau (τ). The value of τ is influenced by sample size and ranges from -1 to $+1$. Tau (τ) is a measure of the strength of a trend (strength of the correlation between discharge and time). For example, if all discharges in a tested time series decrease with time, $\tau = -1$, whereas if all discharges increase with time, $\tau = +1$. The sign of τ thus indicates the direction of the trend. The trend analysis will help select appropriate homogeneous portions of the record, if any, suitable for the low-flow frequency analysis.

As emphasized by the USGS (2012), the reliability of 7Q10 from a gaging station record generally increases with the length of record but the inclusion or exclusion of a significant drought may be a dominant factor. The general relation for estimating the reliability of a point on a frequency curve, in terms of years of record and variability, assumes that the data are completely random. This assumption is not fully acceptable for low flows; a succession of highs or lows is common. Thus, the purely statistical estimate of reliability of a frequency curve is weak. Ordinarily, a fitted theoretical frequency curve should be used, but if that curve fails to fit the data at the lower end, a graphical interpretation should be made on the computer printout and the experimental graphical curve should be considered the basic frequency curve for low flows. It may also be desirable to provide seasonal low-flow frequency curves or curves for a year other than the climatic year (beginning April 1) for some uses and regions.

Extension of low-flow frequency curves to large recurrence intervals generally should be drawn or checked graphically. The low-discharge end of a mathematically fitted frequency curve with a positive skew will be affected by the data points having recurrence intervals of less than 2 years, but the shape of the low-discharge part of a low-flow frequency curve depends on a variety of factors such as

- The distribution of precipitation over the basin in time and space.
- The temperature regimen, which may permit the storage of water as snow for considerable periods and which also influences the evapotranspiration rate.
- The soil and geologic characteristics, which govern the recharge and discharge rates in the basin.

Low-flow characteristics at large recurrence intervals are less reliable than others, not only because of extrapolation but also because these low flows are more likely to

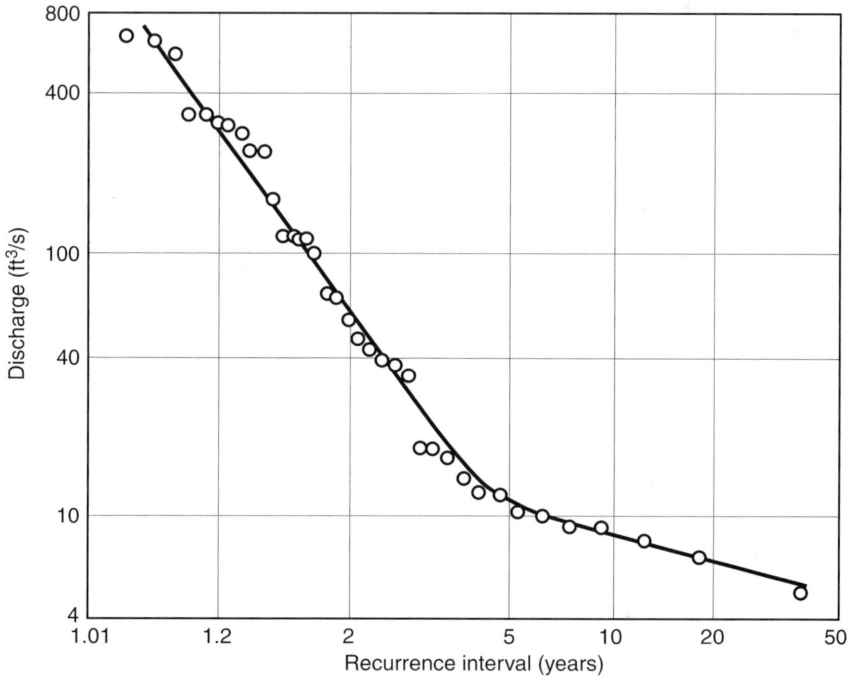

FIGURE 2.58 Low-flow frequency curve of annual minimum 7-day means for Suwannee River at White Springs, FL. (From Riggs, 1972.)

be affected by human activities. In addition, at a given site some of the annual low flows may be derived entirely from groundwater inflow to the stream, whereas others may include some flow from reduction in upstream storage during those years when frequent rains occur during the low-water season. Furthermore, the baseflow of a stream may be derived from several aquifers, not all of which contribute at all times. In other words, annual low flows at a site do not necessarily belong to a single population (Riggs, 1972). For example, the unusual shape of the frequency curve for Suwannee River (Fig. 2.58) reflects geologic differences within the basin. The annual minimum flow of this stream consists of two principal parts: (1) a flow ranging from several hundred cubic feet per second to zero from the headwaters in Okefenokee Swamp and (2) an inflow of 5 to 10 cubic ft per second from a limestone aquifer in a lower reach of the river.

As suggested by the USGS (2012), a low-flow report should include some statements about accuracy of results, even though it is not possible to assess that accuracy closely. On the other hand, the report should not put an accuracy value on each estimate. One can lose credibility by making unwarranted claims about the ability to estimate accuracy as well as by ignoring accuracy.

2.7 References

Adamski, J. C., 2000. Geochemistry of the Springfield Plateau aquifer of the Ozark Plateaus Province in Arkansas, Kansas, Missouri and Oklahoma, USA. *Hydrol Process*, 14:849–866.

Anderholm, S. K., and Heywood, C. E., 2003. Chemistry and age of ground water in the southeastern Hueco Bolson, New Mexico and Texas. U.S. Geological Survey Water-Resources Investigations Report 02-4237, Albuquerque, New Mexico, 16p.

Atkinson, T. C., 1977. Diffuse flow and conduit flow in limestone terrain in Mendip Hills, Somerset (Great Britain). *J Hydrol*, 35:93–100.

Baker, D. B., Richards, R. P., Loftus, T. Y., and Kramer, J. W., 2004. A new flashiness index: Characteristics and applications to midwestern rivers and streams. *J Amer Water Resour Assoc*, 40(2):503–522.

Bonacci, O., 1993. Karst springs hydrographs as indicators of karst aquifers. *Hydrol Sci*, 38(1):51–62.

Bonacci, O., 1995. Ground water behaviour in karst: Example of the Ombla Spring (Croatia). *J Hydrol*, 165(1-4):113–134.

Bonacci, O., 2001. Analysis of the maximum discharge of karst springs. *Hydrogeol J*, 9(4):328–338.

Boussinesq, J., 1904. Recherches théoriques sur l'écoulement des nappes d'eau infiltrées dans le sol et sur les débits des sources. *J Math Pures Appl*, Paris, 10:5–78.

Bureau of Reclamation, 2001. Water Measurement Manual. A water resources technical publication. Water Resources Research Laboratory, U.S. Department of Interior, Washington, D.C., various paging. Available at: http://www.usbr.gov/pmts/hydraulics_lab/pubs/wmm/.

Butt, P., 2003. Making discharge measurements within submerged conduits (Karst Environmental Services, Inc.). Significance of Caves in Watershed Management and Protection in Florida, Workshop Proceedings, April 16th and 17th, 2002, Ocala, FL. Florida Geological Survey Special Publication No. 53.

Clark, I. D., and Fritz, P., 1997. *Environmental Isotopes in Hydrogeology*. Lewis Publishers, New York, NY, 311p.

Coes, A. L., and Pool, D. R., 2005. Ephemeral-stream channel and basin-floor infiltration and recharge in the Sierra Vista subwatershed of the upper San Pedro basin, Southeastern Arizona. U.S. Geological Survey Open-File Report 2005-1023, Reston, VA, 67p.

Dansgaard, W., 1964. Stable isotopes in precipitation. *Tellus*, v. 16, no. 4, pp. 437–468.

Desmarais, K., and Rojstaczer, S., 2002. Inferring source waters from measurements of carbonate spring response to storms. *J Hydrol*, 260:118–134

Dooge, J. C. I., 1973. *Linear Theory of Hydrologic Systems*. Technical Bulletin No. 1468, United States Department of Agriculture, Washington, D.C., various paging.

Dooge, J. C. I., and O'Kane, J. P., 2003. *Deterministic Methods in Systems Hydrology*. IHE Delft Lecture Note Series, A.A. Balkema Publishers, Lisse, The Netherlands, 309p.

Dreiss, S. J., 1989. Regional scale transport in a karst aquifer. 1. Component separation of spring flow hydrographs. *Water Resour Res*, 25(1):117–125.

Drogue, C., 1972. Analyse statistique des hydrogrammes de decrues des sources karstiques. *J Hydrol*, 15:49–68.

Eastoe, C. J., Watts, C. J., Ploughe, M., and Wright, W. E., 2012. Future use of tritium in mapping pre-bomb groundwater volumes. *Ground Water*, 50(1):87–93.

Gat, J. R. and Carmi, I., 1970. Evolution of the isotopic composition of atmospheric waters in the Mediterranean Sea area. *J Geophys Res*, v. 75, pp. 3039–3048.

Geyh, M., 2000. Groundwater, saturated and unsaturated zone. In: Mook, W.G. (ed.), *Environmental Isotopes in the Hydrological Cycle: Principles and Applications*. IHP-V, Technical Documents in Hydrology, No. 39, Vol. IV, UNESCO, Paris, 196p.

Gumbel, E. J., 1941. The return period of flood flows. *Ann Math Statist*, 12(2):163–190.

Hajdin, G., 1981. An example of the hydraulic explanation of the flow out of a karst spring and the piezometric levels in its hinterland. Nas Krs, Bull. Speleological Society, VI(10-11):109–115.

Hajdin, G., and Ivetić, M., 1976. Jedan primer pokušaja objašnjenja hidrauličkih uslova u kraškom podzemnom toku na osnovu opažanja pijezometarskog stanja i izlaznih proticaja. XII jugoslavenski kongres mehanike, Sarajevo, sveske B3/B4.

Halford, K. J., and Mayer, G. C., 2000. Problems associated with estimating ground-water discharge and recharge from stream-discharge records. *Ground Water*, 38(3):331–342.

Hall, L. S., 1943. Open channel flow at high velocities. American Society of Civil Engineers,*Transactions*, Paper No. 2205, vol. 108, 1394p.

Healy, R. W., Winter, T. C., LaBaugh, J. W., and Franke, O. L., 2007. Water budgets: Foundations for effective water-resources and environmental management. U.S. Geological Survey Circular 1308, Reston, VA, 90p.

Helsel, D. R., and Hirsch, R. M., 1992. *Studies in Environmental Science. Statistical Methods in Water Resources*, vol. 49. Elsevier, Amsterdam, 522p.

Herman, E. K., Toran, L., and White, W., 2008. Threshold events in spring discharge: Evidence from sediment and continuous water level measurement. *J Hydrol*, 351:98–106.

Hosking, J. R. M., 1990. L-moments: Analysis and estimation of distributions using linear combinations of order statistics. *J R Stat Soc*, B, 52(1):105–124.

IAEA/WMO (International Atomic Energy Agency/World Meteorological Organization), 2006. Global network of isotopes in precipitation. The GNIP database. Available at: http://www-naweb.iaea.org/napc/ih/GNIP/IHS_GNIP.html.

Imes, J. L., Plummer, L. N., Kleeschulte, M. J., and Schumacher, J. G., 2007. Reacharge area, base-flow and quick-flow discharge rates and ages, and general water quality of Big Spring in Carter County, Missouri. U.S. Geological Survey Scientific Investigations Report 2007–5049, Reston, VA, 80p.

Interagency Advisory Committee on Water Data, 1982. Guidelines for determining flood-flow frequency. U.S. Geological Survey, Office of Water Data Coordination, Hydrology Subcommittee, Bulletin 17B, Reston, VA, variously paged.

Jevdjević, V., 1956. Hidrologija, I deo (Hydrology, Part 1; in Serbian). Hidrotehnički Institut Jaroslav Černi, Beograd, 404p.

Katz, B. G., Hornsby, H. D., Bohlke, J. F., and Mokray, M. F., 1999. Sources and chronology of nitrate contamination in spring waters. U.S. Geological Survey Water-Resources Investigations Report 99-4252, Suwannee River Basin, FL, 54p.

Kauffman, S., and Libby, W. S., 1954. The natural distribution of tritium. *Phys Rev*, 93(6):1337–1344.

Kay, R. T., Bayless, E. R., and Solak, R. A., 2002. Use of isotopes to identify sources of ground water, estimate ground-water-flow rates, and assess aquifer vulnerability in the Calumet Region of Northwestern Indiana and Northeastern Illinois. U.S. Geological Survey Water-Resources Investigation Report 02-4213, Indianapolis, IN, 60p.

Kilpatrick, F. A., 1970. Dosage requirements for slug injections of rhodamine BA and WT dyes. In: Geological Survey Research 1970, U.S. Geological Survey Professional Paper 700-B, pp. 250–253.

Kilpatrick, A., and Cobb, E. D., 1985. Measurement of discharge using tracers. Techniques of Water-Resources Investigations of the United States Geological Survey, Chapter A16, Book 3, Application of Hydraulics, Washington, D.C., 52p.

Kilpatrick, A., and Wilson, J. F., 1989. Measurement of time of travel in streams by dye tracing. Techniques of Water-Resources Investigations of the United States Geological Survey, Chapter A9, Book 3, Application of Hydraulics, Washington, D.C., 27p.

Kresic, N., 1991. Kvantitativna hidrogeologija karsta sa elementima zaštite podzemnih voda (Quantitative karst hydrogeology with elements of groundwater protection, in Serbo-Croatian). Naučna knjiga, Beograd, 196p.

Kresic, N., 1997. *Quantitative Solutions in Hydrogeology and Groundwater Modeling*. CRC Press, Boca Raton, FL, 461p.

Kresic, N., 2007. *Hydrogeology and Groundwater Modeling*, 2nd ed. CRC Press, Taylor & Francis Group, Boca Raton, FL, 807p.

Kresic, N., 2009. *Groundwater Resources. Sustainability, Management, and Restoration*. McGraw Hill, New York, NY, 852p.

Kresic, N., and Mikszewski, A., 2009. Groundwater recharge. In: Kresic, N. (ed.), *Groundwater Resources. Sustainability, Management, and Restoration*, Chapter 3. McGraw Hill, New York, NY, pp. 235–292.

Kresic, N., and Bonacci, O., 2010. Spring discharge hydrograph. In: Kresic, N., and Stevanovic, Z. (eds.), *Groundwater Hydrology of Springs: Engineering, Theory, Management and Sustainability*. Elsevier, New York, NY, pp. 129–163.

Libby, W. F., 1946. Atmospheric helium three and radiocarbon from cosmic radiation. *Phys Rev*, 69: 671–672.

Maillet, E. (ed.), 1905. *Essais díhydraulique souterraine et fluviale*. Herman et Cie, Paris, 1:218.

Maloszewski, P., and Zuber, A., 1996. Lumped parameter models for the interpretation of environmental tracer data. In: Manual on mathematical models in isotope hydrology. IAEA, Vienna, pp. 9–58.

Merlivat L. and Jouzel J., 1979. Global climatic interpretation of the deuterium-oxygen 18 relationship for precipitation. *J Geophys Res*, v. 84, pp. 5029–5033.

Mook, W. G., 1980. Carbon-14 in hydrogeological studies. In: Fritz, P., and Fontes, J. Ch. (eds.), Handbook of Environmental Isotope Geochemistry, Volume 1. The Terrestrial Environment, A, Chapter 2. Elsevier Scientific Publishing Co., New York, NY, pp. 49–74.

NNPSMP (National Nonpoint Source Monitoring Program), 2008. Surface water flow measurements for quality monitoring projects. Technotes 3. Available at: www.bae.ncsu.edu/programs/extension/... /technote3_surface_flow.pdf

Panagopoulos, G., and Lambrakis, N., 2006. The contribution of time series analysis to the study of the hydrodynamic characteristics of the karst systems: Application on two typical karst aquifers of Greece (Trifilia, Almyros Crete). *J Hydrol*, 329(3-4):368–376.

Parkhurst, D. L. and Appelo, C. A. J., 1999. User's guide to PHREEQC (Version 2): A computer program for speciation, batch-reaction, one-dimensional transport, and inverse geochemical calculations. U.S. Geological Survey Water-Resources Investigations Report 99–4259, 310p.

Patton, T., 2005. Giant springs: Climate impact on a first magnitude spring in Montana. Proceeding for Surface Water/Ground Water One Resource, 22nd Annual Meeting of the Montana Section of the American Water Resources Association, Bozeman, Montana, pp. 22–23.

Perry, E. C., Grundl, T., and Gilkeson, R.H., 1982. H, O, and S isotopic study of the ground water in the Cambrian-Ordovician aquifer system of northern Illinois. In: Isotope studies of hydrologic processes: Northern Illinois University Press. DeKalb, Illinois, pp. 35–45.

Plummer, L. N., and Friedman, L. C., 1999. Tracing and dating young ground water. U.S. Geological Survey Fact Sheet 134–99, 4p.

Plummer, L. N., Michel, R. L., Thurman, E. M., and Glynn, P. D., 1993. Environmental tracers for age-dating young ground water. In: Alley, W.M. (ed.), *Regional Ground-Water Quality*. Van Nostrand Reinhold, New York, NY, pp. 255–294.

Plummer, L. N., and Busenberg, E., 2007. Chlorofluorocarbons. In: Cook, P., and Herczeg, A. (eds.), *Excerpt from Environmental Tracers in Subsurface Hydrology*, Kluwer Academic Press. The Reston Chlorofluorocarbon Laboratory, U.S. Geological Survey, various paging.

Poreda, R. J., Cerling, T. E., and Solomon, D. K., 1988. Tritium and helium isotopes as hydrologic tracers in shallow aquifers. *J Hydrol*, 103:1–9.

Raines, T. H., and Asquith, W. H., 1997. Analysis of minimum 7-day discharges and estimation of minimum 7-day, 2-year discharges for streamflow-gaging stations in the Brazos River Basin, Texas. U.S. Geological Survey Water-Resources Investigations Report 97-4117, Austin, TX, 29p.

Riggs, H. C., 1968. Frequency curves. U.S. Geological Survey Techniques of Water Resources Investigations, Book 4, Hydrologic Analysis and Interpretation, Chapter A2, 15p.

Riggs, H. C., 1972. Low-flow investigations. U.S. Geological Survey Techniques of Water Resources Investigations, Book 4, Hydrologic Analysis and Interpretation, Chapter B1, 18p.

Risser, D. W., Gburek, W. J. and Folmar, G. K., 2005. Comparison of methods for estimating ground-water recharge and base flow at a small watershed underlain by fractured bedrock in the eastern United States. U.S. Geological Survey Scientific Investigations Report 2005-5038, Reston, VA, various pages.

Rorabaugh, M. I., 1964. Estimating changes in bank storage and ground-water contribution to streamflow. International Association of Scientific Hydrology Publication 63, IAHS Press, Wallingford, UK, pp. 432–441.

Rowe, G. L. Jr., Shapiro, S. D., and Schlosser, P., 1999. Ground-water age and water-quality trends in a buried-valley aquifer, Dayton area, Southwestern Ohio. U.S. Geological Survey Water-Resources Investigations Report 99-4113. Columbus, OH, 81p.

Rutledge, A. T., 1993. Computer programs for describing the recession of ground-water discharge and for estimating mean ground-water recharge and discharge from streamflow records. *U.S. Geological Survey Water-Resources Investigations Report* 93-4121, 45p.

Rutledge, A. T., 1998. Computer programs for describing the recession of ground-water discharge and for estimating mean ground-water recharge and discharge from streamflow records—Update. U.S. Geological Survey Water-Resources Investigations Report 98-4148, 43p.

Rutledge, A. T., 2000, Considerations for use of the RORA program to estimate ground-water recharge from streamflow records. U.S. Geological Survey Open-File Report 00-156, Reston, VA, 44p.

Ryan, M. and Meiman, J., 1996. An examination of short-term variations in water quality at a karst spring in Kentucky. *Ground Water* 34:23–30.

Schlosser, P., Stute, M., Dorr, H., Sonntag, C., and Oto, K. M., 1988. Tritium/^{3}He dating of shallow ground water. *Earth Planet Sci Lett*, 89:353–362.

Schlosser, P., Stute, M., Sonntag, C., and Muennich, K. O., 1989. Tritiogenic ^{3}He in shallow ground water. *Earth Planet Sci Lett*, 94:245–256.

Shapiro, S.D., Rowe, G., Schlosser, P., Ludin, A., and Stute, M., 1998. Utilization of the ^{3}H-^{3}He dating technique under complex conditions to evaluate hydraulically stressed areas of a buried-valley aquifer. *Water Resour Res*, 34(5):1165–1180.

Sophocleous, M., 2004. Ground-water recharge and water budgets of the Kansas High Plains and related aquifers. Kansas Geological Survey Bulletin 249, Kansas Geological Survey, the University of Kansas, Lawrence, Kansas, 102p.

Soulios, G., 1991. Contribution à l'étude des courbes de récession des sources karstiques: Exemples du pays Hellénique. *J Hydrol*, 127:29–42.

Stedinger, J. R., Vogel, R. M., and Foufoula-Georgiou, E., 1993. Frequency analysis of extreme events. In: Maidment, D. A. (ed.), *Handbook of Hydrology*, McGraw-Hill, New York, NY, Chapter 1–66.

Stichler, W., Trimborn, P., Maloszewski, P., Rank, D., Papesch, W., and Reichert, B., 1997. Environmental isotope investigations. In: Kranjc, A. (ed.), *Karst Hydrogeological Investigations in South-Western Slovenia*, Acta Carsologica, vol. 26, no. 1, pp. 213–236.

Thomas, C. W., and Dexter, R. B., 1955. Modern equipment for application of salt velocity method of discharge. Proceedings of the Sixth General Meeting, vol. 2, International Association of Hydraulic Research, the Hague, the Netherlands.

Trček, B., 2003. Epikarst zone and the karst aquifer behaviour: A case study of the Hubelj catchment, Slovenia. Geološki zavod Slovenije, Ljubljana, 100p.

Trček, B., 2007. How can the epikarst zone influence the karst aquifer hydraulic behaviour? *Environ Geol*, 51(5):761–765.

Trček, B., and Zojer, H., 2010. Recharge of springs. In: Kresic, N., and Stevanovic, Z. (eds.), *Groundwater Hydrology of Springs: Engineering, Theory, Management and Sustainability*. Elsevier, New York, NY, pp. 87–127.

USACE (U.S. Army Corps of Engineers), 1993. Hydrologic frequency analysis. Engineering manual 1110-2-1415, Washington, D.C., various paging. Available at: http://140.194.76.129/publications/eng-manuals/

USGS (United States Geological Survey), 2005. U.S. Geological Survey Streamgaging. Fact Sheet 2005-3131. Available at: http://ga.water.usgs.gov/edu/measureflow.html

USGS (United States Geological Survey), 2012. Low-flow program guidelines. Available at: http://water.usgs.gov/osw/pubs/memos/sw79.06.attachment.html

Vandike, J. E., 1996. *The Hydrology of Maramec Spring*. Missouri Department of Natural Resources, Division of Geology and Land Survey, Water Resources Report Number 55, Rola, MO, 104p.

White, W. B., 2010. Springwater geochemisatry. In: Kresic, N., and Stevanovic, Z. (eds.), *Groundwater Hydrology of Springs: Engineering, Theory, Management and Sustainability*. Elsevier, New York, NY, pp. 231–268.

CHAPTER 3

Drainage Areas in Karst

3.1 Introduction

When applying the term drainage area to a karst setting, most professionals first think of a likely discrepancy between the topographic and the subsurface (hydrogeologic) areas contributing water to a surface stream or a karst spring. This also often includes a close consideration of a possibility that a surface stream may be losing water. However, the term is not restricted to such regional scales only; it would be equally important to define surface and subsurface areas that are relevant for a local-scale project such as an industrial site spread over several flat-laying acres or hectares. As opposed to intergranular, unconsolidated porous media, for example, defining accurately where is the groundwater coming from and going to at such a site in karst would often not be a simple, straightforward task. In other words, it may not be enough to install five or ten monitoring wells and monitor the hydraulic head quarterly to provide the answers, which is the routine practice at non-karstic project sites. The most important thing to have in mind when delineating drainage boundaries in karst, at any scale and for any purpose, is that the dominant groundwater flow directions may, and usually do, change in three dimensions and in time, not necessarily only seasonally. The groundwater drainage boundaries may change temporarily as some preferential flow paths that are inactive during baseflow conditions become activated during storm flow conditions and high hydraulic heads in the aquifer. It is these changing flow paths and groundwater flow directions that result in transfer of water from one topographic drainage area (basin) to another. This subsurface transfer of water may also be permanent so that, in general, it is more appropriate to use term *karst groundwater basin* instead of karst drainage area as it emphasizes the central role aquifers play in most karst water systems.

Figures 3.1 and 3.2 show several settings illustrating the complexity of defining groundwater basins in karst. As emphasized throughout this book, the only predictable thing when working in karst is that the outcome of many specific investigations may be, and usually is, unpredictable. The role of a hydrogeologist therefore is to design and execute scientifically defensible site investigation and, as appropriate, present the case against endless data collection aimed at finding the last mysterious karst feature at their site. This, by default, includes discussions about what constitutes unacceptable financial risks, and/or risks to human health and the environment as they relate to the project at hand. Ultimately, the only real confirmation that a karst groundwater basin (or the relevant local-scale boundaries) has been fully delineated are the positive results of multiple dye-tracing tests performed at varying hydrologic conditions including baseflow

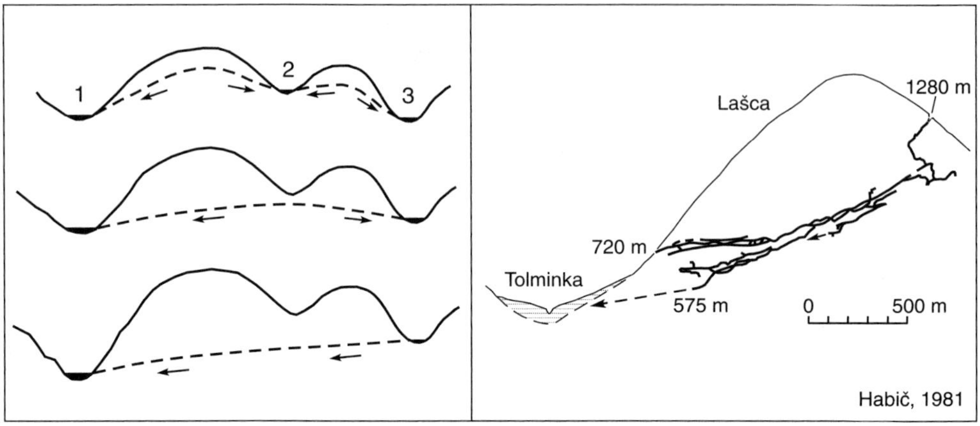

Figure 3.1 *Left*: Example of stream piracy in karst. (1) Permanent stream with the fastest rate of incision. (2) First gaining, then pirated (losing), and finally permanently dry stream. (3) First gaining and then pirated (losing) stream. (From Kresic, 2007.) *Right*: Poloska Jama in Slovenia, typical example of a cave formed by a sinking stream, with inactive upper horizons and hydrologically active lower horizons. (From Habič, 1981.)

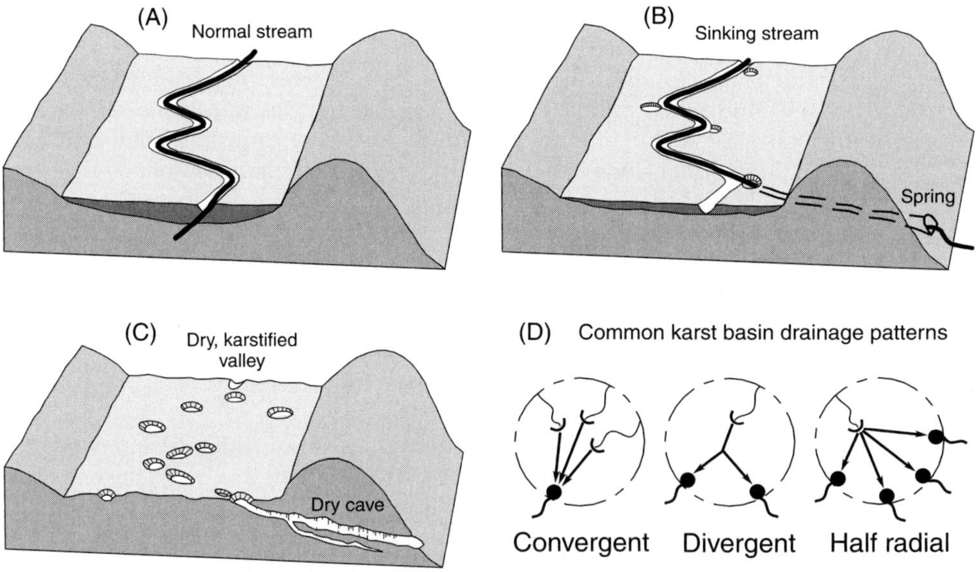

Figure 3.2 (A) to (C) Schematic diagram showing stages of development of a sinking stream initially flowing over a layer of nonsoluble sediments covering soluble carbonates. (A) Initial nonsinking stream, (B) sinking stream losing water to a spring located in different watershed, and (C) karstified dry valley and the former spring converted into a dry inactive cave due to lowering of the base of karstification. (From Kresic, 2010.) (D) Three most common karst drainage patterns in West Virginia, the United States, based on dye-tracing studies. (From Jones, 1977.)

and seasonal storm flows. However, although always advisable in karst, dye tests may be cost-prohibitive or unnecessary in a particular case. With or without dye-tracing tests, the most defensible approach to delineating karst groundwater basins as well as smaller areas of interest should include

- Determination of as many potential preferential flow paths as possible utilizing multiple investigation methods.
- Selection and confirmation of those flow paths pertinent to the issue at hand (e.g., those susceptible to contamination or those likely providing most flow to downgradient receptors/users).
- Detail hydrogeologic investigation of the selected flow paths including flow measurements, aquifer testing, and determination of fate and transport parameters in case of groundwater contamination.

3.2 Preferential Flow Paths

Many times the author of this book wished there was a talented dowser somewhere he could engage for finding preferential flow paths in karst, albeit in secret. The only requirement the dowser would have to meet is that she or he guarantees for the work product. Unfortunately, the search is ongoing. The main reason from the author's experience is that no dowser is willing to provide such a warranty. However, a few hydrogeologists, geophysicists, and some other professionals do not appear hesitant to guarantee they will locate and define all relevant preferential flow paths in a particular karst aquifer. Those professionals that are a lot more experienced may issue this warranty under condition they are given enough funds and time, including possibility to modify their work as needed once they start executing it. As this condition, namely unlimited funds, time, and work change orders, is hard to negotiate with any investor, the consulting world has consolidated into the following:

1. Companies and professionals that do not consider work in karst.
2. Companies and professionals that guarantee for work products such as
   a. accurate determination of preferential flow paths
   b. required water well yield, drawdown, and capture zone delineation
   c. remediation (cleanup) of contaminated groundwater to applicable standards including drinking water standards (e.g., Maximum Contaminant Levels or MCLs in the United States)
   d. Design and construction of nonleaky dams and surface reservoirs.
3. Companies and professionals that work in karst by fully engaging clients and regulators, and by explaining karst intricacies before and as they become apparent. Most importantly, they proactively manage expectations of their clients and other stakeholders but do not make unlimited warranties of the types listed under #2.

Defining preferential flow paths is inseparable from developing a conceptual site model (CSM) in karst; it is also the most important element of such a model in most cases. As discussed by Kresic and Mikszewski (2012), the main purpose of any CSM in hydrogeology is to provide a single product comprising pictures (figures), text, and, if necessary, animations, where all the information about the site is easily accessible and can be used for decision making at any stage of project implementation. At the same time, it is very important to understand that the CSM is a dynamic entity, continuously refined and updated. At the initial stage of the project, it is possible to have two or three competing preliminary CSMs since the readily available information may not lead to a unique concept. As the project progresses and new information is collected, the CSM becomes more detailed and quantitative, helps plan additional investigations, and focuses the project team on feasible solutions. These solutions (such as design of a well field for water supply, or a bioremediation system for aquifer restoration) will be possible only when there is consensus among all involved parties that the final CSM accurately represents the karst hydrogeology of the site at the scale of interest. Although for the convenience of the reader the following sections discuss several common methods of determining preferential flow paths in karst separately, it cannot be emphasized enough that these methods often overlap and the best results are obtained when they are utilized complementarily.

In an informative review article, Worthington and Ford (2009) present results of various laboratory experiments and models simulating formation of preferential flow paths in karst. In soluble carbonates, there is a very strong tendency for developing the so-called self-organized permeability, which results in aquifer drainage by a few or just one large spring, similarly to the hierarchy of surface water drainages. The authors conclude that the invariable result of flow through carbonate aquifers is that the positive feedback between increasing flow and increasing dissolution results in a high-permeability, self-organized network of channels (conduits). If the flux of water through an aquifer is minimal, such as in some confined aquifers, then the formation of channel networks is retarded. Conversely, in confined aquifers where there is a substantial flux of water as well as in unconfined aquifers, channel networks are likely to convey most of the flux of water through the aquifer after periods of 10^3 to 10^6 years following the onset of substantial flow through the aquifer. This range of times is short in comparison to the time that most unconfined limestone aquifers have been functioning, and so it is reasonable to infer that most such aquifers should have well-developed channel networks. The main consequence of the self-organized permeability of karst aquifers is that the drainage areas of karst springs often extend beyond topographic divides, provided the underlying carbonates are relatively thick and extend beyond just one surface water drainage basin. This is also the main reason why karst aquifers give rise to the world's largest springs.

3.2.1 Geomorphologic and Geologic Methods

Distinct surface and subsurface landforms that shape karst landscape also provide various direct and indirect evidences of preferential flow paths. The most obvious are submerged conduits explored by cave divers, a passable cave with a permanent subterranean stream, or a series of karst windows, sinks and resurgences like the ones shown in Fig. 3.3. However, the subsurface karst landforms like these may not be always present or accessible and, in any case, seldom represent all potential pathways in an aquifer. Therefore, rather than immediately mobilizing a team of cavers in search of pothole and cave entrances and passages (Fig. 3.4), the first preliminary

FIGURE 3.3 Oblique winter aerial photograph of karst window at DePriest Cave (foreground) and the resurgence of Charley Run into Culverson Creek (center) near Fort Donnally, West Virginia. (Photograph courtesy of William Jones, Karst Waters Institute.)

CSM of likely flow patterns in the underlying karst aquifer is made using readily available information. This includes topographic maps, geologic maps, aerial photographs, and remote sensing imagery. These sources of visual information are best analyzed when integrated within a common geographic information system (GIS) platform.

FIGURE 3.4 Cavers happy to see the light of day after exploring muddy, wet passages in a cave in southwestern Georgia, the United States.

Thanks to rapid developments in digital cartography, remote sensing technology and easy access to various Internet (online) sources of earth imagery and digital elevation data, it is now possible to perform a very detailed visual analysis of geomorphologic features without even visiting a site. This, however, is not recommended by the author—one should always make every attempt to visit the site she or he is working on. Planning such a visit would also greatly benefit from having a living three-dimensional (3D) image of the topography. Whatever the final decision on visiting the site may be, the site topography can now be easily displayed in three dimensions, rotated, and viewed from different angles by using a variety of commercial and public domain computer programs (see Kresic and Mikszewski, 2012). Aerial and satellite images, geologic maps, and other thematic maps can be easily draped over digital 3D topography and analyzed in the same fashion (Fig. 3.5). In addition, every working professional should have ready access to Google Earth, which is now *the* default, free platform for visualization of land surface features. Although topographic maps printed on paper will continue to be utilized for years to come by default (and in some parts of the world may still be the only available option), a combined use of digital elevation models (DEMs) and remote sensing imagery offers many advantages when developing a CSM for a karst site as illustrated in Fig. 3.6. Another advantage of the 3D DEM visualization is that it is likely easier for nontechnical audiences to understand than two-dimensional contour maps, and it is therefore well suited for presentations and reports prepared for the general public.

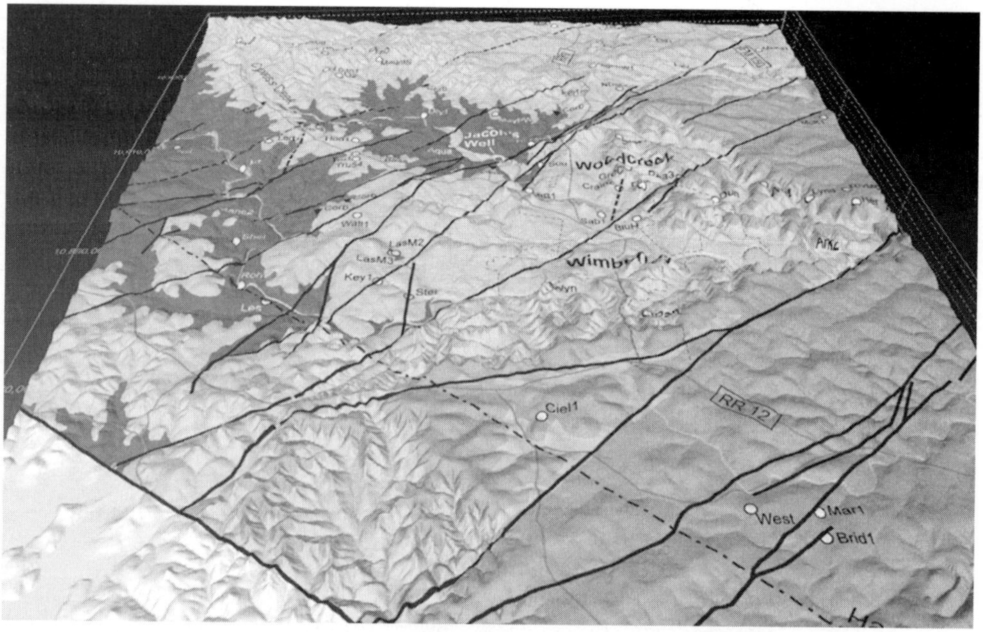

Figure 3.5 Portion of the geologic map from Wierman et al. (2010) draped over a digital elevation model. Thick black lines are faults, and different shaded gray areas (originally in color) are different hydrostratigraphic units. (Courtesy of Gavin Hudgeons, AMEC Environment and Infrastructure, Inc.)

Figure 3.6 Digital elevation model created from high-resolution data collected by airborne LiDAR (Light Detection and Ranging) instrument. Lines of sinkholes like these near Ocala, Florida, developed in surficial sandy sediments, indicate possible preferential flow paths in the underlying karst aquifer. Note extraordinary detail obtained by the LiDAR technology; depth of the wider sinkhole by the road is about 6 m.

FIGURE 3.7 Georeferenced aerial photographs are available for download from GIS Web sites of many state agencies and can be displayed in Google Earth taking full advantage of three-dimensional visualizations. This photo of Bottomless Lakes, large sinkholes developed in gypsum, is available from the New Mexico Resource GIS at http://rgis.unm.edu hosted at the University of New Mexico.

National Elevation Dataset (NED) high-resolution data from the U.S. Geological Survey (USGS) is typically derived from LiDAR (Light Detection and Ranging) technology or digital photogrammetry, and is often break line enforced to account for linear relief features. If collected at a ground sample distance no coarser than 5 m, such data may also be available within the NED at a resolution of 1/9th arc-second, now downloadable for many portions of the United States (e.g., see USGS's Seamless Data Warehouse at http://seamless.usgs.gov/, or The National Map Viewer at http://viewer.nationalmap.gov/viewer/). The new generation of USGS topographic maps at scale 1:24000 (topographic quadrangles) is based on this high-resolution elevation data, incorporates high-resolution photo images, and has various layers of information all available in digital, georeferenced, pdf format. Examples are shown in Figs. 3.7 through 3.9. Both the digital quadrangle maps, with all their layers including aerial photographs, and the accompanying high-resolution NEDs (digital raster files) can be downloaded and analyzed as illustrated. The NED files can be contoured independently and displayed in 3D with some of the commonly used programs such as Surfer, Geostatistical Analyst, or Spatial Analyst (see Kresic and Mikszewski, 2012).

Figure 3.8 shows contours and hydrography layers in portion of the new 2011 USGS Lewisburg, West Virginia quadrangle illustrating unique features of karst topography. Numerous sinkholes and other closed depressions developed in the Greenbrier

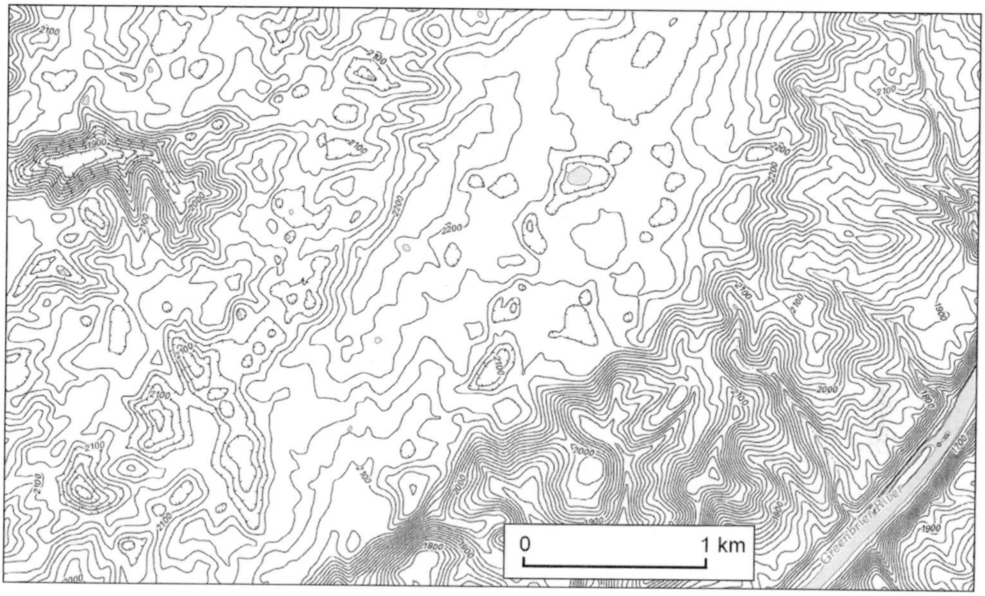

FIGURE 3.8 Contours in part of the new 2011 USGS Lewisburg, West Virginia topographic quadrangle. Contour interval is 20 ft. Former surface drainage is highly disintegrated but parts of it can be inferred based on alignments of sinkholes, shapes of elongated closed depressions, and general orientation of the contour lines.

Limestone are marked with closed contour lines. Dashes on contours point toward centers of depressions. Closed depressions and the absence of surface drainage (flowing streams) are the main characteristic of mature karst terrains. In contrast, less permeable rocks such as shales of the McCrady Formation and Pocono Group in the eastern portion of the map have densely developed surface drainage. Figure 3.9 shows a 3D view of the high-resolution NED for another portion of the same quadrangle and some of the adjacent areas where pronounced lines of sinkholes and other relief lineaments are clearly visible. These linear features, formed by karstification, faulting, and contrasting lithology, are main indicators of potential preferential flow paths within the underlying karst aquifer. The entire karst area of the Lewisburg quadrangle drains at Davis Spring, the largest spring in West Virginia that is located on the adjacent Asbury, West Virginia quadrangle (Kresic and Mikszewski, 2012).

Because of the legibility issues, printed topographic maps are limited by the contour interval even when a finer-resolution elevation data is available. This is illustrated with Fig. 3.10. The map's contour interval of 20 ft is not sufficient to depict smaller sinkholes, which are clearly visible on the aerial photograph when zoomed in. Having the NED file of the same quadrangle enables contouring at any desired interval including 3 ft that results in identifying more sinkholes as shown on the bottom in Fig. 3.10. Better yet is to use a combination of contours and a color-shaded 3D surface map that can be rotated, zoomed in, and displayed with different vertical exaggeration, which all greatly helps in visual identification of various geomorphologic features.

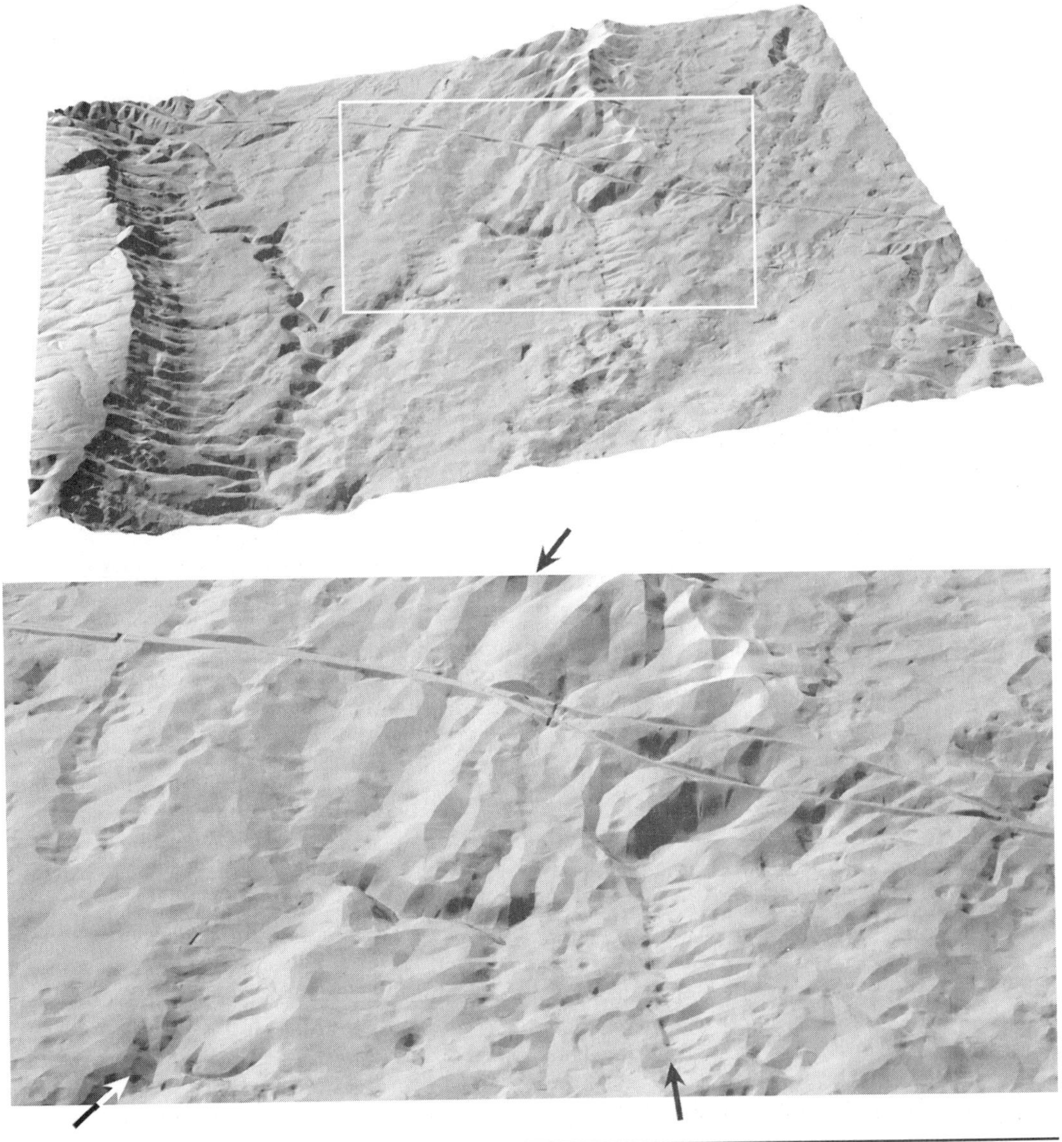

FIGURE 3.9 *Top*: Three-dimensional view of a portion of the Lewisburg, West Virginia topographic quadrangle and the adjacent areas to the west. *Bottom*: Blowup of the three-dimensional surface showing fine detail depicted by the high-resolution National Elevation Dataset (NED). Arrows indicate some of the lineaments. Most of the area is underlain by karstified limestone, dotted with sinkholes and other closed depressions of varying sizes, and with numerous small intermittent streams and dry karstified valleys.

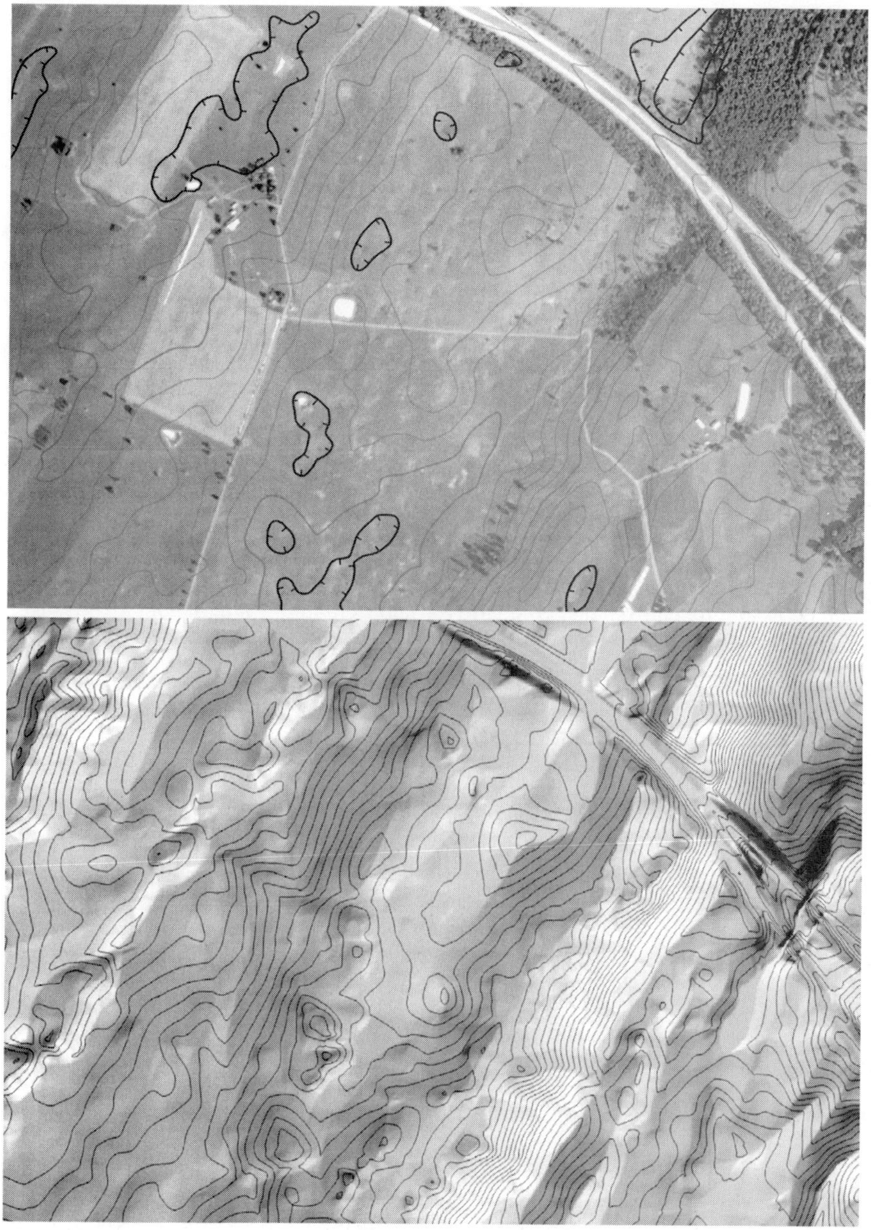

FIGURE 3.10 *Top*: Aerial photograph and original 20-ft contours in a portion of the Lewisburg, West Virginia topographic quadrangle. Contours of closed depressions as indicated by dashes on the original map are emphasized. *Bottom*: 3-ft contours derived from the high-resolution National Elevation Dataset (NED) file using Surfer and displayed together with a shaded relief map also created in Surfer. Note interstate highway for scale.

FIGURE 3.11 Submerged lower entrance to Uvac Cave in Western Serbia and a dry karst valley (to the right) that roughly parallels the cave for approximately 1 km. Almost all passages in this kilometer-long cave are not hydrologically active and the questions are: Where are the current active preferential flow paths in the aquifer? Are they below the explored cave, below the dry valley, in between, or somewhere else in the vicinity?

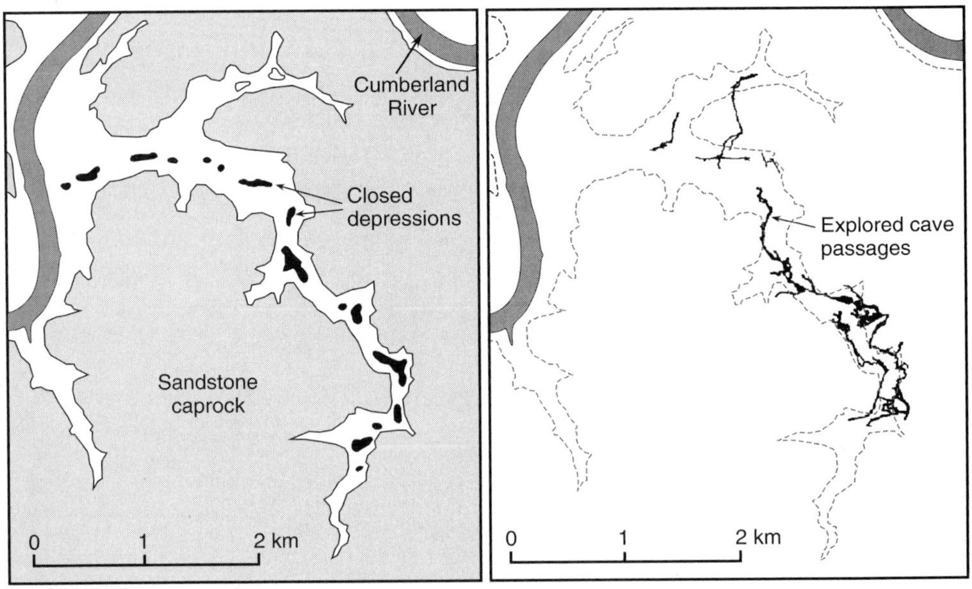

FIGURE 3.12 *Left*: Closed surface depressions in the Cave Creek dry valley, Pulaski County, north-central Kentucky. Stream erosion breached low-permeable caprock of sandstone and shale and exposed underlying limestone in which these depressions are developed. *Right*: Three-dimensional network of cave passages totaling 17 km in length closely follows the valley axis. (Modified from Ewers, 1985; cave information provided courtesy of Louis Simpson.)

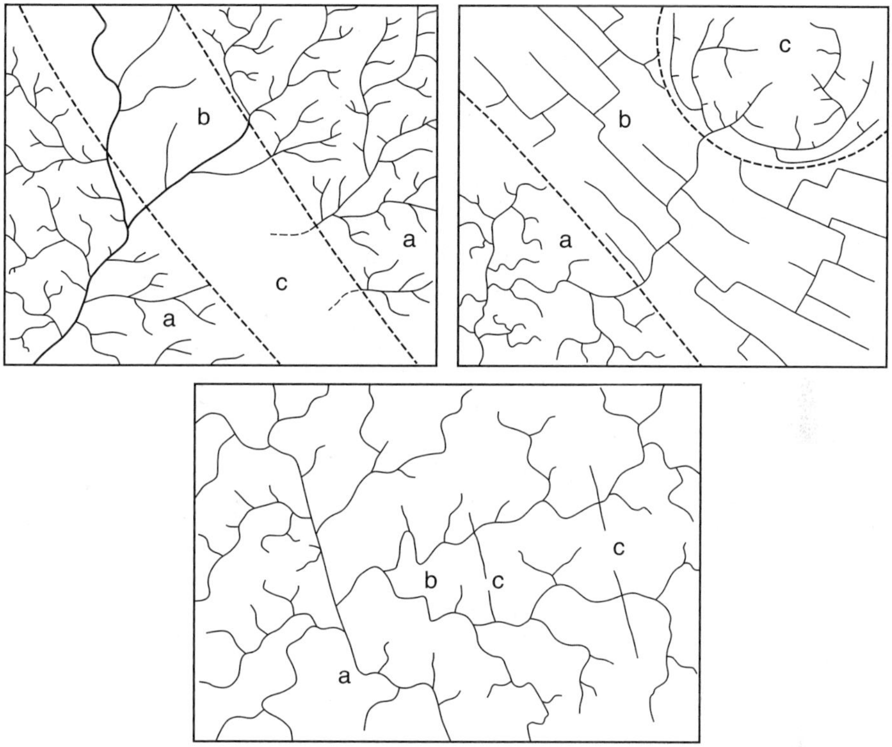

Figure 3.13 Criteria for interpretation of drainage patterns on topographic maps and remote sensing imagery. *Top left*: (a) drainage density is high in less permeable rocks; (b) more permeable rocks have fewer drainage features; (c) surface drainage is disintegrated or missing in karstified rocks. *Top right*: (a) dendritic drainage is characteristics of homogeneous and isotropic geologic terrains; (b) rectangular drainage is common in folded stratified (layered) sedimentary rocks dissected by perpendicular fractures and faults; (c) circular drainage (ring pattern) is characteristic for domes or partially destroyed calderas. *Bottom*: Faults can be inferred from (a) drainage features such as long straight stream segments; (b) aligned segments of the neighboring streams that change flow directions abruptly; and (c) segments of different streams extending (aligning) over the ridges. (Modified from Dimitrijević, 1978.)

In addition to typical karst features shown in Figs. 3.11 and 3.12, which always indicate possible presence of preferential flow paths, the geomorphologic analysis is particularly useful when deciphering geologic reasons for a certain type of surface drainage as illustrated schematically in Fig. 3.13. Sometimes, lineaments indicated this way may represent preferential flow paths transferring groundwater water across different geologic units thus connecting seemingly separate karst aquifers.

For various, and sometimes mystifying reasons, geologic maps do not always show otherwise obvious features that may have an underlying geologic basis and are of critical importance to a hydrogeologic CSM. One such example is the Balcones Fault Zone in Texas responsible for the formation of the geometry of the Edwards Aquifer, one of the most extensive and prolific karst aquifers in the world. Most, if not all, geologic maps of this large area, at varying scales, show numerous predominant faults of the

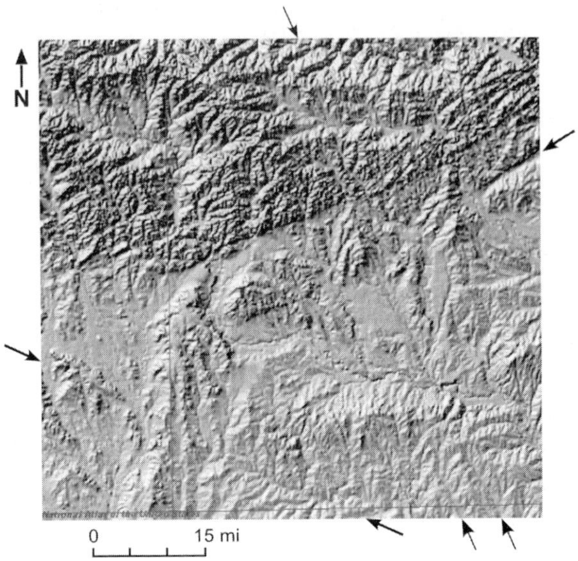

FIGURE 3.14 Shaded relief map of the greater San Antonio, Texas area available from the National Atlas of the United States (www.nationalatlas.gov). Arrows indicate some of the lineaments visible on this small-scale map.

northeast-southwest (NE-SW) strike, with a handful of smaller faults with different orientation. The NE-SW faults are universally interpreted as the most important geologically. However, even very general relief maps at small scales such as the one in Fig. 3.14 clearly show long prominent lineaments trending in other directions, including perpendicular to the main NE-SW system of faults. Even a nongeologist would be able to identify northwest-southeast (NW-SE) striking faults in this figure. In addition to being perfect candidates for preferential groundwater flow paths within the Edwards Aquifer, these faults may also be transferring significant quantities of groundwater from the adjacent Trinity Aquifer to the Edwards Aquifer. Advantages of using high-resolution DEMs to analyze topographic lineaments in the midsection of the Balcones Fault Zone between San Antonio and Austin are illustrated in Kresic and Mikszewski, 2012.

A surprisingly common fallacy when interpreting lithology in the context of karst groundwater basins is to consider noncarbonates and non-karstified rocks as perfect barriers to groundwater flow. Some working professionals and inexperienced karst hydrogeologist may even go so far to differentiate between "less" and "more" karstified rocks and then proceed to conclude that "less" karstified rocks are, in fact, devoid of preferential flow paths and possibly even act as barriers to groundwater flow in the "real" karst. However, nothing can be farther from the truth and also possibly more dangerous if such reasoning is implemented to develop groundwater vulnerability maps (see Sec. 10.3).

Another example from the Edwards Aquifer shown in Figs. 3.15 and 3.16 further illustrates this point. In the geologic map in Fig. 3.15, the darker gray areas are the lower Edwards Limestone and the lighter gray areas are the upper Edwards Limestone. Faults with more than 50% offset of the Edwards Limestone were believed to be barriers to groundwater flow. Prior to tracer testing results, the conventional thought was that groundwater would flow to the southwest around the faults with flow paths of more than 30 miles. More recent tracer tests performed by the Edwards Aquifer Authority clearly

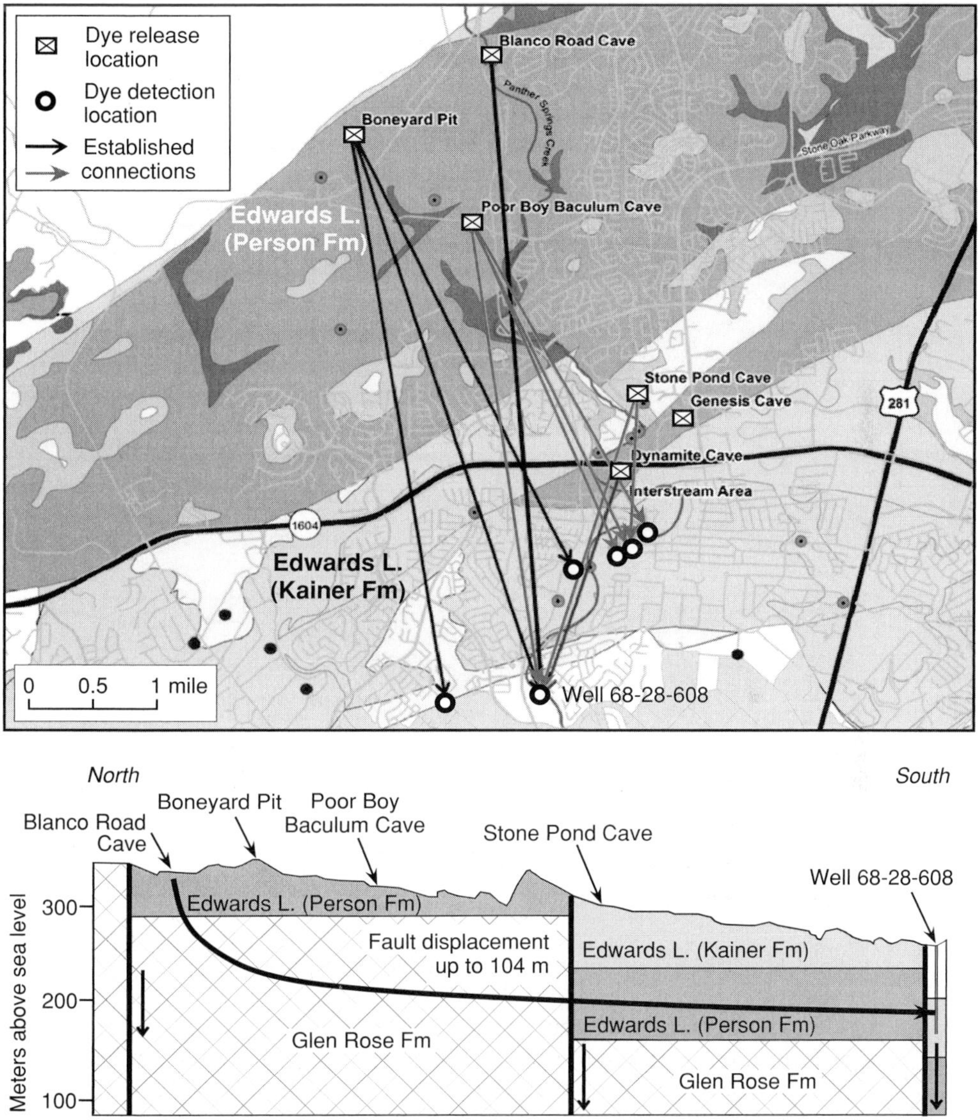

FIGURE 3.15 Geologic map and cross section of an area near San Antonio, Texas, with dye-tracing locations and monitoring. Darker gray areas are the lower Edwards Limestone and the lighter gray areas are the upper Edwards Limestone. The orientation of virtual flow paths is indicated by the arrow lines between the locations of tracer introduction and recovery at monitored wells. (Courtesy of Geary Schindel, Chief Technical Officer of the Edwards Aquifer Authority, San Antonio, Texas.)

showed that faults do not act as barriers. Dyes rapidly traversed more than six faults and one fault with as much as 300 ft of displacement, with tracer test velocities ranging from 80 ft per day to >12,000 ft per day (Schindel et al, 2009). The geologic cross section in Fig. 3.15 shows displacement of the major fault and communication between the Upper Glen Rose Limestone and the Edwards Limestone. The Upper Glen Rose Limestone was

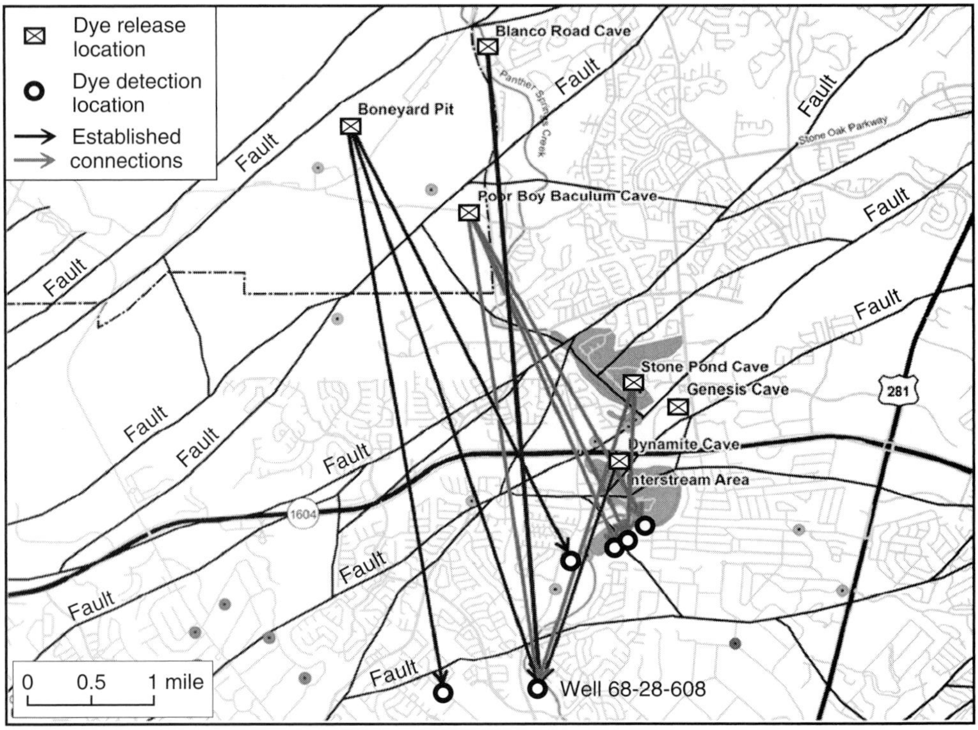

FIGURE 3.16 Faults accompanying the geologic map shown in Fig. 3.15. Faults with more than 50% offset of the Edwards Limestone were believed to be barriers to groundwater flow. The results of dye-tracing tests show dyes rapidly traversed more than 6 faults with velocities ranging from 80 ft per day to >12,000 ft per day. (Courtesy of Geary Schindel, Chief Technical Officer of the Edwards Aquifer Authority, San Antonio, Texas.)

previously thought to be a low-permeable porous medium providing an effective barrier to groundwater flow. Incidentally, the orientation of virtual flow paths indicated by the arrow lines between the locations of tracer introduction and recovery at monitored wells is very similar to the orientation of major topographic lineaments in Fig. 3.14. None of these lineaments is shown on the current geologic maps of the area.

Cvijić recognized very early on that solution cavities develop in all soluble carbonate rocks and, although the shape, size, and other characteristics of such cavities vary based on the rock type, they inevitably interconnect and form passages, however large or small (Fig. 3.17). Cavers and speleologist visiting many caves in The Appalachians, the United States, can also attest that cave passages often traverse layers of noncarbonate, nonsoluble rocks along weakened fracture and fault zones. Although such layers may act as local barriers to groundwater flow in the overlying and the underlying carbonates (Fig. 3.18), one would make a potentially serious conceptual error if such layers, everywhere they are mapped geologically, were considered as perfect flow barriers. An initial breach in a sandwiched noncarbonate layer, perhaps due to faulting as in Fig. 3.18, can rapidly grow because of the mechanical erosion by the flowing water combined with a collapse of the noncarbonates into cavities.

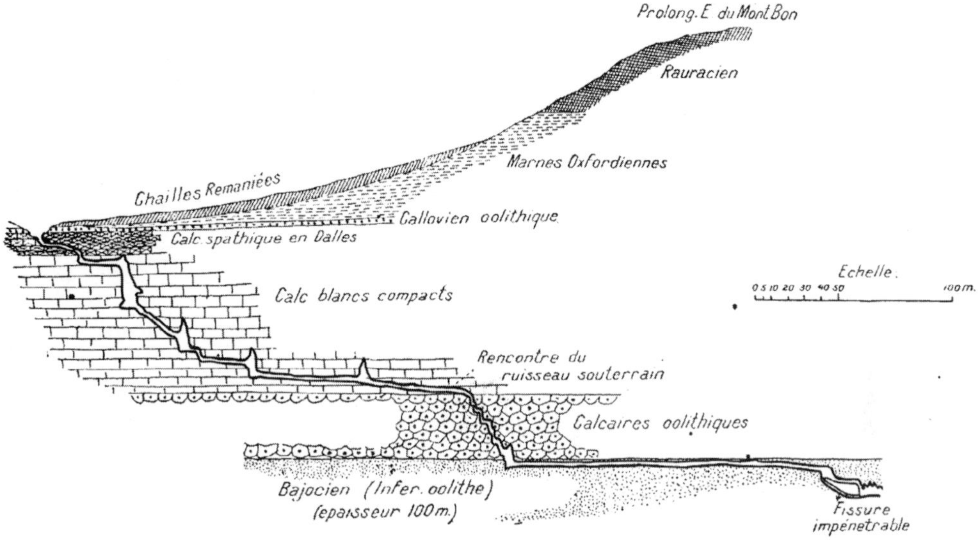

FIGURE 3.17 Cave Paradis in French Jura developed in different types of limestone. (From Cvijic, 1926.)

FIGURE 3.18 Lost Waterfall in the Monongahela National Forest, 6 miles (10 km) west of Elkins, West Virginia. The Taggard shales separate the Greenbrier formation into an upper limestone unit and a lower limestone unit. Water issues from the ground as a spring, cascades over the vertical surface outcrop of the Taggard formation, and immediately sinks back into the ground as it reenters solution channels in the lower limestone units. (Photograph and text courtesy of James van Gundy; accessed at http://epod.usra.edu/blog/2009/10/lost-waterfall.html.)

Figure 3.19 Thick limestone beds in the Peruvian Andes. (Photograph courtesy of Marc Etienne, AMEC Environment and Infrastructure.)

Unfortunately, predicting which features of the geologic fabric in an area underlain by soluble carbonate rocks will have the predominant role in the development of preferential flow paths is half-hazardous at best. Geomorphologic and geologic observations at the land surface or in dry cave passages can only provide hints as to likely preferential flow paths in the saturated zone. For example, one reason for the preferential dissolution of the carbonate layer shown in Fig. 3.19 may be a slightly different chemical composition but other reasons are possible and may not be fully understood just by a simple field observation. This visible discontinuity certainly provides for an enhanced infiltration and thus continuing dissolution with depth. However, it is also certain that at some point it will change its size and orientation for a variety of reasons including folding, faulting, and changes in sedimentary environment. Projecting its 3D orientation and extent in the subsurface therefore cannot be anything but a guess. A more accurate prediction would have to include other methods described further, including more expensive ones such as investigative drilling in case of projects with high financial and other risks. For such projects, which include but are not limited to centralized water supply, mining, dam construction, and tunneling for infrastructure, the investigative drilling is performed by default and in phases.

The best possible approach, although also the most expensive, is a continuous rock coring accompanied with a suite of geophysical logging methods. This would enable collection of confirmatory data for surficial geophysical surveys that, in turn, can gather useful information over large areas at comparably much lower cost than drilling alone. In any case, geologic logging of the rock core will provide an invaluable insight into the

Figure 3.20 Cores of limestone and dolomite retrieved from several borings at varying depths below ground surface (bgs) during drilling campaigns at prospective mining sites. (A and B) Bands of vugs developed along steeply dipping, minute secondary porosity; cores are collected within and adjacent to highly fractured intervals at similar depths (>500 m bgs) in two different borings, indicating a likely preferential flow path; (C) small cavity filled with clay at 208 m bgs indicating a possible preferential flow path in the vicinity that facilitated the transport of clay; and (D) slice through Mesozoic dolomite core showing the effects of dissolution and chemical alteration resulting in high matrix porosity; transmissive intervals like this are very conducive to development of preferential flow paths.

key characteristics of the carbonates such as intervals of enhanced porosity, and presence and orientation of fracture zones and solution cavities at different depths. Interpretation of all this information in the 3D space would often result in a fairly accurate concept of possible preferential flow paths. Figure 3.20 shows rock core from several deep investigative borings at prospective mining sites in the Peruvian Andes. The following discussion based on works of late George Sowers, senior consultant at Law Engineering (now AMEC Environment and Infrastructure, Inc.), provides useful tips for interpretation of carbonate porosity and karst features also applicable to geologic logging of rock cores.

Freshwater circulating through the pores increases solution by substituting fresher water with low saturation for the more saturated water previously in contact with the calcium carbonate. Increased quantities of water from rainfall and increased flow velocities due to increased energy gradients contribute to increased solution. The primary effect of solution is to enlarge all the pores and reduce the proportion of solid calcium carbonate in each unit volume of rock. The consequences are twofold. The increased porosity enhances water circulation and thus aggravates further solution. Secondarily, the increased porosity produces an increase in stress within the remaining solid framework of rock. This directly reduces the strength of the rock as well as possibly indirectly contributing to accelerated solution. Any new tectonic movement (neotectonics) may thus induce new fracturing of this weakened and already brittle limestone. Once the primary (in young carbonates) and secondary (in older carbonates) porosity is enlarged enough to the point of cavity formation, the process of development of preferential flow paths (channels, conduits) accelerates and is self-aggravating; as the channels become larger, the seepage concentrates in them, thus increasing the rate of solution and the concentration of flow.

The solution is most intense at the points of bonding between the individual particles, possibly because of the stress effect on solution, or because the delicate crystal structure of the calcareous bonds is easily damaged. As a result, continued solution partially destroys the induration or cementing of the rock. The indurated materials are converted into a mass of discrete fragments, a homogeneous calcareous sand, or calcareous silt that resembles a talcum powder.

Any change in the seepage environment may induce reprecipitation of the calcium carbonate out of the water. Thus, the calcium carbonate dissolved at one point may add to the cementing of the rock at another point. Sometimes, the calcium carbonate is reprecipitated in sufficiently large openings as beautiful, clear calcite crystals. At other times, it forms an interlocked, crystalline mass of very hard limestone. Finally, solution of the calcium carbonate may be accompanied by chemical precipitation (replacement) by a less soluble material such as magnesium carbonate or silica. As a result, the chemistry of the original rock formation slowly changes. Calcite or calcium carbonate is altered to dolomite (containing magnesium carbonate). Silica accumulates in nodules of chert that tend to grow and force the soft limestone aside.

In conclusion, limestone is not an inert material. It is dynamic, changing with the environment. Moreover, because of the high solubility and chemical activity of calcium carbonate, the changes occur much more rapidly than most other geologic changes; that is, during one's lifetime, the life of a structure, or even a construction project. It is this dynamic character of the rock that leads to engineering problems (Sowers, 1974).

Morphometric Neotectonic Analysis

The analysis of neotectonic activity, or recent tectonic movements, plays an important role in delineating zones of rock mass that are disintegrating due to faulting and fissuring. In the case of karst aquifers, this also means delineation of the zones of higher porosity and karstification developed as a result of the enhanced dissolution of mechanically weakened rock. These zones are thus primary candidates for preferential groundwater flow. In addition, definition of relatively downthrown geologic blocks provides a basis for the delineation of local erosion bases within a karst aquifer: lower blocks are assumed to be subreservoirs collecting groundwater from the higher blocks. This is also the concept of so-called block models of groundwater flow in karst terrains (Kresic, 1991). Common methods of morphometric neotectonic analysis include (1) relief energy, (2) anomaly of partial hydraulic gradient, and (3) difference between the theoretical (attributable to gravity-driven erosion) and the current relief surface (Marković, 1983; Kresic and Tasic, 1984). The latter two methods are often difficult or not feasible to implement in mature karst where surface drainage may be largely disintegrated and solutional karst relief forms dominate.

Relief energy is defined as the potential energy (E) of the earth's surface at a given point ($E = mgh$). At this point, the relief represents a spatial position (h) of a rock mass (m), which has its potential energy. If the surface area under consideration is small enough (unit area), the rock mass within it may be assumed to have constant value. The gravity acceleration (g) within the unit area may also be considered constant. Therefore, the relief energy within the unit area is defined only by its height or, more precisely, by the height difference between the highest and lowest points.

On a local scale, the relief energy is influenced by the lithology and action of exogenous geomorphologic factors, mainly running surface water. Systematic measurements and statistical analysis of the relief energy may locate areas of its highest and lowest values, that is, areas of increased erosion and areas of increased deposition, respectively. On a regional scale, increased erosion and deposition reflect the presence of recent tectonic activity. Regional analysis of relief energy may indicate the position of neotectonic zones and vertical direction and relative intensity of movements, thus providing important information for various earth-related studies.

An example of neotectonic analysis of the Barton Springs, Austin, Texas area, using the relief energy method is provided by Kresic et al. (1994). The basis for the morphometric analysis of relief energy was a DEM created by merging nine USGS DEM 7.5-minute quadrangles. The original data, with 3-second resolution, were resampled to a 50×50 m grid (Fig. 3.21), and the calculations of relief energy and its first two trends were performed with the following procedure:

- Determine elevation difference between the highest and the lowest relief point within a 500×500 m unit area that consists of 100 data points. Assuming that the rock mass is constant, the obtained value corresponds to the potential energy of the relief surface represented by the unit area.
- Determine the reference surface for the study area by averaging all relief energy points.
- Subtract individual values of the relief energy from the reference surface. Positive values represent areas that are relatively uplifting, while negative values represent downthrowing.
- Find first and second trends of the new (relative) relief energy surface, applying a simple one-pass linear digital filter. The procedure removes the influence of local active exogenous processes ("noise") on the relief and enhances the results of recent regional endogenous (tectonic) movements—uplifting and downthrowing.

Figure 3.22 shows the second trend of relief energy for the study area (vertical scale is in feet). Figures A through D show the result of a gradually decreasing contrasting of the digital domains using Image, a public-domain free software package developed at the National Institute of Health, the United States. Although initially designed for the analyses of microphotographs, it has been successfully used for various qualitative and quantitative analyses of remote sensing digital images. The contrasting process with Image allows a fine analysis of the relief energy differences. It is practically irreplaceable when the analysis is performed using the grayscale only, since neotectonically active zones with the highest initial contrasts may be less distinct or even disappear as the contrasting decreases. Two such zones are shown in Figs. 3.22A and 3.22B with white lines for the illustration. Dark/black areas in Fig. 3.22 are neotectonic blocks that are

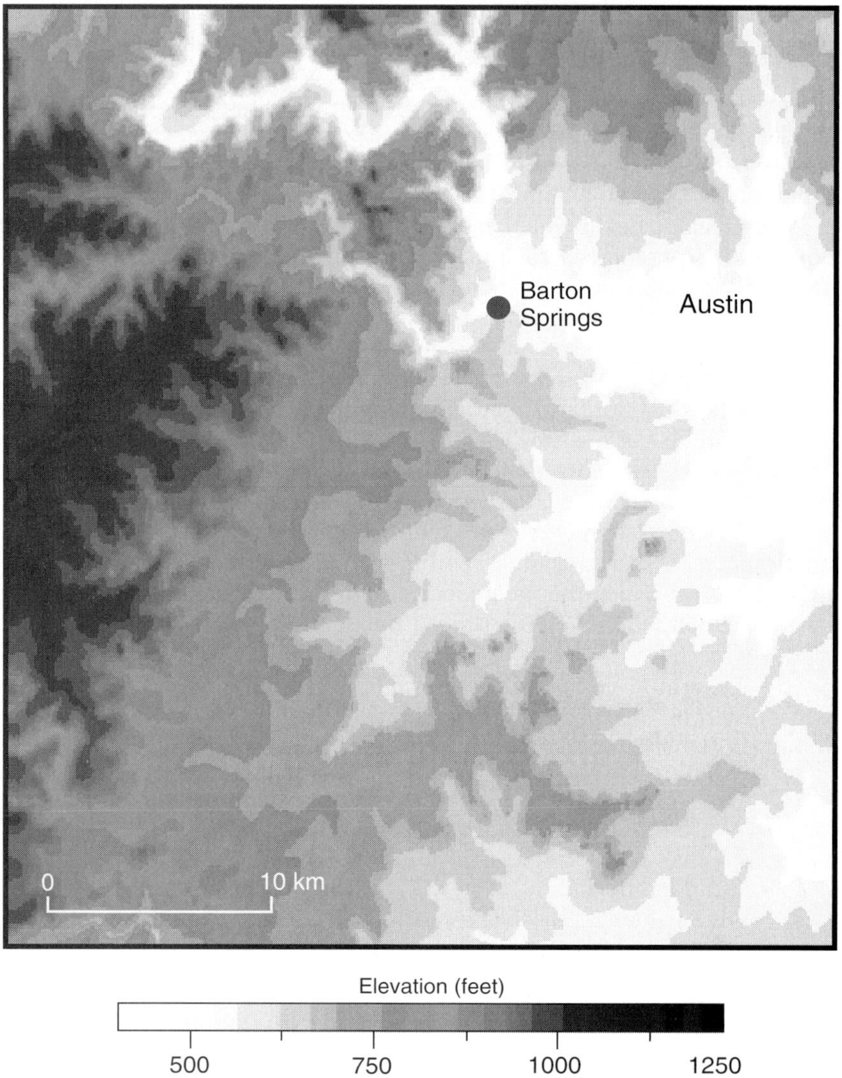

Figure 3.21 Digital elevation model of the Barton Springs, Austin, Texas area. (Modified from Kresic et al., 1994.)

relatively uplifting. The most distinct block is in the northwest and contains incised meanders of the Colorado River. It should be noted that the highest relief in the study area is in its western part (see Fig. 3.21) and that it does not coincide with the highest relief energy.

The narrow zones between the uplifting blocks and the downthrowing blocks (light/white areas) are neotectonically active. They may represent large individual faults or systems of close parallel faults with the predominant vertical component. Figure 3.22D shows the interpreted neotectonically active zones as black/white lines between the

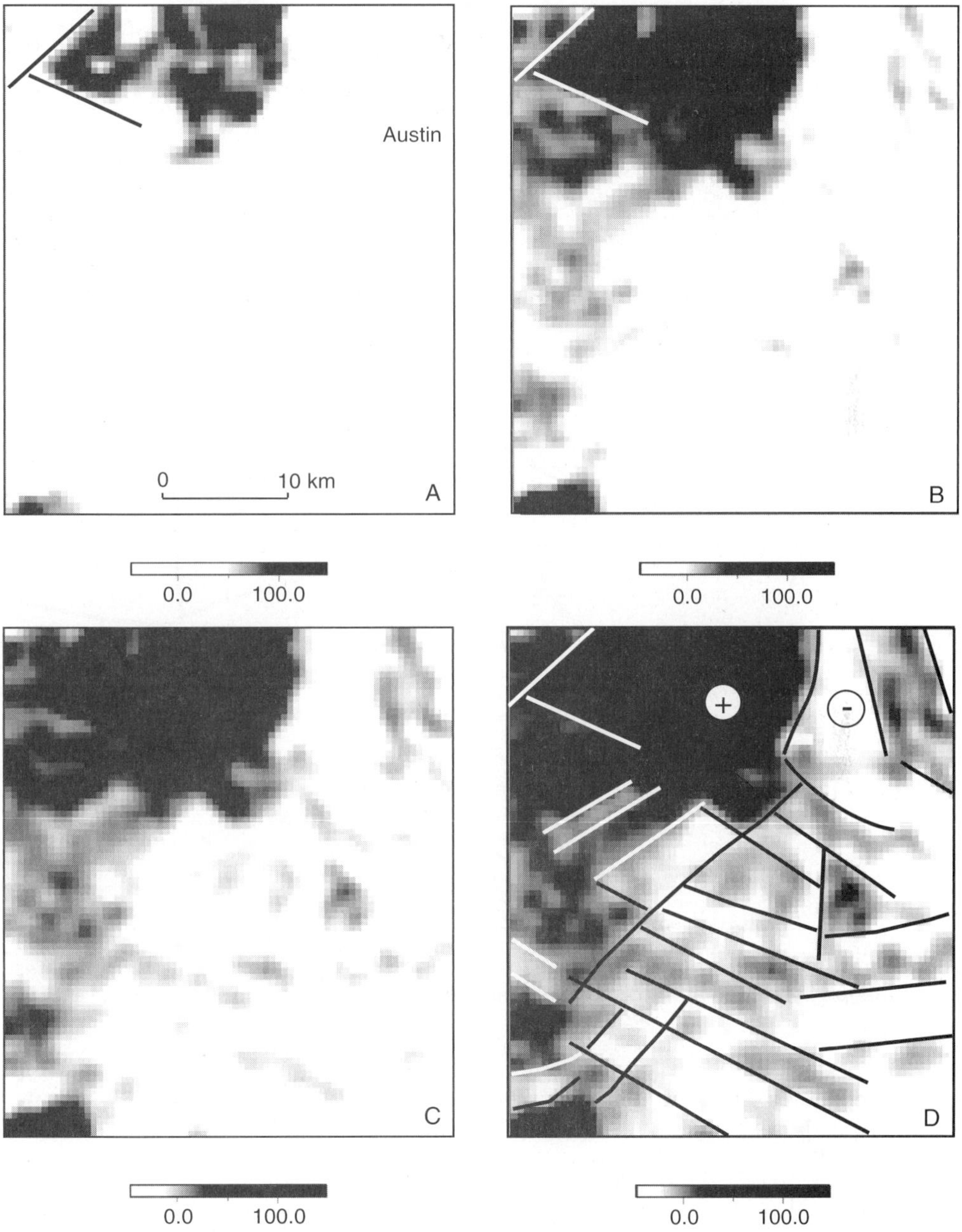

FIGURE 3.22 Map of the second trend of the relief energy with probable neotectonically active zones shown as black/white lines in D. A through C show the result of a gradual contrast decrease between the digital domains. Dark/black areas are relatively uplifted neotectonic blocks (+) and light/white areas are relatively downthrown blocks (−). (Modified from Kresic et al., 1994.)

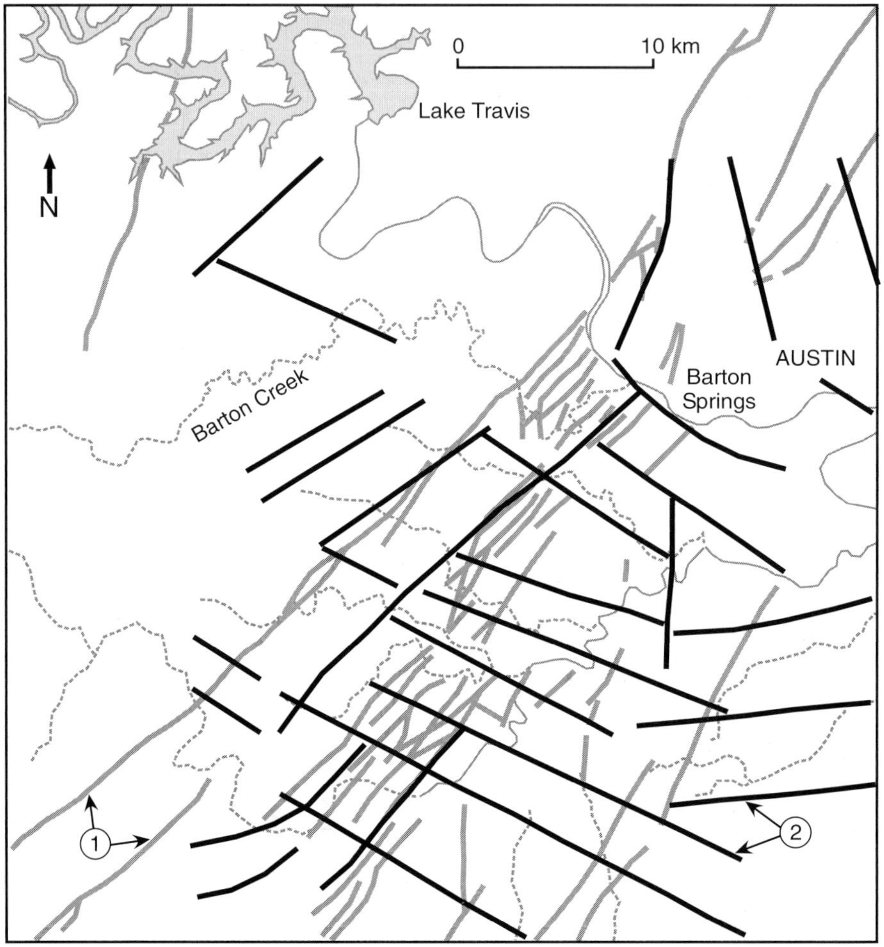

FIGURE 3.23 Geologically mapped faults (1) versus probable neotectonically active zones (2) in the Barton Springs area. (Modified from Kresic et al., 1994.)

neotectonic blocks. The method of relief energy, as well as other quantitative morphometric analyses commonly used in neotectonic studies, is weighted by the interpreter's subjectivism because of its statistical nature and the impact of exogenous agents on the relief evolution (Marković, 1983). Therefore, any morphometric neotectonic analysis should be accompanied by at least geologic and remote sensing studies.

Figure 3.23 is a comparison of the probable neotectonic zones in the Barton Springs area and the faults shown on the published geologic maps (Garner et al., 1976; Proctor et al., 1981). The two most distinct systems of the neotectonic zones are with SE-NW and SW-NE orientations, while practically only the faults with SW-NE strike (the Balcones system) are geologically mapped. The fact that SE-NW faults are much less distinct (or even invisible) in the field or when interpreting boring logs may be explained by their small displacement due to a recent activation. A very good agreement between SW-NE faults and neotectonic zones indicates applicability of the described morphometric method of relief energy analysis. Barton Springs are not accidentally located right at the intersection of two neotectonically active zones of which the one with the Balcones orientation is also the longest in the study area.

Morphometric neotectonic analysis, in conjunction with remote sensing of lineaments, is a fast and inexpensive tool for the preliminary study of possible preferential flow paths in karst aquifers. The utilization of DEMs and various affordable software packages enable application of the method of relief energy for the large areas. The method is very useful for indicating potential recent tectonic activity that is still without a clear geologic reflection at the surface. Delineation of neotectonically active zones and aquifer blocks (subbasins) is also a good basis for an initial conceptualization in groundwater modeling of karst aquifers.

3.2.2 Remote Sensing

As defined in the classic textbook by Floyd Sabins (1987), remote sensing is collection and interpretation of information about a target without being in physical contact with the object. Aircraft and satellites are the common platforms for remote sensing observations. The term is restricted to methods that employ electromagnetic (EM) energy (such as light, heat, and radio waves) as the means of detecting and measuring target characteristics. This definition excludes electrical, magnetic, radioactivity, and gravity surveys that are frequently made from aircraft but are considered airborne geophysical surveys rather than remote sensing.

The original and still the most common form of remote sensing is interpretation of aerial photography including stereo images. These products in the visible portion of the EM spectrum are acquired with optical or digital cameras and are readily available in most countries as part of routine national coverage by various government programs (e.g., for agriculture, forestry, land use planning, and general surveying purposes). Historic aerial photographs are also irreplaceable when studying land use and land cover changes that may have had major impacts on the project issues at hand. For example, as discussed by Panno and Luman (2011), in order to detect and catalog now relict karst features, county-wide historical aerial photography was digitized and orthorectified to create a geometrically accurate image base map of the entire Sinkhole Plain in Illinois, principally within Monroe, St. Clair, and Randolph Counties. Acquired during the summer of 1940 by the U.S. Department of Agriculture, Agricultural Adjustment Administration (USDA-AAA), the AAA aerial photography represents the earliest and most detailed visual record of the landscape prior to the general introduction of mechanized farming practices (Fig. 3.24-top). Because horse-drawn farming equipment was still prevalent in the region, pesticide and fertilizer application mostly unknown, row crops were planted at much lower plant densities, and the near-surface geology can be discriminated through the mature summer crop canopy. Interpretation of the leaf-on AAA aerial photography, augmented with 2005 early spring, (leaf-off) USGS, National Aerial Photography Program (NAPP) imagery (Fig. 3.24-bottom) determined that approximately 30% more sinkholes are present on the land surface than are delineated on the most current USGS 7.5-minute topographic quadrangle maps of the study area. The NAPP aerial photographs were acquired on March 6, 2005, 65 years after the AAA photography. The much wetter nature of the land surface in the NAPP photographs is apparent, with abundant sinkhole ponds that were dry in the AAA photographs. After decades of widespread use of large-scale farming equipment, modification by farm

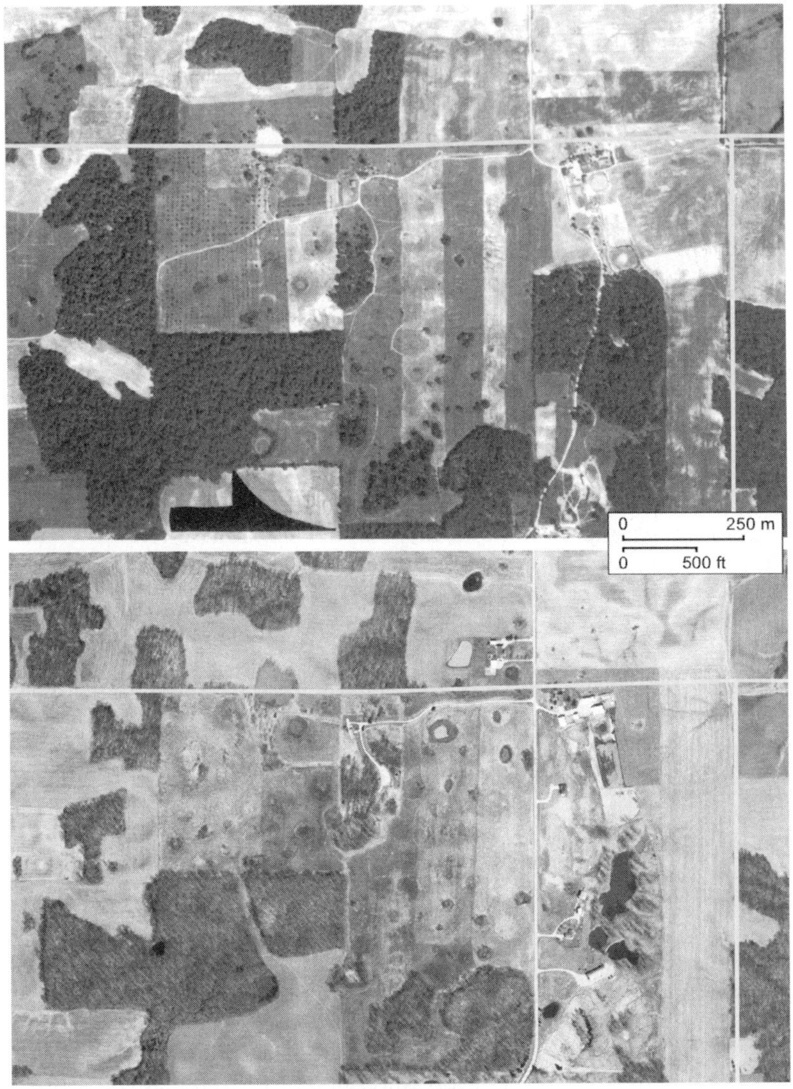

FIGURE 3.24 *Top*: Aerial photograph of a portion of the Illinois Sinkhole Plain in southern St. Clair County acquired in July 1940 by the U.S. Department of Agriculture, Agricultural Adjustment Administration (USDA-AAA). *Bottom*: National Aerial Photography Program (NAPP) photograph acquired on March 6, 2005. (Photographs courtesy of Donald Luman and Sam Panno, Illinois Geological Survey; see also Luman and Panno, 2011.)

operators using drainage stand pipes and fill material, coupled with the widespread adoption of conservation tillage methods beginning in the 1980s, sinkholes apparent on the AAA aerial photography are difficult or impossible to discriminate on recent aerial photography (Luman and Panno, 2011).

Although *photographs* will continue to be an important part of remote sensing studies, the diversity and technological advantages of other *remote sensing imagery* commercially

Figure 3.25 Satellite images of a karst terrain in the Peruvian Andes acquired with GeoEye-1 satellite. This is grayscale version of a color-composite image with ground resolution of approximately 1.6 m; the image is several kilometers across. Steeply bedded limestone is intensely fractured and faulted. See Fig. 3.19 for ground view. (Printed with permission; more detail on GeoEye products is available at www.geoeye.com.)

available today make many of them a preferred choice of working professionals. This includes various high-resolution multispectral satellite images by commercial vendors and government agencies that are acquired at regular intervals for the same area on earth, thus enabling near real-time analysis. Some of these products rival any visible-spectrum aerial photographs in terms of ground resolution and overall image quality. For example, commercial imagery available from GeoEye and acquired with GeoEye-1 satellite has resolution of 0.41 m in panchromatic mode and 1.65 m in multispectral mode (Fig. 3.25). Multispectral imagery offers many advantages of digital signal processing and quantitative analyses not possible with regular photographs. A higher cost of such advanced images compared to common aerial photographs will therefore in many cases quickly pay off multiple times.

Landforms that make karst unique compared to other terrains also make it one of the most photogenic and therefore perfectly suited for remote sensing. Various examples of grayscale (black and white) and color remote sensing imagery are provided further and in the color plates. The main advantage of remote sensing is its ability to combine analyses of images obtained by different sensors, in different spectral regions, and at different scales. Whenever possible, working professionals should perform such comparative analysis as important details may be hidden in one image and clearly visible in another. Finally, the continuous development of new technologies, such as LiDAR, makes the whole task of remote sensing in karst much more interesting as illustrated in Fig. 3.26.

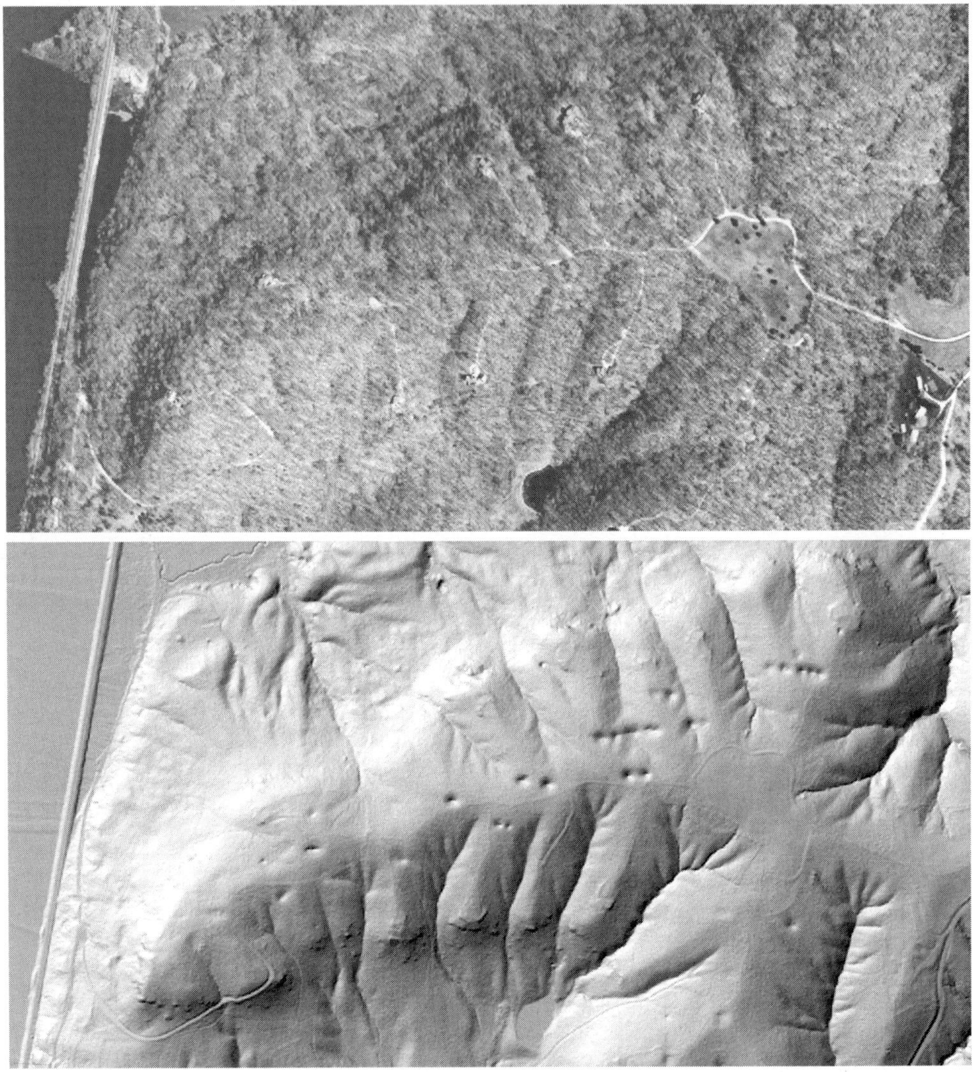

FIGURE 3.26 *Top*: April 2005 aerial photograph in southwestern Jo Daviess County, Illinois, the United States. Dense forest cover obscures the ground surface and prevents detection of geologic features. *Bottom*: LiDAR elevation image of the same geographic area, depicting numerous sinkholes at the ground surface, none of which are visible on the aerial photograph. (Images courtesy of Sam Panno and Donald Luman, Illinois Geological Survey; see also Panno et al., 2012.)

Illinois State Geological Survey scientists are using LiDAR elevation data to better understand the fabric and geometry of the karst aquifers that constitute the groundwater resources of Jo Daviess County in northwestern Illinois. LiDAR is an acronym for **Li**ght **D**etection **A**nd **R**anging and is an imaging technique that incorporates pulsating laser light from a low-flying aircraft to the ground surface. The reflected light from the pulses is detected by instruments that record their location in three dimensions (FGDC, 2006). The imagery produced is a high-resolution characterization of the surface topography that was used by Panno et al. (2012) to better understand the pattern (fabric) and geometry of the karst aquifers. Specifically, it was used to identify the absence or presence of sinkholes, and linear features (lineaments) that were used as guides to locate surface features potentially associated with an underlying karst aquifer.

Nearly one-third of the land area of the Jo Daviess County is dominated by steep slopes with dense forest cover similar to the aerial photograph shown in Fig. 3.26-top. The underlying bedrock is dolomite of Silurian age that is at or near the ground surface. Even though the leaf canopy is absent in this early spring photograph, the landscape obscures the surface topography and makes detection of geologic features difficult or impossible using traditional aerial photography or topographic maps. The second image is of the same geographic area produced from 2008 LiDAR elevation data. When a laser pulse from the LiDAR sensor encounters a soft target such as a forest canopy, a portion of the laser beam continues downward; the last returns recorded by the sensor represent the elevation of the ground, or bare earth surface. Note the distinctive circular features situated along the crests of several of the slopes, which are completely obscured in the aerial photograph. These are cover-collapse sinkholes ranging from approximately 20 to 70 ft in diameter and 10 ft or less in depth. The sinkholes are positioned *en echelon* (in a steplike manner) along nearly east-west—trending lineaments in sediment overlying the Silurian dolomite. Collapse of sediment into crevices creates sinkholes that can be directly observed in the LiDAR elevation data (Panno and Luman, 2011; Panno et al., 2012).

Aerial and Space Photographs

Panchromatic, black and white stereo aerial photographs at scales 1:10,000 to 1:50,000 were until recently the most commonly used images in basic geologic and hydrogeologic remote sensing. They are readily available, affordable, and well suited for structural analysis, inventory of hydrogeologic features such as springs, seeps, sinks, sinkholes with the visible water table (karst windows), analysis of surface drainage, vegetation, and other details. An analysis of photographs of the same area taken in different periods including wet and dry seasons and without leaf cover is highly recommended. Aerial photographs of typical karst terrains are provided in Fig. 3.27. When readily available and affordable or free such as in the United States (e.g., see the Seamless Data Warehouse at http://seamless.usgs.gov/, or The National Map Viewer at http://viewer.nationalmap.gov/viewer/), true-color visible or infrared aerial photographs should always be analyzed. Such photographs can be easily split into three black and white channels (red, green, and blue or RGB) using specialized image processing programs or any consumer program of higher quality such as Adobe Photoshop. The original photograph and the newly created three black and white channels can be separately enhanced, filtered, and otherwise manipulated digitally to reveal various patterns and features that may be less obvious on the original photograph. Legacy paper photographs, both color and black and white, can be scanned at high resolution and also analyzed digitally in the same manner. Examples of such simple but effective analyses are shown in Fig. 3.28. Various linear and other features emphasized by different filters and available in digital form can be further analyzed quantitatively by a computer program including their pattern, lengths and orientation of dominant axes, and areas.

As emphasized earlier, every working professional should have access to Google Earth and take full advantage of 3D visualization of the land surface virtually anywhere in the world. Good quality satellite images including several historic images and, in

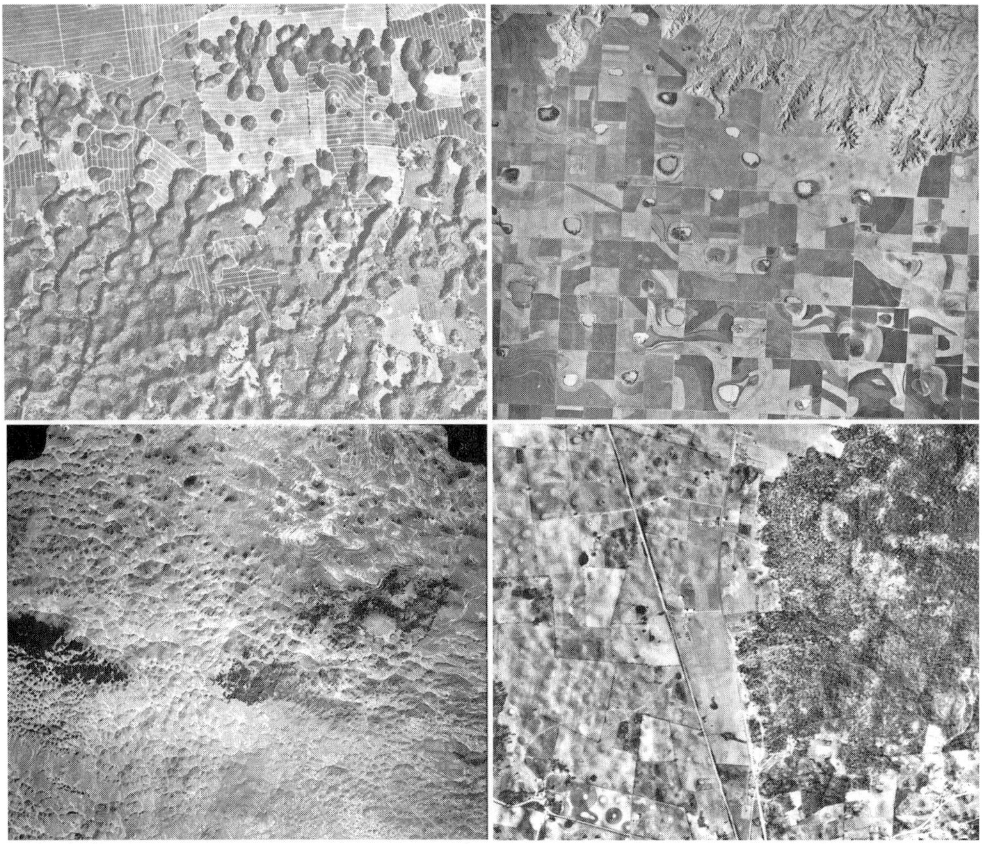

Figure 3.27 *Upper left*: Forested haystack limestone hills or mogotes are clearly distinguishable from flat agricultural fields in alluvial sand on this aerial photograph of the characteristic karst of northern Puerto Rico near Vega Baja (courtesy of USGS). *Upper right*: Shallow sinkholes, some half a mile in diameter, near the High Plains escarpment east of Lubock, Texas (courtesy of USGS). *Lower left*: Sinkholes in Triassic limestone, upper Neretva, Herzegovina; horizontal layers of the overlying dolomite are visible in the upper right portion of photograph (courtesy of Dr. Miroslav Markovic). *Lower right*: Numerous sinkholes developed in St. Louis Limestone near Park City, Kentucky; the wooded hills in the right half of the photograph have sandstone caps (courtesy of USGS).

many cases, high-resolution satellite images are available for anyone to see for free. The user-friendly platform provides for some basic measurements on the images (e.g., lengths and areas) and supports image rotation and zoom in. However, one cannot download the high-resolution images and perform digital processing; this would have to be done at cost, by contacting copyright owners of the images indicated on the Google Earth screen.

The analysis of stereo aerial photographs still offers the best 3D model of the studied terrain as a whole. For example, the aerial photograph in Fig. 3.29 is a typical example of highly dissected relief, inaccessibility, and unfavorable conditions for a field study of structural elements. In such cases, integration of sparse field data into an overall picture may be misleading. However, data on both fractures and folds deciphered from

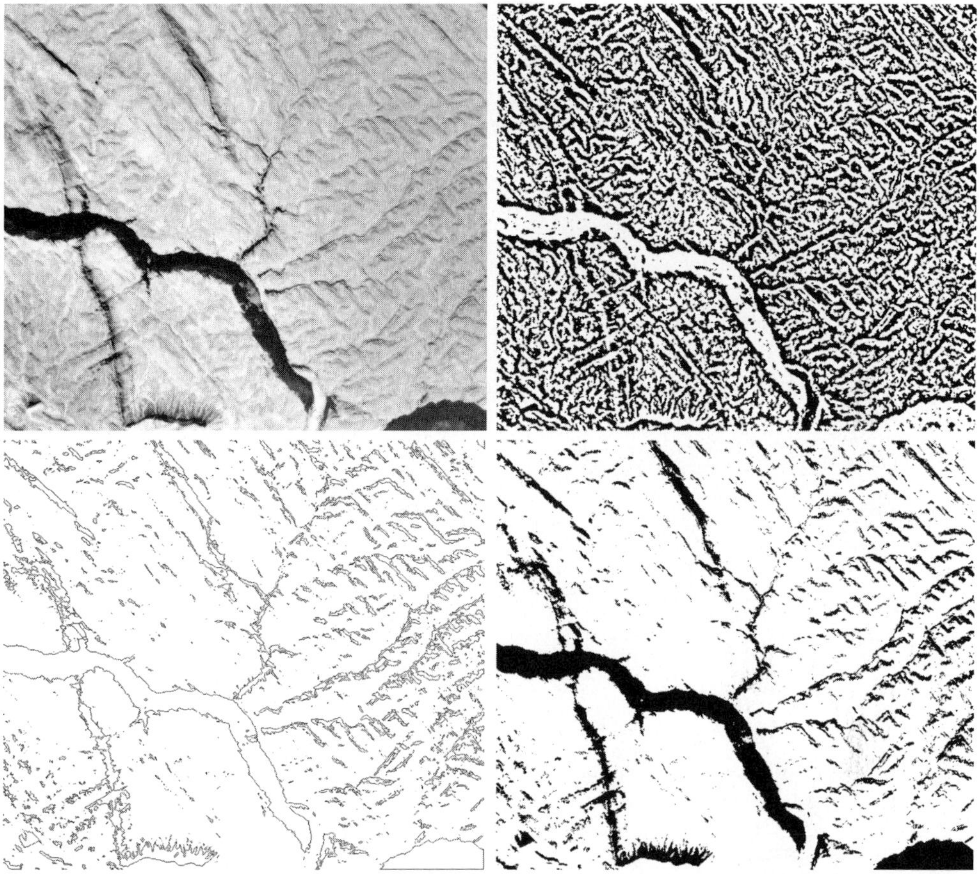

Figure 3.28 *Top left*: Aerial photograph of the limestone plateau above Santa Helena Canyon carved by the Rio Grande, Big Bend National Park, Texas (courtesy of USGS). Image enhancement and filtering is performed in Adobe Photoshop.

stereo aerial photographs are easily and quickly integrated into a preliminary geologic-hydrogeologic map as illustrated in Fig. 3.30. Clearly visible bedding, with the general dip toward the southwest, is dominant in most of the area. Numerous small, tight folds are particularly distinct as well as faults. The system of large faults with SW-NE strike is predominant. Lithological delineation and interpretation of the photographs allow a fairly reliable separation of three large units that have considerable morphological and structural differences: (1) dolomites in the northeast that are covered with dense vegetation, (2) karstified limestones in the central part with numerous dolines (sinkholes), and (3) limestones with clastic sediments (i.e., carbonate flysch) in the southwestern part. The known large permanent and small intermittent springs were clearly visible with the stereoscope. Based on the photogeologic analysis, a hypothesis on groundwater flow directions, shown with arrows in Fig. 3.30, can be formed and later tested in the field (Kresic and Pavlovic, 1990).

Lithological determination using remote sensing images alone is considerably limited, although large and homogeneous areas, especially without vegetative cover, may

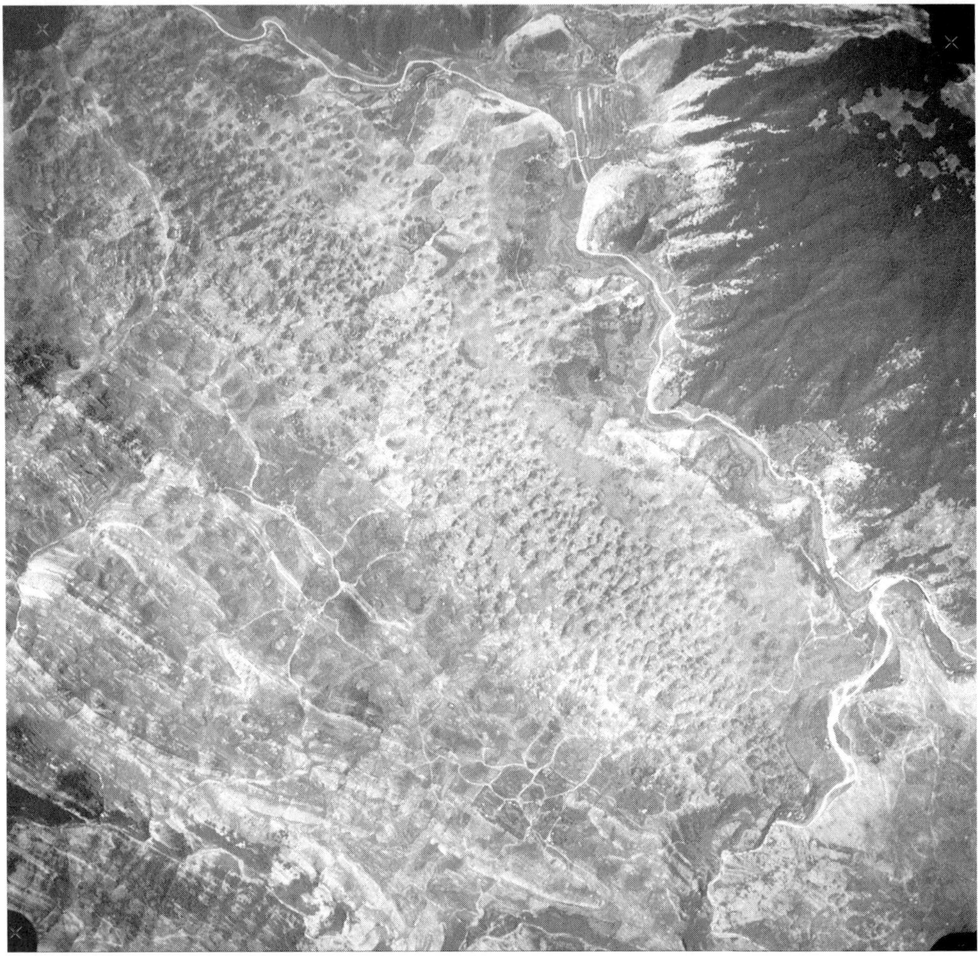

FIGURE 3.29 Panchromatic aerial photograph of the area of a karst spring in the Bregava Valley, Eastern Herzegovina. (From Kresic, 1995.)

sometimes be distinguished from one another fairly accurately. In carbonate terrain, the degree of karstification, the drainage pattern, and the vegetation cover are the most useful for determining lithologic differences. Pure limestones have abundant sinkholes and have a sparse, deeply cut drainage system. Dolomites and carbonate flysch have fewer karst depressions, such as dolines and uvallas, and the vegetation cover usually is richer compared to limestones. Clastic sediments in the karst depressions and sedimentary basins are easily delineated due to a thicker soil cover, its higher moisture content (which results in darker tones on panchromatic and color images), and a relatively mild relief. In general, color-composite multispectral images produced using various combinations of spectral channels are the most useful in analysis of lithology (Oluić, 1983, Sabins, 1987).

Remote sensing of lineaments, interpreted as faults, large fractures, or fault/fracture zones is a well established and practically a mandatory procedure in karst hydrogeology

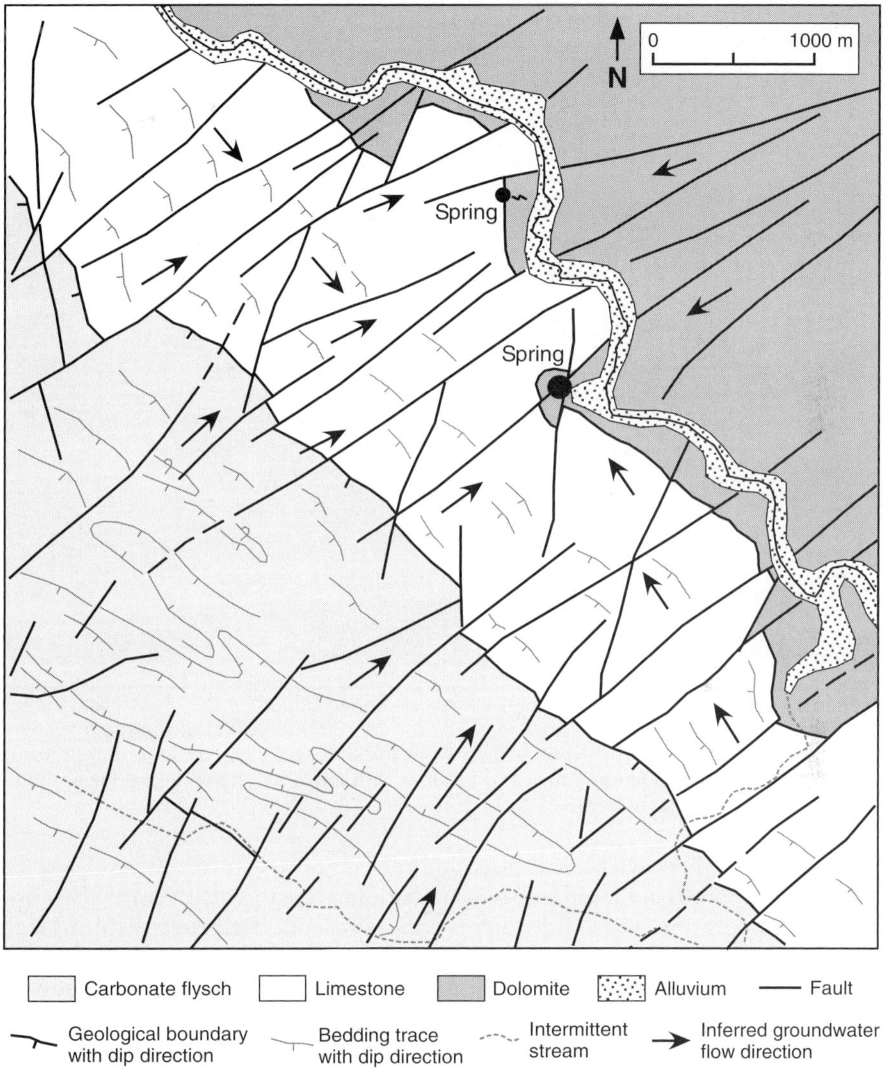

FIGURE 3.30 Preliminary photogeological map of the area shown in Fig. 3.29. (Modified from Kresic and Pavlovic, 1990; Kresic, 1995.)

studies. This is because such lineaments are primary candidates for preferential flow paths (Lattman and Parizek, 1964; Parizek, 1976; Kresic and Pavlovic, 1990; Kresic, 1995). In general, a lineament is defined as a mappable simple or composite linear feature on the surface differing distinctly from the patterns of the adjacent features and presumably reflecting a subsurface phenomenon. Although many lineaments are controlled by structural displacement, they may also represent geomorphic (physiographic) or tonal features caused by contrast difference. A typical geomorphic lineament would be a straight stream valley, whereas a tonal lineament could be caused by differences in vegetation, moisture content, or soil and rock composition (Marković, 1980; Sabins, 1987).

Figure 3.31 Landsat ETM+ Band 4 enhanced image of part of the classic Dinaric karst in Herzegovina, approximately 72 km across. The NW-SE orientation of the main system of lineaments is indicated by white arrows. Note three deeply incised canyons that abruptly terminate at the noted regional lineaments, and an area with mega sinkholes in the lower right. (Landsat TM+ image courtesy of USGS.)

Figure 3.31 shows typical orientation of large structures in the classic Dinaric karst. The most distinct geologic features are regional lineaments that are also clearly reflected in the topography. These lineaments can represent overthrusts, faulting zones, systems of close faults with similar strike, or large individual faults. The NW-SE system with the so-called Dinaric strike (marked by arrows) is the most distinct. The lineaments are either young faults or old ruptures with a recently renewed activity. The Dinaric Mesozoic carbonate platform, which is at the margin between African and Eurasian plates, is one of the most seismically active areas in the world (Ciric, 1984). Many surface streams in the area are intermittent and often abruptly sink into large sinks located at major regional faults.

Caution should be exercised when interpreting various linear features visible on photographs and other remote sensing imagery as many of them do not reflect natural subsurface phenomena. When in doubt deciding which lineament is a potential fault/fracture, that is, for the exclusion of obvious man-made features (roads, power lines, agricultural boundaries), it is best to analyze photographs in 3D using stereoscopes. This, however, is not always feasible. In such cases, an analysis of visible-spectrum photographs combined with an analysis of multispectral satellite images that are readily available in public domain, such as by Landsat, may sometimes help.

The best approach in lineament analysis is to verify their types in the field. For example, the analysis of Panno et al. (2012) has shown that lineaments on LiDAR imagery are quite prominent in the study area in Jo Daviess County, northwestern Illinois. Field evidence from road cuts, quarries, sinkholes, springs, crevice orientations from the field and from references and their relationship with LiDAR-based lineaments suggests that these lineaments are solution-enlarged crevices, clusters of crevices, and/or conduits in the carbonate bedrock and reflect the pattern (fabric) and geometry of the underlying karst aquifer. The lineaments are coincident with a number of sinkholes, springs, linear indentations along stream channels, bedrock-controlled stream channel geometries, and flooding of excavations that approach the carbonate bedrock surface. Orientation of the lineaments follows known crevice directions observed in outcrops, road cuts, and quarries in the study area.

Partially due to high costs involved in data collection, an analysis of high-resolution aerial thermal-infrared images is performed primarily for major projects including construction of dams and reservoirs and centralized water supply. Such images are commonly taken at predawn, at least once during the period of main aquifer recharge (e.g., in April to May in humid temperate climates in the northern hemisphere), and once near the anticipated end of the summer recession period (August to September). Thermal images are particularly useful for a detailed analysis of groundwater discharge sites either at the terrain surface or within the surface water bodies. In some rare favorable cases, thermal-infrared photography also was suitable for discovering shallow karst channels in karst poljes (Milanovic, 1980).

Figure 3.32 illustrates the results of a USGS thermal remote sensing study in coastal waters of the Lower Suwannee River and near-shore tidal marshes on Florida's Gulf Coast. Temperature anomalies were detected with thermal infrared imagery. Imagery included 1.5-m-resolution day and night Thermal Infrared (TIR) and 0.75-m-resolution

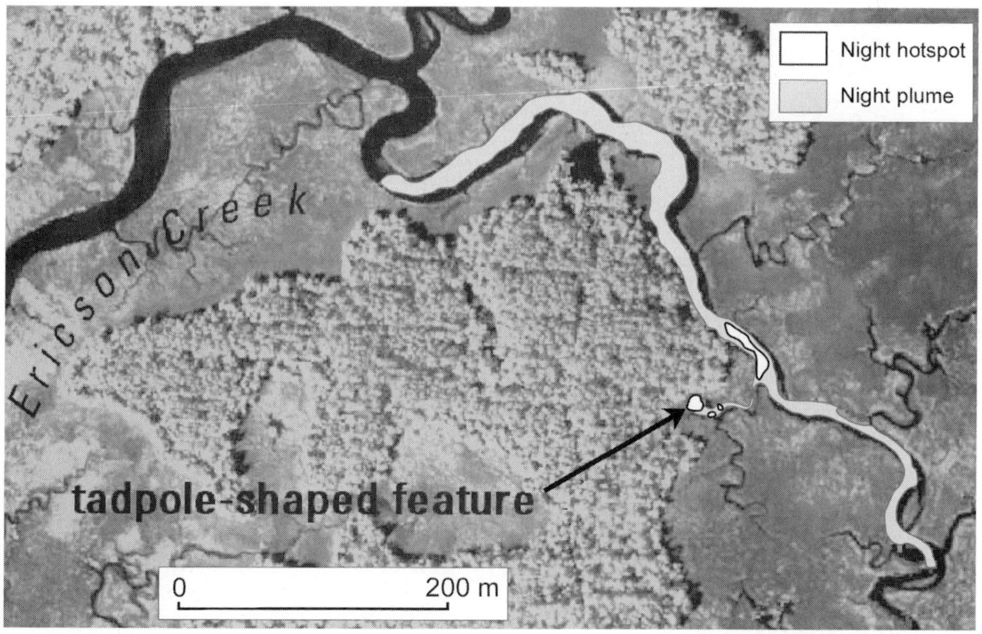

FIGURE 3.32 Temperature anomalies derived from TIR imagery at Ericson Creek. (Modified from Raabe and Bialkowska-Jelinska, 2007.)

Color Infrared (CIR) imagery acquired on March 2-3, 2005, by Verimap, Inc. (Verimap, 2005; www.verimap.com). Coincident temperature readings were collected on the ground and used to calibrate the imagery. The Floridan aquifer is at or near the land surface in this area and bears a constant temperature signature of approximately 22°C. This consistent temperature contrasts sharply with ambient temperatures during winter and summer months. Temperature anomalies identified in the imagery during a late-winter cold spell may be correlated with aquifer seeps. Hot spots were identified as those areas exceeding ambient water temperature by 4°C or more. Warm-water plumes were also mapped for both day and night imagery. Numerous temperature anomalies were identified in small tributaries and tidal creeks from Shired Island to Cedar Key and were confirmed with field reconnaissance. Abundant warm-water features were identified along tidal creeks south of the Suwannee River and near Waccasassa Bay (Raabe and Bialkowska-Jelinska, 2007).

In the USGS study, several tadpole-shaped features were mapped in the intertidal zone. Example is shown in Fig. 3.33. These features, confirmed with summer field reconnaissance, typically consisted of a pond with a small boil connected to the tidal creek by a narrow outlet, creating a tadpole-shaped outline. The boils were small, ranging from 30 to 60 cm across, and shallow, ranging from 0.5 to 2 m deep. Discharge is unknown, but given the small size of the boils, an estimate of 1 cfs is expected. Thermal plumes from each feature extended a considerable distance downstream, 0.1 to 1.5 km, and in some cases mingled with plumes from nearby seeps. Alligator sightings were common at these locations. Temperature anomalies at the boils were confirmed by a temperature difference of 4°C to 10°C lower than the creek itself, during summer high temperature levels (Raabe and Bialkowska-Jelinska, 2007).

FIGURE 3.33 A small boil on Erickson Creek. (From Raabe and Bialkowska-Jelinska, 2007.)

Multispectral Imagery

As expertly described by Dr. Nicholas Short in now discontinued excellent online free tutorial on remote sensing by NASA (the tutorial can be ordered from the NASA Technical Reports Server), spectral measurements involve interactions between illuminating radiation and the atomic/molecular structures of any material, leading to a reflected signal, followed by some further changes to this signal as it returns through the atmosphere. Spectral measurements also depend on the nature of the response of the detector system(s) in the sensor. In practice, objects and features on the earth's surface are described more as *classes* rather than as materials per se. For example, material *concrete* may be found in roadways, parking lots, swimming pools, buildings, and other structural units, each of which might be treated as a separate class. *Vegetation* can be subdivided in a variety of ways: trees, crops, grasslands, lake bloom algae, and others; finer subdivisions are possible, by classifying trees as deciduous or evergreen, or deciduous trees into oak, maple, hickory, poplar, and others.

These various classes, some of which can share the same materials, are distinguished by two additional properties beside their spectral attributes, namely, shape (geometric patterns) and use or context (sometimes including geographical locations). Two features with nearly identical spectral signatures for vegetation might be assigned to the classes "forest" and "crops" depending on whether their areas of similar spectral response, as seen in an image, have irregular or straight (often rectangular) boundaries. The main use of remote sensing data is in classifying the myriad of features in an image into meaningful categories or classes that can be converted to a *thematic* map (the theme is selectable, e.g., land use; geology; and vegetation types). An *unsupervised* classification results when features are separated solely on their spectral properties and a *supervised* classification is made when some prior or acquired knowledge of the classes present in parts of the scene is utilized in setting up *training sites* to estimate and identify the spectral characteristics of each class.

The task of any remote sensing system is simply to detect radiation signals, determine their spectral character, derive appropriate signatures, and interrelate the spatial positions of the classes they represent. This ultimately leads to some type of interpretable display product, be it an image or a map or a numerical data set. Color is often an essential ingredient in most interpretations of remote sensing images. While variations in black and white imagery can be very informative and were the norm in the earlier aerial photographs, the number of different gray tones separable by the eye is limited to about 20 to 30 (out of a maximum of ~200) steps on a contrast scale. By contrast, the eye can pick out and distinguish 20,000 or more color tints, so that small but often important variations related to real differences within the target materials or classes can be discerned. Liberal use of color in the illustrations found throughout this book was not possible; however, examples of color-composite remote sensing images are provided in the color plates. Dr. Short's tutorial (Short, 1982) takes advantage of this capability as do various other educational sites on the Web.

Since 1972, the Landsat series of earth-observing satellites has been imaging earth's land areas. Landsat represents the world's longest continuously acquired collection of space-based moderate resolution land remote sensing data. Virtually all principles of satellite imagery interpretation in agriculture, geology, forestry, regional planning, education, mapping, and global change research are either developed using Landsat images or are based on the vast experience gained from analyzing them. Interested reader can

Band	Portion of Spectrum	Use	Wavelength (μm)	Resolution (m)
1	Blue-green	Monitor sediment in water, map coral reefs, and water depth; can also be used to distinguish forest types	0.45–0.52	30
2	Green	Monitor vegetation health, identify man-made structures	0.52–0.60	30
3	Red	Help identify plant species; can help highlight barren lands, urban areas, and street patterns	0.63–0.69	30
4	Near-infrared	Monitor vegetation (leafy plants reflect this band well)	0.76–0.90	30
5	Mid-infrared	Measure vegetation moisture content and soil moisture	1.55–1.75	30
6	Thermal infrared	Measure surface temperature, identify rock types; detect changes in ground temperature caused by moisture variation	10.4–12.5	60
7	Mid-infrared	Monitor moisture content in soil and vegetation; identify minerals	2.08–2.35	30
8	Panchromatic	Captures visible and near-infrared bands; used to sharpen other bands	0.52–0.90	15

Modified from USGS (2011).

Table 3.1 Landsat 7 Enhanced Thematic Mapper (ETM+) Spectral Bands and Their Common Use

find many useful open-source software and free guidance documents on the Web that can be used to analyze imagery acquired by the Landsat program.

Enhanced Thematic Mapper Plus (ETM+) multispectral sensor on board of Landsat 7, the latest and still orbiting satellite of the Landast program has seven spectral bands (Table 3.1), as well as the panchromatic band (Band 8) that captures visible and infrared light at 15-m resolution (Fig. 3.34) and is often used to sharpen images from the other bands. The resolution of the spectral bands is 30 m (thermal band has resolution of 60 m), that means one pixel is equal to a square 30 m by 30 m; one Landsat 7 image contains about 35 million pixels. The satellite sensors that capture the images record them in grayscale—each pixel is assigned a value of brightness on a scale from 0 to 255. Black is 0, and white is 255, and shades of gray are in between. Grayscale images (bands) can be assigned the color red, green, or blue, which display the brightness of each of these colors. Combining any of the three grayscale images, which correspond to individual spectral bands, results in a false color image. With all of the possible combinations of red,

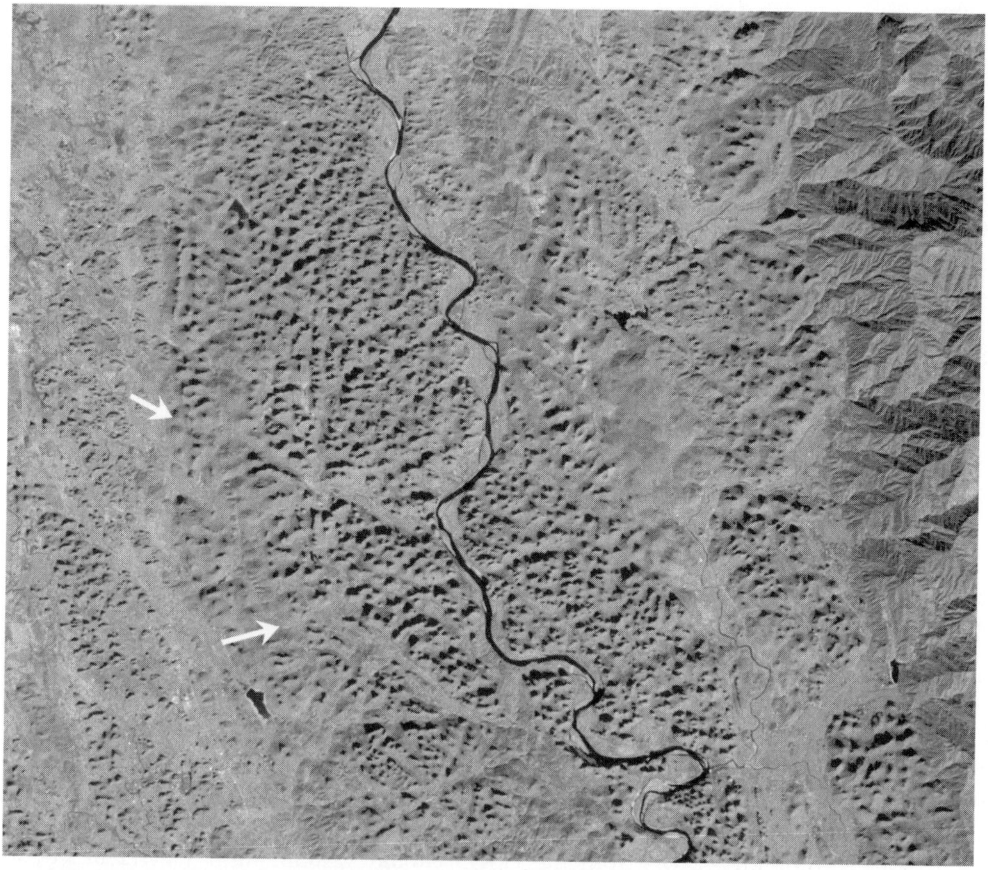

FIGURE 3.34 Enhanced Landsat 7 ETM+ Band 8 image of tower karst along Li River, Guangxi Province, China. Shadows from the limestone tower hills appear black. Note major lineaments across the hill area, two of which are indicated with arrows. See the color plates for a color-composite image of the same area, RGB = 7,4,2. (Raw image courtesy of USGS.)

green, and blue values, this provides for a display system capable of providing over a million different colors.

Different band combinations reveal different details about the surface. Some common combinations are (USGS, 2011):

- 3,2,1 this band combination is the closest to natural color; used in urban studies and to detect sediment in water.
- 4,3,2 commonly used to indicate vegetation, which appears red because band 4 (near-infrared) is assigned the R in RGB (vegetation reflects near-infrared light); urban areas = shades of light blue; soils = dark to light brown.
- 7,4,2 healthy vegetation = bright green; sparsely vegetated areas = oranges and browns; urban areas = varying shades of magenta; dry vegetation = orange; barren soil = pink. An example of using this combination of bands for the same area shown in Fig. 3.34 is provided in the color plates.

- 4,5,1 healthy vegetation = reds, browns, oranges, and yellows; soils = greens and browns; urban areas = white, light blue, and gray.
- 4,5,3 similar to 4,5,1 but offers added definition of land–water boundaries.
- 5,4,3 healthy vegetation = bright green; soils = pale lavender.

Other combinations, including ratios of different bands may be suitable as well, depending largely on the characteristics of the surface analyzed, project goals, but also the preference of the user. For example, the author's experience is that red TM band in many settings appears to be the most suitable for the analysis of lineaments since it carries the least visual noise from the urban and agricultural features. Other researchers may prefer different bands, and certain terrain characteristics may favor one band over another. In any case, it is always advisable to perform remote sensing analyses using a variety of bands and their combinations regardless of the project scope.

Downloaded Landsat images can be combined into color-composites using a variety of commercial and freeware software for image analysis, as well as better consumer programs such as Adobe Photoshop that has "merge channels" function. The favorite free software at many universities is MultiSpec, a multispectral image data analysis system from Purdue University (https://engineering.purdue/edu/~biehl/MultiSpec/), which can be used to perform spectral classification of features on images and produce thematic maps. GRASS (Geographic Resources Analysis Support System) is free GIS software that includes image processing tools (http://grass.fbk.eu/index.php). It was originally developed by the U.S. Army Construction Engineering Research Laboratories and has since evolved into a powerful utility with a wide range of applications. UNESCO-Bilko is free software for learning and teaching remote sensing image analysis skills (http://www.noc.soton.ac.uk/bilko/).

Various satellite images including from several generations of the Landsat satellites are available for free download from the USGS Earth Explorer site at http://earthexplorer.usgs.gov, from the USGS Global Visualization Viewer at http://glovis.usgs.gov/, and from the Global Land Cover Facility at http://www.landcover.org. Search of up-to-date Landsat images and free download of global orthorectified Landsat data is available at http://landsat.org.

Commercial high-resolution multispectral satellite imagery including high-resolution panchromatic products is available from GeoEye (ICONOS and GeoEye-1 satellites, with the ground resolution as fine as 0.5 m; http://www.geoeye.com); from Digital Globe acquired by QuickBird, WorldView-1 and WorldView-2 satellite sensors with 0.5 m panchromatic ground resolution (http://www.digitalglobe.com); and from French SPOT (2.5 m resolution panchromatic images; http://www.astrium-geo.com). Synthetic aperture radar (SAR) high-frequency X-band satellite images from the TerraSAR-X mission are commercially available from EADS Astrium GmbH at ground resolution down to 1 m (http://www.astrium-geo.com).

Quantitative Spatial Analysis

Karst-specific features visible on remote sensing imagery, namely closed depressions such as sinkholes, uvalas, and poljes, can be analyzed quantitatively using a variety of image processing programs including commercial software developed for ESRI's ArcGIS and ERDAS' Imagine. The quantitative analysis commonly includes the orientation of the longer axes, areas, and densities of selected features per unit area of the

FIGURE 3.35 *Left*: Part of the enhanced Landsat 1 image covering the area of Popovo Polje in Eastern Herzegovina (acquisition date October 31, 1973, near-infrared reflectance – Band 7). (1) Karstified dry valley (the black tone is the shadow on the southeastern side of the valley) and (2) flat quaternary deposits of Popovo Polje. *Right*: The result of a density-slice enhancement using the NIH *Image* program. Black areas ("particles") are interpreted as karst depressions—uvallas and dolines. Obvious shadows on the mountainsides are excluded from the analysis and shown with gray tone. (From Kresic, 1995.)

image. The features are identified automatically by the software, based on their spectral signature, but it is always recommended that this selection be interactively guided by the user and modified as needed using best professional judgment. Aerial photographs and new generation satellite images have very fine ground resolution (1 m or less) that enables accurate delineation of even the smallest sinkholes. This greatly aids karst studies since ground measurement of sinkholes in many terrains such as shown in Fig. 3.29 and Fig. 3.35 is not feasible. This is due to great number of sinkholes and difficult access for fieldwork.

The alignment of several karst depressions almost certainly indicates the presence of a fracture or a fault. The statistical maxima of the orientation of all sinkholes and uvallas in an area of interest will reveal the existence of different fracture/fault systems and, consequently, the most probable general directions of groundwater flow within the aquifer. The results of one such analysis using *Image*, free public-domain software developed at National Institutes of Health (NIH), the United States, are shown in Figs. 3.35 and 3.36. The enhancement of karst depressions, which, due to shadowing, are darker than the surrounding terrain, is performed using Density Slice or Threshold options in *Image*. The resulting "particles" (notification in *Image*) are shown as black areas in Fig. 3.35-right.

Both the qualitative observation of aligned features in the photograph (Fig. 3.35-left) and the statistical analysis of sinkhole orientation (Fig. 3.36) show that the NE-SW direction predominates. In general, particles that do not represent obvious karst depressions, such as long shadows along hillsides, can be excluded from the analysis manually. Particles longer than a specified value can also be excluded from the analysis automatically by the program. This is another useful option that allows more input from the user in an otherwise automated procedure.

Although visual interpretation of lineaments on high-quality aerial photographs, preferably using stereoscope, is somewhat subjective, it is still the most reliable method assuming the user is not a complete novice. For example, a geologic analysis of stereo photographs of the terrain shown in Fig. 3.37 will leave little doubt that linear features

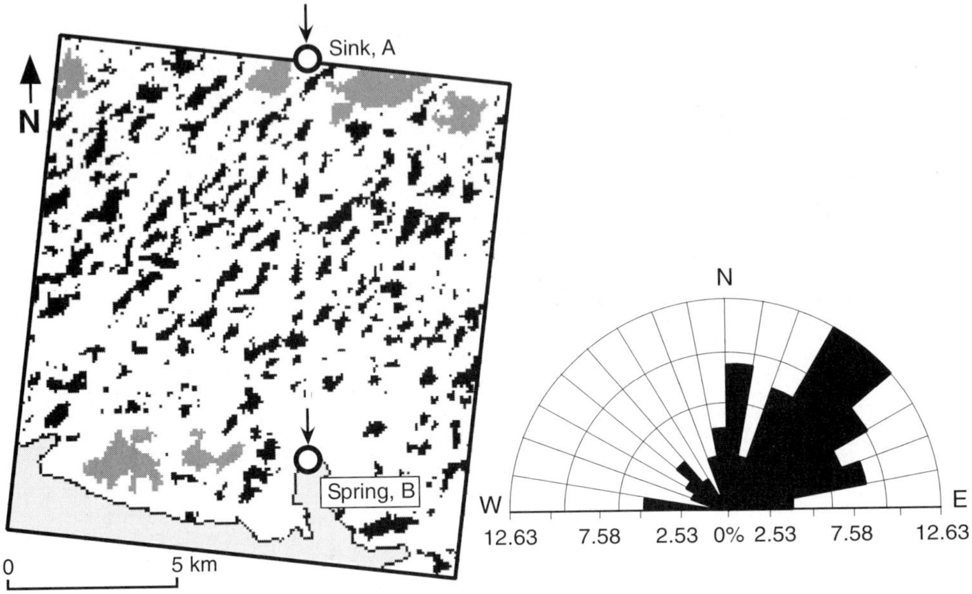

FIGURE 3.36 *Left*: Enlargement of the area outlined in Fig. 3.35. Note north-south aligned dolines between the sink A and the spring B. *Right*: Rose diagram of the orientation of the "particles" (karst depressions) measured with the NIH *Image* program. (From Kresic, 1995.)

are the result of an underlying geologic process that most likely caused formation of sinkholes. Pronounced faults and fractures, and a variety of smaller and larger sinkholes are clearly visible even without a stereoscope. In contrast, an automated selection of lineaments by computer software always results in a myriad of natural and anthropogenic features that would have to be sorted out by the user in any case (see Abdullah et al., 2009). Nevertheless, various computer programs such as LESSA (Lineament Extraction and Stripe Statistical Analysis; http://www.lineament.ru/indese.htm), Geomatica (http://www.pcigeomatics.com/geomatica), and the lineament analysis wizard in ERDAS Imagine (www.erdas.com) have been used in such analyses with varying success.

Figure 3.38 shows a set of lineaments and fracture traces derived from an aerial photograph interpretation performed through on-screen digitizing (Haryono et al., 2005). Total of 5,221 lineaments were identified, with a minimum and maximum length of 90 and 3,234 m, respectively. Three major sinks and one spring in the study area, which were found to be connected by dye tracing, were used in the network modeling. Applying the "find best route" procedure in the ArcView's Network Analyst over the network of lineaments, three hypothetic underground river networks were generated. They show generally the same patterns as those resulted from previous studies that also included geoelectrical sounding in the Sumurup-Baron Spring section (Haryono et al., 2005).

3.2.3 Geophysical Methods

As emphasized throughout the book, investigation of subsurface karst features is always a challenging task. Because of their random, unpredictable distribution, natural cavities or buried sinkholes are notoriously difficult and expensive to accurately locate due to

Figure 3.37 Aerial photograph of a typical karst terrain in Montenegro. The original natural-color photograph is acquired at ground resolution less than 50 cm. The large sinkhole in center-right is approximately 60 m across. See also color plates. (Photograph courtesy of Dr. Radmila Pavlović, Laboratory for Remote Sensing and GIS, School of Mining and Geology, University of Belgrade.)

high costs of drilling numerous borings, the only direct method of confirming karst cavities. However, combining geophysical surveys with the previously described indirect methods of locating possible preferential flow paths significantly increases the likelihood of drilling positive borings. One advantage of geophysical methods is that many of them can be applied both in borings and at the ground surface. This provides an invaluable feedback between the direct and indirect methods and enables calibration of the geophysical results. Performed this way, geophysical surveys applicable to karst will be the most cost effective. However, when geophysical methods are applied individually and, worse yet, without a field confirmation from borings, their results may be misleading; the experience of this author and many other karst hydrogeologists is that such surveys

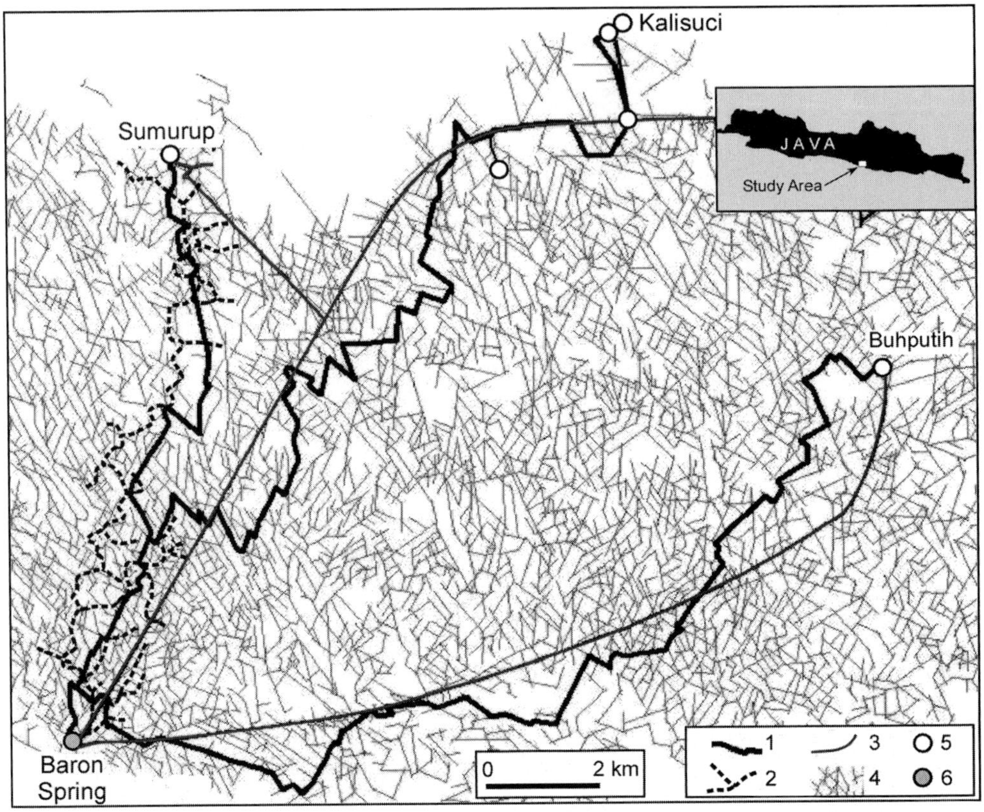

FIGURE 3.38 Underground river network in portion of the Java, Indonesia karst as interpreted by the ArcView 3.2 Network Analyst from manually delineated lineaments visible on aerial photographs. (1) Main networks derived by GIS modeling, (2) networks from Kusumayuda and Santoso (1998), (3) underground connections from MacDonald and Partners (1984), (4) lineaments and fracture traces, (5) sinkhole, and (6) spring. (From Haryono et al., 2005.)

should therefore be avoided altogether. In other words, it cannot be overemphasized enough that same geophysical properties of various karst features that appear together in the subsurface often mask (cancel) each other out so that any interpretation is ambiguous at best. For example, air-filled cavities and clay-filled cavities have opposite resistivity signature so that a mixture of open and filled cavities cancels each other out. In addition, space resolution of any geophysical method rapidly decreases with depth that makes karst terrains with pronounced surface relief and thick karstified carbonates especially difficult to work in. In such cases, only general geophysical assessments are warranted and include estimates of depth to bedrock in case of covered carbonates, or depth of karstification, or zones where the karstification intensity is potentially higher. Attempting to locate cavities (caves) and/or actual (current) preferential flow paths in the saturated zone at depths greater than about 50 m (150 ft) is beyond capabilities of any geophysical method applied at the land surface.

As discussed by Enviroscan (2012), even at relatively shallow depths there is no single geophysical technique capable of detecting all karst-related subsurface features

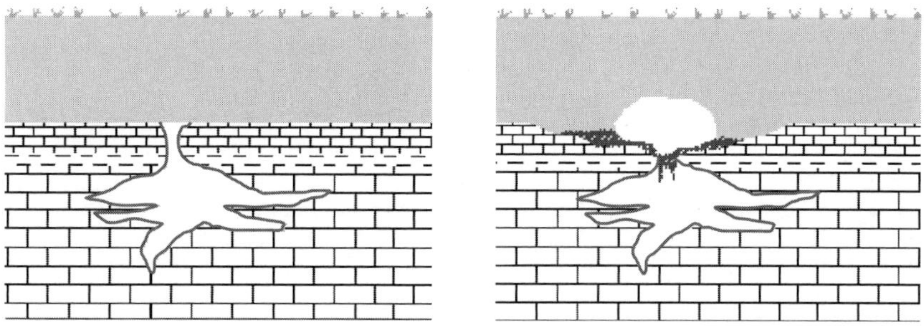

Figure 3.39 Cartoons illustrating some common karst features and their geophysical signatures that can be surveyed by different geophysical methods. *Left*: Open bedrock cavity. *Right*: Open bedrock cavity, soil cavity above it, wet soil with water seeping into the bedrock cavity, and bedrock depression. (Courtesy of Tim Bechtel and Enviroscan; many more similar cartoons, and information are available at http://www.enviroscan.com.)

since they are very diverse and have different physical properties. Figure 3.39 illustrates several karst features and their properties that a geophysical survey using a combination of methods may be able to identify. Other common features and their combinations are provided by Enviroscan with a series of informative cartoons and discussions available at http://www.enviroscan.com.

Geophysics is recommended not as a replacement for borings or other direct testing that can confirm karst-related subsurface features, but as a reconnaissance tool to identify areas that might require additional borings, and areas where fewer borings may be necessary thereby optimizing the number and location of borings, and minimizing the overall investigation cost. The most common reconnaissance geophysical techniques are electrical methods based on rock resistivity and conductivity, EM, spontaneous potential (SP), and seismics. Using these methods, data can be collected relatively rapidly over large areas. In contrast, microgravity surveys are labor-intensive and typically cost two to three times what an electrical, EM, SP, or seismic survey would cost for similar data coverage. Therefore, microgravity surveys are rarely used for reconnaissance. Instead, they are a cost-effective way of providing detailed mapping and dimensions of a cavity or cavity system in a specific area (e.g., where cavities are encountered in a boring, where sinkhole activity suggests their presence, or within the footprint of a critical structure). Table 3.2

Geophysical Method	Void Detection	Cave/Conduit Detection and Mapping	Fault and Fracture Delineation	Sinkhole Hazard Determination	Depth to Rock Profiling	Stratigraphic Profiling or Mapping	Water Table Mapping
Microgravity	A	A	A	A	B	B	NR
Electrical Resistivity Imaging	A	B	A	A	B	B	A
Electrical Resistivity Profiling	B	NR	A	B	B	B	A
Electrical Resistivity Sounding	B	NR	B	NR	NR	NR	A
Spontaneous Potential (SP)	B	NR	B	A	NR	NR	NR
Ground-Penetrating Radar (GPR)	A	NR	B	B	B	B	B
Seismic Surface Wave Analysis	B	A	A	B	B	B	A
Seismic Refraction: P-Wave	B	B	B	B	A	A	A
Seismic Reflection: P-Wave	B	A	A	B	A	A	A
Seismic Refraction: S-Wave	B	A	A	B	A	A	A
Seismic Reflection: S-Wave	B	A	A	B	A	A	A
Magnetic Susceptibility or Field Intensity	B	B	B	NR	NR	NR	NR
Frequency Domain Electromagnetics (FDEM)	B	NR	B	A	B	B	NR
Very Low Frequency Domain Electromagnetics (VLFDEM)	NR	NR	A	B	NR	NR	NR
Time Domain Electromagnetics (TDEM)	NR	A	B	NR	B	B	B
Radio Frequency Electromagnetics (RFEM)	NR	B	NR	NR	NR	NR	NR

From Enviroscan (2012).
A, generally recommended; B, recommended for certain conditions; NR, generally not recommended.

TABLE 3.2 Applicability of Different Geophysical Methods to Karst Studies provides guidance for selection of geophysical methods capable of detecting various karst-related features and subsurface characteristics important for defining position of carbonate (karst) aquifers and preferential flow paths. Many of these methods can be applied at the land surface and directly in borings, and some can be used to rapidly survey large areas with aircraft or maritime vessels (Table 3.3).

A classic book entitled *Geophysics in Karst* by Dušan Arandjelović of the Geophysical Institute in Belgrade is to the best of the author's knowledge still the most comprehensive and informative textbook on the subject. It contains numerous examples of applied geophysics used to help solve hydrogeologic and geotechnical problems encountered during implementation of complex projects in the classic Dinaric karst including dams and reservoirs, water supply, and tunneling for infrastructure, to name just a few. Although written more than 30 years ago (Arandjelović, 1976), it is surprisingly innovative as it contains theoretical and practical aspects of various methods that were either perfected or pioneered in the Dinaric karst; it also has examples from the western literature and numerous examples from the Russian literature still largely unknown in the West. Unfortunately, it is published in Serbian and it is the author's hope that an updated edition will soon be published in English by the Serbian colleagues. The following sections briefly describe geophysical methods most commonly applied in karst and include illustrations courtesy of Tim Bechtel, Enviroscan (http://www.enviroscan.com) and from the work of Arandjelović (1976).

Electrical Resistivity and Conductivity

Electrical resistivity surveys involve injecting current into the ground between two electrodes, and measuring the voltage difference between two other electrodes, usually all at the ground surface. For nearly a century, these surveys have been used to perform one-dimensional profiling (mapping) and sounding. Profiling is detection of lateral changes in underground electrical properties by moving a fixed electrode array along a survey transect, whereas sounding is detection of vertical changes by expanding an electrode array about a fixed location. Within the last decade, development of multi-conductor electrode cables and computer-driven, automated switching, as well as innovative processing of large resistivity data sets, have allowed simultaneous performance of profiling and sounding to produce two- and three-dimensional electrical images depicting detailed variations in subsurface electrical properties. For multiple transects, the data can be directly processed in three-dimensions to produce sliceable, rotatable block images, resulting in maps for discrete depths and fence diagrams for example (Fig. 3.40).

Electrical (or, more correctly, EM) properties of rocks differ and, in karst, it is not enough to consider only the carbonates (primarily, limestone and dolomite) because their karstification results in formation of insoluble clastic sediments such as clay. These sediments can be deposited at the land surface or in the subsurface in various cavities. Karstified carbonates can also be interbedded or in lateral contact with noncarbonates or non-karstified carbonates that additionally complicates the application and interpretation of electrical surveys.

Electrical resistivity (ρ) and its inverse value and electrical conductivity ($\nu = 1/\rho$) are the basic electrical properties of rocks. The resistivity depends on the electrical conductivity of rock matrix (which in general has high resistivity), the conductivity of free water occupying the effective porosity, and the conductivity of water film held by mineral grains. More broadly, the hydrogeologic factors in karst that contribute to different electrical resistivity of rocks are matrix porosity, secondary porosity (fractures and karst

Geophysical Method	Surface Only	Borehole to Surface	Borehole to Borehole	Boreholes and Surface Combined	Borehole Logging	Marine	Aerial
Microgravity	Yes	No	No	No	Yes	Yes	Yes
Electrical Resistivity Imaging	Yes	Yes	Yes	Yes	No	Yes	No
Electrical Resistivity Profiling	Yes	No	No	No	No	Yes	No
Electrical Resistivity Sounding	Yes	No	No	No	Yes	Yes	No
Spontaneous Potential (SP)	Yes	Yes	Yes	Yes	Yes	No	No
Ground-Penetrating Radar (GPR)	Yes	Yes	Yes	Yes	No	Yes	Yes
Seismic Surface Wave Analysis	Yes	No	No	No	No	No	No
Seismic Refraction: P-Wave	Yes	Yes	Yes	Yes	Yes	Yes	No
Seismic Reflection: P-Wave	Yes	Yes	Yes	Yes	No	Yes	No
Seismic Refraction: S-Wave	Yes	Yes	Yes	Yes	Yes	No	No
Seismic Reflection: S-Wave	Yes	Yes	Yes	Yes	No	No	No
Magnetic Susceptibility or Field Intensity	Yes	No	No	No	Yes	Yes	Yes
Frequency Domain Electromagnetics (FDEM)	Yes	Yes	Yes	Yes	Yes	No	Yes
Very Low Frequency Domain Electromagnetics (VLFEM)	Yes	No	No	No	No	Yes	Yes
Time Domain Electromagnetics (TDEM)	Yes	Yes	Yes	Yes	No	Yes	Yes
Radio Frequency Electromagnetics (RFEM)	Yes	No	No	No	No	Yes	Yes

From Enviroscan (2012).

Table 3.3 Locations and Implementation of Different Geophysical Methods

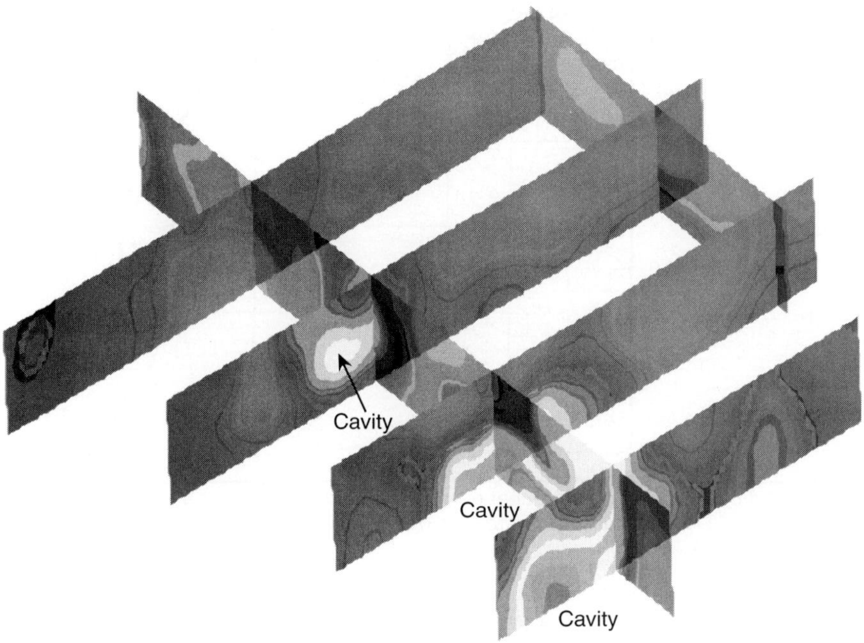

Figure 3.40 Fence diagram created from a three-dimensional electrical imaging data over a "lost" cave (Port Kennedy Bone Cave), known in the 19th century, but later covered by quarry spoils. (Courtesy of Enviroscan.)

voids), water saturation, temperature and mineralization, and rock structure/texture. Several of these relationships typical for the Dinaric karst but also applicable elsewhere are schematically illustrated in Figs. 3.41 and 3.42.

Since great many underground features have electrical properties that differ from their surroundings, electrical imaging (EI) can be used to detect and delineate an incredible array of targets. These include clay layers; sand or gravel lenses; the top-of-rock and rock pinnacles or float blocks; saltwater intrusion; leachate plumes; nonaqueous phase liquid (NAPL) hydrocarbons; dam seepage; mineralized zones; water-bearing fractures; waste disposal pits or trenches; mines, tunnels, and caves; and more. EI specifically can detect variations in subsurface electrical properties related to anomalously thick or wet soils (electrical conductivity highs), or voids in the electrically conductive clay soil mantle (producing electrical conductivity lows). For some targets, recording the induced polarization or chargeability improves detection and discrimination. Each electrical image can depict both the lateral and vertical extents of a target.

Typically, electrical images are one sided, that is, with electrodes only at the ground surface. However, where boreholes are available, images can be collected as two sided (e.g., electrodes in two boreholes) or three sided (electrodes in two boreholes and along the surface between the holes) to increase the image resolution and accuracy.

In some situations, EI may be used to provide cross-sectional imaging of potential voids. Great care must be taken in interpreting electrical anomalies as voids, however, since a void or competent rock (both resistive) may appear identical on an electrical image, but represent very different engineering concerns (Enviroscan, 2012).

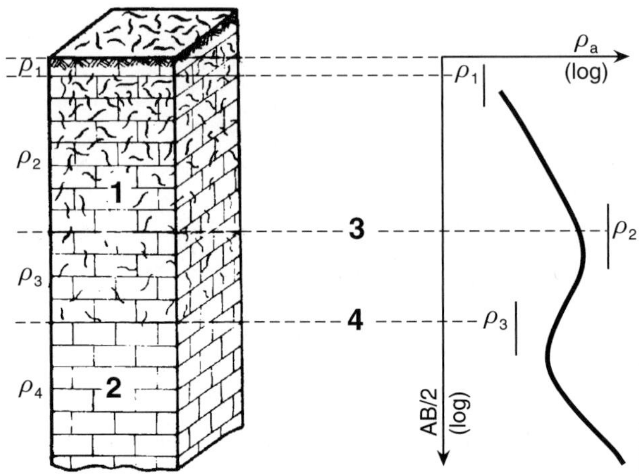

Figure 3.41 Diagram of theoretical electrical sounding curve for a karstified column. (1) Karstified limestone, (2) non-karstified limestone, (3) water table, and (4) base of karstification. (From Arandjelović, 1976.)

Figure 3.43 illustrates application of electrical resistivity surveys for estimating position of the buried carbonates using vertical sounding, and Fig. 3.44 demonstrates profiling along transects that can be used to detect zones of possible preferential flow paths in mature karst terrains. In the first case, the apparent electrical resistivity of the low porosity limestone-dolomite bedrock is higher than the overlying flysch sediments, whereas in the second case the intensely karstified zones in the limestone exhibit a lower resistivity than the surrounding rock.

An application of vertical electrical sounding along radial transects is illustrated in Fig. 3.45. This method can sometimes detect an underlying anisotropy that may be indicative of a more karstified zone and/or a preferential flow path.

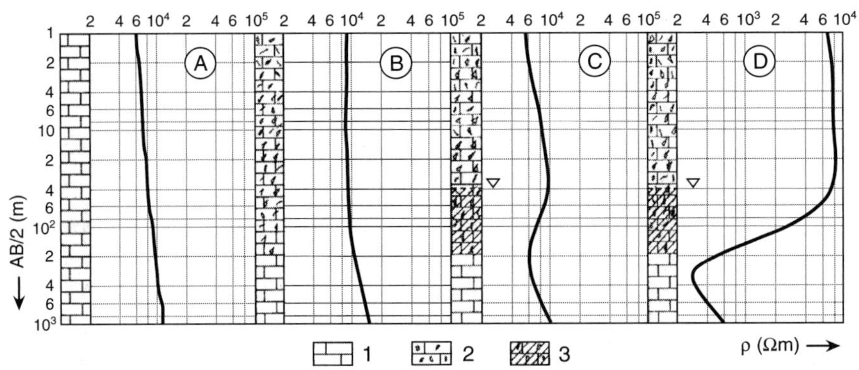

Figure 3.42 Some typical electrical sounding curves in the classic Dinaric karst for (A) non-karstified limestone; (B) unsaturated karstified limestone; (C) saturated karstified limestone; and (D) karstified limestone saturated with seawater, where karstification decreases with depth. (1) Non-karstified limestone, (2) unsaturated karstified limestone, and (3) saturated karstified limestone. (From Arandjelović, 1976.)

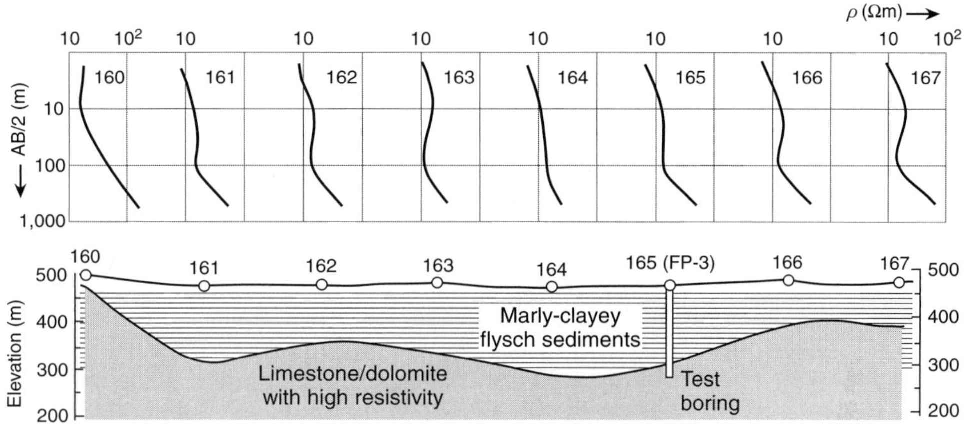

FIGURE 3.43 The limestone-dolomite base of the Fatničko Polje in Eastern Herzegovina determined by electrical sounding. (From Arandjelović, 1976.)

Mise à la Masse

The mise à la masse technique is an electrical resistivity method that has been used in the mining industry since the 1920s for delineating electrically conductive subsurface ore bodies. In that application, a current electrode is placed in the conductive body (either at a surface exposure or in a drillhole), with a second current electrode (to complete the circuit) in the ground at electrical infinity (actually just a distance greater than five times the largest expected dimension of the subsurface body). This causes the entire body to radiate electric current. The equipotential lines from this current flow pattern can be mapped with a voltmeter and two mobile probes. The shape of these equipotentials

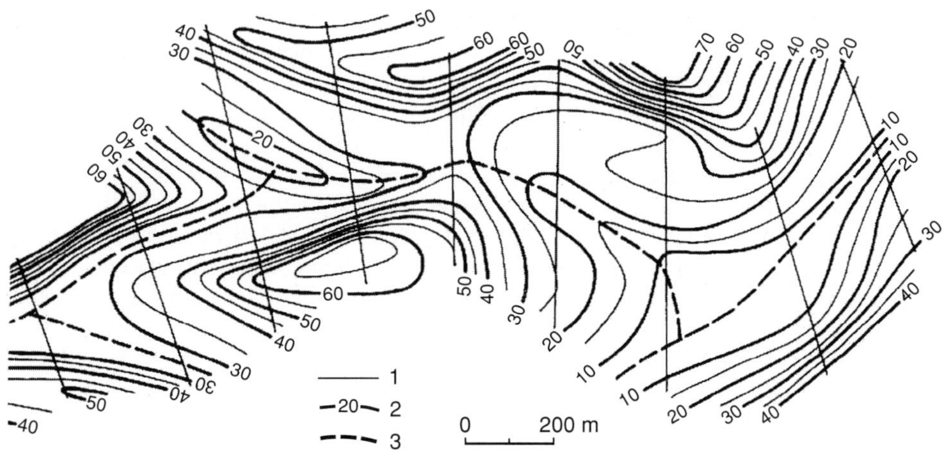

FIGURE 3.44 Apparent resistivity map showing zones of intense karstification and/or tectonically disturbed zones in limestone, Eastern Herzegovina. (1) Trace of electrical mapping profile, (2) contour line of apparent resistivity for AB = 120 m, and (3) axis of zone with lower apparent resistivity. (From Arandjelović, 1976.)

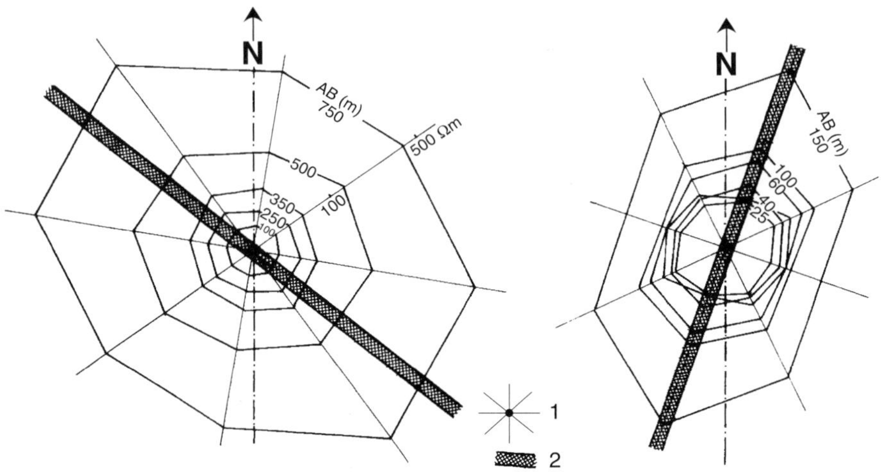

FIGURE 3.45 Diagrams of radial electrical sounding at two locations in Fatničko Polje, Eastern Herzegovina. (1) Azimuths of sounding and (2) main direction of karstification. (From Arandjelović, 1976.)

typically mimics to some degree the footprint of the conductive body. The method has also been successfully employed to map conductive (ionic) leachate plumes, flooded mine workings, mineralized fault zones, and preferential groundwater flow paths since the physics of electric current and water flow are identical (Enviroscan, 2012; see Fig. 3.46).

In karst, the mise à la masse method can in some cases be used to map potential preferential flow paths providing water to a spring. One electrode is placed in the spring and the other one in electrical infinity that should result in good transfer of electrical current to the underground flow. A number of transects is then utilized to measure the gradient of the electric potential at increasing distances from the spring and determine the likely preferential flow path(s). Flow direction of water entering the subsurface via a sink (ponor) can be determined in a similar manner with an added benefit of possibility to introduce an electrolyte such as NaCl into the sinking stream. In time, the electrolyte will create a strong potential field elongated in the direction of the flow that can be mapped with a number of radial transects centered on the ponor as illustrated in Fig. 3.47. The same method can also be used in a single well. As discussed by Arandjelović, the mise à la masse technique using electrolytes has been successful in determining both the directions and velocities of groundwater flow in the Dinaric karst at depths up to 100 m below ground surface. The main limiting factor for the application of the method is presence of steeply dipping layers, and layers with either very high or very low electrical conductivity above the water table.

Spontaneous Potential (SP)

Self-potential or SP methods measure the electrical potential field caused by ambient DC electrical currents in the earth. Electrical currents occur nearly everywhere in the earth due to

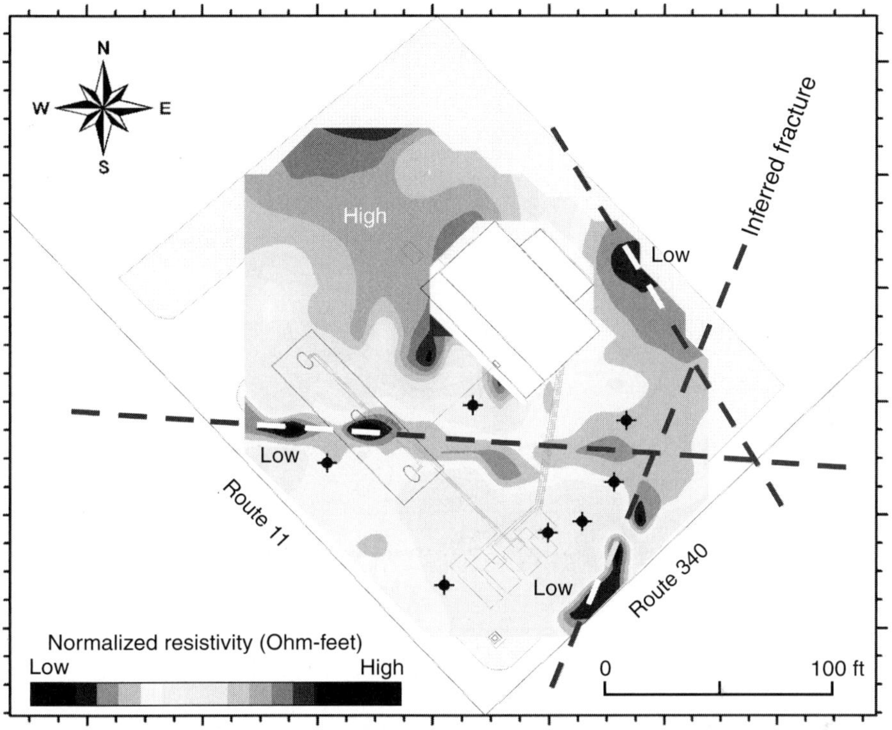

Figure 3.46 Mise-à-la-masse resistivity map: with one current electrode in a well and the other at electrical infinity, potential electrodes can map the preferential flow of current along water-bearing conduits. Note that "infinity" is in reality just a distance greater than five times the largest expected dimension of the subsurface target body. (Courtesy of Enviroscan.)

- Oxidation–reduction reactions around metals or metal ores (e.g., corrosion).
- Movement of ionic fluids across soil or rock contacts (e.g., saltwater intrusion).
- Underground temperature gradients (e.g., near mine fires or geothermal zones).
- Underground chemical gradients (e.g., near sulfides or decomposing petroleum).
- Movement of subsurface water (i.e., seepage, leakage, and flow pathways).

The first four above are generally due to release or capture of charged particles in underground chemical reactions. The fifth arises where a pressure gradient causes any fluid to flow through a permeable medium. This electrofiltration creates a charge separation along the flow path that is measurable as a positive increase in SP in the direction of flow. Thus, SP surveys can locate zones of chemical reaction, as well as detect the actual movement and direction of underground fluid flow, to delineate preferred pathways, or detect leaks or seepage through dams, levees, or earthen liners.

Flowing electrolytic solutions such as acid mine drainage or leachate make dramatic SP targets. The concentrated infiltration of water into incipient sinkholes is also a common source of detectable SP anomalies in karst areas (Fig. 3.48; see the original color figure in the color plates). SP data can be presented as profiles or plan-view maps (based on GPS

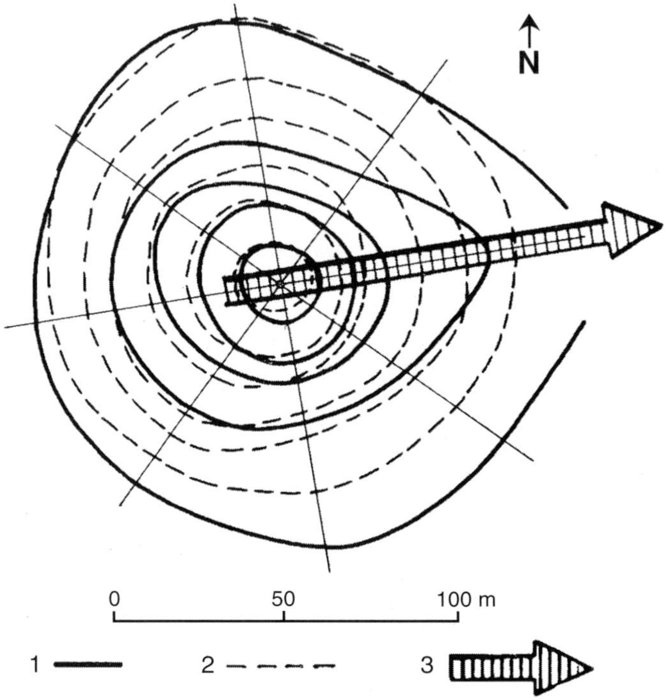

FIGURE 3.47 Determination of the direction of preferential flow originating from a ponor (sink) using mise à la masse method and electrolyte. (1) Equipotential lines after introduction of electrolyte; (2) equipotential lines in natural conditions, before introduction of electrolyte; and (3) direction of preferential flow. (From Arandjelović, 1976, based on Krulc, 1966.)

positioning) that can be interpreted to identify a wide variety of subsurface conditions and features. Often, a charge imbalance created by fluid flow will remain long after the flow ceases—making it possible to detect former flow paths, seepage zones, or infiltration points (e.g., paleosinkholes).

Where there are contaminant plumes or other subsurface phenomena that contain decomposing or reacting chemicals, the release or capture of charged ions creates currents detectable by SP surveys. Acid mine drainage generation zones, as well as underground fires in mines or organic waste masses (e.g., buried leaf mulch or stumps) in particular, produce strong SP anomalies.

As with all geophysics, SP is best interpreted in conjunction with other data. For example, a conductive zone on an electrical image could indicate a water-bearing fracture—an interpretation that would be strongly supported by a coincident SP electrofiltration anomaly. In karst terrains, both microgravity and SP surveys may be used to evaluate sinkhole hazard. However, an SP survey indicates where there is active or former infiltration of the type that could drive soil piping, while microgravity shows where there are bedrock cavities of the type that may not yet be receptors for soil piping, but could accept infiltration if the hydrology of the site is altered such as during regrading or construction (Enviroscan, 2012).

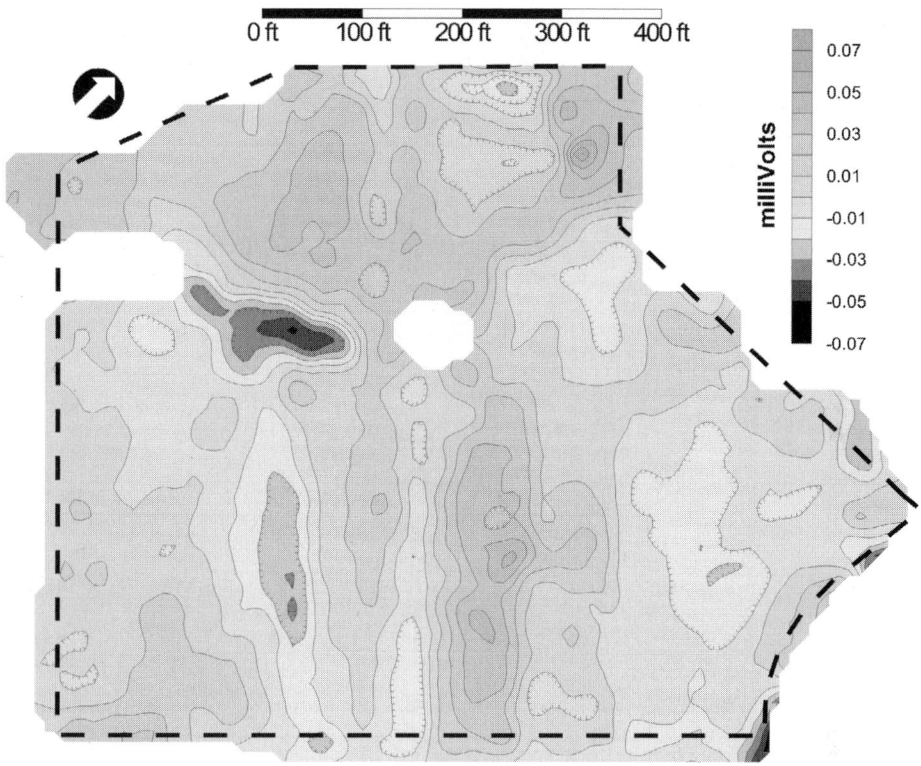

FIGURE 3.48 SP mapping revealed zones of concentrated infiltration (dark gray; red on the original figure—see color plates) that were later confirmed to be deep paleosinkholes covered by more recent sediments, but still piping water down to the karst aquifer system. (Courtesy of Enviroscan.)

Electromagnetic Surveys (EM)

EM array profiling has recently gained popularity in karst studies. The method involves measurements of naturally occurring or induced geomagnetic fields and induced geoelectric (tellyric) fields, collectively called magnetotelluric fields. Where the EM waves encounter electrical conductors in the ground, they induce electrical currents in these conductors, which in turn generate EM waves that can be collected at the surface by an antenna (Fig. 3.49). The method can be useful for mapping karst where voids in bedrock are filled with clay or water. Clay-filled fractures and caves partly filled with water conduct electrical signals far better than does carbonate bedrock, and often produce high-amplitude anomalies. However, air-filled fractures or voids are transparent to EM signals and are difficult to detect (Stierman, 2004). The method is not feasible in urban and industrial areas that typically have a variety of EM disturbance sources.

Microgravity

Large volume voids such as significant washouts, mine workings, tunnels, and karst-related cavities are excellent targets for microgravity surveys. The missing mass of the void creates a measurable disturbance in the earth's gravitational field, with the magnitude of the disturbance directly proportional to the volume of the void (Fig. 3.50).

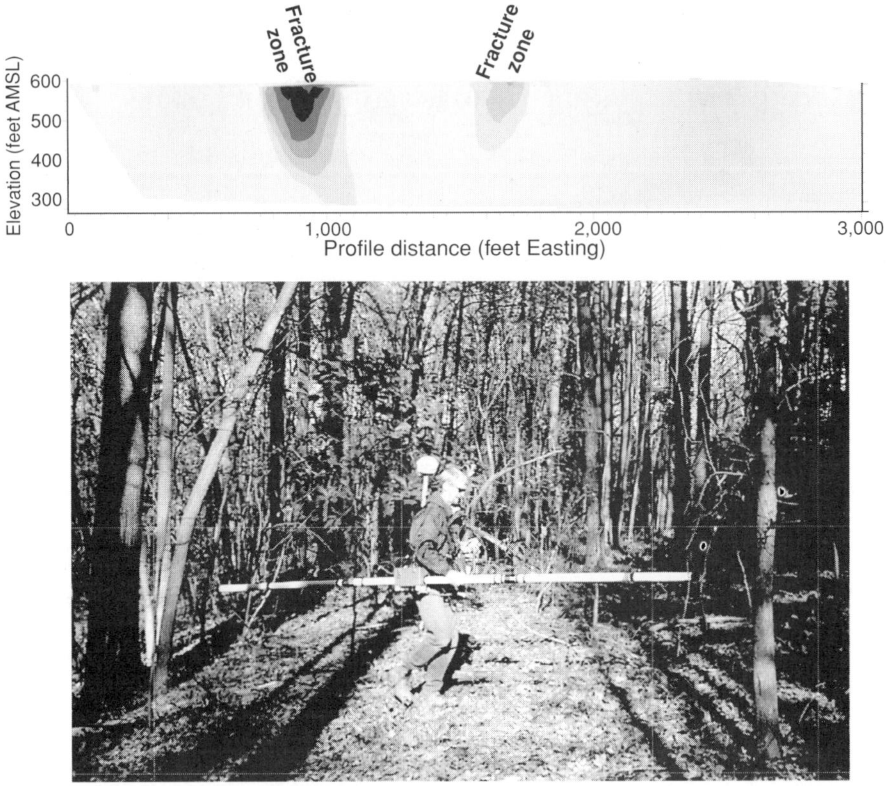

FIGURE 3.49 *Top*: Fracture zones revealed by VLF-EM profile. *Bottom*: Geonics EM-31 terrain conductivity meter. (Courtesy of Enviroscan.)

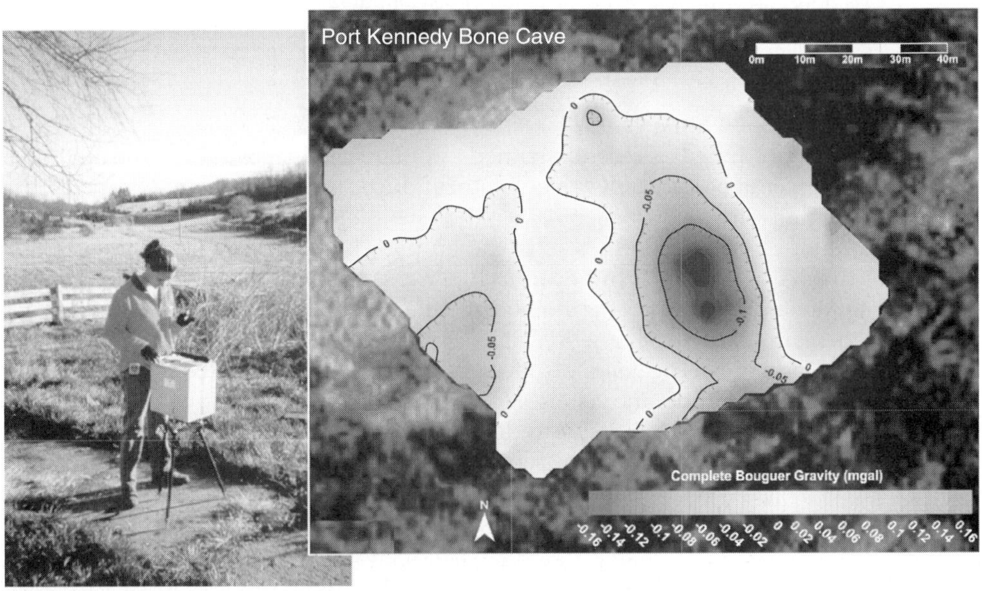

FIGURE 3.50 *Left*: Scintrex CG-3M microgravity meter. *Right*: Gravity map of the "lost cave" shown in Fig. 3.40. (Courtesy of Enviroscan.)

Color Plates

Figure 1 Folded limestone beds in the Durmitor National Park, Montenegro. (Photograph courtesy of Dobrislav Bajović-Bajone.)

Figure 2 Buna Spring near Mostar, Bosnia and Herzegovina, one of the world's largest. This ascending spring issuing from a single cave has been explored by cave divers to a depth of 68 m.

Figure 3 Molnár János cave in Budapest, Hungary. (Photo courtesy of Radoslav Husák.)

Figure 4 Spring of the Sava River, Slovenia. (Photo courtesy of Zoran Stevanovic.)

Figure 5 Havasu Falls with travertine pools in the foreground, Havasu Canyon, part of the Grand Canyon area, in Arizona, the United States. (Photo courtesy of David Price.)

Figure 6 Christy Henson stands in front of a small inlet in Espey Cave, Tennessee, the United States. Espey is 6.6 miles long and was surveyed by the Tennessee Central Basin Grotto in the early 1990s. (Photo by Bob Biddix; copyright Innermost Imagery, printed with permission; more photos available at http://innermostimagery.com.)

Figure 7 Mystery Falls is one of the deepest pits in Tennessee and contains the deepest single free drop in Tennessee at 281 ft. The drop starts by a stream, which issues from a side passage, pours down a pit 100 ft in diameter and 281 ft deep. At the bottom the stream continues flowing to the Tennessee River. (Photo by Bob Biddix; copyright Innermost Imagery, printed with permission.)

Figure 8 Stream formed by a large karst spring, Lukavica mountain, Montenegro. (Photograph courtesy of Dobrislav Bajović-Bajone.)

Figure 9 Dashiwei Tiankeng (collapse doline) in the Leye karst, Guangxi, China. (Photo courtesy of John Gunn.)

Figure 10 Waterfall and pool in travertine, karst spring in Mlini, Croatia.

Figure 11 Historic San Pedro Springs near San Antonio, Texas, the United States, are now dry most of the time due to pumpage from Edwards Aquifer. Shown is one of rare events when the springs are flowing after a major storm. (Photo copyright Gregg Eckhardt, printed with permission.)

Figure 12 Villagers collecting water from a newly drilled flowing well completed in a karst aquifer, northern Iraq. (Photo courtesy of Zoran Stevanovic.)

Figure 13 The late Jim Quinlan is standing next to the remains of a borehole that was drilled through the cave roof to intersect one of the water bodies, the Nullarbor Plain in southern Australia. (Photo courtesy of John Gunn.)

Figure 14 Landsat ETM+ color-composite RGB = 7,4,2 image of the tower karst along Li River, Guangxi Province, China (see also Fig. 3.34).

Figure 15 Land use/land cover maps for the area of Rainbow Springs, Florida, for the years 1995 (left) and 2007 (right). The maps are simplified versions of the original shapefiles provided by the Southwest Florida Water Management District. Red and pink/purple colors indicate residential, industrial, commercial, and transportation developments where aquifer recharge is likely reduced. (From Kresic and Mikszewski, 2012; copyright Taylor & Francis, printed with permission.)

Figure 16 Geophysical SP mapping revealing zones of concentrated infiltration (shown in red) that were later confirmed to be deep paleosinkholes covered by more recent sediments, but still piping water down to the karst aquifer system. (Courtesy of Tim Bechtel, Enviroscan, Inc.)

Figure 17 Typical doline pond ("telaga") in the Gunung Sewu Karst area, Indonesia. (Photo courtesy of Eko Haryono.)

Figure 18 Mystery Falls is one of the deepest pits in Tennessee, the United States. (Photo by Bob Biddix; copyright Innermost Imagery, printed with permission.)

Figure 19 Underground river (maximum flow rate 174 m^3/s) exiting the gigantic cave at the bottom of Xiaozhai Tiankeng (collapse sinkhole) in the Xinlong karst, Chongqing, China. (Photo courtesy of John Gunn.)

Figure 20 Scalloped walls of a gypsum cave near Carlsbad, New Mexico, the United States. (Photo courtesy of Jim Goodbar.)

Figure 21 The larger of Clay Springs in South Snake Valley, Utah, the United States, issues from a deep unexplored pool. The spring is threatened by the proposed expansion of the Southern Nevada Water Authority water supply system driven by the growth of Las Vegas. (Photo courtesy of Gretchen Baker, www.ProtectSnakeValley.com.)

Figure 22 Gypsum chandeliers in Lechuguilla Cave, New Mexico, the United States, formed by evaporation of sulfate-rich water films, which have dissolved gypsum at higher levels. (Photo courtesy of Arthur Palmer.)

Figure 23 One of the Mammoth Springs in Yellowstone National Park, Wyoming, the United States. (Photo courtesy of Miles Kresic.)

Figure 24 Red Lake near Imotski, Croatia. Depth of the sinkhole is 518 m, and the depth of water in the sinkhole during average low conditions is about 250 m. (Photograph courtesy of Croatian National Tourist Board, www.croatia.hr.)

Figure 25 Speleologists following Kali Suci (Clean River), a major sinking stream in the Gunung Sewu Karst, Indonesia. (Photo courtesy of Eko Haryono.)

Figure 26 Dairy Farm sinkhole, Florida, the United States. (Photo by George Sowers, the George Sowers Collection, AMEC Environment and Infrastructure Inc.)

Figure 27 Draperies in the Shenandoah Caverns near New Market, Virginia, the United States.

Figure 28 *Proteus anguinus*, endemic to the Dinaric karst. (Photo by G. Aljani, printed with permission.)

Figure 29 Killeany spring, formerly an important water source, drains a part of the plateau karst of the Burren, Ireland, and is fed by numerous streams generated on surrounding non-limestone rocks which sink at the limestone contact. (Photo courtesy of David Drew.)

Figure 30 Škocjan Caves in Slovenia, protected by UNESCO as a World Natural Heritage Site. (Photos courtesy of Borut Petric, Škocjan Caves National Park.)

Figure 31 Old Ottoman stone bridge over the largest sinking stream in Europe, the Trebišnjica River, Trebinje, Herzegovina. (Photo courtesy of Neno Kukuric.)

Figure 32 Pearlsian Gulf: stalagmites, columns, stalactites, and draperies next to a calcite-laded pool in Lechuguilla Cave, New Mexico, the United States. (Photograph courtesy of Dave Bunnell; more photographs are available at http://www.goodearthgraphics.com/under_earth/.)

Figure 33 High-resolution natural color aerial photograph of a typical karst landscape in classic Dinaric karst of Montenegro. Green vegetated areas with flat bottoms are sinkholes used for agriculture by inhabitants of the small village. (Courtesy of the Center for GIS and Remote Sensing, School of Mining and Geology, University of Belgrade.)

Figure 34 Walled Cities in a Bath County cave, Virginia, the United States. (Photo courtesy of Phil Lukas.)

Figure 35 Forest of soda straws in a cave on New Zealand's North Island. (Photograph courtesy of Dave Bunnell.)

Figure 36 Dye tracing test with uranine in the karst system of Alta Cadena range near Malaga, Spain. (Photo courtesy of Bartolome Andre Navarro, CEHIUMA.)

FIGURE 1

FIGURE 2

FIGURE 3

FIGURE 4

FIGURE 5

FIGURE 6

FIGURE 7

FIGURE 8

FIGURE 9

FIGURE 10

FIGURE 11

FIGURE 12

FIGURE 13

Figure 14

Figure 15

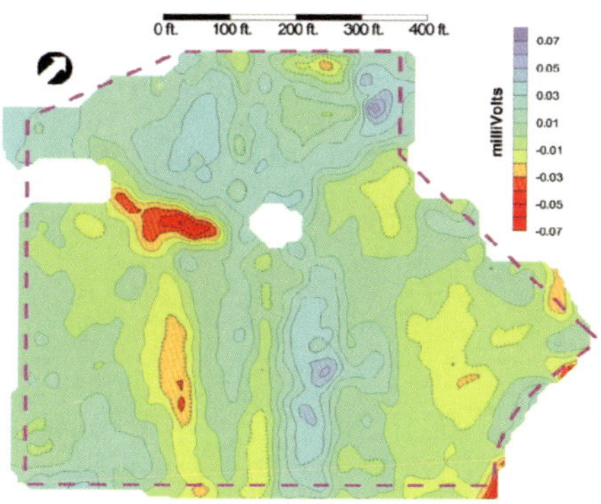

Figure 16

Figure 17

Figure 18

Figure 19

Figure 20

Figure 21

Figure 22

Figure 23

Figure 24

Figure 25

Figure 26

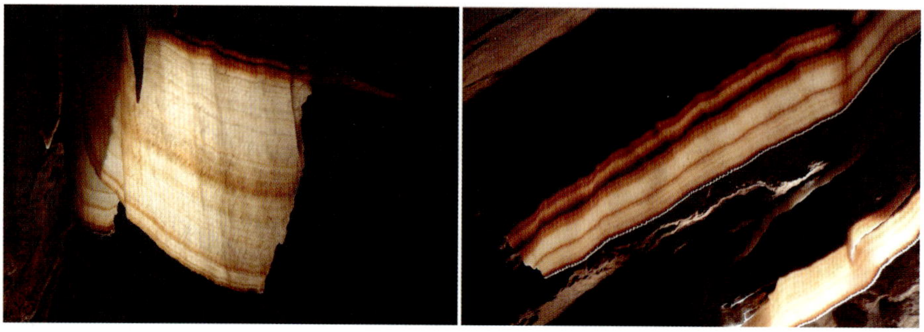

FIGURE 27

FIGURE 28

FIGURE 29

Figure 30

Figure 31

FIGURE 32

Figure 33

Figure 34

FIGURE 35

FIGURE 36

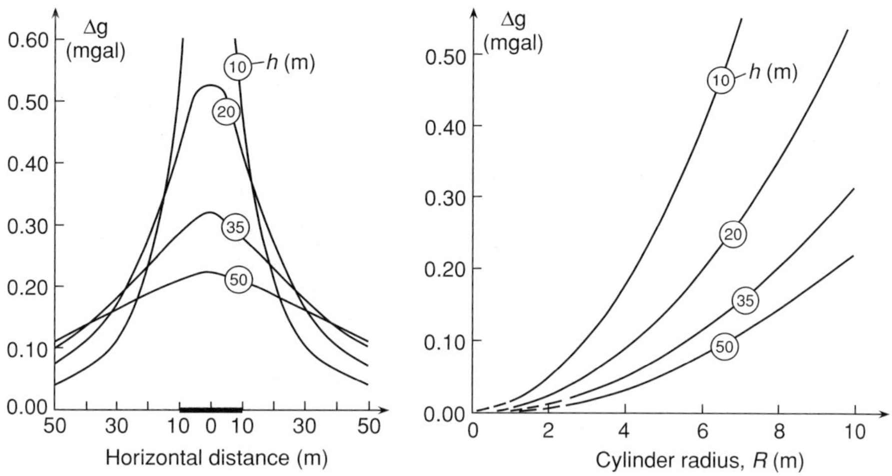

FIGURE 3.51 *Left*: Theoretical curves of Δg above a cylindrical space (underground channel) with a 10-m radius, located at various depths (*h*) in a porous medium with density 2.60 g/cm^3. *Right*: Theoretical curves of changes in Δg for a horizontal air-filled cylinder of varying radius (*R*) and at different depths (*h*) from the land surface. (From Arandjelović, 1976.)

Other examples of application of microgravity include mapping of undocumented mine workings, location of abandoned or clandestine tunnels, and delineation of areas of unsupported floor slab. Since gravity surveys are immune to many common sources of electronic or acoustic noise that can impede many geophysical techniques, they are particularly suited to highly developed or industrial sites (Enviroscan, 2012).

The main limitation of gravity surveys in karst is their shallow effective depths and rapid lateral changes in subsurface geology and rock density such as succession of pinnacles and chimneys for example (see Figs. 1.61 and 1.62). This means that voids may still be present where no gravity anomaly is detected, due to a combination of additive and negative anomalies. Sophisticated computer processing of recorded data can filter some ambiguities but can do little with the surveying precision as illustrated in Fig. 3.51. The graphs show that air-filled cavities have to be very large and at shallow depths to be detected at the precision of about 0.2 mgal or coarser. Smaller cavities filled with water, which significantly reduces the gravity anomaly, at depths greater than 30 to 40 m even in a relatively homogeneous karst environment (if such a thing exists), have little chance of being detected; practical field accuracy of gravity meters in ideal conditions is seldom better than 0.01 mgal (Stierman, 2004). As in other cases of geophysical surveys in karst, microgravity is best utilized in combination with methods that measure a different physical contrast in the subsurface such as electric properties (Fig. 3.52).

Ground-Penetrating Radar

Ground-penetrating radar (GPR) is a relatively inexpensive, nonintrusive, electrical surface geophysical method that can be used to define shallow lithologic contacts and subsurface features. In GPR surveys, a radio-frequency EM signal is transmitted into the ground, and the signal reflected from subsurface lithologic and hydrologic features and boundaries can be interpreted to identify sediment thicknesses, depths to the water

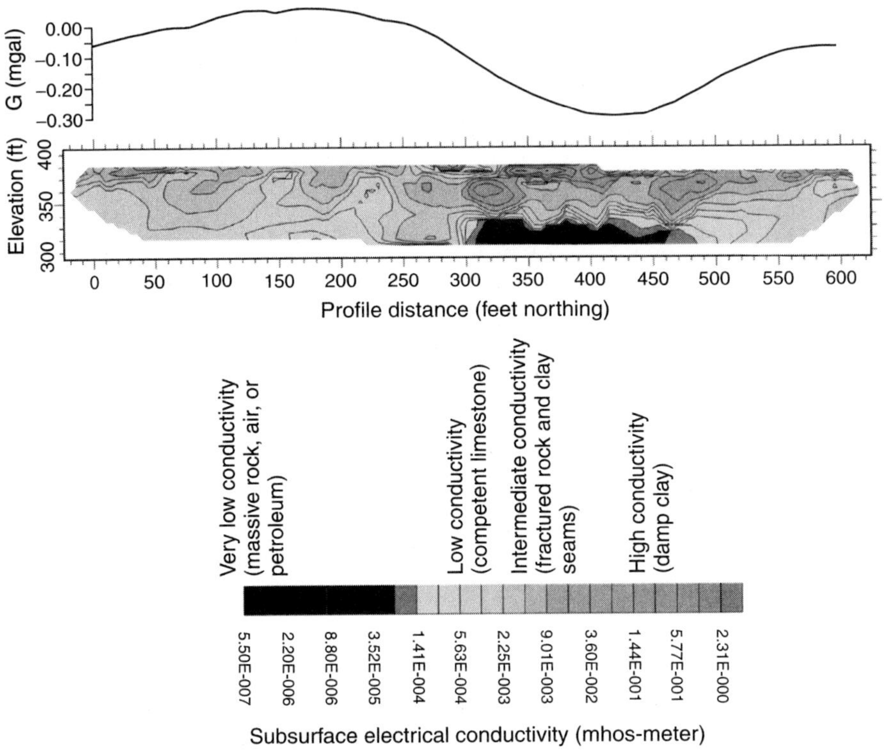

FIGURE 3.52 Two-dimensional electrical image (*bottom*) and matching gravity profile (*top*) showing low gravity over a very resistive zone, perhaps a cavity filled with air. (Courtesy of Enviroscan.)

table and to clay beds, breaches in confining beds, karst development, buried objects, and lake-bottom structure. The effective exploration depth of a GPR survey is the maximum depth at which coherent reflections can be recognized. High subsurface conductivities attenuate the transmitted signal, limiting exploration depths. Barr (1993) reports that effective exploration depths in predominantly unsaturated and saturated sand and clay sediments at the five study sites in west-central Florida ranged from a few feet to more than 50 ft below land surface. Exploration depths were limited as a result of strong signal attenuation when high conductivity clay was encountered, whereas greater exploration depths were possible when low conductivity sand was encountered.

As discussed by Stierman (2004), the selection of radar frequency is probably the most important choice when undertaking a GPR survey, since lower frequencies (10 to 300 MHz) allow depth penetration to several tens of meters, but they will fail to detect small diameter anomalies. Frequencies of 500 MHz and greater provide excellent resolution, but they limit the subsurface depth penetration of the GPR to less than 5 m. Because clay-rich soils attenuate GPR signals and restrict depth interrogation to depths of sometimes just 1 m, GPR surveys have limited overall applicability in most karst terrains.

When conditions are favorable, such as in certain parts of Florida where thick sand overlies clay and carbonate layers, GPR has been successful in identifying sinkhole development hazard, water table position, depth to carbonate bedrock, and presence of

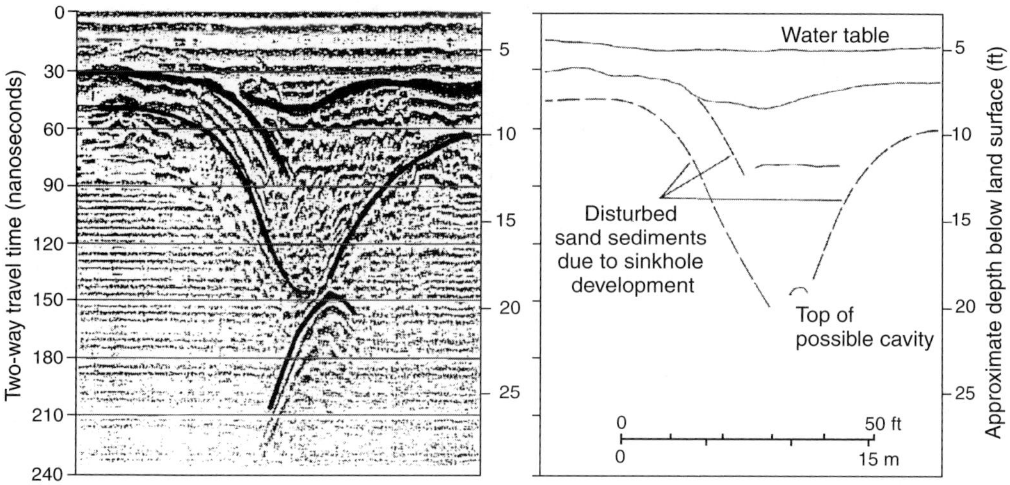

FIGURE 3.53 Ground-penetrating radar profile and hydrogeologic interpretation along a traverse over a subsurface cavity at the Pemberton Creek, Hillsborough County, Florida site. (From Barr, 1993.)

buried karst features as illustrated with the following examples provided by Barr (1993). Figure 3.53 shows a GPR traverse at the Pemberton Creek site in Hillsborough County, Florida. A feature tentatively identified as a cavity below a sinkhole was not verified, but the area has many sinkholes, and other investigators have documented some chevron shapes as cavities. The sediments within the interpreted sinkholes in Fig. 3.54 have chaotic reflection patterns that are characteristic of disturbed or deformed sediments.

One particularly useful feature of GPR surveys is standard equipment that produces a graphic display that allows the user to interpret data while in the field. GPR equipment consists of components that are carried either manually or are towed behind vehicles or boats during a GPR traverse.

Seismic Surveys

Seismic refraction, reflection, and surface wave analysis provide profiles of the top-of-rock that may display conical depressions of the type associated with ancient sinks, or linear deeps that may represent sinkhole-prone faults, fractures, bedding planes, or other lineaments. Seismic rock depths are used to calibrate microgravity results where no boring data are available. As emphasized by Enviroscan (2012), the seismic profiling is also often used to discriminate between EM or EI conductive anomalies (which could represent either a bedrock or wet, soupy soils), or between EM or EI resistive anomalies (which could be caused by either dry, competent rock or air-filled voids).

As discussed by Stierman (2004), single sets of ground-based seismic data have often failed to produce interpretable results, but tomographic sections produced by computer analysis of larger banks of data (obtained from seismic velocities measured on hundreds of ray paths between multiple positions in boreholes) have proved more useful. The development of three-dimensional seismic tomography (3DT) is again dependent on powerful computer software, but has produced some impressive images of complex karstic voids in unseen ground. This is currently the most promising tool in geophysical

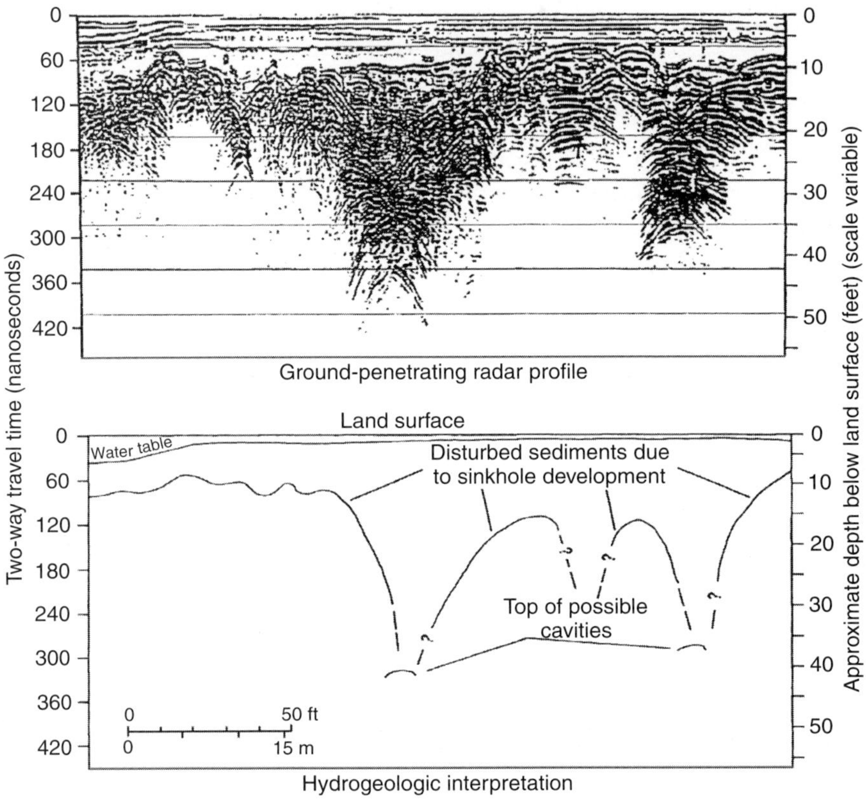

FIGURE 3.54 Ground-penetrating radar profile and hydrogeologic interpretation along a traverse over a sinkhole development at the Northwest Pinellas Sewage Treatment Plant, Pinellas County, Florida site. (From Barr, 1993.)

searches for caves, but does represent a major investigation incorporating boreholes unless underground access is already available (as on a tunnel project).

Case Study: High-Resolution Seismic Reflection

Kindinger et al. (2000) present the data from digital high-resolution seismic profiling (HRSP) used to identify subbottom features from selected lakes in Florida. The objectives of their study were to (1) identify evidence of breaches or discontinuities of the confining units between surficial water bodies and the Floridan aquifer and (2) identify diagnostic features, structure, and geomorphology of sinkhole lakes. The results are also applicable to surface reservoirs in karst terrains and general issues of reservoir leakage/water loss through underlying cavities.

An estimated 95% of the surface waters in northeastern Florida are sinkholes. The formation of lakes is the direct result of chemical and/or mechanical processes. The mechanical processes include (1) slumping or subsidence of underlying clastic or carbonate sediments, (2) clustering of sinkholes, or (3) a combination of the previous two. Sinkhole lakes in Florida occur in areas of thin overburden, typically less than 61 m (100 ft). In areas with an impermeable confining layer and no breaches, a lake might be a perched lake (the lake level is held above the groundwater level) with no communication with subsurface aquifers. Otherwise, a lake will form in conditions where a lack of overburden or permeable confining layer allows for increased karstification of the underlying limestone, producing a depression due to limited fill material. Florida sinkhole lakes are in various stages of development and some of them went from drying out due to flushing of a sinkhole plug to being inundated again. It has been also widely recognized that some of the largest Florida lakes are in fact flooded poljes. A typical example is Orange Lake, which includes active subsidence and transitional features.

Importantly, sinkholes and other karst features underlying the lakes provide vital conduits for surface water recharge to the aquifer as breaches through the semiconfining layer. However, as breaches they also become sources of contaminants to the potable water supply. Consequently, sinkholes and sinkhole lakes need to be characterized by their hydraulic connection potential and subsequently maintained and protected as the rapid population growth and land development in Florida continues unabated.

The Elics Delph2 Geophysical system is used in the study measures and displays two-way travel time (TWTT) of the acoustical pulse in milliseconds (ms). Amplitude and velocity of the signal are affected by variations in lithology of the underlying strata. Laterally consistent amplitude changes (lithologic contacts) are displayed as continuous horizons on the seismic profiles. Depth to horizon is determined from the TWTT, adjusted to the subsurface velocity of the signal. The interbedded nature of the lake-bottom sediments provides good reflecting surfaces for acoustic signals. These layers appear on the seismic records as convergent, divergent, or parallel bands. Folds, faults, and sedimentary changes can be recognized as bands, lateral and vertical discontinuities, and truncations of the bands by other reflections.

Historically, HRSP has been used to determine the regional distribution of stratigraphic units having distinct acoustical characteristics. In this study, the lakes are well distributed and have a relatively small diameter, making stratigraphic correlation difficult. HRSP data has been used primarily to map the shallow subsurface features found beneath selected lakes of northeastern Florida. Subsurface diagnostic features are used to define the structural history and to locate possible breaches in the confining layer that maintains the perched lakes above the Floridan aquifer. In many cases, the acoustical records show fine details of karst features. Compilation of these features from seismic profiles acquired from the lake surveys indicated that certain acoustic patterns reoccur from lake to lake.

Two types of karst features identified in Fig. 3.55-top represent depressions that have been subsequently filled to the present lake bottom. The fill is represented by horizontal reflectors that completely cover the subsided area. Evidence of stress fractures, slumping, faulting, or dissolution fractures around the depression differentiate the two dolines and may indicate more rapid or continuous subsidence, or a more competent overburden. These breaches within the depression may provide a significant hydraulic connection between surface waters and the underlying aquifer.

Other features commonly detected by the survey are high frequency or chaotic reflections interspersed between horizontal reflections (Fig. 3.55-bottom left). These reflectors indicate a disturbance within a relatively intact stratigraphic sequence and may represent solution pipes or fractures through the overburden. The features may connect to dissolution systems in the underlying limestone and could represent direct hydraulic connection through the semiconfining layer to the underlying aquifer. The disturbed reflections indicate areas of potential subsidence or collapse. These features have a high

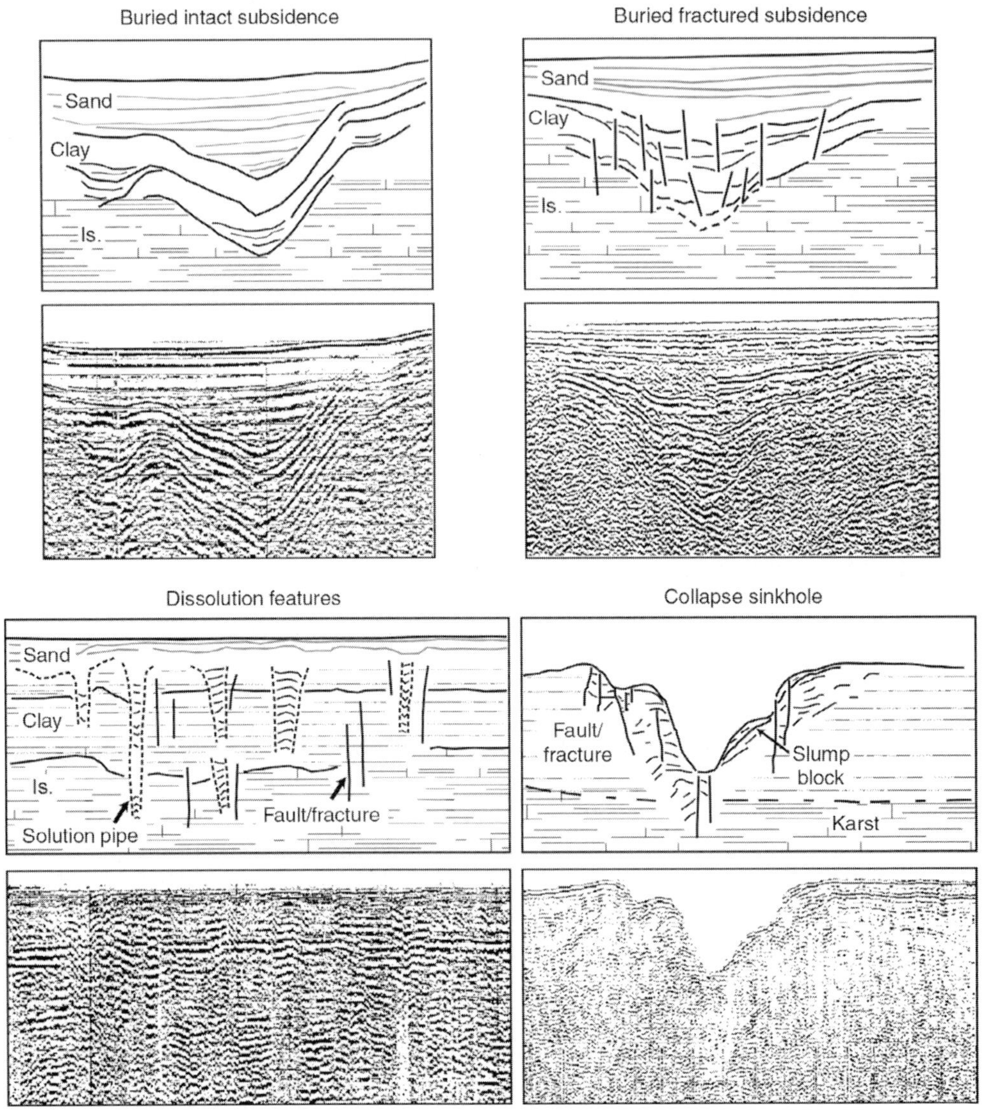

FIGURE 3.55 Seismic profiles with line drawing interpretations of different types of features described from the lakes of northeastern Florida. (From Kindinger et al., 2000.)

potential for reactivation since the plugs that fill solution pipes may dislodge during periods of major rainfall variations. Solution pipes and related features commonly occur in areas where cohesive overburden is moderate to thin. Dissolution is focused and material directly over the cavity is washed into the void during the piping process.

The feature illustrated in the bottom right in Fig. 3.55 represents the classic collapse sinkhole, with steep walls that show evidence of slumping and active development along the periphery of the collapse. Freshwater plumes have been imaged emerging from similar collapse features found in marine environments (e.g., Crescent Beach Spring). In seismic profiles, areas of negligible acoustic return below the collapse have been postulated to represent subterranean cavities. These active phase collapse sinkholes are typically evident at the surface without imaging and occur in areas of minimal overburden. They also indicate areas of internal drainage or discharge depending on the location of the potentiometric surface of the Floridan aquifer (Kindinger et al., 2000).

Case Study: Electrical Resistivity

An informative work on the application of electrical resistivity to examine known karst features and locate water-bearing conduits in the Inner Bluegrass of Kentucky is provided by Graham (2000). As discussed by the author, the method of electrical resistivity practiced today is relatively fast and nondestructive, can be used to locate a variety of subsurface features, and has rapidly gained popularity. Analytical solutions and mathematical models have been derived by various researchers to describe the behavior of an electrical field in different geologic materials and geometries. The most commonly applied of these solutions are (1) the behavior of an electrical field as it is carried laterally across a high-angle discontinuity, and (2) the changes in the apparent resistivity measured at the surface as an electrical field are expanded to impinge on successively deeper layers. The author provides detailed discussion on the intricacies of the method, common misinterpretations of the field data in karst, and a comprehensive list of references.

The work by Graham (2000) took advantage of numerous borings and known karst features in three study areas including conduits at the Marshall Spring in Scott County. Drilling concurrent to the resistivity surveying provided information vital to interpretation of the geophysical data (Fig. 3.56). For example, a drillhole on a very pronounced low-resistivity anomaly encountered a major water-bearing karst conduit, interpreted to feed Royal Spring in nearby Georgetown. Profiles in the Sinking Creek Karst Basin using two different electrode arrays located anomalies interpreted as corresponding to water-bearing conduits and to abandoned, air-filled conduits of that karst system. The author concludes that the electrical resistivity method is applicable to the prospection for water in the karst terrain of the Inner Bluegrass, particularly if there is local drillhole data available. He also suggests that survey and interpretive methods relying on simple geometries in the subsurface can lead to errors in interpretation of the subsurface bodies causing resistivity anomalies. Recommendations for the use of electrical resistivity methods in this region include consideration of topography and its effect on resistivity measurement, use of diagnostic signatures of known geologic features, and use of multiple arrays to locate narrow anomalies accurately. One example of a practical advice given by Graham (2000) is that "The response of the electrical field to near-surface conditions is very important. Wet ground, soil moisture and the ability of different soils to retain moisture, and the presence or absence of water at the soil-rock interface can have a profound impact upon the apparent resistivities, even when the electrode array is arranged to measure the resistivity at greater depths. When taking resistivity soundings it is important to keep in mind that eliminating short-spacing readings can hamper detailed interpretation of sounding data, in spite of field time saved."

Geophysical Logging of Boreholes

Single-borehole hydraulic methods and the geophysical logging methods taking advantage of ambient or introduced fluids (hydrogeophysics) are described in Sec. 1.5.1.

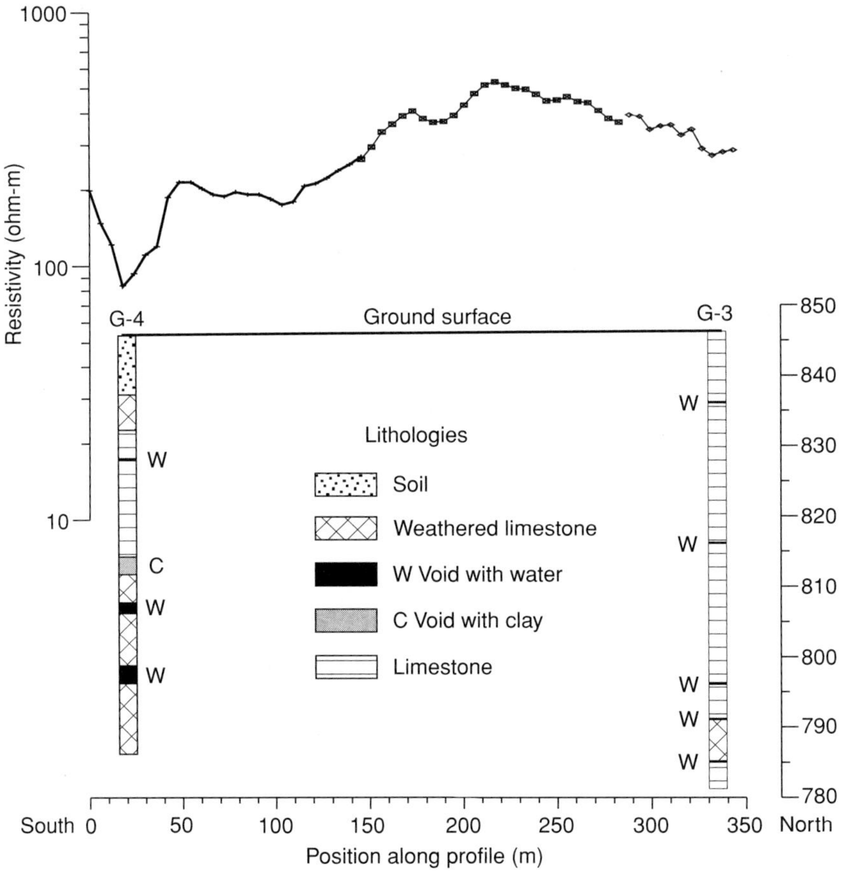

FIGURE 3.56 Resistivity profile and drilling results at Line A, southern and central segments, Kentucky Horse Park. (From Graham, 2000.)

When these methods are applied at multiple boreholes and combined with the surficial geophysical surveying techniques described in the preceding sections, the resulting conceptual site model will be as complete as possible for a karst environment.

In the same way that surface geophysics makes it possible to "see" beneath the ground surface, borehole geophysical logging makes it possible to see beyond the walls of a boring or well. To perform borehole logging, sensors (or sondes) that measure different physical properties of the formation around the boring are lowered down the hole to record continuous data (or logs). A multiconductor cable on a motorized winch controls the sonde, and transmits data back up the hole to a computer and graphic display. Geophysical logging also typically samples a larger volume of material around the borehole providing more representative, in situ and undisturbed data than a core sample. Often, multiple logs (called a suite) are recorded for a single boring, each measuring a different property, to allow more complete knowledge of subsurface conditions (Enviroscan, 2012). Various logging methods used in karst settings and well conditions required for their implementation are provided in Tables 3.4 and 3.5, respectively.

Geophysical Logging Method	Density	Porosity	Clay vs. Sand	Fracture Zones	Fracture Thickness and Dip	Water-Bearing Zones	Low-Flow Water Movement	High-flow Water Movement	Cavity Dimension
Natural Gamma (nγ)	NR	NR	A	B	NR	NR	NR	NR	NR
Short Normal Resistivity	NR	B	A	B	NR	B	NR	NR	NR
Long Normal Resistivity	NR	B	A	B	NR	B	NR	NR	NR
Single Point Resistance (SPR)	NR	NR	B	B	NR	B	NR	NR	NR
Spontaneous Potential (SP)	NR	NR	B	B	NR	NR	NR	NR	NR
Magnetic Susceptibility	NR	NR	B	B	NR	NR	NR	NR	NR
Magnetic Field Vector	NR	NR	NR	B	NR	NR	NR	NR	NR
Caliper	NR	NR	NR	A	A	NR	NR	NR	B
Fluid Temperature	NR	NR	NR	NR	NR	A	B	B	NR
Fluid Conductivity	NR	NR	NR	NR	NR	A	B	B	NR
Electromagnetic Induction	NR	NR	A	B	NR	B	NR	NR	NR
Verticality	NR	NR	NR	NR	NR	NR	NR	NR	NR
Full Waveform Sonic	A	A	B	A	B	NR	NR	NR	NR
Digital Video	NR	NR	B	A	A	A	NR	B	B
Borehole Acoustic Televiewer (BHTV)	NR	NR	NR	A	A	NR	NR	NR	B
Borehole Optical Televiewer (OPTV)	NR	NR	NR	A	A	B	NR	NR	B
Heat Pulse Flowmeter	NR	NR	NR	NR	NR	B	A	B	NR
Impeller Flowmeter	NR	NR	NR	NR	NR	B	B	A	NR
Induced Polarization (IP)	NR	NR	B	B	NR	NR	NR	NR	NR
Borehole Gravity	A	B	B	B	B	NR	NR	NR	A
Gamma Density (γ-γ)	A	NR	B	B	B	NR	NR	NR	NR
Neutron Density	NR	A	B	B	B	A	NR	NR	NR
pH	NR	NR	NR	NR	NR	A	B	B	NR
Dissolved Oxygen (O_2)	NR	NR	NR	NR	NR	A	B	B	NR
Cavity Sonar	NR	NR	NR	NR	NR	NR	NR	NR	A
Cavity Ultrasound	NR	NR	NR	NR	NR	NR	NR	NR	A

From Enviroscan (2012).
A, generally recommended; B, recommended for certain conditions; NR, generally not recommended.

TABLE 3.4 Applicability of Different Geophysical Logging Methods to Karst Studies

Technique	Required Well Conditions in Interval to be Logged		Information Beyond Casing, if Present?
	Casing	Contents	
Natural Gamma ($n\gamma$)	None, steel or PVC—solid or screen	Air or water—clear or turbid	Yes
Short Normal Resistivity	None or PVC screen	Clear or turbid water	Yes
Long Normal Resistivity	None or PVC screen	Clear or turbid water	Yes
Single Point Resistance (SPR)	None or PVC screen	Clear or turbid water	Yes
Spontaneous Potential (SP)	None or PVC screen	Clear or turbid water	Yes
Magnetic Susceptibility	None, or PVC—screen or solid	Air or water—clear or turbid	Yes
Magnetic Field Vector	None, or PVC—Screen or solid	Air or water—clear or turbid	Yes
Caliper	None, steel or PVC—Solid or screen	Air or water—clear or turbid	No
Fluid Temperature	None, steel or PVC—solid or screen	Clear or turbid water—fully equilibrated	No
Fluid Conductivity	None, steel or PVC—solid or screen	Clear or turbid water—fully equilibrated	No
Electromagnetic Induction (Formation Conductivity)	None, or PVC—screen or Solid	Air or water—clear or turbid	Yes
Verticality	None, or PVC—screen or solid	Air or water—clear or turbid	No
Full Waveform Sonic	None	Clear or turbid water	Yes
Digital Video	None, steel or PVC—solid or screen	Air or clear water	No
Borehole Acoustic Televiewer (BHTV)	None, steel or PVC—solid or screen	Clear or turbid water	No
Borehole Optical Televiewer (OPTV)	None, steel or PVC—solid or screen	Air or clear water	No
Heat Pulse Flowmeter	None, steel or PVC—solid or screen	Clear or turbid water—slow moving	No
Impeller Flowmeter	None, steel or PVC—solid or screen	Clear or turbid water—fast moving	No
Induced Polarization (IP)	None or PVC screen	Clear or turbid water	Yes
Borehole Gravity	None, steel or PVC—solid or screen	Air or water—clear or turbid	Yes
Gamma Density (γ-γ)	None, steel or PVC—solid or screen	Air or water—clear or turbid	Yes
Neutron Density	None, steel or PVC—solid or screen	Air or water—clear or turbid	Yes
pH	None, steel or PVC—solid or screen	Clear or turbid water—fully equilibrated	No
Dissolved Oxygen (O_2)	None, steel or PVC—solid or screen	Clear or turbid water—fully equilibrated	No
Cavity Sonar	None	Clear or turbid water	No
Cavity Ultrasound	None	Air	No

From Enviroscan (2012).
A, generally recommended; B, recommended for certain conditions; NR, generally not recommended.

TABLE 3.5 Required Well Conditions for Application of Different Geophysical Logging Methods

Acoustic and optical borehole televiewer logging allows quantitative and statistical measurement of the depth, thickness, and orientation of features such as cavities, fractures, and joints. For this purpose, downhole geophysical measurements can actually be more accurate than core samples since the televiewer depicts in situ conditions that are magnetically referenced to true north (eliminating the need for oriented cores), and are not affected by incomplete core recovery (a common problem in fractured rock). Together with borehole video camera, these methods provide for invaluable direct observations of subsurface conditions (Fig. 3.57).

In some settings, tomographic (tomo = slice, graph = picture) measurements between boreholes or wells (crosshole) or between boreholes and the ground surface (uphole) can provide subsurface images superior to those based only on surface measurements. These settings include targets that are very small relative to their depth, targets that lie beneath buildings or other structures, or targets beneath sites with shallow or surficial interference

FIGURE 3.57 *Left*: Borehole video inspection setup. *Top right*: Still frame from borehole video showing water jetting into the borehole about 50 ft above the local static water table. *Bottom right*: Optical televiewer image of solution-enlarged horizontal and subvertical fractures in limestone wrapped into virtual core. (Courtesy of Enviroscan.)

FIGURE 3.58 Setup for determining presence and three-dimensional orientation of fractures using electrical resistivity tomography between boreholes. (Photograph courtesy of Peter Thompson, AMEC.)

or "noise." Tomography typically involves sending signals between a transmitter in one borehole and a receiver in another borehole. The signal can be a seismic/acoustic pulse, an electrical current, or an EM field or radar pulse. The advantage of tomography over surface geophysics is that signal paths are one-way and much shorter—that is, from one borehole to another rather than down from the surface and back up again. The shorter signal paths mean that each signal samples a smaller volume of subsurface material (creating higher resolution images), and suffers lesser signal loss, thus creating the ability to image more subtle targets (Enviroscan, 2012).

As with any geophysical survey, the exact choice of a tomographic technique (e.g., seismic versus electrical or EM) is dictated by the physical characteristics of the target and its surroundings, as well as any potential sources of interference or noise. An example of crosshole electrical tomography is provided in Figs. 3.58 and 3.59.

3.2.4 Dye Tracing

Tracing of groundwater is the oldest investigative method in karst hydrogeology. Although points of view among hydrogeologists and hydrologists may vary widely (and wildly) on just about every aspect of water in karst, it can be argued that tracing, when correctly executed, is still considered by many as the most accurate method for finding answers to a variety of questions including

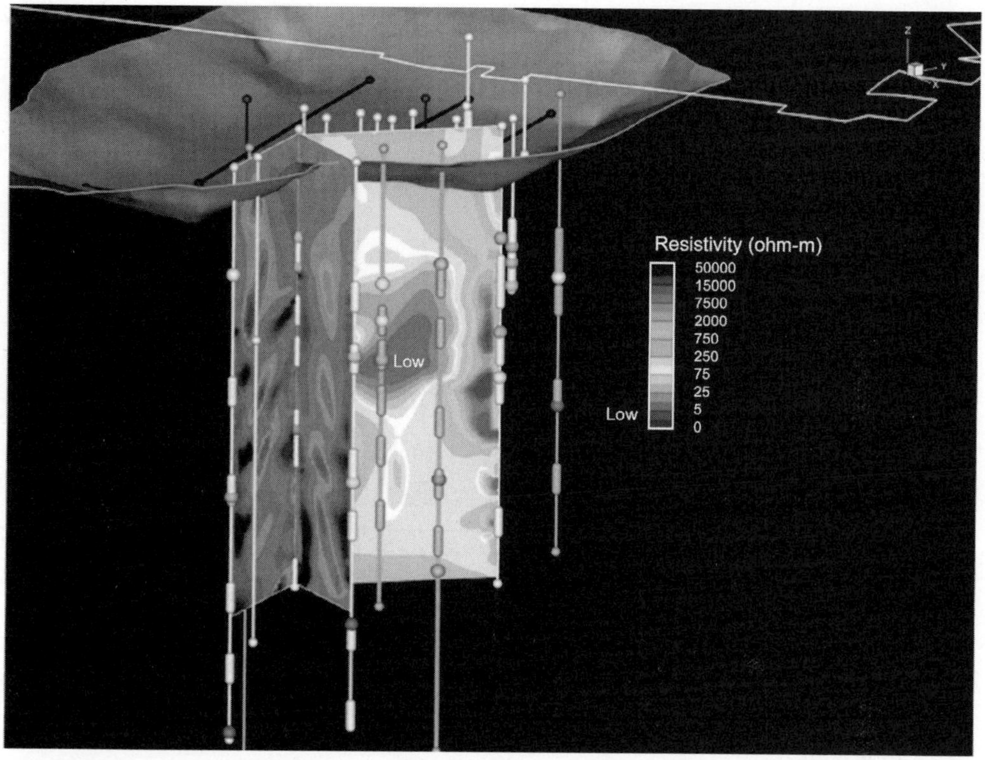

Figure 3.59 Electrical resistivity tomography results shown in panels between boreholes. Fractures and fracture zones are indicated by low resistivity. (Courtesy of Peter Thompson, Scott Calkin, and Rod Rustad, AMEC.)

- Existence of underground connections between the points of interest, such as a sink or an industrial site with groundwater contamination and a spring or water supply well (e.g., see Quinlan et al., 1988).
- Time of tracer travel and therefore groundwater flow velocity between the points of interest.
- Delineation of the hydrogeologic (subsurface) drainage area of a spring or a stream.
- Delineation of the capture zone of a water supply well (wellhead protection zone).
- Assessment of contaminant fate and transport parameters required for a groundwater model.
- Assessment of groundwater flow rates in a conduit or network of conduits.

It is experience of the author and many colleagues from around the world that tracing in karst is increasingly utilized and required by various regulatory agencies as part of decision making. This is driven largely by the growing realization by nonspecialists that

Figure 3.60 *Left*: Injecting 50 mL of fluorescein dye into Monitoring Well 2, Gibsonburg, OH. *Right*: Charcoal passive sample collection method applied at a monitoring well. (From Ohio Environmental Protection Agency, 2010.)

many investigative methods applicable in traditional hydrogeology are not providing adequate information in karst. It is now not unusual to find statements similar to the following:

"This was the first hydrogeologic investigation in Ohio to utilize an artificial injection site and public water system well receptor sites and provides evidence that even though the area lacks surficial karst features, it still exhibits fast groundwater flow rates indicative of a karst aquifer. Data from this study will be used to refine the drinking water source protection area for the Village of Gibsonburg and other public water systems utilizing this karst aquifer." (Ohio Environmental Protection Agency, 2010; see Fig. 3.60.)

There are many common and not so common types of tracers that have been and could be used in karst studies. Since a detailed discussion on the topic of tracer types is beyond the scope of this book, the focus of the following overview is on the application of commercial dyes, the most widely used tracers in hydrogeology today. As a brief sidebar, here are several excerpts from an excellent historic overview of tracing in karst provided by Jones (2012) including arguably the first scientific dye-tracing test performed in karst in 1872 by Hagler.

"A 2000-year-old tracer test is attributed to tetrarch Philippus who supposedly established the source of the springhead of the Jordan River in 10 A.D. Chaff was thrown into Phialo Pond (Berekhat Ram Crater Lake) and reemerged from the cave spring Panium, the head of the Jordan River. Although the results of this test have been questioned, it does establish that the basic idea of tracing undergroundwater is quite old."

"A scientifically planned tracer test was conducted in 1872 by A. Hagler to determine the origin of a typhoid fever outbreak in the village of Lausen in Switzerland. About 800 kg of salt (NaCl) was injected in a sinking stream on a farm south of the village and the water supply spring for the village showed a strong reaction for chloride the following day (Käss, 1998)."

"One novel tracing experiment in Slovenia in 1929 used marked eels. The dorsal fins of 494 eels were notched and the eels were released in the Reka River that sinks into Skocjanske Cave and resurges 34 km to the west at the Timavo Springs in Trieste Bay (Italy). Twenty-nine of the eels were caught in eel pots at the Timavo Springs during the one-year observation period."

Dye tracing can be quantitative and qualitative. The quantitative tracing includes continuous or frequent collection of tracer concentration data and flow rates at the locations of tracer discharge. This data is used to plot tracer breakthrough curves and determine mass recovery (balance) of the tracer. The qualitative dye tracing is performed with the simple goal of finding, if there is a subsurface connection between the points of interest. It often precedes a quantitative test to help its planning including selection of sites for more frequent (or continuous) sampling, thus minimizing cost.

The main problem with performing only a qualitative test is a possibility that an old dye residing in the subsurface and not detected during background sampling may remobilize under new hydrologic conditions, interfere with the newly injected dye, and significantly bias the test interpretation. A carefully designed and executed quantitative dye-tracing test should help eliminate this concern. It is the opinion of the author, shared by many colleagues working in karst, that use of qualitative tests for final decision making, whatever that may be, should be avoided.

The general principles of quantitative dye tracing described previously in Sec. 2.5 are applicable to tracing of both surface streams and groundwater in karst. Four widely cited public-domain publications on groundwater tracing provide a wealth of information and references including real-world examples and practical advice: Mull et al. (1988), Quinlan (1989), and Field (2002, 2003). The first two can be downloaded from http://www.karstwaters.org (Web page of Karst Waters Institute) and all four are available from the U.S. EPA main publication search and download page at http://www.epa.gov/nscep/index.html. Other useful overviews of dye tracing in karst are provided by Atkinson et al. (1973), Benischke et al. (2007), and Trček and Zojer (2010). For an in-depth discussion on various tracing techniques, the reader should consult work of Käss (1998).

Preparation for Dye-Tracing Test

Prior to any tracing efforts, background water samples need to be collected and analyzed for the tracer of interest. If the values obtained are low enough (e.g., few μg/L), then the chosen tracer may be used. If not, then a different tracer should be chosen. Low background concentrations in samples will then need to be averaged and subtracted from every sample of recovered tracer from subsequent tracing efforts. For example, sodium fluorescein is used to color automobile antifreeze. Because there are so many automobiles in existence and so many of them have leaks in their radiators, fluorescein-colored antifreeze is fairly ubiquitous in the environment (Field, 2002). Commercially available dyes commonly used in tracer tests are included in Table 3.6.

The most important aspect of dye tracing in karst is that the project team be prepared for unexpected by not ruling out any possibility, however remote, of dye appearance at multiple locations. It is better to collect sample from one location too many than from one too little. This advice may appear as an unnecessary luxury to some, especially when there are serious budgetary constraints. However, in many cases a combination of (inexpensive) passive and active sampling should be sufficient to help eliminate possible doubts during the analysis of test results. Several examples of complicated subsurface

Dye Type and Common Name	Generic Name and Color Index	CAS No.	Detection Limit (μg/L)	Sorption Tendency
Xanthenes				
Sodium fluorescein (Uranine)	Acid Yellow 73	518-47-8	0.002	Very low
Eosine	Acid Red 87	17372-87-1	0.01	Low
Rhodamines				
Rhodamine B	Basic Violet 10	81-88-9	0.006	Strong
Rhodamine WT	Acid Red 388	37299-86-8	0.006	Moderate
Sulforhodamine G	Acid Red 50	5873-16-5	0.005	Moderate
Sulforhodamine B	Acid Red 52	3520-42-1	0.007	Moderate
Stilbenes				
Tinopal CBS-X	Fluorescent Brightener 351	54351-85-8	0.01	Moderate
Tinopal 5BM GX	Fluorescent Brightener 22	12224-01-0	?	?
Phorwite BBH Pure	Fluorescent Brightener 28	4404-43-7	?	?
Diphenyl brilliant flavine 7GFF	Direct Yellow 96	61725-08-4	?	?
Functionalized polycyclic Aromatic hydrocarbons				
Lissamine flavine FF	Acid Yellow 7	2391-30-2	?	?
Pyranine	Solvent Green 7	6358-69-6	?	?
Amino G acid	—	86-65-7	?	?

From Field (2002).

TABLE 3.6 Commonly Used Fluorescent Dyes and Their Properties. Detection Limits Are Typical Values for Tracer Detection in Clean Water Using Spectrofluorometers. Values Can Be Adversely Affected by Augmented Fluorescence and/or Scattered Light Background

connections in mature karst terrains that were deciphered after multiple dye-tracing tests are shown in Figs. 3.61 and 3.62. Notably, some dye traces to different springs cross each other and all major springs receive water from several topographic watersheds and distances exceeding tens of miles.

A well-developed hydrogeologic conceptual site model (CSM) for the project should serve as the basis for planning the dye-tracing test. A number of CSM elements will help decide where to inject the dye, where to take samples, which type of dye(s) to use, how much dye to inject, and when to start monitoring at selected sampling locations and how often. Important CSM elements include known or suspect locations of surface water sinking, likely preferential flow paths, hydrologic characteristics of the season, flow rates of surface streams and springs, anticipated predominant type of flow at the locations of dye injection (conduit, diffuse/matrix), estimated flow velocities in the subsurface,

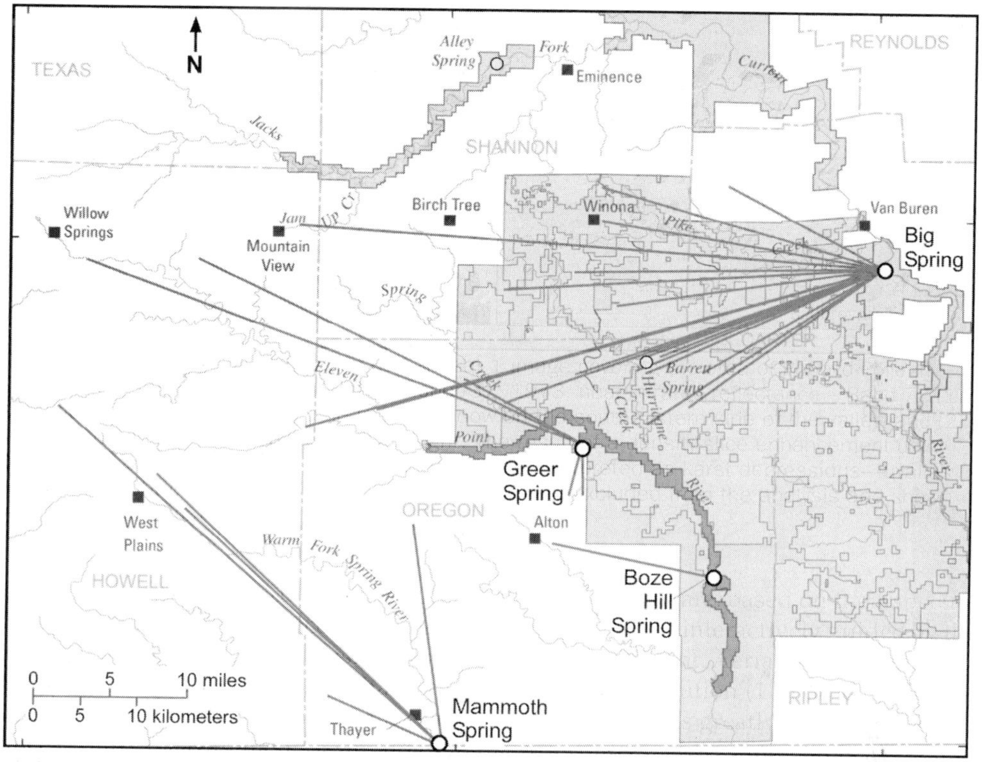

FIGURE 3.61 Results of dye-tracing investigations that successfully identified a hydrologic connection between a sinkhole or a losing stream reach and a major spring, south-central Missouri. Dye traces are long straight lines. Different gray polygons represent various land uses. (Modified from Imes et al., 2007.)

any groundwater withdrawals in the general area of the dye test, and other project-specific factors that may influence transport of dye between the location(s) of injection and sampling. Whenever feasible, sampling should also include selected water supply and/or monitoring wells in the test area.

Below is an example of a qualitative dye-tracing test that may have benefited from developing a CSM and using it to plan the test, thus leaving less room for possible erroneous conclusions. Apparently, the only goal of the test was to confirm if there is a connection between a known sink and a spring used for water supply. Notably, no flow measurements of any kind were performed during the entire test and the site visits only included sampling of the spring. The test was generally performed as follows:

- Place a passive dye sampler consisting of a bag of granular activated charcoal into the spring pool.
- Dump an arbitrary amount of a dye into the sink located several miles away from the spring (incidentally, the stream sinking into the sink is intermittent).

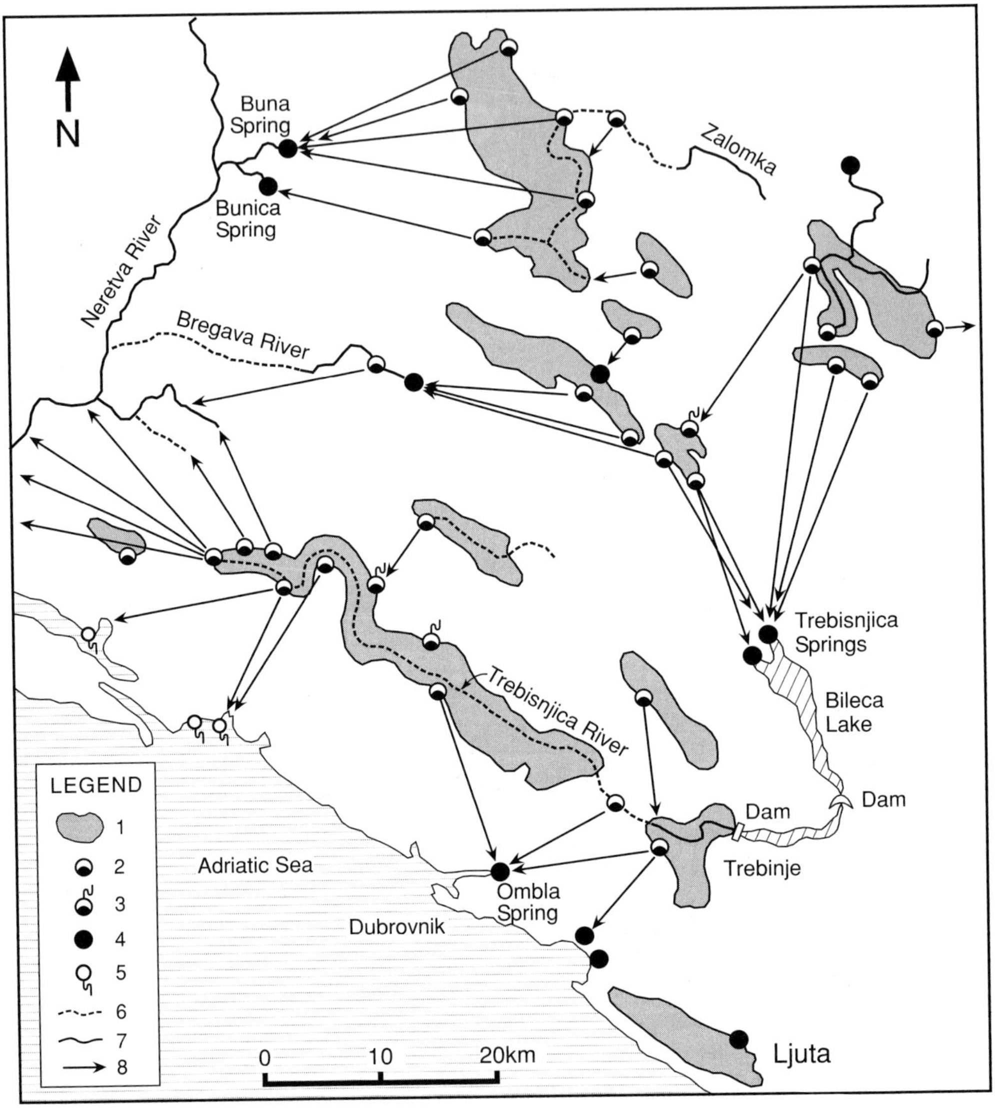

FIGURE 3.62 Dye traces in karst of Eastern Herzegovina. (1) Karst polje, (2) ponor (sink), (3) estavelle, (4) major permanent karst spring, (5) submarine spring, (6) intermittent stream, (7) permanent stream, and (8) dye trace. (Modified from Milanovic, 1979.)

- Visit the spring once a week for 4 weeks after introducing the dye into the sink and replace the charcoal bags; analyze the charcoal samples in the laboratory for any presence of dye.
- Discontinue the sampling after 4 weeks since no dye was detected in the charcoal samples.
- Based on the absence of dye in the samples, conclude that there is no connection between the sink and the spring.

In addition to the most obvious concern regarding the above example, namely that the dye may have arrived at the spring during the fifth week after the injection, a number of other possibilities illustrate how a previously developed CSM would have helped in avoiding the following ambiguities and the possibly erroneous final conclusion:

- The sinking stream may have dried up at any time following the introduction of dye resulting in a considerable slowdown of the tracer movement.
- The flow rate of the sinking stream and/or the spring may have been too high for the mass of dye used resulting in dye dilution to nondetectable concentrations.
- The dye was inadequate for the characteristics of the spring, that is, it may have degraded below the detection limit under the sunlight in the exposed spring pool due to high photochemical decay.
- Although apparently not of importance to the issue at hand (the connection between the sink and one particular spring), the dye may have discharged at unmonitored locations.
- The sink and the spring may be connected during different hydrologic conditions (seasons).

In contrast to the above example, an ill-conceived qualitative test can also result in a false positive result and conclusion that the connection between a sink and a spring exists, when in fact it does not. Reasons for possible false-positive detections with passive detectors are explained in detail by Field (2003). One common problem is that activated carbon enhances background concentrations of fluorescent dyes and other similarly appearing compounds. Available information shows that dye concentration by activated carbon may be as high as 1,000 times the concentration in water (Käss, 1998).

Dye injection locations vary widely based on the goals of the study. Most obvious are sinks and swallow holes, sinkholes, caves, and monitoring wells and borings (Fig. 3.63). Dye can also be sprayed at the land surface or released into trenches excavated down to the bedrock to study recharge mechanisms or possible impacts of point and nonpoint sources of groundwater contamination. In any case, several different dyes may be injected simultaneously so multiple release points may be tested at the same time and under the same flow conditions. When more dyes are planned for injection, the background concentration of each has to be determined, as well as possible limiting factors for their use. This includes interference from other fluorescent substances, quenching, and presence of fine sediments and clay in karst conduits (caves) that can adsorb the dyes. For example, rhodamine B is stable in sunlight, but it is readily adsorbed to sediments in water. Rhodamine WT was developed specifically for water tracing and is recommended for such routine use, but it is particularly susceptible to photochemical decay. Fluorescent substances found in aquatic systems, which might share excitation or emission spectra with rhodamine WT, and thereby source of interference, include, but are not limited to, algae; certain salt compounds; naturally occurring colored organic compounds, and man-made organic pollutants such as dyes, petroleum distillates, and detergents (YSI Environmental, 2001).

Quenching is the suppression of fluorescence, and results from the action of other substances in solution with the dye. Quenching agents may act in any or all of the following manners: (1) absorb excitation energy, (2) absorb emission (fluorescence) energy, (3) degrade excitation-state energy, and (4) chemically alter the dye molecular structure

FIGURE 3.63 Dye injections. *Top left*: Sodium fluorescein injection into collapse sinkhole formed in a pond; stream of water is outflown from a settling pond at a public supply water-treatment plant (photograph by Charles J. Taylor, U.S. Geological Survey). *Bottom left*: Injection of rhodamine WT into a water-level observation well (photograph by Charles J. Taylor, U.S. Geological Survey). *Right*: Uranine (Acid Yellow 73) release into the underground river in Sainte Anne Cave, Belgium. (Photo courtesy of Philippe Meus; from Field, 2002).

(Williams and Bridges, 1964). Chlorine is an example of a quenching agent that changes fluorescent properties of rhodamine WT by altering its molecular structure.

The pH of a sample may also affect fluorescence intensity, therefore the measured concentration of dye. Fluorescence of rhodamine WT is stable in the pH range of 5 to 10 and decreases outside those limits. Decreased fluorescence due to pH variation is generally not a problem except in studies of instantaneous pollutant discharges, industrial waste streams, acidic streams, and certain stages of municipal and industrial waste treatment (YSI Environmental, 2001).

In highly aerated waters, such as in the whitewater reaches of mountain streams or in other situations where super-saturation of dissolved oxygen exists, oxygen can manifest the same effect as chlorine (Wilson et al., 1986). Care should be taken to identify stream reaches where there is potential for this problem and to quantify its effects.

Where applicable, precipitation and flow measurements at the locations of dye injection and sampling should be conducted as part of any dye-tracing test, during the entire duration of the test. This information will be critical when interpreting the results of a quantitative test, and the flow measurements will also be used to plot (calculate) the dye mass recovery curve.

Determining Mass of Dye Required

Field (2003) provides a thorough overview and quantitative analysis of 33 equations for calculating the amount of tracer material to be released into a flow system. The examination of the equations reveals little about each except that most rely to some extent on volumetric flow rate (discharge). Presumably, the reliance on discharge by the originators of the equations was intended to address probable dilution effects. The 33 equations were developed empirically, so maintaining mass balance appears not to have been an important consideration. Sources for the equations were not always specific about which type of tracer a particular equation was meant for in the design or about whether it even matters, but most were probably designed for the fluorescent dye Na-fluorescein. Also, it is not always clear whether an equation was intended for visual tracer detection, instrument detection in water samples, or sample collectors designed to enhance tracer concentrations (e.g., carbon sorption of fluorescent dyes via detectors).

Some equations require a "fudge factoring" constant that takes into consideration the relative detectability of the tracer to be used, the residence time within the aquifer, the type of sampling done, the method of sample collection, and/or the method of analysis. Specifics such as tracer type, methods of sampling and analysis, and fudge factor multipliers should generally be regarded as inappropriate when calculating necessary tracer mass unless a clear scientific basis for such can be established (e.g., solution–conduit sinuosity ≤ 1.5).

Flow discharge is clearly the main factor for the majority of the 33 equations listed, suggesting that tracer dilution estimates were an overriding concern during development. Expected transport distances or transport times are also generally common elements. All other factors included in the equations were intended to address either known complications (e.g., transport distance) or unknown complications (e.g., sorption) that are expected to influence final downstream concentrations. As emphasized by Field (2003), with the exception of two equations, they were all developed on the false assumption that a simple algebraic expression that appears to function adequately for a selected hydrologic setting will then adequately function in all hydrologic settings. In addition, these equations fail to properly account for the important effects of axial dispersion and solute transport theory.

As a result of the above analysis, Field (2003) proposes The Efficient Hydrologic Tracer-Test Design (EHTD) method and includes a computer program that estimates the amount of dye needed for injection. The model also provides forward modeling capability needed to predict tracer-breakthrough curve characteristics and the time intervals needed for effective sampling of the passage of the dye pulse—information that is important to planning quantitative tracer tests. The EHTD method calculates the amount of dye needed for injection by using various forms of the advection-dispersion equation for open-channel flow, closed-conduit flow, and flow through porous equivalent media. The program enables the user to designate the mass of tracer dye to be injected and an injection flow rate. For open-channel and closed-conduit flow conditions, the EHTD method requires the following input values: (1) discharge at the sampling station (spring), (2) estimated longitudinal distance from the dye-injection site to the anticipated resurgence site, (3) estimated cross-sectional area of the discharge point (i.e., spring or stream cross-sectional area), and (4) a sinuosity factor applied to straight-line estimates of the distance between a dye injection and potential dye resurgence site (Taylor and Greene, 2008).

As a very general guidance, Jones (2012) suggests that tests in conduit aquifers involving distances less than 1 km and discharges less than 30 L/s should require less than 200 g of sodium fluorescein (Uranine). Most of the other commonly used tracer dyes probably require two to four times the mass used for a sodium fluorescein test. Less dye is usually needed for qualitative tests than for quantitative tests. If the possible coloration of the springs and receiving streams is not objectionable, the estimated amount of dye should be increased. Although no strict requirements based on toxicological or other health studies have been established for acceptable concentrations of commonly used nontoxic dyes approved for tracers studies, care must be taken to keep maximum tracer recovery concentrations well below the visible threshold (about 30 μg/L for sodium fluorescein) if the tracer could affect a water supply.

Sampling Interval

The most uncertain aspect of any tracing study is the schedule for sample collection (Kilpatrick and Wilson, 1989). Although much effort has gone into estimating tracer mass for a tracing study, very little appears to have been done in terms of determining sample collection frequency. Sampling frequencies are generally based on travel distances, which suggest a direct relationship between travel distance and expected time of arrival. This relationship is obviously correct, but it is ambiguous because transport velocity as a function of residence time is unknown. Transport velocities can achieve extreme ranges, rendering invalid sampling schedules based on transport distances that are devoid of residence time estimates (Field, 2003).

Tracer recovery from qualitative tracing tests (e.g., sorption of fluorescent dyes onto detectors) that are sampled every few hours, days, or weeks sometimes serves as a basis for determining appropriate sampling frequencies for quantitative-tracing tests. However, sampling schedules based on qualitative tracing tests have been found to result in false positive and false negative results and cannot be relied upon for predicting sampling times (Field, 2003).

For all of the above reasons, the author of this book deliberately does not provide any recommendation as to the tracer sampling frequency; for each project site this will have to be determined using the CSM, and the best estimates of flow velocities and travel distances. In any case, the best solution is to have some type of continuous sampling in the field (e.g., using probes) supplemented by laboratory analysis of a number of duplicate samples for quality control.

Dye Injection, Collection, and Sample Analysis

In practice, dye injection is best accomplished at locations that provide rapid, direct transport of the tracer into conduits, thus minimizing loss of dye through photochemical decay, sorption, or other field conditions. This injection can be in form of a slug (instantaneous) or it can be continuous over several or more hours if the test is, in part, designed to estimate subsurface flow rates (Figs. 3.63 and 3.64). Open-throated swalets in sinkholes and the swallow holes of sinking streams are ideal sites. In the absence of naturally occurring runoff (inflow), dyes can be injected into a stream of potable water discharged from a tanker truck or large carboy. In general, 300 to 500 gallons of water are a minimum quantity needed for dye injection, and quantities of 1,000 gallons or more are preferable. Approximately one-half of the water is used to initiate flow into the swallet

Figure 3.64 Continuous injection of dye into a small swallow hole (shown with arrow) using a pump. (Photograph courtesy of Malcolm Field.)

prior to dye injection. This is done to test the swallet's drainage capacity, to initiate flow, and to flush the flow path to minimize losses to sorption. The remainder of the water is discharged after the dye is injected as a "chaser." Under most conditions, this technique does not substantially change the naturally occurring flow conditions or alter hydraulic heads in the aquifer (Taylor and Greene, 2008).

The injection of dye into flooded sinkhole depressions or swallets choked with sediment generally is not advisable, particularly for quantitative tracer tests. Unless there is evidence that drainage through the regolith into the subsurface is relatively rapid, excessive loss of the tracer dye may be incurred. If necessary, swallets that are partly choked with sediment may be cleared out with a shovel or backhoe and pretested for drainage capacity prior to an attempted injection of dye (Taylor and Greene, 2008).

Traditional manual sample collection for laboratory analyses is performed much less frequently today due to high labor costs associated with long stays in the field. Instead, automated samplers are used whenever possible because they can be installed in a variety of conditions (Fig. 3.65) and exclude possibility for cross-contamination of samples that is always a concern when samples are collected manually. In either case, great many samples must be collected, stored, transported, and analyzed in the laboratory (Fig. 3.66) to achieve sufficient resolution required for plotting tracer breakthrough and mass recovery curves.

Mull et al. (1988) recommend that all samples, manual or with automated samplers, should be collected using glass bottles to minimize dye losses. Glass sample bottles are less likely to adsorb the tracer than are plastic sample bottles, which may distort

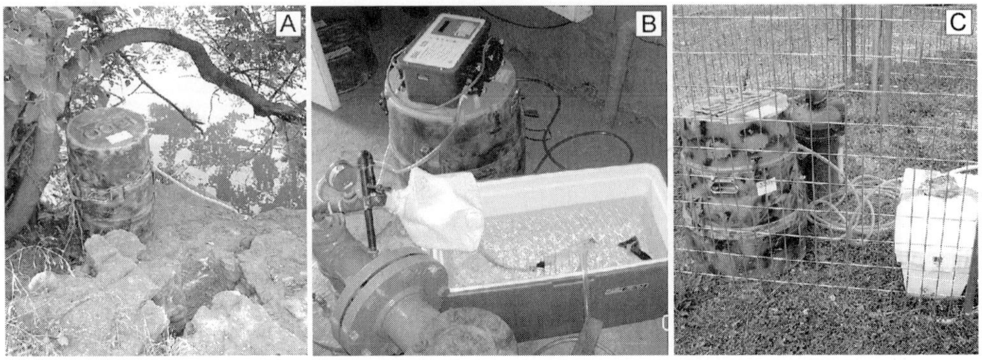

Figure 3.65 Automated water sample collection methods. (A) Sampler at a quarry, (B) sampler at a water supply well (inside well house), and (C) sampler at a monitoring well (inside fenced area). (From Ohio Environmental Protection Agency, 2010.)

sample-analysis results. The sample bottles need only be large enough to hold a maximum of approximately 32 mL of water in most instances. Samples should be stored tightly capped in a cool, dark place. Shipping to the laboratory should be by cooler, with an ice block enclosed (Field, 2002).

Several important issues must be considered when collecting samples with packets of granular activated charcoal if fluorescent dyes are used as tracers. It is believed that activated charcoal will ensure dye recovery because the supposed much lower dye

Figure 3.66 *Left*: Collected samples are prepared for analysis in the laboratory. (From Ohio Environmental Protection Agency, 2010.) *Top right*: Fiber optic fluorometer (Courtesy of Dr. Malcolm Field). *Bottom right*: Tandem of fluorometers for measuring concentrations of Rhodamine WT and fluorescein in samples. (Courtesy of Dr. Malcolm Field.)

concentrations found in water samples may not be detected, or sampling frequencies may not have been adequate. The ability of activated charcoal to continue sorbing and concentrating fluorescent dye provides a sound means for determining fluorescent dye occurrence when water samples are ambiguous. However, at best, activated charcoal will result in a qualitative tracing test only. More importantly, there is considerably more opportunity for sample contamination from handling. Still more serious is the recently considered problem of false positives and false negatives associated with activated charcoal packets (Smart and Karunaratne, 2001; Smart and Simpson, 2001).

Despite the above concerns, the use of passive granular activated carbon detectors is popular for a variety of reasons as explained by Taylor and Greene (2008). The principal advantage of charcoal detectors is their economy and relative ease of use for groundwater reconnaissance studies, for simultaneous monitoring of many potential dye resurgence sites, and for mapping of conduit-flow paths or karst basin boundaries in areas where these are primarily or completely unknown. The detectors are relatively easy to conceal, thus minimizing the potential for disturbance and vandalism; and they are inexpensive.

The detectors typically are constructed of fiberglass screen, nylon netting, or a similar material, fashioned into packets that contain several grams of charcoal sampling media. The size and shape of the packets and the amount of charcoal used in them are not particularly critical—the only requirements are that detectors should be relatively durable, securely retain the activated granular charcoal, and allow water to flow easily and evenly through the packets. The detectors, or "bugs", generally are used in rapidly flowing water in streams and springs suspended above the substrate using a wire-and-concrete or wire-and-brick anchor. The detectors also can be staked or pinned directly into the streambed in very shallow water, and they can be easily suspended in monitoring or water supply wells using, for example, monofilament line, snap-swivels, and steel shot-weights sold as fishing tackle. A common practice is to use detectors at all anticipated resurgences and exchange the detectors at 2- to 10-day intervals throughout the duration of the test (Quinlan, 1989). As a practical matter, it is generally not advisable to leave detectors in the field longer than 10 days because of physical degradation that can diminish the adsorptive capability of the charcoal. Taylor and Greene (2008) provide detailed explanation of how to extract dye from the charcoal samples and perform laboratory analyses.

The use of commonly available fluorometers with a flow-through cell for online analysis of a tracer concentration (Fig. 3.66) has the advantage of a continuous measurement. However, the sampling procedure may sometimes cause a delay and errors in the concentration measurement by mixing processes in the sampling pipe and chamber or by disturbing the natural flow processes, especially at low flow rates. Frequent calibration, temperature corrections, and equipment check will be necessary when fluorometer is used in the field in the flow-through mode. When desired, the restrictions of manual, automated sampler, or fluorometer methods can be overcome by using probes and data loggers for continuous in situ tracer concentration measurements. Advanced instruments, such as in situ fluorometers for the measurement of rhodamine WT (http://www.ysi.com) and fiber optic fluorometer (FOF), can be installed in monitoring wells, springs, or submerged cavities. Simultaneous recording of different tracers at multiple measuring points can be achieved by using multiple probes for individual tracers.

Despite the obvious advantages described by the inventors of the FOF (Barczewski and Marschall, 1992), this method has yet to gain wide acceptance partially because of the high cost of commercially available instruments, and partially because of certain practical limitations. For example, Schmid and Barczewski (1995) caution that sometimes significant differences between laboratory measurements and continuous field measurements with the FOF probe may be caused by the growing of a light absorbing layer on the probe by sedimentation of suspended solids, as was the case after several thunderstorms in the catchment area of the spring they studied.

Regardless of the selected sample collection method, it is of utmost importance that the following recommendations be followed.

- Duplicate samples should be collected for quality control.
- For the analytical laboratory control, several samples should be collected at remote sites where detection of dye is highly unlikely, such as city tap water.
- All samples should have chain of custody.
- Different people should inject dye, different should collect samples, and different should analyze them in the laboratory. Ideally, these three groups should not be in physical contact for the duration of the test to avoid any possibility, however remote, of cross-contaminating the samples.
- A selected number of samples should be analyzed with a different method. For example, if water samples are analyzed with a field fluorometer, a number of samples should also be analyzed in the laboratory using a scanning spectrofluorophotometer.

Interpretation of Results

Breakthrough curves generated from tracer tests of preferential flow paths are typically positively skewed and may be quite simple, with one single sharp peak (Fig. 3.67), or quite complex with several peaks, plateaus, and irregular oscillations (Fig. 3.68).

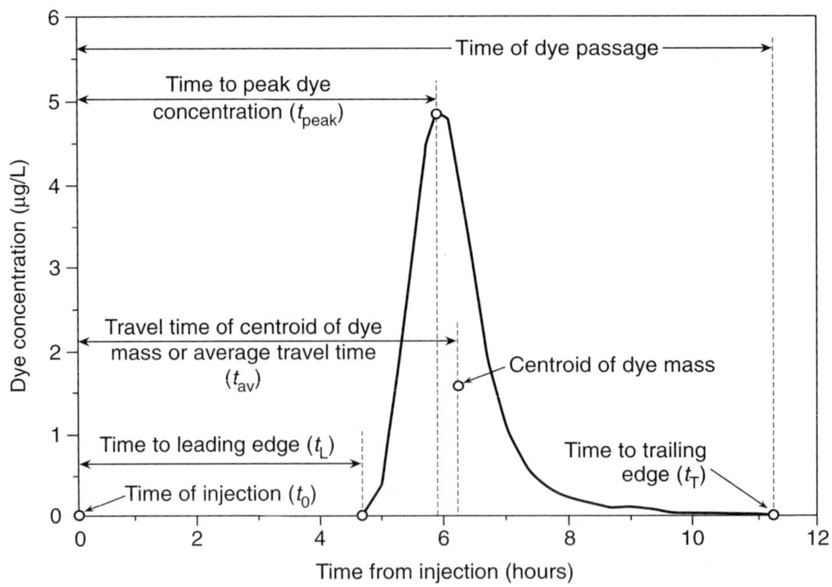

FIGURE 3.67 Characteristics of a simple dye breakthrough curve. (Modified from Mull et al., 1988.)

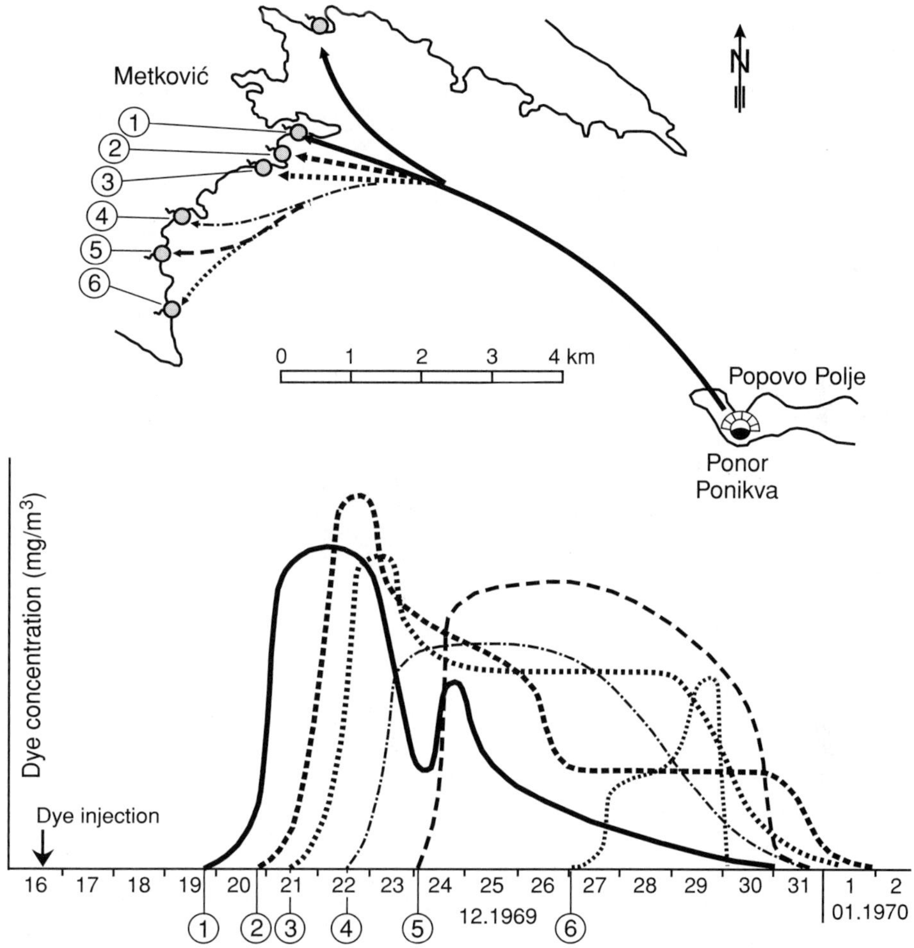

Figure 3.68 Results of a dye test at the Ponikva Ponor (sink) on the Trebišnjica River, Eastern Herzegovina. (Modified from Ramljak et al., 1975.)

Multiple reasons may exist for complex curves and there is no simple book of recipes that can be used to select the right interpretation. The exception, although rare in most situations, is when one can follow the entire path of the tracer through the subsurface and observe how any pooled water, turbulence, and tributary or distributary conduits affect the tracer concentration and transport times. In any case, however, it helps when more detail is known about possible subsurface conduit patterns from cave surveys in the same watershed or in similar geologic settings. As can be seen from a small sample shown in Fig. 3.69, many different cave types and cave systems exist. Without knowing any details about the underlying karst aquifer, it would be just a wild guess as to which specific features in the subsurface are responsible for certain shapes of complex breakthrough curves. For an initial concept, the reader should consult Jones (1984) and Smart (1988) who provide examples of conduit configurations that may result in some types of breakthrough curves.

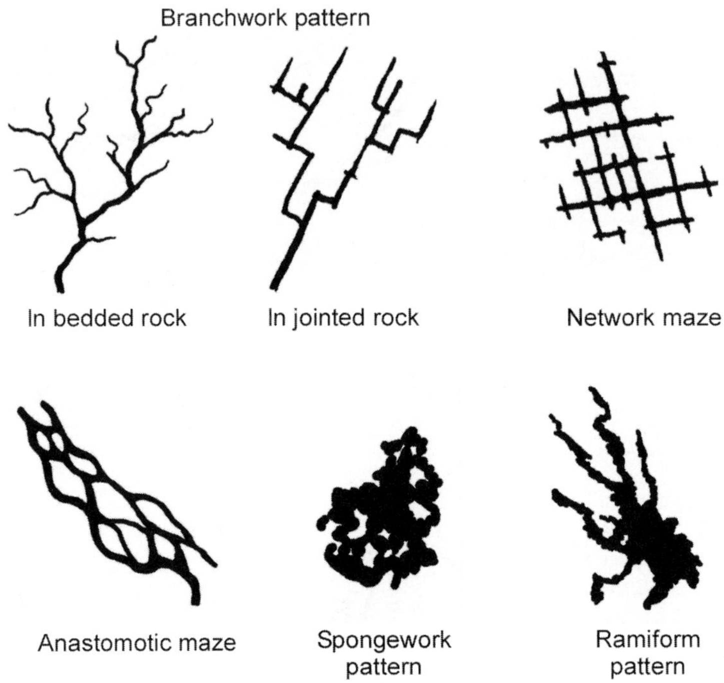

FIGURE 3.69 Common patterns of solution caves. Single-passage caves are simply rudimentary versions of those shown here. (From Palmer, 2007.)

Dye Velocity

Average groundwater flow velocity from the location(s) of dye injection to the location(s) of dye recovery is calculated from the average dye residence time and the distance between the points of interest. The average dye residence time (denoted with t_{av} in Fig. 3.67) is the length of time required for the centroid (gravity mass) of the tracer dye to traverse this distance. The centroid generally is not the same as the peak concentration of the dye mass in the breakthrough curve, but the more the dye plume conforms to Fick's law (the mass of the diffusing substance passing through a given cross section per unit time is proportional to the concentration gradient), the less obvious the difference between the dye centroid and peak concentration will be (Taylor and Greene, 2008).

Average (mean) tracer residence time is estimated by the equation:

$$t_{av} = \int_0^\infty tC(t)Q(t)\mathrm{d}t \Big/ \int_0^\infty C(t)Q(t)\mathrm{d}t \tag{3.1}$$

where t is time of sample collection, $C(t)$ is measured dye concentration of the sample, and $Q(t)$ is the discharge measured at the sampling location.

If QTRACER2 or another suitable mathematical software program is not used, and the sampling frequency was done at regularly spaced intervals, the integration can be done by using a simple summation algorithm as detailed in Field (2002) and Mull et al. (1988).

Tracer Mass Recovery

The accuracy of calculations of mean tracer residence time, flow velocities, and other conduit hydraulic properties from breakthrough curves data is entirely dependent on tracer mass recovery. Few tracing tests result in 100% recovery of dye, but as the percentage of mass recovery decreases, the margin of error in the calculated hydraulic parameters increases and confidence in the values obtained declines. As indicated earlier, both the breakthrough curves and the tracer recovery may be affected by the internal structure of conduit networks. It therefore is important to assess tracer mass recovery as a starting point in the analysis of quantitative dye-tracing tests.

The quality of the tracer experiment may be quantified in terms of the relationship between the mass of dye tracer injected (M_{in}) during the experiment and the total mass of dye tracer recovered (M_{r}). A test accuracy index proposed by Sukhodolov et al. (1997) is calculated by

$$A_{\text{I}} = M_{\text{in}} - M_{\text{r}}/M_{\text{in}} \tag{3.2}$$

This index provides a semiquantitative assessment of the quality of the test. A value $A_{\text{I}} = 0$ indicates a perfect tracing experiment with no loss of tracer dye mass. A positive A_{I} value indicates that more tracer dye mass was injected than was recovered—a common result, whereas a negative value indicates more dye mass was recovered than was injected—an impossibility unless residual tracer dye is present in the aquifer, errors are made in determining the dye concentration in test samples, or initial calculations of the injected dye mass are in error (Taylor and Greene, 2008).

In the previous equation, the value for M_{r}, the total mass of tracer dye recovered is given by the following equation:

$$M_{\text{r}} = \int_0^{\infty} C(t)Q(t)\mathrm{d}t \tag{3.3}$$

This equation assumes that the total dye mass is recovered at a single spring site. If dye has resurged at multiple spring outlets, these calculations are repeated for each site and the results are summed to obtain M_{r}.

If the tracer is conservative (not subject to decay or sorptive losses) and all of the water from the injection point is accounted for, the mass of dye recovered should be equal to the mass injected. None of the fluorescent tracers is completely conservative, but tracer tests with travel times of less than a few days should balance reasonably well if all of the resurgences have been monitored. A significant loss of the tracer from the system suggests that additional resurgences are present or some of the tracer has moved into storage areas along the flow route. Less conservative tracers usually produce a longer travel time estimate due to retardation caused by sorption and desorption along the flow route (Jones, 2012).

Storage of the tracer in the conduit system may be very complex. Storage may just represent very low velocity flow through one section of a branching conduit, called inline storage. A test done under conditions of diminishing flow may leave some of the dye abandoned in upper level pockets to be remobilized when the flow increases at a later date. Some dye may become stored offline in lateral areas along the route such as small voids or in pores in the sediments. Inline storage tends to create dye pulses that correlate with flushing due to storm events. Dye from offline storage tends to drain slowly during groundwater recessions (Jones, 2012).

Solute transport in karst aquifers is of particular concern when it takes place in simple flow-through conduits where transport is rapid, turbulent, and relatively unrestrictive. Breakthrough curves with a single sharp peak like the one shown in Fig. 3.67 are typical of such conditions, indicating a complete mixing of the solute with little or no dispersion. More complex tracer breakthrough curves can be described with models of contaminant fate and transport that account for a variety of possible processes in the subsurface including dispersion, delays due to pooling, sorption, decay, and diffusion into the rock matrix.

In order to understand the circumstances under which multipeaked positively skewed breakthrough curves occur, Field and Leij (2012) performed physical experiments

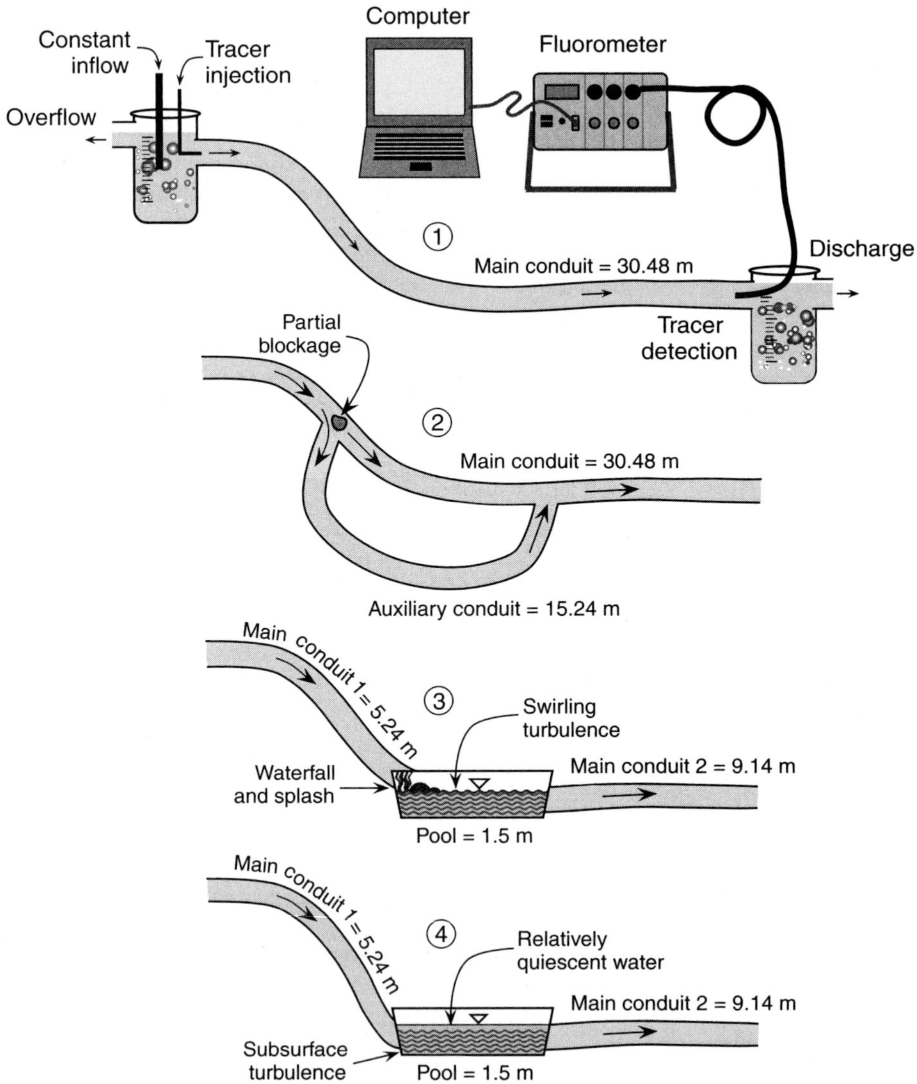

FIGURE 3.70 Schematic diagrams of four laboratory dye-tracing experiments performed to study impacts of different conduit configurations and flow characteristics on tracer breakthrough curves; see Fig. 3.71 for examples of breakthrough curves. (Modified from Field and Leij, 2012.)

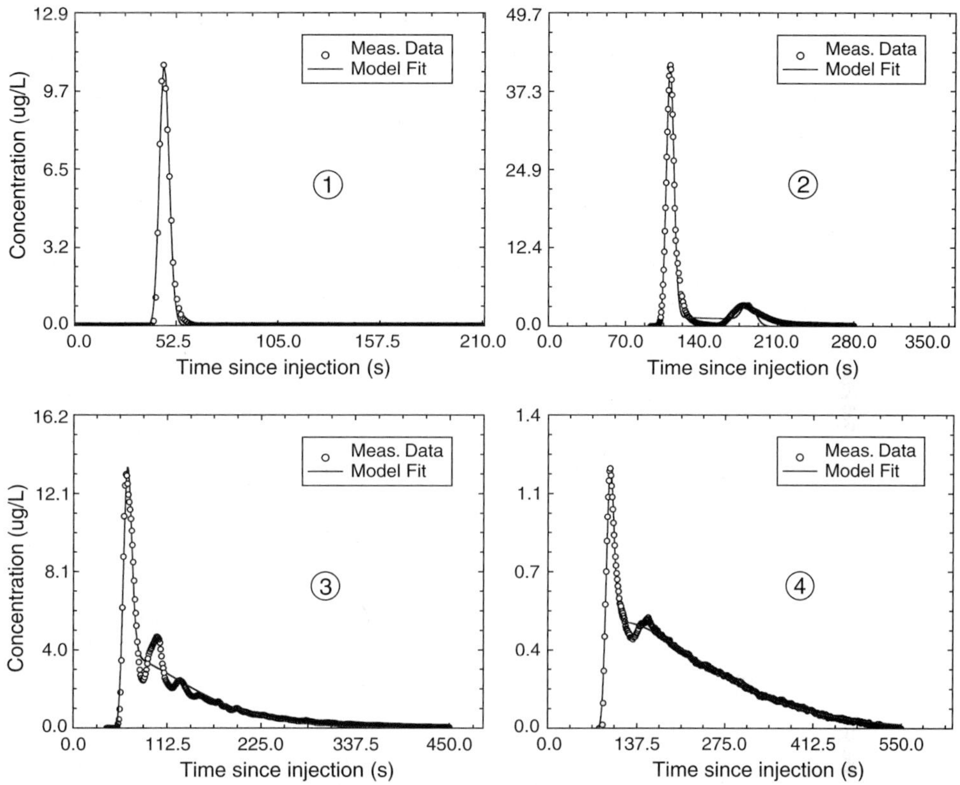

FIGURE 3.71 Tracer breakthrough curves and optimized dual-advection, dispersion equation (DADE) fit to the breakthrough curves for four dye-tracing experiments shown in Fig. 3.70. (Modified from Field and Leij, 2012.)

utilizing single- and multiple-flow channels. Experiments also included waterfalls, short-term solute detention in pools, and flow obstructions (Fig. 3.70). Results demonstrated that breakthrough curve skewness nearly always occurs to some degree but is magnified as immobile-flow regions are encountered. Multipeaked breakthrough curves occurred when flow in the main channel became partially occluded from blockage in the main channel that forced divergence of solute into auxiliary channels and when waterfalls and detention in pools occurred (Fig. 3.71). In this paper, a dual-advection dispersion equation (DADE) with first-order mass transfer between conduits was derived. The DADE was then applied to the multipeaked breakthrough curves obtained from the physical experiments in order to obtain some insight into the operative solute-transport processes through the acquisition of a consequent set of velocities, dispersions, and related parameters. Application of the equation to a tracer test conducted in a real karst aquifer, such as the Upper River Sink (URS) of the Leon Sinks–Wakulla Cave System in Florida, illustrates the value of applying a multidispersion model to a tracer test that exhibits multiple peaks. The model fit to the URS provided a good indication of the transport velocities and dispersivity operative in the Leon Sinks-Wakulla Cave System. Of particular interest was the direct observation by cave divers that the URS system conforms to requirements of the DADE of two independent flow domains (conduits) with limited exchange between them (Field and Leij, 2012).

3.3 References

Abdullah, A., Akhir, J. M., and Abdullah, I., 2009. A comparison of Landsat TM and SPOT data for lineament mapping in Hulu Lepar area, Pahang, Malaysia. *Eur J Sci Res*, 34(3): 406–415.

Arandjelovic, D., 1976. Geophysics in karst (in Serbian, English abstract). Geozavod Special Edition, Book 17, Belgrade, 220p.

Atkinson, T. C., Smith, D. I., Lavis, J. J., and Whitaker, R. J., 1973. Experiments in tracing underground waters in limestones. *J Hydrol*, 19: 323–349.

Barczewski, B., and Marschall, P., 1992. Development and application of a light fibre fluorimeter for tracer tests. In: Hôtzl, H., and Werner, A. (eds.), *Tracer Hydrology*, Balkema, Rotterdam, pp. 33–39.

Barr, G. L., 1993. Application of ground-penetrating radar methods in determining hydrogeologic conditions in a karst area, west-central Florida. U.S. Geological Survey Water-Resources Investigations Report 92–4141, Tallahassee, FL, 26p.

Benischke, R., Goldscheider, N., and Smart, C., 2007. Tracer techniques. In: Goldscheider, N., and Drew, D. (eds.), *Methods in Karst Hydrogeology. International Contributions to Hydrogeology 26, International Association of Hydrogeologists*, Taylor & Francis, London, pp. 147–170.

Ciric, B., 1984. Earth crust in Yugoslavia. In: Mijatovic, B. F. (ed.), *Hydrogeology of the Dinaric Karst, International Contributions to Hydrogeology*, vol. 4, Heise, Hannover, pp. 18–41.

Cvijić, J., 1926. Geomorfologija (Morphologie Terrestre). Knjiga druga (Tome Second). Beograd, 506p.

Dimitrijević, M. D., 1978. Geološko kartiranje. Izdavačko-informativni centar studenata, Beograd, 486p.

Enviroscan, 2012. Geophysics done right. Available at: http://www.enviroscan.com

Ewers, R. O., 1985. Patterns of cavern development along the Cumberland Escarpment in southeastern Kentucky. In: Dougherty, PH. (ed.), *Caves and Karst of Kentucky*, Kentucky Geological Survey Special Publication 12, Series XI, Lexington, KY, pp. 97–118.

FGDC (Federal Geographic Data Committee), 2006. FGDC Digital Cartographic Standard for Geologic Map Symbolization. FGDC Document Number FGDC-STD-013-2006. Available at http://www.fgdc.gov/standards/standards_publications/.

Field, M. S., 2002. The QTRACER2 Program for Tracer-Breakthrough Curve Analysis for Tracer Tests in Karstic Aquifers and Other Hydrologic Systems. U.S. Environmental Protection Agency, Office of Research and Development, EPA/600/R-02/001, Washington, D.C., 179p.

Field, M. S., 2003. Tracer-Test Planning Using the Efficient Hydrologic Tracer-Test Design (EHTD) Program. U.S. Environmental Protection Agency, National Center for Environmental Assessment, EPA/600/R-03/034, Washington, D. C., 175p. Available at http://www.epa.gov/ncea.

Field, M. S., and Leij, F. J., 2012. Solute Transport in Solution Conduits Exhibiting Multi-Peaked Breakthrough Curves. *Journal of Hydrology*, vol. 440–441, pp. 26–35.

Garner, L. E., Young, K. P., Rodda, P. U., Dawe, G. L., and Margaret, A. R., 1976. Geologic Map of the Austin Area, Texas (scale 1:62,500). Bureau of Economic Geology, The University of Texas at Austin, Austin.

Graham, C. D. R., 2000. Electrical resistivity studies in the Inner Bluegrass Karst Region, Kentucky. Thesis Series 1, Series XII, The Kentucky Geological Survey, Lexington, 92p.

Habič, P., 1981. The tectonic influence on water discharge in karst. Naš krš, Bulletin of Speleological Society, Sarajevo, vol. VI, no. 10-11, pp. 37–46.

Haryono, E., Nurcahyo, A. D., Gunawan, T., and Purwanto, T. H., 2005. Underground river network modelling from lineaments and fracture traces by means of remote sensing and geographic information system. In: Stevanovic, Z., and Milanovic, P. (eds.), Water Resources and Environmental Problems in Karst, Proceedings of the International Conference and Field Seminars, Belgrade & Kotor, Serbia & Montenegro, September 13–19, 2005, Institute of Hydrogeology, University of Belgrade, Belgrade, pp. 571–574.

Imes, J. L., Plummer, L. N., Kleeschulte, M. J., and Schumacher, J. G., 2007. Reacharge area, base-flow and quick-flow discharge rates and ages, and general water quality of Big Spring in Carter County, Missouri. U.S. Geological Survey Scientific Investigations Report 2007-5049, Reston, VA, 80p.

Jones, W. K., 1977. *Karst Hydrology Atlas of West Virginia*. Special Publication 4, Karst Waters Institute, Charles Town, WV, 111p.

Jones, W. K., 1984. Analysis and interpretation of data from tracer tests in karst areas. *Nat Speleolog Soc Bull*, 46: 41–47.

Jones, W. K., 2012. Water tracing in karst aquifers. In: White, B.W., and Culver, D.C. (eds.), *Encyclopedia of Caves*. Academic Press, Chennai, pp. 886–896.

Käss, W., 1998. Tracing technique in geohydrology. Balkema, Rotterdam, The Netherlands, 581p.

Kilpatrick, A., and Wilson, J. F., 1989. Measurement of time of travel in streams by dye tracing. Techniques of Water-Resources Investigations of the United States Geological Survey, Chapter A9, Book 3, Application of Hydraulics, Washington, D.C., 27p.

Kindinger, J. L., Davis, J. B., and Flocks, J. G., 2000. Subsurface Characterization of Selected Water Bodies in the St. Johns River Water Management District, Northeast Florida. U.S. Geological Survey Open File Report 00-180, 7p.

Kresic, N., 1991. Kvantitativna hidrogeologija karsta sa elementima zaštite podzemnih voda (Quantitative karst hydrogeology with elements of groundwater protection, in Serbo-Croatian). Naučna knjiga, Beograd, 196p.

Kresic, N., 1995. Remote sensing of tectonic fabric controlling groundwater flow in Dinaric karst. *Remote Sens Environ*, 53(2): 85–90.

Kresic, N., 2007. *Hydrogeology and Groundwater Modeling*, 2nd ed. CRC Press, Taylor & Francis Group, Boca Raton, FL, 807p.

Kresic, N., 2010. Types and classification of springs. In: Kresic, N., and Stevanovic, Z. (eds.), *Groundwater Hydrology of Springs: Engineering, Theory, Management and Sustainability*. Elsevier, New York, pp. 31–85.

Kresic, N., and Tasic, Z., 1984. The significance of neotectonic investigations in hydrogeology of karst (in Serbo-Croatian, English abstract). *Proceedings of Geoinstitute*, Belgrade, no. 17, pp. 145–151.

Kresic, N., and Pavlovic, P., 1990. Remote sensing as a method for determining directions of karst groundwater flow, *Nas Krs*, Sarajevo, XVI, 28-29, pp. 21–34.

Kresic, N., Busbey, A. B., and Morgan, K. M., 1994. Remote sensing and neotectonic analyses of possible groundwater flow paths in karst. Proceeding of the Second International Conference on Ground Water Ecology, American Water Resources Association, Herndon, VA, pp. 289–295.

Kresic, N., and Mikszewski, A., 2012. Hydrogeological Conceptual Site Models: Data Analysis and Visualization. CRC/Taylor & Francis Group, Boca Raton, FL, 552p.

Krulc, Z., 1966. Primene geoelektričnih ispitivanja u sklopu istražnih radova na velikim hidroenergetskim objektima. Saopštenja sa VII Kongresa Jugoslovenskog nacionalnog komiteta za visoke brane, Sarajevo.

Lattman, L. H. and Parizek, R. R., 1964. Relationship between fracture traces and the occurrence of ground-water in carbonate rock. *Journal of Hydrology*, v. 2, p. 73–91.

Luman, D. E. and Panno, S. V., 2011. Mapping palimpsest karst features of Illinois' sinkhole plain using historical aerial photography. Proceedings of the 12th Multidisciplinary Conference on Sinkholes and the Engineering & Environmental Impacts of Karst. January 10–14, 2011, St. Louis, MO, 25p.

MacDonald and Partners, 1984. Greater Yogyakarta Groundwater Study, Vol. 3C Cave Sutvey. Directorate General of Water Resources Development of the Republic of Indonesia.

Marković, M., 1980. Method of remote sensing in geologic investigations (in Serbian), Simpozijum 20 godina LMGK, School of Mining and Geology, University of Belgrade, Belgrade, pp. 143–160.

Marković, M., 1983. Basics of applied geomorphology (in Serbian), *Geoinstitute Special Edition*, Book 8, 172p.

Milanovic, P., 1979. Hidrogeologija karsta i metode istraživanja (in Serbian; Karst hydrogeology and methods of investigations). HE Trebišnjica, Institut za korištenje i zaštitu voda na kršu, Trebinje, 302p.

Milanovic, P., 1980. Possibilities of application of remote sensing in karst hydrogeology (in Serbian), Ph.D. Thesis, University of Belgrade, Belgrade, 213p.

Mull, D. S., Liebermann, T. D., Smoot, J. L., and Woosley Jr., L. H., 1988. Application of dye-tracing techniques for determining solute-transport characteristics of ground water in karst terranes. U.S. Environmental Protection Agency, EPA 904/6-88-001, Atlanta, GA, 103p.

Ohio Environmental Protection Agency, 2010. Gibsonburg karst investigation. Available at: http://www.epa.state.oh.us/portals/28/documents/swap/GibsonburgDyeTracesReport_DRAFT_July2011.pdf

Oluić, M., 1983. Remote sensing in geology (in Serbo-Croatian). In: Donassy, P., et al. (eds.), *Remote Sensing in Earth Sciences, Council for Remote Sensing and Photointerpretation*, Yugoslav Academy of Sciences and Arts, Zagreb, pp. 141–268.

Palmer, A. N, 2007, *Cave Geology*. Cave Books, Trenton, NJ, 454p.

Panno, S. V., and Luman, D. E., 2011. Illinois height modernization program. LiDAR data applications. Detecting karst features in forested terrain. Available at: www.isgs.illinois.edu/maps-data-pub/publications/geobits/geobit7.shtml

Panno, S. V., Luman, D. E., Kelly, W. R., Larson, T. H., and Taylor, S. J., 2012. Mapping karst terrane of northwestern Illinois' unglaciated region, Jo Daviess County using LiDAR elevation data. Illinois State Geological Survey Circular. In review.

Parizek, R. 1976. On the nature and significance of fracture traces and lineaments in carbonate and other terranes. In: Yevjevich, V. (ed.), Proceedings of the U.S.-Yugoslavian Symposium, Dubrovnik, June 2–7, 1975. Karst Hydrology, vol. 1, Water Resources Publications, Fort Collins, CO.

Proctor, C. V., Brown, T. E., McGowen, J. H. and Waechter, N. B., 1981. Austin Sheet, Geologic Atlas of Texas, Scale 1:250,000, Bureau of Economic Geology, The University of Texas at Austin, Austin.

Quinlan, J. F., 1989. Ground-water monitoring in karst terranes. Recommended protocols & implicit assumptions. U.S. Environmental Protection Agency, EPA/600/X-89/050, Environmental Monitoring Systems Laboratory, Las Vegas, NV, 79p.

Quinlan, J. F., Ewers, R. O., and Field, M. S. 1988. How to use ground-water tracing to "prove" that leakage of harmful materials from a site in a karst terrane will not occur. In: Proceedings of 2nd Conference Environmental Problems in Karst Terranes and Their Solutions, Nashville, TN, National Water Well Association, Dublin, OH, pp. 289–301.

Raabe, E. A., and Bialkowska-Jelinska, E., 2007. Temperature anomalies in the Lower Suwannee River and tidal creeks, Florida, 2005. U.S. Geological Survey Open-File Report 2007–1311, Reston, VA, 28p.

Ramljak, P., Filip, A., Milanović, P., and Arandjelović, D., 1975. Establishing karst underground connections and responses by using tracers (in Serbo-Croatian). In: Mikulec, S., Sarić, A., Šunjić, J., and Trumić, A. (eds.), Hidrologija i vodno bogatstvo krša, Hidrotehnički zavod Sarajevskog Univerziteta, Sarajevo.

Sabins, F. F., Jr., 1987. *Remote Sensing Principles and Interpretation*, 2d ed., W.H. Freeman & Co., New York, 449p.

Schindel, G., Johnson, S., Hoyt, J., Green, R. T., Alexander, E. C., and Krietler, C., 2009. Hydrology of the Edwards Group: A Karst Aquifer Under Stress. A Field Trip Guide for the US EPA Groundwater Forum November 19, 2009, San Antonio, TX, 57p.

Schmid, G., and Barczewski, B., 1995. Development and application of a fibre optic fluorimeter for *in situ* tracer concentration measurements in groundwater and soil. In: Tracer Technologies/or Hydrological Systems (Proceedings of a Boulder Symposium, July 1995), Wallingford, UK, IAHS Publ. no. 229.

Short, N. M., 1982. The LANDSAT Tutorial Workbook: Basics of Satellite Remote Sensing (updated in 1998). NASA Goddard Space Flight Center, E83-10001, NASA-RP-1078, 554p.

Smart, C. C., 1988. Artificial tracer techniques for the determination of the structure of conduit aquifers. *Ground Water*, 26: 445–453.

Smart, C. C., and Karunaratne, K. C., 2001. Statistical characterization of natural background fluorescence as an aid to dye tracer test design. In: Beck, B.F., and Herring, J.G. (eds.), Proceedings of the Eighth Multidisciplinary Conference on Sinkholes and the Engineering and Environmental Impacts of Karst, Geotechnical and Environmental Applications of Karst Geology and Hydrology, Balkema, Rotterdam, the Netherlands, pp. 271–276.

Smart, C. C, and Simpson, B., 2001. An evaluation of the performance of activated charcoal in detection of fluorescent compounds in the environment. In: Beck, B.F., and Herring, J.G. (eds.), Proceedings of the Eighth Multidisciplinary Conference on Sinkholes and the Engineering and Environmental Impacts of Karst, Geotechnical and Environmental Applications of Karst Geology and Hydrology, Balkema, Rotterdam, the Netherlands, pp. 265–270.

Sowers, G. F., 1974. Foundation subsidence in soft limestones in tropical and subtropical environments. Special Technical Publication G-7, Law Engineering Testing Company, Atlanta, GA, various pagination.

Stierman, D. J., 2004. Geophysical detection of caves and karstic voids. In: Gunn, J. (ed.), *Encyclopedia of Caves and Karst Science*, Fizroy Dearborn/Taylor & Francis, New York, pp. 377–380.

Sukhodolov, A. N., Nikora, V. I., RowiÒski, P. M., and Czernuszenko, W., 1997. A case study of longitudinal dispersion in small lowland rivers. *Water Environ Resour*, 97: 1246–1253.

Taylor, C. J., and Greene, E. A., 2008. Hydrogeologic characterization and methods used in the investigation of karst hydrology. In: Rosenberry, D. O., and LaBaugh, J. W. (eds.), *Field Techniques for Estimating Water Fluxes between Surface Water and Ground Water*, U.S. Geological Survey Techniques and Methods 4–D2, Chapter 3, 114p.

Trček, B., and Zojer, H., 2010. Recharge of springs. In: Kresic, N., and Stevanovic, Z. (eds.), *Groundwater Hydrology of Springs: Engineering, Theory, Management and Sustainability*. Elsevier, New York, pp. 87–127.

USGS (United States Geological Survey), 2011. Tracking change over time. Teacher guide. Images for education. General Information Product 133, 20p.

Verimap, 2005, Suwannee River project final report. VeriMap PLUS Inc., 30p. Available at: www.verimap.com

Wierman, D. A., Broun, A. S., and Hunt, B. B., 2010. Hydrogeologic Atlas of the Hill Country Trinity Aquifer, Blanco, Hays, and Travis Counties, Central Texas. The Hays-Trinity, Barton Springs/Edwards Aquifer, and Blanco Pedernales Groundwater Conservation Districts, July 2010, 17, Plates + DVD, Austin, TX.

Williams, R. T., and Bridges, J. W., 1964. Fluorescence of solutions—A review. *J Clin Pathol*, 17: 371–394.

Wilson, J. F., Jr, Cobb, E. D., and Kilpatrick, F. A., 1986. Fluorometric procedures for dye tracing. U.S. Geological Survey Techniques of Water Resources Investigations, Reston, VA, Book 3, Chapter A12, 34p.

Worthington, S. R. H., and Ford, D. C., 2009. Self-organized permeability in carbonate aquifers. In: Kresic, N. (guest editor), Theme Issue: Ground Water in Karst. *Ground Water*, 47(3): 326–336.

YSI Environmental, 2001. Water Tracing, *In Situ* Dye Fluorometry and the YSI 6130 Rhodamine WT Sensor. White Paper, 7p. Available at: http://www.ysi.com

PART 2

Management of Water in Karst

CHAPTER 4
General Principles

4.1 Introduction

The general principles of water resources management are with one overriding factor in mind—water supply. At the most basic level, water supply management is a process that secures enough water of suitable quality to meet demand at all times, provided that the demand is reasonable and that there is no waste of water. This definition is centered on a variety of anthropogenic uses and, unfortunately, in many cases it ignores the needs of ecological users, thus putting into question the term reasonable use. More broadly, as pointed out by Rogers and Hall (2003), water supply management is an integral part of water governance, which refers to the range of political, social, economic, and administrative systems that are in place to develop, manage, and deliver water services at different levels of society. Water governance includes the ability to design public policies and institutional frameworks that are socially acceptable, equitable, and environmentally sustainable. Given the complexities of water use within society, effective water governance requires the involvement of all stakeholders and must ensure that disparate voices are heard and respected in decisions on development, allocation, and management of common waters, and the use of financial and human resources. Governance aspects overlap with the technical and economic aspects of water, but include the ability to use political and administrative elements to solve a problem or exploit an opportunity.

In the context of Integrated Water Resources Management (IWRM) and in order to achieve sustainable solutions for future generations, both water governance and water management, in any geologic setting or watershed, must include groundwater, surface water, stormwater, precipitation (rainfall) water, and wastewater (often referred to as used water to minimize negative connotation). These different types of water are all essential elements of the overall hydrologic cycle. For example, groundwater is inseparable from surface water because it provides baseflow to surface streams, and sustains aquatic life in surface water as well as in the aquifer itself (such as in karst cavities). Withdrawal of groundwater may affect surface water flows and quality and vice versa. Surface water may become groundwater at some point, and the same water may again emerge as surface water after flowing through a groundwater system for miles, days, or centuries. Upstream (in surface water streams) and upgradient (in aquifers) water withdrawals and wastewater discharges will affect downstream and downgradient users, water availability, and water quality. In karst, these interactions between different types of water are accentuated, thus presenting a great challenge for successful IWRM.

One example of water governance on a grand political scale is the Groundwater Directive by the European Parliament, which refers to groundwater as "the most sensitive and the largest body of freshwater in the European Union and, in particular, also a main source of public drinking water supplies in many regions" (The European Parliament and the Council of the European Union, 2006). This Directive establishes specific measures in order to prevent and control groundwater pollution, defined as the direct or indirect introduction of pollutants into groundwater as a result of human activity. These measures include (a) criteria for the assessment of good groundwater chemical status and (b) criteria for the identification and reversal of significant and sustained upward trends of contamination and for the definition of starting points for trend reversals. The Directive also requires the "establishment by Member States of groundwater safeguard zones of such size as the competent national body deems necessary to protect drinking water supplies. Such safeguard zones may cover the whole territory of a Member State."

The Directive emphasizes that groundwater is a valuable natural resource and as such should be protected from deterioration and chemical pollution. This is particularly important for groundwater-dependent ecosystems and for the use of groundwater in water supply for human consumption. The protection of groundwater may in some areas require a change in farming or forestry practices, which could entail a loss of income. The Common Agricultural Policy provides for funding mechanisms to implement measures to comply with community standards.

Europe's biggest environmental nongovernmental organization, European Environmental Bureau (EEB), stated the following in response to the Groundwater Directive: "Members of the European Parliament have successfully fought off attempts by governments to renationalise groundwater protection. They ensured that preventing pollution and achieving quality standards is robust and legally binding. Without this kind of EU-wide approach, countries would have been left exposed to pressure from powerful, globalised businesses" (EEB, 2006).

Water management is commonly divided into supply-side management and demand-side management. This division is more for technical and administrative purposes, however, since the two aspects are interdependent. At the same time, the overall water management is not "just" about making sound engineering, scientific, economic, and environmental decisions. Water governance is in many cases disproportionately influenced by policies that favor growth of one or more groups of water users—urban, industrial, or agricultural—without much regard for the sustainability of water use or the environmental impacts. The worst possible outcome of failed policies is an uncontrolled spiral of increasing demand causing increasing groundwater withdrawals, which in turn result in serious aquifer depletion and overall environmental degradation. Where groundwater is viewed as both an economic and a public good, the use of which is influenced and overseen by most if not all stakeholders, it is less likely that this spiral would continue unchecked.

Various aspects of water management in karst are best illustrated with examples of large springs since they often represent end points of integrated surface water and groundwater systems spanning several watersheds. As pointed out by Brune (1975), the study of springs is a borderline discipline, because they represent the transition from groundwater to surface water. Hence, they have been studied to some extent by groundwater specialists (hydrogeologists), and to some extent by surface water specialists (hydrologists). Springs are direct reflection of the state of groundwater in aquifers that feed them, and they directly influence streams and other surface water bodies into which they discharge, including all dependent ecosystems. Spring management

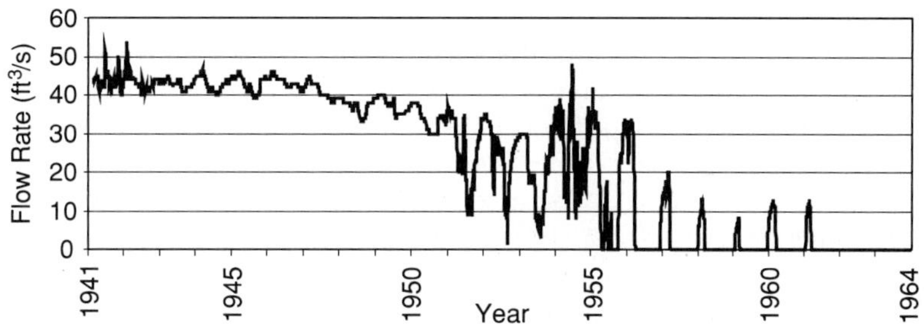

Figure 4.1 Mean daily discharge hydrograph of the historic Comanche Springs, Pecos County, Texas, based on data from the USGS.

therefore includes many of the same principles that guide management of both surface water and groundwater resources. The following examples illustrate how management of karst springs and therefore overall water in karst should not look like.

As discussed by Brune (1975), the artesian Comanche Springs at Fort Stockton, Pecos County, Texas, issuing from a Comanchean limestone groundwater reservoir, formerly flowed as much as 66 ft^3/s (1.7 m^3/s), and served the Comanche and other Indians for uncounted thousands of years. From 1875 on, the spring was the basis for an irrigation district that supplied water to 6,200 acres of cropland. Heavy pumping of the aquifer lowered the water table so that the spring discharge began to fall off in May 1947. The irrigation district sought an injunction in 1954 to restrain pumping that interfered with the normal flow of Comanche Springs. The injunction was denied by the Texas courts, and the springs ceased to flow in March 1961. The historic hydrograph of the Comanche Springs based on flow measurements by the United States Geological Survey (USGS) is shown in Fig. 4.1.

A similar scenario of decreasing discharge of springs in the basin-and-range aquifers in Nevada and Utah is very likely due to a planned expansion of the Southern Nevada Water Authority (SNWA) water supply system driven by the growth of Las Vegas. This is the main reason why various environmental and citizens groups have challenged such expansion in courts and with public outreach. In addition, the State of Utah have initiated a comprehensive groundwater monitoring program in the area and have completed a report summarizing likely impacts of the proposed large-scale pumping from the Snake Valley aquifer on its water resources (Kirby and Hurlow, 2005).

A preview of what may be the ultimate impact of the proposed water diversion by SNWA is destiny of the Needless Point Spring located in the southern part of Snake Valley (Baker Ranches, Inc., 2008). A series of wells were drilled from 1993 to 2000 to supply water for 13 center-pivot irrigation systems. An assessment of the historic and hydrogeologic conditions at the Needle Point Spring, including possible causes of the cessation of spring flow, is provided by Summers (2001–2005). As stated by Summers, information from drillers logs indicated that the wells are 16-inch-diameter wells, designed for high-capacity irrigation pumping between 500 and 1,300 gpm. The U.S. Bureau of Land Management has expressed concern about their prior rights in Needle Point Spring to the responsible regulatory agency, but it did not protest an application for pumping from a new irrigation well. This application was granted by the regulatory agency that, in its ruling, stated: "However, the records of the State Engineer have been reviewed. It appears that the well is about one mile distant from Needle Point Spring which reduces any likelihood that impairment will occur." As a result of this ruling, in summer of 2001, the impairment did occur and the spring dried up. A number of wild horses died because they had imprinted on the spring and did not know where else to look for water. Although irrigation pumping only lasts from April to October of each year, water has never flowed again from Needle Point Spring.

Conclusions of the Utah Geological Survey assessment of likely impacts of the proposed water diversion from Snake Valley by SNWA are as follows (Kirby and Hurlow, 2005):

- Wells proposed by the SNWA will likely adversely affect groundwater conditions in nearby Utah.
- Total drawdown of groundwater near Garrison in western Millard County could be greater than 100 feet (31 m).
- The proposed pumping may change or reverse groundwater flow patterns for much of the east-central Great Basin in Utah and Nevada. The effects may eventually propagate eastward, and impact discharge at important regional springs in Wah Wah Valley and Tule Valley.
- Discharge of agriculturally and ecologically important springs will decrease.

Clay Spring is just one of the many springs in Snake Valley that could be impacted by the proposed large-scale groundwater pumping. It originally attracted some of the first settlers in Snake Valley, the Clay family. Later it was used by the Civilian Conservation Corps as a water source for their camp in the 1930s. The spring pool is very deep and has not been explored yet. The photograph of Clay Spring in the color plates illustrates the spring ecosystem that supports a variety of wildlife including migrating birds, deer, elk, pronghorn, small mammals, and aquatic organisms within the surrounding dry Great Basin desert. A small drawdown in groundwater near this area could reduce or eliminate spring flow and result in the disappearance of this and other important oases (Baker Ranches, Inc., 2008).

With an average annual discharge of 1,219 million cubic meters, or 38.7 cubic meters per second, Ras-el-Ain karst spring in Syria was among the largest of the great springs of the world. The spring, located midway between the Tigris and Euphrates Rivers, just south of the Syrian-Turkish frontier, issued from some 13 locations combined to form the effective head of the Khabour River, tributary of the Euphrates. The aquifers that fed the spring are the Eocene limestones, with subsidiary supplies from the Miocene evaporites and limestones. The major infiltration area is located over basalt and Eocene limestone outcrops in the north and is estimated to extend over some 8,100 square kilometers. Measurements from December 1942 to November 1959 indicated that the average monthly discharge varied from 36 m^3/s in June to 42 m^3/s in November, while the recorded maximum was 107.8 m^3/s. The calculated stored reserves in the aquifer feeding the spring were estimated at about 7,420 million cubic meters. A study carried out as part of the work done in Syria by the Technical Assistance Group sent there by the Food and Agriculture Organization of the United Nations concluded that a more efficient use of the water for irrigation was possible by reducing winter and increasing summer discharge from the springs (Burdon and Safadi, 1963). Nevertheless, the great Ras-el-Ain Spring eventually ceased to flow due to an unchecked overexploitation of the aquifer with numerous irrigation and water supply wells.

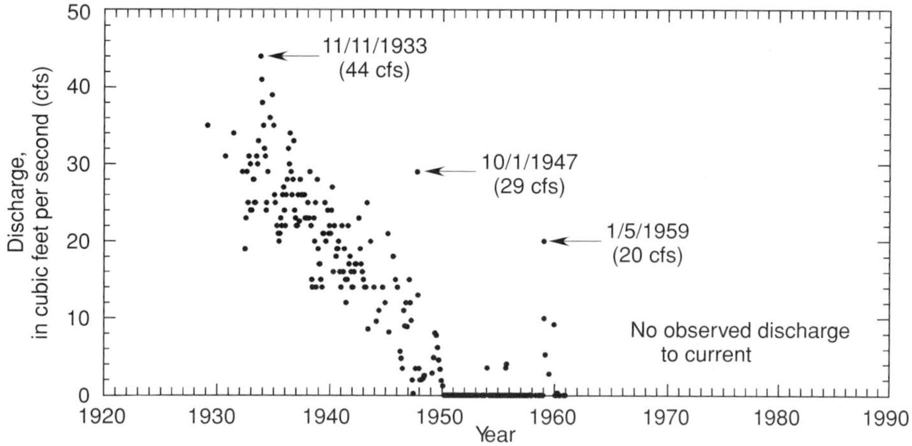

FIGURE 4.2 Periodic discharge measurements at Kissengen Spring in Polk County, central Florida. The spring remains dry through the present day. (Modified from Lewelling et al., 1998.)

Disappearance of karst springs due to large-scale groundwater withdrawals from aquifers is not unique to regions with low rainfall and low natural aquifer recharge. Figure 4.2 shows measured discharge rates of Kissengen Spring in Florida. This historic large spring, located near the city of Bartow in Polk County, served as a popular recreational area for decades (Fig. 4.3) and had average flow of about 29 ft^3/s (0.82 m^3/s).

FIGURE 4.3 Historic photo of Kissengen Spring. (From Peek, 1951.)

The cessation of spring flow, both at Kissengen Spring and at many minor springs in the Upper Peace River basin, was related to the regional lowering of the potentiometric surfaces of the intermediate aquifer system and the Upper Floridan Aquifer from 1937 to 1950 for as much as 60 feet, mainly due to phosphate mining (Peek, 1951; Stewart, 1966; Lewelling et al., 1998). Permanent cessation of the spring flow occurred in April 1960. Although the phosphate mining industry has since improved its water use efficiency and greatly reduced groundwater withdrawals, the Kissengen Spring is still dry mainly because of excessive drawdowns caused by various additional permitted groundwater users. There is little hope that the Kissengen Spring will ever flow again, as the responsible regulatory agencies in Florida are continuously struggling with possible approaches to managing karst aquifers in this and many other parts of the state due to ongoing land development and associated political pressures.

In an ideal situation, when the karst water source is not consumed in any way, its management should be focused on protecting the quality and quantity of water for the ecological users and possible ongoing nonconsumptive uses. Some examples include recreation, fisheries, and power generation. When the water source (a spring, a stream, or an aquifer) is used for consumptive water supply of any type, the additional management requirements include

- Securing clearly defined and reliable quantities of water delivered to all users using various engineering means (see Chaps. 8 and 9).
- Establishing source protection zones as required by the regulations (Sec. 10.3).
- Restoring impaired water sources (Chap. 11).

Three main prerequisites for establishing a realistic, workable management plan are

1. Hydrogeologic and hydrologic characterization of the source drainage area, and its recharge and discharge parameters (Chaps. 1, 2, and 3).
2. Monitoring of water quantity and quality of both the source and the overall water resource (Sec. 4.3).
3. Predictive modeling of the source water quantity and water quality under natural and engineered conditions (Chap. 6).

In addition to the above-listed supply-side management components, all sources of water supply should include demand-side management as well. The concept of water demand management generally refers to initiatives, which have the objective of satisfying existing needs for water with a smaller amount of available resources, normally through increasing the efficiency of water use. Water demand management can be considered a part of water conservation policies, which describe initiatives with the aim of protecting the aquatic environment and ensuring a more rational use of water resources.

Unfortunately, water demand management is often given a low priority or practiced reluctantly, in part due to a false premise that it only involves raising fees. The water sector is heavily subsidized in both developed and developing countries and politicians are usually very hesitant to tackle the issue of water pricing, especially during elections. However, in addition to pricing, which is usually an effective means of demand management, there are many other measures that, when combined, can be as effective. A very detailed discussion on demand management tools, water use, and water conservation,

Figure 4.4 *Top*: Old Town of Dubrovnik, Croatia, built on Mesozoic limestone, has had centralized water supply based on karst springs since its beginnings. *Bottom*: The Old Town is dotted with beautiful public fountains fed by spring water.

with examples from European countries, is presented in the report "Sustainable water use in Europe. Part 2: Demand management" published by the European Environment Agency (EEA) in 1999.

One example of a very efficient demand-side management in the Middle Ages comes from the Republic of Dubrovnik (now part of Croatia) and its capital city of Dubrovnik that still has centralized water supply and beautiful public fountains fed by springs (Fig. 4.4). After the fountain shown in Fig. 4.5 was completed, a new law was enacted

FIGURE 4.5 Great Fountain of Onofria in Dubrovnik, Croatia, designed and built by the Italian architect Onofria della Cave in 1440. Water for the fountain, which includes the water storage reservoir, was conveyed from a group of springs by a 12-km-long system of canals and aqueducts.

stating that "The right hand of everyone caught diverting water from or plugging the conveyance canal will be cut off." and "Public servants will be sent every week to check the entire water conveyance system and determine if anything is damaged or spoiled." (Tušar, 2008). Although hardly imaginable today in the civilized world, this practice shows that those stealing or "spoiling" water designated for the use of others should bear some well-defined consequences.

Water resource management must have a clearly stated objective. This is true for any level of management, starting with a local water agency or water purveyor, and ending at the national (federal) level. The management objective should include the establishment of threshold values for readily measured quantities such as stream and spring flows, groundwater levels in the aquifer, and water-quality parameters. When the threshold level is reached, the rules and regulations require that the amount of captured water diverted from the stream, spring, and/or groundwater extraction from the aquifer should be adjusted or stopped to prevent exceeding that threshold.

Management objectives may range from entirely qualitative to strictly quantitative. For each particular water source, they would have a locally determined threshold value that can vary widely. For example, in establishing a management objective for water quality, one agency may simply choose to establish an average value of turbidity or nitrate concentration as the indicator of whether a management objective is met, whereas another agency may choose to have no constituents exceeding the maximum contaminant level for public drinking water standards. While there is great latitude in establishing management objectives, local water managers should remember that the objectives serve to support the goal of a sustainable supply for the beneficial use of the water in their particular area (DWR, 2003).

Unfortunately, given the enormous pressures that competing water users exert on water agencies, utilities, and managers, in many instances the water resources management is devoid of scientific or engineering principles and stripped down to pure political decisions. In such cases, dissatisfied stakeholders may have a last resort in form of a lawsuit. An illustrative example presented by Eckhardt (2010) comes from Texas where The Lone Star Chapter of the Sierra Club filed a lawsuit against the U.S. Fish and Wildlife Service claiming the Service was not adequately protecting endangered species that depend on the Edwards Aquifer. The Sierra Club argued that Comal and San Marcos Springs (Fig. 4.6) could dry up if unrestricted pumping continued and that would constitute a "taking" as defined by the Endangered Species Act. The Sierra Club asked that the Service be required to ensure minimum spring flows to protect the endangered species. After a 2-year trial, in January 1993, Federal Judge Lucius Bunton of the U.S. District Court in Midland ruled in favor of the Sierra Club and others who had joined the suit along with the Club. The Court found that if unrestricted withdrawals continued, endangered and threatened species would be "taken" as defined by the Act. The Court also found the Fish and Wildlife Service had failed to implement a recovery plan for San Marcos and Comal Springs and had caused risk or jeopardy to the endangered species. Judge Bunton ordered that spring flow must be maintained even during the most severe drought, such

FIGURE 4.6 San Marcos Springs. In San Marcos, Texas, more than 200 springs issue from three large fissures and many smaller openings. In 1849, the springs were inundated when General Edward Burleson, who served as Vice-President of the Texas Republic, built a dam just downstream of the springs to operate a gristmill. About half of the combined discharge of the springs is seen here, leaving Spring Lake over a spillway. (Copyright Gregg Eckhardt, printed with permission.)

as in the 1950s. He directed the Texas Water Commission to prepare and submit a plan to ensure spring flows, and he directed the Fish and Wildlife Service to determine spring flow levels that would result in "take" or "jeopardy" of the species. The Service subsequently determined that level was 150 cubic feet per second at Comal Springs (Eckhardt, 2010).

In the 1993 ruling, Judge Bunton also announced the Texas Legislature had to enact a regulatory plan to limit withdrawals from the Edwards Aquifer or he would implement his own plan. In May 1993, the Texas Legislature passed Senate Bill 1477 that abolished the Edwards Underground Water District, created the Edwards Aquifer Authority, and authorized the new agency to issue permits and regulate groundwater withdrawals from the Edwards Aquifer (Eckhardt, 2010). To ensure adequate spring flows during dry years, new Critical Period Management rules were also adopted that set more stringent trigger levels at the several monitoring wells, called index wells, used to index the hydraulic head in the Edwards Aquifer. For example, everyone living in the San Antonio area knows of J-17, the index well featured daily in the local newspapers and TV news informing them if and how severe the restrictions on water use will be.

Drought restrictions in San Antonio are triggered when the level of the J-17 index well remains below elevation 660 feet for 10 consecutive days, and restrictions stay in place for 30 days after the well rises back above the trigger level (Eckhardt, 2010). Figure 4.7 shows declining levels at J-17 in late spring of 2008, after an extremely dry 10-month period that also caused seriously declining spring flows. As a result, the San Antonio Water System's Twin Oaks aquifer storage and recovery (ASR) facility, which stores treated wastewater in the Carrizo-Wilcox sand aquifer, switched to recovery mode and began producing about 15 cfs (0.42 m^3/s) of stored water, about one-third of its pumping capacity. As seen in Fig. 4.7, the J-17 well level rose in response to reduced pumping from the Edwards Aquifer

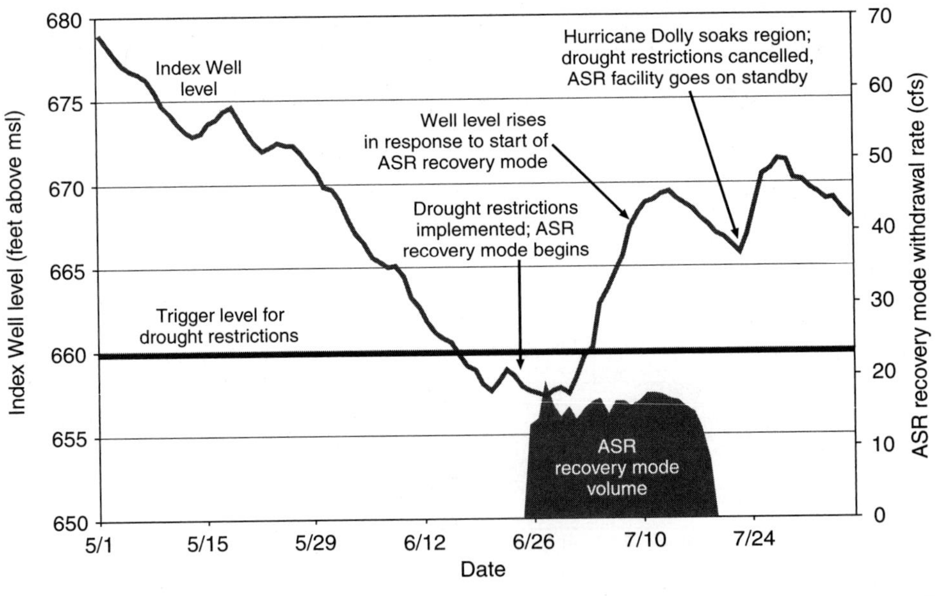

FIGURE 4.7 Index well J-17 and Twin Oaks ASR facility recovery mode, from May to July 2008. (Copyright Gregg Eckhardt, printed with permission.)

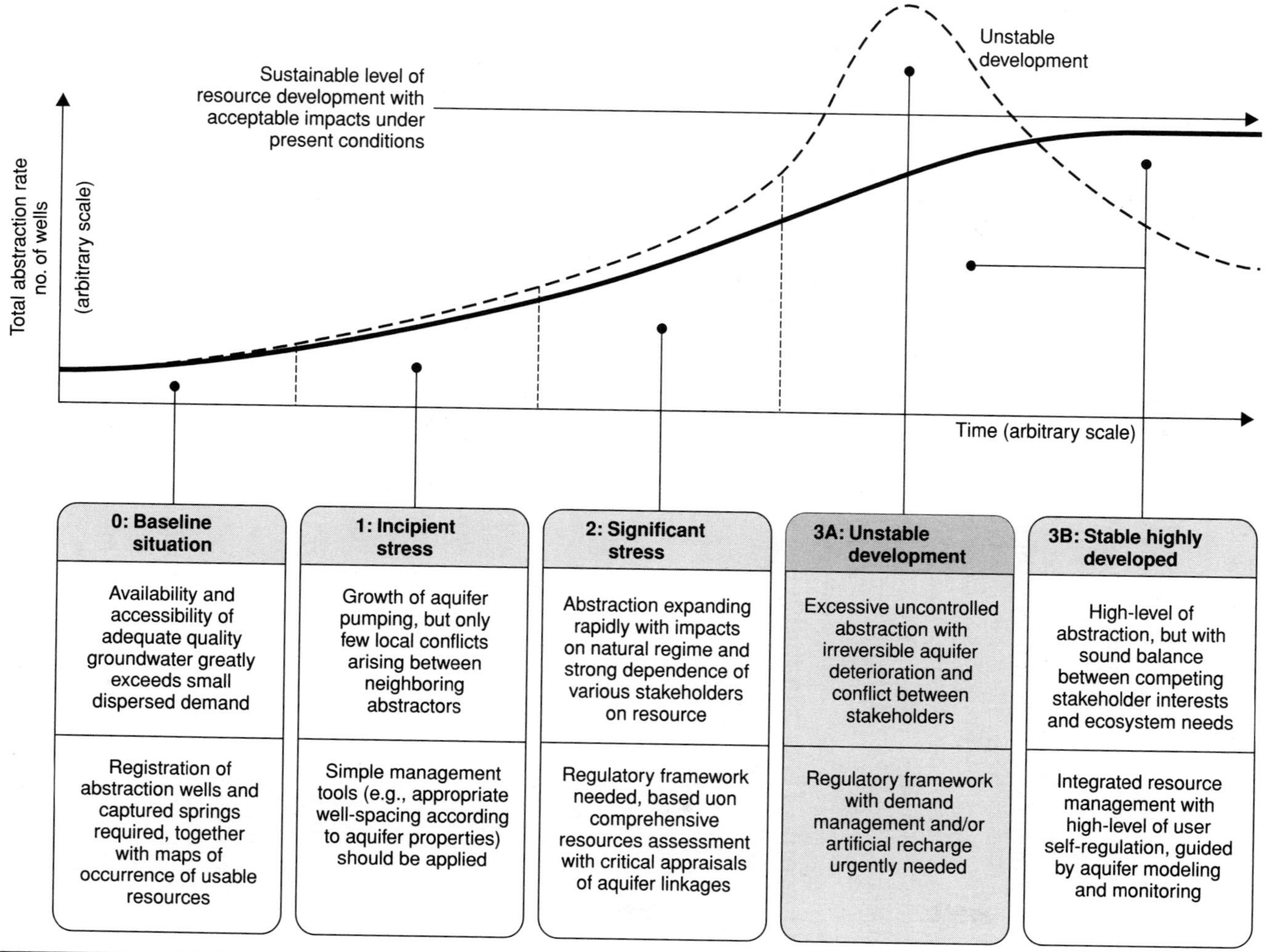

FIGURE 4.8 Stages of groundwater resource development in a major aquifer and their corresponding management needs. (From Tuinhof et al., 2002–2005a.)

and calls for conservation. Extremely heavy demands later in the month caused the well level to begin declining again and facility operators contemplated increasing the recovery (pumping) rate; however, hurricane Dolly soaked the region, drought restrictions were cancelled, and the Twin Oaks facility went on standby, having successfully maintained well levels and spring flows during a critical time (Eckhardt, 2010).

As emphasized by Tuinhof et al. (2002–2005a), water management will in most situations need to keep a balance between the costs and benefits of management activities and interventions. It should take into account the susceptibility to degradation of the water system involved and the legitimate interests of water users, including ecosystems and those dependent on downstream baseflow. Figure 4.8 illustrates a common evolution of groundwater resources development in general and the associated stages based on the impacts of the hydraulic stress (groundwater extraction) on the system.

The condition of excessive and unsustainable extraction (3A—Unstable Development) is also included in Fig. 4.8. For this case, the total abstraction rate (and usually the number of production wells) will eventually fall markedly as a result of near-irreversible degradation of the aquifer system itself.

4.2 Concept of Sustainability

A successful water resource management starts with an agreement of all stakeholders as to what constitutes its sustainable use. This agreement has to be binding and within a clearly defined regulatory framework. The term *sustainable development* was popularized by the World Commission on Environment and Development (WCED) in its 1987 report (published as a book) titled *Our Common Future*. The aim of the report, accepted by the United Nations General Assembly, was to find practical ways of addressing the environmental and developmental problems of the world. In particular, it had three general objectives:

- To reexamine the critical environmental and developmental issues and to formulate realistic proposals for dealing with them.
- To propose new forms of international cooperation on these issues that will influence policies and events in the direction of needed changes.
- To raise the level of understanding and commitment to action of individuals, voluntary organizations, businesses, institutes, and governments.

In various publications, debates, interpretations, and reinterpretations over the course of years, the Commission's report was in many cases stripped down to the following widely cited single sentence: "Sustainable development is development that meets the needs of the present without compromising the ability of future generations to meet their own needs."

One commonly held but inaccurate belief when estimating water availability and developing sustainable water supply strategies is that groundwater use can be sustained if the amount of water removed is equal to recharge—often referred to as *safe yield*. However, there is no volume of groundwater withdrawal that can be truly free of any adverse consequence, especially when time is considered. The safe yield concept is therefore a myth because any groundwater that is artificially extracted from an aquifer will not flow to the natural discharge area any more. The myth falsely assumes that there will be

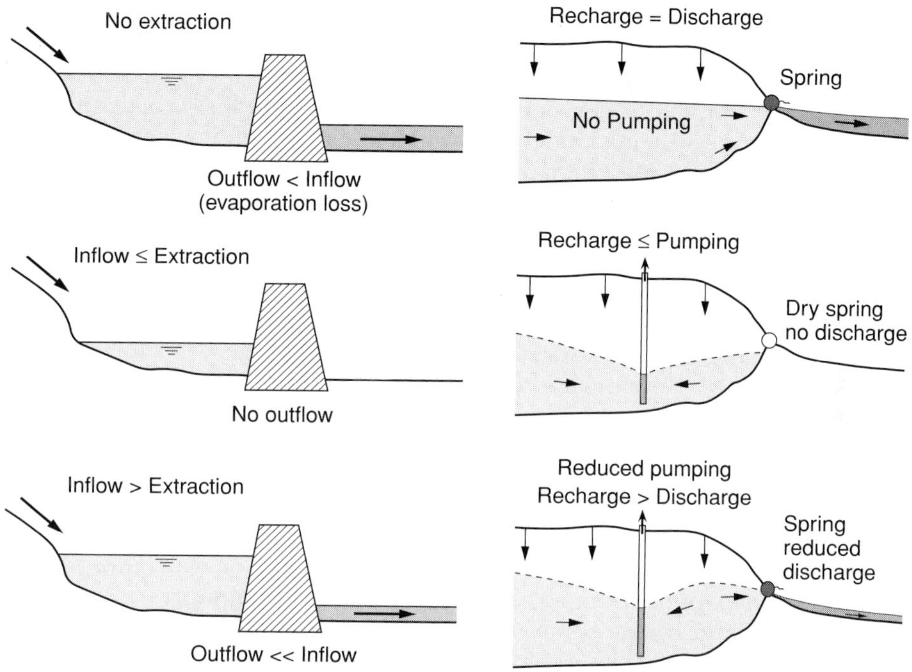

FIGURE 4.9 Illustration of a surface water reservoir (*left*) and an aquifer feeding a spring (*right*). Well yield and water use from the reservoir that are considered sustainable can be achieved only by accepting some consequences of groundwater pumping and water withdrawal, respectively. Safe yield often is referred to as water withdrawal equal to recharge, but as shown, can result in reduced or no outflow and unacceptable consequences. (Modified from Kresic, 2009, 2010.)

no effects on other elements of the overall water budget, such as flow of any spring(s) draining the aquifer or the baseflow of surface stream to which groundwater discharges. Bredehoeft (2002) and Devlin and Sophocleous (2005) provide illustrative discussions about various aspects of the safe yield concept and the related water budget myth.

The false premise of the safe yield myth is illustrated in Fig. 4.9 that compares water withdrawals from a surface water reservoir and from an aquifer drained by a spring. If pumping from the aquifer equals or exceeds all inflows (recharge) to the aquifer, the aquifer discharge via the spring will eventually cease, resulting in some adverse consequence on the surface water-dependent ecosystems and users. Similarly, if most or all water discharging from the aquifer at the spring is captured and diverted to other users, there will inevitably be some adverse consequences. Although this concept is obvious when applied to surface water reservoirs and springs, it still is less comprehensible to some when applied to groundwater reservoirs (aquifers). The direct hydrologic effects in either case will be equal to the volume of water removed from the natural system, but those effects may require some time to manifest in groundwater systems. Because aquifer recharge, groundwater withdrawals, and amount of water captured at the spring or by water wells can vary substantially over time, these changing rates must be taken into consideration when developing both surface water and groundwater management strategies.

With an increased demand for water and pressures on groundwater resources, the decades-long debate among water professionals about what constitutes safe withdrawal of groundwater has now changed into a debate about sustainable use of groundwater. The difference between terms "safe groundwater withdrawal" and "sustainable groundwater use" is not only semantic, and confusion has occasionally resulted. For example, there are attempts to introduce a concept of sustainable pumping that is defined as the pumping rate that can be sustained indefinitely without mining or dewatering the aquifer. What sometimes appears most difficult to explain to inexperienced water managers is that the karst water system is a very dynamic one—any change in one portion of the system will often rapidly and certainly ultimately affect its other parts as well. As groundwater moves from the recharge area toward the discharge area (e.g., a spring or a river), it constantly flows through the saturated zone that acts as the groundwater storage (reservoir). If another discharge area (such as wells for water supply) is created, less water will flow toward the old discharge area (spring or river). This fact seems to be ignored by those who argue that groundwater withdrawals may actually increase aquifer recharge by inducing inflow from recharge boundaries (such as surface water bodies) and therefore result in sustainable pumping rates. Although such groundwater management strategy may be preferable for the intended use, another question is if it has any consequences for the sustainable use of the surface water that is now losing water to rather than gaining it from the groundwater system.

Another argument for sustainable pumping is based on managing groundwater storage. This management strategy adjusts withdrawal (pumping) rates to take advantage of natural recharge cycles. For example, during periods of high demand, some water may be withdrawn from the storage by greatly increasing pumping rates and lowering the hydraulic heads (water table) in the aquifer. During periods of low demand and high natural recharge, this depleted storage would then be replenished (this is also one of the principles of engineered spring use discussed in Sec. 8.3). However, the same question of the sustainability of such approach remains. Any portion of the natural recharge that does not contribute to the natural (nonanthropogenic) discharge will have some consequences for the water users and water uses that rely on it. Depending on the volumes and rates of the denied groundwater discharge, the affected users may or may not be able to adapt to the new reality. For these reasons, the IWRM in many parts of the world relies increasingly on various strategies of artificial aquifer recharge including treated wastewater.

The term sustainability in the case of nonrenewable groundwater systems has an entirely social rather than engineering or scientific context. It implies that full consideration must be given not only to the immediate benefits but also to the negative socioeconomic impacts of development and to the "what comes after" question—and thus to time horizons longer than 100 years. As discussed by Foster et al. (2002–2005), groundwater resources are hardly ever strictly nonrenewable. However, in certain cases, the period needed for replenishment (100s to 1,000s of years) is very long in relation to the normal time frame of human activity in general and water resources planning in particular. For this reason, it is valid in such cases to talk of the utilization of nonrenewable groundwater or aquifer mining. Two general groups of groundwater systems, both of which involve the extraction of groundwater that originated as recharge in a distant past, fall into the category of nonrenewable:

1. Unconfined aquifers in areas where contemporary recharge is very infrequent and of small volume.
2. Confined portions of large aquifer systems, where groundwater development intercepts or induces little active recharge, and the hydraulic head falls continuously with groundwater extraction.

Saudi Arabia is a good example of two main stages in the exploitation of nonrenewable groundwater resources including karst aquifers: initial very rapid, large-scale, and unrestricted development for all uses, subsequently supplemented by desalinated water and treated wastewater. Saudi Arabia has become the largest desalinated water producer in the world. The present production presents about 50% of the total current domestic and industrial demands, with the rest met from groundwater resources (Abderrahman, 2006). However, irrigated agriculture is still the largest user of the nonrenewable groundwater in Saudi Arabia where food security concerns have the highest priority (Fig. 4.10).

Public awareness campaigns on the nature, uniqueness and value of nonrenewable groundwater will be necessary to create social conditions conducive to aquifer

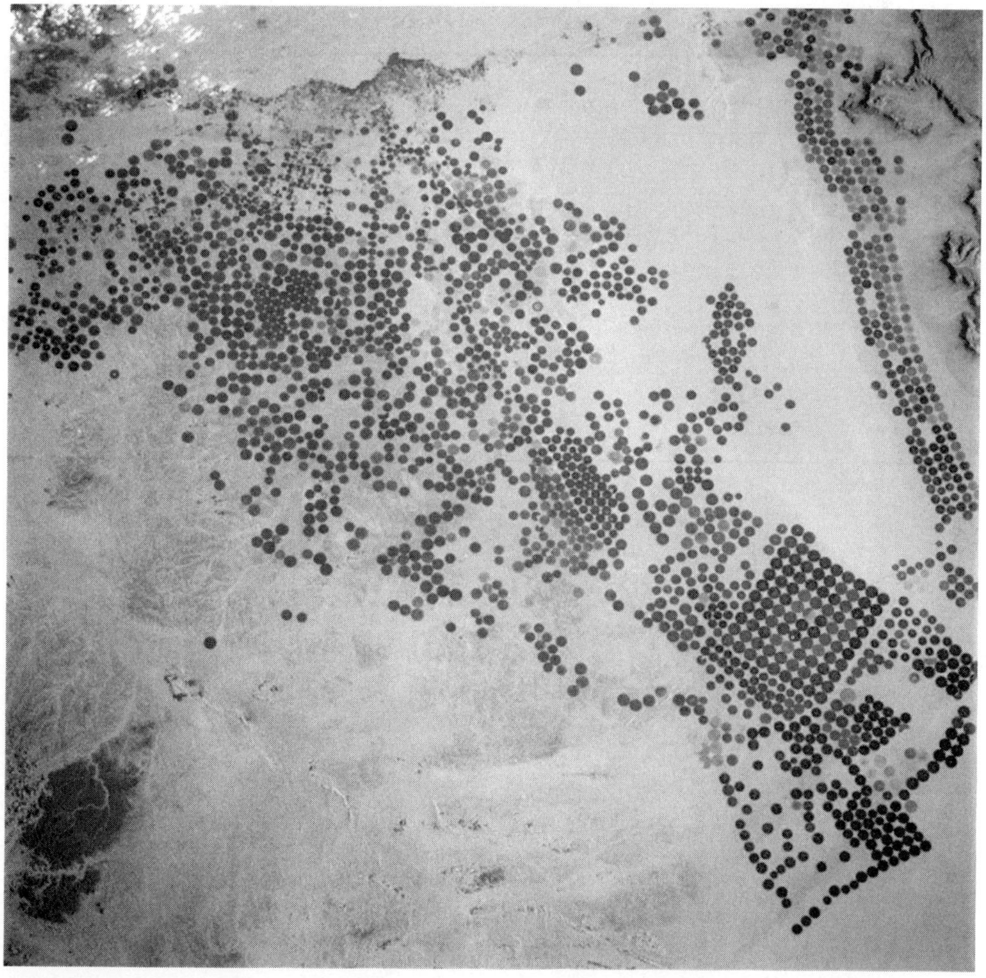

FIGURE 4.10 Space Shuttle astronaut photograph taken in January 1990 showing area of Jabal Tuwayq, southwest of Riyadh in Saudi Arabia. Circles are fields with center-pivot irrigation systems, drawing groundwater via drilled wells in the centers. The diameter of circular fields varies from several hundred feet to over a mile (2 km). (Photograph courtesy of the Image Science & Analysis Laboratory, NASA Johnson Space Center, http://eol.jsc.nasa.gov.)

Figure 4.11 *Top*: Initial abundance provided by a prolific, nonrenewable aquifer. *Bottom*: Possible final outcome of groundwater withdrawal from the nonrenewable aquifer. (Courtesy of Marin Kresic.)

management, including full user participation wherever possible. In this context, all groundwater data (reliably and independently synthesized) should be made regularly available to stakeholders and local communities. This transparency is essential because any depletion of nonrenewable groundwater resources is not sustainable in the long term by definition. Although it may sustain certain socioeconomic development and policies for one or several generations, this depletion will ultimately lead to an inevitable change of the socioeconomic and political environments it helped create in the first place. It is also important that the political entity engaged in mining of nonrenewable groundwater and the society as a whole clearly understand its negative impacts (Fig. 4.11) that are certain to occur without appropriate management plans including artificial aquifer recharge. Environmental impacts will ultimately result in disappearance of springs, surface streams, marshes, and oases fed by the mined groundwater system.

In order to sustain valued ecosystems and endangered species, segments of societies worldwide expect water to be made available, in volumes not easily quantified, to meet key habitat requirements. This relatively recent trend is accompanied by actions of environmental groups that include legal challenges and lawsuits against various government agencies incharge of water governance as illustrated with the example of Edwards Aquifer. Prior to 1980s, similar involvement of nongovernmental groups or the public was virtually nonexistent as shown with the examples of Kissengen Spring in Florida and Comanche Springs in Texas.

4.3 Monitoring

In order for both to be effective or even possible, the water source management and the IWRM must rely on monitoring of water quantity, quality, and their spatial and temporal changes for all parts of the water cycle, that is, surface water, groundwater, precipitation, stormwater, and wastewater (used water). All monitoring data, as well as all data generated during water resource evaluation, development, and exploitation, should be stored and organized within an interactive, geographic information systems (GIS)–based database. Making the products of such database available via the Internet will enable various stakeholders to visualize the historic and current monitoring data and trends, and make informed decisions (see Kresic and Mikszewski, 2012).

Applicable regulations should make a provision for monitoring of surface water and groundwater use and status by assigning facets of this task to the water resources administration and to water users. To be effective, this legislation should set realistic requirements that take account of existing institutional capacity. A typical division of responsibilities is (modified from Tuinhof et al., 2002–2005b):

- Central government/national water authority—ambient (basic reference) monitoring network.
- Regional/basin/aquifer water resource agency—monitoring as a function of resource regulation and protection.
- Water-well contractors/drilling companies—obligations for well logs and aquifer testing data.
- Large groundwater developers (utilities)—records of metered groundwater extraction and groundwater levels; early-warning groundwater quality monitoring at sentinel wells near water supply wells (wellfields).

- Small groundwater developers (small utilities, water purveyors, farms)—general feedback on well characteristics and performance.
- Potential groundwater polluters—compliance and early-warning groundwater quality monitoring at site level.

There are three basic types of monitoring applicable to both groundwater and surface water resources: (1) ambient, (2) compliance, and (3) performance monitoring. All three

Figure 4.12 USGS monitoring well located at the headquarters of the National Ground Water Association in Columbus, Ohio. Water level data are recorded in real-time and transmitted via satellite to the USGS processing center. The data are available online 15 minutes after recording. (From Kresic and Mikszewski, 2012.)

types can monitor both water quantity and quality, and can be short-term or long-term, depending on the management goals.

4.3.1 Ambient Monitoring

Ambient water monitoring programs measure water quality and water quantity such as stream and spring flows, lake levels and water levels in monitoring wells. This type of monitoring is designed to collect long-term data on regional water resources, but can also be implemented for short periods in order to collect site-specific information required by a particular project. An example of short-term ambient monitoring is quarterly concentration of certain water-quality parameters (constituents) in background monitoring wells assumed not to be impacted by anthropogenic contamination at a site. Another example is measurement of spring flows, stream flows, or water levels in monitoring wells for several years. This short-term, site-specific ambient information can then be correlated with long-term data series collected at closest monitoring sites that are part of a regional network.

Long-term, regional ambient monitoring of water quality and quantity is the most important foundation of water resources management at any scale. Because of this importance, as well as associated costs and complexity, such monitoring programs are implemented and managed by state and federal government agencies. The USGS maintains one of the most extensive and sophisticated water resources monitoring networks in the world. Data on stream and spring flows, lake levels, groundwater levels, and physical and chemical water parameters are collected for all major watersheds and aquifers in the country at approximately 1.5 million sites. All historic and current data, including sites with real-time monitoring (Fig. 4.12), are in the public domain and available online (http://waterdata.usgs.gov/nwis/). Users can browse the USGS National Water Information System online database, download the data, or choose to visualize the data on screen by selecting various options for plotting time-series graphs of selected parameters (Figs. 4.13 and 4.14).

One of the most important factors in deciding monitoring priorities at various levels of government is economic constraints. Characterizing and monitoring groundwater quality is expensive. Even in the United States, few states have the economic resources to assess groundwater quality across an entire state (USEPA, 2000). Therefore, states are applying different approaches to groundwater monitoring. These approaches are based on each state's individual challenges and economic constraints. Approaches range from implementing statewide ambient groundwater monitoring networks to monitoring selected aquifers on a rotating basis. States determine the approach based on the use of the resource, vulnerability to contamination, and state management decisions. Large water management districts dependent on groundwater use also have their own monitoring programs that may be better funded and more comprehensive than monitoring programs in many states. Examples include water management districts in Florida.

To protect its groundwater resources, Pennsylvania developed a groundwater monitoring system that accomplishes the following goals:

- Measures ambient groundwater quality.
- Provides an indication of long-term groundwater quality trends resulting from land use practices.
- Assesses the success or failure of land management practices.

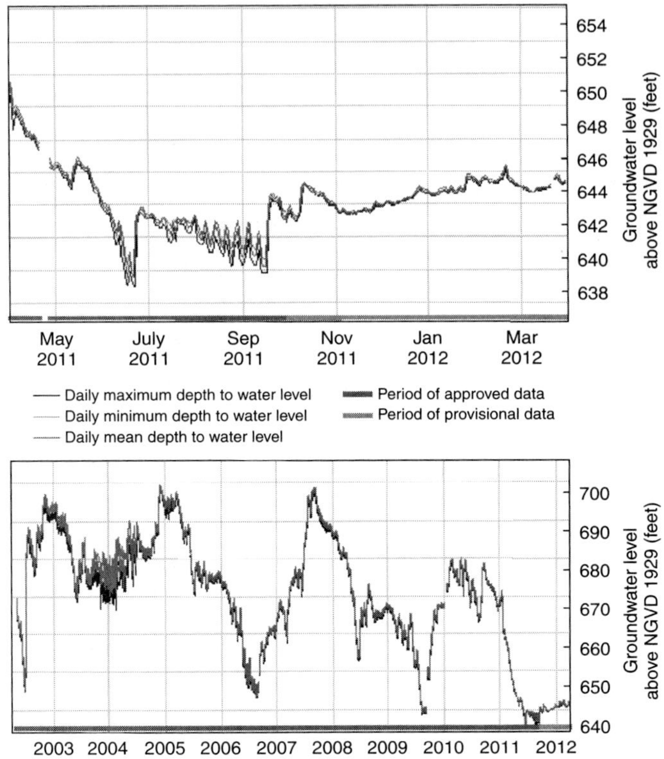

FIGURE 4.13 *Top*: USGS monitoring well # 292556098260701 AY-68-37-526 near San Antonio, TX, showing impact of the severe ongoing drought and groundwater extraction from the Edwards Aquifer. *Bottom*: Well graph for the period of record. (Modified from USGS; available at http://waterdata.usgs.gov/nwis/.)

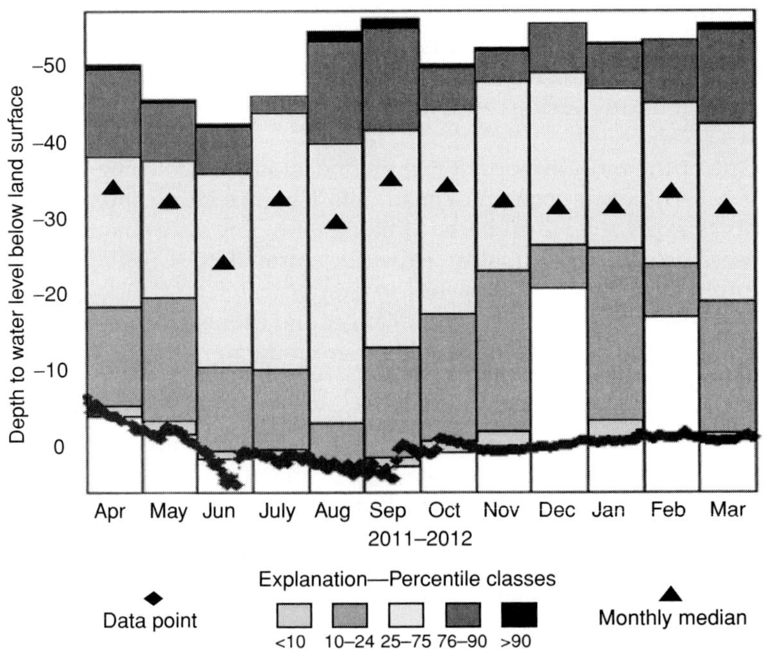

FIGURE 4.14 Statistics for the USGS monitoring well shown in Fig. 4.13. The severity of the drought is illustrated by the fact that the water level during the last 12 months is at or below the 10th percentile for the period of record.

The groundwater monitoring program was developed following a division of the state into 478 groundwater basins. Although the basins are not true hydrogeologic units, each basin considers similarities in hydrogelogic, hydrologic, and physical features. The basins were prioritized for monitoring purposes in 1985 according to three main factors:

- Groundwater use.
- Potential unmonitored sources of groundwater contamination.
- Environmental sensitivity.

In spite of the overall good quality of groundwater, exceedances of drinking water standards were detected in Pennsylvania (USEPA, 2000). Some exceedances result from naturally elevated concentrations of substances such as iron, total dissolved solids (TDS), manganese, or low pH. However, trend analyses of nitrate, sodium, chloride, and total hardness suggest that groundwater quality in Pennsylvania is undergoing some change that likely results from human activities. Downward trends in nitrate and sulfate at many monitoring points may reflect a reduction in sources of nitrate from agricultural areas (fertilizers), septic systems, and atmospheric deposition. Increasing trends in TDS, chloride, calcium, potassium, total hardness, and sodium at many monitoring points may result from increased nonpoint source pollution such as road salting and sprawling paved developments and suburbs (USEPA, 2000).

In contrast to surface water monitoring, a worldwide inventory of groundwater monitoring compiled by the International Groundwater Resources Assessment Centre (IGRAC) shows that in many countries systematic monitoring of groundwater quantity or quality at a regional scale is minimal or nonexistent (Jousma and Roelofsen, 2004). This lack of monitoring may result in undiscovered degradation of water resources due to either overexploitation or contamination, leading to the following scenarios:

- Declining groundwater levels and depletion of groundwater reserves.
- Reductions in stream/spring baseflows or flows to sensitive ecosystems such as wetlands.
- Reduced access to groundwater for drinking water supply and irrigation.
- Use restrictions due to deterioration of groundwater quality.
- Increased costs for pumping and treatment.
- Subsidence and foundation damage.

A number of factors contribute to the lack of groundwater monitoring. Insufficient financial resources and lack of technical capacity to implement monitoring are perhaps the major factors. Other factors that may contribute are a lack of clear institutional responsibilities and legal requirements for monitoring. Even where monitoring programs are operating, they may fail to provide adequate information to support effective management because

- The objectives are not properly defined.
- The program is established with insufficient knowledge of the groundwater system.

- There is inadequate planning of sample collection, handling, and storage.
- Data are poorly archived and not readily accessible in the type of formats that can be used to support management and inform other stakeholders.

In order to improve this situation, IGRAC assembled an international working group of groundwater professionals and assigned it the task of developing a groundwater monitoring guideline for countries with limited financial resources. This document is the result of concerted action by the working group. It focuses on the first stage of groundwater monitoring for general reference, a prerequisite for sound groundwater management. The guideline includes a detailed discussion of monitoring principles and objectives, institutional requirements, design, methods, implementation, and data management (Jousma, 2006). An overview of national strategies for groundwater monitoring in England and Wales, European Union countries, and the United States and a review of current availability of field chemistry methods, measurements techniques, novel analytical tools, and the use of chemical indicators are provided by Stuart et al. (2003).

Groundwater monitoring is a scientifically designed, continuing measurement, and observation of the groundwater characteristics. It also includes data evaluation and reporting procedures. Within a monitoring program, data should be collected at set locations and regular time intervals as much as possible. Although on the regulatory basis institutional framework and funding situation will impose their own objectives and constraints, still the underlying scientific or technical objective is to describe groundwater characteristics in space and time. Meeting this objective in karst environments, namely selecting monitoring locations and frequency, is often challenging even when the monitoring program is well funded. As illustrated throughout the book with various graphs, karst aquifers in open hydrogeologic structures and with well-developed preferential flow paths will react rapidly to recharge episodes. In such cases, sampling frequency becomes the most critical parameter for the success of a monitoring program.

Notwithstanding the limitations associated with the high-frequency (e.g., daily) or continuous (real-time) monitoring, such as relatively high cost and feasibility of implementation, a site-specific situation may deem any other sampling interval entirely inadequate. For example, Fig. 4.15 shows discharge, rainfall, temperature, pH, specific conductance, and dissolved oxygen measured at 10- or 15-minute intervals at the Wilson Spring in Central Basin karst region of Tennessee, the United States. Nonisokinetic dip-sampling methods were used to periodically (mostly during baseflow conditions) collect samples of volatile organic compounds (VOCs) from the spring. During selected storms, automatic samplers were used to collect samples that were analyzed using a portable gas chromatograph (GC). Quality-control samples included trip blanks, equipment blanks, replicates, and field-matrix spike samples. Significant changes in water quality and discharge were detected with rapid changes observed during storms. Specific conductance ranged from 81 to 663 μS/cm, and chloroform concentrations ranged from 0.073 to about 34 mg/L. The greatest change was observed during the first storm during fall 2000, when chloroform concentrations increased from about 0.5 to about 34 mg/L. From these results, it is apparent that a sampling interval of 1 week or 1 month would not provide any meaningful information as to the actual volumes of water discharging from the aquifer, response of the aquifer to precipitation events, and the actual range of concentrations (fluctuation) of a possible contaminant.

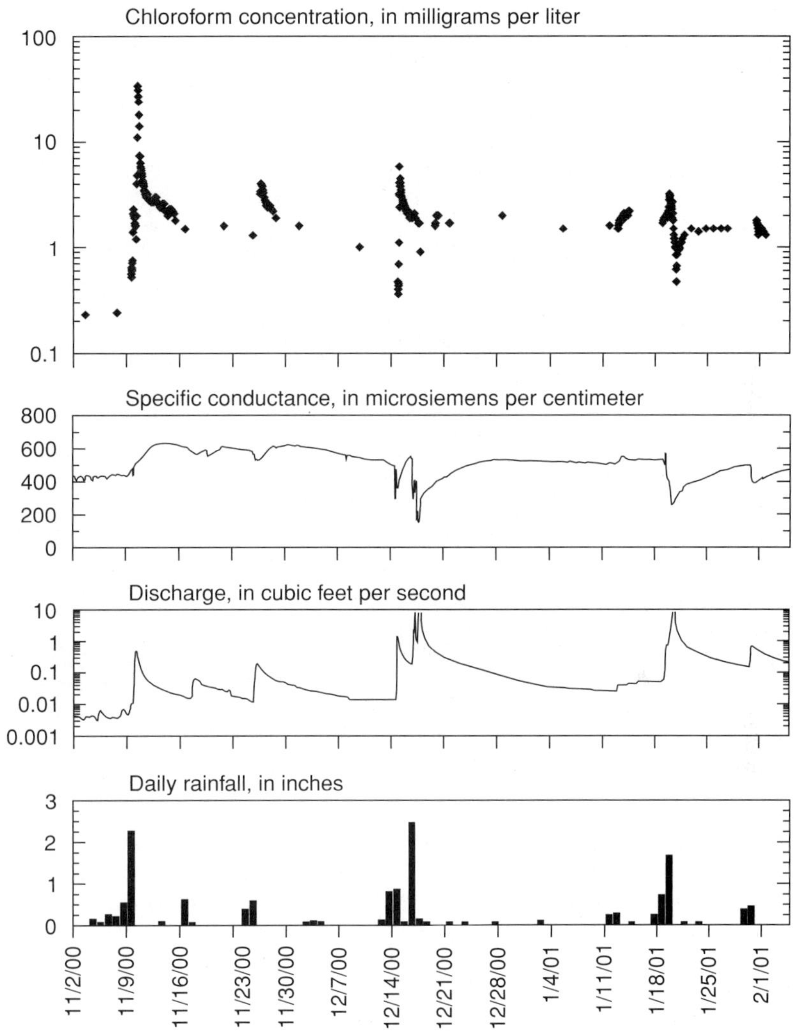

FIGURE 4.15 Results of high-frequency measurements at Wilson Spring in Central Basin karst region of Tennessee, the United States. (Modified from Williams and Farmer, 2003.)

In any case, and regardless of the specific karst type, monitoring of groundwater levels, spring flows, and surface flows (stages) should be automated and accompanied with the similar recording of precipitation (Fig. 4.16), since the associated costs are lower compared to manual measurements while the accuracy and overall benefits are much higher. Once the relationship between the water input and output (e.g., rainfall and spring flow, or groundwater levels and well pumpage) is established based on high-frequency or continuous recordings, it will be possible to adequately scale the monitoring requirements for water-quality parameters of interest that are typically more expensive to analyze. For example, Fig. 4.17 shows that the well water level is not responding to most rainfall events with the exception of three episodes in February 1992. The steady

Figure 4.16 Solar-powered automated rain gauge with data logger deployed as part of the water budget monitoring program at a mine site in karst, The Andes, Peru. (Courtesy of Marc Etienne, AMEC Environment and Infrastructure.)

7-month water level increase is caused by one major weeklong rainfall series in December 1991. Analyzing for synthetic organic chemicals every day at this well would therefore be hardly justified as opposed to the example shown in Fig. 4.15.

Based on analytical data for nitrate and triazines from 566 samples collected over a 3-year period at Pleasant Grove Spring, Logan County, Kentucky, Currens (1999)

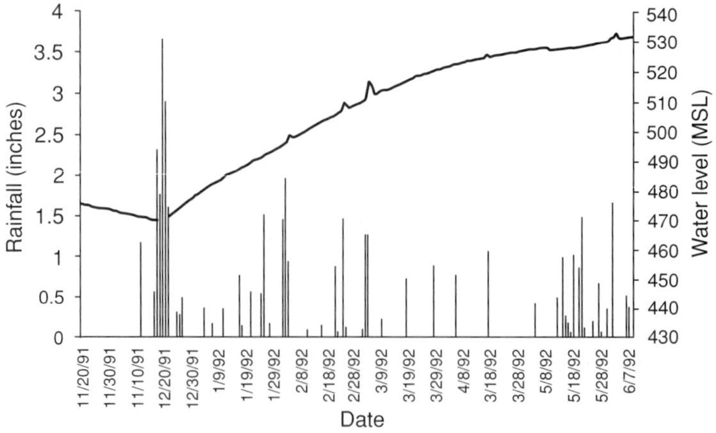

Figure 4.17 Edwards Aquifer Lovelady well levels and rainfall events 1991–1992. (Modified from Hauwert and Vickers, 1994.)

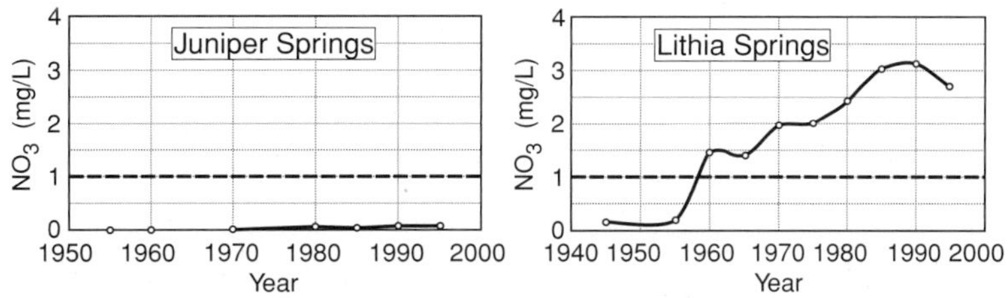

FIGURE 4.18 Nitrate concentration (in mg/L) in two Florida springs over 40 to 45 years. (From The Florida Springs Task Force, 2000.)

concludes that a biweekly sampling schedule augmented with bihourly samples from high-flow events will provide meaningful suspended-constituent and dissolved-constituent statistics. The collected data were statistically analyzed to determine the minimum data set needed to calculate meaningful yearly averages for a conduit-flow karst spring.

Figure 4.18 illustrates the importance of long-term ambient monitoring. The bold lines on the graph show nitrate concentrations (in mg/L) in two Florida springs over 45 to 50 years. Juniper Springs is located within the Ocala National Forest, with a recharge basin that is mainly in conservation lands. Land uses within the Lithia Springs recharge basin are mainly agricultural. Field observations indicate that as nitrate levels approach 1 mg/L (doted line), spring biological communities become degraded by out-of control plant growth. The exact concentration that causes impairment is still not known.

4.3.2 Compliance and Performance Monitoring

Compliance monitoring programs of groundwater quality are required by federal or state regulations at and near facilities where groundwater contamination has occurred or where there is a potential for release of contaminants (such as at gas stations with underground storage tanks or at landfills). These programs measure concentrations of specific constituents of concern (COCs) in groundwater to ensure that there is no contaminant migration that would pose unacceptable risk to human health and the environment.

Groundwater monitoring networks at hydrogeologically complex sites with significant point source contamination are quite elaborate and may include multiple monitoring wells screened at different depths and in different hydrogeologic units. Figure 4.19 illustrates one such network where well clusters are arranged in transects perpendicular to the main groundwater flow direction. In non-karstic porous media, configurations like this enable determination of the contaminant concentrations and mass flux, and the effects of remedial actions undertaken in the source zone to reduce the contaminant mass and flux. In karst aquifers, however, even much more elaborate monitoring well configurations often fail to fully characterize the fate and transport of contaminants due to the specific porosity of karst media as discussed throughout in Chap. 1 (see also Fig. 4.20). The monitoring costs associated with characterization, delineation, and remediation of contaminant plumes are typically very high, and often exasperated by long investigation and remediation times spanning multiple years. Depending on the nature of the aquifer, the contamination being monitored, and the goals of the specific project, multiple monitoring wells may have to be located and screened in both the preferential flow paths and the rock matrix.

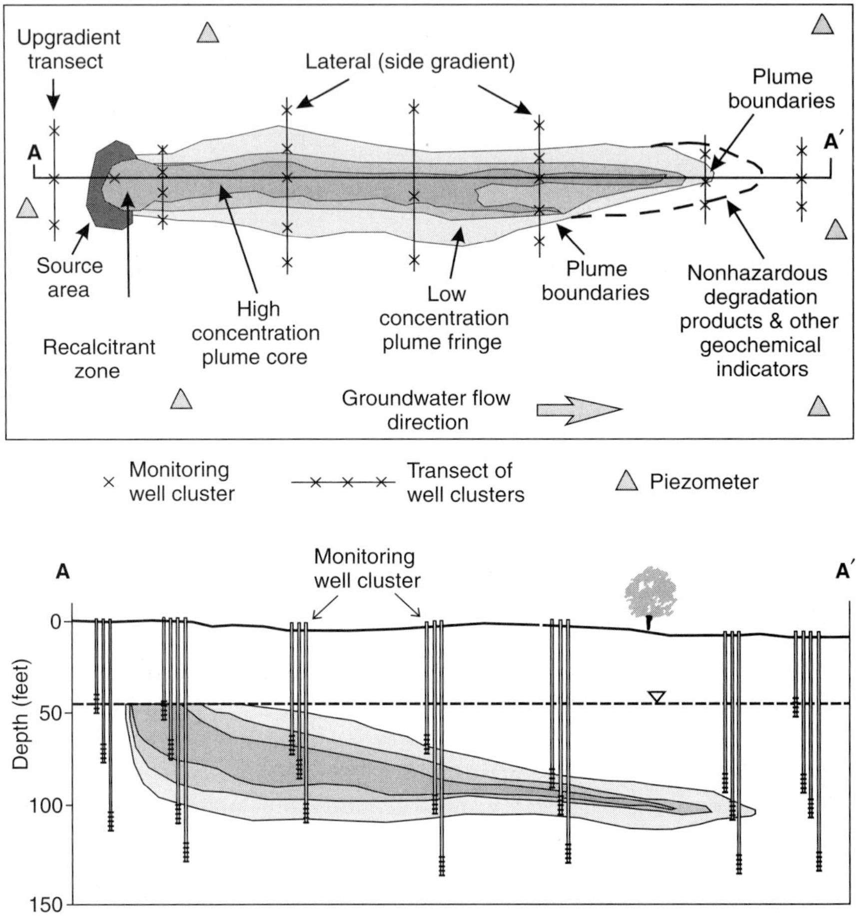

Figure 4.19 Example of a network design for performance monitoring, including target zones for monitoring effectiveness with respect to specific remedial objectives. *Top*: map view. *Bottom*: cross-sectional view AA′. (From Ford et al., 2007.)

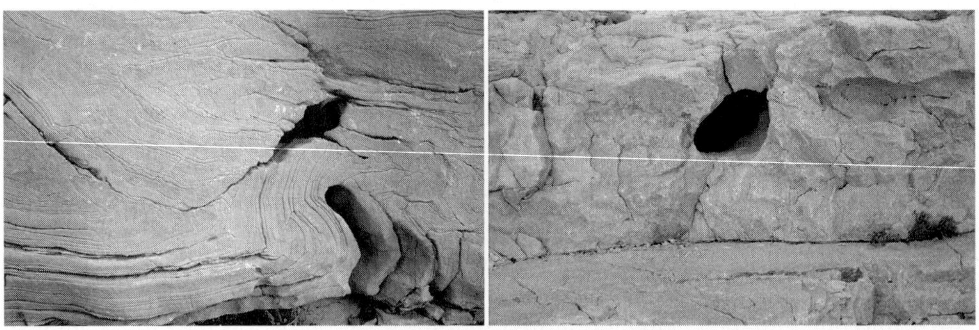

Figure 4.20 Dissolution features in Mesozoic limestone, Dalmatian coast, Croatia. Left photo is approximately 1 foot (30 cm) across and right photo is 2 feet across.

Adding to the troubles when dealing with complex nature of karst aquifers are unsubstantiated guidance documents and recommendations occasionally published by various groups and institutions. One example is the following mystifying statement from a standard published by American Society for Testing and Materials (ASTM, 1995) that contradicts quite a few other statements in the same publication and the experience of many karst hydrogeologists throughout the world: "Fracture lineaments and the intersection of vertical fractures are potential sites for monitoring wells, especially in crystalline rocks. However, in carbonate rocks, most conduits and high-permeability zones are developed along bedding planes; monitoring wells located on the basis of fracture-trace and lineament analysis are not likely to intercept major conduits."

It is the hope of the author that this and similar statements will not be taken seriously by karst specialists. Unfortunately, there is a high probability that novice hydrogeologists and many regulators will be confused when reading this guide and then comparing it with a conceptual site model proposed by experienced karst professionals for their particular site.

Performance monitoring refers to the collection, on a regular basis, of quality and quantity data associated with water supply pumping or groundwater remediation. Monitoring at individual groundwater extraction locations is one of the key elements for successful groundwater management. It provides crucial information needed for the accurate determination of local and regional water budgets, interactions between individual pumping locations, optimization of pumping, assessment of impacts of groundwater extraction on surface water, groundwater levels, and water quality.

Performance monitoring of large-production wells is especially important. These wells should be equipped with a flow or volume-metering device at a minimum. Ideally, monitoring should also include frequent (preferably continuous) recording of water levels and basic physical and chemical parameters such as water temperature, pH, and electrical conductance. At least once a year, or as required by the applicable regulations for drinking water quality, water analysis of the full suite of regulated chemicals should be performed. If the well has been impacted or is threatened by certain natural or anthropogenic contaminants, constituent-specific analyses would have to be performed more frequently.

In the case of irrigation wells operated by individual farmers, or water supply wells for individual households, it is unlikely that any information regarding actual groundwater extraction rates or even quality would be available. In such cases, groundwater extraction (e.g., total volume per time or pumping rates) would have to be estimated based on energy consumption for pump operation, irrigated area, and water requirements of individual crops, or typical water use patterns per person in the given climatic and socioeconomic conditions (Kresic, 2009).

The general principles of groundwater monitoring in karst are described at length by Quinlan (1989); this publication can be downloaded at http://www.karstwaters.org.

4.3.3 Detecting Trends

A trend is defined as a monotonic change in a particular measured parameter (e.g., a chemical constituent in groundwater, or water level) with time. Trends can be examined visually (qualitatively), by plotting the observed parameter versus time. If a trend is apparent, simple linear regression analysis can be performed in an attempt to quantitatively describe the relationship between the two variables (e.g., concentration and time). This analysis is useful even when the data has a high degree of scatter and the coefficient of linear regression is small. The linear relationship depicted by the regression line slope can be negative or positive indicating a decreasing or an increasing trend, respectively. Parameter used to estimate the confidence in the slope is the standard error of the slope (Aziz et al., 2003).

Another, more robust approach to trend analysis is based on probability analysis. In statistical terms, this refers to a determination of whether the probability distribution from which the random variable (e.g., concentration) arises has changed over time. Also of interest is the description of the amount or rate of that change, in terms of changes in some central value of the distribution such as a mean or median. Of the statistical tests for trends available, the nonparametric Mann-Kendall test is the most robust and simple. It indicates whether a particular random variable (concentration) has a statistically significant increasing or decreasing trend at an individual location (e.g., a monitoring well). The test will also indicate the absence of a statistically significant trend. Mann-Kendall is particularly useful since it has very few limitations. For example, sample data need not be in a particular distribution and missing values are allowed. Mann-Kendall can also use data reported as either a trace value or one less than the method detection limit; such values are usually assigned a common value that is less than the lowest measured value (e.g., half detection limit). This is allowed because Mann-Kendall uses the relative magnitude of the values to determine a trend and not the measured values (Gilbert, 1987). The Mann-Kendall test analyzes the sign difference between later-measured data and earlier-measured data. Each later-measured datum is compared with all data measured earlier. An increasing trend is identified if later-measured values tend to be larger than earlier-measured values (this difference, denoted by τ, is positive: $\tau > 0$). Conversely, a decreasing trend is identified if later-measured values tend to be smaller than earlier-measured values ($\tau < 0$).

An example of using the Mann-Kendall test to evaluate apparent nitrate-N trends observed in water supply wells sampled in Hays, Kansas, from 1995 to 2000 is discussed by Townsend and Macko (2007). In west-central Kansas, nitrate concentration has become increasingly problematic in an area with a limited groundwater supply. This is illustrated by the elevated nitrate-N concentrations in Hays' municipal supply wells (Fig. 4.21) that

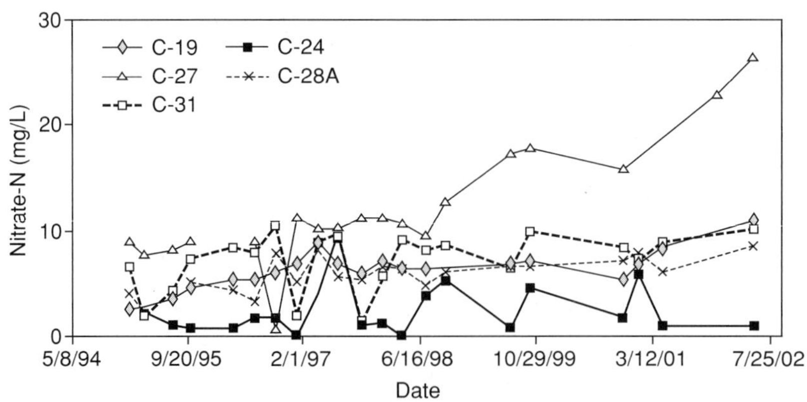

Figure 4.21 Nitrate-N concentration versus time at the City of Hays wells. (Modified from Townsend and Macko, 2007.)

Well	tau (τ)	p-value
C-19	0.658	**0.0001**
C-24	0.164	0.3445
C-27	0.673	**0.0001**
C-28A	0.521	**0.0115**
C-31	0.314	**0.0496**

Modified from Townsend and Macko (2007).

TABLE 4.1 Kendall Test for Nitrogen-N Trend for City of Hays Wells (1995–2002). Statistically Significant *p* Values Are Shown in Bold

show an overall increasing trend, particularly in well C-27 (this well is surrounded on the west and south by farmed land). In well C-24, nitrate-N values are generally lower than in the other wells, likely due to a reducing water quality (the well is not routinely used). The δ15N value for this well also indicates that denitrification may have occurred.

As mentioned earlier, the Mann-Kendall test uses tau (τ) values to evaluate the monotonic correlation between concentration and time. If the concentration of a constituent (nitrate) increases with time, then τ will have a positive value and the parameter p will show a significant value based on the α value assigned to the test. In this example, α is selected to be 0.10 so that $p < \alpha = 0.10$ shows significance. Table 4.1 shows the results of the test for increasing or decreasing trends of nitrate-N values in groundwater samples from the wells shown in Fig. 4.21. All wells, except for well C-24, experience a statistically significant increasing trend for nitrate-N as indicated by a positive τ value (Townsend and Macko, 2007).

Groundwater quality data possess unique characteristics that may require specialized approaches to statistical testing of trends. Data sets have a base limit of zero due to positive values only and can contain censored values (e.g., less than detection limit or <DL) values, outliers, multiple detection limits, missing values, and serial correlation (autocorrelation). These characteristics commonly present problems in the use of conventional parametric statistics based on normally distributed data sets. The presence of censored data, nonnegative values, and outliers may lead to an asymmetric or nonnormal distribution instead of a normal, symmetric, or bell-shaped (Gaussian) distribution, which is common for many data sets. In order to avoid bias and unreasonable trend-test results, these skewed data sets may require use of specific nonparametric statistical procedures for their analysis. The use of nonparametric statistical procedures also is preferred when determining trends of many constituents at multiple locations. Additionally, nonparametric statistical tests are more powerful when applied to nonnormally distributed data, and almost as powerful (under certain conditions) as parametric tests when applied to normally distributed data (Helsel and Hirsch, 1992, 2002).

There are quite a few possible causes of seasonal variation in groundwater quality data such as the effects of varying natural recharge or applied irrigation rates, and varying groundwater flow rates. Such factors need to be taken into consideration in order to discern the specific anthropogenic or natural processes that affect groundwater quality over time. Any effects of seasonality, as well as methods of dealing with both censored and noncensored data, have to be included in trend analysis.

The statistical approach that is used to compensate for seasonal variability in water-quality data is the distribution-free, nonparametric Seasonal Kendall (SK) trend test. The SK test was developed by the USGS in the 1980s to analyze trends in surface-water quality throughout the United States (Hirsch et al., 1982; Hirsch and Slack, 1984; Helsel and Hirsch, 1992, 2002). This test has since become the most frequently used test for trend in the environmental sciences and has been applied to a variety of media at many different locations. The SK test compares relative ranks of data values from the same season (specified by the user). For example, January values are compared to January values, February values are compared to February values, and so on. No comparisons are made across seasonal boundaries. Plus values are recorded if the subsequent value in time is higher, and a minus is recorded if the subsequent value in time is lower. If pluses predominate, a positive trend exists and if minuses predominate, a negative trend results. No trend is the result of pluses and minuses being equal. The null hypothesis is that the concentration of the water-quality constituent is independent of time (Smith et al., 1982). The test assumes that the data are independent and from the same statistical distribution. The SK test statistic is the summation of the Mann-Kendall test results from all the seasons. The attained significance level (or p-value) is the probability of incorrectly rejecting the null hypothesis of no trend when actually there is a trend. The SK slope estimator is computed according to the method of Sen (1968); it is the median slope of all the pair-wise comparisons from all of the seasons expressed as rate of change per year in original units (usually in milligrams per liter depending on the constituent) and in percent per year.

The SK test was modified by Hirsch and Slack (1984) to account for serial correlation (autocorrelation) in the time series. Serial correlation violates the assumption of the independence of data—that there is no short-term correlation between observations. When serial correlation occurs, p-values for significance tests will be too small, and trends may be incorrectly identified. Hirsch and Slack (1984) suggest using the adjusted p-value whenever there are more than 10 years of data, as it commonly takes this much data to detect serial correlation, if present.

The Regional Kendall test extends this concept to spatial locations rather than seasons. A Mann-Kendall test is computed for individual locations and results are combined into one overall test for consistent regional trend (Helsel and Frans, 2006). The test is applicable to data where observations have been made annually (seasonally) at numerous locations, such as at water wells, and one overall test is desired to determine whether the same trend is evident across those locations.

A public-domain computer program that can be downloaded from the USGS software Web page performs the Seasonal Kendall, Regional Kendall, and Mann-Kendall tests for trend. Two other public-domain programs, MAROS and ProUCL, also have modules for Mann-Kendall tests. They are available at http://www.gsi-net.com/en/software/free-software/maros.html and http://www.epa.gov/osp/hstl/tsc/software.htm, respectively.

4.4 References

Abderrahman, W. A., 2006. Saudi Arabia aquifers. In: Foster, S. and Loucks, D. P. (eds.), *Non-Renewable Groundwater Resources: A Guidebook on Socially-Sustainable Management for Water-Policy Makers*. IHP-VI, Series on Groundwater No. 10, UNESCO, Paris, pp. 63–67.

ASTM (American Society for Testing and Materials), 1995. Standard Guide for Design of Ground-Water Monitoring Systems in Karst and Fractured-Rock Aquifers. D 5717-95, West Conshohocken, PA, 17p.

Aziz, J. A., Newell, C. J., Ling, M., Rifai, H. S., and Gonzales, J., 2003. MAROS: A decision support system for optimizing monitoring plans. *Ground Water* 41(3): 355–367.

Baker Ranches, Inc., 2008. Protect Snake Valley, Water. Available at: http://protectsnakevalley.com/water.html. Accessed December 2008.

Bredehoeft, J. D., 2002. The water budget myth revisited: Why hydrogeologists model. *Ground Water*, 40(4): 340–354.

Brune, G., 1975. Major and historical springs of Texas. Texas Water Development Board, Report 189, Austin, TX, 95p.

Burdon, D. J., and Safadi, C., 1965. Ras-el-Ain: The great karst spring of Mesopotamia: An hydrogeological study. *J Hydrol*, 1(Issue 1): 58–64.

Currens, J. C., 1999. A sampling plan for conduit flow karst spring. Minimizing sampling cost and maximizing statistical utility. *Eng Geol*, 52(1-2): 121–128.

Devlin, J. F., and Sophocleous, M., 2005. The persistence of the water budget myth and its relationship to sustainability. *Hydrogeol J*, 13: 549–554.

DWR (Department of Water Resources), 2003. California's groundwater. Bulletin 118, Update 2003. State of California, The Resources Agency, Department of Water Resources, 246p.

Eckhardt, G., 2010. Case study: Protection of Edwards Aquifer, the United States. In: Kresic, N., and Stevanovic, Z. (eds.), *Groundwater Hydrology of Springs: Engineering, Theory, Management and Sustainability*. Elsevier, New York, pp. 527–542.

EEB (European Environmental Bureau), 2006. Protecting Europe's Groundwater: Parliament achieves precautionary safeguards. Press Release, 18 October 2006, Brussels, Belgium. Available at: http://www.eeb.org/press/pr_groundwater_181006.htm. Accessed October 26, 2006.

Ford, R. G., Wilkin, R. T., and Puls, R. W. (eds.), 2007. *Monitored Natural Attenuation of Inorganic Contaminants in Ground Water. Volume 1—Technical Basis for Assessment*. EPA/600/R-07/139, the U.S. Environmental Protection Agency, National Risk Management Research Laboratory, Ada, Oklahoma, 78p.

Foster, S., Nanni, M., Kemper, K., Garduño, H., and Tuinhof, A., 2002–2005. Utilization of non-renewable groundwater: A socially-sustainable approach to resource management. Sustainable Groundwater Management: Concepts and Tools, Briefing Note Series Note 11, GW MATE (Groundwater Management Advisory Team), the World Bank, Washington, D.C., 6p.

Gilbert, R. O., 1987. *Statistical Methods for Environmental Pollution Monitoring*. Van Nostrand Reinhold, New York, 320p.

Hauwert, N. M., and Vickers, S., 1994. Barton Springs/Edwards Aquifer hydrogeology and groundwater quality. Barton Springs/Edwards Aquifer Conservation District, Austin, TX, various paging.

Helsel, D. R., and Frans, L. M., 2006. Regional Kendall test for trend. *Environ Sci & Technol*, 40(13): 4066–4073.

Helsel, D. R., and Hirsch, R. M., 1992. Statistical methods in water resources. Elsevier, Amsterdam, 529p.

Helsel, D. R., and Hirsch, R. M., 2002. Statistical methods in water resources. U.S. Geological Survey Techniques of Water-Resources Investigations, Book 4, Chapter A3, 524p. Available at: http://water.usgs.gov/pubs/twri/twri4a3/

Hirsch, R. M., and Slack, J. R., 1984. A nonparametric trend test for seasonal data with serial dependence. *Water Resour Res*, 20(6): 727–732.

Hirsch, R. M., Slack, J. R., and Smith, R. A., 1982. Techniques of trend analysis for monthly water-quality data. *Water Resour Res*, 18(1): 107–121.

Jousma, G. (ed.), 2006. Guideline on: Groundwater monitoring for general reference purposes. International Working Group I, International Groundwater Resources Assessment Centre (IGRAC), Report no. GP 2006-1, Utrecht, the Netherlands, various paging.

Jousma, G., and Roelofsen, F. J., 2004. World-wide inventory on groundwater monitoring. IGRAC, Report no. GP 2004-1. International Groundwater Resources Assessment Centre, Utrecht, the Netherlands, various paging.

Kirby, S., and Hurlow, H., 2005. Hydrogeologic setting of the Snake Valley hydrologic basin, Millard County, Utah, and White Pine and Lincoln Counties, Nevada—Implications for possible effects of proposed water wells. Report of Investigation 254, Utah Geological Survey, Utah Department of Natural Resources, 39p.

Kresic, N., 2009. *Groundwater Resources: Sustainability, Management, and Restoration*. McGraw-Hill, New York, 852p.

Kresic, N., 2010. Sustainability and management of springs. In: Kresic, N., and Stevanovic, Z. (eds.), *Groundwater Hydrology of Springs: Engineering, Theory, Management and Sustainability*. Elsevier, New York, pp. 1–29.

Kresic, N., and Mikszewski, A., 2012. *Hydrogeological Conceptual Site Models: Data Analysis and Visualization*. CRC/Taylor & Francis Group, Boca Raton, FL, 552p.

Lewelling, B. R., Tihansky, A. B., and Kindinger, J. L., 1998. Assessment of the hydraulic connection between ground water and the Peace River, West-Central Florida. U.S. Geological Survey Water-Resources Investigations Report 97–4211, Tallahassee, 96p.

Peek, H. M., 1951. Cessation of flow of Kissengen Spring in Polk County, Florida. In: Water resource studies, Florida Geological Survey Report of Investigations No. 7, Tallahassee, 73p.

Quinlan, J. F., 1989. Ground-Water Monitoring in Karst Terranes. Recommended Protocols & Implicit Assumptions. U.S. Environmental Protection Agency, EPA/600/X-89/050, Environmental Monitoring Systems Laboratory, Las Vegas, NV, 79p.

Rogers, P., and Hall, A. W., 2003. Effective water governance. TEC Background Papers No. 7, Global Water Partnership Technical Committee (TEC), Global Water Partnership, Stockholm, Sweden, 44p.

Sen, P. K., 1968. Estimates of the regression coefficient based on Kendall's Tau. *J Am Stat Assoc*, 63: 1379–1389.

Smith, R. A., Hirsch, R. M., and Slack, J. R. 1982. A study of trends in total phosphorus measurements at NASQAN stations. U.S. Geological Survey Water-Supply Paper 2190, 34p.

Stewart, H. G., Jr., 1966. Ground-water resources of Polk County, Florida. Florida Geological Survey Report of Investigations No. 44, Tallahassee, 170p.

Stuart, M. E., Gaus, I., Chilton, P. J., and Milne, C. J., 2003. Development of a methodology for selection of determinand suites and sampling frequency for groundwater quality monitoring. National Groundwater and Contaminated Land Centre Project NC/00/35, Environment Agency, Olton, West Midlands, 89p.

Summers, P., 2001–2005. *Hydrogeologic Analysis of Needle Point Spring (Revised Final)*. Bureau of Land Management, Fillmore Field Office, Utah.

The European Parliament and the Council of the European Union, 2006. Directive 2006/118/EC on the protection of groundwater against pollution and deterioration. *Off J Eur Union*, 2006: L 372/19-31.

The Florida Springs Task Force, 2000. Florida's springs. Strategies for protection & restoration. 58p. Available at: http://www.dep.state.fl.us/springs/reports/files/SpringTaskForceReport.pdf

Townsend, M. A., and Macko, S. A., 2007. Preliminary identification of ground-water nitrate sources using nitrogen and carbon stable isotopes, Kansas. Current Research in Earth Sciences, Kansas Geological Survey, Bulletin 253, part 3. Available at: http://www.kgs.ku.edu/Current/2007/Townsend/index.html. Accessed January 2008.

Tuinhof, A., Dumars, C., Foster, S., Kemper, K., Garduño, H., and Nanni, M., 2002–2005a. Groundwater resource management: an introduction to its scope and practice. Sustainable Groundwater Management: Concepts and Tools, Briefing Note Series Note 1, GW MATE (Groundwater Management Advisory Team), the World Bank, Washington, D.C., 6p.

Tuinhof, A., Foster, S., Kemper, K., Garduño, H., and Nanni, M., 2002–2005b. Groundwater monitoring requirements for managing aquifer response and quality threats. Sustainable Groundwater Management: Concepts and Tools, Briefing Note Series Note 9, GW MATE (Groundwater Management Advisory Team), the World Bank, Washington, D.C., 10p.

Tušar, B., 2008. Vodoopskrba u Dubrovniku (Water Supply in Dubrovnik, in Croatian). *Obrada vode*, 4/2008, pp. 54-59.

USEPA, 2000. National Water Quality Inventory: 1998 Report to Congress: Ground water and drinking water chapters. EPA 816-R-00-013, Office of Water, Washington, D.C., various paging.

Williams, S. D., and Farmer, J. J., 2003. Volatile organic compound data from three karst springs in middle Tennessee, February 2000 to May 2001. U.S. Geological Survey Open-File Report 03-355, Nashville, TN, 69p.

CHAPTER 5

Regulations and Education

5.1 Introduction

In the United States, the management of water and the system of issuing water rights has historically been an undenied responsibility of the States, rather than the Federal Government. However, many aspects of Federal law intrude today into this State-based system of water management. The Endangered Species Act, the Clean Water Act (CWA), and the Wild and Scenic Rivers Act are just a few of the Federal laws that impinge upon State authority. As an example, the legal concept of Federal-reserved water rights was thought to apply only to Indian reservations until the mid-1970s when the U.S. Supreme Court issued its ruling in the case of *Cappaert* v. *United States*, 1976 (from Anderson and Woosley, 2005):

> This Court has long held that when the Federal Government withdraws its lands from the public domain and reserves it for a Federal purpose, the Government, by implication, reserves appurtenant water then unappropriated to the extent needed to accomplish the purpose of the reservation. In doing so the United States acquires a reserved water right in unappropriated water which vests on the date of the reservation and is superior to the rights of future appropriators.

The quantity of the water is limited to the quantity needed to accomplish the purpose(s) of the reservation. A significant challenge today is to determine the amount of water required to sustain native peoples, a riparian system, or an endangered species. Under this ruling, the Federal Government can claim a volume of water, required to sustain the lands set aside (reserved) from the public domain for a particular purpose, with an early priority date. National forests, national parks, national wildlife refuges, and wild and scenic rivers have a water right that goes along with the land.

Western water law historically has placed a higher value on water being used off-stream. As an example, the Constitution of the State of Colorado states the following: "The right to divert the unappropriated waters of any natural stream to beneficial uses shall never be denied." Notice the requirement that the water be put to beneficial use. If a diverter fails to use his full allocation of water, after some period of time, he can be forced to forfeit some or all of his right. In this way, the prior appropriation doctrine discourages conservation, for there is a serious disincentive to conserve water on the part of the water-right holder. Such provisions, established in law and set by historical precedent, make it difficult to change the allocation of water to other uses, such as instream use for aquatic life and habitat maintenance and enhancement (Anderson and Woosley, 2005).

The common-law rule regarding groundwater is the rule of capture or the English rule, which essentially provides that, absent malice or willful waste, landowners have

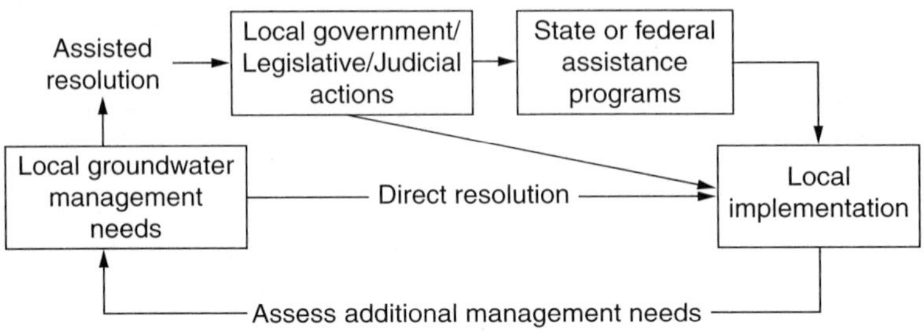

Figure 5.1 Process of addressing groundwater management needs in California. (From DWR, 2003.)

the right to take all the water they can capture under their land and do with it what they please, and they will not be liable to neighboring landowners even if in so doing they deprive their neighbors of the water's use. The rule of capture is in contrast to "reasonable use" or the "American rule," which provides that the right of a landowner to withdraw groundwater is not absolute, but limited to the amount necessary for the reasonable use of his land, and that the rights of adjoining landowners are correlative and limited to reasonable use. Except for Texas, all other states have some form of the American rule, usually left for interpretation to local regulators (Potter, 2004). For example, although the regulation of groundwater has been considered on several occasions, the California Legislature has repeatedly held that groundwater management should remain a local responsibility. Figure 5.1 depicts the general process by which groundwater management needs are addressed under existing law. They are identified at the local water agency level and may be directly resolved at the local level. If groundwater management needs cannot be directly resolved at the local agency level, additional actions such as enactment of ordinances by local governments, passage of laws by the Legislature, or decisions by the courts may be necessary to resolve the issues. Upon implementation, local agencies evaluate program success and identify additional management needs. The State's role is to provide technical and financial assistance to local agencies for their groundwater management efforts, such as through the Local Groundwater Assistance grant program (DWR, 2003).

Approaches to water resources protection from contamination differ at various levels of government and may mean different things to different stakeholders, thus emphasizing the need for public education and dialogue. For an individual household that has just discovered (or was told) that the water they were drinking was contaminated with a carcinogenic substance for years, it would be impossible not to state that their government has failed them. Unfortunately, similar cases are occurring daily, and worldwide. On the other hand, in many developed countries, there are quite a few governmental programs and regulations aimed at water resources protection.

In the United States, some examples of programs that fully or in part address pollution prevention include CWA, Source Water Assessment Program, Pollution Prevention Program, Wellhead Protection Program, Pesticide State Management Plan, Underground Injection Control Program, and Superfund Amendments and Reauthorization Act Title III Program.

For many years following the passage of CWA in 1972, the United States Environmental Protection Agency (USEPA), States, and Indian tribes focused mainly on the chemical aspects of surface water protection. During the last two decades, however, more attention has been given to physical and biological integrity. Also, in the early years of the Act's implementation, efforts focused on regulating discharges from traditional point-source facilities, such as municipal sewage plants and industrial facilities, with little attention paid to runoff from streets, construction sites, farms, and other wet-weather sources. Starting in the late 1980s, efforts to address polluted runoff have increased significantly. For nonpoint runoff, voluntary programs, including cost sharing with landowners, are the key tool. For wet-weather point sources like urban storm sewer systems and construction sites, a regulatory approach is implemented.

Evolution of CWA programs also included something of a shift from a program-by-program, source-by-source, pollutant-by-pollutant approach to more holistic watershed-based strategies. Under the watershed approach, equal emphasis is placed on protecting healthy waters and restoring impaired ones. A full array of issues are addressed, not just those subject to CWA regulatory authority. Involvement of stakeholder groups in the development and implementation of strategies for achieving and maintaining state water quality and other environmental goals is another hallmark of this approach.

The key elements of the CWA are establishment of water quality standards (WQSs) and monitoring and, if WQSs are not met, developing strategies for meeting them. If all WQSs are met, then antidegradation policies and programs are employed to keep the water quality at acceptable levels. Ambient monitoring is also implemented to ensure that this is the case. The most common strategy of meeting WQS for surface streams is the development of a Total Maximum Daily Load (TMDL). TMDLs determine what level of pollutant load would be consistent with meeting WQS. TMDLs also allocate acceptable loads among sources of the relevant pollutants.

As opposed to surface water, an overall protection of groundwater resources still remains a rather vague concept in the United States where it is not subject to any legally enforced, overall regulation by the States or the Federal Government. As noted by the USEPA (1999):

> Ground water management in this country is highly fragmented, with responsibilities distributed among a large number of federal, state, and local programs. At each level of government, unique legal authorities allow for the control of one or more of the ground water threats described in Section 3.0. These authorities need to complement one another and allow for comprehensive management of the ground water resource.

The main reason for this lack of comprehensive, overall groundwater protection regulation is that any such regulation would require a lot of political will because it would have to include land-use controls and significant resources to monitor and enforce land-use practices. This, by definition, includes any agricultural, industrial, or other activity that uses land. As this is not feasible in the near future, in most cases, the real protection of both the existing sources and the resource is, in the end, left to the local communities and public water systems (PWSs). They have to develop management plans, involving local stakeholders and the public, which would minimize risks to their water supplies and their resource. Simply put, if the community (all stakeholders included) believes that a certain industry or land-use activity will not threaten the quantity and quality of their groundwater sources and the resource as a whole, everyone should be satisfied. If this is not the case, there are usually four options: (1) public education and outreach that may resolve the issue by voluntarily changing questionable land uses; (2) enactment of local regulations and ordinances that may leave some stakeholders still unhappy; (3) land acquisitions and land conservation; and (4) a lawsuit (Kresic, 2009).

As discussed in Chap. 10, there are many sources, past and present, of potential or actual groundwater contamination, and an infinite variety of natural and anthropogenic contaminants. It is often very difficult to pinpoint every single source of groundwater contamination, and even more difficult to quickly restore the resource to beneficial use. In most cases, the quality of the resource cannot be directly controlled by the end user due to legal, financial, and other constraints. A simple example is contamination that is (has been) occurring miles away, outside the jurisdiction of the utility that extracts groundwater, and is affecting its well field for public water supply. Even when the sources of contamination are well defined and the legal authority for groundwater restoration is clearly established, it may take years before any measures to mitigate the situation are taken. One common reason is the high cost of groundwater remediation, which can prohibit small and large users alike from attempting to solve the problem on their own. This is the main reason why in some societies such as the United States, where the legal rights of both water users and alleged polluters are protected, exorbitant sums of money are spent each year on litigation over groundwater contamination rather than on its direct mitigation.

5.2 Groundwater Rule

The USEPA promulgated the final Ground Water Rule (GWR) in October 2006 to reduce the risk of exposure to fecal contamination that may be present in PWSs that use groundwater sources. The rule establishes a risk-targeted strategy to identify groundwater systems that are at high risk for fecal contamination and specifically addresses karst and high-permeability gravel aquifers. The GWR also specifies when corrective action (which may include disinfection) is required to protect consumers supplied by water from groundwater systems at risk (USEPA, 2006, 2008).

PWSs are defined as systems for providing water for human consumption through pipes or other constructed conveyances, if such systems have at least 15 service connections or regularly serve at least 25 people at least 60 days a year. A system includes any collection, treatment, storage, and distribution facilities under control of the operator of such system and used primarily in connection with such a system, and any collection or treatment facilities not under such control that are used primarily in connection with such a system (USEPA, 2008).

In the explanation of its ruling, the USEPA points out that groundwater occurrence studies and recent outbreak data show that pathogenic viruses and bacteria can occur in PWSs using groundwater and that people may become ill due to exposure to contaminated groundwater. Most cases of waterborne disease are characterized by gastrointestinal symptoms (e.g., diarrhea, vomiting) that are frequently self-limiting in healthy individuals and rarely require medical treatment. However, these same symptoms are much more serious and can be fatal for persons in sensitive subpopulations (such as young children, the elderly, and persons with compromised immune systems). Viral and bacterial pathogens are present in human and animal feces, which can, in turn, contaminate drinking water. Fecal contamination can reach groundwater sources, including drinking water wells, from failed septic systems, leaking sewer lines, animal farms, and large animal feeding operations.

Although data indicate that relatively small percentage of public water supply systems using groundwater experience fecal contamination, the severity of health impacts and the number of people potentially exposed to microbial pathogens in groundwater prompted the USEPA to promulgate the regulation. The GWR applies to PWSs that use groundwater (more than 147,000 as of 2003). It also applies to any system that mixes surface water and groundwater if the groundwater is added directly to the distribution system and provided to consumers without treatment equivalent to surface water treatment. In total, these systems provide drinking water to more than 100 million consumers.

The GWR addresses risks through a risk-targeting approach that relies on four major components:

1. Periodic sanitary surveys of groundwater systems that require the evaluation of eight critical elements and the identification of significant deficiencies (e.g., a well located near a leaking septic system). States must complete the initial survey by December 31, 2012, for most community water systems (CWSs) and by December 31, 2014, for CWSs with outstanding performance and for all non-CWSs.
2. Source water monitoring to test for the presence of *Escherichia coli*, enterococci, or coliphage in the sample. There are two monitoring provisions: (a) *Triggered monitoring* for systems that do not already provide treatment that achieves at least 99.99% (4-log) inactivation or removal of viruses and that have a total coliform-positive routine sample under Total Coliform Rule sampling in the distribution system; (b) *Assessment monitoring* as a complement to triggered monitoring, a state has the option to require systems, at any time, to conduct source water assessment monitoring to help identify high-risk systems.
3. Corrective actions required for any system with a significant deficiency or source water fecal contamination. The system must implement one or more of the following correction action options: (a) correct all significant deficiencies; (b) eliminate the source of contamination; (c) provide an alternate source of water; or (d) provide treatment that reliably achieves 99.99% (4-log) inactivation or removal of viruses.
4. Compliance monitoring to ensure that treatment technology installed to treat drinking water reliably achieves at least 99.99% (4-log) inactivation or removal of viruses.

More details on the GWR can be found at www.epa.gov/safewater.

5.3 Groundwater Under Influence

Some pathogens, such as parasites *Giardia* and *Cryptosporidium*, are naturally present in surface water bodies and are not necessarily associated with poor sanitation practices. In addition, the parasites are not killed by simple disinfection aimed at bacteria. For this reason, the USEPA has instituted specific water treatment requirements for public supply systems using surface water, and groundwater under direct influence of surface water or GWUDI (see USEPA 2001). This by default includes springs and infiltration galleries. However, groundwater extraction from karst aquifers is not directly mentioned, although it is another logical target of these requirements. As a result, many PWSs relying on water wells pumping from karst aquifers are not even considering threats from *Giardia* and *Cryptosporidium*. Adding to the inadequacy of interpreting GWUDI in karst by various state and local water agencies is the USPA's statement that groundwater sources less than 200 ft from surface water are more susceptible. It then follows, by some interpretations, that if this distance is greater than 200 ft, the groundwater withdrawn by a water supply well is *not* considered to be under influence of surface water. However, surface streams may lose water or sink many miles away from a public supply well and this water can enter the capture zone of the well via preferential flow paths rather rapidly. Any pathogens present in such losing or sinking streams may therefore still be alive when extracted by the well. In addition to sinking streams, there are many other pathogen sources that can quickly impact wells in karst aquifers with unfavorable conditions. For example, common sources of *Giardia* include livestock, on-site sewage systems, feedlots, injection wells, and sewage sludge pits.

5.4 Public Education and Outreach

Probably, the single most important aspect of karst water management is public education and outreach. Unfortunately, it is also often the most underfunded or completely disregarded. There are many simple means of educating the public and the regulators that can pay off many times more than the investment made. Some examples include public outreach with programs describing septic tank maintenance and proper use (e.g., see Riordan, 2007), disposal of toxic wastes generated in households (e.g., paints, solvents, garden pesticides), and proper disposal of unused pharmaceuticals. In terms of protecting the availability (quantity) of groundwater in the areas where it is used for water supply, public outreach programs on water conservation and land use practices are irreplaceable.

Perhaps the most receptive audience to groundwater education programs are the numerous visitors of national, state, and local parks, which have been established because of various karst features such as caves, springs, and unique surface landforms (Fig. 5.2).

FIGURE 5.2 Recreational swimming pools built directly over discharge outlets of two large limestone springs in Texas, the United States. Areas of both springs are designated as state parks. *Left*: San Solomon Spring, one of the Balmorhea Springs in Balmorhea, Texas; between 22 and 28 million gallons of groundwater flow through the pool each day. *Right*: Barton Spring in Austin, Texas, with the main spring on the far bank. On this day in the early 1990s, the pool was closed and drained because of water quality concerns. (Photograph copyright Gregg Eckhardt, printed with permission.)

These parks should be used as role models for the importance of water and environmental protection, and featured in media and school programs as much as possible. One such park is the Barton Springs in Austin, Texas (Fig. 5.2, *right*). In 2006, the U.S. Geological Survey (USGS) published a scientific investigations report that summarized water quality sampling performed at the springs from 2003 to 2005. The water was found to be contaminated by persistent low concentrations of atrazine (an herbicide), chloroform (a by-product of drinking water disinfection), and tetrachloroethane (a synthetic organic solvent chemical). In 2008, the fight to preserve Barton Springs was the subject of *The Unforeseen*, a documentary coproduced by Robert Redford, who learned to swim there as a child. The movie uses the struggle over development in the Barton Creek watershed to illustrate the many clashes between private property rights and resource protection that are occurring across the country. The film drew great reviews, but some developers said it went too far and portrays them unfairly. Environmentalists said the movie is not hard enough on those who would develop lands at the expense of common resources like Barton Springs. (For more details, see http://www.edwardsaquifer.net/barton.html.)

The following words of wisdom and a summary of the importance of public education and outreach are provided by The Florida Springs Task Force, which produced a comprehensive analysis of the state of Florida karst springs and suggested management strategies for their protection.

> Public sentiment is everything. With it nothing can fail. Without it nothing can succeed.
>
> Abraham Lincoln

Local officials cannot make informed decisions about land uses in spring recharge basins if they do not understand the relationship between land uses within spring basins and the quality and quantity of spring water. Homeowners, farmers, golf course managers, and public works officials will not reduce their use of fertilizers if they do not understand the impacts of their actions on the groundwater that feeds springs. The successful resolution of many other threats to springs is dependent on the actions of an educated populace.

Education can nurture appreciation of Florida's springs and bring about cooperation and voluntary compliance. Educational activities carried out on behalf of Wekiwa Springs, Silver Springs, and Ichetucknee Springs have resulted in substantial public support and increased protection. Education can build public support for land acquisition, restoration, and other protective actions for springs. In the absence of public support, lawmakers and policymakers are unlikely to provide funding that will benefit Florida's springs. The public must be informed that Florida's springs are in trouble. A positive message of hope must be delivered—we *can* protect Florida's springs if we work together (The Florida Springs Task Force, 2000).

5.5 References

Anderson, M. T., and Woosley, L. H. Jr., 2005. Water availability for the Western United States—Key scientific challenges. U.S. Geological Survey Circular 1261, Reston, Virginia, 85p.

DWR (Department of Water Resources), 2003. California's groundwater. Bulletin 118, Update 2003. State of California, The Resources Agency, Department of Water Resources, 246p.

Kresic, N., 2009. *Groundwater Resources: Sustainability, Management, and Restoration*. McGraw-Hill, New York, 852p.

Potter, H. G., 2004. History and evolution of the Rule of Capture. In: Mullican, W. F., and Schwartz, S. (eds.), *100 Years of Rule of Capture from East to Groundwater Management*. Texas Water Development Board Report 361, Austin, TX, pp. 1–9.

Riordan, M. J., 2007. Septic system checkup: The Rhode Island Handbook for Inspection. Rhode Island Department of Environmental Management, Office of Water Resources. Available at: www.dem.ri.gov/pubs/regs/regs/water/isdsbook.pdf

The Florida Springs Task Force, 2000. *Florida's Springs. Strategies for Protection and Restoration*. Florida Department of Environmental Protection, Tallahassee, FL, 58p.

USEPA, 1999. Safe Drinking Water Act, Section 1429, Ground Water Report to Congress. EPA-816-R-99–016, United States Environmental Protection Agency, Office of Water, Washington, D.C., various paging.

USEPA, 2001. Implementation Guidance for the Interim Enhanced Surface Water Treatment Rule. EPA 816-R-01–011, United States Environmental Agency, Office of Water, Washington, D.C., various paging.

USEPA, 2006. Final Ground Water Rule. Fact Sheet, EPA 815-F-06–003, United States Environmental Agency, Office of Water, Washington, D.C., 2p.

USEPA, 2008. Sanitary Survey Guidance Manual for Ground Water Systems. EPA 815-R-08–015, United States Environmental Agency, Office of Water, Washington, D.C., various paging.

CHAPTER 6
Predictive Models

6.1 Introduction

Predictive modeling is the cornerstone of decision support systems (DSS) for water resources management. As water resources-related projects and their management vary in scope and complexity, so do the models. Figure 6.1 illustrates application of high-level models that serve to optimize water supply for multisectoral, and often competing, demands on a regional, watershed (basin) scale. Such models perform operational simulations of existing water systems as well as conceptual alternatives, include any number of decision variables, and have automated sensitivity analyses and optimization algorithms. They can operate with a variety of management constraints set by regulations or agreed upon by stakeholders. In the most complex cases, an integrated management of water resources (IMWR) in a large watershed may require several hierarchic models connected with scenario-decision-feedback loops. These models can be used in real time to make short-term and long-term predictions based on estimated climatic input. They can also be used to evaluate various engineering projects for development, augmentation, protection, and restoration of both surface water and groundwater. Finally, they can be used in support of new regulations aimed at balancing demand and supply and competing interests of various water users.

The basin-scale management models are periodically updated by the output from physical models developed for smaller watersheds and groundwater systems, which, at the same time, are updated using input parameters measured in the field at various monitoring locations. Some examples of output from local models are predicted stream water quality changes due to application of fertilizers, turbidity loads after storm events, or changes in flows caused by controlled releases of water from reservoirs. Groundwater models may provide predictions of baseflow and incorporate expected withdrawal from aquifers for water supply and irrigation based on seasonal weather or soil moisture conditions. In short, the level of complexity of integrated modeling systems will depend on the management goals set by end users and stakeholders.

As both water resources and their users are spatially distributed, all models are naturally best integrated in a geographic information system (GIS) that links databases, models, and model outputs in a user friendly, graphics-based visual environment (e.g., see Kresic and Mikszewski, 2012). Monitoring systems, models, and GIS are the three main pillars of DSS that is quickly becoming a standard in water resources management. Visual and multimedia tools are now irreplaceable in transferring technical knowledge to decision- and policy-makers and the public. They are also the most efficient means of

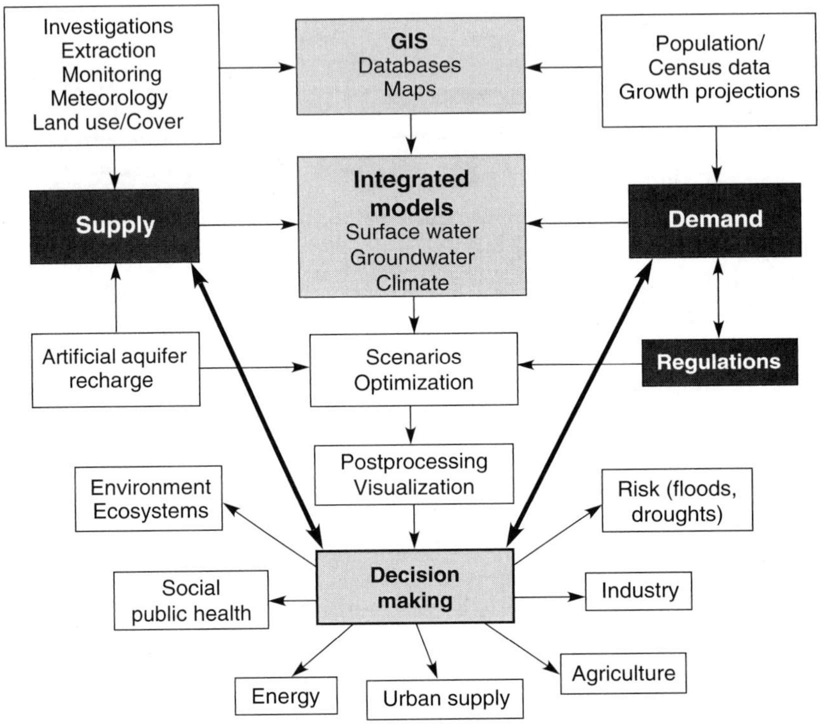

FIGURE 6.1 Resource management and decision support system (DSS) in the context of water sector. Decisions can be made based on supply, demand, modeling, and regulations.

identifying the economic, environmental, and social impacts of different management alternatives.

Karst water systems present ongoing challenges for application of physically based numeric models. The main explanation for these challenges is straightforward: in aquifers with intergranular porosity, the equations of groundwater flow are mostly based on the relatively simple Darcy's law. However, the nature of porosity in karst aquifers requires the application of different sets of equations for distinct porous media: (1) rock matrix; (2) rock discontinuities such as faults, fractures, and bedding planes; and (3) solutionally enlarged voids such as channels and conduits developed from the initial discontinuities. Any meaningful quantitative integration of various equations describing these distinct flow regimes is further complicated by the uncertainties associated with the field distribution and identification of different porosity types.

Unfortunately, the results of the related efforts have generally lagged behind the advances in hydrogeology of intergranular (nonfractured, non-karstic) aquifers. As a consequence, the majority of modeling approaches in karst are still based on various applications of time-series analyses, as well as general statistical and probabilistic methods developed in surface water hydrology. Such methods have one common thread: the necessity for a relatively long time series of data on aquifer recharge and spring discharge and various input–output relationships. This, however, is also the key limiting factor for many practical engineering projects with short execution times. It is therefore not surprising that quite a few practicing hydrogeologists still elect to characterize a karst aquifer as an equivalent porous medium (EPM) and then describe groundwater flow or even fate and transport of contaminants with popular deterministic (physically based) models based on Darcian equations, thus arriving at conclusions that are often questionable (Kresic, 2009).

In any case, with the continued improvement in field and laboratory methods for fractured-rock and karst aquifer characterization, including the development of new analytical and numerical solutions for groundwater flow, it is likely that the successful management of karst water systems will in many cases require simultaneous application of both time-series and deterministic models thus facilitating close cooperation between traditional surface water hydrologists and traditional hydrogeologists.

To be useful as an engineering or water resources management tool in karst, any model should be able to answer some of the following or similar questions:

- What will the spring flow rate be at the end of this month if there were no rain? What would it be if there is average rainfall this month?
- If there is future well pumping of 200 L/s 3 km from the spring, how will it affect the spring?
- What would water availability from the aquifer be if there was catastrophic drought of historic proportions?
- Can we regulate the aquifer (spring) and withdraw more water in August-September than what naturally discharges? Can we count on being able to do that every year? How will that affect the stream?
- Can we count on good water quality from that spring? All the time? How often do we have to treat water?
- Can we use that spring for a trout hatchery?
- What capacity hydroelectric plant could we build at the spring to be profitable at current energy prices?
- How long would it take the contaminant to reach the well (spring)? What would the contaminant concentration be once it reaches it?

It is obvious that the ability to quantitatively (i.e., using a model) answer these questions will depend primarily on the available field data and the funds required to obtain such data including the time involved. However, regardless of the selected modeling approach and the available data, any answer would be more or less probable by default. For example, even though deterministic models are based on some established law(s) of groundwater flow in the subsurface, every prediction about future state of the aquifer (or the spring) will to some degree depend on future precipitation that is a random periodic process.

Models that use mathematical equations to describe the elements of groundwater flow are called mathematical. Depending on the nature of equations involved, these models can be empirical (experimental), probabilistic, and deterministic. Empirical models are derived from experimental data that are fitted to some mathematical function. A good example is Darcy's law. (Note that Darcy's law was later found to be theoretically grounded and actually became a physical or deterministic law.) Although empirical models are limited in scope, they can be an important part of a more complex modeling effort. For example, tracer studies in a spring drainage area can lead to development of mathematical expressions describing the tracer(s) transport toward the spring, including fast (conduit) and slow (diffuse) flow components where applicable; these functions can then be used in stochastic time-series modeling or as calibration targets in numeric deterministic models.

Probabilistic models are based on laws of probability and statistics. They can have various forms and complexity starting with a simple probability distribution of a spring discharge or precipitation, and ending with complicated stochastic time-series models. Such models do not attempt to describe the physics of how various water inputs are transformed into stream or spring discharge and are therefore often called black-box models. Their main goal is to find an appropriate statistical–mathematical expression that can reasonably accurately perform this transformation of one (or more) time series into position of the hydraulic head in the aquifer, spring discharge, or some other component of discharge (e.g., water turbidity or coliform count). The main limitation of time-series models in hydrogeology is that they cannot be used for some common tasks such as predicting the effects of planned new pumping from an aquifer or artificial aquifer recharge.

6.2 Regression

Regression between different variables that are correlated in some manner and represented with a sufficient number of data is probably the most common quantitative method in most scientific fields. In hydrology and hydrogeology, this usually refers to finding a simple or multiple regression equation describing the flow rate, the hydraulic head, or some other dependent variable using the observed time series of independent variables known to influence them. The independent variables may include various water budget components and parameters such as rainfall, soil moisture, and many others.

As with all other models developed from time-series data, regression is performed between variables observed with the same time interval and represented by the same number of data points. In other words, daily flows are correlated with daily rainfall, monthly flows are correlated with monthly rainfall, and so on. Regression is included by default in all commercial statistical computer programs, most spreadsheet programs, and many public-domain programs available for free download on Internet. Accompanied by various plots (graphs), it can provide an invaluable insight into the karst water system.

Figures 6.2 through 6.6 illustrate one such analysis of daily flows at Comal and San Marcos Springs draining Edwards Aquifer in Texas. The daily pumpage from the aquifer in the San Antonio area, the aquifer levels at the Bexar County Index Well J17, and the precipitation in San Antonio are plotted against the daily flow rates at the two springs. As can be seen without any quantitative analysis, Comal Springs, which is closer to well J17 and San Antonio, shows a better overall correlation with the pumpage and the aquifer levels. The simple regression model of the Comal Springs discharge based on the J17 water levels (Fig. 6.3) appears almost perfect judging from the model correlation coefficient ($r = 0.978$). This value is very close to 1 that would correspond to a mathematical function. However, when looking at the graph of model residuals (Fig. 6.4), it is apparent that the model cannot accurately describe an unknown cyclic component that may be caused by pumping. However, adding pumping from the aquifer to the regression does not significantly improve the model.

Since both springs are ascending artesian springs, there is no rapid reaction to daily precipitation at either one, except for the Comal Spring in early 2005. This is confirmed

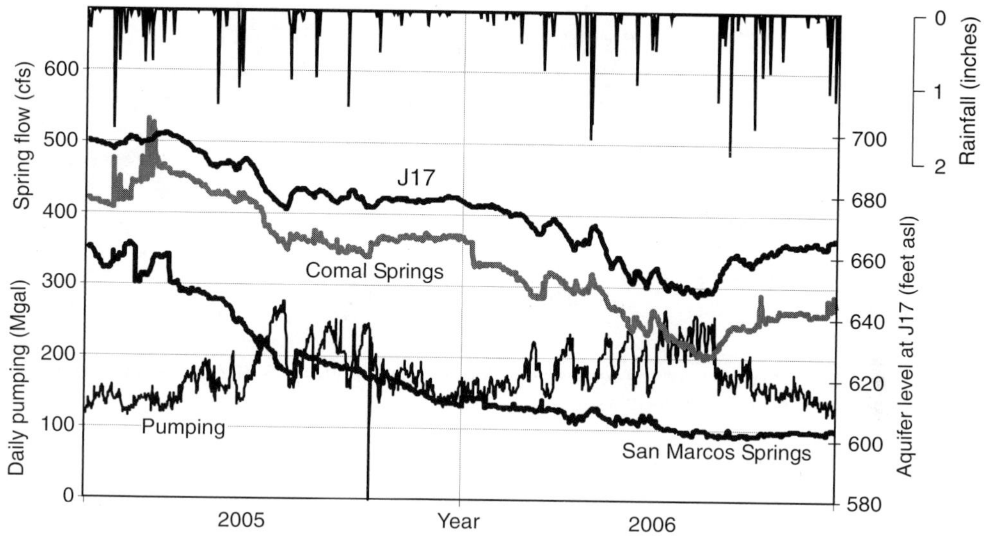

FIGURE 6.2 Daily flows at Comal and San Marcos Springs (in cubic feet per second) versus pumpage from the Edwards Aquifer in the San Antonio area (in million gallons per day), daily aquifer level at the Bexar County Index Well J17, and daily precipitation in San Antonio. (From Kresic, 2010; copyright Elsevier, reprinted with permission.)

by the complete absence of correlation as shown in Fig. 6.5. Even for springs that more readily react to daily or weekly rainfall events, the simple linear regression between flows and total precipitation in the spring drainage area is always week. In addition to the fact that there would be many days with zero precipitation during a hydrologic year, there are other reasons for this: the same amount of rainfall will produce different seasonal response at the spring depending on the antecedent rains, soil moisture, and evapotranspiration rates, and there will usually be some delay in the spring response as the newly infiltrated water travels to the water table and then to the spring.

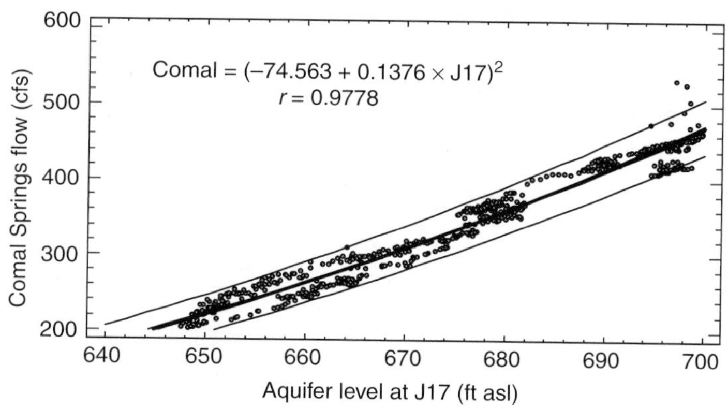

FIGURE 6.3 Simple square root regression model of the Comal Springs flow on aquifer level at the Bexar County Index Well J17, with the 95% prediction limits. (From Kresic, 2010; copyright Elsevier, reprinted with permission.)

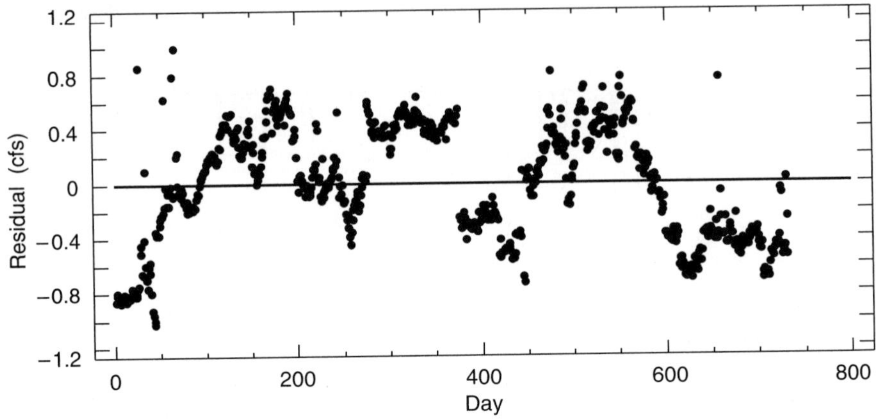

Figure 6.4 Residuals of the regression model shown in Fig. 6.3. (From Kresic, 2010; copyright Elsevier, reprinted with permission.)

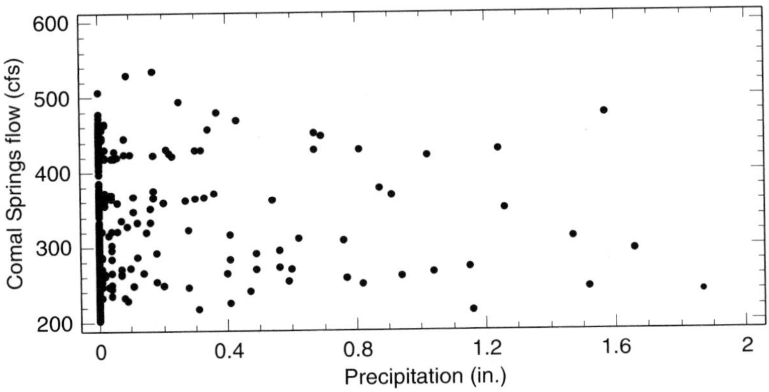

Figure 6.5 Comal Springs flow versus precipitation. (From Kresic, 2010; copyright Elsevier, reprinted with permission.)

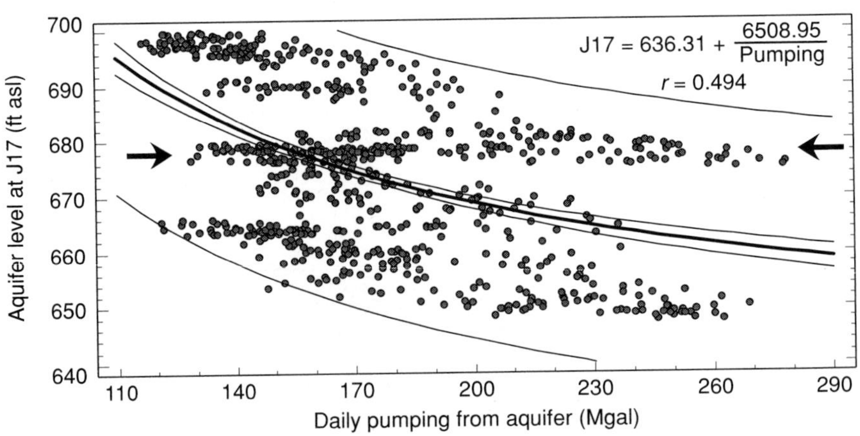

Figure 6.6 Regression model of the Comal Springs flow rate and the aquifer pumpage, with the 95% confidence and prediction limits. (From Kresic, 2010; copyright Elsevier, reprinted with permission.)

The analysis also demonstrates that some correlations may be linear or close to linear (Fig. 6.3), whereas others may show strong nonlinearity (Fig. 6.6). One possible explanation for the zigzag pattern visible in Fig. 6.6 is that there may be some threshold aquifer levels (hydraulic heads) in certain portions of the aquifer that are important for the pumping rate. As the pumping increases, the aquifer levels remain stable (indicated with arrows), then slowly decrease only to drop significantly to the next lower level when the threshold is reached.

One convenient way to improve regression of spring discharge on precipitation is to use the so-called index of antecedent precipitation (*IAP*) that indirectly accounts for the influence of soil moisture and aquifer recharge due to precipitation that fell prior to the present day. The general form of IAP for springs with rapid response to rainfall is

$$IAP = \sum_{t=1}^{i} (C_t \times P_t) \tag{6.1}$$

where t is time interval (e.g., 1 day), i is total number of time intervals for the calculation of *IAP*, C_t is empirical coefficient, and P_t is precipitation at time t. Figure 6.7-top shows some common expressions for C_t and the corresponding decrease in relative weights given to precipitation that fell during 10 preceding days.

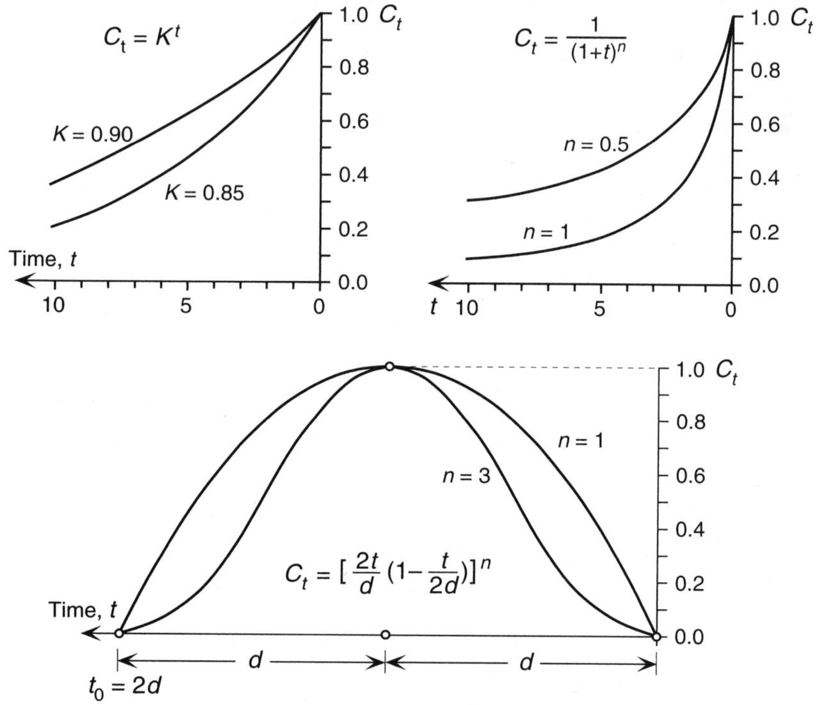

FIGURE 6.7 *Top*: Some common expressions for coefficient C_t used to define index of antecedent precipitation (*IAP*) when spring shows rapid response to rainfall. *Bottom*: Coefficient C_t for springs with delayed response to rainfall. (From Kresic, 2010; copyright Elsevier, reprinted with permission.)

When the spring has a delayed response to rainfall (recharge), *IAP* has peak that is shifted *d* days to the past so that the precipitation that fell on that day has the highest influence on the present-day discharge (Fig. 6.7-bottom). The delay (lag) time can be determined by trial based on the highest value of the regression model coefficient that incorporates *IAP*. It can also be selected from the cross-correlogram of spring discharge and precipitation (see next section). Any number of asymmetrical mathematical expressions can be tested as well and may sometimes provide a superior fit.

6.2.1 Autocorrelation and Cross-Correlation

Regression model of a karst water system output such as spring discharge can be significantly improved if one of the variables used to describe it is the discharge itself—the so-called autoregression. It is evident that the present-day discharge is influenced by the discharge of the preceding day, of the discharge 2 days ago, and so on. This is true for both the descending (recession) and the rising limbs of spring hydrographs. Some springs will exhibit a higher degree of this interdependence of successive flow rates, whereas others will have shorter memory. The flow rate, say, 10 and more days ago may have a statistically insignificant impact on the present-day flow. This memory of the spring is evaluated by the autocorrelation coefficient (also called serial correlation coefficient) of the time series described earlier in Sec. 1.7.1 [see Eqs. (1.81) through (1.85)]. If there is some predictability of the time series' present value based on its past values, the series is autocorrelated. Terms that are also often used to describe an autocorrelated series are persistence and memory. If a series is not autocorrelated (it does not have persistence), it is without memory.

Mangin (1982) proposed that the time required for the correlogram to drop below 0.2 is called memory effect. According to the author, in case of karst aquifers, a high memory of a spring indicates a poorly developed karst network with large groundwater flow reserves (storage). In contrast, a low memory is believed to reflect low storage in a highly karstified aquifer. However, Grasso and Jeannin (1994) analyzed the autocorrelograms of a synthetic, regular discharge time series and demonstrated that the increase in the frequency of flood events resulted in a steeper decreasing limb in the associated correlogram. They also pointed out that the sharper the peak of the flood event, the steeper a decreasing limb of the correlogram. Similarly, the decrease in the recession coefficient entails a steeper decreasing limb of the correlogram. Numerical forward simulation of spring hydrographs by Eisenlohr et al. (1997) confirmed that the shape of the resulting correlogram strongly depends on the frequency of precipitation events. These authors also showed that the spatial and temporal distribution of rainfall and the ratio between diffuse and concentrated infiltration had a strong influence on the shape of the hydrograph and subsequently on the correlogram. Consequently, the shape of the correlogram and the derived memory effect depend not only on the state of maturity of the karst system but also on the frequency and distribution of the precipitation events under consideration (Kresic, 1995; Kovács and Sauter, 2007).

Figure 6.8 shows hydrographs of the same spring for two characteristic years (wet and dry—note the difference in the precipitation axis scale) and Fig. 6.9 shows their autocorrelograms and cross-correlograms. As can be seen on the autocorrelograms, the limits of confidence diverge slightly since the number of pairs in correlation (i.e., the accuracy of correlation) decreases with the increasing lag. For the dry year, the autocorrelogram exceeds the confidence limits for approximately 60 days indicating that the

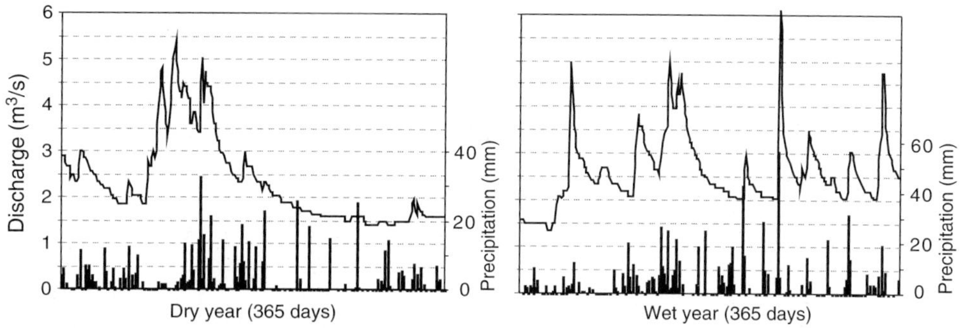

FIGURE 6.8 Characteristic discharge hydrographs of a large karst spring for typical dry and wet years. (From Kresic, 1997; copyright Taylor & Francis, reprinted with permission.)

autocorrelation coefficients for this period are significantly different from zero and that the process of spring discharge is not independent. In other words, the system has a long memory and, generally speaking, the flows for up to 60 preceding days are important for the present-day flow. Possible explanation is that the aquifer storage, including matrix porosity, is significant and that it releases water gradually.

For the wet year the picture is quite different. The system memory of about 20 days is considerably shorter which could be explained by the fact that the inflow from precipitation is frequent enough to cause a dominant role of groundwater flow through large open fissures/fractures or karst conduits. These fast-draining pathways have high flow amplitudes that can be seen from the spring hydrographs in Fig. 6.8. The frequent flow changes in these pathways mask the influence of the surrounding matrix during the dry year.

As discussed in Chap. 1, the aquifer hydraulic head and/or spring flow are more or less influenced by precipitation (or any other water input), and this influence may be delayed for a variety of reasons. In the cross-correlation analysis, the time-dependent relationship between an output (the dependent variable) such as daily spring flow, and an input (the independent variable) such as daily precipitation, is analyzed by computing coefficients of cross-correlation for various time lags and plotting the corresponding

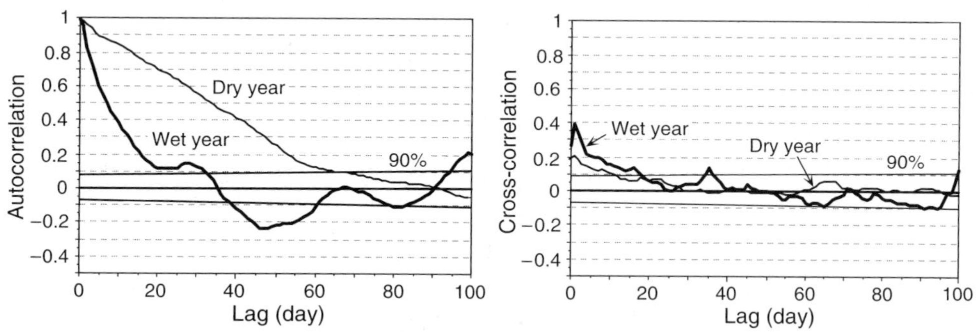

FIGURE 6.9 *Left*: Autocorrelograms of the spring hydrographs shown in Fig. 6.8 for typical wet and dry years. *Right*: Cross-correlograms of the spring flow and precipitation in its drainage area for the same years. (From Kresic, 1997; copyright Taylor & Francis, reprinted with permission.)

cross-correlogram. The cross-correlation coefficient for any lag k is given as follows:

$$r_k = \frac{\mathrm{COV}(x_i, y_{i+k})}{(\mathrm{VAR}x_i \cdot \mathrm{VAR}y_i)^{1/2}} \tag{6.2}$$

where COV is the covariance between the two series, x_i and y_i are the observed daily precipitation and flow, respectively, and VAR is the variance of each series. In practice, the coefficient of cross-correlation for lag k is estimated from the sample using the following equation:

$$r_k = \frac{\sum_{i=1}^{n-k} x_i \cdot y_{i+k} - \frac{1}{n-k}\left(\sum_{i=1}^{n-k} x_i\right)\left(\sum_{i=1}^{n-k} y_{i+k}\right)}{\left[\sum_{i=1}^{n-k} x_i^2 - \frac{1}{n-k}\left(\sum_{i=1}^{n-k} x_i\right)^2\right]^{1/2} \cdot \left[\sum_{i=1}^{n-k} y_{i+k}^2 - \frac{1}{n-k}\left(\sum_{i=1}^{n-k} y_{i+k}\right)^2\right]^{1/2}} \tag{6.3}$$

Figure 6.9 shows that the cross-correlation coefficients are statistically insignificant after about 12 and 17 days for the dry and wet years, respectively, which indicates that the antecedent precipitation beyond this period has no significant influence on spring discharge. A distinct peak at lag 1 for the wet year shows an important, 1-day-delayed response of large transmitters in the aquifer to significant rainfall events. In a typical dry year, this response is much less pronounced, but it has the same lag (1 day) that indicates the same discharge mechanism after major rainfall. An interesting secondary peak at 35 days is evident on the wet year cross-correlogram. This peak may indicate slower contribution from another set of conduits and/or from a distant point source recharge. It may also indicate delayed recharge through the thicker vadose zone overlying less permeable residuum, or non-karstic sediments.

The following examples illustrate possible applications of the autocorrelation and cross-correlation analyses when developing conceptual models of karst aquifers and spring discharge. Ombla Spring (Fig. 6.10), tapped for the water supply of the Croatian coastal city of Dubrovnik, drains more than 600 square kilometers of pure mature classic karst terrain of the Dinarides. The ratio of maximum to minimum flow (coefficient of

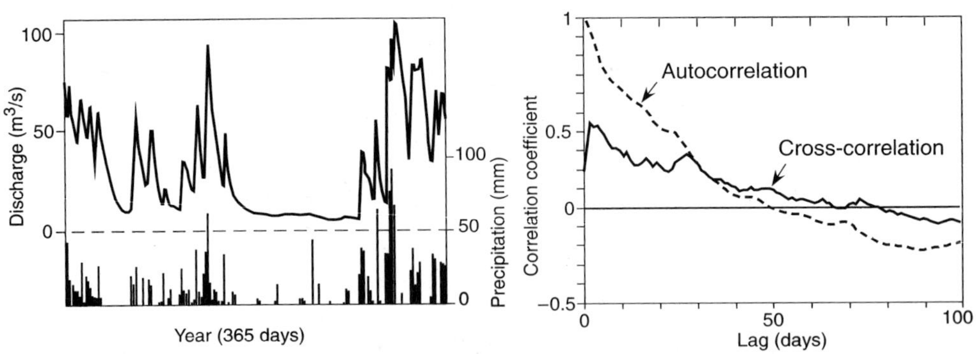

FIGURE 6.10 Autocorrelaton analysis of the Ombla Spring flow, and cross-correlation of the spring flow and precipitation in its drainage area developed in mature classic karst of the Dinarides, Croatia. (From Kresic, 1995; copyright American Institute of Hydrology; reprinted with permission.)

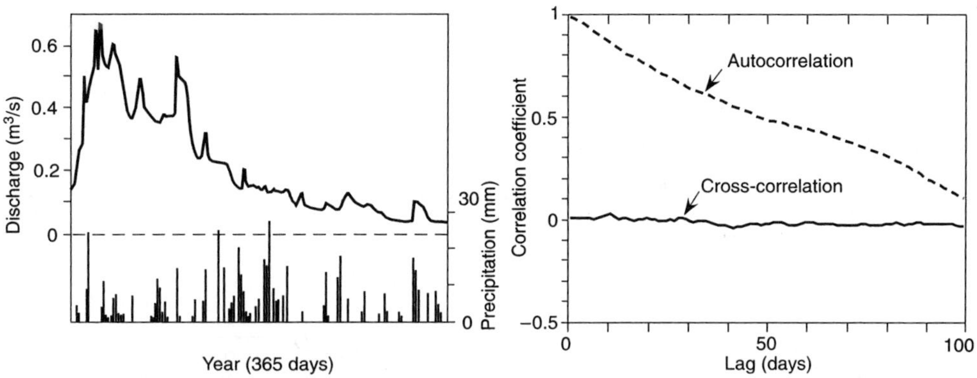

FIGURE 6.11 Autocorrelaton analysis of the Grza Spring flow, and cross-correlation analysis of the spring flow and precipitation in its drainage area developed in semicovered karst of Eastern Serbia. (From Kresic, 1995; copyright American Institute of Hydrology; reprinted with permission.)

spring nonuniformity) is more than 10 for most years, and a very rapid response to major rain events is evidenced by a short time lag of 2 to 3 days and the corresponding high coefficient of cross-correlation close to 0.5 (the peak on the cross-correlogram in Fig. 6.10). Statistically significant autocorrelation of flow according to Mangin ($r_k > 0.2$) lasts over 30 days due to both frequent precipitation and stable (although low) baseflow during summer months. These facts provide for a quick preliminary assessment: the flow takes place mainly through large conduits capable of rapidly transmitting equally rapidly infiltrated rainfall. The conduit network, however, drains quickly and does not have significant storage. Other types of porosity contribute to a very uniform regional baseflow (between, 6 and 7 m^3/s) during long summer periods. However, knowing that the spring drainage area is probably more than 600 square kilometers, it follows that the effective matrix porosity (storage) of the aquifer is quite low.

Grza Spring (Fig. 6.11), located in the semicovered karst of Eastern Serbia, has a very high coefficient of nonuniformity of 22.5 but at the same time a significantly higher and longer autocorrelation. The cross-correlation is statistically insignificant, although the precipitation in the drainage area is frequent and relatively uniformly distributed throughout the year. Preliminary assessment is that the infiltration is quite slow for a karst terrain. The conduit flow is not predominant, and other nonconduit types of effective porosity/storage are more significant. It also helps to know that the drainage area is a mountainous terrain with a significant snowcap that melts relatively quickly during spring. This snowmelt contributes to peak flows that are not directly related to the ongoing precipitation events (Kresic, 1995, 2010).

6.2.2 Autoregressive-Cross-Regressive (ARCR) Models

A separate analysis of simple autoregressive (AR) and cross-regressive (CR) models of spring discharge provides useful information on the aquifer structure. Figure 6.12 shows comparison of these models for typical wet and dry years (spring hydrographs are shown in Fig. 6.8). The AR model of order p is (Kresic, 1997):

$$Q_t = a + b_1 Q_{t-1} + b_2 Q_{t-2} + \cdots + b_p Q_{t-p} \tag{6.4}$$

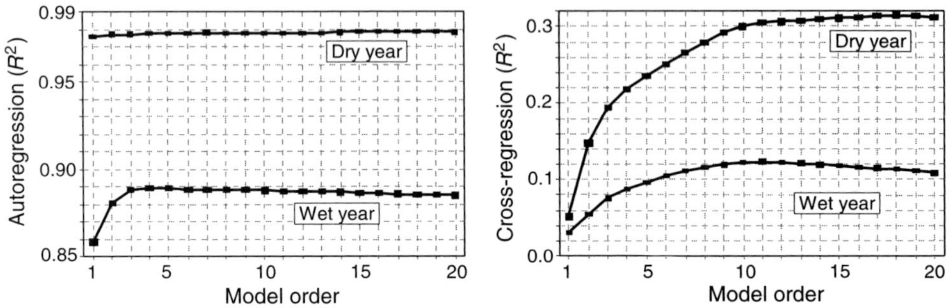

FIGURE 6.12 Comparison of simple autoregressive (*left*) and cross-regressive (*right*) models of the spring discharge shown in Fig. 6.8. The goodness of the models is measured by the squared coefficient of multiple regression (R^2). (From Kresic, 1997; copyright Taylor & Francis, reprinted with permission.)

where Q_t is the predicted discharge at time t; $Q_{t-1}, \ldots, Q_{t-p}$ is the discharge for $1, \ldots, p$ preceding days; and $a, b_1, \ldots, b_p$ are the model parameters.

The CR model of order q is

$$Q_t = a + c_1 P_{t-1} + c_2 P_{t-2} + \cdots + c_q P_{t-q} \tag{6.5}$$

where Q_t is the predicted discharge at time t; $P_{t-1}, \ldots, P_{t-q}$ is the precipitation for $1, \ldots, q$ preceding days; and $a, c_1, \ldots, c_q$ are the model parameters.

As can be seen in Fig. 6.12, AR flow models have much higher coefficients of multiple regression (which is the measure of the model goodness) than CR models. This is mainly because the system has a long internal memory resulting from the gradual release of groundwater from storage. The same can be concluded from the shape of the autocorrelograms in Fig. 6.12. In addition, the CR model uses only total precipitation and the magnitude and time distribution of the effective infiltration remain unknown. This introduces a lot of zero values into the multiple cross-regression and significantly decreases its coefficient.

The autoregression model coefficients remain consistently high for both years. For the dry year, they are higher than for the wet year and change insignificantly for all tested model orders. This again can be explained by the fact that the flow generated by the slow draining of small fissures, filled cavities, and/or rock matrix is statistically predominant. In the wet year, however, flow through karst conduits for the preceding 3 to 4 days is more important (see the maximum on the autoregression graph in Fig. 6.12). The cross-regression graph for the wet year changes its slope considerably for the order of 3 and, less abruptly, again for the order of 10, after which it remains steady. This also indicates that, in the wet year, discharge of the spring is statistically influenced by the precipitation of the preceding 10 days, with the first 3 days being the most important.

Because of their statistical insignificance, simple CR models cannot be used for the prediction of spring discharge: the coefficient of multiple regression for the wet and dry years for all orders is less than 0.32 and 0.13, respectively. However, the AR models for both years have a very high coefficient of regression that for the dry year approaches almost a functional dependence (R^2 is close to 0.98). They are therefore good predictive models and can be used for forecasting when the spring discharge is not under the direct influence of precipitation, that is, in the periods of recession. AR models, however, cannot be used for forecasting based on some expected precipitation because they cannot cause change in the direction of the hydrograph (e.g., from falling to rising). They also cannot be used for generating historic flows based on historic precipitation.

ARCR models belong to the category of multivariate time-series models and are similar to ARMAX models, that is, autoregressive moving average (ARMA) models with exogenous variables (Salas, 1993). However, they are less complex since the error term is not included in the parameter estimation. The order of the models is generally low—discharge and precipitation for up to 3 or 4 preceding days are included. Higher orders do increase multiple correlation coefficients (R) as a measure of the model goodness, but these increases are usually statistically insignificant. However, when a drainage area is large (50 or more square kilometers) and/or there is a significant delay in the influence of precipitation on the spring discharge, an increase in its CR part may considerably improve the validity of an ARCR model.

In this case, the combination of AR and CR models significantly improves validity of the new ARCR model for the wet year. For the dry year, this merging increases the coefficient of multiple regression insignificantly, since its value for the simple AR model of all tested orders is already close to 0.98. The ARCR model is given as

$$Q_t = a + b_1 Q_{t-1} + \cdots + b_p Q_{t-p} + c_1 P_{t-1} + \cdots + c_q P_{t-q} \tag{6.6}$$

where the notification is the same as in Eqs. (6.4) and (6.5).

The model order (p, q) can be chosen from various combinations of its autoregressive (p) and cross-regressive (q) parts as shown in Fig. 6.13. For the wet year, the model coefficient (R^2) has maximum for $p = 3$ and $q = 3$. A further increase in the orders of AR and CR parts results in slight negative changes of R^2 that remains close to 0.925. The model for the dry year does not have a single maximum of the multiple regression coefficient. The AR part has the first abrupt positive change for the order of 2 and then for 7 when it also reaches the first maximum. R^2 increases again for $p = 14$ and 15. However, these changes of the coefficient of multiple regression are with an insignificant magnitude, that is, less than 0.001. Therefore, introducing more variables into the autoregression by increasing the AR order does not increase the model validity. Generally speaking, when a time series is highly autocorrelated (as in this case), AR models of first and second orders are considered sufficient in hydrologic practice.

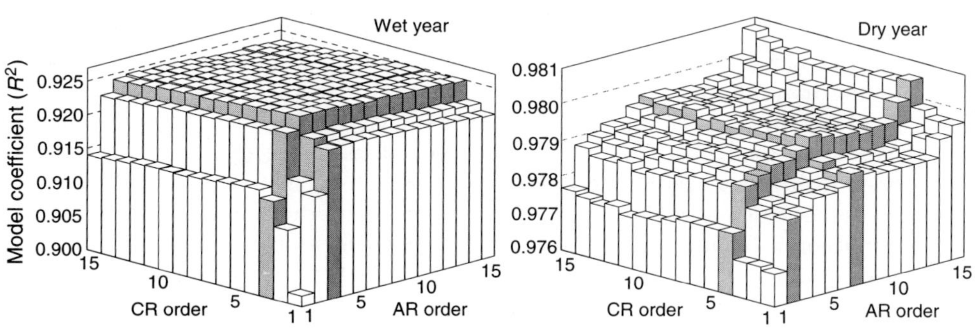

FIGURE 6.13 Dependence of the ARCR model coefficient (R^2) on the order of cross-regressive and autoregressive parts. (From Kresic, 1997; copyright Taylor & Francis, reprinted with permission.)

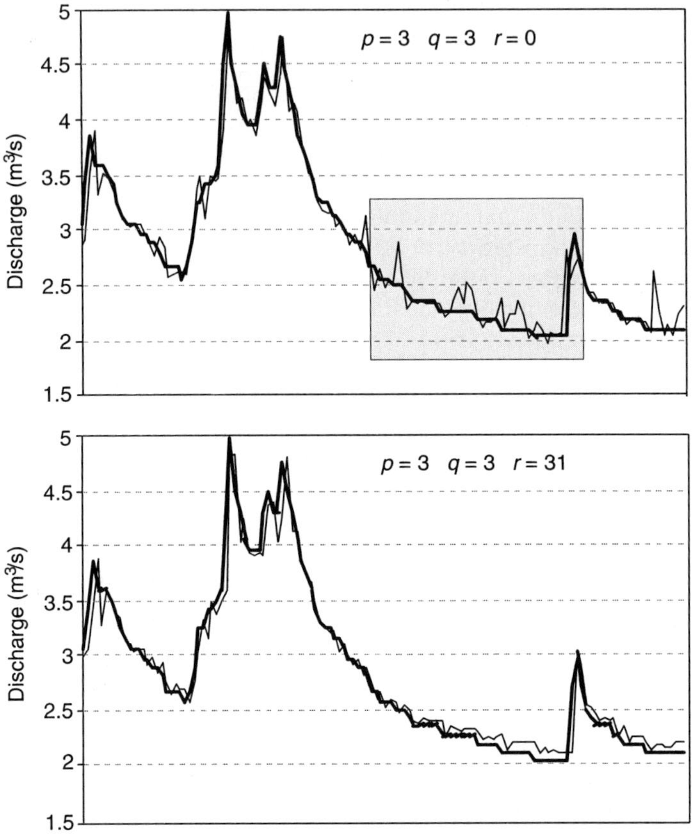

Figure 6.14 Recorded (*bold line*) versus predicted discharge of the spring for the summer period of a typical wet year. *Top*: Result of the ARCR model without the moving average filter ($r = 0$). Both the autoregressive and the cross-regressive parts are of third order. The outlined area is also shown in Fig. 6.15-left. *Bottom*: Result of the application of a 31-day moving average linear filter on the time series of total precipitation in the spring drainage area. (From Kresic, 1997; copyright Taylor & Francis, reprinted with permission.)

The model's CR part for the dry year is consistent and its order of 4 has the most significant increase throughout the analyzed domain (see Fig. 6.13). The ARCR model order for the dry year can be chosen as $p = 2$, and $q = 4$, that is (2,4). The model-predicted versus recorded spring discharge for a part of the wet year is shown in Fig. 6.14. When the model uses actual total precipitation, the predicted values differ significantly from the measured ones during the recession period (top hydrograph in Fig. 6.14). This is because the model gives the same weight to all precipitation data including isolated summer storms. It does not account for increased evapotranspiration and moisture deficit in the unsaturated zone or other factors during the summer months that reduce effective precipitation.

Correct estimation of the effective (net) precipitation is a major difficulty in hydrologic modeling of rainfall–runoff relationship. This is particularly true for the groundwater systems where the rainfall infiltration may be significantly attenuated by the porous media. The response of aquifers to recharge episodes varies throughout the year

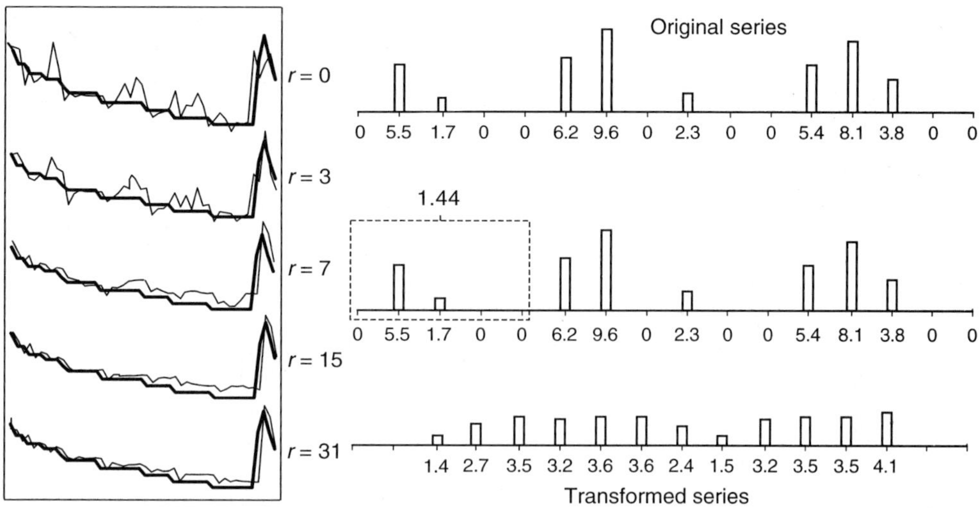

FIGURE 6.15 *Left*: Results of the increasing (*top to bottom*) moving average window on the predicted discharge of the spring (*thin line*). The thick line is the recorded spring discharge. *Right*: Application of a linear moving average filter for the transformation of precipitation series. Note that the total number of data in the transformed series is reduced by $m - 1$, where m is the length of the filter window ($m = 5$ in this case). (From Kresic, 1997; copyright Taylor & Francis, reprinted with permission.)

depending on the conditions and thickness of the unsaturated zone, water table levels, position of the discharge areas, and other factors.

A simple way to mathematically describe the transformation of total precipitation is to transform data using a moving-average linear filter. Figure 6.15 shows the results of such transformation with different moving averages, and the principle of transformation with the moving-average window of 5.

The ARCR model incorporating the modified precipitation input data is given as

$$Q_t = a + b_1 Q_{t-1} + \cdots + b_p Q_{t-p} + c_1 P_{r,t-1} + \cdots + c_q P_{r,t-q} \tag{6.7}$$

where r is the length of the linear filter window used to transform the precipitation series. The other notification is as in Eqs. (6.4) and (6.5).

The application of the moving-average filter actually slightly decreases the model's coefficient of multiple regression: the reason why the visual determination of the filter length is preferred. It can be seen in Fig. 6.14 that for the periods of recharge both total and transformed precipitation give good predictions. The difference between the two is significant in the case of recession when the transformed precipitation produces a much better result.

A simple ARCR model can be built by preparing data using any spreadsheet program and multiple regression programs; the discharge and precipitation time series are arranged in a number of columns corresponding to the total number of ARCR orders and then shifted one lag at the time. For an ARCR(3,3) model, the spring flow is predicted based on precipitation recorded that day, and flows and precipitation for the preceding 3 days:

$$Q_t = a + b_1 Q_{t-1} + b_2 Q_{t-2} + b_3 Q_{t-3} + c_0 P_t + c_1 P_{t-1} + c_2 P_{t-2} + c_3 P_{t-3}$$

As mentioned earlier, the CR models generally have a low coefficient of multiple regression that is one measure of their validity. This is because the continuous daily spring flow is regressed on daily precipitation that is often zero. One way to transform the precipitation series is to apply a simple linear moving-average filter as shown in Fig. 6.15. In this example, the filter window is 5 which means that five data at the time are summed and their average is assigned to the central point in the window. The window then moves one position and the averaging is repeated. The resulting series is smoother: the high precipitation values decrease and the low values (including zeros) increase. The new transformed series is then included in the ARCR model instead of the total precipitation. There are various other mathematical and statistical filters that can be tested and used to produce a better fit (e.g., see Fig. 6.7).

ARCR models give valid future predictions for only one step at the time (e.g., next day) because for the second and following intervals the true values of the output (spring discharge) for the preceding days are not known, and the model confidence and prediction limits rapidly deteriorate. This is also true for any other type of time-series models.

6.3 Convolution and Transfer Functions

Application of convolution and various mathematical functions that describe transformation of an input (such as rainfall) into an output (runoff) has been popular among surface water hydrologists since the late 1960s. As described briefly in Sec. 2.6 with Figs. 2.20 and 2.21, this approach divides precipitation into (1) precipitation excess and (2) infiltration and other losses. The precipitation excess produces direct storm runoff. In karst hydrogeology, the precipitation excess is equivalent to rapid infiltration to the saturated zone that produces the first response at the water table or a spring. Any excess infiltration after soil moisture is satisfied forms recharge to groundwater, which eventually emerges as baseflow of surface streams or discharge of springs.

The presence of the threshold in the soil storage phase of the water system makes it impossible to treat the whole system as linear, even where the evaporation and transpiration are completely known. In surface water hydrology, the development of the unit hydrograph theory as a linear relationship between precipitation excess and storm runoff avoided this difficulty by the elimination of the baseflow and the infiltration. It is the existence of this threshold—rather than the difference in response time between the surface response and the groundwater response—that necessitates the separation. In applied hydrology, the full model shown in Fig. 2.21 (Sec. 2.6) is seldom used. Rather, the baseflow is separated from the total hydrograph in some arbitrary fashion, and the precipitation excess is then taken so as to be equal in value (volume) to the storm runoff. If soil moisture accounting is included in the analysis, then the inherent threshold effect in soil moisture storage must be defined.

As emphasized by Dooge (1973), if there is desire to consider the whole system shown in either Fig. 2.20 or 2.21, then one would of necessity have to deal with a nonlinear system. This brings all the difficulties of nonlinear mathematics. It is not surprising, therefore, that the concentration in both surface water and groundwater hydrology has been on individual components shown in Fig. 2.21. Various unit hydrographs techniques (impulse response functions) have been developed for dealing with the direct response in runoff, and they are all based on the assumption of linear behavior. It is only relatively recently that the nonlinear system modeling of spring response to recharge has taken the lead (see Jukić and Denić-Jukić, 2006). The unsaturated phase involving soil moisture storage remains the most difficult part of the hydrologic cycle to handle. Not only does a threshold exist but also there is a feedback mechanism because the state of the soil moisture determines the amount of infiltration.

The foundation of the most commonly applied methods of runoff and spring discharge modeling using linear system approach are the following three principles: (1) superposition, (2) time-invariance, and (3) lumping of the system inputs and outputs. In reality, none of the three principles is valid but they greatly simplify the procedure and may produce satisfactory results (Dooge, 1973). Superposition means that any number of inputs (x) can be added together so that the output (y) will be the sum of the individual corresponding outputs. The system linearity defined by the principle of superposition must be distinguished from the existence of a general linear (i.e., a straight line) functional relationship between input and output. A system is said to be time-invariant when its parameters do not change with time. For such a system, the form of the output depends only on the form of the input and not on the time at which the input is applied.

The problems of systems analysis and synthesis are also greatly simplified if the input and output of a system are assumed to be lumped. In a lumped system with a single input and a single output, the behavior of the system would be described by a set of differential equations. If the inputs and outputs are not lumped, then the system behavior must be described by partial differential equations that are much more difficult to handle than ordinary differential equations.

The mathematical formulation for a lumped linear time-invariant system is based on the concept of an impulse (input) function and the impulse response (output) as follows (Dooge, 1973):

$$y(t) = \int_{-\infty}^{\infty} h(t-\tau)x(\tau)d\tau \tag{6.8}$$

where $y(t)$ is the output in time t, $x(\tau)$ is the input in time τ represented by the orthogonal coefficients of the impulse or delta function, and $h(t-\tau)$ is the impulse response function.

The right-hand side of Eq. (6.8) represents the well-known mathematical operation of convolution, which is often expressed by an asterisk:

$$y(t) = h(t)^* x(t) \tag{6.9}$$

The impulse response function in Eq. (6.9) is called instantaneous unit hydrograph (IUH) in hydrology; in mathematics, it is called a *kernel function*.

For two discrete finite series that are related (e.g., rainfall and spring discharge), the convolution equation becomes (Dooge, 1973; Dreiss, 1982, 1989a):

$$y_i = \Delta t \sum_{j=0}^{i} x_j h_{i-j} + \varepsilon_i \quad i = 0, 1, 2, \ldots N \tag{6.10}$$

where N is the number of sampling intervals of equal length Δt, y_i is the mean value of the output during the interval i, x_j is the mean value of the input during the interval

j, h_{i-j} is the kernel function during the interval $i - j$, and ε_i is error due to inherent nonlinearities in the system and measurements errors.

If x_j and h_{i-j} are known, then y_i can be determined directly by convolution. If x_j and y_i can be identified, then h_{i-j} (kernel function) can be determined through deconvolution (Dooge, 1973; Dreiss, 1982, 1989a). The solution of Eq. (6.10) is obtained by minimizing the sum of the square errors

$$\sum_{i=0}^{N} \varepsilon_i^2 \Rightarrow \min \tag{6.11}$$

assuming that the discrete kernel function is nonnegative

$$h_k \geq 0 \qquad k = 0, 1, 2, \ldots M \tag{6.12}$$

where M is the system memory or the length of the period in which the impulse (input) has the influence on the output. The solution is obtained when the total volumes of the input and the output series are equal to each other, that is, when the area under the kernel function is equal to 1:

$$\Delta t \sum_{k=0}^{M} h_k = 1 \tag{6.13}$$

System identification can also be performed using correlation methods as described by Dooge (1973) who gives the following form of the convolution equation connecting two discrete time series and optimized in the least squares sense:

$$\phi_{xy}(k) = \sum_{j=0}^{j=\infty} h_{\text{opt}}(j)\phi_{xx}(k - j);\ \text{when } k > 0 \tag{6.14}$$

where ϕ_{xy} is cross-correlation of the two series that is defined as the limit:

$$\phi_{xy}(k) = \frac{1}{n} \sum_{i=-p}^{p} x(i)y(i + k) \tag{6.15}$$

as p tends to infinity. ϕ_{xx} is the autocorrelation of a time series defined as the limit:

$$\phi_{xx}(k) = \frac{1}{n} \sum_{i=-p}^{i=p} x(i)x(i + k) \tag{6.16}$$

where $n = 2p + 1$ is the number of data points as p tends to infinity.

As illustrated in Fig. 6.16, the impulse response function (kernel) describes the transfer of the fast discrete input such as fast infiltration, into the output such as spring discharge. The principle of superposition allows that the simulated unit hydrographs from isolated rainfall episodes (unit impulses) be combined into the real (complex) spring

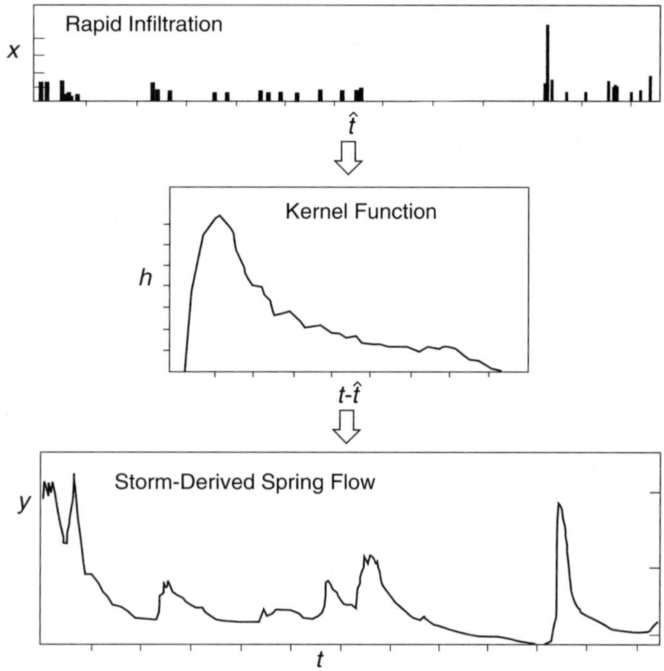

FIGURE 6.16 Schematic diagram of linear systems analysis of karst spring flow. (From Dreiss, 1989a; copyright American Geophysical Union.)

hydrograph. The kernel identification should be performed on as many isolated hydrographs as possible for which the precipitation episodes that caused them can be defined. It may be possible to apply one average kernel or try with a selective application of several kernels for different seasons if they are notably different.

One of the main limitations of convolution in simulating spring discharge is that long periods of baseflow (without precipitation) longer than the base of kernel(s), that is, longer than the system memory, cannot be simulated because the kernel will calculate zero flows. In addition, most commonly applied numeric methods generally have problems in accurately describing tails of recession hydrographs. In such cases, Snyder (1968) suggests that unit hydrographs be added to the previously separated baseflow so that the beginning of the output hydrograph does not start from zero. In order to avoid complicated algorithms, Canceill (1974) proposes adding a simple threshold with a decreasing trend into the convolution:

$$y(t) = \int_{-\infty}^{t} h(t-\tau)x(\tau)d\tau + C_k e^{-\infty(t-t_k)} \tag{6.17}$$

where t_k is the date of the last effective (nonzero) rainfall and C_k is parameter to be optimized. The main idea behind Eq. (6.17) is to separate two types of discharge: the linear fast flow component (impulse response) and the slow, nonlinear baseflow during recession as described by Maillet [see Eq. (2.23)].

Dreiss (1989b) identifies kernels of different moments by combining the results of a tracer test performed at Maramec Spring in Missouri, the United States with a long-term, 12-month sequence of naturally occurring regional tracer events and, three short-term subsets of the record.

A new form of the transfer functions for karst aquifers, the so-called composite transfer function (CTF) has been proposed by Denić-Jukić and Jukić (2003). The CTF simulates discharges by two transfer functions adapted for the quick flow and the slow flow hydrograph component modeling. A nonparametric transfer function (NTF) is used for the quick flow component. The slow flow component is modeled by a parametric transfer function that is an IUH mathematically formulated and defined from a conceptual model. By using the CTF, the irregular shape of the tail of the identified transfer function can be avoided, and the simulation of long recession periods as well as the simulation of a complete hydrograph becomes more successful. The NTF, the Nash model, the Zoch model, and other similar conceptual models can be considered separately as simplified forms of the CTF. The rainfall–runoff model based on the convolution between rainfall rates and the CTF was successfully tested on the Jadro Spring in Croatia. The results of the application were compared with the results obtained by applying NTFs independently.

6.4 Time-Series Models

A time series is a series of time-dependent hydrologic variable such as the flow rate in a surface stream or at a spring. When analyzing a time series, one deals with a limited amount of recorded data—a sample. This sample, regardless of its size, consists of a limited number of realizations of the same hydrologic process. All possible realizations of that process constitute a population. The goal of most hydrologic and hydrogeologic studies is to understand and quantitatively describe the population, as well as the process that generates it, based on a limited number of samples measured in the field (Gabric and Kresic, 2009; Kresic, 1997).

A time series can be continuous (such as the flow rate of a spring) or discrete (such as daily precipitation). For practical and computational purposes, most continuous time series are converted into discrete time series by introducing the recording (or modeling) time interval such as 1 day, 1 week, or 1 month. When a time series is described with statistical and probabilistic parameters, it represents a probability of occurrence (realization) of one of its possible stages. A good example is a time series of monthly precipitation at a certain location in a moderate climate. Our long-term experience can tell us that, for example, April through June is the wet period, and July through September is the dry period of the year. Accordingly, it can be expected, with a high probability, that in the near future (say, next year), these two periods will again last about the same time. However, no one can state with 100% accuracy that this will indeed happen (e.g., June may be an unusually dry month next year), because it is impossible to accurately predict the annual or monthly amount of precipitation using some physical laws of nature.

One can only apply tools of statistics and make predictions about the future using probabilistic models based on past data. A time series studied in this way is called a stochastic time series. In contrast, the stage of a deterministic process at time t is defined with certainty knowing its stage at some earlier time t_0. In other words, a deterministic process is described with physical laws rather than laws of probability. An example is the flow of groundwater from point A to point B when described with equations such as the Dupuit equation, the Laplace equation, or the Theis equation, to name just a few. Quantitative hydrogeology is based on the physical laws of groundwater flow as are the traditional numeric models presented in the next section.

Strictly speaking, most time series in hydrologic and hydrogeologic studies are stochastic since they depend on at least one random variable, with precipitation often being the most important one. They can be described mathematically (statistically) in time domain or in spectral (frequency) domain as explained in detail in the subsection "Precipitation" in Sec. 1.7.1 [see Eqs. (1.81) through (1.94)]. The most widely cited general reference on time-series analysis is the classic book by Box and Jenkins (1976), and the first comprehensive reference on the stochastic time-series analysis in hydrology is the classic book by Yevjevich (1972). Both are highly recommended by the author to readers interested in application of time-series models in karst. An informative public-domain publication by the Unites States Geological Survey (USGS), available for free download at the Survey's Web site, explains in detail the fundamentals of stochastic signal processing and its application in hydrology (Westlake and Dracup, 1980).

In general, a time series has the following five components, all of which may or may not be present (adapted from McCuen and Snyder, 1986):

- *Trend*, which is a tendency to increase or decrease continuously for an extended period of time in a systematic manner. This component can often be described by fitting a functional form such as a line or polynomial. The coefficients of the equation are commonly evaluated using regression analysis. The trend is also referred to as a deterministic component even though its physical explanation may not always be clear.
- *Periodicity*, which is very common in hydrologic time series: annual and seasonal periodicity of precipitation, temperature, flows, pumping from the aquifer. The period(s) in time series can be identified using a moving-average analysis, an autocorrelation analysis, or a spectral analysis, after which it is described by one or more trigonometric functions.
- *Cycle*, which occurs with an irregular period and is hard to detect (e.g., hydrometeorological time series are thought to be influenced by sunspot activity that has an irregular period).
- *Episodic variation*, which results from extremely rare or one-time events such as hurricanes. Identification of this component requires supplementary information.
- *Random fluctuations*, which are often a dominant source of the variation in time series and are the main target of a probabilistic identification.

Figure 6.17 illustrates various time-series components with a small sample of actual spring hydrographs created from long-term records kept by the USGS.

A generalized framework for a time-series model development consists of three phases: (1) identification, (2) parameter estimation, and (3) verification/diagnostic checking. Model identification is not a standardized, automated procedure but it is rather heuristic. The usual approach is an iterative trial-and-error procedure. The first step is to investigate if the time-series data is stationary and if there is any significant seasonality that needs to be modeled. A visual inspection of the time plot of the historic times

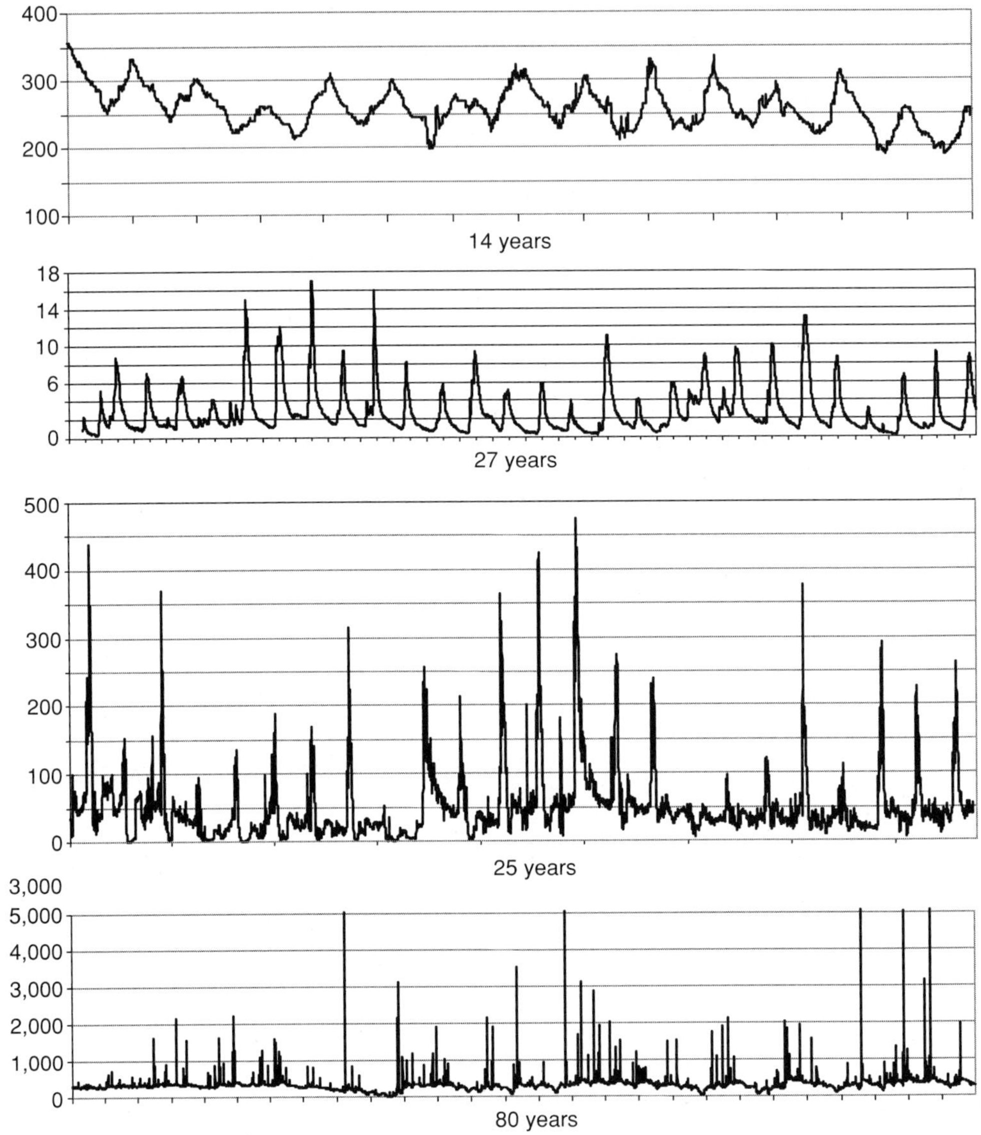

FIGURE 6.17 Hydrographs of springs showing various components of time series—trends, periodicity, cycles, episodic variation, and random fluctuations. All flow rates are in cubic feet per second. (Source of data USGS.)

series can help in deciding between seasonal and nonseasonal models, whether local differentiation is needed to produce stationarity, and to acquire a general feeling about the order of possible models. A further identification process is to examine the shapes of autocorrelation and the partial autocorrelation functions of the historic time series. To allow for the possible identification errors, a set of several models with a close structure should be considered. It is advisable to always select the simplest acceptable model.

Stochastic models describe time series formally (as black boxes) and do not consider their physical nature. Simply stated, they statistically (mathematically) analyze the past of the time series, as system input, and then predict the present or the future as the system output (so-called AR models). They can also analyze the past of one time series and use it to predict the present and future of some other time-dependent series proven to be correlated with the first one. Stochastic models can also combine several inputs and give one or several outputs. Examples would be a model that predicts water table elevation based on its position in the past, a model that predicts spring discharge based on its own past and the antecedent effective precipitation, or a model that also includes past stages of a nearby river.

Because of their relative simplicity and an abundance of mathematical approaches and computer programs developed for general applications in many different scientific and engineering fields, various time-series models in karst have been very popular for decades. Two main applications of the time-series models are the generation of synthetic samples and short-term forecasting of hydrologic events. Generated time series, which are statistically indistinguishable from historic time series, also often serve as input to the analysis of complex water resources systems. They can be used to provide a probabilistic framework for engineering analyses and design. Generated series show many possible hydrologic conditions that do not explicitly appear in the historic record. Consequently, using synthetic time series, different designs and operational schemes can be tested under many different conditions contained in these time series. Data generated from known historic observations can help in evaluating options for a real-time system operation. However, as emphasized earlier, the main limitation of time-series models is that they cannot provide definitive answers to common questions associated with water management decisions such as installation of new well fields, artificial recharge, or development of a new subdivision, to name a few. This is because these new conditions have not been observed yet and there is no data to identify the required model parameters.

The general form of an input–output stochastic model of a discrete time series (or a continuous time series transformed into a discrete one) with the same recording time interval is

$$y_t = f(x_t, x_{t-1}, x_{t-2}, \ldots; y_{t-1}, y_{t-2}, \ldots; \theta_1, \theta_2, \ldots) + \varepsilon_t \tag{6.18}$$

where f is the selected mathematical function; y_t is the predicted output at time t; y_{t-1}, $y_{t-2}\ldots$ are the successive members of the output time series recorded at corresponding time intervals t, t–1, t–2; x_t, x_{t-1}, $t_{t-2}\ldots$ are the successive members of the input time series recorded at time intervals t, t–1, t–2; θ_1, θ_2,... are the model parameters found by mathematically minimizing the differences between estimated (calculated) and observed y_t values; and ε_t is the model error (residual) given as the difference between the calculated and the recorded value of the output series at time t.

Stochastic modeling generally follows the approach proposed by Box and Jenkins (1976), who introduced ARMA models. The mathematical formulation of ARMA models is

$$z_t = \sum_{j=1}^{p} \phi_j z_{t-j} + \sum_{j=0}^{q} \theta_j \varepsilon_{t-j} + \varepsilon_t \tag{6.19}$$

where z_t represents the time-dependent series with mean zero and variance one; $\theta_1, \ldots \theta_p$ are time varying AR coefficients; $\theta_0, \ldots \theta_q$ are time varying moving-average coefficients; and ε_t is an independent normal variable.

Time-series models used to generate synthetic time series can be classified into autoregressive models [AR(p)], moving-average models [MA(q)], and their combination, [ARMA(p,q)] with variations such as autoregressive integrated moving-average models [ARIMA(p,d,q)] and others, where p and q are the orders of autoregressive and moving-average terms, respectively, and d is the differentiation order.

An AR model estimates values for the dependent variable Z_t as a regression function of previous values $Z_{t-1}, Z_{t-2} \ldots . Z_{t-n}$. A moving-average model is conceptually a linear regression of the current value of the series against the white noise or random shocks of one or more prior values of the series. PureAR model, commonly called Thomas–Fiering model, has been extensively applied in hydrology for modeling annual and periodic hydrologic time series. Because the parsimony ("the less the better") in the number of parameters is very desirable (since the parameters are estimated from data), the second order of these models is usually the highest lag necessary in representing hydrologic time series. A parsimonious model can be achieved using a mixed ARMA model as combination of a moving-average process and an AR process rather than a pure AR or MA model. Therefore, low-order models have been widely used in hydrologic practice (Graupe et al., 1976; Armbruster, 1979; Salas et al., 1982; Salas, 1993; Weeks and Boughton, 1987; Padilla et al., 1996; Montanari et al., 2000). More sophisticated stochastic models can include various linear and nonlinear transformations of input series, consideration of periodicity (seasons), and filtering (minimizing) of the model residuals. One widely used linear filter for the residuals is the Kalman filter that usually significantly improves ARMA models (Kalman, 1960; Bierkens et al., 2001).

When an aditional input time series is included in the model, such as rainfall, the models are called ARMAX. Application of an ARMAX model for estimating probability of the minimum daily spring discharge for each month of the year is discussed by Kresic et al. (1993). On the basis of 4-year observation of daily flow rates and a 30-year observation of daily precipitation in the spring drainage area, the ARMAX model was used to simulate 30 years of spring discharge. Different orders of the model were tested for reliability as illustrated in Fig. 6.18. For all 4 years, the model of the fourth order for both the internal (spring discharge) and the external (rainfall) inputs had the highest regression coefficient; the reliability of the model increased with the decreasing precipitation

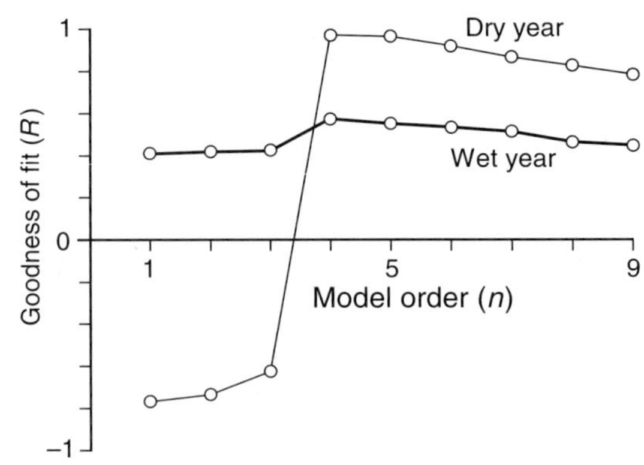

FIGURE 6.18 Dependence of ARMAX model goodness of fit (expressed through regression coefficient R) on the model order, n. (From Kresic et al., 1993.)

Year	Deviation from Average Annual Precipitation (%)	Model Goodness of Fit (*R*)	Model No.
1980	17	0.59	1
1981	−3	0.72	2
1982	−22	0.92	3
1983	−24	0.96	4

From Kresic et al. (1993).

TABLE 6.1 Dependence of ARMAX Model Reliability on Annual Precipitation in the Grza Spring Drainage Area

as shown in Table 6.1. The simulation of the spring discharge for a 30-year period was performed with different ARMAX models. Model no. 2 was used for years with precipitation within 10% of the average. Dry years with precipitation lower by 10% or more than the average were simulated with Model no. 4, and wet years with precipitation exceeding the average by 10% or more were simulated with Model no. 1. Thirty calculated values of the minimum daily discharge for each month were fitted to six different probability distributions. The results for May (generally the month with highest spring discharge) and August (month with generally lowest discharge) are shown in Fig. 6.19.

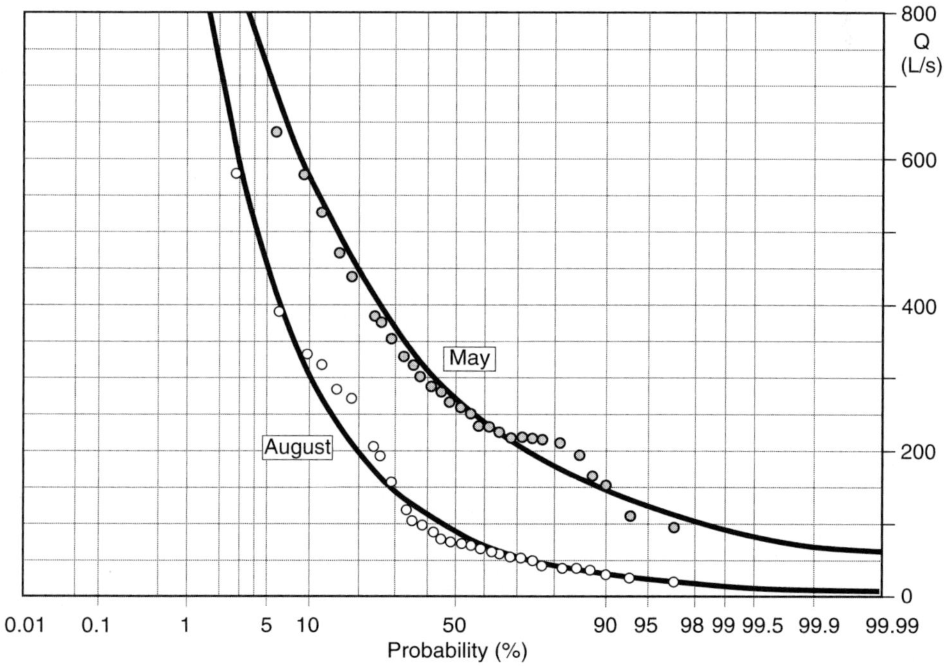

FIGURE 6.19 Theoretical probability distributions (log-Pearson III) of the spring minimum discharges calculated by the ARMAX models. (From Kresic et al., 1993.)

The best fitting probability distribution for all months, except for January and December, is log-Pearson III.

An important aspect of stochastic modeling is the problem of nonstationarity in hydrologic time series. Stationarity is usually assumed when modeling annual time series. When dealing with monthly or weekly time series, seasonal nonstationarity is present and it may be necessary to use a model that has seasonally varying properties. Significant contributions in developing periodic models were made by Hirsch (1979), Salas et al. (1985), and Salas and Obeysekera (1992). For modeling seasonal time series, two approaches can be used. The first is a direct approach in which a model with periodic parameters is fitted directly to the seasonal flows. This method requires a considerable number of years of data. The number of coefficients involved can be very large. If available historical data are limited, the parameters are poorly estimated. Consequently, the main problem of all seasonal models with time varying coefficients is the lack of parsimony. The second approach is decomposition ("disaggregation") in which the seasonal flows are generated at two or more levels. For instance, the first level is modeling and generating annual flows, and the second level is their decomposition into seasonal flows based on a linear model. However, if the autocorrelation structure of a historical time series shows a significant periodicity, then seasonal models that explicitly incorporate a periodic structure must be used. If the seasonality of time series under consideration is in the mean and the variance, then such seasonality can be removed by simple seasonal standardization, and a stationary model can be applied. Another peculiarity of hydrologic processes is the skewed distribution functions observed in most cases. Therefore, attempts have been made to adapt standard models to enable treatment of skewness (Bras and Rodriguez-Iturbe, 1994).

Figure 6.20 illustrates the results of stochastic modeling of surface stream flow at different timescales in the Sangsoorakh karst basin, Iran. The models are based on a seasonal ARIMA and a deseasonalized autoregressive moving average (DARMA). For this study, stochastic approaches were used to develop weekly, monthly, and bimonthly

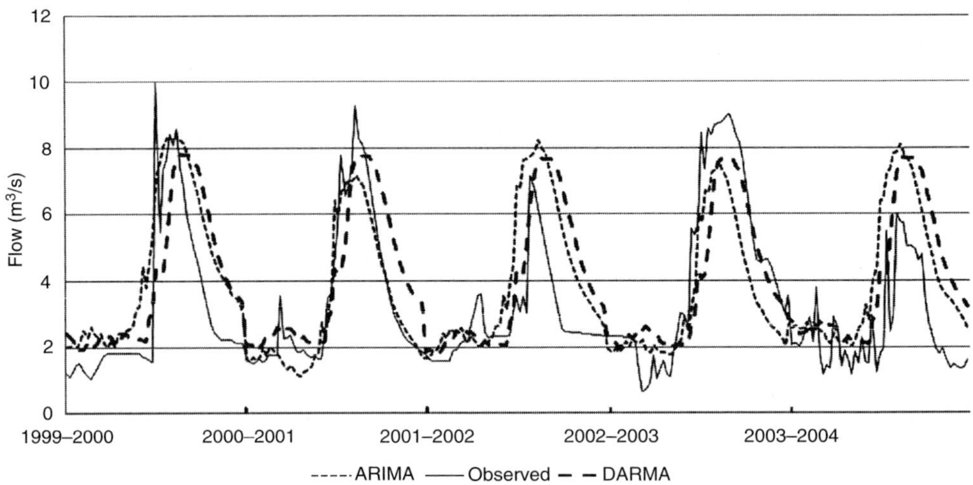

FIGURE 6.20 Comparison of observed and forecasted weekly flow using ARIMA and DARMA models for validation period 1999–2000 to 2003–2004. (From Ghanbarpour et al., 2010.)

flow simulation and forecasting models. As concluded by the authors, ARIMA models perform better in this particular case than the DARMA models (Ghanbarpour et al., 2010).

Time-series models in spectral domain or mixed time-spectral domains are also very popular in karst hydrologic and hydrogeologic studies. One advantage of using such models is their ability to incorporate multiple cyclic (periodic) input components at different frequencies such as well pumpage and operation of locks and dams on surface streams. Parameter identification performed for spectral-domain models may also reveal superimposed and therefore masked input and/or output components that may or may not be fully known. Quite a few papers have been published on the topic including Mangin (1981), Padilla and Pulido-Bosch (1995) Larocque et al. (1998), Labat et al. (2000), and Rahnemaei et al. (2005).

When the selected mathematical function f in Eq. (6.18) ignores any physical laws that govern the transformation of input(s) into output(s), the model is a pure stochastic one. If, in any form, the mathematical function incorporates physical laws, the model is called a stochastic-conceptual model or gray model. The knowledge of various physical processes and relationships related to the system of interest is invaluable and provides the physical background for stochastic modeling (Klemeš, 1978; Koch, 1985; Salas and Obeysekera, 1992; Kresic et al., 1993; Kresic and Busbey, 1995; Lee and Lee, 2000; Petrič, 2002; Jemcov and Petric, 2009). It is always preferable to make detailed structural and physical analyses of the hydrologic processes involved before performing stochastic modeling.

As mentioned earlier, there are many possible uses of time-series models. Typical applications of synthetic time series in surface water engineering are reservoir design, risk and reliability assessment, planning of hydropower production, and flood and drought hazard analysis. In groundwater studies, stochastic models can be used to analyze and forecast the hydraulic head fluctuations, fill-in data gaps, and detect and quantify trends (Ahn, 2000; Knotteres and Bierkens, 2000; Bierkens et al., 2001; Kim et al., 2005). Time-series models of spring discharge rates are becoming increasingly popular. They use observed flows and incorporate factors that influence it, such as groundwater levels, precipitation, evapotranspiration, and anthropogenic hydrologic disturbances such as pumpage from the aquifer.

6.4.1 Case Study: Jadro Spring, Split, Croatia

The following example, courtesy of Ivana Gabrić, Croatian Waters, illustrates the application of synthetic hydrologic time series in karst water resources management. Jadro Spring (Fig. 6.21), with an average discharge rate of 9.82 m^3/s, provides water supply for the Croatian port city of Split and its 300,000 inhabitants. The main characteristic of the spring discharge is large fluctuation in response to rainfall. During high discharge rates, there is more intense flushing of the soil and sediment accumulated in the subsurface, resulting in sudden, short-lived changes in water quality. Consequently, the springwater is characterized by the occasional occurrence of high turbidity exceeding regulatory standards. Turbidity is the key problem in water quality management of this spring as well as many other karst springs. For management purposes, it is important to know the nature of the turbidity and predict its occurrence as early as possible, because elevated turbidity is often associated with high bacteria counts indicating possible contamination. An accurate prediction of elevated turbidity can help optimize sampling strategies. In the case of Jadro Spring, the turbidity monitoring was not systematic, leaving insufficient

Figure 6.21 Jadro Spring used for water supply of the Croatian port city of Split. The spring was initially tapped in fourth century AD for the summer residence of Roman Emperor Diocletian and water was transferred via a viaduct. (Photographs courtesy of Split Waterworks; available at http://www.vodovod-st.ht; see also Fig. I.10 in Introduction.)

information for a reliable water supply system management. A stochastic time-series modeling can therefore provide for a more comprehensive understanding of the turbidity, using available short-term measurements of the turbidity and long-term recording of the spring discharge (Margeta and Fistanić, 2004; Rubinić and Fistanić, 2005).

Time-series analyses of discharge and turbidity show that the turbidity is higher during the first large rainfall following a dry period. Similar discharge rates following the first rainy period produce generally lower turbidities. Consequently, a reliable prognosis of turbidity can be carried out using different regression functions of discharge and turbidity for different parts of the year. A stochastic model was built based on 3 years of daily measurements of turbidity and discharge, and 28 years of daily discharge measurements, as illustrated in Fig. 6.22. Block A presents identification, estimation, and verification phases of the Jadro Spring stochastic modeling. Synthetic time series of monthly spring discharge were generated using a Thomas–Fiering AR(2) model with constant coefficients, which provided the best preservation of historic statistical characteristics of the spring discharge—seasonal means, variances, and correlations of the processes. Block B describes the procedure of the turbidity time-series generation. Based on the functional dependence between daily turbidity and discharge, the synthetic time series of the average monthly discharge (box A) are used to generate up to 100 time series of the average monthly turbidity. The time series of maximum daily turbidity are then generated using the time series of the average monthly turbidity.

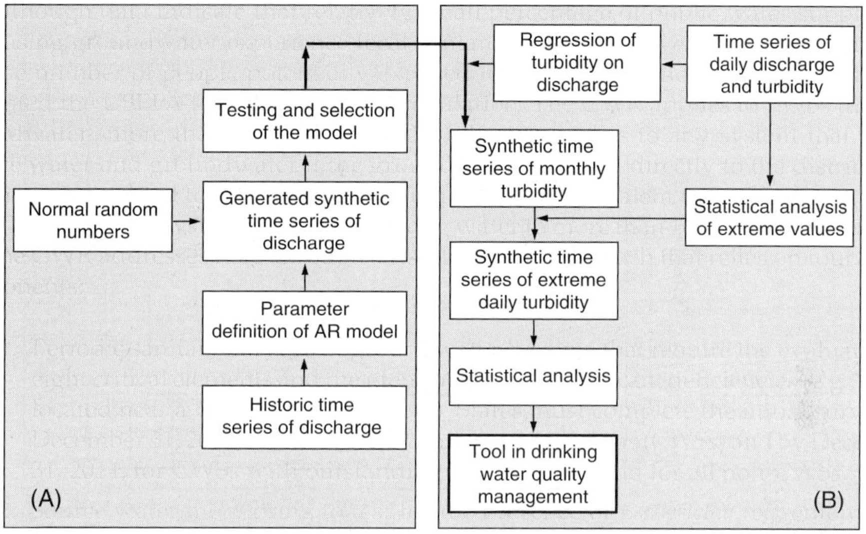

FIGURE 6.22 Building steps of the stochastic time-series model of daily turbidity at Jadro Spring. Explanation in text. (Courtesy of Ivana Gabrić, Croatian Waters.)

Figure 6.23 shows the maximum daily values of the generated turbidity at Jadro Spring as a function of the number and length of the synthetic time series. The analysis of the high turbidity occurrences indicates that their seasonal behavior is preserved in the model as shown in Fig. 6.24. The highest averages, as well as maximum turbidity values, usually appear during late autumn months, when rainfall after long summer droughts mobilizes fine sediments deposited in the subsurface.

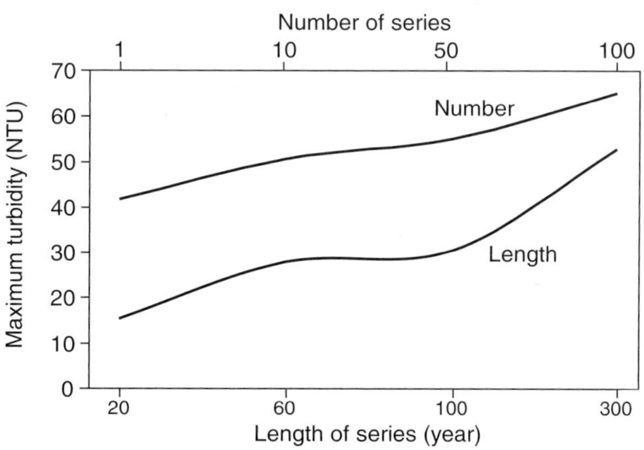

FIGURE 6.23 Maximum values of synthetic turbidity at Jadro Spring versus number and length of synthetic time series. (Courtesy of Ivana Gabrić, Croatian Waters.)

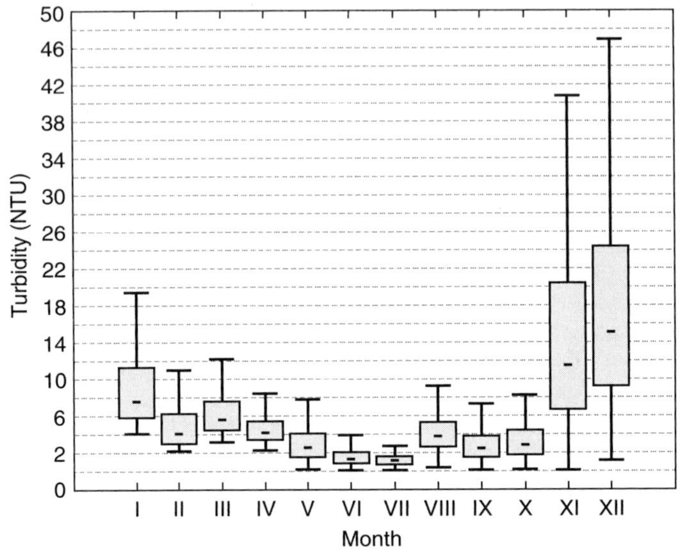

FIGURE 6.24 Box-whisker diagram of synthetic daily values of turbidity, without extremes and outliers. (Courtesy of Ivana Gabrić, Croatian Waters.)

6.5 Numeric Models

Numeric models describe the entire flow field of interest at the same time, providing solutions for as many data points as specified by the user. The area of interest is subdivided into many small areas (referred to as cells or elements) and a basic flow equation is solved for each model cell considering its water balance (water inputs and outputs). The solution of a numeric model is the distribution of hydraulic heads at points representing individual cells. These points can be placed in the center of the cell, at intersections between adjacent cells, or elsewhere. The basic differential flow equation for each cell is replaced (approximated) by an algebraic equation so that the entire flow field is represented by x equations with x unknowns where x is the number of cells. This system of algebraic equations is solved numerically, through an iterative process, thus the name *numeric models*. Based on various methods of approximating differential flow equations, and methods used for numerically solving the resulting system of new algebraic equations, numeric models are divided into several groups. The two most widely applied groups are finite differences (numeric models) and finite elements (numeric models).

Both types of models have their advantages and disadvantages, and for certain problems one may be more appropriate than the other. However, because they are easier to design and understand, and require less mathematical involvement, finite-difference models have prevailed in hydrogeologic practice. In addition, several excellent finite-difference modeling programs have been developed by the USGS and are in public domain, which ensures their widest possible use. One of these is MODFLOW (http://water.usgs.gov/nrp/gwsoftware/modflow.html), probably the most widely used, tested, and verified modeling program today, and it has become the industry standard thanks to its versatility and open structure: independent subroutines called modules are grouped into packages that simulate specific hydrologic features. New modules and packages can be easily added to the program without modifying the existing packages or the main code. USGS has also significantly upgraded its finite-element model SUTRA, now capable of simulating three-dimensional (3D) flow (Voss and Provost, 2002). This computer program can simulate both unsaturated and saturated flow, heat and contaminant transport, as well as variable density flow. Unfortunately, unlike MODFLOW, SUTRA3D is not yet part of any of the popular user friendly, graphical user interface (GUI) commercial programs for processing model input- and output data, which severely limits its greater application. In 2008, the USGS released a highly anticipated addition to MODFLOW-2005 called Conduit Flow Process (CFP) (Shoemaker et al., 2008) capable of simulating various flow types in karst aquifers explicitly using physically based approach.

The success of translating a conceptual site model (CSM) for a karst site into a numerical model will depend on both the experience of the user and the capabilities and limitations of the selected computer program. The main advantage of classic MODFLOW and the companion fate and transport models such as MT3D and RT3D is their broad user base and the continuous updates including frequent introduction of new modules and numerical solvers. Most modeling concepts today across various computer programs are therefore indirectly or directly based on those first implemented in MODFLOW such that an experienced MODFLOW user usually has little trouble when transitioning to another program. What also generally makes groundwater modeling more efficient than ever before is that modern-day computers and computer operating systems do not pose model size limitations for most practical applications. Models can have literally hundreds of layers and millions of cells and still be solved relatively quickly with desktop PCs.

However, despite its numerous advantages, even classic MODFLOW has some serious conceptual limitations that many practitioners have to deal with on a routine basis. Four limitations that immediately come to mind are (1) requirement that all model layers have to be continuous throughout the model domain; (2) model instability when trying to simulate large vertical displacements caused by faults or artificial structures; (3) all cells in the model must be rectangular, with rows and columns extending from one edge of the model to the other; and (4) model instability when simulating contacts between porous media with highly contrasting hydraulic conductivities. These limitations seriously affect attempts to simulate complex 3D geologic relationships and discontinuities characteristic of karst, as well as water fluxes between cells.

Fortunately, at the time of this writing, a true breakthrough in groundwater modeling has taken place in the form of a fundamentally new program called MODFLOW-USG (see Sec. 6.5.3 for a detailed description). MODFLOW-USG was developed by Dr. Sorab Panday of AMEC, and retains full compatibility with previous versions of MODFLOW while taking advantage of Un-Structured Grids and finite-volumes numerical solutions. The program is released in the public domain; data input and visualization of model results are at the time of this writing supported exclusively by the latest version of GUI program Groundwater Vistas. It enables hydrogeologists to accurately translate even the most complex CSMs into a numerical environment, thus eliminating the need for various surrogate-modeling solutions. This includes flow in fractured rock and karst aquifers.

Predictive numeric models are divided into two main groups: (1) models of groundwater flow and (2) models of contaminant fate and transport. The later ones cannot be developed without first solving the groundwater flow field of the system studied, that is, they use the solution of the groundwater flow model as the base for fate and transport calculations.

Regardless of the intended use or type, it is essential that for any groundwater model to be interpreted and used properly, its limitations should be clearly understood. In addition to strictly technical limitations, such as accuracy of computations (hardware/software), the following is true for any model:

- It is based on various assumptions regarding the real natural system being modeled.
- Hydrogeologic and hydrologic parameters used by the model are always just an approximation of their actual field distribution that can never be determined with 100% accuracy.
- Theoretical differential equations describing groundwater flow are replaced with systems of algebraic equations that are more or less accurate.

It is therefore obvious that a model will have a varying degree of reliability, and that it could not be misused as long as all the limitations involved are clearly stated, the modeling process follows industry-established procedures and standards, and the modeling documentation and any generated reports are transparent, also following the industry standards (Kresic, 2007).

For a thorough explanation of finite-element and finite-difference models, and their various applications, the reader should consult the excellent work by Anderson and Woessner (1992).

6.5.1 Equivalent Porous Media (EPM) Models

Until very recently, most if not all widely used computer programs for numeric groundwater modeling were lacking capabilities for a physically based simulation of the characteristic triple porosity of karst aquifers. Consequently, hydrogeologists and groundwater modelers were either avoiding numeric modeling of karst aquifers altogether or were engaged in searching for an EPM approach that may work. One such approach when using MODFLOW and similar codes based on Darcy's law includes assigning very high values of the hydraulic conductivity to those model cell known to contain or suspected to contain highly transmissive conduits. Figure 6.25 illustrates how this procedure may result in a reasonable representation of the preferential pathways and the field-observed hydraulic head distribution.

However, none of the EPM models can accurately reproduce the fast (conduit) flow component of the spring discharge after major rainfall events and the very important hydraulic interaction between the conduits and the surrounding matrix. Namely, as the hydraulic head rapidly rises in the conduits, there may be a very significant transfer of water into the surrounding matrix and storage of water provided that the matrix has some effective porosity. At the same time, the spring may react rapidly (sometimes in a matter of hours) to the propagation of pressure through the conduits caused by the newly infiltrated water. Finally, this newly infiltrated water may start discharging at the spring only several days after a localized rainfall in a remote portion of the spring drainage area.

In an attempt to simulate the apparent high groundwater velocity along the virtual conduit cells and the related spring response, some EPM modelers select to assign a very low effective porosity to those model cells, much lower than in the surrounding matrix cells, which is nonsensical (see Fig. 1.40 and the related discussion in Chap. 1). Even if this approach produces the desired effect (fast flow of groundwater), it will not result

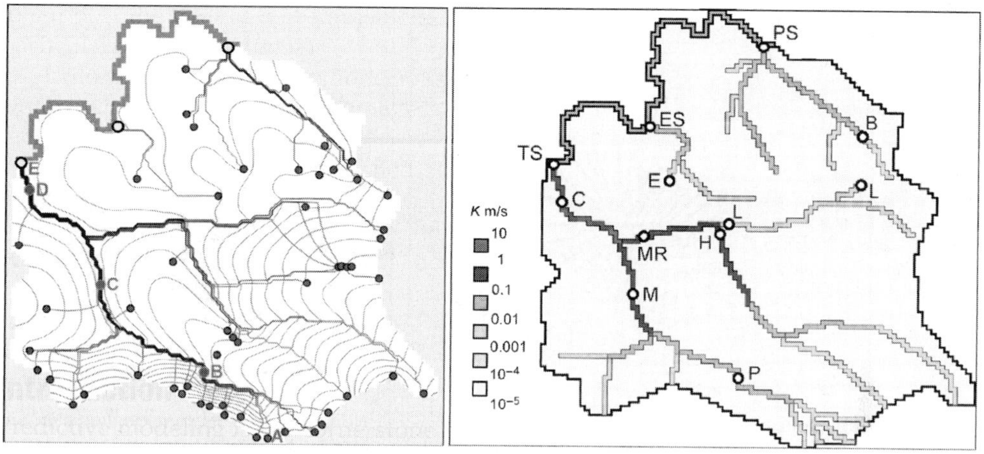

FIGURE 6.25 *Left*: Modeled hydraulic heads with an EPM numeric model of a karst aquifer showing good agreement between actual tracer paths and the hydraulic gradients. Small dark circles are dye-tracing locations. All tracer paths go to correct springs, shown with three larger circles along the river boundary (gray model cells) in the top left. *Right*: Hydraulic conductivity distribution that resulted in the map shown in the left figure. Note that portions of the virtual conduits simulated by the EPM model cells have the assigned hydraulic conductivity as high as 10 cm/s or 8640 m/day. (From Worthington, 2003.)

in a higher hydraulic head in the virtual conduit cells at any given time by default; the EPM model will keep the heads in these cells lower than in the surrounding cells since they are much more transmissive than the rest of the model area.

Figure 6.26 shows some of the results of contributing area delineation for a well field in the Central Swamp region near Tampa, Florida, using an EPM MODFLOW-based model with particle tracking. As stated by the authors of this USGS report (Knochenmus and Robinson, 1996), time-related areas of contribution were analyzed for six carbonate aquifer system types: an isotropic and homogeneous single-layer system, an anisotropic in a horizontal plane single-layer system, a discrete vertically fractured single-layer system, a multilayered system, a double porosity single-layer system, and a vertically and horizontally interconnected heterogeneous system. It is important to note that fractures simulated in the model were virtually represented by the model cells in which the flow equation was still for intergranular porous media (i.e., Darcian continuum).

The simulated aquifer anisotropy was 5:1 and was determined from program TENSOR2D results. The simulated virtual vertical discrete fracture network represents locations of photolineaments. The simulated enhanced flow zones were determined from borehole video and geophysical logs. The authors conclude that the distribution and nature of aquifer heterogeneities affect the size, shape, and orientation of areas of contribution in a karst carbonate aquifer system. The size of the 50-year time-related areas of contribution ranged from 8.2 to 39.1 square miles. Simulations showed that the size of areas of contribution is primarily affected by simulated withdrawal rates, effective porosity of the carbonate rock, and transmissivity. The shape and orientation of the simulated areas of contribution primarily result from aquifer anisotropy, well distribution, flow along virtual solution-enhanced zones, and short-circuiting of flow through model cells simulating fracture networks. Flow velocities and particle path length respond to

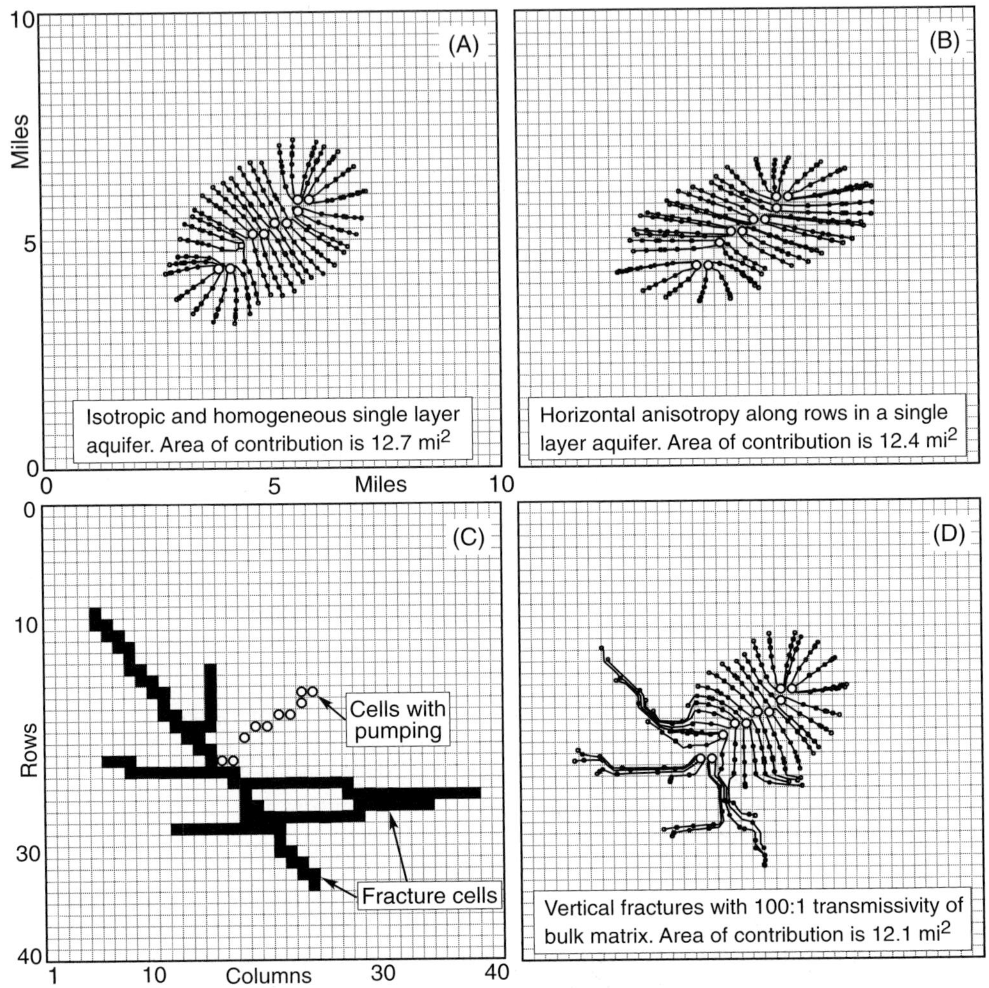

FIGURE 6.26 Influence of aquifer anisotropy and heterogeneity on the modeled capture zone for the Central Swamp region, Cypress Creek well field near Tampa, Florida. (A) Isotropic and homogeneous one-layer aquifer, (B) anisotropic hydraulic conductivity with five times greater value along rows, and (C and D) simulation of vertical fractures with "fracture" cells where transmissivity is 100 times greater than in the surrounding "matrix" cells. (Modified from Knochenmus and Robinson, 1996.)

withdrawal rates, simulated effective porosity, short-circuiting of flow by virtual fractures, and transmissivity. Results of simulations incorporating aquifer heterogeneity indicate that oversimplification of the flow system may result in erroneous definition of flow fields. For example, aquifer anisotropy typical of many carbonate aquifers creates elliptical flow fields. Circular protection zones do not adequately characterize the elliptical areas of contribution. Also, areas of contribution were larger when the carbonate aquifer system was simulated as a multilayered system with discrete flow zones (Knochenmus and Robinson, 1996).

Although numerous other examples illustrating many limitations of the EPM models in karst abound, the practice is still widely spread. In some cases, authors justify the EPM approach by stating that the porous media is relatively uniformly karstified such that the key model parameters can be assigned throughout the modeled layer(s). This includes common procedures for their spatial interpolation such as contouring with kriging. Floridan aquifer is a good example of literally dozens of EPM models created for various purposes and at various scales. In some cases, the EPM approach is defended by stating that the model is regional and therefore suitable for estimating various general water budget elements and not necessarily for accurately simulating groundwater flow velocities, local flow directions, and local drawdowns caused by pumping for example.

The problems with using EPM models in karst for water management usually arise when limitations of a model are not clearly stated and/or are ignored in a desire to have a quantitative management tool of some kind. Unfortunately, once a model is produced, often times it takes life of its own and is used by various stakeholders for various purposes and agendas, especially if it was created by an entity without apparent stakes in the related project or by an independent government agency such as USGS. However, in all such cases, the model documentation should be readily available for an independent review and prepared according to industry standards. This includes discussion on the CSM and karst development, model limitations, assumptions, purpose, and model errors (residuals), to name just a few mandatory sections. The following discussion provided by Putnam and Long (2009) illustrates these points.

The city of Rapid City, South Dakota, obtains more than one-half of its municipal water supplies from the Minnelusa Aquifer and from the karstified limestone Madison Aquifer (Fig. 6.27) through deep wells and springs. Numerous additional users in the Rapid City area obtain water from these two aquifers for domestic, commercial, industrial, and irrigation usage. The USGS, as part of a long-term cooperative study with the city of Rapid City, has compiled numerous datasets designed to better understand groundwater flow in the Minnelusa and Madison hydrogeologic units. A numerical EPM groundwater flow model was developed for a 1000-square-mile study area that includes these two units in the Rapid City and surrounding areas. The purposes of the modeling report prepared by the USGS are to (1) document the development of a numerical groundwater flow model of the Minnelusa and Madison hydrogeologic units, which contain the Minnelusa and Madison Aquifers, in the Rapid City area in South Dakota, and (2) present simulated responses to stress and describe model limitations.

As stated by the USGS, the model is limited by numerous simplifying assumptions. One of the most important simplifying assumptions is representation of a hydrogeologic system that contains substantial secondary porosity as a porous media. A bulk parameter that represents the average for a given model cell area provides only a generalized approximation of hydraulic properties, especially for the karstic Madison hydrogeologic unit. These simplifications probably result in a simulated hydraulic head response to stress that is more damped than observed responses. Site-specific responses in some areas could be substantially in error (Putnam and Long (2009).

Figures 6.28 and 6.29 illustrate some of the model calibration results presented in the USGS report. As can be seen, the model residuals (difference between the model calculated and the observed hydraulic heads) show substantial bias, especially for the Minnelusa Aquifer; the data points are within a narrow vertical band with an excessive vertical scatter and values exceeding 100 feet in either direction from zero. This indicates a rather poor calibration as it has been widely accepted in the industry that the model

FIGURE 6.27 Swallow hole in Madison Limestone in Spring Creek Canyon. (Photograph by Ingrid E. Arlton; from Putnam and Long, 2009.)

residuals should be randomly distributed (unbiased). The graph of simulated versus observed transient discharge at the Jackson-Cleghorn Springs (Fig. 6.29-left) is somewhat puzzling since the observed flow rate at the springs does not vary in time at all, which is quite unusual for any spring, not just in karst.

In any case, presumably in part based on the model calibration results, the USGS states that, with additional data, further refinement of the model would be possible, which could improve the accuracy of model estimates of the effects of additional stresses on the system, such as increased withdrawals or drought. The USGS concludes that the model can guide management decisions and planning and provides a useful tool for general characterization of the effects of stresses and management alternatives on a regional basis (Putnam and Long, 2009).

In addition to regional assessments such as the above example of Madison Aquifer, it may be equally or more important for a water management agency to understand model uncertainties and how they may impact future planning and decisions that will affect users at a local scale. For example, according to Putnam and Long (2009), site-specific responses predicted by the Madison karst aquifer model in some areas could be *substantially in error* (emphasis added). It is therefore unclear what the current utility of this model would be in predicting site-specific responses. This, for example, includes a reduced discharge or even drying of a particular spring due to an increased permitted groundwater withdrawal from the aquifer. If this uncertainty is not quantified in some

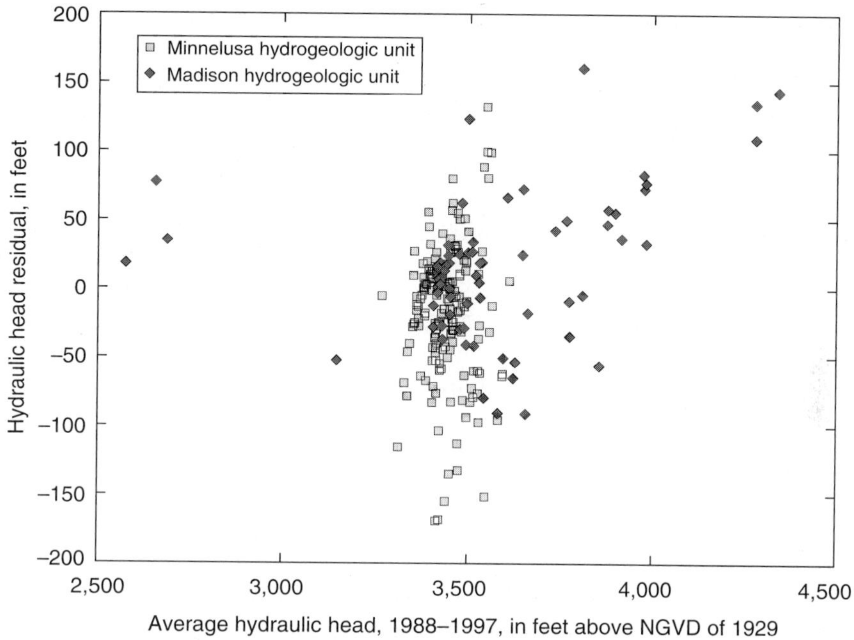

Figure 6.28 Residuals between simulated steady-state and observed hydraulic head values and their relation to average (water years 1988 to 1997) hydraulic head values. (Modified from Putnam and Long, 2009.)

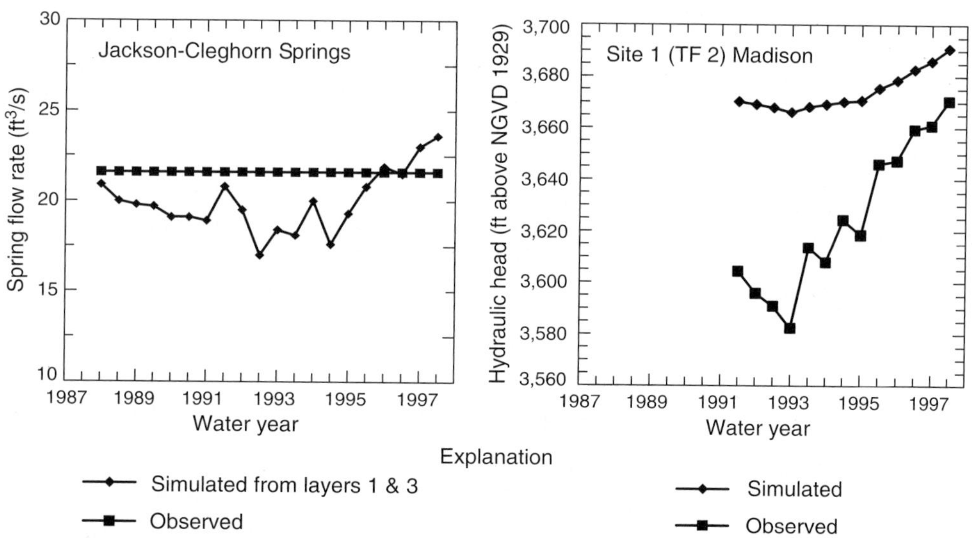

Figure 6.29 *Left*: Simulated and observed or estimated spring flow values. *Right*: Simulated and observed hydraulic heads. (Modified from Putnam and Long, 2009.)

manner, including limitations of an EPM model, the water management agency would not have any basis for asserting risks associated with management decisions.

As discussed by Kresic and Mikszewski (2012), a common problem with the application of complex numerical models by practitioners not well versed in either groundwater modeling or principles of groundwater flow in karst is not realizing that the numerical model may be producing results that contradict the CSM and/or basic physical principles. In other words, the numerical model will always do exactly what it was asked to do by its developer including producing impossible results. This is true even when the model developer firmly believes his or her CSM is correct and is properly translated into the numerical environment (computer program). The model developer may also firmly believe that the computer program is producing correct results because the conceptual model is correct. Unfortunately, if the whole process is not checked by independent, knowledgeable hydrogeologists and groundwater modelers, it would sometimes be impossible to find real reasons for questionable results created by a model if one were to rely only on a published modeling report and some colorful figures. A disconnect between the CSM, the numerical model developed from it, and the presentation of modeling results is illustrated with the following example.

A numerical groundwater flow model that incorporates important components of the latest information and conceptualization of the karstic Edwards Aquifer was developed by the USGS in cooperation with the U.S. Department of Defense and Edwards Aquifer Authority (Lindgren et al., 2004). This one-layer model includes both the San Antonio and Barton Springs segments of the Edwards Aquifer in the San Antonio region, Texas, and was calibrated for steady-state and transient conditions. Transient simulations were conducted using monthly recharge and groundwater pumpage (withdrawal) data. This EPM model was developed using MODFLOW and incorporates virtual conduits simulated as one-cell-wide zones with very large hydraulic conductivity values of as much as 30,00,00 (three hundred thousands) feet/day. The flow in the conduit cells is simulated with Darcy's equation, just like in all other cells in the model. As stated by the model authors, the locations of the virtual 1,320 feet (400 m) wide, and tens of miles long conduits were based on a number of factors, including major potentiometric surface troughs in the aquifer, the presence of sinking streams, geochemical information, and geologic structures (Lindgren et al., 2004).

Figure 6.30A shows a portion of the model domain with the simulated steady-state potentiometric surface of the Edwards Aquifer. The model's no-flow boundary is emphasized by annotation added to the original figure of Lindgren et al. (2004). This no-flow boundary behaves as expected based on general principles of groundwater flow described in most hydrogeology textbooks. In other words, the simulated potentiometric surface lines are perpendicular to the no-flow boundary. As a reminder, a no-flow boundary in hydrogeology and groundwater modeling is equivalent to an impermeable boundary—water cannot enter or leave the active model domain across this type of boundary in either steady-state or transient conditions.

Figure 6.30B shows the same portion of the model domain with the simulated potentiometric surface for transient, drought conditions in August 1956. The situation shown is completely different than in Fig. 6.30A. The potentiometric lines are parallel to the no-flow boundary. If one were to draw groundwater flow lines perpendicular to the potentiometric surface lines, it would appear that the no-flow boundary is now providing water to this one-layer model. This is emphasized by the thick black arrows and question marks added to the original figure of Lindgren et al. (2004).

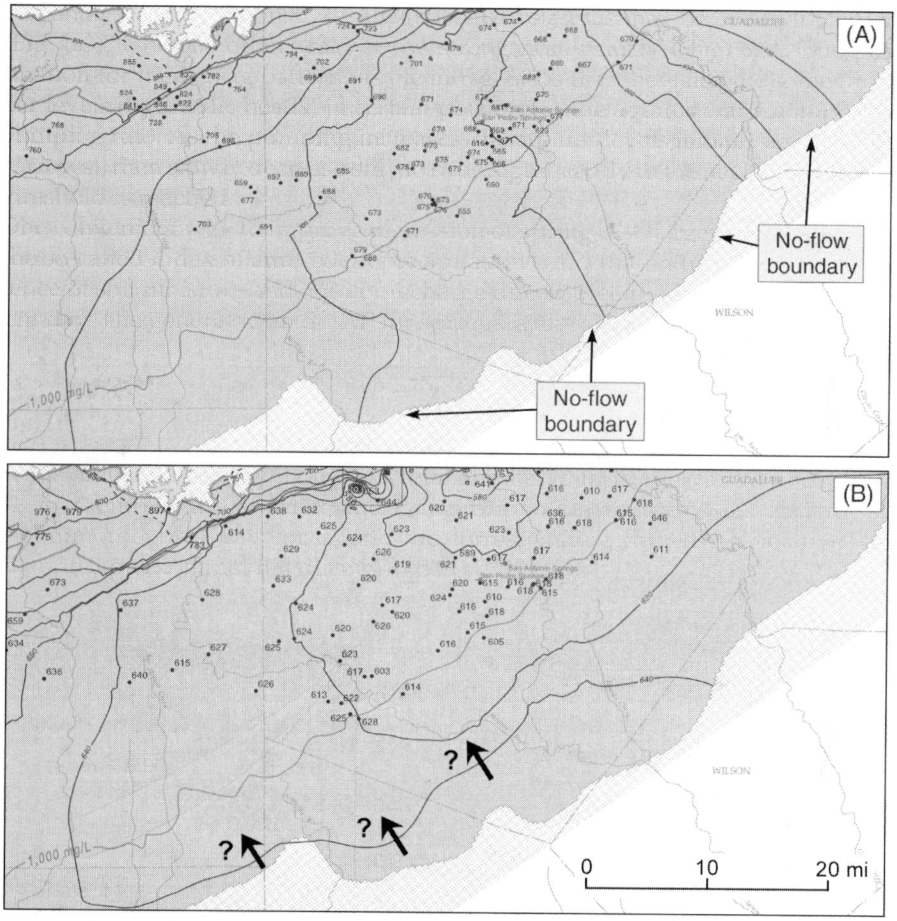

FIGURE 6.30 Portion of the model domain showing results of a numerical groundwater flow model developed for karstic Edwards Aquifer in Texas. (From Lindgren et al., 2004.) (A) Calibrated steady-state model with the model-predicted potentiometric surface contours; annotation added for emphasis. (B) Calibrated transient model for drought conditions of August 1956; thick black arrows and question marks are added for emphasis.

Figure 6.31 shows the same steady-state and transient conditions depicted in Fig. 6.30 when the virtual conduits are in the model (the conduits are represented by the jagged gray lines, originally red in the USGS report). Notably, the simulated potentiometric contour lines for the conduit conditions are omitted from the two figures. Instead, the authors of the report have chosen to schematically show groundwater directions using hand-drawn arrows. (Note that all arrows on Figs. 6.31A and 6.31B are original to the USGS publication.) The USGS arrows showing groundwater flow directions are perpendicular to the no-flow boundary as emphasized with added question marks in Fig. 6.31B, indicating that flow from the no-flow boundary is occurring.

The author of this book is not aware of any plausible explanation for the strange, hydraulically impossible behavior of the no-flow boundary in this one-layer EPM

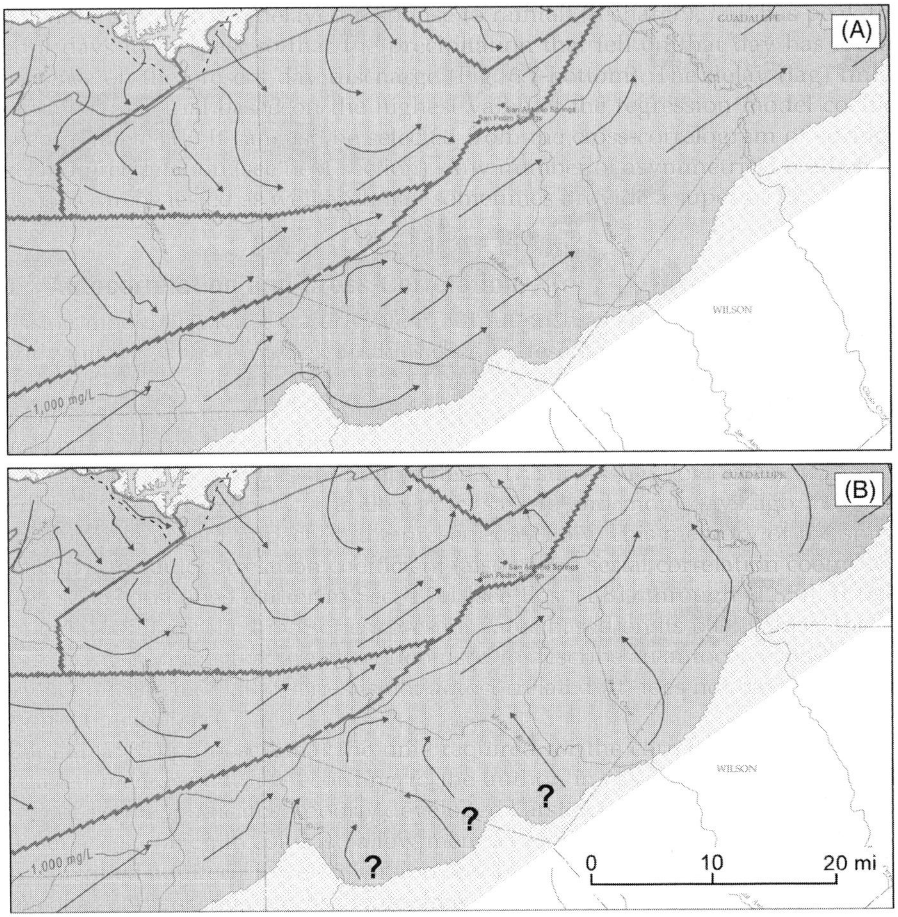

FIGURE 6.31 Portion of the model domain showing results of a numeric groundwater flow model developed for karstic Edwards Aquifer in Texas when virtual conduits (jagged gray lines) are included in the model. (From Lindgren et al., 2004.) Hand-drawn arrows showing groundwater flow directions are original to this USGS report. Note the absence of model-predicted potentiometric surface contour lines. (A) Calibrated steady-state model. (B) Calibrated transient model for drought conditions of August 1956; question marks are added for emphasis.

numerical model of the Edwards Aquifer. In addition to the USGS, multiple public and private professionals are listed as authors of the model or members of the Ground Water Model Advisory Panel (GWMAP) that endorsed it. Consequently, it may appear that both the numerical groundwater model and the conceptual hydrogeologic model it is supposed to simulate must be correct, since they were endorsed by numerous institutions and consulting companies.

The above example illustrates that one should not accept previously published, peer-reviewed information at face value without independent, critical analysis. While the USGS produces many useful and accurate reports, the institution is not infallible. It is often desirable for a hydrogeologist to directly cite the USGS or other agencies such as USEPA as there will generally be less resistance from regulators in accepting the related concepts and conclusions. However, if the assumptions and results of the study in question are wrong, it can lead to the rapid propagation of conceptual errors that can become entrenched in professional practice (Kresic and Mikszewski, 2012).

6.5.2 Coupled Continuum Conduit Flow (CCCF) Models

As discussed earlier, the nature of porosity in karst aquifers requires the application of different sets of equations for three distinct porous media: (1) rock matrix; (2) rock discontinuities such as faults, fractures, and bedding planes; and (3) solutionally enlarged voids such as channels and conduits developed from the initial discontinuities. Within large-enough discontinuities, such as large fractures and conduits (pipes), groundwater flow can become turbulent and therefore has to be described by nonlinear (non-Darcian) equations. Numerical groundwater flow models of karst aquifers that explicitly account for pipe hydraulics by coupling the Darcian porous continuum model with a conduit (pipe) network representing flow in karst conduits and channels are commonly called Coupled Continuum Conduit Flow (CCCF) models. An illustrative overview of the concept's historic development including its mathematical basis is provided by Hu (2009).

Reimann et al. (2011) have recently introduced a research-oriented karst model (ModBraC) that accounts for unsteady and nonuniform discrete flow in variably saturated conduits employing the Saint-Venant equations. According to the authors, model performance tests indicate that ModBraC is able to simulate (1) unsteady and nonuniform flow in variably filled conduits, (2) draining and refilling of conduits with stable transition between free-surface and pressurized flow and correct storage representation, (3) water exchange between matrix and variably filled conduits, and (4) discharge routing through branched and intermeshed conduit networks. ModBraC is applied to an idealized catchment to investigate the significance of free-surface flow representation. A parameter study is conducted with two different initial conditions: (1) pressurized flow and (2) free-surface flow. If free-surface flow prevails, the systems is characterized by (1) a time lag for signal transmission, (2) a typical spring discharge pattern representing the transition from pressurized to free-surface flow, and (3) a reduced conduit-matrix interaction during free-surface flow.

A step in the right direction of simulating true nature of groundwater flow in karst aquifers with well-established and widely available programs is the CFP module for the USGS' MODFLOW-2005 (Shoemaker et al., 2008). Many research-oriented computer programs are available for simulating dual porosity aquifers, but have not been fully documented for wider use (e.g., Clemens et al., 1996; Kiraly, 1998; Bauer, 2002; Birk, 2002). Additionally, the structure of MODFLOW has changed with MODFLOW-2005 (Harbaugh, 2005), making the groundwater flow computer code even more modular and allowing easier addition of new processes to the code.

CFP has the ability to simulate turbulent groundwater flow conditions by (1) coupling the traditional groundwater flow equation with formulations for a discrete network of cylindrical pipes (Mode 1 or CFPM1), (2) inserting a high-conductivity flow layer that can switch between laminar and turbulent flow (CFPM2), or (3) simultaneously coupling a discrete pipe network while inserting a high-conductivity flow layer that can switch between laminar and turbulent flow (CFPM3). As explained by the USGS, CFPM1 may represent dissolution or biologic burrowing features in carbonate aquifers, voids in fractured rock, and/or lava tubes in basaltic aquifers and can be fully or partially saturated under laminar or turbulent flow conditions. Preferential flow layers (CFPM 2)

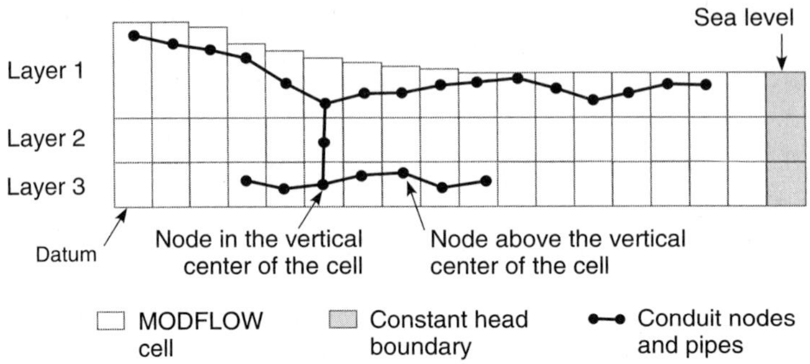

FIGURE 6.32 Possible variations in elevation of conduit nodes in MODFLOW cells. (From Shoemaker et al., 2008.)

may represent: (1) a porous media where turbulent flow is suspected to occur under the observed hydraulic gradients; (2) a single secondary porosity subsurface feature, such as a well-defined laterally extensive underground cave; or (3) a horizontal preferential flow layer consisting of many interconnected voids.

As stated by the USGS, the CFP was designed to be flexible enough for use in locations with both limited and abundant field data. In some geologic environments, such as Mammoth Cave, Kentucky, detailed information is available (or could be derived) on the location, diameter, tortuosity, and roughness of the subsurface caverns. CFPM1 was designed with these locations in mind. In other locations, such as the Biscayne Aquifer of southern Florida, void connections and distributions are so complicated within preferential flow layers that a complete characterization is not possible. CFPM2 was designed with these locations in mind; specifically, laminar and turbulent flow through complicated void connections is represented with a limited number of effective or bulk layer parameters.

One of the options in the CFP is that, in cases with abundant field data on the void architecture and hydraulic behavior, complex two- or three-dimensional networks of conduit flow pipes and nodes can be designed to represent interconnected or dead-end voids in the subsurface. Flow calculations assume pipe nodes are located in the center of MODFLOW cells. An exception is in the vertical direction, for which there are two options. First, pipe nodes can be assigned elevations above a datum, and therefore, are not restricted to center elevations of MODFLOW cells. Second, pipe nodes can be assigned a distance above or below the center of the MODFLOW cell (Fig. 6.32). With this second option, if the distance is set to zero, pipe nodes are assumed to exist at the vertical center of the MODFLOW cell. Pipes can connect diagonally between two finite-difference cells within the same layer or within and adjacent model layers (Shoemaker et al., 2008).

Data requirements for simulations with CFPM1 are more complex than using CFPM2; they include conduit pipe locations, length, pipe diameter, water temperature, turtuosity, internal roughness, critical Reynolds' numbers, and exchange conductances between pipes and the matrix. The parameters for CFPM2 simulations include water temperature, mean void diameter, and critical Reynolds' number. Simulations with CFPM3 include the combination of CFPM1 and CFPM2 input parameters. Boundary and initial conditions (for transient simulations) are the same as for classic porous media MODFLOW models. Concentrated aquifer recharge, such as via sinks, can be assigned directly to pipe nodes.

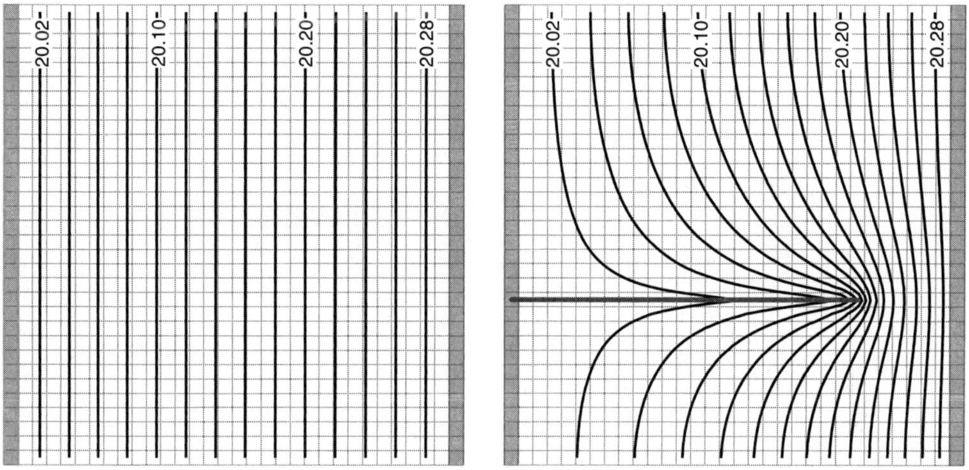

Figure 6.33 Comparison of EPM (*left*) and CCCF (*right*) model-calculated hydraulic head contours for the same boundary conditions and same hydraulic conductivity of the matrix. Created with MODFLOW 2005 and the Conduit Flow Process (CFP) module for the example on the right. (From Mikszewski and Kresic, 2012.)

Figures 6.33 through 6.35 illustrate several conceptual models developed with the MODFLOW CFP module using Groundwater Vistas GUI. A comparison of an EPM and a coupled continuum pipe-flow (CCPF) model-calculated hydraulic head contours is shown in Fig. 6.33. The flow direction is from the right to the left as prescribed with two constant head boundaries. The hydraulic conductivity assigned to all finite-difference cells is the same in both cases. The CCPF example includes a conduit (thick line) for

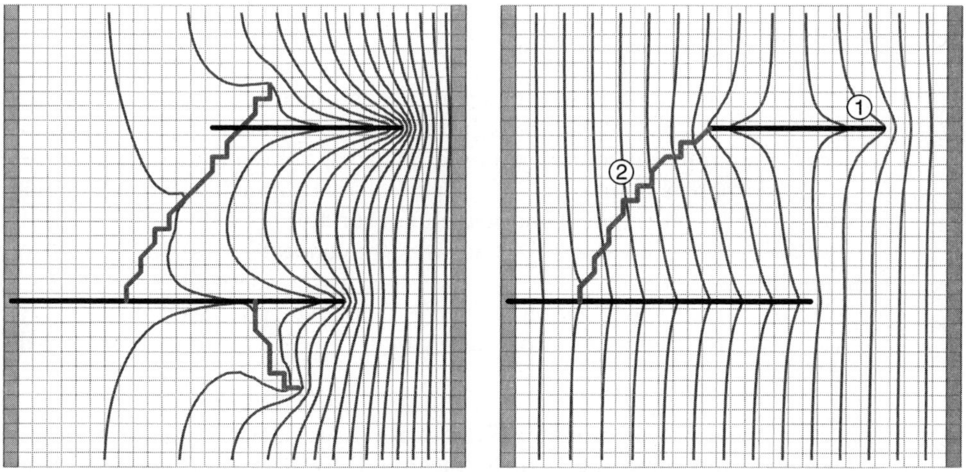

Figure 6.34 Influence of a network of conduits on the hydraulic head contours in the surrounding aquifer matrix. The example on the right shows influence of a focused recharge to conduit no. 2 that is more transmissive (has larger diameter) than conduit no. 1. (From Mikszewski and Kresic, 2012.)

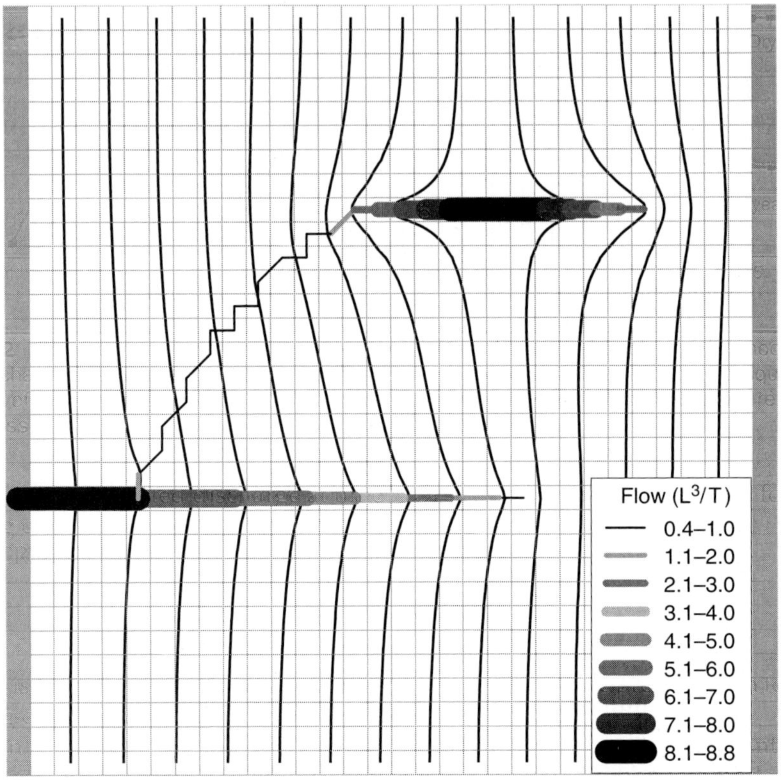

FIGURE 6.35 Model-calculated flow rates in the conduit network shown in Fig. 6.34-right. Processing of model output results for display of the flow rates is performed externally to MODFLOW-2005. (From Mikszewski and Kresic, 2012.)

which the flow equation is solved separately and then coupled with the Darcian flow solution for the finite-difference cells. Influence of a network of conduits on the hydraulic head contours in the surrounding aquifer matrix is shown in Fig. 6.34. The example in Fig. 6.34-right shows influence of a focused recharge applied via a conduit node directly to conduit no. 2 that is more conductive than conduit no. 1. Because of this difference, the hydraulic head in conduit no. 1 increases, the conduit backs up and discharges to the surrounding porous media matrix as indicated by the shape of the contours. The flow mass balance results saved by MODFLOW-2005 can be processed externally to the GUI and visualized for either the matrix portion of the model or for the conduits as illustrated in Fig. 6.35.

The MODFLOW-2005 CFP module is still a work in progress and some useful add-ons available for traditional porous media simulations are not yet available. This includes fate and transport modeling and particle tracking. For example, particle tracks and related groundwater flow velocities, which can be optionally marked by the Groundwater Vistas GUI using tickmarks, stop short of conduits and completely ignore them. The user must process the model results externally in order to analyze and visually present these features. More serious, however, are various problems associated with the model instability and long runtimes due to quite a few additional model parameters required to simulate turbulent flow in conduits. For example, Ashok and Sophocleous (2008) tested the CFP with various model setups and found a number of inconsistencies. In one of the analyses, CFPM2 was tested by increasing and decreasing the diameter of voids by one order of magnitude, but the results were still the same. It was even tested by increasing Ks in the layer-property flow (LPF) package by three orders of magnitude, increasing pumping to –200 m^3/day, and increasing recharge by one order of magnitude. The Reynolds' numbers did not seem to have any impact on simulation no matter what the ranges of values between upper and lower Reynolds' numbers were (Ashok and Sophocleous, 2008.)

Mikszewski and Kresic (2012) report that, when desire is to create laminar flow in the conduits (either based on field observations or well-established hydraulic principles of flow in pipes) by increasing the critical Reynolds' number, the model appears to rebel and apparently puts more water into the conduit than exists in the model so that it can maintain turbulent flow.

In any case, encouraging continuing efforts of improving the CFP by both USGS and non-USGS researchers will hopefully increase its utility and facilitate its adoption by the modeling community. For example, Reimann et al. (2012) discuss recent adjustments to the existing CFPM2 made by implementing a generalized power function with a user-defined, nonphysical exponent. This allows for matching turbulence in porous media or pipes and eliminates the time steps required for iterative adjustment of hydraulic conductivity. The original CFPM2 approximates turbulent flow by reducing the hydraulic conductivity within the existing linear head gradient of the MODFLOW continuum model. This approach was selected because it reduces the practical as well as numerical efforts for simulating turbulence. The original formulation was for large pore aquifers where the onset of turbulence is assumed to be at low Reynolds' numbers (1 to 100) and not for conduits or pipes. In addition, the existing code requires multiple time steps for convergence due to iterative adjustment of the hydraulic conductivity.

6.5.3 MODFLOW-USG

The description of this new groundwater-modeling program is provided courtesy of Dr. Sorab Panday, the Groundwater Modeling Practice Leader at AMEC Environment and Infrastructure, Inc.

As described earlier, the USGS' finite-difference model MODFLOW (Harbaugh, 2005) is the most popular groundwater code worldwide. MODFLOW solves the groundwater flow equation on a rectangular finite-difference grid. Although there are options available for blocking out portions of a rectangular grid that are outside of the simulation domain and for local grid refinement, a rectangular grid structure is generally less effective or efficient in fitting irregular domain geometries or for having refined grid structures within areas of high activity or interest. The MODFLOW-USG code extends the MODFLOW simulation capabilities to irregular domains using unstructured grids and retains full compatibility with previous versions of MODFLOW. The program is released in the public domain; data input and visualization of model results is at the time of this writing supported exclusively by the latest version of the GUI program Groundwater Vistas.

Unstructured grids provide a high level of flexibility of discretization by implementing different shaped grid-block geometries and by use of nested grid structures. Figure 6.36 shows some unstructured grid geometries that may be used by MODFLOW-USG. The grid may be created as a combination of nested grids and polygons of different

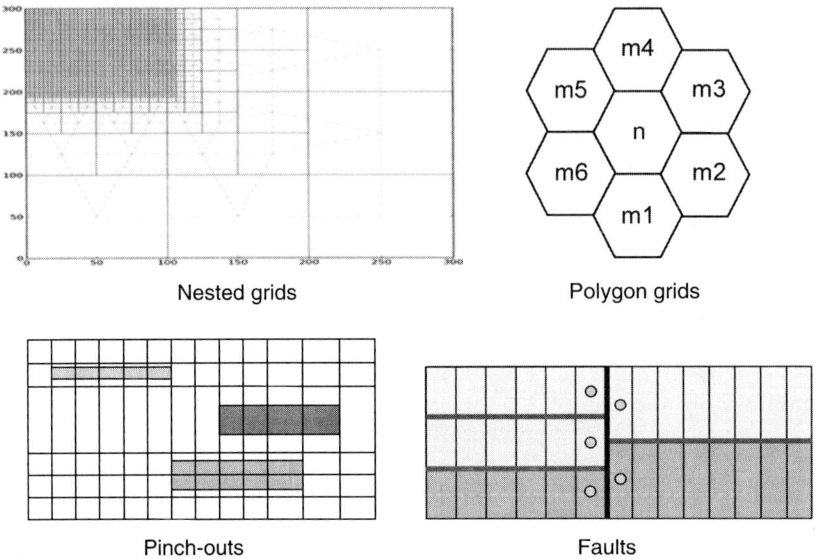

FIGURE 6.36 Examples of unstructured grid geometries supported by MODFLOW-USG. (Courtesy of Dr. Sorab Panday, AMEC Environment and Infrastructure, Inc.)

geometries. Sublayering and vertical displacements along faults may be directly accommodated without excessive discretization. The grid is however limited to a prismatic shape in the vertical direction.

Description of Method

A finite-volume formulation is implemented to discretize the groundwater flow equations for an irregular grid structure. Advantages of the finite-volume scheme include

- Finite-volume discretizations provide gridding flexibility and are easy to understand and implement.
- Finite-volume discretizations provide mass-conserved solutions that are a topologic extension of the finite-difference formulation (Peaceman, 1977; Moridis and Pruess, 1992).
- Finite-volume schemes are robust and efficient with the flexibility of finite-elements but without the computationally intensive numerical integration, elemental assembly schemes, and expanded matrix connectivity.
- The finite-volume formulation provides flexibility with various grid-block shapes and connectivities without the need for extensive elemental catalogues as required by the finite-element schemes.

The Integrated Finite Difference (IFD) formulation is applied to the groundwater flow equation. The discretized formulation is identical to the finite-difference approximation, but with generalized computations for volumes, flow areas, flow lengths, and connections. The cell-by-cell conductance term may be computed using any of the averaging options provided by the block-centered flow (BCF) and LPF packages of MODFLOW-2005 including harmonic, arithmetic, and logarithmic averaging. The upstream weighting formulation of Niswonger et al. (2011) is also included along with Newton Raphson linearization to provide added robustness for difficult nonlinear problems. The conductance term may further be modified for presence of a horizontal flow barrier conceptualized using the HFB package of MODFLOW-2005.

Use of an unstructured formulation with arbitrary flow connections between nodes allows for a natural extension of the basic groundwater flow solution to include various other connected flow processes in a fully implicit manner. Therefore, there exists a framework for incorporating additional flow domains and connections to the subsurface porous medium system. An important flow process significant in several situations includes flow through long-screened wells, conduits, and fracture networks. A Conduit Domain Flow (CDF) package has been included with the current version of MODFLOW-USG to simulate flow through conduit networks and between the porous medium and conduits. Conduits may be horizontal, vertical, or angled in any direction. Exchange of water between the conduit and the porous medium is expressed via a linear leakance term, or via the Thiem equation to simulate the net head loss between the conduit node and the porous-medium grid-block. Therefore, the CDF package provides most of the combined functionalities of the CFP (Shoemaker et al., 2008) and the Multi-Node Well (MNW) (Halford and Hanson, 2002; Konikow et al., 2009) packages of MODFLOW-2005 in a fully implicit manner.

For finite-volume grids, the formulation is a second-order approximation only when the perpendicular from the cell centroid to a face between two nodes coincides with the midpoint of the face (Dehotin et al., 2011)—as is the case for isosceles triangles, rectangles, and regular higher-order polygons. There is an error in the flux computation, however, for irregular polygon shapes or nested grids where the cell center does not coincide with the perpendicular bisector of the face. A ghost node correction (GNC) module based on the work of Edwards (1996) and Dickinson et al. (2007) has also been introduced with MODFLOW-USG to maintain higher-order accuracy for irregular grids. The concept of the GNC is that the nodal value of the head at an irregular grid-block is not representative for flux across the interface between grid-blocks, and that instead a more representative value is obtained by interpolating the head value to a "ghost node" location that does lie along the perpendicular bisector of the face. As illustrated in Fig. 6.37, the ghost node location interpolation can be performed in one or multiple dimensions. The GNC is also applicable to conduits that are located off-center from the porous matrix grid-block, to provide subgrid scale location adjustments without grid refinement.

MODFLOW-2005 boundary packages that have been currently converted for use with unstructured grids include the recharge package (RCH), the evapotranspiration package (EVT), the transient flow and head boundary package (FHB), the well package (WEL), the drain package (DRN), the river package (RIV), the head-dependent flux (general head boundary) package (GHB), the stream package (STR7), the stream flow routing package with unsaturated flow beneath streams (SFR2), the stream-gage monitoring package (GAGE), and the lake package (LAK3). Boundary nodes are identified within these packages by indexing the global node number instead of the structured layer, row, and column classification used by MODFLOW-2005.

Unstructured grids cannot be accommodated by the structured, symmetric solvers present in MODFLOW-2005. Therefore, the MODFLOW-USG code incorporates its own suite of unstructured solvers for symmetric and nonsymmetric matrices.

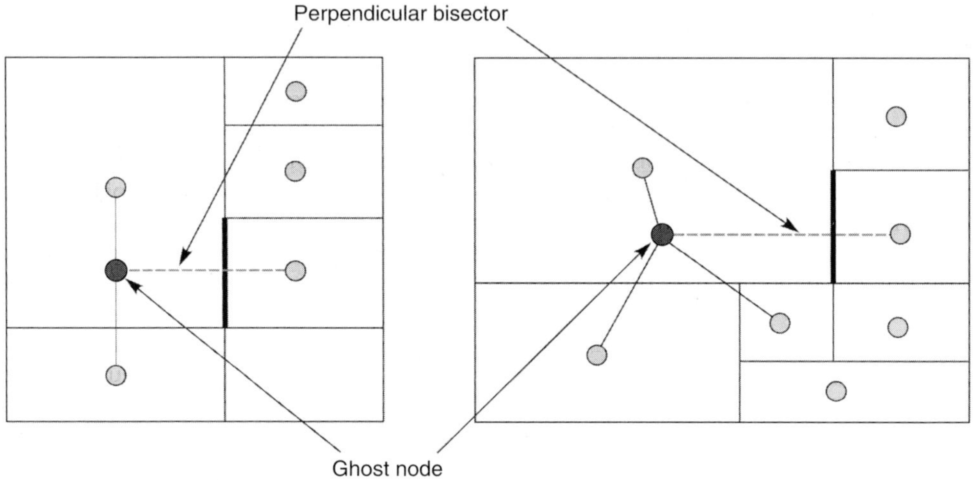

FIGURE 6.37 Ghost node conceptualization in MODFLOW-USG. (Courtesy of Dr. Sorab Panday, AMEC Environment and Infrastructure, Inc.)

Input and Output

The input/output (I/O) formats for MODFLOW-USG follow the MODFLOW-2005 conventions. A name file is read by MODFLOW during problem initialization, which provides all other file names that are opened by MODFLOW-USG. Input for a structured or an unstructured grid is accommodated. If the structured grid I/O option is selected, the code uses the finite-difference input data files of MODFLOW-2005, with minimal changes such as switching to one of the unstructured solver routines. Output from a structured grid is consistent with MODFLOW-2005 ASCII and binary formats, for easy postprocessing by tools available for MODFLOW.

For unstructured grids, the code provides various levels of generalization to ease I/O burden. Furthermore, I/O is based on a global node number convention rather than using the "layer-row-column" formats. Nodal outputs are provided in layer lists, to accommodate easier processing of results from each model layer. Flow outputs are provided for all connections to a node. Only the upper triangular portion of the cell-by-cell flow matrix is saved to avoid redundant output (because $Q_{nm} = -Q_{nm}$).

Benchmarking and Testing

Several test problems have been simulated to verify and benchmark the code. First, the suite of test problems available from the USGS for MODFLOW-2005 was run to ensure that the new code and solvers provide the same results for rectangular finite-different problems, as the MODFLOW-2005 code. These same problems were also simulated with unstructured grid inputs, to again verify that the code behaves appropriately for unstructured grid inputs. Further testing was done using field-scale applications on finite-difference grids to compare performance of the new code and solvers against MODFLOW-2005. Simulated water levels and mass balances and execution speeds are similar depending on the solver options selected. Additional testing was then performed using various nested grid structures to evaluate code performance for unstructured, rectangular grids. These tests are discussed by Langevin et al. (2011). Finally, the CDF package

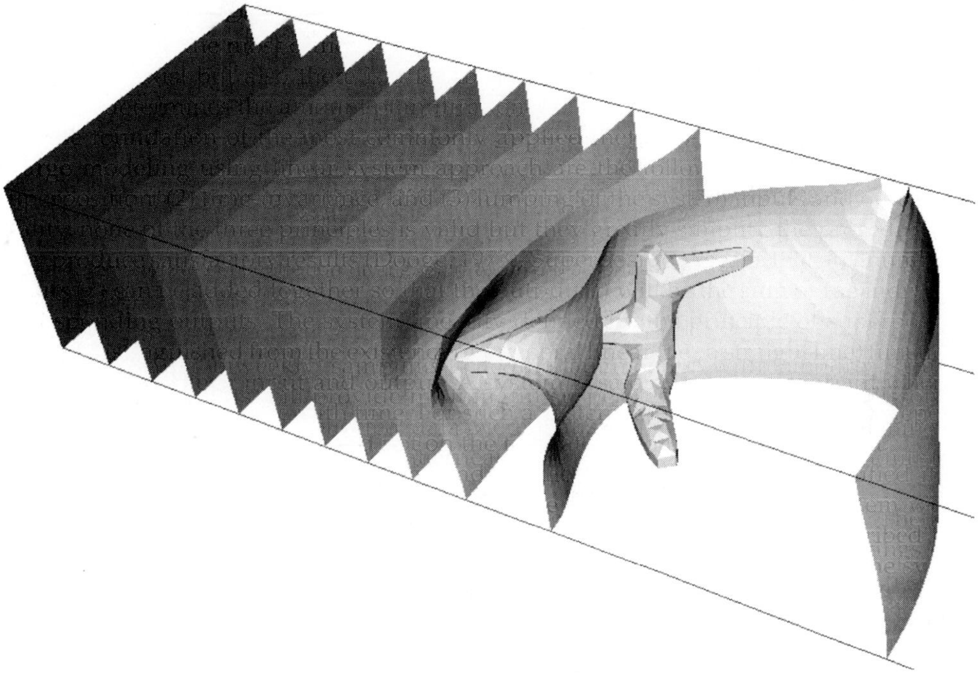

Figure 6.38 Modeled iso-surfaces of the hydraulic head in aquifer matrix at 0.5-m intervals reflecting groundwater extraction from a central shaft and three radial adits. This exploratory model is created with MODFLOW-USG. (Courtesy of Tim Lewis, AMEC Environment and Infrastructure, Inc.)

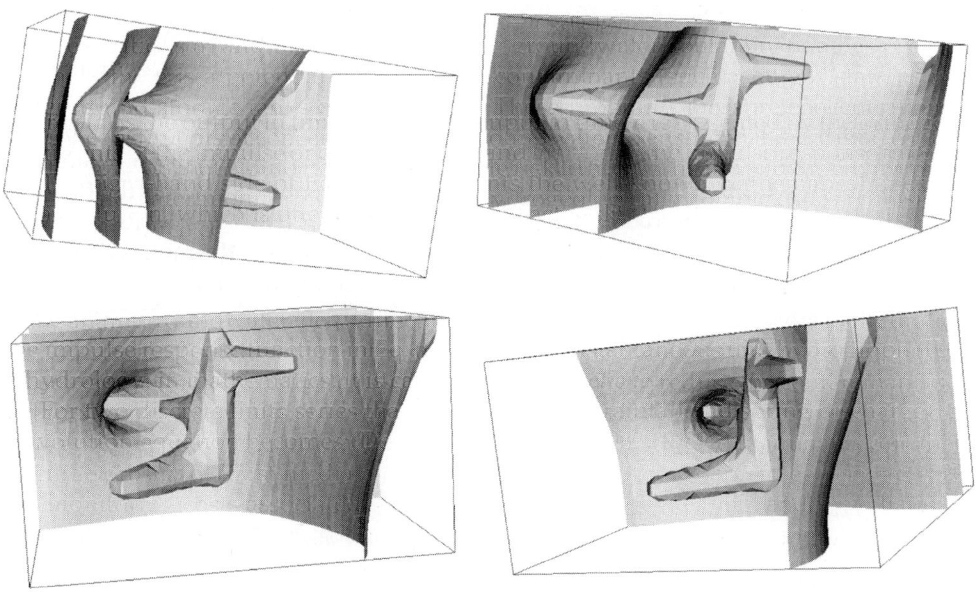

Figure 6.39 Detailed 3D views of the modeled iso-surfaces shown in Fig. 2.61. (Courtesy of Tim Lewis, AMEC Environment and Infrastructure, Inc.)

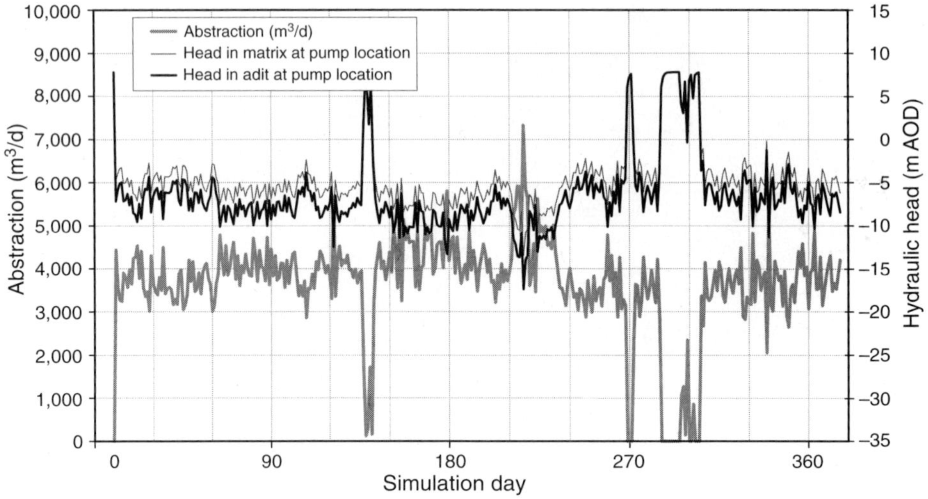

FIGURE 6.40 Hydraulic head in the aquifer matrix adjacent to the location of pump extracting water from the adit system shown in Fig. 2.61 versus the hydraulic head in the adit at the pump location and the pumping rate. (Courtesy of Tim Lewis, AMEC Environment and Infrastructure, Inc.)

and GNC module were also tested using simple and complex examples and comparing the results with more comprehensive codes or refined solutions. The code has also been tested on multiple field examples.

Figures 6.38 through 6.40, provided courtesy of Tim Lewis, AMEC Environment and Infrastructure, Inc., show results of an exploratory model that simulates an adit system typical of that used for public water supply in parts of England. From a central shaft, three adits radiate in different directions and at different levels. In this model, the maximum extent of the adits is around 150 m. Using the conduit domain functionality, water is abstracted from the shaft at a rate that varies daily (see Fig. 6.40). The model was constructed with 20 layers to permit sufficient vertical resolution. There is very little head loss within the adit system itself, with consequent distortion of the flow field in the aquifer matrix compared to the case of a single borehole without adits. The illustrations in Figs. 6.38 and 6.39 show iso-surfaces of hydraulic head in the matrix at 0.5 m intervals; the elongation of these iso-surfaces along the adits is clear. An example of the difference in hydraulic head between the adit system and the immediately adjacent matrix at the pump location is shown with the graph in Fig. 6.40.

6.6 References

Ahn, H., 2000. Modeling of groundwater heads based on second-order difference time series models. *J Hydrol*, 234: 82–94.

Anderson, M. P., and Woessner, W. W., 1992. *Applied Ground Water Modeling: Simulation of Flow and Advective Transport*. Academic Press, San Diego, CA, 381p.

Armbruster, J. T., 1979. Regional stochastic generation of streamflows using an ARIMA (1,0,1) process and disaggregation. U.S. Geological Survey Water-Resources Investigations Report 79-3, 54p.

Ashok, K. C., and Sophocleous, M., 2008. Recent MODFLOW developments for groundwater modeling. Kansas Geological Survey Open-File Report 2009-4, Lawrence, KS, 139p.

Bauer, S., 2002. Simulation of the genesis of karst aquifers in carbonate rocks. Vol. 62 of Tübinger Geowissenschaftliche Arbeiten: Tübingen, Germany, Reihe C. Institut und Museum für Geologie und Paläontologie der Universität Tübingen.

Bierkens, M. F. P., Knotters, M., and Hoogland, T., 2001. Space-time modeling of water table depth using a regionalized time series model and the Kalman filter. *Water Resour Res*, 37(5): 1277–1290.

Birk, S., 2002. Characterization of karst systems by simulating aquifer genesis and spring responses: Model development and application to gypsum karst: Vol. 60 of Tübinger Geowissenschaftliche Arbeiten: Tübingen, Germany, Reihe C. Institut und Museum für Geologie und Paläontologie der Universität Tübingen.

Box, G. E. P., and Jenkins, G. M., 1976. *Time Series Analysis: Forecasting and Control.* Holden Day, San Francisco, CA, 575p.

Bras, R. L., and Rodriguez-Iturbe, I., 1994. *Random Functions and Hydrology.* Dover Publications, New York, 559p.

Canceill, M. F., 1974. Convolution et tarissement dans l'etude des relations pluie-debit. Pub. Com. Fr. AIH, Congress de Montpellier, AIH Memoires, Tome X, Paris, pp. 181–184.

Clemens, T., Hückinghaus, D., Sauter, M., Liedl, R., and Teutsch, G., 1996. A combined continuum and discrete network reactive transport model for the simulation of karst development. In: Calibration and reliability in groundwater modelling—Proceedings of the ModelCARE 96 Conference, Golden, Colorado, September 1996, International Association of Hydrological Sciences Publication 237, pp. 309–318.

Dehotin, J., Vazquez, R. F., Braud, I., Debionne, S., and Viallet, P., 2011. Modeling of hydrological processes using unstructured and irregular grids: 2D groundwater application. *J Hydrol Eng*, 16(2): 108–125.

Denić-Jukić, V., and Jukić, D., 2003. Composite transfer functions for karst aquifers. *J Hydrol*, 274: 80–94.

Dickinson, J. E., James, S. C., Mehl, S., et al. 2007. A new ghost-node method for linking different models and initial investigations of heterogeneity and nonmatching grids. *Adv Water Resour,* 30: 1722–1736.

Dooge, J. C. I., 1973. Linear Theory of Hydrologic Systems. Technical Bulletin No. 1468, United States Department of Agriculture, Washington, D.C., various paging.

Dreiss, S. J., 1982. Linear kernels for karst aquifers. *Water Resour Res*, 18(4): 865–876.

Dreiss, S. J., 1989a. Regional scale transport in a karst aquifer. 1. Component separation of spring flow hydrographs. *Water Resour Res*, 25(1): 117–125.

Dreiss, S. J., 1989b. Regional scale transport in a karst aquifer. 2. Linear systems and time moment analysis. *Water Resour Res*, 25(1): 126–134.

Edwards, M. G., 1996. Elimination of adaptive grid interface errors in the discrete cell centered pressure equation. *J Comput Phys,* 126: 356–372.

Eisenlohr, L., Kiraly, L., Bouzelboudjen, M., and Rossier, I., 1997. Numerical versus statistical modeling of natural response of a karst hydrogeological system. *J Hydrol*, 202: 244–262.

Gabric, I., and Kresic, N., 2009. Time series models. In: Kresic, N. (ed.), *Groundwater Resources: Sustainability, Management, and Restoration*. McGraw Hill, New York, pp. 657–661.

Ghanbarpour, M. R., Abbaspour, K. C., Jalalvand, G., and Moghaddam, G. A., 2010. Stochastic modeling of surface stream flow at different time scales: Sangsoorakh karst basin, Iran. *J Cave Karst Stud*, 72(1): 1–10.

Grasso, D. A., and Jeannin, P. Y., 1994. Etude critique des methodes d'analyse de la reponse globale des systemes karstiques. Application au site de Bure (JU, Suisse). *Bulletin d'Hydrogéologie (Neuchatel)*, 13: 87–113.

Graupe, D., Isailovic, D., and Yevjevich, V., 1976. Prediction model for runoff from karstified catchments, in Proceedings of the U.S.-Yugoslavian Symposium on Karst Hydrology and Water Resources, Dubrovnik, June 2–7, 1975, pp. 277–300.

Halford, K. J., and Hanson, R. T., 2002. User Guide for the Drawdown-Limited, Multi-Node Well (MNW) Package for the U.S. Geological Survey's Modular Three-Dimensional Finite-Difference Ground-Water Flow Model, Versions MODFLOW-96 and MODFLOW-2000, U.S. Geological Survey Open-File Report 02–293.

Harbaugh, A. W., 2005. MODFLOW-2005, the U.S. Geological Survey modular ground-water model-the Ground-Water Flow Process. U.S. Geological Survey Techniques and Methods 6-A16, Reston, VA, variously paginated.

Hirsch, R. M., 1979. Synthetic hydrology and water supply reliability. *Water Resour Res*, 15(6): 1603–1615.

Hu, B. X., 2009. Examining a coupled continuum pipe-flow model for groundwater flow and solute transport in a karst aquifer. *Acta Carsologica*, 39(2): 347–359.

Jemcov, I., and Petric, M., 2009. Measured precipitation vs. effective infiltration and their influence on the assessment of karst systems based on results of the time series analysis. *J Hydrol*, 379: 304–314.

Jukić, D., and Denić-Jukić, V., 2006. Nonlinear kernel functions for karst aquifers. *J Hydrol*, 328: 360–374.

Kalman, R. E., 1960. A new approach to linear filtering and prediction problems. *Trans. ASME, J. Basic Eng*, 82: 35–43.

Kim, S-J., Hyun, Y., and Lee, K-K., 2005. Time series modeling for evaluation of groundwater discharge rates into an urban subway system. Geosciences Journal, v. 9, no. 1, pp. 15–22.

Kiraly, L., 1998. Modelling karst aquifers by the combined discrete channel and continuum approach. *Bulletin d'Hydrogeologie*, 16: 77–98.

Klemeš, V., 1978. Physically based stochastic hydrologic analysis. In: Chow, V. T. (ed.), *Advances in Hydroscience*, Vol. 11. Academic Press, New York, pp. 285–352.

Knochenmus, L. A., and Robinson, J. L., 1996. Descriptions of anisotropy and heterogeneity and their effect on ground-water flow and areas of contribution to public supply wells in a karst carbonate aquifer system. U.S. Geological Survey Water-Supply Paper 2475, Washington, D.C., 47p.

Knotteres, M., and Bierkens, M. F. P., 2000. Physical basis of time series models for water table depths. *Water Resour Res*, 36(1): 181–188.

Koch, R. W., 1985. A stochastic streamflow model based on physical principles. *Water Resour Res*, 21(4): 545–553.

Konikow, L. F., Hornberger, G. Z., Halford, K. J., and Hanson, R. T., 2009. Revised multi-node well (MNW2) package for MODFLOW ground-water flow model. U.S. Geological Survey Techniques and Methods 6-A30, 67p.

Kovács, A., and Sauter, M., 2007. Modelling karst hydrodynamics. In: Goldscheider, N., and Drew, D. (eds.), *Methods in Karst Hydrogeology*. International Contributions to Hydrogeology 26, International Association of Hydrogeologists. Taylor & Francis, London, pp. 201–222.

Kresic, N., 1995. Stochastic properties of spring discharge. In: Dutton, A. R. (ed.), *Toxic Substances and the Hydrologic Sciences*. American Institute of Hydrology, Minneapolis, MN, pp. 582–590.

Kresic, N., 1997. Quantitative solutions in hydrogeology and groundwater modeling. Lewis Publishers/CRC Press, Boca Raton, Florida, 461p.

Kresic, N., 2007. *Hydrogeology and Groundwater Modeling*. 2nd ed. CRC Press, Taylor & Francis Group, Boca Raton, FL, 807p.

Kresic, N., 2009. *Groundwater Resources: Sustainability, Management, and Restoration.* McGraw-Hill, New York, 852p.

Kresic, N., 2010. Modeling. In: Kresic, N., and Stevanovic, Z. (eds.), *Groundwater Hydrology of Springs: Engineering, Theory, Management and Sustainability*. Elsevier, New York, pp. 165–230.

Kresic, N., and Mikszewski, A., 2012. *Hydrogeological Conceptual Site Models: Data Analysis and Visualization*. CRC/Taylor & Francis Group, Boca Raton, FL, 584p.

Kresic, N., Kukuric, N. and Zlokolica, M., 1993. Numeric versus stochastic modeling of water balance and minimum discharge of a karst hydrogeological system. In: Günay, G., Johnson, I., and Back, W. (eds.), *Hydrogeological Processes in Karst Terranes (Proceedings of the Antalya Symposium and Field Seminar, October 1990), IAHS Publ. No. 207*. IAH Press, Wallingford, CT, pp. 253–259.

Kresic, N. A., and Busbey, A. B., 1995. Transformation of precipitation series for stochastic modeling of karst aquifer discharge. Proc. Solutions '95 Congress of International Association of Hydrogeologists, Edmonton, Canada, 6p.

Labat, D., Ababou, R., and Mangin, A., 2000. Rainfall-runoff relations for karstic springs. Part I: Convolution and spectral analyses. *J Hydrol*, 238: 123–148.

Langevin, C. D., Panday, S., Niswonger, R. G., and Hughes, J. D., 2011. Evaluation of mesh alternatives for an unstructured grid version of MODFLOW, MODFLOW and More 2011, Golden, CO.

Larocque, M., Mangin, A., Razack, M., and Banton, O., 1998. Contribution of correlation and spectral analyses to the regional study of a large karst aquifer (Charente, France). *J Hydrol*, 205: 217–231.

Lee, J. -Y., and Lee, K. -K., 2000. Use of hydrologic time series date for identification of recharge mechanism in a fractured bedrock aquifer system. *J. Hydrol*, 229(3–4): 190–201.

Lindgren, R. J., Dutton, A. R., Hovorka, S. D., Worthington, S. R. H., and Painter, S., 2004. Conceptualization and simulation of the Edwards Aquifer, San Antonio region, Texas. U.S. Geological Survey Scientific Investigations Report 2004–5277, 143p.

Mangin, A., 1981. Apports des analyses correlatoire et spectral crois dans la connaissance des systems hydrologiques. *Comptes Rendus de L Academie Des Sciences, Paris*, 293(11): 1011–1114.

Mangin, A., 1982. L'approche systémique du karst, conséquences conceptuelles et méthodologiques. Proc. Réunion Monographica sobre el karst, Larra, pp. 141–157.

Margeta, J., and Fistanić, I., 2004. Water quality modelling of Jadro Spring. *Water Sci Technol*, 50(11): 59–66.

McCuen, R. H., and Snyder, W. M., 1986. *Hydrologic Modeling: Statistical Methods and Applications.* Prentice-Hall, Englewood Cliffs, NJ, 568p.

Mikszewski, A., and Kresic, N., 2012. Numeric modeling of karst aquifers: Comparison of EPM and CFP models. Proc. IAH 2012 Congress, Niagara Falls, Canada.

Montanari, A., Rosso, R., and Taqqu, M. S., 2000. A seasonal fractional ARIMA model applied to the Nile River monthly flows at Aswan. *Water Resour Res*, 36(5): 1249–1259.

Moridis, G. and Pruess, P., 1992. TOUGH simulations of Updegraff's set of fluid and heat flow problems.Lawrence Berkeley Laboratory Report LBL-32611, Berkeley, CA, November 1992.

Niswonger, R. G., Panday, S., and Ibaraki, M., 2011. MODFLOW-NWT, A Newton Formulation for MODFLOW-2005. U.S. Geological Survey Techniques and Methods 6-A37, 44p.

Padilla, A., and Pulido-Bosch, A., 1995. Study hydrographs of karstic aquifers by means of correlation and cross-spectral analysis. *J Hydrol*, 168: 73–89.

Padilla, A., Pulido-Bosch, A., Calvache, M. L., and Vallejos, A., 1996. The ARMA models applied to the flow of karstic springs. *Water Resour Res*, 32(5): 917–928.

Peaceman, D. W., 1977. *Fundamentals of Numerical Reservoir Simulation*. Elsevier, Amsterdam, 176p.

Petrič, M., 2002. *Characteristics of Recharge-Discharge Relations in Karst Aquifer*. ZRC Publishing, Postojna-Ljubljana, Slovenia, 154p.

Putnam, L. D., and Long, A. J., 2009. Numerical Groundwater-Flow Model of the Minnelusa and Madison Hydrogeologic Units in the Rapid City Area, South Dakota. U.S. Geological Survey Scientific Investigations Report 2009–5205, Reston, VA, 81p.

Rahnemaei, M., Zareb, M., Nematollahic, A. R., and Sedghid, H., 2005. Application of spectral analysis of daily water level and spring discharge hydrographs data for comparing physical characteristics of karstic aquifers. *J Hydrol*, 311: 106–116.

Reimann, T., Birk, S., Rehrl, C., and Shoemaker, W. B., 2012. Modifications to the conduit flow process mode 2 for MODFLOW-2005. *Ground Water*, 50(1): 144–148.

Reimann, T., Geyer, T., Shoemaker, W. B., Liedl, R., and Sauter, M.M., 2011. Effects of dynamically variable saturation and matrix-conduit coupling of flow in karst aquifers. *Water Resour Res*, 47, W11503, doi:10.1029/2011WR010446.

Rubinić, J., and Fistanić, I., 2005. Application of time series modeling in karst water management. In: Stevanovic, Z., and Milanovic, P. (eds.), *Water Resources and Environmental Problems in Karst, Proceeding of the International Conference and Field Seminar*. Belgrade & Kotor, September 13–19, 2005, Institute of Hydrogeology, University of Belgrade, Belgrade pp. 417–422.

Salas, J. D., 1993. Analysis and modeling of hydrologic time series. In: Maidment, D. R., (ed.), *Handbook of Hydrology*, McGraw-Hill, New York, pp. 19.1–19.72.

Salas, J. D., and Obeysekera, J. T. B., 1992. Conceptual basis of seasonal streamflow time series models. *J Hydraulic Eng*, 118(8): 1186–1194.

Salas, J. D., Boes, D. C., and Smith, R. A., 1982. Estimation of ARMA models with seasonal parameters. *Water Resour Res*, 18(4): 1006–1010.

Salas, J. D., Delleur, J. W., Yevjevich, V., and Lane, W. L., 1985. *Applied Modeling of Hydrologic Time Series.* Water Resources Research Publication, Littleton, CO.

Shoemaker, W. B., Kuniansky, E.L., Birk, S., Bauer, S., and Swain, E. D., 2008. Documentation of a Conduit Flow Process (CFP) for MODFLOW-2005. U.S. Geological Survey Techniques and Methods, Book 6, Chapter A24, Reston, VA, 50p.

Snyder, W. M., 1968. Subsurface implications from surface hydrograph analysis. In: *Proceedings of Second Seepage Symposium, Phoenix, AZ*. U.S. Department of Agriculture, Washington, D.C., pp. 35–45.

Voss, C. I., and Provost, A. M., 2002. SUTRA: A model for saturated-unsaturated, variable-density ground-water flow with solute or energy transport. U.S. Geological Survey Water-Resources Investigations Report 02–4231. Reston, VA, 250p.

Weeks, W. D., and Boughton, W. C., 1987. Tests of ARMA model forms for rainfall-runoff modeling. *J Hydrol*, 91: 29–47.

Westlake, P. R., and Dracup, J. A., 1980. Stochastic signal processing and analysis of water level data. U.S. Geological Survey Open File Report 80–1156, 226p.

Worthington, S., 2003. Characterization of the mammoth cave aquifer. In: *Significance of Caves in Watershed Management and Protection in Florida, Workshop Proceedings*, April 16th and 17th, 2003, Ocala, FL, Florida Geological Survey Special Publication No. 53.

Yevjevich, V., 1972. *Stochastic Processes in Hydrology*. Water Resources Publications, Fort Collins, CO, 302p.

CHAPTER 7

Floods, Droughts, and Climate Change

7.1 Introduction

Although not all karst water systems and aquifers are equally susceptible to precipitation fluctuations, when compared to other porous media they generally show faster responses and are therefore more vulnerable to extreme weather events such as floods and droughts. Karst communities in many parts of the world inhabited for centuries or millennia have collective memory that enables them to cope with these extreme events and sometimes even take advantage of them. Figure 7.1 illustrates this point. More recently, however, the population growth, together with the accompanying land use changes and large-scale engineering projects, is disturbing the delicate water balance in karst areas worldwide. These new practices in many cases disregard the existing collective experience, thus increasing the vulnerability of karst environments to extreme weather events and likely impacts of the long-term climate change.

Karst aquifers in open hydrogeologic structures directly exposed to recharge from precipitation and with low matrix porosity are especially vulnerable. During high-intensity rainfall events, the low aquifer porosity cannot store and transmit large volumes of infiltrating rainwater that causes rapid buildup of hydraulic pressure in the conduit network. Consequently, the hydraulic heads in the conduits can sometimes rise tens of feet or meters in less than 24 hours (see Fig. 1.146). This can result in emergence of temporary springs at the land surface and flooding via activated conduits. At the same time, the accompanying general rise of water table within the more weathered surficial portion of the aquifer can cause diffuse flooding by groundwater and impact large contiguous areas. Communities established around periodically flooded karst depressions are well aware of these mechanisms such that no permanent residences or vulnerable infrastructure are built in the affected areas; agriculture is usually the only activity left to seasonal risks (see Fig. 7.2 and Fig. 1.76). In order to facilitate uninterrupted natural draining of flooded karst depressions and maximize time available for agriculture, key ponors and swallets should be regularly cleared of debris and sediment or protected from clogging with various engineered structures including circular dams as illustrated in Fig. 7.3. Drainage tunnels and galleries may also be constructed to evacuate floodwater from the depressions faster and into lower catchments where feasible.

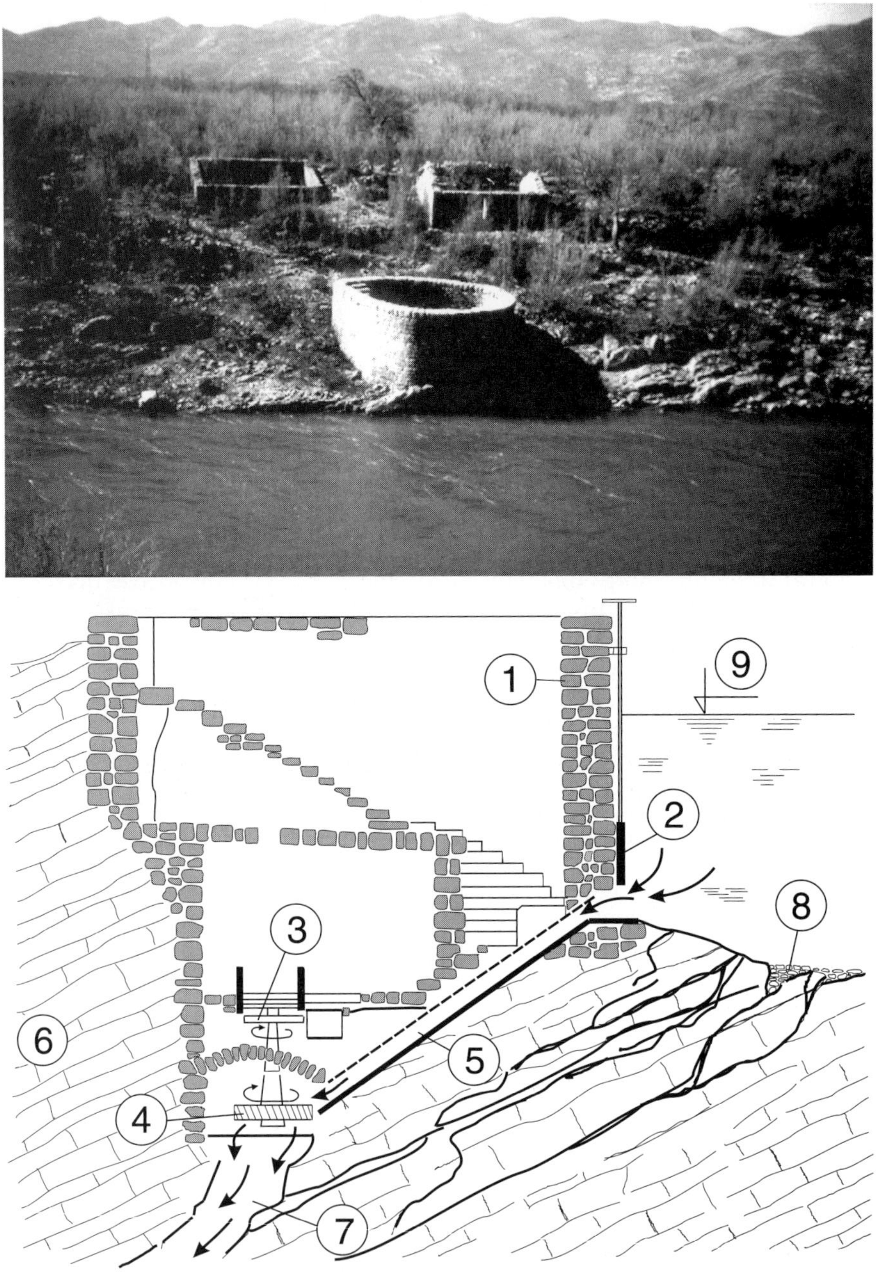

Figure 7.1 *Top*: Abandoned water mill at the Trebisnjica Riverbank, Herzegovina, built over a sink (ponor). *Bottom*: Cross section of typical water mill in Popovo Polje. (1) Cylindrical stone wall, (2) water gate, (3) millstone, (4) wooden turbine, (5) water inlet, (6) karstified limestone, (7) karst channel between ponor and water turbine, (8) riverbed, and (9) river water level. (Courtesy of Dr. Petar Milanovic.)

Figure 7.2 Small patches of soil tilled to grow maize in a doline (sinkhole) with soil erosion caused by stormwater during the summer season. The doline is near Jiangzhou Township in northern Fengshan County, Guangxi, China. (The upper photo taken in July 2008, courtesy of Zhang Cheng; the lower photo taken in July 2010, courtesy of Pei Jianguo.)

Surface streams in karst are vulnerable to flooding for the same reasons described above. The infiltration rate is much higher and the evapotranspiration loss comparably much lower in the exposed karst compared to non-karst areas. Large volumes of water can therefore be quickly routed toward the streams through the karst subsurface during and after intense rainfall and/or rapid snowmelt. The flooding in a stream may be caused by a single large spring, or by any combination of distributed groundwater discharge via spring(s) and directly into the stream. Figure 7.4 shows one of more extreme floods on a surface stream in eastern Herzegovina, the area in classic Dinaric karst where both floods and droughts are common and routinely dealt with.

FIGURE 7.3 Protection of ponor in a periodically flooded karst polje, Peloponnesus, Greece. (Courtesy of Dr. Petar Milanovic; see also Milanovic, 2004.)

Some land use practices that can exasperate flooding include urban sprawl and paving that may concentrate stormwater runoff into smaller areas and increase subsequent localized infiltration into the subsurface that can then overwhelm the conduit network. Improperly scaled ongoing agricultural irrigation can maintain relatively high water table and/or moisture content in the residuum sediments. This creates favorable conditions for enhanced recharge and quick rise of the water table after major recharge episodes. Inadequate operation of dams and locks can also contribute to downstream flooding when the reservoir levels are kept high and water inflow after heavy rainfall is greater than what can be safely released from the reservoir.

In these and similar situations, the key for minimizing negative impacts are operational models that fully take into account specifics of karst and are continuously updated based on monitoring of precipitation, surface flows, and water levels in monitoring wells. Hydrologic and weather forecasts including precipitation amount are becoming ever more reliable and are available online in some countries (e.g., see http://water.weather.gov/ahps/). Precipitation forecasts for 1-day, 5-day, 7-day, and longer periods can be incorporated into numeric and time-series models developed for index monitoring wells and flow gauging stations. By accounting for antecedent precipitation and hydraulic conditions in the aquifer, such models issue real-time, short-term prognosis of the expected hydraulic heads in the aquifer or flows at springs and surface streams.

All water bodies in karst terrains are particularly impacted by droughts. Even during average years, there is general shortage of surface flows in sparse drainage networks due to various characteristics of karst explained in detail in Chap. 1. Limited volumes of

Figure 7.4 Bridge over Zalomka near Nevesinje, Herzegovina. Upper photo taken during normal low flow conditions, lower photo taken during an exceptional flood in fall 2010. (Photographs courtesy of Željko Zubac.)

water stored in the rock matrix cannot provide enough baseflow to many streams for any longer periods, and this condition only worsens during seasons and years with below-average precipitation. In areas where the base of karstification is deep, longer periods without precipitation can quickly translate into lowering of the aquifer hydraulic heads well below stream channels and drying of many semipermanent springs (Fig. 7.5). In mature karst terrains with thick carbonate deposits, the saturated zone is often very deep and the disintegrated surface drainage is dry most of the year even in regions with abundant precipitation. Many karst terrains are therefore in a state of permanent virtual drought compared to all other geologic environments. Nevertheless, succession of dry

FIGURE 7.5 Water mill on an intermittent karst spring, Croatia.

years can have devastating long-term effects on karst water resources in all regions where water supply is largely dependent on groundwater withdrawal via wells. An increased pumping triggered by the drought can quickly drain the limited matrix storage of older compacted limestones as the unsaturated zone thickens due to drawdown. Refilling of this storage may require long periods of both increased precipitation and lowered pumpage so that the persistently higher hydraulic heads in the conduit network can drive water back into the rock matrix. However, due to possibly increasing demands in the affected area regardless of the climatic conditions, natural restoration of the matrix storage may not be feasible in many such cases.

Long-term ambient monitoring is the key for successful management of extreme weather events in all environments. Without it, it would be impossible to accurately assess trends and magnitude of droughts and floods. The probability analysis of floods and droughts based on the existing measurements of precipitation and flows is described in detail in Chap. 1, section on Precipitation, and Chap. 2, section on Probability of High and Low Flows, respectively. The same methodology is applicable to other extreme event indicators such as hydraulic heads in the aquifer measured at monitoring wells. When the period of record is too short, the synthetic time series of sufficient length can be generated using some of the time-series models described in Sec. 6.4. The probability analysis can then be performed for the synthetic time series in order to develop water management plans and obtain required design parameters for various engineering projects such as construction of dams and reservoirs, spring capture, development of well fields, or artificial aquifer recharge.

The ongoing challenge of water management in karst is achieving balance between floods and droughts by storing surplus water for times of increased demand. The most common means of achieving this goal worldwide has been construction of dams and reservoirs, although storing water in aquifers is now an increasing trend. This is mainly because surface waters in many countries are largely developed, with little opportunity available to increase storage along main rivers because few suitable sites remain for dams, and there is general concern about the environmental impacts of artificial reservoirs. Unfortunately (again), karst presents additional challenges for construction of surface reservoirs and for aquifer recharge as explained in more detail in Sec. 10.4.2.

Classic Dinaric karst showcases many complex water resources projects including some of the world's highest arch dams built in karst, largest reservoirs located in pure mature karst, and longest hydrotechnical tunnels built exclusively for karst water resources management. Among such projects, the Trebisnjica River water system in Eastern Herzegovina stands out (Milanovic, 1979, 2004, 2006). Probably the most fascinating aspect of the system is flood and drought control of several large karst poljes including paving of a 62.5-km-long section of the Trebisnjica River, formerly the longest and largest sinking stream in Europe (Fig. 7.6). Before paving the river channel with sprayed concrete (shotcrete), the cumulative water losses along its losing stretch were estimated at 63 m^3/s during dry periods; the river ultimately disappeared into the karst subsurface via several large ponors at the western end of Popovo Polje. The maximum flow of the Trebisnjica River exceeds 1,400 m^3/s. In the spring, following heavy rainfall and snowmelt in the higher elevations of the Dinarides, the river flow is much higher than the combined capacity of all its sinks and sinks along the polje perimeters. Popovo Polje was therefore regularly flooded in the spring and the retention lasted for up to 5 months. After completion of the regulation works in the river drainage basin, small areas of the polje have been flooded only several times for few weeks since the late 1970s.

FIGURE 7.6 Popovo Polje in Herzegovina after completion of massive engineering regulations in the Trebisnjica River drainage area. The river channel, visible in the right, is paved with shotcrete, the polje does not flood anymore and is used for intensive agriculture, and portion of the river water is utilized for irrigation during dry season.

7.2 Influence of Climate Change

Climate is defined as an aggregate of weather conditions, representing a general pattern of weather variations at a location or in a region. It includes average weather conditions, as well as the variability of elements and information on the occurrence of extreme events (Lutgens and Tarbuck, 1995). The nature of both weather and climate is expressed in terms of basic elements, the most important of which are (1) the temperature of the air, (2) the humidity of the air, (3) the type and amount of cloudiness, (4) the type and amount of precipitation, (5) the pressure exerted by the air, and (6) the speed and direction of the wind. These elements constitute the variables by which weather patterns and climatic types are characterized (Lutgens and Tarbuck, 1995). They also all influence the water budget of an area primarily by affecting natural processes of aquifer recharge.

The main difference between weather and climate is the timescale at which these basic elements change. Weather is constantly changing, sometimes from hour to hour, and these changes create almost an infinite variety of weather conditions at any given time and place. In comparison, climate changes are more subtle and were, until relatively recently, considered important for timescales of hundreds of years or more, and usually only discussed in academic circles. A broader definition of climate is that it represents the long-term behavior of the interactive climate system, which consists of the atmosphere, hydrosphere, lithosphere, biosphere, and cryosphere or ice and snow that are accumulated on the Earth's surface (Lutgens and Tarbuck, 1995).

The most significant theory relating Earth motions and long-term climate change, later confirmed with geologic and paleoclimatic evidence collected from around the globe, was developed in the 1930s by the Serbian mathematician and astrophysicist Milutin Milankovitch, professor at the University of Belgrade. His work was published in German in 1941 by the Royal Serbian Academy, under the title *Kanon der Erdbestrahlung und seine Anwendung auf das Eiszeitenproblem (Canon of Insolation of the Earth and Its Application to the Problem of the Ice Ages)*, but it was largely ignored by the international scientific community (Milankovitch, 1941). In 1969, it was translated into English and published with the title *Canon of Insolation of the Ice-Age Problem* by the U.S. Department of Commerce and the National Science Foundation, Washington, D.C. In 1976, a study published in the journal *Science* examined deep-sea sediment cores and found that Milankovitch's theory did in fact correspond to periods of climate change (Hays et al., 1976). Specifically, the authors were able to analyze the record of temperature change going back 450,000 years and found that major variations in climate were closely associated with changes in the geometry (eccentricity, obliquity, and precession) of Earth's orbit; ice ages had indeed occurred when the Earth was going through different stages of orbital variation. Since this study, the National Research Council of the U.S. National Academy of Sciences has embraced the Milankovitch Cycle model (NRC, 1982): ". . . orbital variations remain the most thoroughly examined mechanism of climatic change on timescales of tens of thousands of years and are by far the clearest case of a direct effect of changing insolation on the lower atmosphere of Earth."

Milankovitch was intrigued by the puzzle of climate change and studied climate records, noting differences over time. He theorized that global climate change was brought about by regular changes in Earth's axis, tilt, and orbit that altered the planet's relationship to the Sun, triggering ice ages. Milankovitch determined that the Earth wobbles in its orbit and calculated the slow changes in the Earth's orbit by careful measurement of the position of the stars and using the gravitational pull of other planets and stars. The three variables quantified by Milankovitch are now known as Milankovitch Cycles:

1. Eccentricity cycle of the Earth's orbit; every 90,000 to 100,000 years there is a change in Earth's orbit about the Sun. Its almost circular orbit becomes more elliptical, taking Earth farther from the Sun.
2. The tilt of the Earth's axis or obliquity cycle; on average, every 40,000 years there is a change in the tilt of the Earth's equatorial plane in relation to its orbital plane, moving either the Northern or Southern Hemisphere farther from the Sun.
3. Precession, or orientation of the Earth's rotational axis; on average every 22,000 years there is a slight change in its wobble (Earth does not rotate perfectly like a wheel about an axis; it spins like a wobbling top).

These cycles mean that during certain periods there is less solar energy arriving to Earth, resulting in less melting of snow and ice. Instead of melting, these cold expanses of frozen water grow. The snow and ice last longer and, over many seasons, begin to accumulate. Snow and ice reflect some sunlight back into space, which also contributes to cooling. Temperatures drop, and glaciers begin to advance (Tesla Memorial Society of New York, 2007).

The climate is influenced by all three cycles that can combine in a number of different ways, sometimes strongly reinforcing each other, and sometimes working against each other. The general influence of the Milankovitch Cycles on the long-term climate and their current state is presented below based on NASA (2007; see also Fig. 7.7).

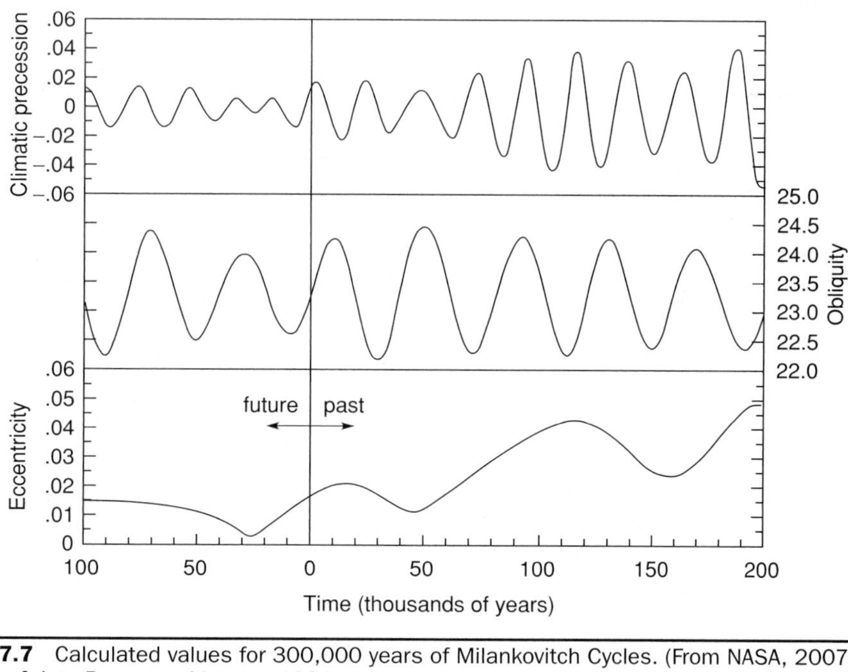

FIGURE 7.7 Calculated values for 300,000 years of Milankovitch Cycles. (From NASA, 2007; source of data Berger and Loutre, 1991.)

The eccentricity of the Earth's orbit changes slowly over time from nearly zero (circular) to 0.07 (eccentric). As the orbit becomes more eccentric (oval), the difference between the distance from the Sun to the Earth at perihelion (closest approach) and aphelion (furthest away) becomes greater and greater. Currently, a difference of only 3% (5 million kilometers) exists between perihelion, which occurs on or about January 3, and aphelion, which occurs on or about July 4. This difference in distance amounts to about a 6% increase in incoming solar radiation (insolation) from July to January. The current trend of eccentricity is decreasing. When the orbit is highly elliptical, the amount of insolation received at perihelion would be on the order of 20% to 30% greater than at aphelion, resulting in a substantially different climate from what we experience today.

Today, the Earth's axis is tilted 23.5 degrees from the plane of its orbit around the Sun. During a cycle that averages about 40,000 years, the tilt of the axis varies between 22.1 and 24.5 degrees. Because of tilt changes, the seasons as we know them can become exaggerated. More tilt means more severe seasons—warmer summers and colder winters; less tilt means less severe seasons—cooler summers and milder winters. It is the cool summers that allow snow and ice to last from year-to-year in high latitudes, eventually building up into massive ice sheets. An Earth covered with more snow reflects more of the Sun's energy into space, causing additional cooling. The current trend in the Earth's axis tilt is decreasing.

Changes in axial precession alter the dates of perihelion and aphelion, and therefore increase the seasonal contrast in one hemisphere and decrease the seasonal contrast in the other hemisphere. If a hemisphere is pointed toward the Sun at perihelion, then hemisphere will be pointing away at aphelion, and the difference in seasons will be more extreme. This seasonal effect is reversed for the opposite hemisphere. Currently, the Northern Hemisphere summer occurs near aphelion which means that the Northern Hemisphere should have somewhat less extremes between the seasons. The climatic precession is close to its peak and shows a decreasing trend.

Although the Milankovitch Cycles can explain long-term climatic changes on geologic timescales (on the order of tens of thousands of years or more), their long duration makes them ineffective tools to explain or predict changes that are of significance for water resources evaluation and planning, namely at timescales of decades to centuries. However, what we can learn from the well-established science of long-term climate change and the geologic evidence of it occurring in the past is that it will inevitably occur in the future as well. A fourth cycle, not addressed by Milankovitch, may accelerate natural climate change—human activity on Earth.

The phrases *climate change* and *global warming* became household words across the globe as the Intergovernmental Panel on Climate Change (IPCC) and the former vice-president of the United States, Mr. Al Gore, both received the 2007 Nobel Peace Prize for their efforts in researching and explaining the many aspects and impacts of climate change. There are few scientific graphs, if any, that have became so widely reproduced in the media and discussed worldwide as the two shown here in Fig. 7.8. They illustrate the connection between an increasing concentration of carbon dioxide in the atmosphere caused by human activity such as burning of fossil fuels and an increasing global temperature. As if determined to convert a few remaining skeptics, 2007 was also the driest year on record in Southern California, with a catastrophic drought affecting the southeastern United States as well. At the time of this writing (summer of 2012) large areas of the United States are affected by an exceptional drought of record. Exceptional droughts tend to galvanize politicians, economists, water professionals, and the general public in searching for answers to many drought-related problems, water

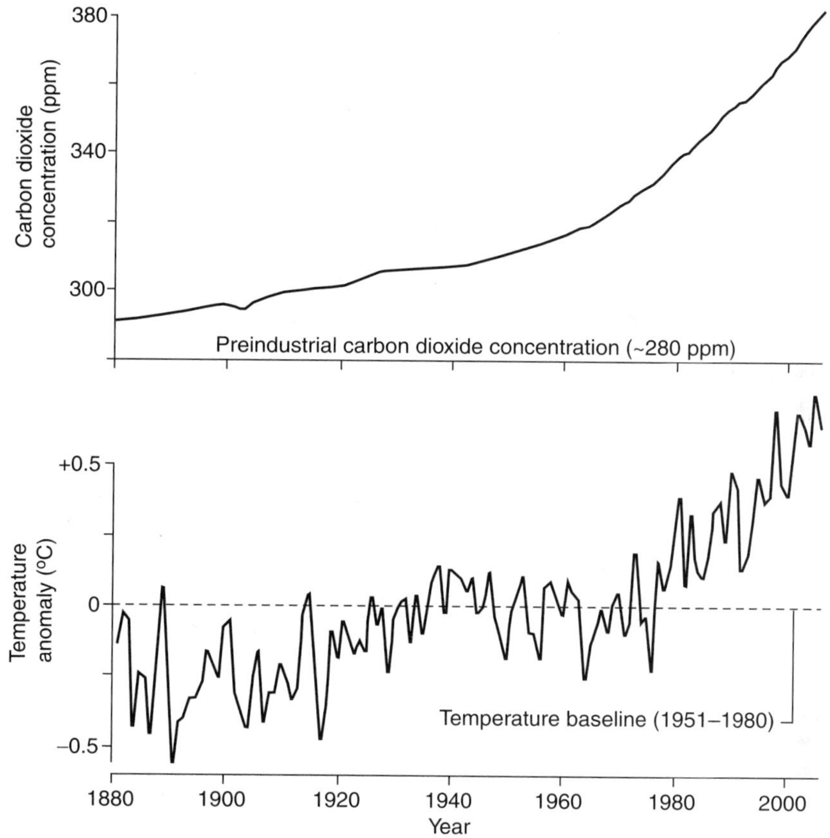

Figure 7.8 Global temperature anomaly and carbon dioxide concentration in parts per trillion since 1880. (From Riebek, 2007. NASA graphs by Robert Simmon, based on carbon dioxide data from Dr. Pieter Tans, NOAA/ESRL and temperature data from NASA Goddard Institute for Space Studies.)

supply being the most important. In that sense, 2007 and 2012 may be called perfect years for rethinking societal approaches to both water management and climate change.

Accurate and systematic measurements of weather and climate elements are paramount to fully understanding the climate of a region and anticipating future climatic changes that may impact water supplies. Unfortunately, records of air temperature and precipitation, the most important direct measures of climate, go back only several hundred years in Europe, and less than that in other parts of the world. The situation is even worse with hydrologic measurements of stream or spring flows and worse yet with records of groundwater levels, the three most important direct measures of freshwater budget. Even though the time record of direct climatic and hydrologic measurements is increasing, it is becoming more and more evident that one hundred years or so is still too short to capture the statistics necessary for a more accurate probability analysis of the extreme climate events such as floods and droughts.

Studies in the last several decades have revealed that some climatic fluctuations once thought to be local phenomena are part of a large-scale atmospheric circulation that periodically affects global weather and contributes to long-term climate characteristics of different world regions. The best known and the most studied is ENSO (El Niño-Southern Oscillation). Centuries ago, the local residents on the coasts of Ecuador and Peru named a regular annual weather event El Niño ("the child") after the Christ child, because it usually appeared during the Christmas season. During this event that lasts a few weeks, a weak, warm countercurrent flows southward along the coasts of Ecuador and Peru replacing the cold Peruvian current. However, every 3 to 7 years, this countercurrent is unusually warm and strong and is accompanied by a pool of warm ocean surface water in the central and eastern Pacific that influences weather worldwide (Lutgens and Tarbuck, 1995).

The second strongest El Niño on record occurred in 1982 and 1983 (Fig. 7.9) and was blamed for weather extremes of a variety of types in many parts of the world. Heavy rains and flooding affected normally dry portions of Peru and Ecuador. Australia, Indonesia, and the Philippines experienced severe droughts, while one of the warmest winters on

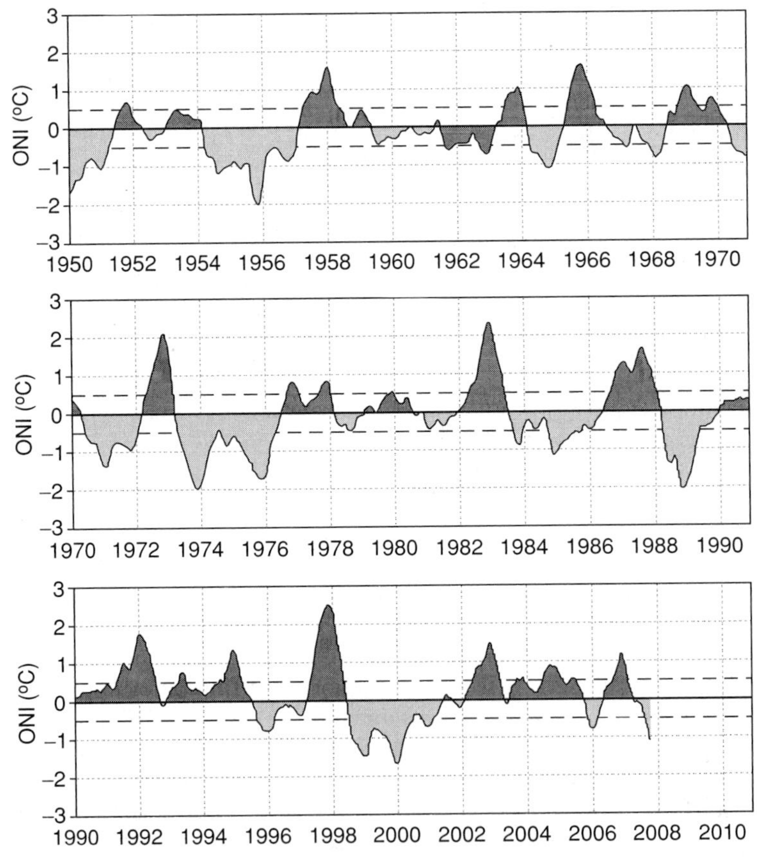

Figure 7.9 Oceanic Niño Index, ONI (°C), evolution since 1950. ONI is principal measure for monitoring, assessing, and predicting ENSO. Positive values greater than +0.5 generally indicate El Niño conditions, and negative values less than −0.5 generally indicate La Niña conditions. (From CPC, 2007.)

record was followed by one of the wettest springs for much of the United States. Heavy snows in the Sierra Nevada and the mountains of Utah and Colorado led to mudflows and flooding in Utah, Nevada, and along the Colorado River in the spring of 1983. The unusual rains brought floods to the Gulf States and Cuba. Unfortunately, as discussed by Lutgens and Tarbuck (1995), the effects of El Niño are highly variable depending in part on the temperatures and size of the warm pools in the Pacific. During one El Niño, an area may experience flooding, only to be hit by drought during the next event. It is such extreme events that water managers both fear and are constantly preparing for. Climate Prediction Center (CPC) of the National Weather Service, National Aeronautic and Oceanic Administration (NOAA), maintains a Web page dedicated to the research and weather predictions associated with El Niño and La Niña events (CPC, 2012).

In addition to El Niño and La Niña, several other large-scale interconnected atmospheric circulation patterns affect global climate including Pacific Decadal Oscillation (PDO) that changes sign about every 25 years or longer. As discussed by Wang et al. (2009), during the past four decades, precipitation near the central Intermountain Region of the western United States has experienced a pronounced increase in temporal variability, exhibiting an amplified cycle with a frequency of about 15 years, a timescale that falls between the 2 to 7 years ENSO cycle and the ≥25 years PDO variation. There is a consistent time lag between the phases of the PDO and the Intermountain rainfall, with the PDO cycle leading rainfall by about 3 to 4 years. Cross-correlation function (CCF) of the bandpass filtered PDO and Intermountain rainfall depicts a clear oscillation pattern repeating about every 12 years. The ~3-year lead time of the PDO appears to be a quarter phase of this ~12-year cycle. CCF of the 18-month low-pass filtered PDO and rainfall shows significant oscillations with the same periodicity and phase lag, supporting the ~12-year cycle (Wang et al., 2009). Interestingly, the analysis of annual precipitation in the drainage area of the largest karst spring in France, Fontaine de Vaucluse, presented in Sec. 1.7.1 (Precipitation) also indicates a 12-year cycle (see Figs. 1.170 and 1.171).

According to Chen (2006), during the last 40 years, the air temperature has increased about 0.24°C every decade in the area north of 35 degrees North in China and it will continue to increase. Climate change will greatly affect China's availability of water resources, ecosystems, and agricultural production. The author presents an analysis of precipitation cycles and probability at three locations with long-term data availability and in different climate regions, humid, subarid, and arid. At the humid tropical climate site, there is a general trend of increasing precipitation as opposed to the other two sites where the precipitation is decreasing. Spectral analysis showed that at all three sites the cycles of extreme precipitation fluctuations (droughts and floods) during growing seasons have shortened, thus leading to potentially decreased primary productivity of biomass (crops).

The occurrences of extreme precipitation (maximum and minimum) in the growing seasons at the three locations have strong agreement with ENSO events. Higher precipitation has a higher probability of coinciding with ENSO warm phase; lower precipitation has a coincidence with ENSO cold phase. The average coefficient of fluctuation at the arid and subarid sites is significantly higher after 1980, whereas at the humid climate location it is slightly lower. The precipitation variation in China is also impacted by monsoons, but the interaction of the monsoon with other large-scale climatologic features, such as ENSO, is very complex and therefore difficult to interpret. One possible explanation of the precipitation trends in China over the past several decades might be related to the increased anthropogenic black carbon aerosols and land and water use policies (Chen, 2006).

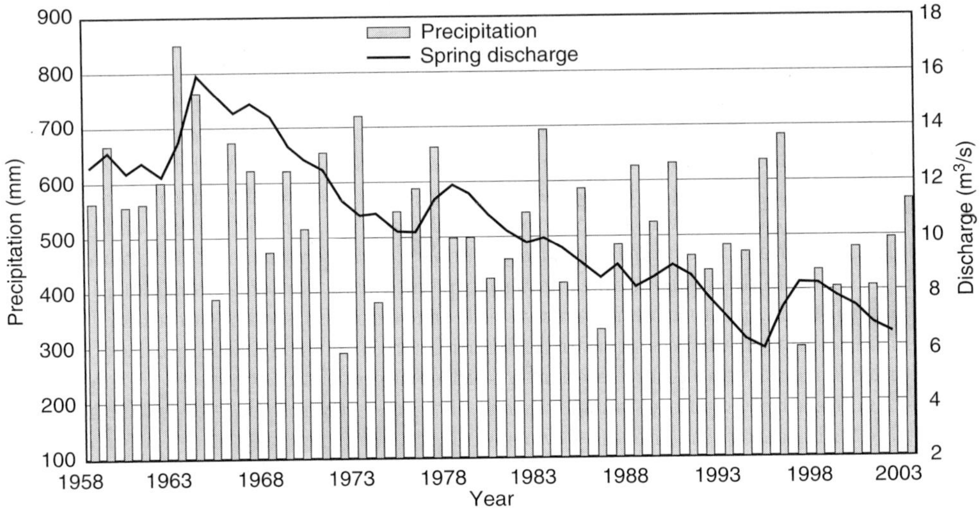

FIGURE 7.10 Discharge hydrograph of Niangziguan Spring, the largest karst spring in northern China, and precipitation in its drainage area both show a decreasing trend. (From Wu et al., 2010; copyright Elsevier Inc., printed with permission.)

Figure 7.10 illustrates negative impacts of both the decreasing precipitation trend and the land use policies in the drainage basin of Niangziguan Spring, the largest karst spring in North China. Since the early 1970s, karst groundwater in the Niangziguan Basin has been exploited for irrigation, municipal use, and industrial water supply. Water usage has accelerated since the 1980s. Nowadays, the karst groundwater is a major water source for the Niangziguan Springs Basin, one of the heavily industrialized regions in China with major coal mining, power generation, chemical engineering, and metallurgy. Karst groundwater in the Niangziguan Springs Basin is utilized from spring discharge and from wells. In 1998, the total volume of karst groundwater use was 149 million m^3, 82% of which came from spring outlets, and 18% from wells. From 1980 to 2001, the annual volume of abstracted water from wells increased from 6.43 million to 30.8 million m^3. This enormous exploitation of groundwater has reaped large socioeconomic benefits for the area, but has caused water quality and quantity problems. Overexploitation, uncontrolled urban and industrial discharges, and agricultural intensification are causing increasingly widespread degradation of the aquifers and reducing the discharge of karst springs. The discharge of Niangziguan Springs has steadily decreased since 1956. This has become a major concern for local economic development. The spring system is further impacted by dewatering for coal mining and lack of land use control (Wu et al., 2010).

According to IPCC (2007), there is high confidence that some hydrologic systems have been already affected by ongoing climate change through increased runoff and earlier spring peak discharge in many glacier- and snow-fed rivers. Global warming has also affected the thermal structure and water quality of warming rivers and lakes. IPCC projects that the mechanism of decreasing snowpack, receding glaciers, and increasing early spring runoff will affect mountainous regions across the globe and cause shortages of surface water supplies in regions at lower elevations. Most notably, the water supply of some of the most populated regions of India and China depends on large rivers originating in the high mountain ranges of Himalayas.

The reduced winter snowpack will also have consequences for groundwater recharge of intermountain basins; less and shorter snowmelt means that the surface flow will last for shorter periods of time that directly translates into less groundwater recharge at basin margins and via bedrock. A greater percentage of precipitation falling on mountain ranges and at basin margins will be in the form of rainfall immediately becoming surface water runoff. Similarly unfavorable for groundwater recharge conditions is the IPCC's projection that an increase in rainfall in some regions will come in the form of bigger, wetter storms, rather than in the form of more rainy days. In between, those larger storms will be longer periods of light or no rain, so the frequency of drought will increase.

Rising temperatures and heat waves will result in an increased demand for water, and together with droughts will put additional pressure on both surface water and groundwater resources. Warming of reservoirs and surface streams will result in water quality problems such as algal blooms. Higher temperatures will cause higher water losses from surface water reservoirs due to increased evapotranspiration. Increased evapotranspiration will also affect shallow subsurface soil causing higher soil moisture deficits. A larger fraction of the infiltrated precipitation will therefore be spent on satisfying this soil moisture deficit before deep circulation and actual groundwater recharge can take place. In areas with irrigated agriculture, higher temperatures, heat waves, and droughts will put additional demand on water resources because more water will have to be diverted from streams and reservoirs or pumped from aquifers to offset higher evapotranspiration and higher soil moisture deficit.

More frequent episodes of heavy precipitation will cause deterioration of water quality in surface streams due to increased turbidity and runoff from nonpoint sources of contamination. A related increased incidence of floods will put pressure on water supply infrastructure and increase the risks of waterborne diseases.

A sea-level rise will cause salinization of irrigation water in coastal areas, estuaries and freshwater systems, and saltwater intrusion in shallow aquifers underlying low coastal areas. In addition, population relocation from the semiarid and arid regions severely affected by prolonged droughts, and from the low-lying coastal areas affected by the sea-level rise will put pressure on water resources in other regions that otherwise may not be significantly impacted by global warming.

Some of the projected continental-scale impacts of global warming are given below (IPCC, 2007).

Africa

- By 2020, between 75 and 250 million people are projected to be exposed to increased water stress due to climate change.
- By 2020, in some countries, yields from rain-fed agriculture could be reduced by up to 50%. Agricultural production, including access to food, in many African countries is projected to be severely compromised. This would further adversely affect the predictability of food production and exacerbate malnutrition.
- Toward the end of the 21st century, the projected sea-level rise will affect low-lying coastal areas with large populations. The cost of adaptation could amount to at least 5% to10% of gross domestic product (GDP).

- By 2080, arid and semiarid land in Africa is projected to increase by 5% to 8% of under a range of climate scenarios.

Asia

- By 2050, freshwater availability in Central, South, East and South-East Asia, particularly in large river basins, is projected to decrease.
- Coastal areas, especially heavily populated megadelta regions in South, East and South-East Asia, will be at greatest risk due to increased flooding from the sea and, in some megadeltas, flooding from the rivers.
- Climate change is projected to compound pressures on natural resources and the environment, along with rapid urbanization, industrialization, and economic development.
- Endemic morbidity and mortality due to diarrhoeal disease primarily associated with floods and droughts are expected to rise in East, South, and South-East Asia due to projected changes in the hydrologic cycle.

Australia and New Zealand

- By 2020, a significant loss of biodiversity is projected to occur in some ecologically rich sites including the Great Barrier Reef and Queensland Wet Tropics.
- By 2030, water availability problems are projected to intensify in southern and eastern Australia and, in New Zealand, in Northland and some eastern regions.
- By 2030, production from agriculture and forestry is projected to decline over much of southern and eastern Australia, and over parts of eastern New Zealand, due to increased drought and fire. However, in New Zealand, initial benefits are projected for some other regions.
- By 2050, ongoing coastal development and population growth in some areas of Australia and New Zealand are projected to exacerbate risks from sea-level rise and increases in the severity and frequency of storms and coastal flooding.

Europe

- Climate change is expected to magnify regional differences in Europe's natural resources and assets. Negative impacts will include an increased risk of inland flash floods, more frequent coastal flooding, and increased erosion (due to an increase in storms and sea-level rise).
- Mountainous areas will experience glacier retreat, reduced snow cover and winter tourism, and extensive species losses (in some areas up to 60% by 2080).
- In Southern Europe, climate change is projected to worsen conditions (high temperatures and drought) in a region already vulnerable to climate variability, and to reduce water availability, hydropower potential, summer tourism and, in general, crop productivity.
- Climate change is also projected to increase the health risks due to heat waves and the frequency of wildfires.

Latin America

- By mid-century, increases in temperature and associated decreases in soil water are projected to lead to gradual replacement of tropical forest by savanna in eastern Amazonia. Semiarid vegetation will be slowly replaced by arid-land vegetation.
- There is a risk of significant biodiversity loss through species extinction in many areas of tropical Latin America.
- Productivity of some important crops is projected to decrease and livestock productivity to decline, with adverse consequences for food security. In temperate zones, soybean yields are projected to increase. Overall, the number of people at risk of hunger is projected to increase.
- Changes in precipitation patterns and the disappearance of glaciers are projected to significantly affect water availability for human consumption, agriculture, and energy generation.

North America

- Warming in the mountains of the American west is projected to cause decreased snowpack, more winter flooding, and reduced summer flows, exacerbating competition for overallocated water resources.
- In the early decades of the century, moderate climate change is projected to increase aggregate yields of rain-fed agriculture by 5% to 20%, but with important variability among regions. Major challenges are projected for crops that are near the warm end of their suitable range or which depend on highly utilized water resources.
- During the course of this century, cities that currently experience heat waves are expected to be further challenged by increased number, intensity, and duration of heat waves during the course of the century, with the potential for adverse health impacts.
- Coastal communities and habitats will be increasingly stressed by the changing climate.

Small Islands

- Sea-level rise is expected to exacerbate inundation, storm surge, erosion, and other coastal hazards, thus threatening vital infrastructure, settlements, and facilities that support the livelihood of island communities.
- By mid-century, climate change is expected to reduce water resources in many small islands, such as in the Caribbean and Pacific, to the point where they become insufficient to meet demand during low-rainfall periods.

7.3 References

Berger, A., and Loutre, M. F., 1991. Insolation values for the climate of the last 10 million years. *Quat Sci Rev*, 10 :297–317. Data available at: http://www.ncdc.noaa.gov/paleo/forcing.html#orbital

Chen, X., 2006. Comparison of the recent precipitation variation at three locations in China. *J App Sci*, 6(1): 144–150.

CPC (Climate Prediction Center), 2007. ENSO cycle: Recent evolution, current status and predictions, update prepared by Climate Prediction Center/NCEP, December 24, 2007. National Weather Service. Available at: http://www.cpc.ncep.noaa.gov/products/analysis_monitoring/lanina/index.html. Accessed 28th December 2007.

CPC (Climate Prediction Center), 2012. El Niño/La Niña Home. National Weather Service. Available at: http://www.cpc.ncep.noaa.gov/products/analysis_monitoring/lanina/index.html

Hays, J. D., Imbrie, J., and Shackleton, N. J., 1976. Variations in the earth's orbit: Pacemaker of the ice ages. *Science*, 194 (4270): 1121–1132.

IPCC (Intergovernmental Panel on Climate Change), 2007. Summary for policymakers of the Synthesis Report of the IPCC Fourth Assessment Report; Draft Copy 16 November 2007 – subject to final copyedit, 23p. Available at: http://www.ipcc.ch/.

Lutgens, F. K. and Tarbuck, E. J., 1995. *The Atmosphere*. Prentice-Hall, Inc. Englewood Cliffs, NJ, 462p.

Milanovic, P., 1979. Hidrogeologija karsta i metode istraživanja (in Serbian; Karst hydrogeology and methods of investigations). HE Trebišnjica, Institut za korištenje i zaštitu voda na kršu, Trebinje, 302p.

Milanovic, P. T., 2004. *Water Resources Engineering in Karst*. CRC Press, Boca Raton, FL, 312p.

Milanovic, P. T., 2006. Karst istočne Hercegovine i dubrovačkog priobalja (Karst of eastern Herzegovina and Dubrovnik litoral; in Serbian). ZUHRA, Belgrade, 362p.

NASA (National Aeronautic and Space Agency), 2007. On the shoulders of giants. Milutin Milankovitch (1879–1958). NASA Earth Observatory. Available at: http://earthobservatory.nasa.gov/Library/Giants/Milankovitch/

NRC (National Research Council), 1982. *Solar Variability, Weather, and Climate*. National Academy Press, Washington, D.C., p. 7.

Riebek, H., 2007. Global warming. NASA Earth Observatory. Available at: www.earthobservatory.nasa.gov/library/globalwarmingupdate/

Tesla Memorial Society of New York, 2007. Milutin Milankovitch. Available at: http://www.teslasociety.com/milankovic.htm

Wang, S.-Y., R. R. Gillies, J. Jin, and L. E. Hipps, 2009: Recent rainfall cycle in the Intermountain Region as a quadrature amplitude modulation from the Pacific Decadal Oscillation, Geophys. Res. Lett., 36, L02705.

Wu, Q., Xing, L., and Zhou, W., 2010. Utilization and protection of large karst springs in China. In: Kresic, N., and Stevanovic, Z. (eds.), *Groundwater Hydrology of Springs: Engineering, Theory, Management and Sustainability*. Elsevier, New York, pp. 543–565.

CHAPTER 8

Groundwater Extraction

8.1 Introduction

Groundwater extraction in karst is generally accomplished by the following: (1) installation of individual water wells or wellfields, (2) construction of underground dams (reservoirs), and (3) tapping of springs including engineered regulation of spring discharge. In addition to classical vertical wells, groundwater extraction can be also accomplished with slanted wells and drainage galleries. Some other methods common to aquifers developed in unconsolidated sediments, such as drainage trenches, drains, or collector wells, are often not feasible in karst. In any case, when selecting locations and method(s) of groundwater extraction, the following general factors should be considered:

- Capital cost.
- Vicinity to future users.
- Existing groundwater users and groundwater permits.
- Hydrogeologic characteristics and depth to different water-bearing zones (aquifers).
- Required flow rate of the water-supply system and expected yield of individual wells.
- Well drawdown and radius of well (wellfield) influence.
- Interference between wells in the wellfield.
- Water treatment requirements.
- Energy cost for pumping and water treatment, and general operations and maintenance (O&M) costs.
- Aquifer vulnerability and risks associated with the existing or potential sources of contamination.
- Interactions with other parts of the groundwater system, and with surface water.
- Options for artificial aquifer recharge, including storage and recovery.
- Societal (political) requirements.
- Existence or possibility of an open water market.

The above factors are not all-inclusive and are not listed in order of importance; sometimes just one or two are all that is needed or feasible for proceeding with the final design, including various compromises that often have to be made. However, as the development and use of groundwater resources is becoming increasingly regulated in the United States and in many other countries, it is likely that most of these factors will have to be addressed as part of a groundwater extraction permitting process. Even in cases where permitting requirements are absent, it is prudent to consider most, if not all, of the listed factors since they ultimately define the long-term sustainability of the project.

8.2 Water Wells in Karst

Probably the first thought that comes to mind to many people when discussing groundwater development and use, in general, is a water well. For nonhydrogeologists and those who are not in a related water-supply profession, a well usually means a nondescript hole in the ground that somehow produces water; this may include an image of a fenced well house, or a picturesque countryside image of a dug well with a rotating wooden wheel and a bucket. In any case, a relatively small number of people fully understand the complexity, importance, and cost of a properly constructed well used for public water supply. The same is true in many developed countries where modern drilling technologies have been routinely used for a long time to construct wells for both the public and domestic supply (Fig. 8.1). The end users usually leave this "well business" to well drillers and do not care to learn much about their own hole in the ground. However, hydrogeologists and groundwater professionals think of wells in many different contexts, and some of them spend lifetimes trying to better understand and design them.

Selecting the "best" site for a well or wellfield in karst is considered by some to be an art or a matter of luck, by some to be as simple as finding a recommended dowser, and by some (such as well drillers) as a natural part of collecting fees for drilling holes into the

FIGURE 8.1 Small-diameter private water supply well intersecting a submerged passage of Little River Spring Cave, Florida. (From The Florida Springs Task Force, 2000.)

subsurface as deep as possible. Although a few groundwater professionals may agree with some of these statements (except, arguably, relying on a dowser), sitting and then designing wells for public supply or large irrigation projects is very complex and should result from thorough considerations of multiple design elements. The most important, however, is to first consider all indirect and direct methods for locating preferential flow paths described in Sec. 3.2. A well located in or adjacent to a preferential flow path in karst will have much higher probability of being successful than a well located at random. The exceptions are wells drilled in younger carbonates with high matrix porosity such as Floridan aquifer in the United States. In such aquifers, virtually all wells will have significant yields, while those close to or in preferential flow paths will be exceptionally productive.

Well depth, diameter, and construction methods vary widely and there is no such thing as a "one size fits all" approach to well design. Answers to just about any question regarding well design can be found in the classic 1,000-page book *Groundwater and Wells* by Driscoll (1986). Another exhaustive reference book on well design is *Water Well Technology* by Campbell and Lehr (1973). Various public-domain publications by U.S. government agencies provide useful information on the design and installation of water supply and monitoring wells (e.g., USEPA, 1975, 1991; U.S. Bureau of Reclamation, USBR, 1977; Aller et al., 1991; Lapham et al., 1997). Well design, installation, and well construction materials should conform to applicable standards. In the United States, the most widely used water well standard is the ANSI/AWWA A100 standard, but the authority to regulate products for use of, or in contact with, drinking water, rests with individual states that may have their own standard requirements. Local agencies may choose to impose requirements more stringent that those required by the state (AWWA, 1998).

The general design elements of water wells in any porous media include the following: (1) drilling method, (2) boring (drilling) and casing diameter, (3) depth, (4) well screen, (5) gravel pack, (6) well development, (7) well testing, and (8) selection and installation of the permanent pump. In most karst aquifers, however, wells can often be installed without screens and gravel packs and completed as open borings, which significantly reduces well costs and eliminates many issues associated with well screen design, well efficiency/loss, and overall well maintenance. In cases of large public supply wells, it still is desirable to install coarse screens that will maintain borehole stability and prevent mechanical damage to the pump caused by collapsing cavities and dislodged rocks.

Figures 8.2 and 8.3 illustrate some of the more common well designs that can vary widely based on project-specific requirements and can be combined in a single well. Continuing advances in drilling and well installation technology allow for elaborate designs such as underreaming (widening of borehole below already installed and grouted casings), use of temporary casings for drilling in unstable conditions, telescopic screens, multiple screen intervals with or without continuous gravel packs, and slanted wells.

The expected well yield, the well depth, and the geologic and hydrogeologic characteristics of the porous media (rock) all play an important role in selecting the drilling diameter and the drilling method. Deep wells, or thick stratification of permeable and low-permeable porous media, may require drilling with several diameters, and the installation of several casings of progressively smaller diameter, called telescoping casing (Fig. 8.2A). This is done to provide stable and plumb boreholes in deep wells, and to bridge difficult or undesirable intervals (e.g., flowing sands, highly fractured and unstable walls prone to caving, and thick sequences of swelling clay). The cost of drilling increases progressively with the drilling diameter and it is important to balance this

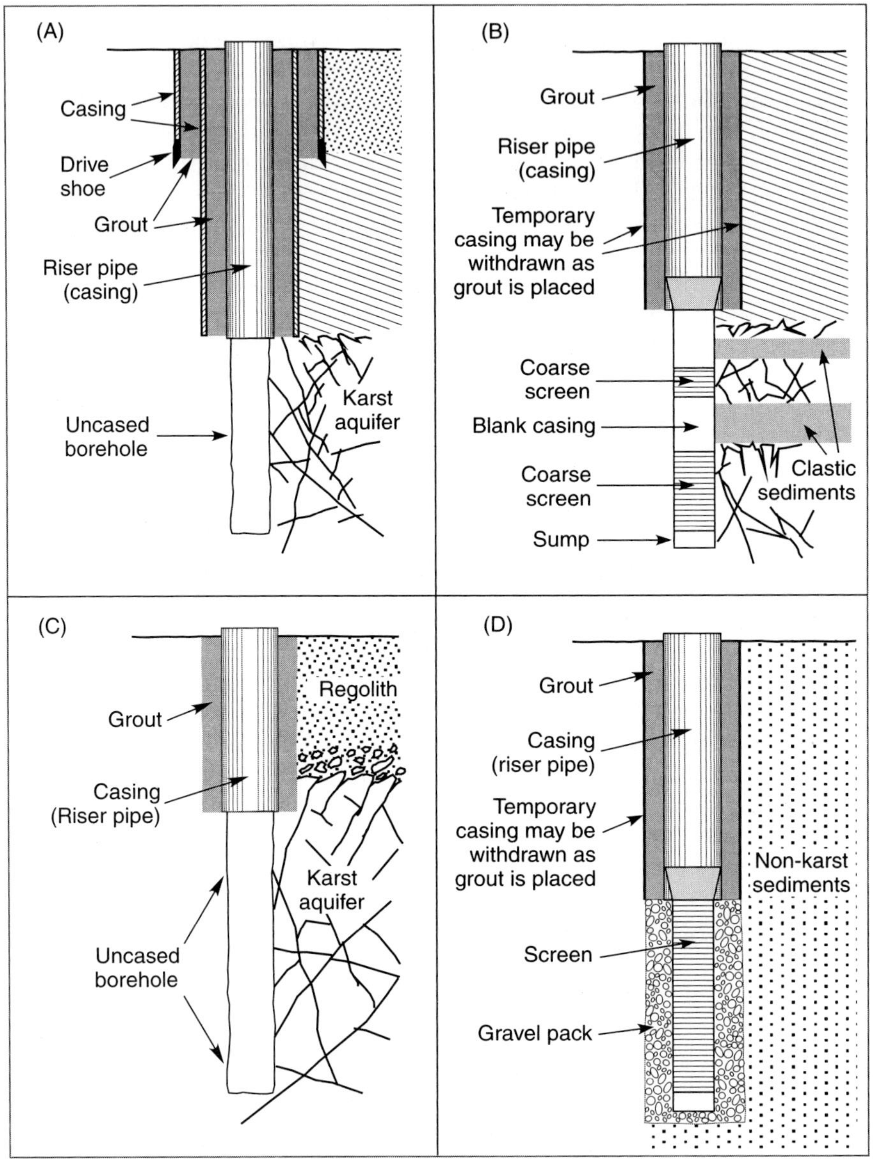

FIGURE 8.2 Some basic well types. (A) Deep well with multiple cemented casing for bridging unstable and undesired formations, completed as open borehole in the deeper karst aquifer; (B) Well cased through overlying unconsolidated sediments, and with multiple coarse screen intervals in the karstified aquifer separated by blank casing to isolate unconsolidated sediments and unstable intervals; (C) Well completed as open borehole in stable karstified bedrock and with casing cemented in place, extending through regolith and upper portion of the bedrock. (D) Well in unconsolidated (non-karstic) sediments with telescoped screen and gravel pack, and with riser pipe cemented in place.

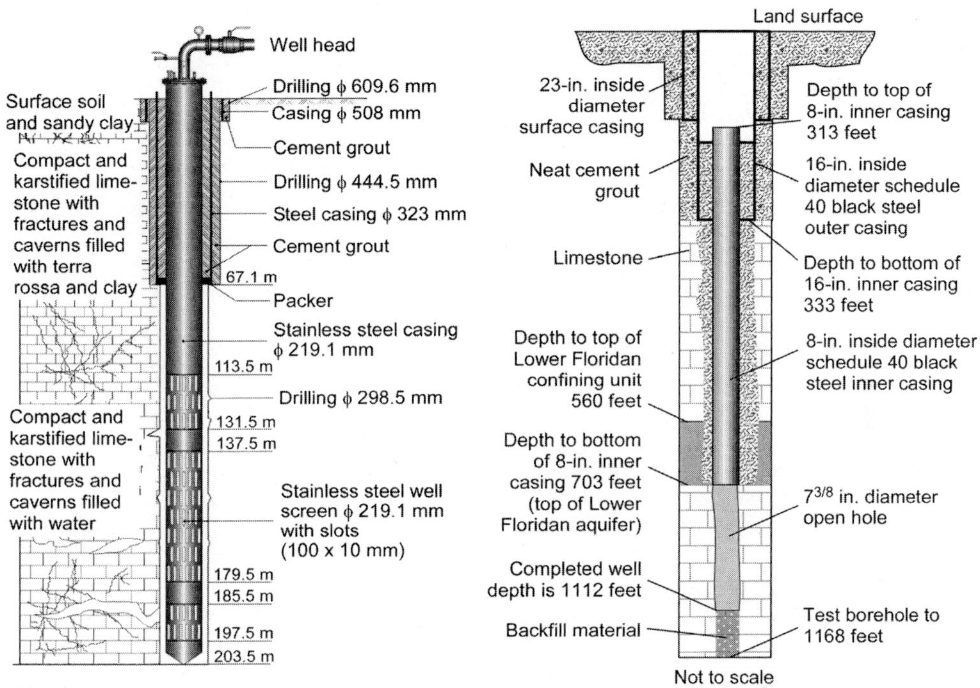

FIGURE 8.3 *Left*: Construction of well RAJ-2 in the village of Rajkovic near Mionica, western Serbia yielding 2 L/s with drawdown of 112 m. (From Djokić et al., 2005.) *Right*: Well-construction diagram for well HAAF 11 (36Q392) completed in the Lower Floridan aquifer, Hunter Army Airfield, Chatham County, Georgia. (From Clarke et al., 2010.)

cost with other design requirements, some of which may be desirable but not always necessary. For perspective, a public supply, high capacity well, that is several thousands feet deep (e.g., 1,000 m), may easily cost well over 1 million U.S. dollars. Such wells are drilled with large drill rigs and may utilize special large-diameter drilling bits and include large-diameter casing as shown in Fig. 8.4.

Whenever possible, a well design should be based on information obtained by a pilot boring drilled prior to the main well bore. Geophysical logging, coring, and sample collection in the pilot boring provide important information on depth, thickness, and permeability of the water-bearing intervals, and physical and chemical characteristics of the porous media and groundwater. Unknown geology and hydrogeology of the formation(s) to be drilled may result in the selection of an improper drilling technology, sometimes leading to a complete abandonment of the drilling location due to various unforeseen difficulties such as flowing unconsolidated materials (sands), collapse of boring walls, or loss of drilling equipment in karst cavities.

All permanent well casings have to be continuous and watertight from top to bottom. Grouting prevents possible short-circuiting of groundwater along the boring walls and between various untargeted water-bearing zones (aquifers), as well as contamination from the land surface. In the United States, most states require that the upper casing be grouted a minimum depth from the land surface, usually 50 feet. Casing material must be compatible with the groundwater chemistry to prevent corrosion or other failures.

Figure 8.4 *Left*: Roller cone-type bits (pointed out by arrows in the center and bottom center) are preferable when drilling deep, large-diameter wells. They are constructed in configurations such as this one to maintain a plumb borehole and enlarge its diameter in stages. (Photo courtesy of Adrian Ezeagy.) *Right*: 48-inch-diameter surface casing being placed in a 54-inch-diameter borehole. The target depth for the well is 1,300 feet with an open interval from 1,000 to 1,300 feet in the Floridan aquifer. The final casing inside diameter is 20 inches. The well is drilled for water supply of the City of Miramar in Florida, the United States. (Photo courtesy of Richard Crowles.)

The selection of materials for well casing is critical in locations where there is likelihood of its exposure to significant concentrations of contaminants comprising low molecular weight petroleum products or organic solvents and their vapors. Casing materials such as polyethylene, polybutylene, polyvinyl chloride (PVC), and elastomers, such as used in jointing gaskets and packing glands, may be subject to permeation by lower molecular weight organic solvents or petroleum products (AWWA, 1998). If the well casing extends through such a contaminated area or an area subject to contamination, the well casing material should be selected accordingly. Casing has to be strong and thick enough to provide for structural stability during its installation, well development, and use. This is especially important in deep wells where high formation pressures may cause casing to collapse if undersized (under designed). In general, selecting an inferior casing material will initially reduce the capital cost, but may result in irreparable damage to the casing and premature loss of a well. ANSI/AWWA A100 standard provides specifications for casing materials and diameters, and calculations of minimum acceptable casing strengths.

As mentioned earlier, wells in stable bedrock are in many cases completed as open borehole intersecting as many fractures and cavities as possible to maximize well yield and minimize construction cost. Such wells should still have an appropriately grouted casing in the top portion to prevent caving-in of regolith material and pieces of weathered rock, and contamination of the well from the land surface. It is highly recommended that the grouted casing extends through the entire thickness of the regolith and the highly weathered rock, and for some distance into the more competent bedrock. This prevents deterioration of the well due to inflow of fine particles and it extends the life of the well pump. Although the final boring diameter of unlined, open-borehole wells still has to accommodate the pump assembly and provide for easy maintenance access, it is not limited by the well screen diameter and gravel pack thickness required for the wells screened in unconsolidated sediments.

The selection of drilling methods based on geologic formation and well depth and the optimum and minimum size of well inner casing (riser pipe) for various pumping rates are presented in Tables 8.1 and 8.2, respectively.

8.2.1 Well Development and Completion

Proper well development will improve almost any well regardless of type and size, whereas without development an otherwise excellent well may never be satisfactory. As discussed by USEPA (1975) and Drsicoll (1986), in any well drilling technology the permeability around the borehole is reduced. Compaction, clay smearing, and driving fines into the wall of the borehole occur in the cable tool drilling method. Drilling fluid invasion into the aquifer and formation of a mud cake on the borehole walls are caused by direct rotary method. Silty and dirty water often clog the aquifer in the reverse rotary drilling method. In consolidated formations, compaction may occur in some poorly cemented rocks, where cuttings, fines, and mud are forced into fractures, bedding planes, and other openings, and a mud cake forms on the wall of the borehole.

Proper well development breaks down the compacted borehole wall, liquefies jelled mud, and draws it and other fines that have penetrated the formation (aquifer) into the well, from which they are removed by bailing or pumping. Well development also removes smaller grains initially present in the formation and creates a more permeable and stable zone adjacent to the well screen. Development of new wells is based on the mechanical action of water or air and should involve backwashing, that is, movement of water in both directions through the well screen openings. Chemicals should be used in well development only exceptionally, in small quantities, and with prior approval of both the well owner and the regulatory agency where applicable. Such chemicals include mud-dispersing agents such as crystalline and glassy polyphosphates, and acids for wells completed in limestone. Various problems can arise from improper use of chemicals leading to an inferior well or sometimes to a complete screen clogging. For example, an over rich solution of sodium hexametaphosphate (SHMP) can precipitate glassy phosphates on contact with the cold groundwater. These glassy precipitates are gelatinous masses that are extremely difficult to remove because no effective solvents exist (Driscoll, 1986).

There are various methods of well development, and their selection depends primarily on the applied drilling technology and the formation characteristics. However, availability of the equipment and driller's preference in many cases play unjustifiably more important roles. It is often impossible to anticipate how a well will respond to certain types of development and how long it will take to achieve adequate development. Since a lump-sump basis for well development may result in unsatisfactory work, it is

				Drilled			
					Rotary		
Characteristics	Dug	Bored	Driven	Percussion	Hydraulic	Air	Jetted
General range of common depths	0–50 feet	0–100 feet	0–50 feet	0–1,000 feet	0–1,000 feet	0–750 feet	0–100 feet
Diameter	3–20 feet	2–30 in.	$1^{1/4}$–2 in.	4–18 in.	4–24 in.	4–10 in.	2–12 in.
Type of geologic formation							
Clay	Yes	Yes	Yes	Yes	Yes	No	Yes
Silt	Yes	Yes	Yes	Yes	Yes	No	Yes
Sand	Yes	Yes	Yes	Yes	Yes	No	Yes
Gravel	Yes	Yes	Finer size	Yes	Yes	No	1/4 in. pea gravel
Cemented gravel	Yes	No	No	Yes	Yes	No	No
Boulders	Yes	Yes if < than well diameter	No	Yes when in firm bedding	Difficult	No	No
Sandstone	Yes, if soft or fractured	Yes, if soft or fractured	Thin layers	Yes	Yes	Yes	No
Limestone	No	No	No	Yes	Yes	Yes	No
Dense igneous rock	No	No	No	Yes	Yes	Yes	No

From USEPA (1991).

TABLE 8.1 Applicable Drilling Method for Different Types of Geologic Formations

Anticipated Well Yield		Optimum Casing Size		Smallest Casing Size	
gpm	l/s	in.	mm	in.	mm
Less than 100	Less than 5	6 ID	152 ID	5 ID	127 ID
75–175	5 to 10	8 ID	203 ID	6 ID	152 ID
150–350	10 to 20	10 ID	254 ID	8 ID	203 ID
300–700	20 to 45	12 ID	305 ID	10 ID	254 ID
500–1,000	30 to 60	14 OD	356 OD	12 ID	305 ID
800–1,800	50 to 110	16 OD	406 OD	14 OD	356 OD
1,200–3,000	75 to 190	20 OD	508 OD	16 OD	406 OD
2,000–3,800	125 to 240	24 OD	610 OD	20 OD	508 OD
3,000–6,000	190 to 380	30 OD	762 OD	24 OD	610 OD

Modified from Driscoll, 1986; Reprinted by Permission of Johnson Screens/a Weatherford Company.
ID, inside diameter; OD outside diameter.

TABLE 8.2 Recommended Well Diameters for Various Pumping Rates

better to provide for development on a unit price per hour basis and continue until the following conditions have been met (modified from AWWA, 1998):

1. Content of sand and fines should average not more than 5 mg/L for a complete pumping cycle of 2-hour duration when pumping at the design discharge capacity.
2. No less than 10 measurements should be taken at equal intervals to permit plotting of sand content as a function of time and production rate and to determine the average sand content for each cycle.
3. There should be no significant variation in specific capacity during at least 24 hours of development.

General methods of well development are pumping, surging, fracturing, and washing, each of which has several variations (USEPA, 1975). It is recommended that at least two methods be applied for best results. One of the less effective—but commonly used—methods is overpumping the well. Here, water flows in one direction only—toward the well. The flow velocities are generally not fast enough to remove much of the fine material plugging the formation. During overpumping, a surging action can be created by periodically shutting off the pump and allowing the water in the pump column to flow back into the well. This is more effective than overpumping. However, water will reenter the most permeable parts of the formation or those that have been least damaged during well construction. Thus, the portion of the formation that requires the most active development is largely excluded (Johnson Screens, 2007). Injecting water into the well and then pumping it back from the well (backwashing) will result in movement of water through the wells screen and borehole walls in both directions, thus increasing the effectiveness of development.

Pumping with compressed air or airlift is probably the most common method of well development (Fig. 8.5). Driscoll (1986) provides a very detailed discussion on various

Figure 8.5 Development of a high-capacity water supply well completed in Floridan aquifer. (Photograph courtesy of Dr. Gerald Seaburn.)

airlift techniques, including determinations of quantitative parameters required for proper airlift design. Unfortunately, many well drillers and contractors always apply the same airlift method (one they are familiar with) without regard for site-specific conditions.

As with unconsolidated formations, all drilling methods including air-drilling cause some plugging of fractures and other openings in consolidated sediments and hard rock formations. Therefore, any material that clogs openings in such formations should be removed during well development. In many cases, the best method is the water jetting combined with airlift pumping. Inflatable packers can isolate the productive zones (fractures) supplying water to the well and increase efficiency of their development.

Well yield in carbonate rocks can sometimes be increased considerably if one or more well stimulation methods are applied. Well stimulation is considered as a second level of development which can increase well performance beyond that obtained through traditional methods. Hydrofracturing is used to stimulate both new and old wells in consolidated rock formations. In hydrofracturing, water at extreme pressures can be injected into the entire well or into discrete intervals sealed by packers. The injected water removes sediment from the existing fractures and cavities and creates new fractures resulting in an increased permeability of the formation adjacent to the well.

Blasting with explosive charges lowered in an uncased borehole in consolidated rock is sometimes used to increase well specific capacity. Similarly to hydrofracturing, this method enlarges the existing fractures and creates new ones resulting in an increased hydraulic conductivity of the formation. However, blasting with explosives should be applied with utmost care and only after considering many factors including legal requirements and environmental impacts.

Acid can be used for well stimulation and formation development in limestone and dolomite aquifers. Acid dissolves carbonate minerals and enlarges voids and small fissures in the formation adjacent to the borehole. Acid can also be forced into discontinuities away from the well resulting in dissolution and removal of a larger volume of the native material. This increases the overall hydraulic conductivity of the aquifer around the well and may result in a significant increase in the specific capacity of the well.

Well completion includes disinfection of the entire casing and the screen if present, installation of the permanent well pump, and construction of the well housing (wellhead) with all its sanitary requirements. It is highly recommended that a dedicated small-diameter perforated pipe (sounding line) be installed in the well for water level measurements and depth-discrete sampling.

Probably the only design element of the entire well that cannot be guaranteed by any well driller, hydrogeologist, or engineer is the well yield and its long-term sustainability. It is not uncommon that, for various reasons, a well that costs several hundred thousand dollars to complete, disappoints all stakeholders by actually producing just a fraction of its designed capacity. However, many such surprises can be avoided by following well-established hydrogeologic principles of karst aquifer evaluation and testing, and, of course, well design and installation itself. And the best-kept secret is that a little bit of luck is always needed when installing a well in karst for any use.

8.3 Extraction from Cavities

Natural karst cavities, such as sinkholes and cave passages filled with water, often are attractive targets for groundwater withdrawal. As opposed to drilled wells, where finding preferential flow paths that are actually transmitting water is always uncertain to varying extent, known submerged karst cavities are the best indicators of transmissive aquifer zones or are part of such zones themselves. When of sufficient size, they can accommodate large-capacity pumps and sustain high yields when connected with an extensive conduit network. Stewart (1977) provides a very detailed example of pumping from 200-feet deep Morris Bridge Road Sink near Tampa, Florida (Fig. 8.6). The sink was pumped for 25 days at an average rate of 4,000 gallons per minute (250 L/s) to test its possible use as a supplemental water source for municipal supply. Drawdowns at the end of the test were 5.8 feet (1.8 m) in the sink, 1.5 feet in an observation well 655 feet (200 m) from the sink, and no detectable drawdown in an observation well 2,500 feet (760 m) from the sink. Based on drawdown data from the observation wells near the sink, the lumped aquifer transmissivity was estimated at 1.3×10^5 ft^2/day (1.2×10^4 m^2/day). The test indicated that the sink would be capable of yielding 15 million gallons per day (0.7 m^3/s) on a sustainable basis, with a maximum drawdown likely not to exceed about 23 feet (7 m).

One of the most productive wells in Alabama, the Williams Well, which can sustain pumping rate of over 5,000 gpm (316 L/s) penetrates a cavity 3 m in diameter, 30 m below land surface. The well is located in an area approximately 7 miles southwest of Huntsville, where a grouping of high-yield wells is attributed to large-scale lineaments (Fig. 8.7) clearly visible on LANDSAT imagery (Moore et al., 1977). Of this cluster, one more well has yield greater than 5,000 gpm, two wells yield more than 1,000 gpm (63 L/s),

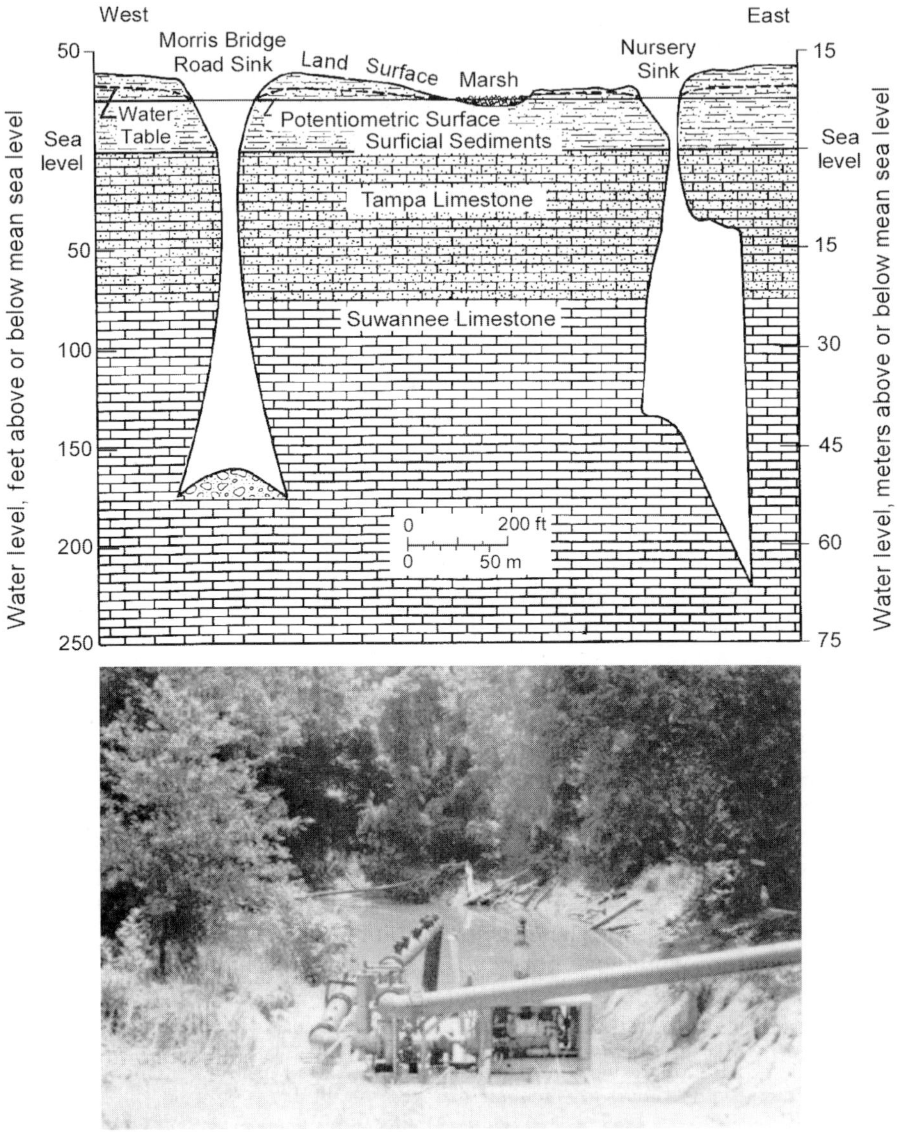

FIGURE 8.6 *Top*: Generalized section through Morris Bridge Road Sink and Nursery Sink, based in part on reports of scuba divers. *Bottom*: Morris Bridge Road Sink during pumping at an average rate of about 4,000 gal/min or 250 L/s. (From Stewart, 1977.)

and two springs also have yields greater than 1,000 gpm. Compared to the average yield of approximately 20 gpm (1 L/s) for wells completed in low matrix porosity Mississippian carbonate rocks in this part of Alabama, Moore et al. (1977) qualify the Williams Well group as an anomalous hydrologic condition. One likely explanation for this "anomaly" is that the wells are connected to an extensive regional network of karst conduits as indicated by the large cavity at Williams Well.

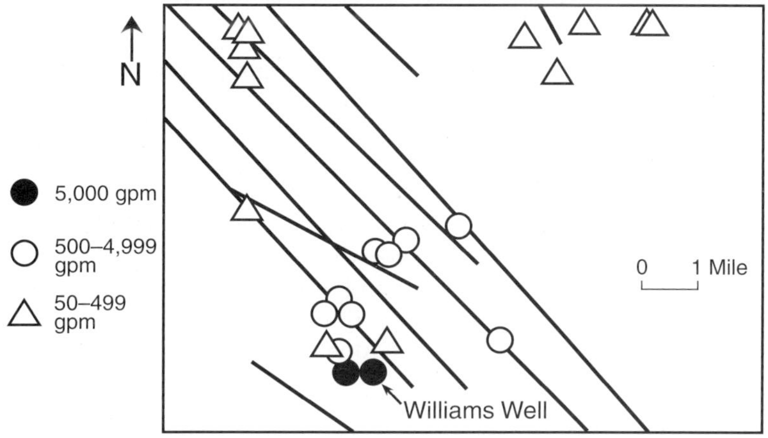

Figure 8.7 Large-scale lineaments in the Williams Well area with high-capicity wells (denoted with circles) and springs (triangles). (From Moore et al., 1977.)

Figure 8.8 shows what may be an ultimate dream by well drillers: a well that did not have to be drilled and just appeared at wish. This is what is written on a board by the well: "The 0–9 well is a Texas historical landmark. The date of discovery of this natural cavern is unknown. It has played a major role in the development of the northern part of Crockett County by providing an endless supply of fresh water for the cattle and sheep

Figure 8.8 0-9 Well, a natural 120-foot-deep pit near Ozona, Texas, on the Edwards Plateau. Wagon trains used to stop here for water as they could lower a bucket on a rope and get water from the pit. (Photograph courtesy of Geary Schindel, Edwards Aquifer Authority.)

industry. As early as the mid-1800s, cattle were watered at the well on their long journey north from Mexico to Kansas. The 'natural well,' also served as a regular stop for the stage route between Ozona and San Angelo until the early twentieth century. The well and surrounding property was granted to the University of Texas by the state of Texas in 1876 and remains a part of the U.L.S. system today. 0–9 well is protected under the Texas Cave Protection Act and under the management of the Texas Cave Management Association. No unauthorized entry."

Cavities, vertical conduits, and siphons in the vicinity of and supplying water to large vauclusian springs can be tapped directly using a variety of engineered structures including vertical and slanted wells, and drainage galleries (adits, tunnels). As opposed to aquifer pumping away from springs, this form of groundwater extraction is considered spring regulation that is explained in more detail in the following sections.

8.4 Capture of Karst Springs

The following analysis by Sun et al. (1963) illustrates the importance of springs in karst terrains, as well as some limitations for their wider utilization as reliable sources of water supply. Springs are important sources of water for municipal, domestic, and farm use in east Tennessee; in 1959, 39 of the 95 municipal water-supply systems used water derived solely from springs, and 15 others used springwater as a supplemental source. Many of the springs are not developed, because of their inaccessibility or the lack of information regarding their adequacy and dependable low flows available for small-industry and community supplies. A spring discharging 450 gpm (gallons per minute), or about 1 cfs (cubic foot per second), is capable of supplying a town of 6,500 population, if one assumes a per capita consumption of 100 gpd (gallons per day). A number of large springs have been identified as excellent sources of water for future development in east Tennessee. However, the variability of their flow may prevent full utilization unless adequate storage facilities are provided. If there is no storage to provide water during periods of peak use and minimum yield, which usually are concurrent, a spring may be developed only to the extent of its lowest dependable flow. If adequate storage facilities were provided, however, development could approach the average annual flow.

The subject of tapping, engineering, utilization, and management of springs for water supply is covered in great detail in Kresic and Stevanovic (2010). Case studies from around the world and engineering design principles described in this book include numerous karst springs. As pointed out by Stevanovic (2010), tapping of springwater is an ancient art. Historically, to have easy access to water, cities were often situated near large karst springs, while those cities without reliable water supply were destroyed or abandoned because they could not survive long sieges. As a rule, the cities with plentiful water drawn from successfully constructed spring intakes provided a base for prosperous development and a safe haven for their citizens. For example, in the narrow historical centre of Rome, there were 23 springs that initially supplied water throughout the city, while at the height of the Roman Empire, 11 long aqueducts delivered more than 13 m^3/s of water to the city from karst springs at distances ranging from 16 to 91 km (Lombardi and Corazza, 2008).

Springs can meet water demands on a wide scale: from the supply of large cities at the regional level to the supply of just one or several households. In the latter case, small springs with yields of 0.1 L/s or less will often suffice even without construction of a

FIGURE 8.9 *Left*: Part of Kläffer Spring during a flood in August 2002. Waterfall coming out of the spring cave is visible in the upper right. The spring flows into Salza River shown at the bottom. *Right*: Partly flooded gallery that receives direct inflow of large Kläffer Spring. The gallery was excavated into the limestone beneath natural spring orifices and intercepts two water-bearing cavities. (Photographs courtesy of Dr. Lukas Plan; see also Plan et al., 2010.)

large storage chamber (reservoir). However, major consumers require prolific springs with high-sustained flows. A long tradition of prioritizing springwaters exists in Central and Southeastern Europe where as many as five capital cities obtain water from karst springs (Stevanovic, 2010). Details on one of the most famous such springs, the Kläffer Spring (Fig. 8.9) utilized by the Vienna Waterworks, are provided by Plan et al. (2010). The increased demands have, however, caused many cities to substitute or enhance their primary water system based on springwater with surface waters or groundwater from other aquifers, most commonly from alluvial ones. Population growth and land use changes are the main reasons why many previously utilized springs around the world are now abandoned (Fig. 8.10). Orientation away from springs and toward aquifers, river water, or water from reservoirs has been a common trend for decades.

FIGURE 8.10 Capture of this karst spring in Winchester, Virginia, formerly utilized for public water supply, is now abandoned. Large intake pipe is visible in the left photo. The outflow from the spring basin, shown in the right photo, now sustains a recreational lake in the town's main park. Winchester switched to surface water source after the spring's yield became insufficient to support its population growth.

Figure 8.11 *Left*: Clauton Mill Spring, Bath County, Virginia. (Courtesy of Phil Lukas). *Right*: Part of a karst spring captured for water supply of an industrial facility near Nova Varoš, Serbia.

Before deciding on the exact engineering means of spring capture, several key characteristics of the spring should be analyzed in detail including its hydrogeological and hydraulic characteristics, discharge pattern (flow magnitude, fluctuation, seasonality, probability of minimum and maximum flows), and the physical and chemical properties of water including during and following major storms. However, even when all these characteristics are favorable, in many cases the most important limiting factor for spring development is the (in)ability to control land use in the spring drainage area and its vulnerability to contamination (see Chap. 10).

The main desire in most cases is to capture a permanent spring that can secure a required flow rate during season(s) of highest demand. However, if there are no other options, even capture of an intermittent spring or control of seasonal flow may be reasonable. Such control can be achieved by tapping deeper parts of the aquifer with wells and galleries, or by using subsurface dams to store water for later use.

The first item for consideration when capturing a spring is feasibility of storing water. For small springs supplying few individual users sometimes a very simple structure such as a spring box is sufficient (Fig. 8.11, left). If properly developed, even a very small spring flow during dry season can supply needed water. In cases of major springs, other structures such as tanks, reinforced concrete basins (Fig. 8.11, right), or dams (see Fig. 6.21) may be required to store water. In general, the main task of any spring capture design is to try to control as much additional water as possible. Only then it would be possible to manage the water by using the amount needed, and by enabling the surplus water to flow out freely downstream.

Large permanent karst springs usually discharge at topographically low contacts with less permeable formations, and along surface streams that act as regional erosion bases of karst aquifers (Fig. 8.12). Because of the diversity of karst formations, the complexity of karst processes, and the role of geology and tectonics in directing groundwater flow, karst springs can be of any type: ascending, descending, cold, thermal, with uniform discharge, or with discharge varying between 0 and >300 m^3/s (see Fig. 1.163). For these reasons, the capture of karst springs can vary widely.

There are three basic types of spring capture: (1) with minimal or no artificial intervention when the spring is utilized as is, (2) with some form of engineering intervention

FIGURE 8.12 Photograph of the outflow from the spring box of Hughes Spring near Zack, Arkansas, used for public water supply. (Photo courtesy of Joel Galloway, U.S. Geological Survey; see also Galloway, 2004.)

aimed at securing the spring for reliable use and protecting it from immediate surface contamination, and (3) with engineering aimed at artificially increasing the spring discharge rate. When used for potable water supply, a spring should be completely enclosed, protected from contamination and equipped with fixtures for easy access, cleanup, and water distribution. One such design showing a typical capture of a contact gravity spring with a watertight basin ("spring box") constructed in place with reinforced concrete is illustrated in Fig. 8.13. The basin has one side open to the aquifer, allowing inflow of

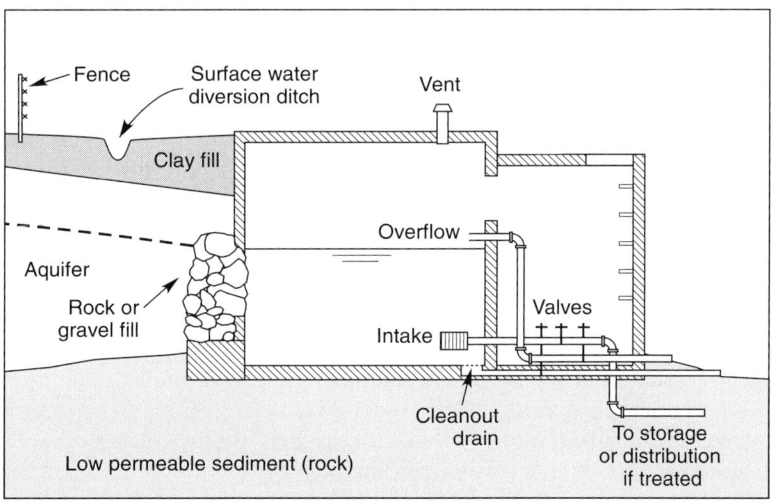

FIGURE 8.13 Typical capture of contact gravity spring using spring box. (From Kresic, 2007; copyright Taylor & Francis Group; used with permission.)

FIGURE 8.14 Lakomica Spring captured for water supply of Nova Varoš, Western Serbia.

water. In case of ascending springs, the basin is open at the bottom. In either case, the side of the basin open to the inflow of groundwater should be stabilized with rock fragments, gabions, or gravel. The basin should be vented to the surface and have easy access for maintenance. Three pipes equipped with valves should allow for (1) water overflow, (2) complete basin drainage for the cleanout and maintenance, and (3) transfer of water to users or storage. All pipes should have screens on either end to prevent entry of rodents, small animals, and insects.

Sanitary protection of the spring is achieved by fencing (Fig. 8.14) and, if needed, placement of an impermeable clay fill and surface drainage ditches located uphill from the spring to intercept surface-water runoff and carry it away from the source. Springs are often contaminated with bacteria during construction or maintenance. All new and repaired springwater systems should be disinfected using shock chlorination. If bacterial contamination occurs on a regular basis, continuous disinfection using chlorination or some other method may be necessary. Shock chlorination requires concentration of at least 200 parts per million (ppm) chlorine (Jennings, 1996).

Depending on site-specific conditions, the basic configuration of spring capture shown in Fig. 8.13 may include additional features such as drainage galleries extending into the saturated zone behind the spring for intercepting more flow (see Fig. 8.9-right). Most importantly, a spring capture should be built at the location of primary discharge of water from the subsurface since secondary springs may move in time. If the discharge is from colluvium and other types of rock debris, there is high probability that the spring is secondary and located away from the primary discharge location that may not be visible. In such cases, every attempt should be made to clear the rock debris and locate the primary spring.

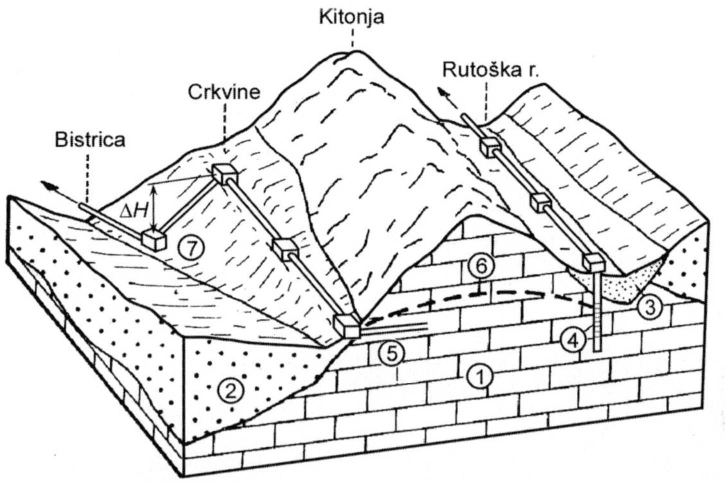

Figure 8.15 Schematic of groundwater utilization in the Crkvine-Rutoši karst system. (1) Triassic limestone, (2) the Diabase-Chert Formation, (3) Neogene clastic sediments, (4) drilled well, (5) drainage gallery, (6) groundwater level during exploitation, and (7) mini hydropower plant. (From Kresic, 1988.)

In terrains with pronounced relief differences, springs are often located at higher elevations than settlements enabling gravity water transport with no expenditure of energy. Additional advantage in such cases is that the high hydraulic head of springs can be utilized for both water supply and hydropower generation. This concept is illustrated in Fig. 8.15. A series of captured overflow springs located at the contact between Mesozoic limestone and the Diabase-Chert Formation is connected with a pipeline that leads to a mini hydropower plant in the river valley. The energy potential (E) of the plant is about 50 kW for the average combined spring flow $Q = 0.065$ m^3/s and the hydraulic head difference of $\Delta H = 78$ m: $E = Q \times \Delta H \times g$ where g is acceleration of gravity (9.81 m/s^2). For the average maximum flow $Q = 0.2$ m^3/s, the energy potential is 0.15 MW.

Graziadei and Zötl (1984) describe a spring regulation and hydropower project completed in Austria to supply water and energy for Innsbruck, the famous Olympic ski resort and capital of Tyrol. The first intake of Mühlau springs for Innsbruck was built in 1887, and reconstructed in the 1950s. Noteworthy are the conditions and difficulties of construction of the new Rumerstollen tunnel through the karstic massif of Triassic limestones overlain by younger sediments. The collector galleries are 564 m long, with additional 1,159 m of branches. Two pipelines, 600 mm in diameter, convey the water to two turbines in the power station. About 25% of the inhabitants of Innsbruck obtain energy from this plant. In addition, the water, which receives no treatment, is of excellent quality. The water temperature is constant at 5°C. The annual discharge ratio is around 1:2. The minimum, during early spring months, is 560 L/s, while the maximum, during summer, is over 1,600 L/s. The relatively steady discharge is due to delayed infiltration of rainfall and snowmelt. There is also a longer residence time within fissured limestone aquifer, which consists more of narrow joints than of large caverns (Graziadei and Zötl 1984, from Stevanovic, 2010).

8.4.1 Case Study: Rockhead Spring

This case study is provided courtesy of Dr. John Gunn. Rockhead Spring (also referred to as Cowdale Spring; Fig. 8.16) was once the potable supply for Rockhead hamlet and Cowdale village near Buxton in the English Peak District. In common with many other villages on the limestone, a piped mains supply was introduced in the 1970s and the spring was unused for around 30 years.

During this time, it was studied by a number of researchers who found that over a period of 20 years the major ions showed only a small variability and suggested that it should be recognized as a type site for springs fed by dispersed autogenic recharge. In this

FIGURE 8.16 Rockhead Spring, the only spring in the English Peak District that drains solely from limestone to be officially recognized by the European Union as a Natural Mineral Water. (Courtesy of Dr. John Gunn.)

respect, it differs from most of the surrounding springs that have a variable component of allogenic recharge and consequently a more variable flow and chemistry. The flow from the spring varies from about 4 to 20 L/s, but the variability is thought to be due to a "piston" effect whereby new rain pushes out old water at an increased rate.

Over the period 1997 to 2000 further water quality testing (including bacteriological testing and analysis for pesticides, herbicides, surfactants, and other trace pollutants) established that Rockhead Spring met all the European Union requirements for recognition as a Natural Mineral Water. In common with most entirely autogenic-fed karst springs, it was not possible to determine the catchment in detail but an approximate boundary was drawn based on geology and a water balance. This was subsequently included in a large Source Protection Zone drawn to protect the catchment of Rockhead, two nearby boreholes, and St Ann's Well, a thermal spring that has a part limestone and part gritstone source.

An environmental assessment showed that abstraction from Rockhead at a rate of 480 cubic meters/day (175,200 cubic meters/year) would have no adverse impacts on water quality or flow in the River Wye and a license to abstract up to these amounts of water for bottling was granted by the Environment Agency.

A pipeline has been constructed linking the spring to a small bottling factory on an industrial estate where glass bottles of Rockhead Mineral Water are to be produced. This will only use a small proportion of the available yield, and a planning application has been made for a large new factory in a disused limestone quarry near the spring.

8.5 References

Aller, L. T., Bennett, T. W., Hackett, G., et al. 1991. Handbook of suggested practices for the design and installation of ground-water monitoring wells. EPA 160014–891034, Environmental Monitoring Systems Laboratory, Office of Research and Development. Las Vegas, Nevada, 221p.

AWWA (American Water Works Association), 1998. AWWA standard for water wells. *American National Standard.* ANSI/AWWA A100–97, AWWA, Denver, Colorado, various paging.

Campbell, M. D., and Lehr, J. H., 1973. *Water Well Technology*. McGraw-Hill Book Company, New York, 681p.

Clarke, J. S., Williams, L. J., and Cherry, G. C., 2010. Hydrogeology and water quality of the Floridan aquifer system and effect of lower Floridan aquifer pumping on the upper Floridan aquifer at Hunter Army Airfield. U.S. Geological Survey Scientific Investigations Report 2010–5080, Chatham County, Georgia, 56p.

Djokić, I., Canić, V., and Cekić, M., 2005. Results of hydrogeological investigations in the Valjevo-Mionica karst region, Rajković-Kjluč area (Western Serbia). In: Stevanović, Z., and Milanović, P. (eds.), *Water Resources and Environmental Problems in Karst – Cvijić 2005, Proceedings of the International Conference and Field Seminars, Belgrade & Kotor, 13–19 September 2005, Institute of Hydrogeology*. University of Belgrade, Belgrade, pp. 597–602.

Driscoll, F. G., 1986. *Groundwater and Wells*. Johnson Filtration Systems Inc., St. Paul, MN, 1089p.

Galloway, J. M., 2004. Hydrogeologic characteristics of four public drinking-water supply springs in northern Arkansas. U.S. Geological Survey Water-Resources Investigations Report 03–4307. Little Rock, Arkansas, 68p.

Jennings, G. D., 1996. Protecting Water Supply Springs. North Carolina Cooperative Extension Service, Publication no. AG 473–15. Available at: http://www.bae.ncsu.edu/programs/extension/publicat/wqwm/ag473–15.html

Johnson Screens, 2007. Well screens and well efficiency. Johnson Screens a Weatherford Company. Accessed in November 2007 at: www.weatherford.com/weatherford/groups/public/documents/general/wft029882.pdf.

Kresic, N., 1988. Karst aquifers of the Lim catchment (SR Serbia). *Des Comptes Rendus des Séances de la Société Serbe de Géologie pour l'année 1985–1986*. Belgrade, pp. 217–223.

Kresic, N., 2007. *Hydrogeology and Groundwater Modeling*. 2nd ed. CRC Press, Taylor & Francis Group, Boca Raton, FL, 807p.

Kresic, N., and Stevanovic, Z. (eds.), 2010. *Groundwater Hydrology of Springs: Engineering, Theory, Management and Sustainability*. Elsevier, New York, 573p.

Lapham, W. W., Franceska, W. D., and Koterba, M. T., 1997. Guidelines and standard procedures for studies of ground-water quality: Selection and installation of wells, and supporting documentation. U.S. Geological Survey Water-Resources Investigations Report 96–4233. Reston, VA, 110p.

Lombardi, L., and Corazza, A., 2008. L'acqua e la città in epoca antica. In: La Geologia di Roma, dal centro storico alla periferia, Part I, Memoire Serv. Geol. d'Italia, Vol LXXX, S.E.L.C.A, Firenze, pp. 189–219.

Moore, J. D., Hinkle, F., and Moravec, G. P., 1977. High-yielding wells and springs along lineaments interpreted from LANDSAT imagery in Madison County, Alabama, U.S.A. In: Tolson, J. S., and Doyle, F. L. (eds.), *Karst Hydrogeology, Proceedings of the 12th Congress of the International Association of Hydrogeologists*. UAH Press, Huntsville, Alabama, pp. 477–486.

Plan, L., Kuschnig, G., and Stadler, H., 2010. Case study: Klaffer Spring—The major spring of the Vienna water supply (Austria). In: Kresic, N., and Stevanovic, Z. (eds.), *Groundwater Hydrology of Springs: Engineering, Theory, Management and Sustainability*. Elsevier, New York, pp. 411–427.

Stevanovic, Z., 2010. Utilization and regulation of springs. In: Kresic, N., and Stevanovic, Z. (eds.), *Groundwater Hydrology of Springs: Engineering, Theory, Management and Sustainability*. Elsevier, New York, pp. 339–388.

Stewart, J. W., 1977. Hydrologic effects of pumping a deep limestone sink near Tampa, Florida, U.S.A. In: Tolson, J. S., and Doyle, F. L. (eds.), *Karst Hydrogeology, Proceedings of the 12th Congress of the International Association of Hydrogeologists*. UAH Press, Huntsville, Alabama, pp. 195–211.

Sun, P-C. P., Criner, J. H., and Poole, J. L., 1963. *Large Springs of East Tennessee*. Geological Survey Water-Supply Paper 1755, 52p.

The Florida Springs Task Force, 2000. Florida's springs. Strategies for protection & restoration. 58p.

USBR, 1977. *Ground Water Manual*. U.S. Department of the Interior. Bureau of Reclamation, Washington, D.C., 480p.

USEPA, 1975. *Manual of Water Well Construction Practices*. EPA-570/9–75-001, Office of Water Supply, Washington, D.C., 156p.

USEPA, 1991. *Manual of Small Public Water Supply Systems*. EPA 570/9–91-003, Office of Water, Washington, D.C., 211p.

CHAPTER 9

Engineering Regulation of Karst Aquifers and Springs

9.1 Introduction

The rate of natural discharge has often been a limiting factor when considering the use of a particular spring for public water supply. Springs with discharge hydrographs similar to the one shown in Fig. 9.1 may have potential for regulation by artificially increasing their minimum and average annual flows. The basic idea is to take advantage of the fact that the spring is capable of discharging large quantities of water during periods of less demand such as in spring or late fall when natural aquifer recharge is the highest. This volume of surplus water may be regulated in two basic ways: (1) by using it to naturally recharge the volume of the aquifer drained by overpumping during periods of peak demand (e.g., summer-early fall) and (2) by storing it in the aquifer above the elevation of natural spring discharge, that is, by creating a surface or underground dam and groundwater impoundment. These two concepts are illustrated in Figs. 9.2 and 9.3.

In either case, the main prerequisite is that the aquifer has adequate storage capacity, below or above the spring elevation, respectively. In case of a dam, for any feasible design the spring must issue from a "V" shaped land-surface contact between the aquifer and the impermeable barrier. The dam is keyed into the impermeable barrier laterally and vertically and, in combination with drainage gallery, is used to control the hydraulic head in the aquifer and the flow rate. The most important requirement for implementing this concept, with either surface or subsurface dams, is that there should not be an uncontrollable water loss around or below the dam. This means that extensive field investigations including drilling, geophysical surveys, borehole testing, and dye tracing must be implemented to accurately map the three-dimensional extent of the karst aquifer and, most importantly, any potential preferential flow paths that may become active after the impoundment.

Figures 9.4 and 9.5 show setup of Ombla Spring tapped for water supply of the Croatian coastal city of Dubrovnik. The spring is soon to become construction site for the world's largest underground dam in karst. Interestingly, one of the first students the author of this book mentored while at the University of Belgrade, wrote his graduation thesis on the Ombla Underground Dam more than 20 years ago. It appears that the initial feasibility study and preliminary dam design, completed by Energoprojekt Hidroinženjering from Belgrade, Serbia, years ago, will finally be brought to fruition as preparatory work on the dam construction has commenced at the time of this writing.

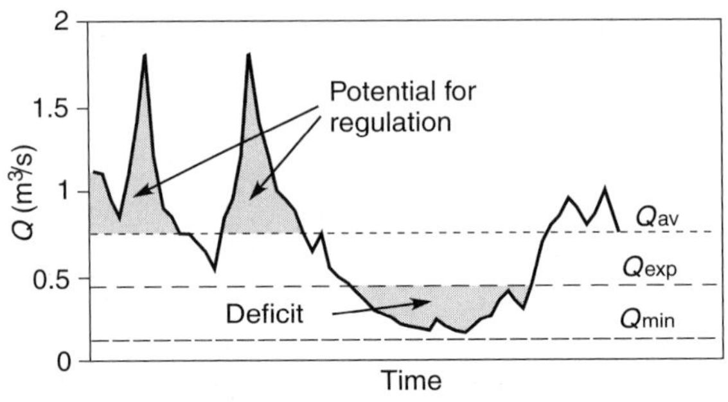

FIGURE 9.1 Hydrograph of a spring with potentially exploitable reserves higher than the minimum spring discharge. Q_{av}, average spring discharge; Q_{min}, minimum spring discharge, Q_{expl}, potentially secure exploitable reserves. (From Stevanovic, 2010.)

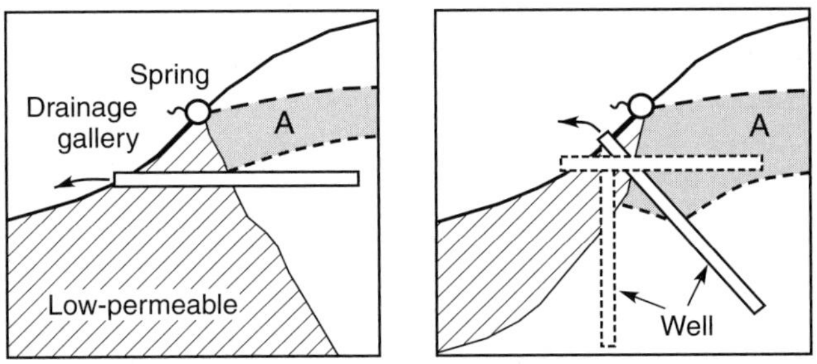

FIGURE 9.2 Potentially favorable conditions for spring regulation using drainage galleries or wells for overpumping. An additional volume of water that may be extracted from aquifer storage during peak demand, assuming recovery during main periods of natural aquifer recharge, is shaded.

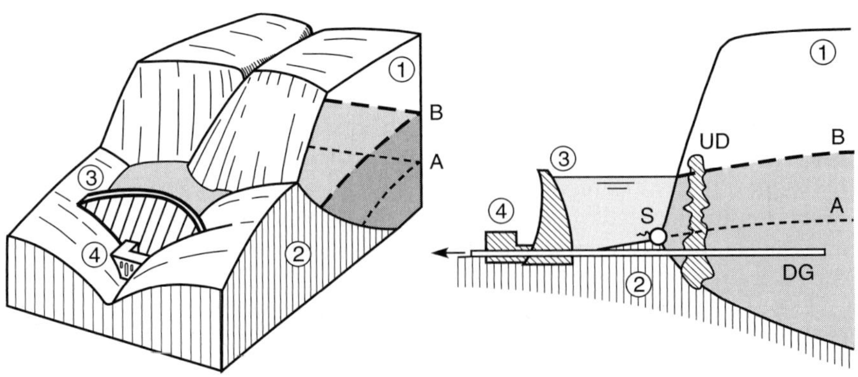

FIGURE 9.3 Possible spring regulation using surface or subsurface dam in the case of favorable geologic and geomorphologic conditions. (1) Aquifer, (2) impermeable base, (3) surface dam with impoundment, and (4) hydropower plant and water treatment. S, original spring; A original natural water table before impoundment; B water table after impoundment; DG, drainage gallery; UD, underground dam, such as grout curtain, instead of surface dam. (From Kresic, 1991).

Figure 9.4 *Top:* Panoramic view of the karst plateau above Ombla Spring. (Courtesy of Dr. Neno Kukuric.) *Bottom*: Close-up of the spring with an abandoned mill building. The underground dam will be constructed within the carbonates visible in the background and keyed laterally and vertically into the natural impermeable barrier composed of flysch sediments.

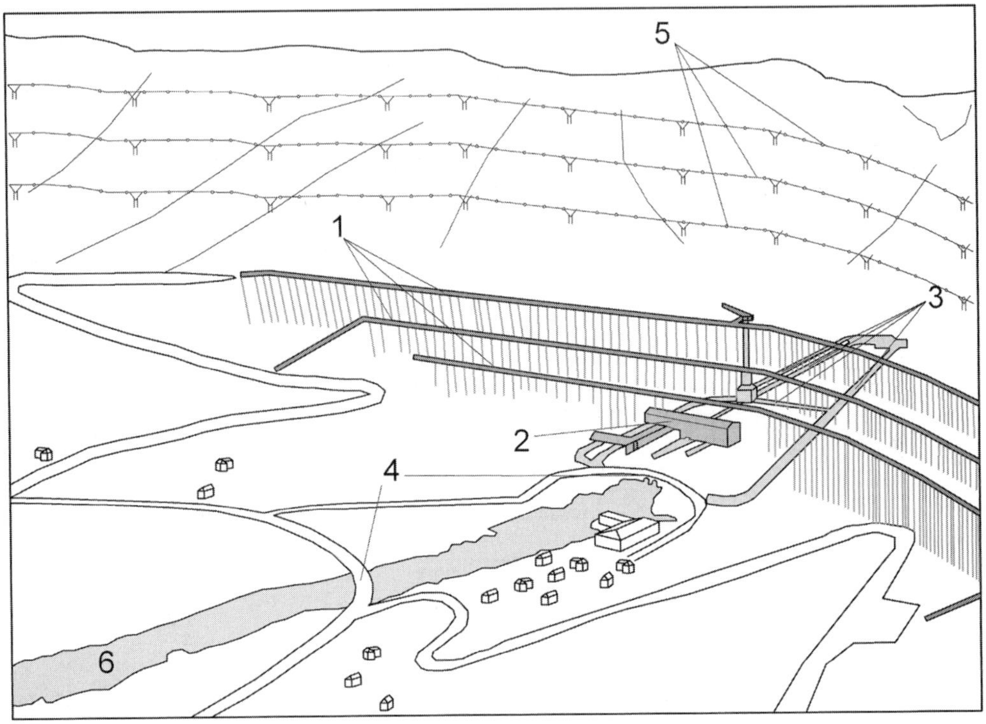

FIGURE 9.5 Ombla underground reservoir. (1) Grout curtains, (2) hydro power plant, (3) pipelines, (4) bridge, (5) electrical transmission lines, and (6) Rijeka Dubrovačka River formed from Ombla Spring. (Modified from original technical documentation of Elektroproject, Zagreb 1999; from Stevanovic, 2010.)

The dam will include three-level grout curtain supported by the existing natural barrier of Eocene Flysch, and plugging of the main natural spring conduit. The underground hydropower plant is proposed to be installed in a large cavern downgradient of the curtains (Milanovic, 1996.)

Large barrier karst springs are the best candidates for flow regulation, since the nature of karst porosity allows rapid recharge and transmittal of groundwater. For example, overpumping of the Krupac Spring (Fig. 9.6) presently secures additional 200 L/s for water supply of Niš, the second largest city in Serbia, during two summer months of peak demand. Based on the initial pumping test conducted in the 1970s that confirmed the presence of a deeper and large underground reservoir that drains a karstic basin of over 70 km^2, an investigation program for spring regulation was implemented in the early 2000s. It included exploratory diving to a depth of 86 m and a new pumping test by two pumps installed at a depth of about 30 m. For a pumping rate of 150 to 350 L/s, the maximum drawdown at the spring was 21 m. Based on the pumping test results, it was concluded that the spring could provide additional 260 to 280 L/s of water constantly for 2 months, with drawdown in the karst channel of 50 to 60 m. If the average discharge of 50 to 150 L/s during the dry season is taken as a reference, the additional amount of water represents a significant increase in meeting the water supply needs (Jevtic et al., 2005). Two new 720-mm-diameter wells, each 80 m deep, were drilled above the spring

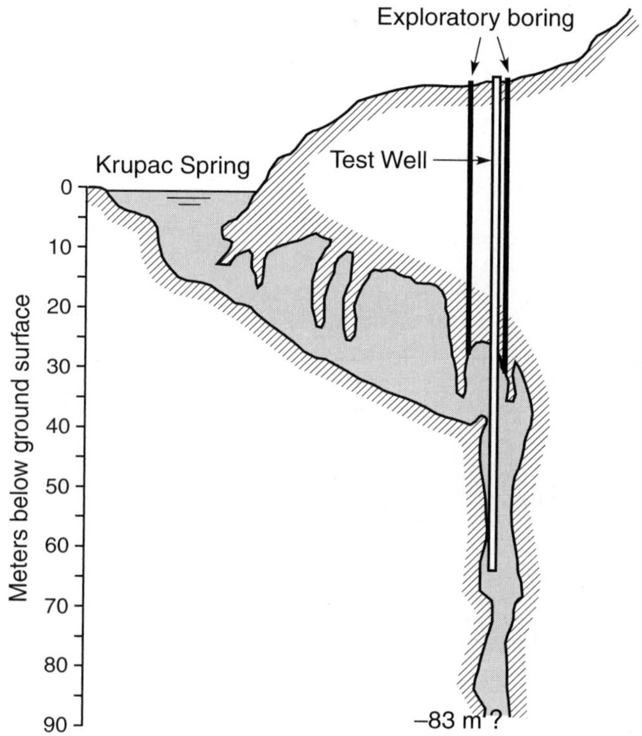

Figure 9.6 Overpumping of Krupac Spring used for water supply of the City of Niš, Serbia. (From Milanovic, S., 2005.)

and completed in the vertical channel. Continuous monitoring has since confirmed that water abstracted from the karst aquifer during dry seasons (static reserves) is regularly replenished during periods of intensive aquifer recharge in winter and early spring (Stevanovic, 2010).

The overpumping of Krupac Spring was in part inspired by the well-known example from the littoral karst of southern France where Lez Spring (Fig. 9.7) supplies water for the city of Montpellier (Avias, 1984). Water demands of Montpellier have increased more than 70-fold during the last 150 years, from just 25 L/s to over 1,800 L/s. As mentioned earlier, it is common in water practice to abandon supply based on springwaters due to large expansion of the cities and rapid industrialization. This, however, was not the case in Montpellier thanks to the engineering idea to tap additional water deep inside the aquifer. Exploration of deep submerged cave channels and their precise mapping allowed for the construction of an extraction shaft, a gallery, and the placement of high-capacity pumps. Diving into channels to more than 500 m from the submerged cave entrance and to depths of over 100 m was a delicate task, but now serves as a proof of concept worldwide. Groundwater is withdrawn by powerful pumps (Fig. 9.7-bottom) from three wells placed in the main drain of this ascending spring, 400 m upstream from the main pool and 48 m below the overflow level. The average pumping rate for the 1997 to 2005 period was 1.1 m^3/s (Fleury et al., 2008). When pumping begins to withdraw the reserves, the water levels in the drain and in the spring pool drop to the level of the pool overflow and the spring dries up. Pumping then causes a drawdown within the drain that can reach several tens of meters at the end of the low-water period. These reserves

FIGURE 9.7 *Top*: Overflow of the Lez Spring near Montpellier, France. *Bottom*: Replacement of large capacity pumps at the Lez Spring. (Photograph courtesy of Arnaud Vestier, Ville de Montpellier.)

are recharged during the autumn and winter. The hydraulic water level in the drain rises above that of the pool, and the spring begins to flow again. During the active pumpage, when the Lez River flow is lower than the pumped flow, part of the pumped flow is channelled to the river. This return flow, set by the Declaration of Public Utility of June 5, 1981, is 160 L/s.

9.2 Subsurface Dams

The benefits of storing water in the subsurface have been recognized in the United States since the late 1890s. For example, Slichter (1902) gives an account of one of the first subsurface dams built in the United States. This dam on Pacoima Creek, Los Angeles County, California, was constructed in 1887 to 1890 and it enabled the owners to use the bedrock flow of water for the three dry years, 1898 to 1900, and thereby to successfully carry through the orange, lemon, and olive growing in Fernando Valley.

Compared with surface water reservoirs formed by conventional dams, use of groundwater impoundments behind underground dams has the following general advantages:

- Very limited or negligible evaporation loss.
- Land use above the groundwater reservoir can continue without change (there is no submergence of houses, infrastructure, and property, in general).
- There is general improvement in water quality because of the porous media filtration of airborne and surface runoff contaminants and pathogens, as well as longer residence times. On conduit-dominated karst with fast flow, subsurface dams and reservoirs can greatly increase residence time of newly infiltrated water that helps with attenuation of various contaminants and also significantly reduces or eliminates turbidity issues at strategically selected locations of groundwater withdrawals or at springs.
- Functioning of underground reservoirs is longer since there is no accumulation of sediment (in case of intergranular aquifers); although a potential concern, such accumulation is much more limited in case of karst aquifer compared to surface reservoirs where it is the main reason for their limited life span.
- There is no danger of dam failure and catastrophic loss of life and property.
- Overall impact on the environment and natural habitat of plants and animals is of much lower magnitude, although the full impact on cave/conduit biota in karst may not be fully understood.

The main disadvantage of subsurface reservoirs in any porous media is that virtually none of the feasible (cost-effective) construction methods can guarantee complete dam impermeability, although practical impermeability, that is, a tolerable level of dam leakage is often possible to achieve. In addition, storage capacity of the underground reservoir cannot be accurately determined and has to be estimated based on more or less limited field data on heterogeneous porous media. These two factors are accentuated in karst, just as in case of surface dams that are explained in more detail in Sec. 10.4.2. Similarly to surface dams and reservoirs, underground dams and groundwater storage basins cannot be constructed "everywhere" and require certain favorable hydrogeologic conditions such as sufficient thickness of the unsaturated zone that will accommodate the rising water table, and natural lateral and vertical containment of the groundwater flow by low-permeable formations.

Subsurface dams can be defined as any artificial structures that intercept or obstruct the natural flow of groundwater and provide storage for water underground. Most commonly, they have been used in unconsolidated sediments in India, Africa, and Brazil, in areas where flows of groundwater vary considerably during the course of the year, from very high flows following rain to negligible flows during the dry season. Groundwater dams can be divided in two types: subsurface dams and sand storage dams. A subsurface dam is built entirely under the ground using low-permeable natural materials such as clay, or impermeable materials such as concrete and concrete grouts. The underground dam causes water table upgradient of the dam to rise, stores additional volumes of groundwater, and reduces water table fluctuations by controlling the outflow (see Fig. 9.3). The ability to control outflow is very important for minimizing negative impacts on downgradient users. Potential salinization in the groundwater reservoir area due to evaporation from the shallow water table can also be avoided by an appropriate outflow control design that can keep the water table from reaching critical depth of evaporation.

Figures 9.8 and 9.9 illustrate the general concept and layout of several large subsurface dams successfully built on Miyakojima Island in Japan. As explained in detail by Osuga (1997), the Shimajiri mudstone layer of bedrock has a low average hydraulic conductivity of 2×10^{-6} cm/s. The highly permeable Lyukyu limestone layer with the hydraulic conductivity of 3.5×10^{-1} cm/s forms the overlying aquifer with the thickness ranging between 10 and 70 m. The effective porosity of the Lyukyu limestone is between 10% and 15%. Tectonic movements along several major faults have formed underground valleys, with groundwater flow parallel to the structural highs (ridges) between them. About 40% of the abundant rainfall on Miyakojima recharges the limestone aquifers but quickly flows out into the ocean unused. In addition, the subtropical climate results in

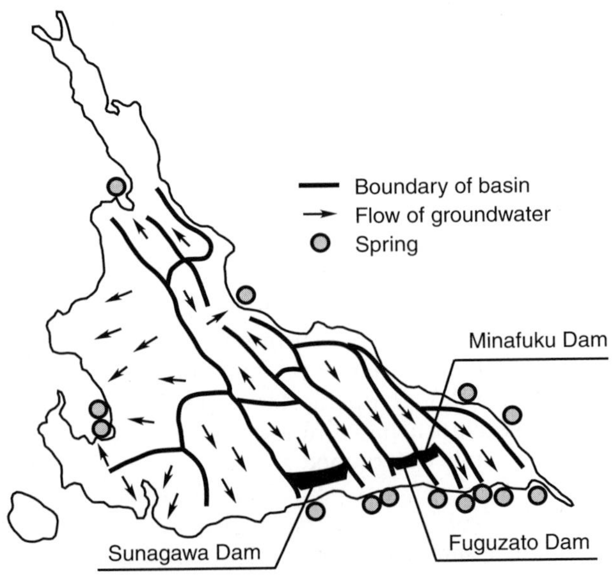

FIGURE 9.8 Location of the groundwater basins and subsurface dams on the Miyakojima Islands, Japan. (Osuga, 1997; copyright United Nations University.)

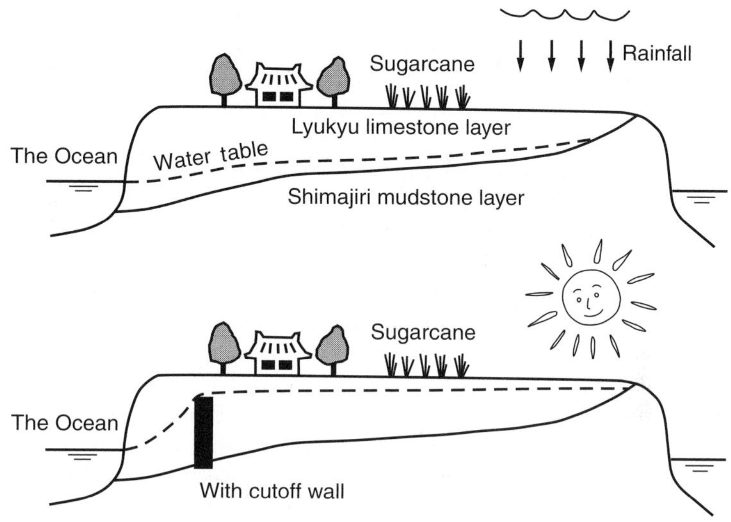

Figure 9.9 The concept of the underground dams on the Miyakojima Islands, Japan. (Osuga, 1997; copyright United Nations University.)

evaporation of as much as 50% of the rainfall. The result is that only 10% of the precipitation ends up as surface flows on the island. Figure 9.10 shows changes in groundwater levels after the Sunagawa Dam cutoff wall was completed in November 1993. The full underground reservoir capacity was reached in September 1995, or less than 2 years after completion of the dam.

Sunagawa Dam has capacity of about 10 million cubic meters of stored groundwater. This mega-dam is 1,677 m long and was built with a mixed-in-place slurry wall method (Ishida et al., 2003). Since the beginning of groundwater withdrawal with vertical wells

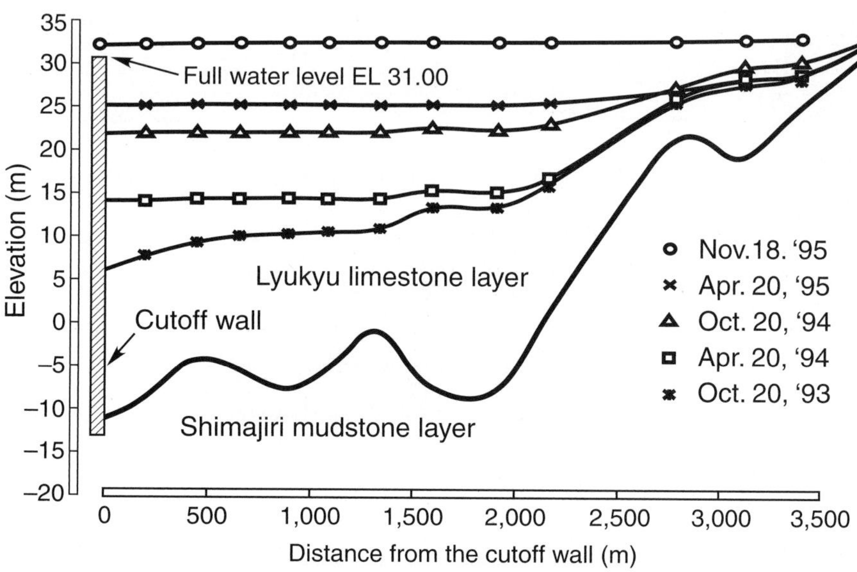

Figure 9.10 Changes in groundwater levels at the Sunagawa Subsurface Dam, Miyakojima Islands, Japan. (Osuga, 1997; copyright United Nations University.)

in 2001, groundwater levels were somewhat higher than the designed full underground reservoir water level, indicating that the rates of groundwater withdrawal were slightly lower than the rate of natural aquifer recharge. The Sunagawa Dam provided 2.7, 4.8, and 5 million cubic meters of water a year for irrigation in 2001, 2002, and 2003, respectively (Ishida et al., 2005). An analysis of nitrate concentrations in groundwater before and after the dam completion and operation for 3 years showed that the dam resulted in their overall decrease. After 3 years, the nitrates were also more uniformly distributed throughout the reservoir due to longer retention of nitrate-free natural recharge, mixing and circulation of water caused by pumping (Ishida et al., 2005).

An example of using underground dam to prevent saltwater intrusion and utilize a brackish karst spring is presented by Potier et al. (2005). The existence of submarine freshwater springs in the Calanques range of hills on the coastline between Marseilles and Cassis has been known for many years. The largest two springs, Port-Miou and Bestouan, discharge a combined 3 m^3/s through submerged karst channels. In 1964, in order to find an appropriate response to the fast-growing water demands of the city of Marseilles, two French institutions, BRGM (Bureau de Recherches Géologiques et Minières) and a local water company from Marseille (SEM), united to become the Syndicat de Recherches

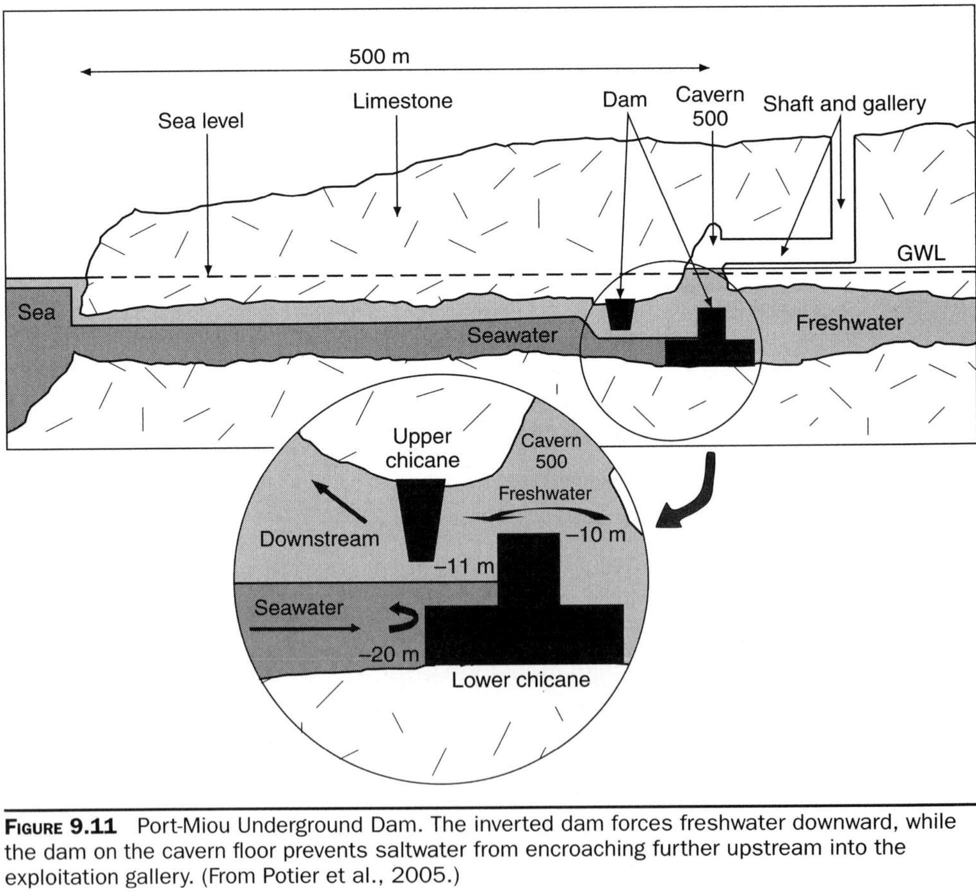

FIGURE 9.11 Port-Miou Underground Dam. The inverted dam forces freshwater downward, while the dam on the cavern floor prevents saltwater from encroaching further upstream into the exploitation gallery. (From Potier et al., 2005.)

de Port-Miou association and conducted an extensive survey including geologic and geophysical field exploration, topographic survey, flow measurements, dye testing, and diving exploration in Port-Miou (to depth of 147 m below sea level and to a distance of 2,230 m from the entrance) and in Bestouan (to depth of 31 m below sea level and to a distance of approximately 3 km from the entrance).

The survey of Port-Miou resulted in the construction of the first underground dam in 1972 (Fig. 9.11) and a complete closure of the natural cave gallery with a second underground dam. The objective of the first dam was to prevent saltwater intrusion without modifying pressure in the aquifer. The principle of a "chicane" dam was adopted. This involved building a pair of dams 530 m from the sea outlet: an upstream chicane dam on the gallery floor and a downstream chicane dam on the gallery ceiling.

Construction of chicane dams did not entirely stop saltwater from contaminating fresh groundwater, but it had a significant positive effect: after construction of the dam the salinity of the water dropped from 4–5 g/L to 2–3 g/L at the surface, and from 18–20 g/L to 3–4 g/L upstream of the dam at a depth of 20 m below sea level. A few years later, it was decided to modify the initial project by completely closing the gallery and creating a spillway. Several other measures have also been undertaken in order to control the pressure in the aquifer during flood episodes and at the same time to reduce saline intrusion. However, a complete prevention of salinization was not possible (Potier et al., 2005).

The main practical experience on underground dams in karst has been obtained from projects implemented in China (Yaoru et al., 1973; Yaoru, 1986). Tens of underground reservoirs have been completed for different purposes (water supply, irrigation, and hydropower) and enabled water storage ranging from 1×10^5 m^3 to 1×10^7 m^3 in individual reservoirs. Although these projects vary in design, the solutions described earlier with a dam in front of the spring and with the plugging of the discharge channel were among the most frequently implemented (Fig. 9.12). Quibei in Yunnan Province is one of the largest such reservoirs formed behind a 25-m-high underground dam; the hydraulic head difference between the underground reservoir (dam) and a downgradient surface power station is 109 m generating 25 MW of electric power (Yaoru, 1986).

9.3 Artificial Aquifer Recharge

The main purpose of artificial aquifer recharge in water resources management is to store water underground for later use. There have been a growing number of artificial aquifer projects around the world as many countries and regions are under increasing pressure from population growth and water shortages. The main benefits and limitations from storing water underground as opposed to surface water reservoirs are the same as listed in Sec. 9.2 on subsurface dams.

The important factors that have to be considered for any scheme of artificial aquifer recharge are (Kresic, 2009) as follows:

- Regulatory requirements.
- The availability of an adequate source of recharge water of suitable chemical and physical quality.
- Geochemical compatibility between recharge water and the existing groundwater (e.g., possible carbonate precipitation, iron hydroxide formation, and mobilization of trace elements).

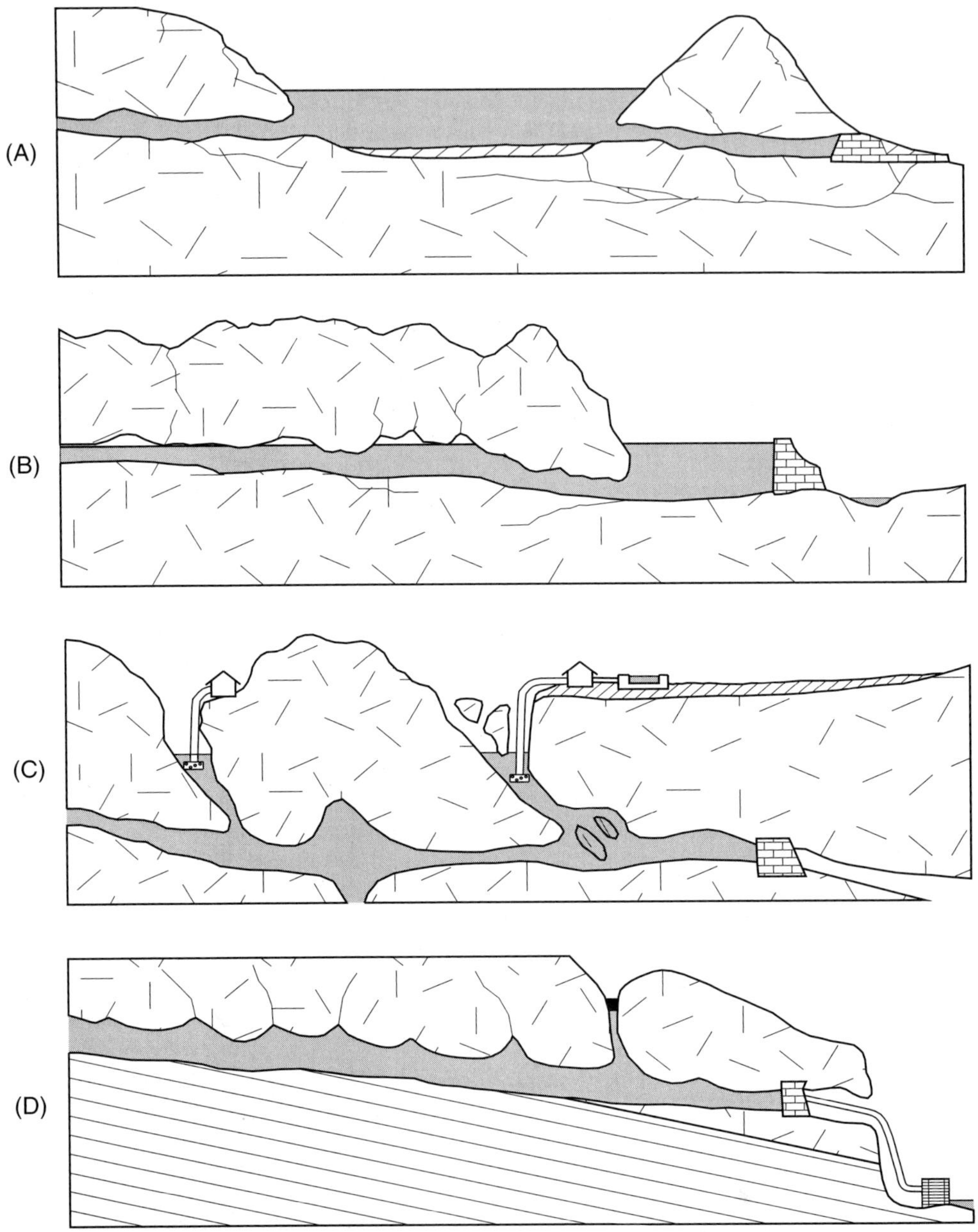

FIGURE 9.12 Some examples of karst aquifer regulation in China using underground dams. (A and B) With underground and surface reservoirs, (C) with underground dam and pumpage from deep sinkholes, and (D) with underground dam and gravity transport of water to water tank. (Modified from Yaoru, 1986.)

- The hydrogeologic properties of the porous media (soil and aquifer) must facilitate desirable infiltration rates and allow direct aquifer recharge. For example, existence of extensive low-permeable clays in the unsaturated (vadose) zone may exclude a potential recharge site from future consideration.
- The water-bearing deposits must be able to store the recharged water in a reasonable amount of time, and allow its lateral movement toward extraction locations at acceptable rates. In other words, the specific yield (storage) and the hydraulic conductivity of the aquifer porous media must be adequate.
- Presence of fine-grained sediments may have an advantage of improving the quality of recharged water due to their high filtration and sorption capacities. Other geochemical reactions in the vadose zone below recharge facilities may also influence water quality.
- Engineering solution should be designed to facilitate efficient recharge when there is an available surplus of water, and efficient recovery when the water is most needed.
- The proposed solution must be cost-efficient, environmentally sound, and competitive to other water resource development options.

Aquifers that can store large quantities of water and do not transmit them away quickly are best suited for artificial recharge. For example, karst aquifers may accept large quantities of recharged water but may also transmit them very quickly away from the recharge area. This may still be beneficial for the overall balance of the system and the availability of groundwater considerable distance downgradient from the immediate recharge sites, as illustrated with examples from the Edwards Aquifer in Texas (discussed later in this chapter). Alluvial aquifers are usually the most suited for artificial recharge because of the generally shallow water table and vicinity to source water (surface stream). Sandstone aquifers are in many cases very good candidates due to their high storage capacity and moderate hydraulic conductivity that does not facilitate rapid transport.

Three common methods of artificial aquifer recharge are (1) spreading water over the land surface, (2) delivering it to the unsaturated zone below the land surface, and (3) injecting water directly into the aquifer. A variety of engineering solutions and combinations of these three methods have been used to accomplish a simple goal—to deliver more water to the aquifer (see Figs. 9.13 and 9.14).

Table 9.1 provides a comparison of major engineering factors that should be considered when installing a groundwater recharge system, including the availability and cost of land for recharge basins (Fox, 1999). If such costs are excessive, the ability to implement injection wells adjacent to the reclaimed water source tends to decrease the cost of conveyance systems for injection wells. Surface spreading basins require the lowest degree of pretreatment, while direct injection systems require water quality comparable to drinking water, if potable aquifers are affected. Low-technology treatment options for surface spreading basins include primary and secondary wastewater treatment with the possible use of lagoons and natural systems. Reverse osmosis is commonly used for direct injection systems to prevent clogging; however, some aquifer storage and recovery (ASR) systems have been operating successfully without membrane treatment when water was stored for irrigation. The cost of direct injection systems can be greatly reduced from the numbers presented in Table 9.1 if the aquifer is shallow and nonpotable (USEPA, 2004).

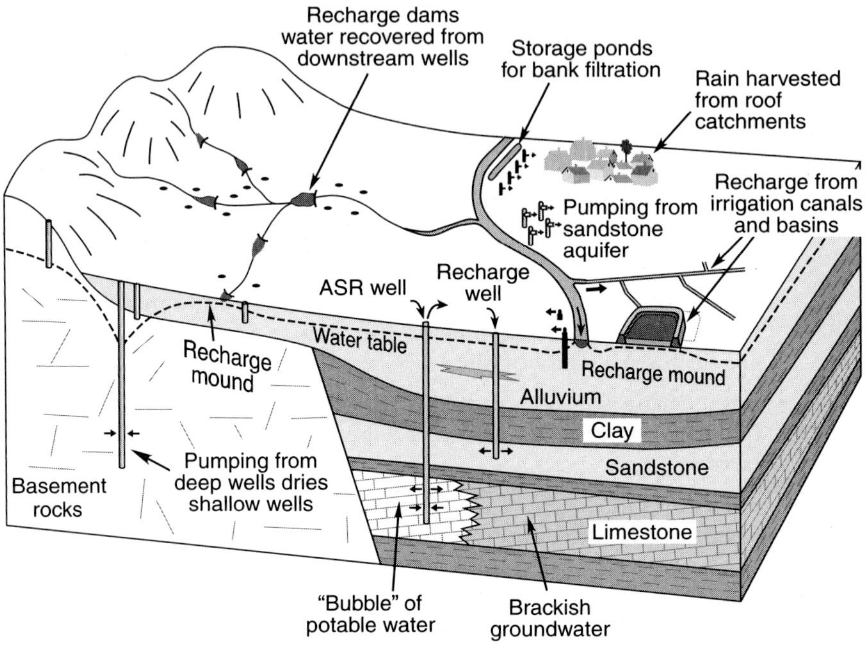

FIGURE 9.13 Possible options for managed aquifer recharge or MAR. (Modified from IAH, 2002.)

Detailed discussion on various general methods for artificial aquifer recharge is provided by Kresic (2009) and reproduced here briefly as related to karst aquifers.

9.3.1 Spreading Basins

Artificial recharge by spreading is the most common technique of artificial aquifer recharge. In karst terrains where carbonates are covered by thick residuum or other

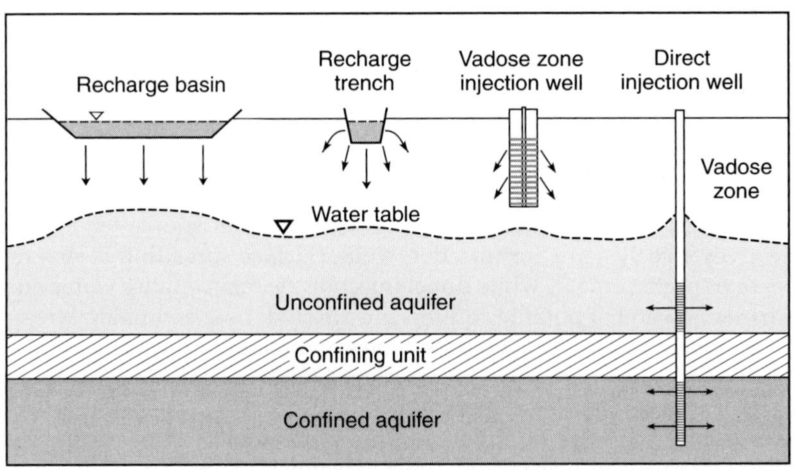

FIGURE 9.14 Engineering methods for artificial aquifer recharge. (Modified from USEPA, 2004.)

	Recharge Basins	Vadose Zone Injection Wells	Direct Injection Wells
Aquifer type	Unconfined	Unconfined	Unconfined or confined
Pretreatment requirements	Low technology	Removal of solids	High technology
Estimated major capital costs	Land and distribution system	\$25,000–75,000 per well	\$500,000–1,500,000 per well
Capacity	100–20,000 m^3/hectare/day	1,000–3,000 m^3/day per well	2,000–6,000 m^3/day per well
Maintenance requirements	Drying and scraping	Drying and disinfection	Disinfection and flow reversal
Estimated life cycle	>100 years	5–20 years	25–50 years
Soil aquifer treatment	Vadose and saturated zones	Vadose and saturated zones	Saturated zone

From USEPA (2004).

Table 9.1 Comparison of Major Engineering Factors for Engineered Groundwater Recharge

relatively thick unconsolidated sediments, this method is preferable mainly because the non-karstic portion of the system will significantly contribute to attenuation of various constituents of concern (COCs) in the recharge water and also provide for steady distribution of recharge.

Recharge is accomplished by spreading water over the ground surface or by conveying the raw water to infiltration basins and ditches. The operational efficiency of the spreading depends on the following factors (modified from Pereira et al., 2002):

- Presence of sufficiently pervious layers between the ground surface and the water table (aquifer).
- Enough thickness and storage capacity of the unsaturated layers above the water table.
- Appropriate transmissivity of the aquifer horizons.
- Surface water without excessive particulate matter (low turbidity) to avoid clogging.

The quantity of water that can enter the aquifer from spreading grounds depends on three basic factors:

- The infiltration rate at which the water enters the subsurface.
- The percolation rate, that is, the rate at which water can move downward through the unsaturated zone until it reaches the water table (saturated zone).
- The capacity for horizontal movement of water in the aquifer, which depends on the hydraulic conductivity and thickness of the saturated zone.

The infiltration rate tends to reduce over time due to the clogging of soil pores by sediments carried in the raw water, growth of algae, colloidal swelling, soil dispersion, and microbial activity. A spreading basin is constructed with a flat bottom that is covered evenly by shallow water. This requires the availability of large surfaces of land for meaningful size recharge works. Several basins may be arranged in line so that excess water runs between the basins. Retaining basins may be used for settlement of suspended sediments before water enters the spreading basins. The settling of sediments may also be assisted with the addition of coagulation agents. In addition to clogging, another major disadvantage of shallow infiltration basins is evaporation loss, which may be rather significant in the case of lover infiltration rates, caused either by low-permeable natural soils or clogging. Proper operations and regular maintenance of infiltration basins are therefore very important for the overall efficiency of spreading grounds and infiltration basins.

Variations in the spreading ground technique consist of the use of trenches that are often easier to handle and for which clogging is a lesser problem since a major part of the sediment is carried out of the ditches by the slow-flowing water (Pereira et al., 2002). The main advantage of recharge trenches is that they are relatively inexpensive, and generally have higher infiltration rates per area than recharge basins because water also infiltrates through the trench sides. The disadvantage is that they eventually do clog up at their infiltrating surface because of accumulation of suspended soils and/or biomass. This, however, is the matter of maintenance where trenches may offer greater flexibility over basins in terms of taking them off line individually to minimize interruptions to the recharge operations.

9.3.2 Recharge Dams

Widely applied artificial recharge structures are floodwater retention dams, the purpose of which is to delay the runoff of water in surface streams and provide the time needed for recharge into the local aquifer. The structures usually consist of low dams, including earth walls and gabions built to be toppled by floods. For example, a number of such dams have been built in the Edwards Aquifer recharge zone (Fig. 9.15) and more are in various planning phases. As discussed earlier, Edwards Aquifer is the principal source of water supply for the City of San Antonio, as well as numerous other communities and agricultural interests throughout south-central Texas. The aquifer also supplies Leona, San Pedro, San Antonio, Comal, and San Marcos Springs, creating unique environments and recreational opportunities while providing baseflow to the Leona, San Antonio, Guadalupe, and San Marcos Rivers. Over the past several decades, increasing water demands on the Edwards Aquifer have raised concerns about the ability of the aquifer to meet these demands without causing social, economic, and environmental problems. Estimated benefits of the proposed four new recharge projects are additional pumpage of 21,440 acft/year (approximately 26.4 billion cubic meters per year) from the aquifer, additional spring flows of 15,240 acft/year, and additional aquifer storage (in initial years) and then additional spring flows of 11,320 acft/year. In the Uvalde County, the additional artificial recharge would raise aquifer levels 11 feet on average, and result in 25% reductions in time the County is in drought management conditions (Nueces River Authority, 2007, from HDR Engineering, Inc.).

When geologic conditions are favorable, surface water reservoirs constructed upgradient from major sinks, and located in non-karstic rocks, can become part of the springwater treatment: turbid water flowing into the reservoir after major rainfall events

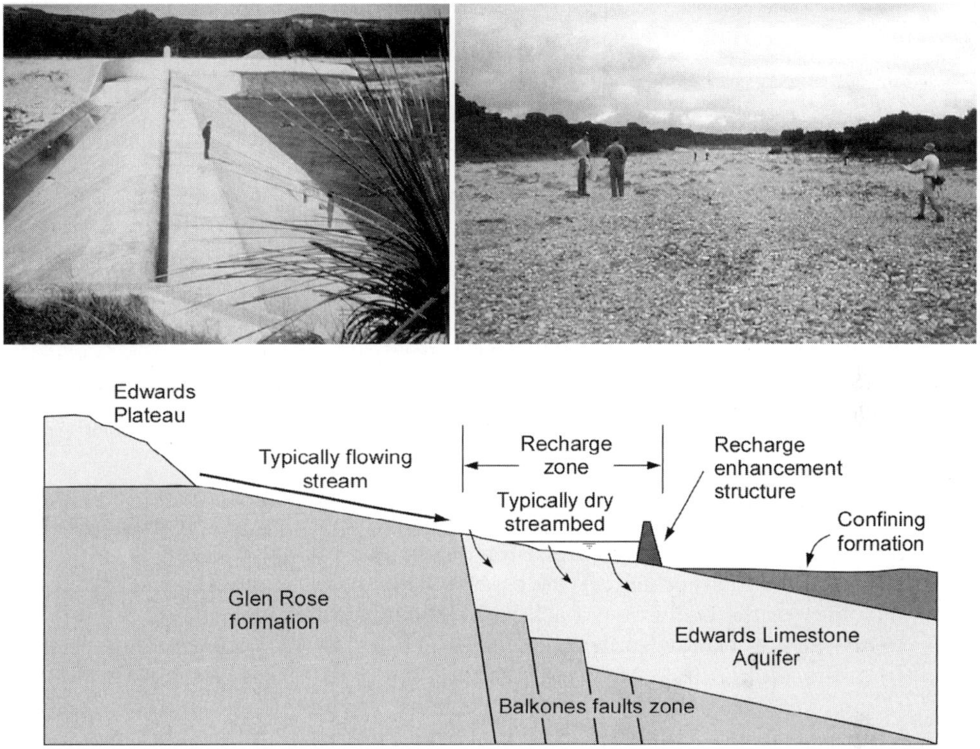

Figure 9.15 *Top*: View of concrete recharge dam at Middle Verde Creek Recharge Project site, Edwards Aquifer, Texas, and typical, highly permeable, streambed at recharge sites. *Bottom*: Principle of artificial recharge of Edwards Aquifer with recharge dams. (Courtesy of Nueces River Authority, 2007, and HDR Engineering, Inc.)

is allowed to settle before it is released and enters the karst aquifer via sinks (Fig. 9.16). Such dams also have added benefit of controlling the recharge rate, thus managing both water supply and demand.

9.3.3 Injection Wells

The advantage of the saturated-zone injection wells is that they can be used to recharge any type of aquifer, at any depth, thus eliminating problems associated with low-permeable surficial soils and low-permeable layers in general. Direct injection requires water of higher quality than for surface spreading because of the absence of vadose zone and/or shallow soil matrix treatment afforded by surface spreading and the need to maintain the hydraulic capacity of the injection wells, which are prone to physical, biologic, and chemical clogging. Treatment processes beyond secondary treatment that are used prior to injection include disinfection, filtration, air stripping, ion exchange, granular activated carbon, and reverse osmosis or other membrane separation processes. By using these processes or various subsets in appropriate combinations, it is possible to satisfy all present water quality requirements for water injection, including with reclaimed

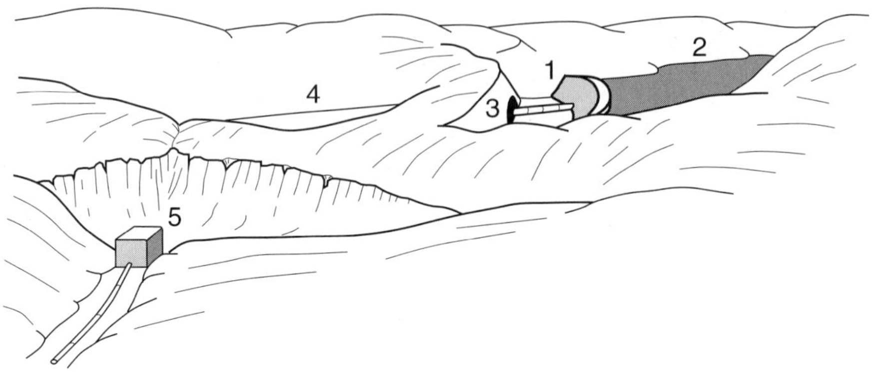

FIGURE 9.16 Schematic of the Bucje Aquifer regulation, Western Serbia and Montenegro. (1) Dam, (2) reservoir, (3) Tmusa River cave sink, (4) dry valley, and (5) Bucje karstic spring. (From Kresic, 1988.)

water. However, even when such high-quality water is used for recharge, trouble-free operations cannot be guaranteed for longer periods of well operation due to various geochemical and mechanical processes, which tend to clog well screens and reduce permeability of the well gravel pack and adjacent aquifer material. A common practice is to inject water via dual-purpose wells, which are used to occasionally pump water back from the aquifer thus removing screen-clogging materials ("backwashing"). The most frequent causes of clogging are accumulation of organic and inorganic solids, biologic and chemical contaminants, and dissolved air and gases from turbulence. Very low concentrations of suspended solids, on the order of 1 mg/L, can clog an injection well. Even low concentrations of organic contaminants can cause clogging due to bacteriological growth near the point of injection (USEPA, 2004).

In many cases, the wells used for injection and recovery are classified by the USEPA as Class V injection wells. Some states require that the injected water must meet drinking water standards prior to injection into a Class V well.

For both surface spreading and direct injection, locating the extraction wells as great a distance as possible from the recharge site increases the flow path length and residence time in the underground, as well as the mixing of the recharged water with the natural groundwater.

9.3.4 Aquifer Storage and Recovery (ASR)

ASR has been widely applied in the coastal areas of the eastern United States, particularly in Florida, where population growth and demand for water resources are of great concern. Figure 9.17 illustrates the basic principle of ASR: potable water is injected (stored) into a portion of the karstic Floridan aquifer with brackish, nonpotable water during periods of low water demand and/or high water availability, and then extracted (recovered) during periods of increased demand. The success of such systems is measured by recovery efficiency, expressed as percentage of the potable water meeting a preset criterion (e.g., chloride concentration less than 250 mg/L) that can be recovered relative to the quantity injected, over one full cycle that includes injection, storage, and recovery. Recovery efficiency in most cases increases with the number of full cycles, since most of the injected potable water remains in the aquifer after each cycle. Recovery of the stored

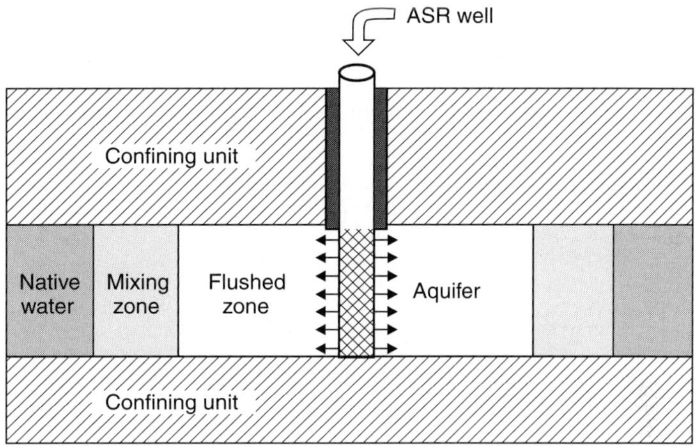

FIGURE 9.17 Aquifer storage and recovery well in a confined aquifer depicting idealized flushed and mixing (transition) zones created by recharge. Flushed zone contains mostly recharged water. (From Reese, 2002.)

water is dependent on the effective placement of a relatively stable, thick lens or bubble of low-density recharge water during the injection phase. To form this lens, enough water must be injected to displace a large volume of saline water, the mixing of the injected and native waters must not be significant, and confinement must be sufficiently tight to prevent rapid vertical migration of the less dense recharged water (Rosenshein and Hickey, 1977, from Yobbi, 1997).

Among various problems associated with ASR operations in brackish aquifers is the formation of lenses and bubbles of freshwater floating on, or embedded into, the denser native water. The lenses may have formed both because of a density contrast between the recharge water and the native groundwater and because of the possible presence of different transmissive zones in the aquifer.

9.4 References

Avias, J., 1984. Captage des sources karstiques avec pompage en periode d'etiage. L'example de la source du Lez. In: Burger, A.,and Dubertet, L.(eds.), *Hydrogeology of Karstic Terrains, Case Histories, International Contributions to Hydrogeology*. Vol. 1., IAH, Heise, pp. 117–119.

Fleury, P., Ladouche, B., Conroux, Y., Jourde, H., and Dörfliger, N., 2008. Modelling the hydrologic functions of a karst aquifer under active water management—The Lez Spring. *J. Hydrol.*, doi: 10.1016/j.hydrol.2008.11.037.

Fox, P. 1999. Advantages of aquifer recharge for a sustainable water supply. United Nations Environmental Programme/International Environmental Technology Centre, International Symposium on Efficient Water Use in Urban Areas, Kobe, Japan, June 8–10, pp. 163–172.

IAH (International Association of Hydrogeologists), 2002. Managing aquifer recharge. Commission on Management of Aquifer Recharge. IAH-MAR, 12 p. Available at http://www.iah.org/recharge/MAR_reports.htm

Ishida, S., Kotoku, M., Abe, E., Fazal, M. A., Tsuchihara, T., and Masayuki, I., 2003. Construction of subsurface dams and their impact on the environment. *RMZ Materials and Geoenvironment*, 50(1): 149–152.

Jevtic, G., Dimkic, D., Dimkic, M., and Josipovic, J., 2005. Regulation of the Krupac spring outflow regime. In: Stevanovic, Z., and Milanovic, P. (eds.), *Water Resources and Environmental Problems in Karst — Cvijić* 2005, Proceeding of International Symposium, University of Belgrade, Institute of Hydrogeology, Belgrade, pp. 321–326.

Ishida, S., Tscuchihara, T., and Imaizumi, M., 2005. Evaluation of impact of an irrigation project with a mega-subsurface dam on nitrate concentration in groundwater from the Ryukyu limestone aquifer,

Miyako Island, Okinawa, Japan. In: Stevanović, Z., and Milanović, P. (eds.), *Water Resources and Environmental Problems in Karst—Cvijić 2005, Proceedings of the International Conference and Field Seminars, Belgrade & Kotor, September 13–19, 2005, Institute of Hydrogeology*. University of Belgrade, Belgrade, pp. 121–126.

Kresic, N., 1988. Karst aquifers of the Lim catchment (SR Serbia). *Des Comptes Rendus des Séances de la Société Serbe de Géologie pour l'année 1985–1986*, Belgrade, pp. 217–223.

Kresic, N., 1991. *Kvantitativna Hidrogeologija Karsta Sa Elementima Zaštite Podzemnih Voda (Quantitative Karst Hydrogeology With Elements Of Groundwater Protection, in Serbo-Croatian)*. Naučna knjiga, Beograd, 196p.

Kresic, N., 2009. *Groundwater Resources: Sustainability, Management, and Restoration*. McGraw-Hill, New York, 852p.

Milanovic, P., 1996. Ombla Spring, Croatia. *Environ Geol*, 27(2): 105–107.

Milanovic, S., 2005. Hydrogeological characteristics of some deep siphonal springs in Serbia and Montenegro karst. In: Stevanovic, Z., and Milanovic, P. (eds.), *Water Resources and Environmental Problems in Karst — Cvijić* 2005, Proceeding of International Symposium, University of Belgrade, Institute of Hydrogeology, Belgrade, pp. 451–458.

Nueces River Authority, 2007. Edwards Aquifer recharge dams. Available at: http://www.nueces-ra.org/II/recharge/; accessed in 14th December 2007.

Osuga, K., 1997. The development of groundwater resources on the Miyakojima Islands. In: Uitto, J. I., and Schneider, J. (eds.), *Freshwater Resources in Arid Lands. UNU Global Environmental Forum V*. United Nations University Press, Tokyo. Available at: http://www.unu.edu/unupress/unupbooks/uu02fe/uu02fe02.htm, accessed October 27, 2005.

Pereira, L. S., Cordery, I., and Iacovides, I., 2002. Coping with water scarcity. *International Hydrological Programme VI, Technical Documents in Hydrology No. 58*. UNESCO, Paris, 269p.

Potier, L., Ricour, J., and Tardieu, B., 2005. Port-Miou and Bestouan freshwater submarine springs (Cassis—France) investigations and works (1964–1978). In: Stevanovic, Z., and Milanovic, P. (eds.), *Water Resources and Environmental Problems in Karst, Proceedings of the International Conference and Field Seminars, Belgrade & Kotor, Serbia & Montenegro, September 13–19, 2005, Institute of Hydrogeology*, University of Belgrade, Belgrade, pp. 267–274.

Reese, R. S., 2002. Inventory and review of aquifer storage and recovery in southern Florida. U. S. Geological Survey Water-Resources Investigations Report 02-4036, Tallahassee, Florida, 56p.

Rosenshein, J. S., and Hickey, J. J., 1977. Storage of treated sewage effluent and storm water in a saline aquifer, Pinellas Peninsula, Florida. *Ground Water*, 15(4): 289–293.

Slichter, C. S., 1902. The motions of underground waters. *U.S. Geological Survey Water-Supply and Irrigation Papers 67*. Washington, D.C., 106p.

Stevanovic, Z., 2010. Utilization and regulation of springs. In: Kresic, N., and Stevanovic, Z. (eds.), *Groundwater Hydrology of Springs; Engineering, Theory, Management and Sustainability*. Elsevier, New York, pp. 339–388.

USEPA, 2004. Guidelines for water reuse. EPA/625/R-04/108. *U.S. Environmental Protection Agency, Office of Wastewater Management, Office of Water*. Washington, D.C., 460p.

Yaoru, L., 1986. Karst in China. Landscapes, types, rules (in Chinese). Spec. Ed. of Geol. Publ. House. Beijing, 288p.

Yaoru, L., Jie, X. A., and Zhang, S. H., 1973. The development of karst in China and some of its hydrogeological and engenireeng geological conditions. *Acta Geologica Sinica*, 1: 121–136.

Yobbi, D. K., 1997. Simulation of subsurface storage and recovery of effluent using multiple wells, St. Petersburg, Florida. U.S. Geological Survey Water-Resources Investigations Report 97-4024. Tallahassee, Florida, 30p.

PART 3

Vulnerability and Restoration of Water in Karst

CHAPTER 10
Vulnerability of Water in Karst

10.1 Introduction

Just like in any other porous media and environmental setting, water in karst is vulnerable to various natural and anthropogenic factors that can negatively impact its quality and quantity. Numerous examples and related discussions are provided in previous chapters in various contexts. In few cases, however, the unique nature of karst vulnerability may have desirable outcome, at least to some. The following narrative provided by Jochen Duckeck and featured on the Web page maintained by Showcaves (2012) illustrates this point:

> This is not a page about a place, but it is one about the funniest accident based on karst geology. At the end of the 19th and the beginning of the 20th century a drink called Absinth was very famous in France. It was a liqueur with many ingredients, including anis. Like all those anis liqueurs (e.g. Ouzo, Raki,...) it is a clear liquid until it is mixed with water. Then it becomes milky.
>
> As a matter of fact the original Absinth contained some slightly poisonous ingredients. It would have been hard to get a lethal dose without being killed by the alcohol much earlier, but this poisonous touch was part of the Absinth myth. And there was a rumor of it being an aphrodisiac. Unfortunately those peculiarities were the reason why it was forbidden some years later.
>
> On 11 August 1901 a lightning hit the plant of the Pernod company in Pontarlier. It was set on fire, and the high temperature made the huge tanks with alcohol explode. Some one million liters of Absinth poured across the plant, down the sewage system and straight into the river Doubs. The river turned white and smelled of anis.
>
> Two days later, André Berthelot visited the Source de la Loue (shown on the front cover of this book) more than 15 km from Pontarlier, on the other side of a high mountain ridge. While enjoying the romantic atmosphere at the clear spring, he suddenly noticed a change. The water turned milky, and a smell of anis filled the air. He knew this smell from the pub, but he had to take a mouthful, to believe it: The water of the big spring turned into a free aperitif! It is not handed down, if he filled his water-bottle...
>
> This story was heard by the founder of modern speleology, E. A. Martel, and nine years later he made a dye tracing experiment. He found a swallow hole in the bed of river Doubs and added a strong, non-toxic, green color. 64 hours later, the Source de la Loue turned green. The connection of the Loue and the Doubs across the mountain ridge was proven and the dye tracing of groundwater invented.
>
> But now the next act in this play started: the owners of some small mills along the Doubs had problems with a strange lack of water for many years now. As they heard of the swallow

> holes, they started to look for them and filled them with concrete when they found them. So the water stayed in the Doubs.
>
> But now the residents along the Loue had problems with missing water. And so the case came before court. The decision: the already sealed sinks may stay as they are, but new changes of the course of the water will be punished.

As discussed by Hötzl (1996), it is practical to distinguish between resource and source protection, although both concepts are closely related to each other—it is impossible to protect a source without protecting the resource. In European countries, for example, groundwater is considered a valuable resource that must be protected. Activities endangering its quality are forbidden by law (see, e.g., German regulation described in WHG, 1996). The European Water Framework Directive (The European Parliament and the Council of the European Union, 2000) emphasizes that water is not a commercial product like any other but a heritage that must be protected, defended, and treated as such. Thus, the directive demands for the protection of groundwater and surface water resources. The highest priority is to protect the groundwater used for drinking water supply. The source may be a captured spring, a pumping well, or any other groundwater extraction point. The European Groundwater Directive of 2006 extends the concept of overall resource protection in detail (The European Parliament and the Council of the European Union, 2006).

Protection of karst water sources, including surface streams, aquifers, and springs, is achieved by preventing possible contamination, by remediating already contaminated water, and by detecting and preventing unsustainable water withdrawal. The prevention aspect includes pollution prevention programs, control measures at potential contaminant sources, land-use control, and public education. Some examples of prevention measures include (Kresic, 2009)

- Land-use controls that prevent obvious introduction of contaminants from industrial, agricultural, and urban sources into the streams and the subsurface.
- Controls of pesticide use in sensitive aquifer recharge areas including their complete banning when warranted such as in open hydrogeologic structures where karst is covered only by residuum sediments or directly exposed to the land surface.
- Land-use controls that minimize the interruption of natural aquifer recharge, such as the paving of large urban areas (urban sprawl).
- Management of urban runoff that can contaminate both surface water and groundwater resources.
- Mandatory installation of devices and monitoring wells for early detection of contaminant releases such as leaks from underground storage tanks at gas stations and landfill leachate migration.

Figure 10.1 shows a prevention measure implemented in the city of Austin, Texas, where the underlying karstic Edwards Aquifer is directly exposed at the land surface or covered by a thin layer of residuum. Surface runoff from state highways is directed toward specially designed concrete catch basins where oil and fine particulate matter are removed before water is allowed to infiltrate into the subsurface.

Unfortunately, almost every human activity has the potential to directly or indirectly impact karst groundwater to a certain extent. Figure 10.2 illustrates just some of the land-use activities that can result in surface water, groundwater, and ultimately karst spring

Figure 10.1 Concrete catch basin by a state highway in Austin, Texas, designed to strip oil and fines from surface runoff before water is allowed to infiltrate into the underlying Edwards Aquifer. (From Kresic, 2009.)

contamination. An exponential advancement of analytical laboratory techniques in the last decade has demonstrated that many synthetic organic chemicals (SOCs) are widely distributed in the environment, including surface water and groundwater, and that a considerable number of them can now be found in human tissue and organs of people living across the globe.

Notwithstanding assertions that any water can be treated to become drinkable (potable), it is not surprising that multinational companies selling bottled water are frantically looking for new sources of springwater across the globe. This is understandable since many consumers worldwide are ready to pay a premium for brands marketed as

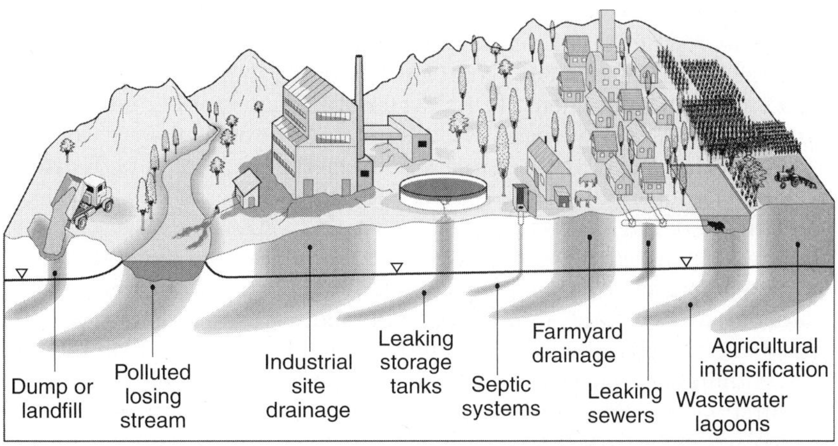

Figure 10.2 Land-use activities commonly generating a groundwater pollution threat. (From Foster et al., 2002–2005.)

"pure spring water." In contrast, as mentioned in Sec. 8.4, quite a few European countries, where karst springs of high water quality are abundant, utilize them as preferred sources of public water supply.

As discussed by Brune (1975), in Texas and the United States as a whole, interest in groundwater initially also focused on springs, particularly in the arid West. As the eastern part of the country developed and most of the land became privately owned, the groundwater-based public water supply shifted from springs to wells. Advances in well drilling, pump technology, and rural electrification made possible the broad-scale development of groundwater in the West beginning in the early 1900s. Large-scale irrigation with groundwater, especially after World War II, spread rapidly throughout the West resulting in the cessation of spring flows at many locations. As a consequence, in both the Eastern and Western United States, the overall utilization of springs for centralized water supply is minor compared to other parts of the world (Kresic, 2009). Where there are numerous large springs, as in the karst regions of Florida, Texas, and Missouri for example, in many cases such springs are located on private land or public parkland, and preserved for other uses including recreation. A paragraph from Meinzer's publication on large springs in the United States (Meinzer, 1927) illustrates this point for Florida springs: "Some of the springs have become well-known resorts, but otherwise not much use is made of their water. The fascinating character of these springs is indicated by the following vivid description of Silver Spring, abbreviated from a description given in a booklet published by the Marion County Chamber of Commerce" (see pl. 1.). "The deep, cool water of Silver Spring, clear as air, flows in great volume out of immense basins and caverns in the midst of a subtropical forest. Seen through the glass-bottom boats, with the rocks, under-water vegetation, and fish of many varieties swimming below as if suspended in mid-air, the basins and caverns are unsurpassed in beauty. Bright objects in the water catch the sunlight, and the effects are truly magical. The springs form a natural aquarium, with 32 species of fish. The fish are protected and have become so tame that they feed from one's hand. At the call of the guides, hundreds of them, of various glistening colors, gather beneath the glass-bottom boats."

Unfortunately, as illustrated by the example of Florida's Kissengen Spring in Sec. 4.1, the preservation of large karst springs for nonconsumptive beneficial uses is becoming increasingly difficult and requires a sustained, coordinated effort by the water utilities, regulatory agencies, legislature, and the public. In addition to quantity, water quality of a once-pristine spring can be diminished by unsustainable land-use practices in its drainage area to the point of a complete loss of the source. As emphasized in Chap. 5, *the* most important component of karst water protection is public education. In that respect, various government agencies with mandate to protect the environment and enforce the related laws play the key role. The following example of educating the public by explaining vulnerability of karst in the New River watershed is provided by the Virginia Department of Environmental Quality (VA DEC, 2008a).

The intimate connection between surface water and groundwater in karst makes for an environment that is easily contaminated. In most non-karst settings, surface water slowly percolates into the ground, allowing time for filtration and natural bioremediation, such as the uptake of nitrates by plants. In karst, on the other hand, voids and conduits allow surface water to enter groundwater quickly, often without ample time for filtration or biologic breakdown of nutrients and contaminants. Consequently, what takes place on the surface largely determines the nature and proportions of contaminants that enter karst waters.

In the New River watershed, three classes of land use dominate forest, agricultural, and urban. Forests cover approximately 58% of the watershed, concentrated along ridgetops and slopes. Much of the forested land in the watershed lays up-slope from the karst. Forest streams commonly sink into the subsurface shortly after flowing onto karst. Streams that drain mature forest areas are generally very clean except for occasional low-level bacterial contamination from wildlife. However, logging operations may increase erosion and runoff, potentially producing large volumes of soil and organic debris, or result in accidentally spilled fuel or other contaminants, thus forest best management practices (BMPs) are essential.

Agricultural land, mostly pasture, crop, and orchard land, covers 37% of the New River watershed. Agriculture is a significant potential source of water quality degradation in the watershed. Fertilizer, pesticides, and manure drain from cropland, active livestock areas, and pastures into streams and sinkholes.

Disposal of wastes in sinkholes, once a common practice, is illegal and directly impacts karst. Implementation of agricultural BMPs, as specified by Virginia's soil and water conservation districts, can greatly reduce the potential impacts of agriculture on both surface water quality and karst resources.

Urban and residential land use, though accounting for only about 4% of the watershed, can result in significant surface and groundwater degradation. Examples of chemical contaminants commonly released include petroleum, antifreeze, and runoff from parking lots and roads, fertilizer and pesticides from landscaping activities and termite treatments, sediment from construction sites, chemical spills, and leaking underground storage tanks (LUSTs). Bacterial and chemical contamination from leaking sewer lines or malfunctioning septic systems can further degrade water quality and is difficult to detect because these structures are buried.

Virginia Department of Conservation and Recreation (VA DCR, 2012) has Natural Heritage Karst Program that emphasizes identification and protection of sensitive karst areas through appropriate land-use planning to help ensure a safe, clean water supply for everyone. These actions also protect rare species and natural communities for the benefit of all. The DCR Karst Program and the governor-appointed Virginia Cave Board assist state and federal agencies, local governments, private organizations, and individuals in matters relating to the management of caves and the protection of karst resources in the commonwealth. The Virginia Cave Protection Act (http://www.dcr.virginia.gov/natural_heritage/vcbprotact.shtml#) acknowledges the unique and irreplaceable nature of caves, their special habitats, their historic and cultural value, and their other special attributes. It recognizes the role caves play as conduits for groundwater as well as their vulnerability to pollution (VA DCR, 2012).

DCR's Natural Heritage Program supports protection of Virginia's caves and karst through inventory and information management, land conservation, and land management. The program is a part of NatureServe, an international network of natural heritage programs that tracks rare biologic elements for environmental planning and in support of state environmental review process. Every year, DCR reviews and comments on hundreds of projects on karst terrain, working with local governments, citizens, and state and federal agencies to find solutions that balance competing demands on these resources. DCR proactively engages in projects to define karst groundwater basins, perform biologic investigations and monitoring in caves, and help develop and promote BMPs that protect Virginia's caves and karst resources. DCR provides karst-related technical assistance to various state, federal, and local government agencies and to organizations, professionals, and private citizens.

Project Underground (http://www.dcr.virginia.gov/underground.shtml), an educational program that promotes a better understanding of caves and karst, is a cornerstone of DCR's karst conservation effort. Through this nationally recognized program, thousands of students and citizens throughout the commonwealth each year learn the special values and sensitive nature of karst. DCR also has stormwater runoff, erosion, sediment, and nutrient management programs to reduce nonpoint source pollution that may be detrimental to karst resources. The agency works with local soil and water conservation districts to help landowners use BMPs that conserve natural resources and protect the environment. DCR has the following advice to visitors of its karst Web site: "Here are a few ways to protect fragile karst in the area. Do not dump trash into sinkholes. Manage stormwater runoff to prevent sediment pollution, and improve water quality by using vegetated filter strips and stream buffers. Develop alternate water sources for livestock and fence cattle away from streams and sinkhole ponds. Develop a nutrient management plan to apply only the fertilizer your plants need. Help friends and neighbors better understand and appreciate their stake in maintaining a sustainable, quality environment, and to protect caves and other karst resources in the upper James and upper Roanoke River basins now and for future Virginians."

Sometimes, with best intentions, various government agencies and other interested groups create guidance documents and educational pamphlets on karst vulnerability that are confusing at best. To its credit, the U.S. Department of Agriculture states the following in the same document where the convoluted flowchart shown in Fig. 10.3 comes from: "A geologic investigation of the potential impacts of the treatment on groundwater, surface water, and the karst features is required and must be conducted by a qualified geologist."

Sometimes, a few suggested rules of thumb by "official" sources may be misleading and potentially harmful. Typical example is a recommended minimum distance for which there should be concerns regarding certain practices; if such distance is greater than the rule of thumb, the concern somehow disappears (e.g., if the distance between points A and B is greater than 1 mile, than there is no concern). Whatever the karst-related concern may be, it is best to perform a site-specific assessment and not rely on some rules of thumb, even if they come from seemingly official sources.

Many speleological and caving clubs and associations throughout the world have excellent educational programs and public outreach. They are also the best ambassadors of karst as well as the most frequent visitors witnessing many unique and fascinating karst features in constant need of protection. This includes karst biota that is especially vulnerable to environmental degradation of water (Fig. 10.4; see also color plates). Nowhere else except in deep oceans can one find such exceptional diversity of organisms many of which are endemic species living only in certain localities and nowhere else. For example, the Madison Cave Isopod (*Antrolana lira*), a blind, free-swimming crustacean is known only from the limestone aquifers of the Great Valley of Virginia and West Virginia and has so far being found in 11 caves and 7 wells (Fig. 10.4-top right). Blind and pigmentless, fully adapted to life underground and underwater, it is believed to have evolved long ago from marine ancestors stranded when ocean water was in contact with limestone of the Shenandoah Valley. Listed as threatened over its entire range under the federal and Virginia Endangered Species Acts, the Madison Cave Isopod is legally protected from impacts due to human activities.

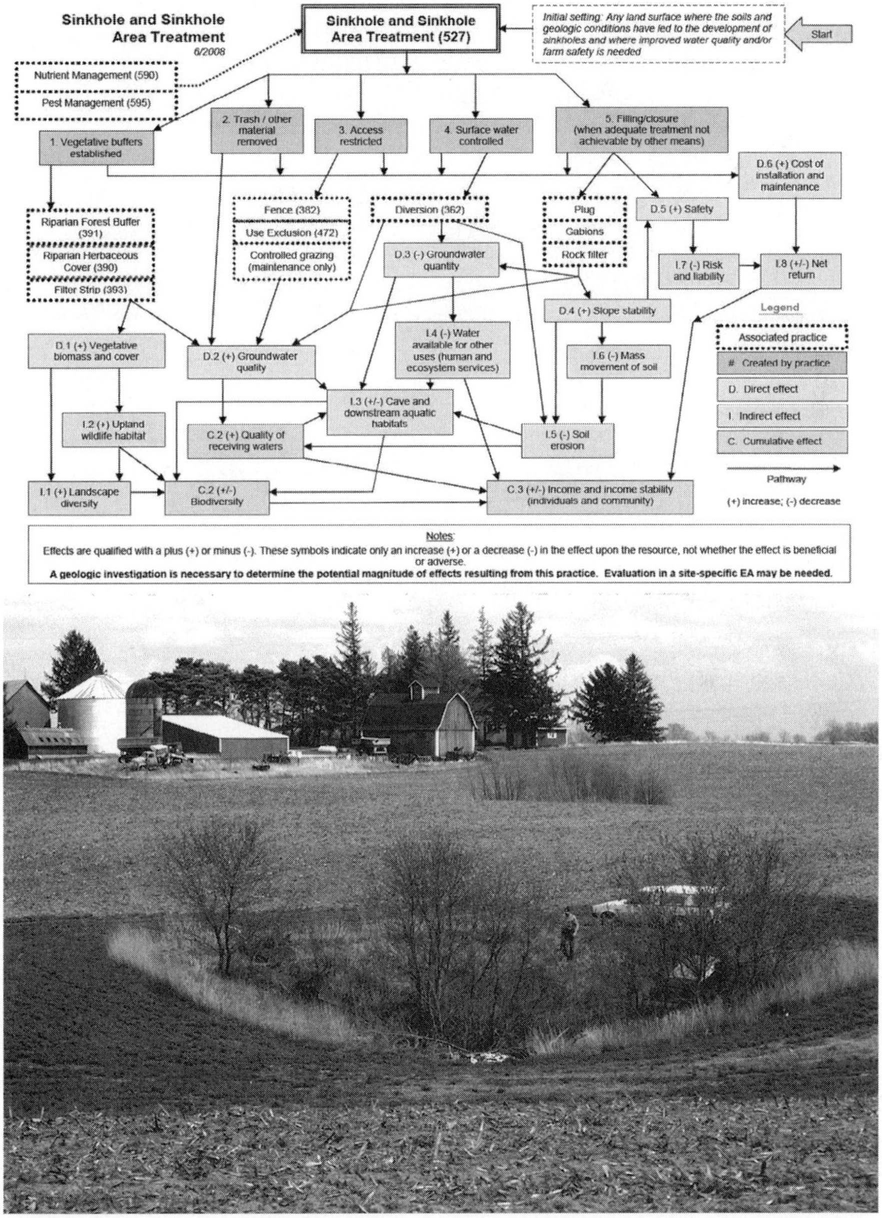

Figure 10.3 *Top*: The original USDA caption states that this rather convoluted diagram identifies the effects expected to occur when this practice (Sinkhole and Sinkhole Area Treatment) is applied according to National Resource and Conservation Service (NRCS) practice standards and specifications. Users are cautioned that these effects are estimates that may or may not apply to a specific site. (From USDA, 2008.) *Bottom*: A sinkhole in the karst topography of northeast Iowa funnels runoff waters directly into an underground aquifer. The conservation buffer surrounding the sinkhole improves water quality by filtering the water from croplands. (Courtesy of NRCS, photograph by Tim McCabe.)

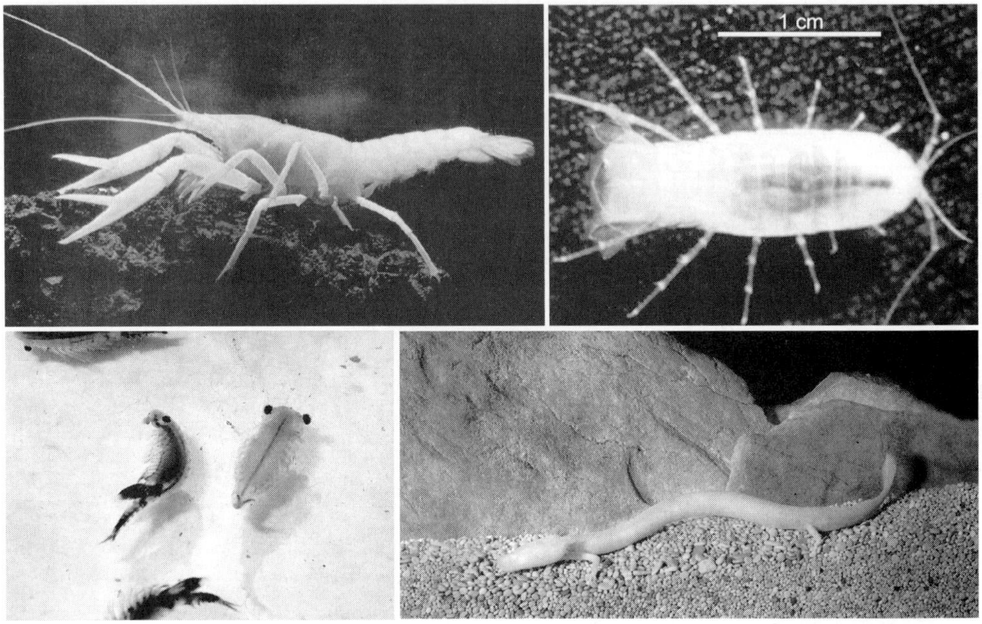

FIGURE 10.4 *Top left*: *Orconectes pellucidus* (Tellkampf), the troglobitic crayfish of the Western Mississippian Plateau (From Barr, 1985.). *Top right*: The Madison Cave Isopod, *Antrolana lira*. (Courtesy of Virginia Division of Natural Heritage Karst Program, Virginia Department of Conservation and Recreation.). *Bottom left*: The fairy shrimp (Anostraca), *Chirocephalus croaticus*, a rare species limited to temporary karst waters in the Croatian Dinaric karst. (Photo courtesy of S. Polak.) *Bottom right*: Completely aquatic amphibian with undeveloped eyes covered with skin, *Proteus anguinus* is endemic to Dinaric karst aquifers in Slovenia, Croatia, and Herzegovina. (Photo courtesy of G. Aljančič.)

Famous *Proteus anguinus* shown in Fig. 10.4-bottom right is the only European exclusively cave-dwelling chordate. Locals call it "human fish" (čovečija ribica) due to its skin color (see color plates). Because of its pinkish-red external gills, few people that ever saw Proteus in the Middle Ages thought of it as a baby dragon hiding in caves. This folk tale persisted until second half of the eighteenth century when it was finally described scientifically. Proteus is protected in Slovenia and Croatia.

Every so often a new karst species is discovered somewhere in the world reminding all of us of this enormous underground treasure that is increasingly under stress from various land-use changes and human activities.

10.2 Water Contamination

Water contamination in karst results from the following activities, with the note that the list is far from being all-inclusive (modified from www.wrds.uwyo.edu/wrds/deq/whp/):

- Illegal dumping or abandonment of household, commercial, or industrial waste and chemicals, particularly in karst depressions (Fig. 10.5).

Figure 10.5 *Top*: Trash disposed in a sinkhole, Laclede County, Missouri. Dye tracing shows this sinkhole to provide recharge to Ha Ha Tonka Spring. (Photograph courtesy of Missouri Department of Natural Resources; available at: http://www.dnr.mo.gov/env/wrc/springsandcaves.htm.) *Bottom left*: Photograph and text used to illustrate vulnerability of groundwater to contamination in an early USGS publication for public education: "Continuous stream of sewage from a college building discharging into a sink from which it finds its way to the underground water channels." (From Fuller, 1910.) *Bottom right*: Spelunker examines trash around a flowstone waterfall at the bottom of the entrance to Midnight Cave near Austin, Texas, on November 20, 1993. The trash includes household garbage, used oil filters, corroded 55-gallon drums, glass pesticide bottles, partially filled turpentine cans, and automobile parts. Note the trash on the higher ledges of the cave. (From Hauwert and Vickers, 1994.)

- Misuse and improper disposal of liquid and solid wastes and chemicals at commercial, industrial, agricultural, and governmental facilities, and in households (Fig. 10.5).
- Accidental spilling of chemicals from trucks, railways, aircraft, handling facilities, and storage tanks.
- Use of road salt in winter.
- Land application (spraying of wastewater treatment effluent on land and its disposal through infiltration basins; spreading of solid sludge from wastewater treatment facilities on land and farmland).
- Urban runoff from parking lots, streets, and construction sites.
- Improper location, design, construction, operation, maintenance, or deterioration of agricultural, residential, municipal, commercial, and industrial drinking water wells and liquid and solid waste disposal facilities.
- Application of natural (manure) and artificial fertilizers, pesticides, and insecticides in agriculture, on household loans and gardens, and on golf courses.
- Livestock feeding operations.
- Atmospheric pollutants, such as airborne sulfur and nitrogen compounds, which are created by smoke, flue dust, aerosols, and automobile emissions, fall as acid rain and percolate through the soil.
- Contaminated water used for aquifer recharge, and geochemical reactions between otherwise clean recharge water and aquifer porous media (Fig. 10.6).

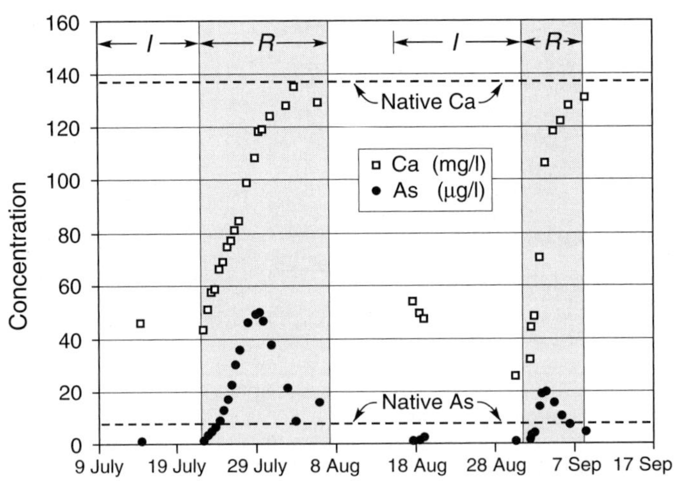

FIGURE 10.6 Changes in arsenic (As) and calcium (Ca) concentrations during Punta Gorda ASR project cycle tests, Floridan karst aquifer; *I*: injection phase; *R*: recovery phase. During recovery, mixing of low-Ca injected waters and higher-Ca native groundwater is observed. Arsenic concentrations in recharge and native groundwaters are less than 10 micrograms per liter (μg/L). An arsenic peak (up to 50 micrograms per liter) is observed indicating water–rock interaction. (Arthur et al., 2002.)

Contaminants can reach surface water and groundwater from nonpoint and point sources. The former includes forms of diffuse contamination caused by sediment, nutrients, organic, and toxic substances originating from land-use activities such as agriculture, urban development, and spraying of wastewater treatment effluent on land. Rainwater, snowmelt, or irrigation water can wash off these substances together with soil particles and carry them with surface runoff to surface streams. This contaminant load dissolved in water can also infiltrate into the subsurface and eventually contaminate groundwater. Contaminated surface streams in karst also present nonpoint sources of groundwater contamination if they lose water over long stream segments.

Point sources result from both unintentional and intentional hazardous waste disposal practices, spills, leaks, or otherwise limited in extent introduction of contaminants into the subsurface. They also include contaminated surface streams sinking at well-defined locations (sinks), sources below the land surface but above the water table such as septic systems, structures beneath the water table such as compromised or improperly constructed wells, and contaminated water used for artificial aquifer recharge.

Many RCRA and Superfund sites in the United States are examples of multiple point sources of groundwater contamination. These sources may form individual plumes of individual contaminants, individual plumes of mixed contaminants from identifiable sources, or, in the most complicated cases, commingled (merged) plumes of various contaminants from multiple sources, some of which are not easily, or not at all identifiable. Sites on military installations, large industrial complexes, and chemical manufacturing plants are likely to have groundwater contaminated by multiple constituents, which may be distributed at various depths in the underlying aquifer(s) and form plumes with complicated shapes. Complex sites of groundwater contamination are often a nightmare for groundwater professionals trying to characterize possible contaminant sources and "attach" to them their own plumes. This, however, is the favorite topic of attorneys working for various Potentially Responsible Parties (PRPs). Heavy involvement of attorneys in groundwater contamination and remediation issues in the United States is understandable since costs associated with groundwater remediation may be astronomical, and the question who is responsible for the plume is of an utmost importance. In addition, remediation of certain contaminants in karst settings may not be technically or economically feasible, which is often very hard to convey to various stakeholders (see Sec. 11.3.2).

Results of a nationwide study of potential contaminant sources conducted by the United States Environmental Protection Agency (USEPA, 2002a) and state environmental protection agencies are shown in Fig. 10.7. Each state was requested to indicate the 10 top sources that potentially threaten their groundwater resources. States added sources as was necessary based on state-specific concerns. When selecting sources, states considered numerous factors, including

- The number of each type of contaminant source in the state.
- The location relative to groundwater sources used for drinking water purposes.
- The size of the population at risk from contaminated drinking water.
- The risk posed to human health and/or the environment from releases.
- Hydrogeologic sensitivity (the ease with which contaminants enter and travel through soil and reach aquifers).
- The findings of the state's groundwater assessments and/or related studies.

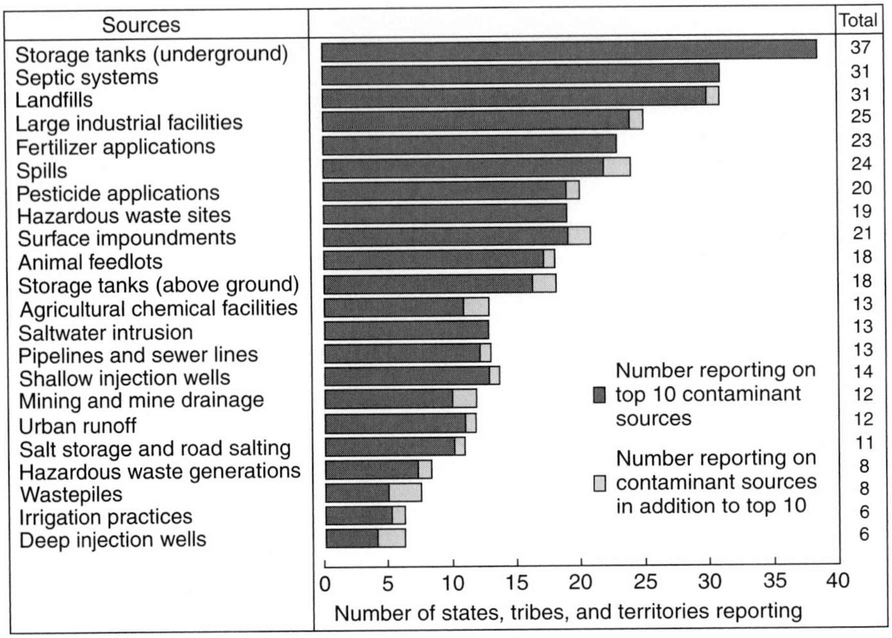

FIGURE 10.7 Major sources of groundwater contamination in the United States. (From USEPA, 2000.)

For each of the 10 top sources, states identified the specific contaminants that may impact groundwater quality. As seen in Fig. 10.7, the sources most frequently cited by states as a potential threat to groundwater quality are LUSTs. Septic systems, landfills, industrial facilities, and fertilizer applications are the next most frequently cited sources of concern. If similar sources are combined, four broad categories emerge as the most important potential sources of water contamination: (1) fuel storage practices, (2) waste disposal practices, (3) agricultural practices, (4) industrial practices, and (5) mining operations.

Case studies detailing various impacts of human activities on karst groundwater are presented in Drew and Hotzl (1999a).

10.2.1 Detecting Water Contamination in Karst

Water contamination may be detected in a number of ways. The most obvious is with human senses: water that looks bad, smells bad, and tastes bad is probably bad. It is also true that relying only on human senses to detect contamination does not make sense because water that looks clear, does not smell, and has no taste may still be contaminated and its ingestion may cause serious health problems or death. Unfortunately, very few individuals, even in most developed countries, can afford testing their drinking water for the full suite of regulated chemicals, let alone doing it on a regular basis (one such analysis can cost well over 3,000 US dollars). Public water supply systems in most countries are required by law to regularly test their raw source water for drinking water standards, thus protecting their customers. Individual well owners are not subject to such requirements and are therefore most at risk from drinking contaminated groundwater, especially if their wells are getting water from shallow unconfined aquifers or anywhere in karst aquifers.

In the United States, the public and regulators alike are increasingly recognizing the vulnerability of private wells to groundwater contamination. For example, anyone thinking of buying one of the 20,000 homes in the New York's Westchester County, served by a private well, will enjoy safer drinking water thanks to a new law that took effect in November 2007. The law requires that a water test be conducted upon signing a contract of sale for any property served by a private drinking water well. The law also requires that new private wells and private wells that have not been used to supply drinking water for a period of 5 years be tested before use. Private wells on leased properties must also be tested regularly. The test ensures that the well water is safe for human consumption through analysis for the presence of coliform bacteria and chemical contaminants. Under the new law, only certified laboratories are authorized to collect and test the water samples. Test results must be submitted to the Westchester County Health Department, as well as to the person requesting the test. The law also makes clear the responsibilities of home sellers and buyers and other parties to ensure that drinking water quality problems are corrected. It also establishes penalties for noncompliance. Information on the law is available on the Westchester County Health Department's Web site at www.westchestergov.com/health.

Monitoring for and detecting groundwater contamination at sites that have confirmed releases of contaminants to the subsurface, or have potential for releases, is not an easy task. As illustrated by the following excerpts from the State of Texas regulations for landfills, it requires proper planning, execution and reporting, and a significant investment (Texas Administrative Code, 2006):

- A groundwater monitoring system must be installed and consist of a sufficient number of monitoring wells, installed at appropriate locations and depths, to yield representative groundwater samples from the uppermost aquifer as defined in §330.3 of this title (relating to definitions).
- Background monitoring wells shall be installed to allow the determination of the quality of background groundwater that has not been affected by leakage from a unit.
- A groundwater monitoring system, including the number, spacing, and depths of monitoring wells or other sampling points, shall be designed and certified by a qualified groundwater scientist. Within 14 days of the certification, the owner or operator shall submit the certification to the executive director and place a copy of the certification in the operating record. The plan for the monitoring system and all supporting data must be submitted to the executive director for review and approval prior to construction.
- The design of a monitoring system shall be based on site-specific technical information that must include a thorough characterization of aquifer thickness; groundwater flow rate; groundwater flow direction, including seasonal and temporal fluctuations in flow; effect of site construction and operations on groundwater flow direction and rates; and thickness, stratigraphy, lithology, and hydraulic characteristics of saturated and unsaturated geologic units and fill materials overlying the uppermost aquifer, materials of the uppermost aquifer, and materials of

the lower confining unit of the uppermost aquifer. A geologic unit is any distinct or definable native rock or soil stratum.

- The owner or operator may use an applicable multidimensional fate and transport numerical flow model to supplement the determination of the spacing of monitoring wells or other sampling points and shall consider site-specific characteristics of groundwater flow as well as dispersion and diffusion of possible contaminants in the materials of the uppermost aquifer.

In addition to the requirement that the plan is certified by a qualified groundwater scientist, several other important points applicable to any site including landfills are

- Determination of background concentrations of various physical and chemical parameters, and potential constituents of concern (COCs).
- Determination of sampling frequency by taking into account seasonal and temporal fluctuations in groundwater flow such that contaminants are not missed.
- Determination of the number and spacing of wells, and screen intervals (depths of monitoring), by taking into account fate and transport characteristics of the COCs, and hydrogeologic characteristics of the karstic porous media including heterogeneity and anisotropy.

Probably the most important factor when monitoring for possible contamination in karst is understanding that most karst water systems are very dynamic and quickly respond to various natural and anthropogenic water inputs. Additionally, contaminants can be introduced in a variety of ways from sources that can be intermittent or continuous; they can be periodically flushed through the vadose zone or out of storage features in the karst subsurface due to varying recharge (precipitation) pattern; they can also reside above the water table and be periodically stripped by the rising water levels. These and other factors should always be considered as they can greatly influence contaminant concentrations at monitoring wells and springs.

Decision as to the number of physical and chemical water parameters and COCs to be monitored will be site-specific depending on the type of contaminant release or potential sources of contamination. Table 10.1 lists types of sites with potential groundwater contamination and related COCs, and Table 10.2 lists analytical laboratory methods for detection of various COCs.

Once the monitoring data is collected and its quality verified, it can be evaluated for contamination detection. Any detection of COCs that are not naturally occurring, such as anthropogenic, synthetic organic chemicals, would constitute groundwater contamination. The same is true for contamination with microorganisms regardless of their origin as they can be either naturally occurring or the result of anthropogenic activities (e.g., wastewater discharges, animal feeding operations, and farming).

When the COCs are not detected in the background monitoring wells (wells located upgradient from the suspected source), and there is a clearly defined and confirmed contaminant release from the suspected source, this contamination may still not be definitively linked to that particular source. Determination of who is responsible for contamination at this point depends on the skills of attorneys, consultants, and representatives if the case were to become the subject of a lawsuit. If the background wells are contaminated with the same COCs, which indicate the presence of another potential source,

Type of Site	Potential Contaminants of Concern (COCs) in Groundwater
Small gas station (with ASTs and USTs)	TPH, PAH, and metals
Heating oil tank (AST or UST)	PAHs
Dry cleaners	VOC
Landfill (Class C and D)	TDS, TPH, PAH, PCB, VOC, SVOC, metals, Chloride
Airport	TPH, PAH
Rail switching yard	TPH, PAH, VOC
Livestock farm (i.e., dairy farm)	Ammonia, nitrate/nitrite, CFU
Crop farm	OP, OC, and herbicides
Orchards	OP, OC, metals (As), and herbicides
Open pit or strip mine	Metals, sulfate, sulfide, pH
Quarry	VOC, HMX, RDX, TNT, perchlorate
Nuclear power plant	Tritium, strontium, cesium
Power plants (thermal)	PAH, TPH, metals
Small arms firing range	Metals (lead and tungsten) PAH (clay targets)
Military ranges	HMX, RDX, TNT, perchlorate, metals
Machine shop (i.e., plating facilities, airline parts manufactures, auto mechanic)	VOC, SVOC, metals
Solid propellant manufacturer (i.e., fireworks, rocket motors)	Perchlorate, metals
Wood treatment plants	PAH, VOC, SVOC
Paper mill	PAH, metals, dioxin
Municipal wastewater treatment plant	Nitrate/nitrite, CFU, TOC, DBP, pharmaceuticals, surfactant (detergent)
Leach fields and septic systems	Ammonia, nitrate/nitrite, CFU, chloride, surfactant (detergent)
Industrial wastewater treatment plant	Metals, VOC
Automated car washes	TPH, PAH, VOC
Chemical manufacturing plant	Chemical specific + VOC, SVOC
Manufactured gas plant (MGP)	PAH, metals, TPH

From Kresic (2009).
Notes: TPH, total petroleum hydrocarbons; PAH, polycyclic aromatic hydrocarbons; OP, organophosphorous pesticides; OC, organochlorine pesticides; PCB, polychlorinated biphenyls; VOC, volatile organic compounds; SVOC, semivolatile organic compounds; CFU, colony-forming unit; AST, above ground storage tank; UST, underground storage tank; DBP, disinfection by-products.

Table 10.1 Types of Sites with Potential Groundwater Contamination and Related Contaminants of Concern

Contaminant of Concern (COC)	Laboratory Analytical Method
General Chemistry	
Alkalinity	EPA 310
Bromide	EPA 300/320
Chloride	EPA 300/325
Conductivity	EPA 120.1
Cyanide	EPA 335
Flouride	EPA 300/340
Ammonia	EPA 350
Nitrate	EPA 300/352/353
Nitrite	EPA 300/354
Nitrate + Nitrite	EPA 353
Ortho phosphate	EPA 365
Perchlorate	EPA 314, EPA 332, EPA 6850
Total phosphorous	EPA 365
Total dissolved solids	EPA 160.1
Total suspended solids	EPA 160.2
Sulfate	EPA 300/375
Sulfide	EPA 300/376
Total organic carbon	EPA 415.1
Turbidity	EPA 180.1
Microbiology	
Total coliform (most probable number—MPN)	SM 9221B
Fecal coliform (colony-forming units—CFU)	SM 9222D
Metals	
Metals (total or dissolved)—all but mercury	EPA 6010/6020
Mercury (total or dissolved)	EPA 7470
Toxicity characteristic leaching procedure (TCLP) metals	EPA 1311
Organics	
PAH	SW 8310
VOC	SW 8260
SVOC	SW 8270
OP	SW 8141
OC	SW 8081
TPH	EPA 418.1
Explosives	EP 8330
Herbicides	SW 8151
PCBs	SW 8082
Detergents—anionic surfactants	EPA 425
Pharmaceuticals	No single standard; varies

From Kresic (2009).

Table 10.2 Commonly Used Analytical Methods for Detection of Potential Contaminants in Groundwater

groundwater professionals involved in the project may have a hard time deciphering contaminant contribution from different source(s) even after spending sometimes exorbitant resources on investigation.

10.2.2 Turbidity and Pathogens

Increased turbidity after recharge episodes of both autogenic (from precipitation) and allogenic (from sinking streams) recharge is the most frequent water quality problem in karst aquifers with significant fast (conduit) flow component. Although this recharge water can bring a great variety of contaminants into the aquifer, it is usually pathogens (bacteria, viruses, and parasites) in the "muddy" water that are of most concern when the karst aquifer is utilized for public water supply (Fig. 10.8). Where proper sanitation facilities are lacking, waterborne diseases caused by pathogens can spread rapidly in karst. Globally, diarrheal disease, the major waterborne disease, is prevalent in many countries where sewage treatment is inadequate—human wastes are disposed of in open latrines, ditches, canals, and water courses or they are spread on cropland. An estimated 4 billion cases of diarrheal disease occur every year, causing 3 million to 4 million deaths, mostly among children (Hinrichsen et al., 1997).

Although surface water is the primary recipient and host of pathogen contamination, karst groundwater is also greatly affected in many regions with poor or nonexistent sanitation where surface water and groundwater are directly connected. However, some pathogens, such as parasites *Giardia* (Fig. 10.9) and *Cryptosporidium*, are naturally present in surface water bodies and are not necessarily associated with poor sanitation practices. For this reason, the USEPA has instituted specific water treatment requirements for public supply systems using groundwater under direct influence (GWUDI) of surface water (see Sec. 5.3) and has instituted Ground Water Rule applied specifically to karst aquifers and addressing contamination with all pathogens (Sec. 5.2).

The increased efforts to reclaim and reuse wastewaters and gray waters pose additional public health concerns. These concerns are heightened by the fact that the indicators of the sanitary quality of water, that is, the total coliform and fecal coliform bacteria, are unreliable indicators of the presence of a number of key pathogenic agents including enteric viruses and cyst-forming protozoans. Wastewater reclamation does

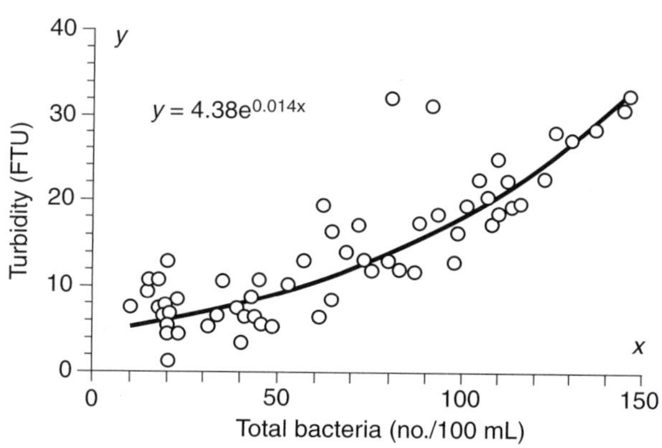

FIGURE 10.8 Total coliform bacteria (no./100 mL) versus turbidity at the Petnica karstic spring near Valjevo, western Serbia. (From Kresic et al., 1992; copyright Springer-Verlag New York Inc.)

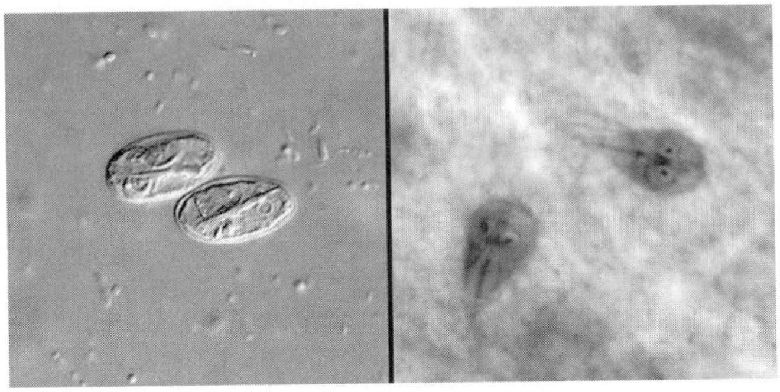

Figure 10.9 *Left*: two *Giardia intestinalis* cysts in a wet mount under DIC microscopy; image taken at 1,000× magnification. The cysts are oval to ellipsoid and measure 8 to 19 μm (average 10 to 14 μm). *Right*: *Giardia intestinalis* trophozoites are pear-shaped and measure 10 to 20 μm in length. (Photographs courtesy of the Center for Disease Control (CDC) Parasite Image Library.)

not have a specific meaning in terms of the degree of treatment for different reclamation projects. Some reuses of wastewater are allowed with very little additional treatment beyond the conventional sewage treatment, of addressing enteric viruses and cyst-forming protozoans. Even after what is considered to be good conventional domestic wastewater treatment and chlorination (chloramination), domestic wastewaters released to surface waters can still contain large numbers of enteroviruses and pathogenic protozoans that can readily cause human disease upon ingestion and, to a lesser extent, body contact with these waters (Lee and Jones-Lee, 1993).

Once in the subsurface (aquifer), the survival and transport of pathogens will depend on various biogeochemical interactions with native groundwater, porous media, and native microorganisms. Whereas some bacteria and parasites cannot survive more than several weeks in the saturated zone regardless of the native conditions, some are known to survive for months or even years. Vesper et al. (2003) describe mechanisms of contaminant transport in karst aquifers including pathogens and provide results of investigations of cave microbiology that have revealed a rich diversity of organisms that seem to thrive very well deep within karst systems. An investigation of total bacteria and coliform bacteria in the Butler Cave—Sinking Creek System of Virginia (Chess, 1987) showed no evidence of die-off anywhere in the system. Roughly, the same populations of organisms were found in streams deep within the system and at the discharging spring as occurred in the sinking surface streams.

A virus is the smallest form of microorganism capable of causing disease. A virus of fecal origin that is infectious to humans by waterborne transmission is of special concern for drinking water regulators. More than 120 different types of potentially harmful enteric viruses are excreted in human feces, and are widely distributed in type and number in domestic sewage, agricultural wastes, and septic drainage systems (Gerba, 1988; from Banks and Batiggelli, 2002). Many of these viruses are stable in natural waters and have long survival times with half-lives ranging from weeks to months. Because they may cause disease even when just a few virus particles are ingested, low levels of environmental contamination may affect water consumers. From 1971 to 1979, approximately 57,974 people in the United States were affected by outbreaks of waterborne pathogens (Craun, 1986; from Banks and Batiggelli, 2002). Outbreaks of waterborne disease attributed to enteric viruses are poorly documented, even though viruses are commonplace in natural waters contaminated with human feces. Illnesses in humans caused by waterborne viruses range from severe infections such as myocarditis, hepatitis, diabetes, and paralysis to relatively mild conditions such as self-limiting gastroenteritis. Currently, enteric viruses are included in the National Primary Drinking Water Standards issued by the USEPA, while several other groups are on the contaminant candidate list (CCL).

In carbonate aquifers where even a relatively small overall volume of rock has developed conduits with fast flow, pathogens pose a permanent threat. They can cause an adverse effect or even death after an acute (short-term) exposure such as ingestion of just one glass of water. Detailed discussions on drinking water treatment technologies for turbidity and pathogens as they relate to groundwater, springs, and water in general are provided by Franchi (2009) and Fotouhi and Kresic (2010).

Because of the great implications they have on water supply management, fate and transport of pathogens in karst are continuous subjects of academic research. From the practical point of view, however, all water managers need to know is if their source (and resource) is karstic. If it is, they should have an adequate water treatment addressing *all* pathogens by default, without trying to somehow differentiate them in a false hope to save money by avoiding "unnecessary" treatment. A study by Boyer and Kuczynska (2003) is very illustrative in that respect. The authors compare the temporal variability in storm flow of fecal coliform bacteria densities and *Cryptosporidium parvum* oocyst densities in agriculturally impacted karst groundwater. *Cryptosporidium parvum* oocyst densities ranged from 0 to 1,050 oocysts/L, and mean storm densities ranged from 3.5 to 156.8 oocysts/L. Fecal coliform densities ranged from less than 1 CFU/100 mL to more than 40,000 CFU/100 mL, and geometric mean storm densities ranged from 1.7 CFU/100 mL to more than 7,000 CFU/100 mL. Fecal coliform densities correlated well with flow during storms, but *Cryptosporidium* oocyst densities exhibited a great deal of sample to sample variability and were not correlated with flow. Fecal coliform densities did not correlate positively with *Cryptosporidium* oocyst densities. Fecal coliform densities were greatest at storm peaks, when sediment loads were also greatest. As concluded by the authors, multiple transport mechanisms for fecal coliform bacteria and *Cryptosporidium parvum* oocysts may necessitate various agricultural land management and livestock health maintenance practices to control movement of pathogens to karst groundwater.

Laroche et al. (2010) present a particularly disturbing study of the conditions promoting the transport of antibiotic-resistant fecal bacteria, in a rural karst system in Northern France providing drinking water. For this purpose, the resistance of the *Escherichia coli* population (436 isolates) to 17 antibiotics was investigated by analyzing water samples from four representative interconnected sites: a creek, a swallow hole, a spring, and a well. The samples were collected during four contrasting hydrologic and grazing periods. This study demonstrated that multiantibiotic-resistant *E. coli*, representing 23% of the total population, infiltrated the karst aquifer at the peak of a rainfall event, with some isolates being resistant to up to eight antibiotics. No *intI1* gene was detected. The search for the origin of the resistant *E. coli* isolated during the study demonstrated that they were of both animal and human origin. The authors conclude that drinking water sources based on groundwater in a rural karst terrain are vulnerable to contamination by antibiotic-resistant bacteria.

Notwithstanding the earlier advice regarding drinking water treatment in karst environments, certain types of karst aquifers and springs are indeed less vulnerable to turbidity increases and associated contamination by pathogens. This includes

Figure 10.10 *Left*: Valdina Farm sink on Seco Creek, Texas, when in flood, receiving sediment-loaded muddy water. *Right*: After a 145-ft vertical drop, the Valdina Farm sink continues into this horizontal passage, dry when the sink is not active. (Photographs courtesy of Geary Schindel, Edwards Aquifer Authority.)

confined aquifers with thick low-permeable cover that consists of noncarbonate sediments, and where ascending springs and well fields are remote from aquifer outcrop areas and locations of allogenic recharge. For example, as discussed by Eckhardt (2012), the Edwards Aquifer in Texas acts as a large settling tank. Rapidly moving surface water, which carries large turbid loads of sediments and debris during storm events, sinks into the subsurface through a variety of karst features, small and large such as the one shown in Fig. 10.10. Once this recharge water reaches the confined portion of the aquifer, it slows down and the sediment and organic material it carries begin to settle out. Divers in the Edwards have reported that underwater cave floors are covered with a thick layer of silt that is easily stirred up, causing a blinding silt-out in the normally crystal clear waters. As a result of this natural clarification taking place in the aquifer, water coming out of the main ascending springs draining the aquifer is always clear and without harmful pathogens.

10.2.3 Agricultural Contaminants and Nitrate

Agricultural activities have had direct and indirect effects on the rates and compositions of groundwater recharge and aquifer biogeochemistry. Direct effects include dissolution and transport of excess quantities of pesticides, fertilizers and associated materials, and hydrologic alterations related to irrigation and drainage. Some indirect effects include changes in water–rock reactions in soils and aquifers caused by increased concentrations of dissolved oxidants, protons, and major ions. Agricultural activities have directly or indirectly affected the concentrations of a large number of inorganic chemicals in groundwater, such as NO_3^-, N_2, Cl, SO_4^{2-}, H^+, P, C, K, Mg, Ca, Sr, Ba, Ra, and As, as well as a wide variety of pesticides and other organic compounds (Böhlke, 2002).

Agricultural practices that have the potential to contaminate groundwater include animal feedlots, fertilizer and pesticide applications, irrigation, agricultural chemical facilities, and drainage wells. Water contamination can be a result of routine applications, spillage, or misuse of pesticides and fertilizers during handling and storage, manure storage and spreading, improper storage of chemicals, and irrigation return drains serving as a direct conduit to groundwater. Fields with overapplied and misapplied fertilizers and pesticides can introduce nitrogen, pesticides, cadmium, chloride, mercury, arsenic, and selenium into the groundwater. As indicated by the USEPA (2000), states report that agricultural practices continue to be a major source of groundwater contamination.

The word pesticide refers to any chemical that kills pests and includes insecticides, fungicides, and nematocides; it also generally includes herbicides. Importantly, extensive use of pesticides is not confined to rural agricultural areas only. They are commonly used in both urban and suburban settings on lawns, parks, and golf courses.

Livestock is an integral component of many states' economies. As a consequence, concentrated animal feeding operations (CAFOs) where animals are kept and raised in confined areas occur in many states. CAFOs congregate animals, feed, manure and urine, dead animals, and production operations on a small land area. Such operations can pose a number of risks to water quality and public health, mainly because of the amount of animal manure and wastewater they generate. Animal feedlots often have impoundments from which wastes may infiltrate to groundwater. Livestock waste is a source of nitrate, bacteria, total dissolved solids, and sulfates.

Shallow unconfined aquifers and karst aquifers in many states have become contaminated from the application of fertilizer. Crop fertilization is the most important agricultural practice contributing nitrate to the environment that is considered by many to be the most widespread groundwater contaminant. Pesticide use and application practices are also of great concern for water quality nationwide and especially in karst areas. The primary route of pesticide transport to groundwater is by leaching through the vadose zone or by spills and direct infiltration through various karst features such as sinkholes and sinks. Pesticide infiltration is generally greatest when rainfall is intense and occurs shortly after the pesticide is applied. Within sensitive areas, groundwater monitoring has shown fairly widespread detections of pesticides, specifically the pesticide atrazine.

The following examples of nonpoint source groundwater contamination require both regulatory, and local land-use changes in order to restore aquifers to their natural condition. The United Kingdom Environment Agency reported that pesticides were found in over a quarter of groundwater monitoring sites in England and Wales in 2004, and in some cases, they exceeded the drinking water limit. Atrazine is a weed killer used mainly to protect maize (corn) crops and it was used in the past to maintain roads and railways. A complete ban on all use of atrazine (and simazine, another pesticide) in the United Kingdom was planned to be phased in between 2005 and 2007, but has not been implemented as of April 2012. As noted by the Agency, banned pesticides can remain a problem for many years after they were last used (Environment Agency, 2007). Some other European countries have banned the use of atrazine: France, Sweden, Norway,

Denmark, Finland, Germany, Austria, Slovenia, and Italy. In contrast, the USEPA has concluded that the risks from atrazine for approximately 10,000 community drinking water systems using surface water are low and did not ban this pesticide, which continues to be the most widely used pesticide in the United States. Incidentally, as stated by the Agency, 40,000 community drinking water systems using groundwater were not included in the related study, and private wells used for water supply were not mentioned in the Agency's decision to allow continuous use of atrazine (USEPA, 2003a).

The United Kingdom Environment Agency also reported that in 2004, almost 15% of monitoring sites in England (none in Wales) had an average nitrate concentration that exceeded 50 mg/L, the upper limit for nitrate in drinking water (for comparison, groundwater naturally contains only a few mg/L of nitrate). Water with high nitrate levels has to be treated or diluted with cleaner water to reduce concentrations. More than two-thirds of the nitrate in groundwater comes from past and present agriculture, mostly from chemical fertilizers and organic materials. It is estimated that over 10 million tons per year of organic material is spread on the land in the United Kingdom. More than 90% of this is animal manure; the rest is treated sewage sludge, green waste compost, paper sludge, and organic industrial wastes. Other major sources of nitrate are leaking sewers, septic tanks, water mains, and atmospheric deposition. Atmospheric deposition of nitrogen makes a significant contribution to nitrate inputs to groundwater (Papic et al., 1991). A study in the Midlands concluded that around 15% of the nitrogen leached from soils came from the atmosphere. The Agency estimates that 60% of groundwater bodies in England and 11% in Wales are at risk of failing Water Framework Directive objectives because of high nitrate concentrations (Environment Agency, 2007).

An illustrative study of nonpoint source contamination from agricultural activities was performed by the Kentucky Geological Survey in 1990s (Currens, 1999). The Pleasant Grove Spring Basin in southern Logan County, Kentucky, was selected for the study because it is largely free of nonagricultural pollution sources. About 70% of the watershed is in crop production and 22% is pasture. The area is underlain by karst geology and the groundwater flow in the basin is divided into a diffuse (slow) flow regime and a conduit (fast) flow regime. The diffuse and conduit-flow regimes have a major influence on the timing of contaminant maxima and minima in the spring during and after major rainfall events. Nitrate is the most widespread, persistent contaminant in the basin, but concentrations average 5.2 mg/L basin-wide and generally do not exceed the drinking water maximum contaminant level (MCL) of 10 mg/L set by the USEPA. Atrazine has been consistently, and other pesticides occasionally, detected. Concentrations of triazines (including atrazine) and alachlor have exceeded drinking water MCLs during peak spring flows, as illustrated in Fig. 10.11. Maximum concentrations of triazines, carbofuran, metolachlor, and alachlor in samples from Pleasant Grove Spring were 44.0, 7.4, 9.6, and 6.1 μg/L, respectively. Flow-weighted average concentrations for 1992–1993 were 4.91 μg/L for atrazine-equivalent triazines and 5.0 mg/L for nitrate-nitrogen. In comparison, the maximum allowed concentration of any individual pesticide in drinking water in the European Union is 0.1 μg/L, and of all pesticides combined it is 0.5 μg/L.

Bacteria counts always exceed standards for drinking water and occasionally exceed standards for drinking water supplies. Basin-wide samples averaged 465 fecal coliform colony-forming units per 100 mL (col/100 mL) and 1891 fecal streptococci col/100 mL; maximum counts were 14,000 and 24,000 col/100 mL, respectively. Bacteriological speciation failed to identify the source of high bacteria counts at Pleasant Grove Spring, but showed that the bacteria are not indigenous to the natural environment of the basin.

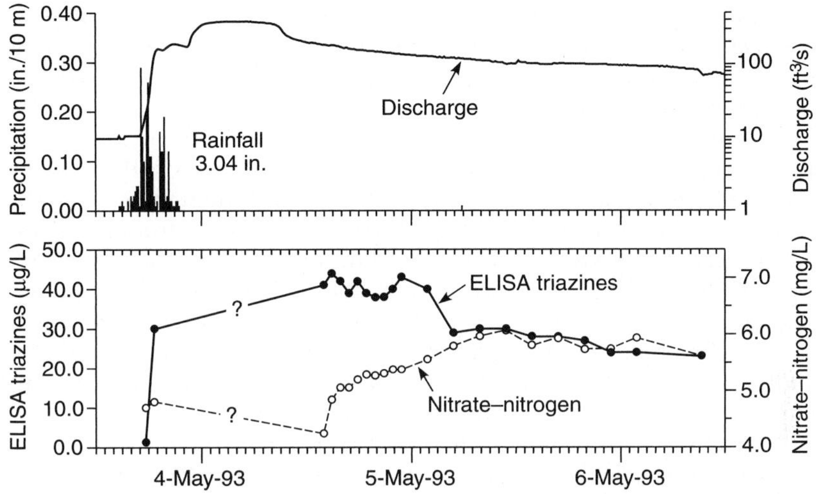

FIGURE 10.11 Chemographs and discharge hydrograph for the May 1993 high-flow event at Pleasant Grove Spring, illustrating highest triazine concentrations determined to date. (From Currens, 1999.)

The hydrogeology of the basin is a significant controlling influence on the temporal variation in contaminant concentrations. The fast-flow conduit region is characterized by intermediate concentrations of nitrate and pesticides during low flow but substantially higher concentrations of triazines and lower concentrations of nitrate during high flow. The diffuse (slow) flow regime, which is estimated to represent slightly less than half of the basin, drains into the area dominated by conduit flow. The diffuse-flow region has persistently higher concentrations of nitrate but lower, less variable concentrations of triazines. The diffuse, slow-flow area is acting as a reservoir of agricultural chemicals, maintaining a background level of triazines and nitrate during low flow in the conduit-flow regime. Triazine concentrations are significantly higher during high flow, while nitrate concentrations are diluted.

As concluded by Currens (1999), both municipal and domestic water supplies derived from groundwater can be adversely affected. BMPs implementation in the basin should focus on controlling animal waste, controlling crop field runoff with associated sediment and pesticide loss, and using more efficient methods of applying nutrients. A strong education program on groundwater protection is highly recommended.

Panno and Kelly (2004) report on an investigation designed to estimate the mass loading of nitrate (NO_3^-) and herbicides in springwater discharging from groundwater basins in an agriculturally dominated, mantled karst terrain. The loading was normalized to land use and NO_3^- and herbicide losses were compared to estimated losses in other agricultural areas of the Midwestern, United States. The study area consisted of two large karst springs that drain two adjoining groundwater basins (total area of 37.7 km^2) in southwestern Illinois' sinkhole plain, the United States. The springs and stream that they form were monitored for almost 2 years. Nitrate–nitrogen (NO_3^-N) concentrations at three monitoring sites were almost always above the background concentration (1.9 mg/L). NO_3^-N concentrations at the two springs ranged from 1.08 to 6.08 with a median concentration of 3.61 mg/L. Atrazine and alachlor concentrations ranged from

<0.01 to 34 μg/L and <0.01 to 0.98 μg/L, respectively, with median concentrations of 0.48 and 0.12 μg/L, respectively (again, note that both these concentrations would constitute groundwater contamination in many European countries where drinking water standard for any individual pesticide is 0.1 μg/L).

Approximately 100,000 kg/year of NO_3^-N, 39 kg/year of atrazine, and 2.8 kg/year of alachlor were discharged from the two springs. Slightly more than half of the discharged NO_3^- came from background sources and most of the remainder probably came from fertilizer. This represents a 21% to 31% loss of fertilizer N from the groundwater basins. The pesticide losses were 3.8% to 5.8% of the applied atrazine, and 0.05% to 0.08% of the applied alachlor. The loss of atrazine adsorbed to the suspended solid fraction was about 2 kg/year, only about 5% of the total mass of atrazine discharged from the springs.

Peterson et al. (2002) describe a quantitative approach used to monitor movement of nitrate through a karst spring basin. The research followed nitrate along flow paths, from application as fertilizer, through the soil matrix, into the bedrock, and ultimately as discharge from a spring. On the basis of data from analysis of soil cores and springwater samples, it was calculated that for the period of 1 year, 22,000 kg of nitrate as nitrogen was discharged from a spring with a drainage basin area of 43.85 km^2. The authors concluded that after application of animal waste as fertilizer, the nitrate leaches through the soil primarily as piston flow in the matrix, unless macropores are encountered. Cores collected from fertilized fields show that the nitrate moves at a rate of approximately 65 cm/year through the silt and clay loams. The cores also revealed the presence of nitrate maxima with depth that were associated with timing of fertilizer application. Soil sampling and spring monitoring suggest that the soil serves as a reservoir for nitrate. NO_3^-N concentrations in the soil range from 0.012 to 0.036 g/kg of soil, which would be equivalent to 11 to 106 mg of NO_3^-N/L water. After reaching the bedrock surface, nitrate enters the aquifer in fractures and solution-enlarged joints, eventually discharging from the spring. Spring discharge data indicate that 74% of the nitrate exits from the basin through the spring during baseflow conditions, and the remaining 26% is discharged as a result of rapid recharge associated with precipitation storm pulses. During precipitation events, overland flow and runoff moves primarily through macropores in the soil carrying mainly surface-derived nitrate, and displaces little of the soil nitrate.

Results of not so recent research regarding the dual effects and risks of multiple agricultural contaminants in drinking water are of particular concern. For example, in a study by the University of Wisconsin-Madison, researchers noted that common mixtures of pesticides and fertilizers can have biologic effects at the current concentrations measured in groundwater. Specifically, the combination of aldicarb, atrazine, and nitrate, which are the most common contaminants detected in groundwater in agricultural areas, can influence the immune and endocrine systems as well as affect neurologic health. Changes in the ability to learn and in patterns of aggression were observed. Effects are most noticeable when a single pesticide is combined with nitrate fertilizer. Research shows that children and developing fetuses are most at risk (Porter et al., 1999; from USEPA, 2000).

Excessive consumption of nitrate in drinking water has been associated with the risk of methemoglobinemia or "blue baby syndrome," an acute effect that is accentuated under poor sanitary conditions such as sewage contamination or dirty drinking vessels (Buss et al., 2005). If left untreated, methemoglobinemia can be fatal for affected infants. The World Health Organization (WHO) and the European Union have set the standard for nitrate in drinking water at 11.3 mg/L measured as nitrogen (mg N/L) that corresponds to 50 mg NO_3/L. The standard in the United States, Canada, and Australia is 10 mg N/L.

10.2.4 Synthetic Organic Chemicals

SOCs are man-made (anthropogenic) compounds that are used for a variety of industrial and agricultural purposes and include organic pesticides. SOCs can be divided into two groups: volatile organic compounds (VOCs) and nonvolatile (semivolatile, SVOCs) compounds.

VOCs are used for a variety of industrial and manufacturing purposes. Among the most common VOCs are degreasers and solvents such as benzene, toluene, and trichloroethylene (TCE); insulators and conductors such as polychlorinated biphenyls (PCBs); dry-cleaning agents such as tetrachloroethylene (PCE); and gasoline compounds. VOCs have the potential to cause chromosome aberrations, cancer, nervous system disorders, and liver and kidney damage (USEPA, 2003b, 2003c).

The U.S. Geological Survey National Water-Quality Assessment (NAWQA) Program collected samples from wells and springs in 12 carbonate aquifers across the country during 1993 to 2005; water quality results for 1,042 samples were available to assess the factors affecting groundwater quality. These aquifers represent a wide range of climate, land-use types, degrees of confinement, and other characteristics that were compared and evaluated to assess the effect of those factors on water quality. Differences and similarities among the aquifers were also identified. Samples were analyzed for major ions, radon, nutrients, 47 pesticides, and 54 VOCs. Altogether, 7 of the 12 assessed carbonate aquifers or aquifer systems had more than 20% of samples with one or more VOCs per sample. At the 0.2 μg/L assessment level, the percent of samples with one or more VOC detections was similar between unconfined (22% of samples) and confined areas (19%).

The finding that one or more VOCs were detected in nearly 70% of the samples analyzed using the low-level method demonstrates the vulnerability of both confined and unconfined carbonate aquifers to VOC contamination (Lindsey et al., 2009).

SVOCs are operationally defined as solvent-extractable organic compounds that can be determined by gas chromatography/mass spectrometry (GC/MS). They include polycyclic aromatic hydrocarbons (PAHs), azaarenes, nitrogenated compounds, phenols, phthalates, ketones, and quinones. Many of these SVOCs have been designated as priority pollutants by the USEPA because of their toxicity and association with industrial activities and processes. SVOCs that are priority pollutants include phthalates used in plastics, phenols used as disinfectants and in manufacturing chemicals, and PAHs.

PAHs and azaarenes contain fused carbon rings that form during the incomplete combustion of organic matter, including wood and fossil fuels (such as gasoline, oil, and coal). Azaarenes are distinguished from PAHs by having a nitrogen atom substituted for a carbon in the fused ring structure. Azaarenes tend to occur in association with PAHs in affected soils and streambed sediment samples because fossil fuel combustion is the primary source of both PAHs and azaarenes. However, additional sources of PAHs include natural or anthropogenic introduction of uncombusted coal and oil and industrial use of PAHs in the dye and plastic industries (Nowell and Capel, 2003; from Lopes and Furlong, 2001). Most SVOCs are moderately to strongly hydrophobic (i.e., they have fairly low water solubility and fairly high octanol–water partition coefficients). Consequently, they tend to sorb to soil and sediment and partition to organic matter in water.

Disinfection of drinking water is one of the major public health advances of the twentieth century. Disinfection is a major factor in reducing the typhoid and cholera epidemics that were common in American and European cities in the nineteenth century and the beginning of the twentieth century. While disinfectants are effective in controlling many microorganisms, certain disinfectants (notably, chlorine) react with natural organic and inorganic matter in source water and distribution systems to form disinfection by-products (DBPs) that are almost all organic chemicals (chromate and bromate are notable exceptions). A large portion of the U.S. population is potentially exposed to DBPs through its drinking water. More than 240 million people in the United States are served by public water systems that apply a disinfectant to water to protect against microbial contaminants. Results from toxicology studies have shown several DBPs (e.g., bromodichloromethane, bromoform, chloroform, dichloroacetic acid, and bromate) to be carcinogenic in laboratory animals. Other DBPs (e.g., chlorite, bromodichloromethane, and certain haloacetic acids) have also been shown to cause adverse reproductive or developmental effects in laboratory animals. Epidemiologic and toxicologic studies involving DBPs have provided indications that these substances may have a variety of adverse effects across the spectrum of reproductive and developmental toxicity: early-term miscarriage, stillbirth, low birth weight, premature babies, and congenital birth defects (USEPA, 2003c). DBPs are of special concern when studying the potential for artificial aquifer recharge using treated wastewater.

Nonaqueous Phase Liquids

Nonaqueous phase liquids (NAPLs) are hydrocarbons that exist as a separate, immiscible phase when in contact with water and/or air. Differences in the physical and chemical properties of water and NAPL result in the formation of a physical interface between the liquids that prevents the two fluids from mixing. NAPLs are typically classified as either light nonaqueous phase liquids (LNAPLs) that have densities less than that of water, or dense nonaqueous phase liquids (DNAPLs) that have densities greater than that of water. It is very important to make the distinction between the actual NAPL liquid in free phase, and the chemical of the same name dissolved in water. For example, most common organic contaminants such as PCE, TCE, and benzene can enter the subsurface as both free-phase NAPL and dissolved in percolating water. However, the fate and transport of free-phase NAPL and the same chemical dissolved in groundwater are quite different (Kresic, 2009).

LNAPLs affect groundwater quality at a variety of sites. The most common contamination problems result from the release of petroleum products. Leaking of underground storage tanks (USTs) at gas stations and other facilities is arguably the most widespread point source contamination of groundwater in developed countries (see Fig. 10.7). Gasoline products are typically multicomponent organic mixtures composed of chemicals with varying degrees of water solubility. Some gasoline additives (e.g., methyl tertiary-butyl ether or MTBE, and alcohols such as ethanol) are highly soluble. Other components (e.g., benzene, toluene, ethylbenzene, and xylenes, collectively known as BTEX) are slightly soluble.

The major types of DNAPLs are halogenated solvents, coal tar, creosote-based wood-treatment oils, PCBs, and pesticides. As a result of widespread production, transportation, utilization, and disposal practices, particularly since 1940s, there are numerous DNAPL contamination sites in North America and Europe. The potential for serious long-term contamination of groundwater by some DNAPL chemicals at many sites is high due to their toxicity, limited solubility (but much higher than drinking water limits), and significant migration potential in soil gas, groundwater, and/or as a separate phase. DNAPL chemicals, especially chlorinated solvents, are among the most prevalent groundwater contaminants identified in groundwater supplies and at waste disposal sites.

Coal tar and creosote are complex chemical mixture DNAPLs derived from the destructive distillation of coal in coke ovens and retorts. Historically, coal tar has been produced by coal tar distillation plants and as a by-product of manufactured gas plant and steel industry coking operations. Creosote blends are used to treat wood alone or diluted with coal tar, petroleum, or, to a very limited extent, pentachlorophenol. In addition to wood-preservation, coal tar is used for road, roofing, and waterproofing solutions. Considerable use of coal tar is also made for fuels (Cohen and Mercer, 1993).

PCBs are extremely stable, nonflammable, dense, and viscous liquids that are formed by substituting chlorine atoms for hydrogen atoms on a biphenyl (double benzene ring) molecule. PCBs, which are not produced any more in the United States, now have a very restricted regulated use, were historically used in oil-filled switches, electromagnets, voltage regulators, heat transfer media, fire retardants, hydraulic fluids, lubricants, plasticizers, carbonless copy paper, dedusting agents, and other products. PCBs were frequently mixed with carrier fluids prior to use. Because of their widespread historic use and persistence, PCBs are often detected in the environment at very low concentrations. The potential for DNAPL migration is greatest at sites where PCBs were produced, utilized in manufacturing processes, stored, reprocessed, and/or disposed of in quantity (Cohen and Mercer, 1993).

Pharmaceuticals and Personal Care Products

With the advancement of analytical methods that can detect concentrations at parts per trillion (ng/L) or lower, a large picture of the numerous chemicals present in water supplies has emerged recently. Certain pharmaceutically active compounds (e.g., caffeine, aspirin, and nicotine), which have been known for over 20 years to occur in the environment, are now joined by a broad group of chemicals collectively referred to as pharmaceuticals and personal care products (PPCPs). It seems that this term has prevailed in practice as a synonym for emerging contaminants, although water and wastewater treatment industry prefers to use term microconstituents.

PPCPs are a diverse group of chemicals comprising all human and veterinary drugs (available by prescription or over-the-counter; including the new genre of "biologics"); diagnostic agents (e.g., X-ray contrast media); "nutraceuticals" (bioactive food supplements such as huperzine A); and other consumer chemicals, such as fragrances (e.g., musks) and sunscreen agents (e.g., methylbenzylidene camphor); also included are "excipients," so-called "inert" ingredients used in PPCP manufacturing and formulation (Daughton, 2007). Nanomaterials are an emerging subgroup of microconstituents considered by many as the next industrial wonder. They are already present in cosmetics, sunscreens, wrinkle-free clothing, and food products. Because of their small size, nanomaterials pose a challenge in terms of detection and treatment. Also, because of their size they can enter all human organs including the brain, but very little is known of their fate and transport in the environment.

Only a subset of PPCPs, such as synthetic steroids, are known to be direct-acting endocrine disruptors. However, little is known about the individual and combined effects of long-term exposure to most PPCPs and their degradation products at very low concentrations.

The widespread use of PPCPs in the environment is a result of their unavoidable, collective discharge by humans as well as animals. Some pharmaceuticals are not completely metabolized after consumption by humans or animals and are excreted in their original form, while others are transformed into different compounds (conjugates).

Domestic sewage is a major source of PPCPs, and CAFOs are a major source of antibiotics and possibly steroids (Daughton, 2007).

Free excreted drugs and derivatives can escape degradation in municipal sewage treatment facilities where their removal efficiency is a function of the drug's structure and treatment technology employed. Some conjugates can also be hydrolyzed back to the free parent drug during the treatment process. After going through the wastewater treatment plant, PPCPs and their degradation products are discharged to receiving surface waters and can find their way to groundwater, including by direct artificial aquifer recharge. The full extent, magnitude, and ramifications of their presence in the aquatic environment are largely unknown (Daughton, 2007). Releases of PPCPs to the environment are likely to continue as the human population increases and ages; the pharmaceutical industry formulates new prescription and nonprescription drugs, and promotes their use; and more wastewater is generated that enters the hydrologic cycle and may impact groundwater resources (Masters et al., 2004).

Einsiedl et al. (2010) report on the occurrence of two pharmaceuticals, ibuprofen and diclofenac, in a vulnerable karst groundwater system. The hydrogeology of the karst system was identified by collecting ^3H samples in groundwater over 27 years and by performing tracer tests. The isotopes and tracer data were interpreted by mathematical modeling to estimate the mean transit time of water and to characterize the hydrogeologic flow paths in the groundwater system. By this approach, a mean ^3H transit time of 4.6 years for the fissured-porous karst aquifer was determined, whereas the fast flowing water in the conduit system showed a mean transit time of days. Both pharmaceuticals that infiltrated along sinkholes and small streams into the karst system were detected in concentrations of up to approximately 1 μg/L in effluent water of the wastewater treatment plants. Diclofenac was present in most samples collected from four springs discharging the karst groundwater to the Rivers Altmühl and Anlauter in concentrations between 3.6 and 15.4 ng/L. In contrast, ibuprofen was rarely detected in groundwater. As concluded by the authors, the results of this study suggest that both pharmaceuticals move into the fractured system of the karst system and go into storage. They also conclude that dilution processes are the dominant control on the concentrations of both pharmaceuticals in the fractured system, whereas biodegradation is likely less important.

As mentioned before, many of the PPCPs are endocrine disruptors as are other synthetic compounds. This list is very long and it is constantly growing as new research results become available. Endocrine disruptors encompass a variety of chemical classes, including natural and synthetic hormones, pesticides, and compounds used in the plastics industry and in consumer products. Endocrine disruptors are often pervasive and dispersed in the environment, including in groundwater. Here are a few from the major groups of synthetic chemicals: persistent organohalogens (1,2-dibromoethane, dioxins, and furans, PBBs, PCBs, pentachlorophenol), food antioxidants (BHA), pesticides (majority, if not all; alachlor, aldrin, atrazine, chlordane, DDT, dieldrin, heptachlor, lindane, mirex, zineb, ziram), and phthalates. Heavy metals, arsenic, cadmium, lead and mercury, are also endocrine disruptors in addition to their toxic effects. In general, the health effects associated with endocrine disrupting compounds include a range of reproductive problems (reduced fertility, male and female reproductive tract abnormalities, and skewed male/female sex ratios, loss of fetus, menstrual problems), changes in hormone levels, early puberty, brain and behavior problems, impaired immune functions, and various cancers (Wikipedia, 2007).

A book by Colborn et al. (1997) *Our Stolen Future* examines mechanisms with which certain synthetic chemicals interfere with hormonal messages involved in the control of growth and development, especially in the fetus. The associated Web site (http://www.ourstolenfuture.org/NewScience/newscience.htm.) discusses scientific findings of the impacts of endocrine disruptors at low doses, emphasizing that new research on endocrine disrupting compounds is revealing that these compounds have impacts at levels dramatically lower than thought relevant to traditional toxicology. The site also includes numerous recent research examples with full reference details.

10.3 Vulnerability and Risk Maps

Two complementary approaches to the protection of quality of groundwater resources are protection of the existing water supplies sources, such as water wells and springs, and protection of the resource as a whole, for all present and future users. In different countries and in different parts of the same country (such as in the United States), the emphasis may be given to one or both approaches, depending on the resource development situation, the prevailing hydrogeologic conditions, and the prevailing political environment. Source-oriented approaches are based on delineation of source protection zones. Such zones serve for setting priorities for groundwater quality monitoring, environmental audit of industrial premises, determining priorities for the cleanup of historically contaminated land, and determining which land-use activities within which zones have acceptable risk. This approach is best suited to relatively uniform, unconsolidated aquifers exploited only by a small number of high-yielding municipal wells with stable pumping regimes. It is not so readily applied where there are a large and growing number of localized groundwater extraction centers, including individual households (Foster et al., 2002–2005).

Resource-oriented strategies are more universally applicable, since they are aimed at achieving protection for the entire groundwater resource and for all groundwater users. They involve aquifer pollution vulnerability mapping over extensive areas, and may include one or more important aquifers. Such mapping would normally be followed by an inventory of potential or existing sources of contamination, and hazards and risks associated with various land-use activities, at least in the more vulnerable areas.

The fundamental concept of groundwater vulnerability is that some areas are more vulnerable to contamination than others. The goal of a vulnerability map is subdivision of an area into several smaller areas that have different degrees of vulnerability (Goldscheider, 2002). Vrba and Zaporozec (1994) emphasize that the vulnerability of groundwater is a relative, nonmeasurable, dimensionless property, and they make the distinction between intrinsic (natural) and specific vulnerability. The intrinsic vulnerability depends only on the natural properties of an area, such as characteristics of the porous media, and recharge. It is independent of any particular contaminant. Specific vulnerability takes into account fate and transport properties of a contaminant. Simplified, this means that, for example, an aquifer may be vulnerable to an improper disposal or spill of chlorinated solvents at the land surface even though the groundwater flow directions and the presence of a low-permeable overlying aquitard may be protective enough in case of a nonpoint source contamination with nitrates. Another example is a thick unsaturated zone (e.g., >300 ft) in arid climates that may be highly protective of the underlying unconfined aquifer simply because of the insignificant present-day aquifer recharge that cannot facilitate migration of a contaminant through such thick vadose zone, all the way down to the water table. However, if there were some land-use practices, such as waste disposal in ponds, which can facilitate contaminant migration, these aquifers would then be considered vulnerable.

Although most definitions and methods for mapping groundwater vulnerability consider only contamination aspects, there are also quantitative aspects of groundwater protection and vulnerability, such as overexploration and aquifer mining (Vrba and Zaporozec, 1994). Maps depicting time-dependent quantities of groundwater available for extraction, and the associated development of drawdown (decrease in water levels), are a very useful tool for groundwater management of nonrenewable groundwater resources, as well as other stressed aquifer systems.

Although vulnerability maps and maps of source zone protection areas are closely related, they usually involve different scales and have different objectives. A wellhead (springhead) protection area (described in more detail in the next section) is an area within which there is a complete pathway between any given location at the water table (top of the unconfined aquifer) and the groundwater extraction point such as a water well. In general, this area is usually defined by the length of time required for all water particles to be extracted, that is, captured by the well. In non-karstic settings, such areas commonly include 5- or 10-year capture zone. In contrast, groundwater vulnerability, which is a qualitative concept, involves a vaguely expressed probability that a theoretical contamination at the land surface is more or less likely to reach the water table. In some cases, the objectives of resource and source protections are merged by creating a map, or several map overlays, that can be used to present both the general vulnerability of an area and the wellhead protection zones for the existing groundwater supplies.

When vulnerability maps include potential sources of contamination and their associated contaminants, possibly ranked by the hazard they pose on groundwater resources (supplies), they are referred to as groundwater contamination hazard or risk maps. Hazards can be grouped into different categories such as from infrastructure development, industrial and agricultural activities. They are commonly evaluated using a hazard index that takes into account the "harmfulness" of a hazard to groundwater, the quantity of relevant substances that can be released in case of an accident, and the probability for a contaminant release to occur. The highest risk is present in situations where a dangerous hazard (high probability of large quantities of harmful substances to be released) is located in a highly vulnerable zone. The risk map shows areas where engineering, legislative (political), or management response is required (Drew and Hötzl, 1999b).

Notably, however, none of the above-described concepts of vulnerability, hazard, and risk of groundwater contamination addresses a three-dimensional likely pathway of any particular contaminant, and its time of travel, between the point of release at the land surface and the receptor (a well or a spring). Note that this pathway must include the vadose zone because, in some cases, a contaminant released to the subsurface may never reach the water table due to various attenuation mechanisms in the vadose zone. For these reasons, some authors criticize that a purely descriptive (qualitative) definition of groundwater vulnerability, which is not quantitatively defined in terms of physics, may be the source of misinterpretations. Andersen and Gosk (1989) point out that it is impossible to create a "general" vulnerability map that expresses in a comparable way permanent protective properties of an area. They emphasize that a vulnerability map cannot at the same time be applicable for both conservative and reactive contaminants, for both instantaneous and long-term contaminant releases, and for both point and diffuse contamination scenarios. They consequently conclude that vulnerability maps should be prepared for well-defined specific situations only. The author of this book fully agrees with this position by Andersen and Gosk (1989), and especially when applied to karst environments.

As discussed by Goldscheider (2002), it is debatable if a descriptive, general definition and the lack of physical precision in vulnerability concepts should be considered as an advantage or as a disadvantage. The advantage of a descriptive definition is that the term vulnerability is often intuitively understood, particularly by decision makers in the planning process (Hötzl et al., 1995). A vulnerability map showing areas of different color symbolizing different degrees of vulnerability (or natural protection) is easy to interpret and can be a practical and applicable tool for land-use planning, protection zoning, and qualitative risk assessment. There are also disadvantages to a purely descriptive definition. A property, which is not precisely defined in terms of physics, cannot be derived unambiguously from measurable physical quantities. Therefore, every method of vulnerability mapping is based on the individual point of view and experience of the person who developed it and is thus subjective (Goldscheider, 2002). It is difficult to compare different vulnerability methods or maps and to decide which one is the best. If different methods are tested in one area, the resulting maps are always different and sometimes contradictory (Gogu, 2000). Another important consequence of the lack of a physical definition is that it is difficult to validate (verify or negate) a vulnerability assessment or mapping (Broyère et al., 2001). The same authors suggest that the three practical questions to which a vulnerability assessment has to answer are the following: If pollution occurs, (1) when will it reach the target, (2) at which concentration level, and (3) for how long will the target be polluted? However, these three questions are more related to protection or vulnerability of a particular water supply source (target), such as a water supply well or a spring, than to groundwater resource protection in general.

Among various methods of qualitative groundwater vulnerability mapping, the so-called index (or parametric) methods are the most common (Fig. 10.12). For example,

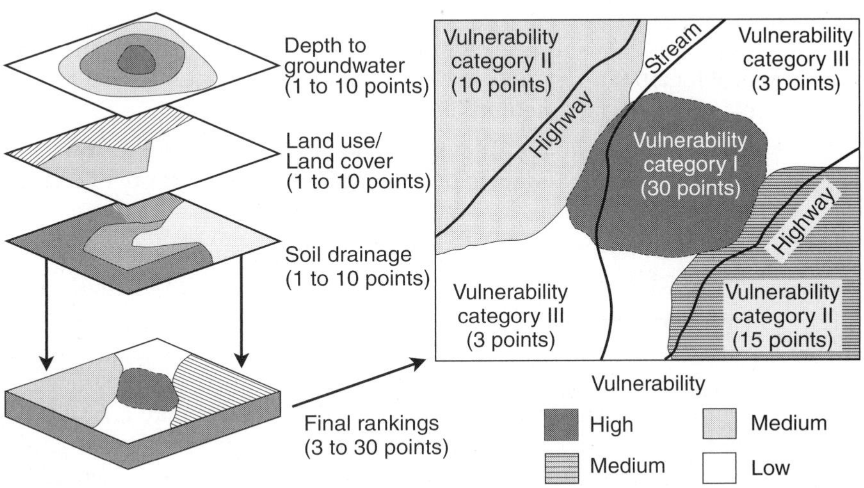

FIGURE 10.12 Schematic of index mapping of groundwater vulnerability. Some indices may be based on quantitative data, but the individual and the final rankings are subjective. (Modified from Focazio et al., 2002.)

Magiera (2000) counted 34 different methods and the author of this book could add a few more, including one with an amazing acronym: GOD (G is for groundwater hydraulic confinement, O is for overlying strata, and D is for depth to groundwater table or strike; see Foster et al., 2002–2005). However, the overall procedure for the various index methods is the same. The first step is the selection of factors (parameters) assumed to be significant for vulnerability. Each factor has a natural range that is subdivided into discrete intervals and each interval is assigned a value reflecting the relative degree of sensitivity to contamination. The vulnerability of an area is determined by combining the values for the different factors using a rating (index) system.

The most widely used index method is DRASTIC, named for the seven factors considered: **D**epth to water, **R**echarge, **A**quifer media, **S**oil media, **T**opography, **I**mpact of the vadose zone, and the hydraulic **C**onductivity of the aquifer (Aller et al., 1985). These seven factors are incorporated into a relative ranking scheme that uses a combination of ratings and weights to produce a numerical value, called the DRASTIC index. Each of the factors is ranked between 1 and 10 and the rank is multiplied by an assigned weighting, which ranges between 1 and 5. The weighted ranks are summed to give a score for the particular hydrogeologic unit with the higher scores indicating greater vulnerability to contamination. The DRASTIC method has been used to develop groundwater vulnerability maps in many parts of the United States, as well as worldwide. However, the effectiveness of the method has met with mixed success due to its subjectivity since the maps are not calibrated to any measured contaminant concentrations or specific contaminants (USEPA, 1993; Rupert, 1999, 2001). DRASTIC does not take into account unique features of fractured rock and karst groundwater systems and therefore *should not be used* for the purposes of mapping any vulnerability in karst.

The following characteristics of karst systems are significant with respect to groundwater vulnerability and should consequently be taken into account when attempting to create vulnerability maps (compiled and modified from Kresic et al., 1992; Hötzl, 1996; Leibundgut, 1998; Trimmel 1998; Drew and Hötzl, 1999b; Goldscheider, 2002; Zwahlen, 2004; Kresic, 2007a, 2007b):

- Each karst system has its individual characteristics and any generalization is problematic; a detailed hydrogeologic investigation of a karst system is irreplaceable for any method of vulnerability mapping.
- Karst systems are highly heterogeneous and anisotropic. Interpolation and extrapolation of field data is problematic and the reliability of a vulnerability map is significantly lower for karst than for other areas.
- Karst aquifers are recharged both by diffuse infiltration and by concentrated point recharge via sinkholes and sinks (swallow holes in surface streams).
- The overlying layers above the karst aquifer, such as top soil, residuum (regolith), and other unconsolidated sediments such as fluvial deposits, may provide a limited form of protection depending on their thickness; however, surface runoff over these areas intersected by sinkholes or sinks may enhance aquifer vulnerability. Additionally, presence of invisible collapse-prone sinkholes below land surface and within surficial clastic deposits increases vulnerability.
- An epikarst zone may have to be taken into consideration where present. In some settings, the main functions of the epikarst are water storage and concentration of flow. The structure and the hydrologic function of the epikarst are difficult to

assess and vary greatly from site to site. Epikarst in itself should not be considered as a protective portion of karst in any respect.

- Karst aquifers are characterized by a dual or triple porosity due to presence of intergranular pores and microfissures in the rock matrix, and fractures and dissolutional voids (conduits). Groundwater storage takes place in the pores and fractures, while conduits act as drains. Consequently, there are both very fast and slow flow components within a karst system. Contaminants can be transported very fast and/or stored and slowly transported for a very long time.
- Karst systems are often characterized by fast and strong hydraulic reaction to hydrologic events. The temporal variations in the hydraulic head in aquifers often reach several tens of feet and sometimes more. In many karst systems, the water table is discontinuous and difficult to determine.
- Karst drainage areas are often extremely large and hydraulically connected over long distances. Watersheds are often difficult to determine and are variable in time, depending on the seasonal hydrologic conditions. Drainage areas of karst springs often overlap and the flow paths proved by tracer tests often cross each other.
- There are possible transitions between less and more intensely karstified portions in the same carbonate aquifer, and these transitions are difficult to define.
- There may be several types of hydraulically connected aquifers in one area, such as a granular aquifer overlying or in lateral contact with a karst aquifer.

In anticipation of the European Union regulations for groundwater protection (The European Parliament and the Council of the European Union, 2006), and especially after they became effective, there has been an amazing proliferation of various index methods proposed for karst environments. It sometimes appears that academic researchers in Europe are competing in the number and relative (qualitative) significance of parameters, subparameters, sub-sub-parameters, and acronyms. Among the first proposed qualitative methods of vulnerability mapping applicable to alpine karst settings is EPIK (Doerfliger, 1996; Doerfliger and Zwahlen, 1998). Brief description of COST 620, developed for a variety of karst settings in Europe, is provided here for reference.

COST (European cooperation in the field of scientific and technical research) initiative no. 620 brought together experts from various countries and several disciplines including hydrogeology, karst geomorphology, environmental chemistry, and microbiology to establish a methodology for producing intrinsic and specific vulnerability maps of karstic environments. As explained in great detail in Zwahlen (2004), COST 620 approach to vulnerability, hazard, and risk mapping is based on an origin-pathway-target model, which applies for both groundwater resource and source protection. Origin is the term used to describe the location of a potential contaminant release. The target is the water, which has to be protected. For resource protection, the target is the groundwater surface, and for source protection it is the water in the well or spring. The pathway includes everything between the origin and the target. For resource protection, the pathway consists of the mostly vertical passage within the protective cover, and for source protection it also includes horizontal flow in the aquifer.

A major task of the COST 620 was to develop a general, nonprescriptive approach to intrinsic vulnerability mapping, which could be adapted into methods appropriate

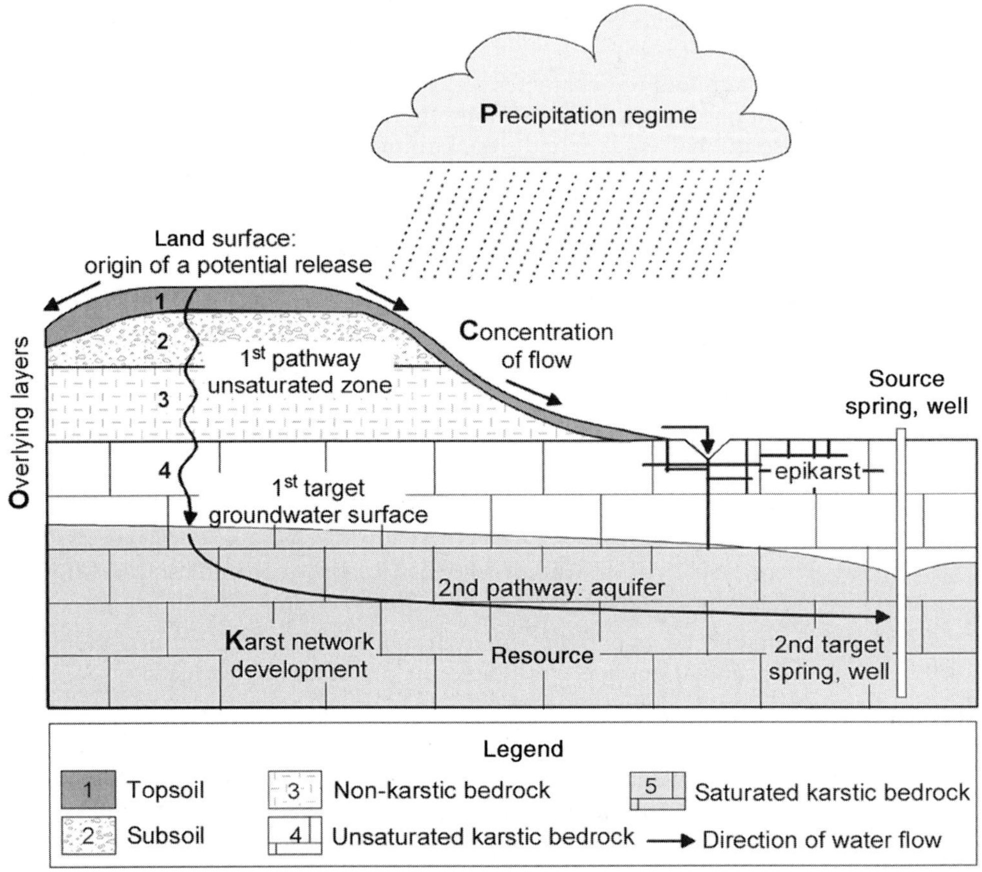

FIGURE 10.13 COST 620, or the European Approach to groundwater vulnerability mapping is based on an origin-pathway-target conceptual model. The main factors for vulnerability assessment are the **P**recipitation regime, the **O**verlying layers, the lateral **C**oncentration of flow, and the **K**arst network development. (1) Topsoil, (2) subsoil, (3) non-karstic bedrock, (4) unsaturated karstic bedrock, and (5) saturated karstic bedrock. (From Zwahlen, 2004.)

for use in individual karst areas of Europe. European karsts include alpine and lowland, Mediterranean and continental and management of karst groundwater also varies from country to country. Hence, flexibility had to be an essential attribute of the approach (Zwahlen, 2004).

The European approach uses four factors in assessing intrinsic vulnerability (Fig. 10.13): Overlying layers (O), concentration of flow (C), precipitation regime (P), and karst network development (K). The factors O, C, and K represent the internal characteristics of the system, while the P factor is an external stress applied to the system. The O factor may comprise up to four layers—soil, subsoil, non-karst rock, and unsaturated karst rock. The C factor recognizes that in karst areas the overlying protective layers may be bypassed by runoff, which is concentrated at or near the surface of the ground and which then enters the groundwater system via a doline (sinkhole) or a stream sink. For resource vulnerability mapping, where the target is the top of the saturated zone, the factors O, C, and P should be taken into consideration, while, in addition, the K factor should be taken into account for source vulnerability mapping where the target is a karst water supply such as a borehole or a spring (Zwahlen, 2004).

COST 620 does not specify how the component factors should be measured or categorized or how vulnerability ratings should be established and thus it is not a methodology. The final aim was that the approach, though karst sensitive, should not be completely karst centered to the extent that it could not be used in other groundwater environments. As stated by Zwahlen, this aim has been realized to the extent that the P and O factors have universal applicability in assessing vulnerability, whereas the C and K factors relate to the particular characteristics of karst aquifer systems. Notably, the intrinsic groundwater vulnerability assessment accounts only for the hydrogeologic characteristics of the system but is, by definition, independent of the properties of specific contaminants. However, each contaminant or group of contaminants behaves differently in the different layers. These contaminants, due to their own physical and chemical properties, can be retarded or degraded during their transit in subsurface. Conduits are excluded from the ranking based on layer characteristics as stated by Zwahlen (2004): "Of course, water flowing in the conductive drains is only slightly submitted to the attenuation processes, due to its generally insufficient residence time in the layer."

In conclusion, the qualitative vulnerability maps (source or resource, intrinsic or specific) in karst show relative sensitivity of different areas to groundwater contamination generated by human activities. They could be used as one of the tools for land-use planning and groundwater management if the limitations of the concept are clarified and fully described to decision makers. Such maps should always be made for a well-defined, specific purpose and should not be a stand-alone element but an integrated part of an overall groundwater protection plan. When issuing specific environmental and land-use permits, a karst vulnerability map cannot be used as a substitute for comprehensive hydrogeologic mapping and field investigations, including dye tracing where applicable. This is because common qualifications inherent to vulnerability maps such as "less sensitive," "more sensitive," and "somewhat sensitive" have no quantifiable risks and therefore have no meaning to water resources managers and regulators.

10.3.1 Delineation of Source Protection Zones

In the United States, public agencies such as the USGS started educating citizens, rural homeowners, and farmers on groundwater contamination issues at the beginning of the twentieth century. These efforts, illustrated by the following excerpts from Fuller (1910) and Figs. 10.14 and 10.15, can be viewed as early examples of the concept of source protection areas.

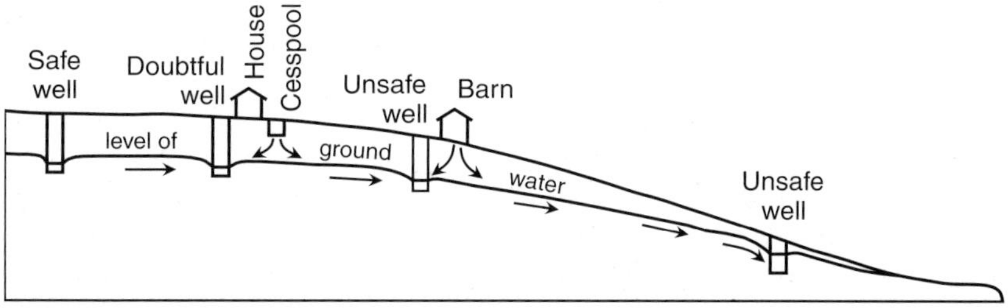

FIGURE 10.14 Diagram showing ordinary location of farm wells. (From Fuller, 1910.)

Figure 10.15 Spring in center of city street. (From Fuller, 1910.)

> Farms, which are generally remote from towns, cities, or other areas of congested population, seem to be almost ideally situated for obtaining pure and wholesome water. In reality, however, polluted water is exceedingly common on them (Fig. 10.14 in this book) and typhoid-fever rates are usually greater in country districts than in cities. Typhoid fever is now almost universally believed to be transmitted solely through drink or food taken into the stomach, and is especially liable to be communicated by polluted waters obtained from shallow wells near spots where the discharges of typhoid patients have been thrown upon the ground and subsequently carried down through the soil and into the wells, and it is doubtless principally this fact that makes the disease so common in farming regions.
>
> An example of danger from refuse of a more disgusting type is shown in Plate X, A (Fig. 3.15 in this book). Located in the middle of a well-traveled street, only a few inches above a gutter filled with paper and refuse, a part of which is sure to enter whenever a heavy rain occurs; open to the rain that washes into it from the steps leading down to it such dirt from the street as is brought in by the feet of the users; subject to the dipping of all sorts of more or less dirty buckets and utensils; receiving the underground drainage and presumably more or less sewage from the buildings on the slopes above; and containing in its bottom several inches of decaying paper and other refuse, this spring is on the whole one of the worst and most dangerously located sources of drinking water in the United States.

As discussed by Kraemer et al. (2005) of the USEPA, the main steps of the source water protection process involve the assessment of the area contributing water to the well or well field or spring, a survey of potential contaminant sources within this area, and an evaluation of the susceptibility of the source to these contaminants. This includes the possibility of contaminant release and the likelihood of transport through the soil

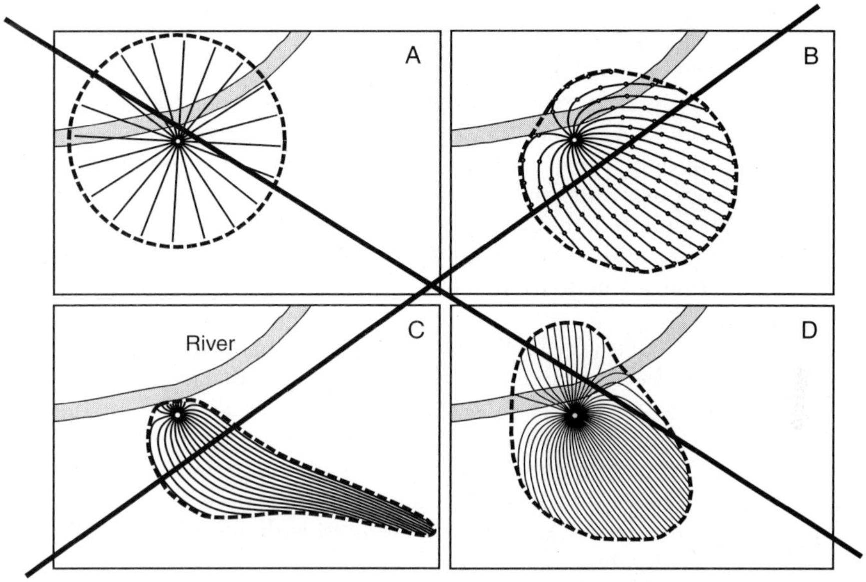

Figure 10.16 Five-year capture zone of a single well near a permanent river delineated using different options in the WhAEM2000 computer program. General groundwater flow direction is from the southeast toward the river. (A) Fixed-radius volumetric method. (B) Uniform flow field method; small circles correspond to 1-year intervals. (C) Zone delineated taking into consideration hydrologic boundaries (not shown). (D) Zone delineated taking into consideration possible resistance of the riverbed sediments (modified from Kraemer et al., 2005). The main reason why the figure is crossed by the author of this book is because any of the results shown, as well as the computer program used to create the maps, are not applicable to karst settings and their application in karst should be avoided at any cost.

and aquifer to the well screen. The designation of the wellhead protection area is a commitment by the community to source area management. USEPA states that delineation of the wellhead (source) protection area is often a compromise between scientific and technical understanding of hydrogeology and contaminant transport, and practical implementation for public safety (Kraemer et al., 2005).

In its efforts to facilitate capture zone delineation and protection area mapping in support of the State's Wellhead Protection Programs (WHPP) and Source Water Assessment Planning (SWAP) for public water supply wells in the United States, the USEPA released an updated version of W*h*AEM2000, a public domain and open source general-purpose groundwater flow modeling program. W*h*AEM2000 is analytic element model limited to steady-state and two-dimensional flow conditions, and includes options for simulating the influence of flow boundaries, such as rivers, recharge, and no-flow contacts (Kraemer et al., 2005). Figure 10.16 illustrates ambiguities associated with the delineation of a wellhead protection area using several approaches outlined by the USEPA. As can be seen, the shapes of the four zones are quite different reflecting critically different assumptions, all of which would be valid based on the USEPA guidance. In all four cases, the same (single) well near a relatively large permanent river operates at a constant (steady state) pumping rate, pumping water from an unconfined aquifer in which the groundwater flow is assumed to be horizontal, and with a regional hydraulic gradient from the southeast to the northwest (toward the river). Most importantly, the methods and the computer program itself are not at all applicable to karst settings and their use should be *avoided at any cost and for any purpose in karst*.

Unfortunately, most states in the United States were confronted with the task of developing wellhead protection programs for thousands of public drinking water supply systems in only a few years time. According to Kraemer et al. (2005), conducting an extensive groundwater modeling campaign for each individual drinking water well (or well field) was out of the question, both in view of the time involved and the cost. "The USEPA recognized this reality from the start and proposed a series of simplified capture zone delineation methods to facilitate a timely implementation of the States wellhead protection programs."

Consequently, it is very likely that many public water supply systems have delineated wellhead protection areas that are not based on any hydrogeologic reality, not just in case of karst aquifers, and may not be protecting them at all, thus creating a false sense of security. In the United States, the majority of community water supply systems serving fewer than 3,300 people use groundwater, while many larger systems also depend on groundwater. Small systems use groundwater as a source because groundwater usually requires less treatment than surface water and is therefore more affordable. This is an important consideration since many small systems without a large, rate-paying base cannot afford appropriate hydrogeologic studies for wellhead (source water) protection purposes regardless of the aquifer type. At the same time, wellhead protection efforts are often among the most cost-effective ways to ensure safe drinking water. These efforts prevent contamination from occurring rather than treating contamination after it has occurred.

Most methods used to delineate source protection zones in all porous media address the residence-time criterion. This criterion is based on the assumption that

- Nonconservative contaminants, subject to various fate and transport processes (e.g., sorption, diffusion, and degradation), may be attenuated after a given time in the subsurface.
- Detection of conservative contaminants (not subject to attenuation) entering the wellhead protection area will give enough lead time for the public water supply entity to take necessary action, including groundwater remediation and/or development of a new (alternative) water supply.
- Detection of any contaminants already in the wellhead protection area would require an immediate remedial action.

The most critical decision regarding an appropriate residence time is made by the stakeholders in each individual case, although 5-year, 10-year, and 20-year wellhead capture zones have been most widely used to delineate certain subzones of various land-use restrictions within the main wellhead capture zone. This concept for a single well in a homogeneous isotropic aquifer, pumping with the same pumping rate, and assuming strictly horizontal flow with the same hydraulic gradient that does not change in time, is illustrated in Fig. 10.17.

Commonly, water utilities implement (or regulatory agencies require) delineation of the following three wellhead protection zones, although the number may vary based on local hydrogeologic conditions and regulatory requirements:

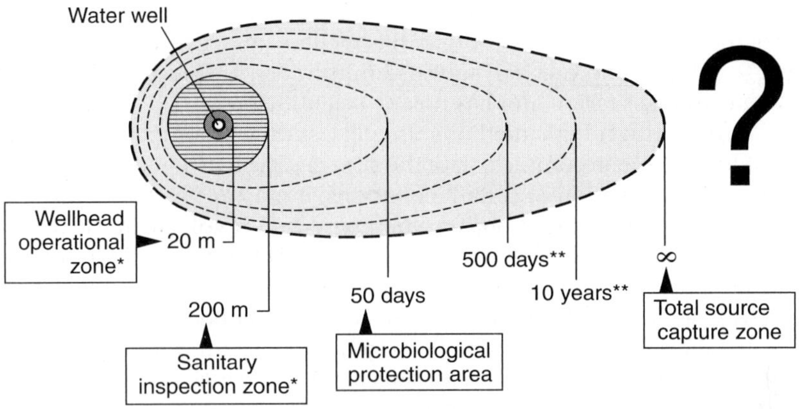

Figure 10.17 Idealized scheme of surface sanitary zones and groundwater flow perimeters for the protection of a water well in an unconfined aquifer. *Empirical fixed-radius area; **Intermediate flow-time perimeters sometimes used (from Foster et al., 2002–2005). The main reason why there is a big question mark added to the figure by the author of this book is because the concept shown is not applicable to most karst aquifers as discussed in the text.

- Zone 1 (sanitary zone or zone of strict protection) is essentially an administrative zone of physical protection of the water source, such as fencing and restricted access measures. Its purpose is to prevent accidental or deliberate damage and/or contamination of the source itself including contamination of the aquifer through the source. The size of Zone 1 may vary between several tens and several hundreds of feet (meters) for simple systems consisting of one well for example, or it may be larger in case of several closely spaced springs or multiple wells.
- Zone 2, sometimes called the inner protection zone, is based on time-of-travel analysis. It reflects the site-specific understanding of the minimum time of potential contaminant travel between its introduction into the subsurface and the groundwater withdrawal locations during which it would be possible to initiate remediation activities and execute contingency plans for water supply if necessary. In some states, Zone 2 is considered an attenuation zone in which pathogens entering the zone from septic systems or surface water bodies would be attenuated before reaching the well (e.g., see Wyoming DEQ, 2007). It is obvious that Zone 2, in general, may vary in shape and size widely, depending on aquifer type and hydraulic conductivity of the most permeable zones. Glacial gravel deposits and karst aquifers are examples of porous media where contaminant travel times may be extremely short (e.g., several days or less) over distances of miles.
- Zone 3, also called the contributing area, includes the entire aquifer volume within the groundwater divide from which all groundwater will eventually discharge through the well(s) or spring(s), regardless of time of travel. This zone may be subdivided based on various criteria, including time of travel. Its main importance is for long-term planning of groundwater resources management, and the related land-use and aquifer protection regulations.

Wellhead protection zone delineation methods can be divided into the following three categories (Kresic, 2007a, 2009): (1) nonhydrogeologic, (2) quasi-hydrogeologic, and

(3) hydrogeologic. The nonhydrogeologic method is a selection of an arbitrary fixed-radius or fixed-shape area around the well(s) in which authorized personnel implement some type of protection such as limited access. This method does not consider the residence-time criterion and has no any scientific basis.

Quasi-hydrogeologic methods use very simple assumptions, which, in many cases, do not have much in common with the site-specific hydrogeologic conditions. Since they include the application of certain equations, it may appear, to nonhydrogeologists, that such methods must have some credibility. When involved in the application of quasi-hydrogeologic methods, hydrogeologists should clearly explain the limitations of various unrealistic assumptions and their implications on the final wellhead zone delineation. For example, these methods do not consider aquifer heterogeneity and anisotropy, presence of confining layers, vertical flow components, interference between multiple wells screened at different or the same depths in the same well field, or depth to screen of individual wells. Three common quasi-hydrogeologic methods include (1) calculated fixed radius, (2) well in a uniform flow field, and (3) groundwater modeling not based on hydrogeologic mapping (Kresic, 2007a, 2009). None of these methods should be applied in any of the situations listed above, and never in any karst environment.

As mentioned repeatedly, karst aquifers are particularly vulnerable to contamination. Contaminants can easily enter underground and may be transported rapidly over large distances in the aquifer. Processes of contaminant retardation and attenuation often do not work effectively in karst systems. Therefore, a detailed knowledge of karst hydrogeology is a precondition for the delineation of source protection areas in karst. However, karst drainage areas contributing water to a single spring or a large supply well can be very extensive and convoluted due to many karst characteristics described in Chap. 1. It is therefore often unrealistic to designate maximum protection for the entire karst water system as the resulting land-use restrictions would not be acceptable to some stakeholders. In addition, many public water systems do not have financial and other resources to "do it right" and embark on an adequate characterization of karst hydrogeology in their service area. Unfortunately, this gamble may sometimes prove fatal, as illustrated with the following example provided by Worthington et al., 2003 (see also Goldscheider et al., 2007).

Walkerton is a rural town in Ontario, Canada, with a population of some 5,000 people. In May 2,000, about 2,300 of them became ill and 7 died from bacterial contamination of the municipal water supply. The principal pathogens were *E. coli O157:H7* (a pathogenic strain of *E. coli*) and *Campylobacter jejuni*. Subsequent epidemiologic investigations indicated that most of the contamination of the water supply has occurred within hours or days at most, after a heavy rain. Three municipal wells had been in use at the time of the outbreak. Soon after the outbreak, a hydrogeologic investigation was carried out that included the drilling of 38 boreholes, surface and downhole geophysics, pumping tests, and testing of numerous water samples for both bacteriologic and physical parameters. The aquifer at Walkerton consists of 70-m-thick flat-bedded Paleozoic limestones and dolostones, which are overlain by 3 to 30 m of till. A numerical model of groundwater flow (using MODFLOW and Darcian approach) indicated that the 30-day time of travel capture zones extended 290 m from Well 5, 150 from Well 6, and 200 m from Well 7. These results suggested that if a groundwater pathway was implicated in the contamination, then the source must have been very close to one of the wells.

A public inquiry (the Walkerton Inquiry) was held to investigate the causes of what came to be known as the Walkerton Tragedy. During the inquiry, the question was raised as to whether the aquifer might be karstic and thus have rapid groundwater flow. The original hydrogeologic investigation, carried out after the outbreak, had not mentioned the possibility of karstic groundwater flow. Subsequent investigations by karst experts found that there were many indications that the aquifer is karstic (Worthington et al., 2003). These included a correlation between bacterial contamination in wells and antecedent rain, demonstrating rapid recharge and flow to wells; localized inflows to wells that video images showed to be dissolutionally enlarged elliptical openings on bedding planes; the presence of springs with discharges up to 40 L/s; rapid changes in discharge and chemistry at these springs following rain; and rapid, localized changes to electrical conductance in a well during a pumping test.

All these tests strongly suggested that the aquifer is karstic, but the most persuasive evidence were the results of aquifer tracing tests. Earlier numerical modeling had suggested that groundwater velocities were typically in the range of a few meters per day, but tracer tests demonstrated that actual velocities were some 100 times faster (Fig. 10.18). In conclusion, the investigations by the karst experts demonstrated that the source for the pathogenic bacteria could have been much further from the wells than the earlier investigations and modeling had indicated (Goldscheider et al., 2007).

Vulnerability mapping has sometimes been used to derive source protection zones in karst. The main potential problem with this method is relying exclusively on desktop studies by using published geologic maps, land-use maps, soil cover maps, and other general maps and documents. In most cases, these general and readily available documents were not prepared with any specific focus on various karst characteristics

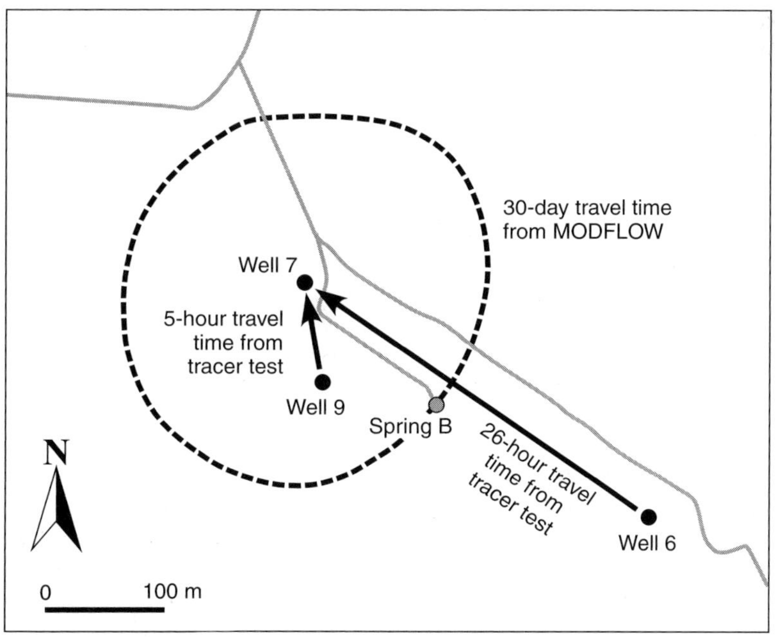

FIGURE 10.18 Trajectories and travel times for tracers injected in monitoring wells 6 and 9 and recovered in pumping Well 7 showing velocities >300 m/day, compared to a 30-day capture zone for Well 7 predicted using MODFLOW. (Modified from Worthington et al., 2003.)

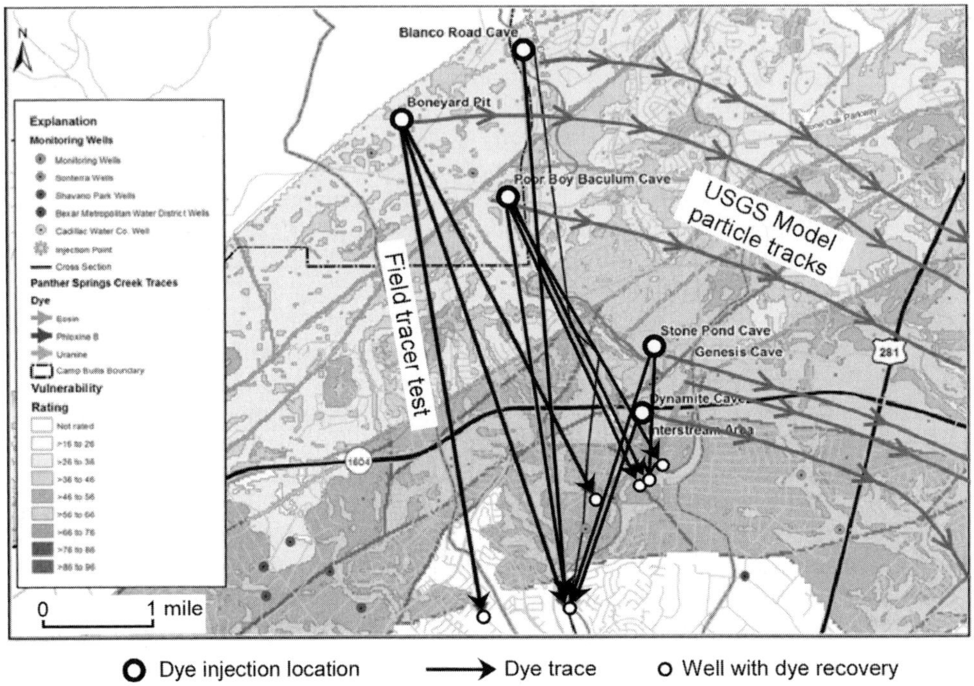

FIGURE 10.19 Map showing tracer test results and flow paths modeled with MODFLOW and Darcian-equivalent porous media (EPM) approach. Tracer test velocities range from 80 ft/day to over 12,000 ft/day. The groundwater travel velocity between arrows placed on the modeled flow paths, as calculated from the USGS' numeric model of Edwards Aquifer, Texas, is 1 mile/year. (Courtesy of Geary Schindel, Edwards Aquifer Authority; dye tracing results presented in Schindel et al., 2009.)

and specific features that are of critical importance when attempting to delineate source water protection zones. It is also erroneous to assign low sensitivity to areas and rocks that are "less" karstified, and then draw conclusions that such areas need not have land-use restrictions applicable to karst in general.

Figure 10.19 illustrates some of the above points. The locations of and the areas surrounding the three northernmost caves shown on the map, Boneyard Pit, Poor Boy Baculum Cave, and Blanco Road Cave, were assigned low relative sensitivity on a vulnerability map prepared by USGS for the San Antonio Water Systems. However, the results of multiple tracer tests performed by the Edwards Aquifer Authority show groundwater velocities ranging from 80 ft/day to >12,000 ft/day between different locations of dye injection and dye recovery at quite a few wells to the south. Tracer testing also clearly shows dyes rapidly traversing one fault with as much at 300 ft of displacement and six other faults. Previously, for unknown reasons, faults with more than 50% offset of the Edwards Limestone were thought to be barriers to groundwater flow. Interestingly, the map also shows groundwater flow paths modeled with MODFLOW and Darcian-equivalent porous media (EPM) approach. The groundwater travel velocity between arrows placed on the modeled flow paths, as calculated from the USGS' numeric model of Edwards Aquifer, Texas, is 1 mile/year. (Courtesy of Geary Schindel, Edwards Aquifer Authority.)

In conclusion, karst-specific hydrogeologic mapping described in detail in Chap. 3 is the only method recommended by the author of this book for delineation of source protection zones. Unfortunately, every so often karst hydrogeologists are faced with a challenging choice: either perform the work knowing that the funding is insufficient to meet the karst challenges or leave the work to someone else. As always, the choice is individual.

10.4 Water Hazards in Karst

10.4.1 Sinkhole Development

Every so often a sinkhole develops somewhere in karst terrains around the world to surprise citizens and make their lives miserable, sometimes including serious consequences (Fig. 10.20). Public agencies on such occasions may issue press releases to reassure the public that something will be done about it as illustrated by the following statement from the USGS and the accompanying photograph by Doug Gouzie (Fig. 10.21): "A sinkhole collapse early Sunday morning August 13, 2006 in Nixa, Missouri is responsible for the disappearance of a garage and the Chevy Cavalier once parked inside. The sinkhole was initially estimated to be approximately 60 ft in diameter and 75 ft deep. While sinkholes are common in karst areas such as the Missouri Ozarks, the sudden collapse of one beneath a home is not a common occurrence. A team of researchers from the U.S. Geological Survey and Missouri State University will be cooperating on a study to better understand the spatial distribution of sinkholes and the potential for the formation of additional karst features in southwest Missouri." (http://mcmcweb.er.usgs.gov/mcgsc/news_nixa_sinkhole.html)

Figure 10.20 *Left*: On September 7, 2008, a large-scale collapse took place on the Jinshazhou Road in Guizhou City, Guangdong Province, China, which was almost 20 m wide and more than 10 m deep. It swallowed a car that happened to drive by at that time. The collapse was caused by an underground tunnel construction and fluctuating water table. *Right*: On December 20, 2008, a collapse induced by water table fluctuation during the foundation construction for nearby residential buildings occurred in XiaMao Village, Guangzhou City, Guangdong Province, China. During this event, 6 buildings collapsed, 3 buildings slanted out of shape, the walls and floors of more than 10 buildings cracked and deformed, within a disaster area of about 5,000 m^2. (Photograph courtesy of Lei Mingtang.)

Figure 10.21 A sinkhole collapse in Nixa, Missouri. (Photograph courtesy of Doug Gouzie.)

Unfortunately, sinkholes in karst will continue to develop including where not expected by the public, both naturally and because of various anthropogenic impacts. The latter accelerate the process, sometimes dramatically, and may cause widespread damage to water resources and infrastructure. The hazards of sinkhole development are well understood and described in numerous publications, and there is no excuse for not considering them for any land-use planning or land-use activity in karst terrains.

As discussed by Tihansky (1999) in a very informative publication on sinkhole development in west-central Florida, induced sinkholes are generally cover-collapse type sinkholes and tend to occur abruptly (Fig. 10.22). They have been forming at increasing rates during the past several decades and pose continuing hazards in developed and developing areas of west-central Florida. The increasing incidence of induced sinkholes is expected to continue as demand for groundwater and land resources increases. Regional declines of groundwater levels increase sinkhole occurrence in sinkhole-prone regions. This becomes more apparent during the natural, recurring periods of low annual rainfall and drought, as well as during extensive groundwater development for public water supply and agricultural irrigation. Beck and Sinclair (1986), and Tihansky (1999) provide the following dramatic examples to illustrate these points.

- The city of St. Petersburg developed a new well field north of Tampa in 1963 (Section 21). Within 1 month of increasing the pumping rate in May 1964, 64 new sinkholes formed within a 1-mile radius of the well field. Most of the sinkholes were formed in the vicinity of Well 21–10, which was pumping at nearly twice the rate of the other wells. Neighboring areas also noticed dramatic declines in lake levels and dewatering of wetland areas. Although the wells only tapped the limestone, the water level in the overlying sands and clays also declined

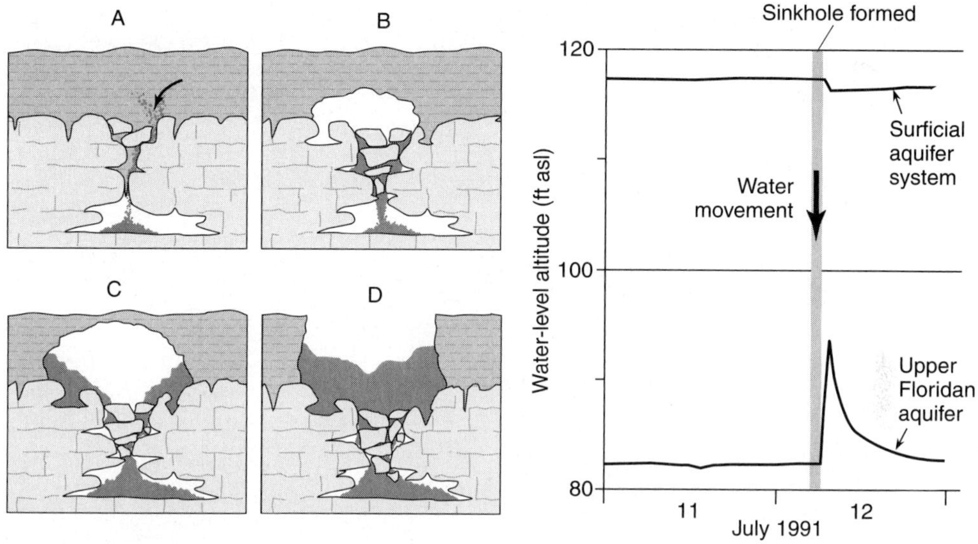

FIGURE 10.22 *Left*: Schematic of cover-collapse sinkhole formation. (A) Sediments spall into a cavity. (B) As spalling continues, the cohesive covering sediments form a structural arch. (C) The cavity migrates upward by progressive roof collapse. (D) The cavity eventually breaches the ground surface, creating sudden and dramatic sinkholes. *Right*: A sinkhole that breached a confining clay layer that separates the Floridan aquifer and the surficial aquifer illustrates the interconnectivity of the aquifers. The water level drop in the surficial aquifer system and the coincident rise in the Upper Floridan aquifer occurred as the sinkhole drained. The water levels were recorded at a well site less than 1,000 ft from the sinkhole. (From Tihansky, 1999.)

similarly, demonstrating that increased downward leakage was occurring to the limestone.

- Hundreds of sinkholes ranging in diameter from 1 ft to more than 150 ft formed within a 6-hour period on February 25, 1998, during the development of a newly drilled irrigation well. Unconsolidated sand overburden collapsed into numerous cavities within an approximately 20-acre area as pumping and surging operations took place in the well. The affected land is located near the coast in an upland region that straddles parts of Pasco and Hernando Counties. A 20-ft-thick sediment cover composed primarily of sand with little clay is underlain by cavernous limestone bedrock. The well was drilled through 140 ft of limestone, and a cavity was reported in the interval from 148 to 160 ft depth, where drilling was terminated. Very shortly after development began, two small sinkholes formed near the drill rig. As well development continued, additional new sinkholes of varying sizes began to appear throughout the area. Trees were uprooted and toppled as sediment collapse and slumping took place, and concentric extensional cracks and crevices formed throughout the landscape. The unconsolidated sandy material slumped and caved along the margins of the larger sinkholes as they continued to expand. The first two sinkholes to form eventually expanded to become the largest of the hundreds that formed during the 6-hour development period. They swallowed numerous 60-ft-tall pine trees and more than 20 acres of forest, and left the well standing on a small bridge of land.

FIGURE 10.23 Sudden collapse of U.S. 19 highway bridge near Tarpon Springs, Florida, caused drowning of one person, while several others were injured as cars could not stop short of the gap. (From the George Sowers Collection, courtesy of AMEC Environment and Infrastructure, Inc.)

- During the disastrous freeze of December 1983, ferneries near Pierson, Florida, were irrigated for freeze protection. The drop in water level was severe enough to cause shallower domestic wells in the area to go dry, and some homes were without water for several days. During this period, a large sinkhole formed in the fernery area. It was 80 ft in diameter and estimated at 35 ft deep. It formed next to a small lake that immediately drained into the sinkhole where the water disappeared. Previous sinkholes that have been documented in this same area due to freeze protection have damaged homes, yards, and highways.

Other common mechanisms that can trigger catastrophic sinkhole collapses include

- Intercepted drainage from a large construction area, or runoff from building roofs and parking lots that are concentrated into one spot.
- Leaky sewers and water lines.
- Land application (spraying) of treated wastewater.
- Increased surface loading due to new buildings and other construction (Fig. 10.23), solid waste and other disposal, and standing water such as artificial reservoirs.
- Natural lakes formed by flooding of an old sinkhole basin when sinkholes that served as drains of the natural depression are plugged. Such karst depression lakes may persist for years, only to empty again when the old drain becomes unplugged or a new collapse sinkhole forms.

In addition to structural damage, sinkhole collapse incidents in urban and industrialized areas can cause serious groundwater contamination by triggering spills of hazardous materials from ruptured pipelines, tanks, and other damaged infrastructure. Buried infrastructure in both urban and rural areas may also be damaged by the carbonate rock dissolution-related settling mechanisms other than catastrophic sinkhole collapse; leaks formed that way may not be noticed for long periods of time resulting in serious groundwater contamination and extensive contaminant plumes.

In conclusion, as discussed by Tihansky (1999), in areas underlain by cavernous limestone with thin-to-moderate thickness of overburden, increased sinkhole development and property loss are strongly correlated to human activity and cultural development. There are several reasons for this correlation. First, rapid growth and development makes it more likely that new sinkholes will be reported, and the construction of roads (Fig. 10.24) and industrial or residential buildings increases exposure to the risk of property damage. Second, land-use changes in rapidly developing areas are often loosely controlled and include altered drainage, new impoundments for surface water, and new construction in sinkhole-prone areas. Finally, the changing land use is often associated with population increases and increasing demands for water supplies, which may lead to increases in groundwater pumpage and the lowering of local and regional groundwater levels.

Although one cannot adequately predict sinkhole development, it may be possible to prevent or minimize the effects of sinkholes or reduce their rate of occurrence. Well-documented episodes of accelerated sinkhole activity are directly related to groundwater pumping events that lower groundwater levels. It is likely that many induced sinkholes can be prevented by controlling fluctuations in groundwater levels. The overall regional decline in water levels in the Upper Floridan aquifer has been a long-standing concern of water resource managers. Local declines around municipal well fields, often much greater than the regional declines, have led to dewatering of lakes and wetlands, upconing of poorer-quality water, saltwater intrusion, and accelerated sinkhole development. Developing plans for minimum groundwater levels can help minimize sinkhole impacts by ameliorating some of the conditions that cause them (Tihansky, 1999).

Figure 10.24 Repaired highway damage caused by the active sinkhole. (From the George Sowers Collection, courtesy of AMEC Environment and Infrastructure, Inc.)

Figures 10.25 through 10.27 show several of numerous sinkhole photographs from the United States and from around the world taken by George Sowers while on different engineering assignments for Law Companies Group, now AMEC Environment and Infrastructure, Inc. The definitive book on the subject of sinkholes by Sowers (1996) entitled *Building on Sinkholes: Design and Construction of Foundations in Karst Terrain* is highly recommended further reading. Another informative reference is book by Waltham et al. (2004).

10.4.2 Dams and Reservoirs

The following description of dam problems provided by Turkmen et al. (2002) is very common in karst. To prevent seepage at the right abutment of a 77-m-high rock-fill dam in Turkey, a 200-m long and 60-m deep grout curtain along the dam axis was constructed. After the impoundment, some springs occurred downstream, and for this reason, extra grouting was performed. However, the seepage problem could not be solved. In order to determine the seepage direction and karstification pattern, additional field investigations were performed including drilling and dye tracing. As a result of these hydrogeologic studies, seepage paths were detected in the karstic limestone between the dam and the spillway. As concluded by the authors, the seepage problems are certain to continue because the carbonate rocks extend beneath the spillway.

Literary hundreds of similar cases of dam leakage in karst are described in literature, with many examples of dam abandonment including some very large, expensive dams (Fig. 10.28). Roughly two periods of dam construction in karst can be identified:

Figure 10.25 Winter Park sinkhole occurred during the spring of 1981. It severed two city streets, truncated a municipal swimming pool, swallowed a house, and damaged an automobile repair shop together with several automobiles parked behind it. The final hole diameter was more than 300 ft (90 m) with the cone-shaped opening approximately 50 ft (15 m) deep through clayey sands. Below was a 60 ft (18 m) diameter cylindrical opening in a soft, porous limestone, possibly an ancient pit or old collapse sinkhole in the limestone. Movement was apparently triggered by wide fluctuations in the groundwater level from well pumping and possibly aggravated by leaking of the swimming pool. (From the George Sowers Collection, courtesy of AMEC Environment and Infrastructure, Inc.)

(1) the first half of the twentieth century when dams were often built without fully understanding the nature of karst and (2) the period after the initial dam construction boom revealed many problems. Unfortunately, the second period is also characterized by gambling with karst in cases where investors, although fully aware of the inherent risks, were trying to save on the field and design investigation phases. Examples abound where this gambling did not pay off, and the "saved" money resulted in tens of millions US dollars or more spent on the postconstruction remedy. Unfortunately, even when all karst-specific concerns are addressed with textbook field investigations, desktop studies, and the subsequent design and construction, there still will remain an inevitable risk of a dam not being fully leak-proof. In any case, when planning a dam construction in karst, investors and other stakeholders should weigh all benefits of the project against the risks and engage karst specialists during all phases of the project.

The world's tallest arch dam built entirely in mature karst, Piva Dam in Montenegro (Fig. 10.29), is an example where the benefits of power production outweigh the costs of grouting that continues several decades after the dam completion. Some of the first practical karst experiences the author of this book had as a young engineer of hydrogeology were from the initial hydrogeology and grouting work on the Piva Dam. During

Figure 10.26 *Top*: A large sinkhole near Johannesburg, South Africa, that swallowed an ore processing plant and several homes, killing 29 persons. *Bottom*: A gaping sinkhole formed abruptly on June 27, 1994, within a 400-acre, 220-ft high gypsum stack at a phosphate mine in south-central Florida. In addition to the weight of the stack, the sinkhole development may have been accelerated by the aggressive chemical properties of the acidic waste slurry. Note trucks in the upper right and the drilling rig for scale. (From the George Sowers Collection, courtesy of AMEC Environment and Infrastructure, Inc.)

FIGURE 10.27 Dairy farm sinkhole southeast of Bartow, Florida. *Right*: Aerial view; note vehicle tracks for scale. *Left*: Looking into the cylindrical rock throat and funnel-shaped overburden slope from the sinkhole rim. (From Sowers, 1974; courtesy of AMEC Environment and Infrastructure, Inc.)

a recent visit to Belgrade, Serbia, he was happy to learn that the dam is still holding water and the grouting and related dye tracing tests are still being performed for the benefit of young hydrogeologists gaining experience in karst (Fig. 10.30). As discussed by Milanovic (2004), the dam site is heavily karstified with numerous karst features encountered in both canyon abutments. Fractures enlarged by dissolution and cavities of all sizes were frequently intersected by exploration galleries (adits) and the main grouting galleries. One of the unusual features was that many large karst cavities were completely filled with lithified clay; the largest such cavity had dimensions of 100 × 100 m. The dam curtain was tied to the low-permeability rock mass on both flanks and reached 250 m below the riverbed. The curtain consists of three rows close to the dam, laterally and below, and one row further into the rock mass. All encountered cavities, 66 in total, were sealed with grout mix using between 5 and 98 tons per cavity. The performance of this very complex grout curtain is being continuously monitored by tens of piezometers inside, upgradient and downgradient of the curtain. As needed, the curtain is repaired by new grouting based on the monitoring results and dye tracing tests.

There are also examples of arch dams built in pure Dinaric karst with minimal leakage rates and no need for postconstruction grouting (Fig. 10.31). In all such cases, complex geotechnical improvements and grouting were performed based on extensive, multidisciplinary field investigations before building the actual dam. Similar examples can be found throughout the classic Dinaric karst where concrete gravity dams and arch dams

FIGURE 10.28 Montejaque Dam in the Sierra de Grazalema, Spain, was abandoned before the Second World War because of large losses of water from the reservoir and at the dam site through numerous sinks and other karst features. This arc dam on the river Compobuche-Guadares, a tributary of the Guadiaro River, is 83.75 m high with an 84-m long crest. After heavy rains, the reservoir behind the dam fills with water only to lose it soon after. (Photograph courtesy of Dr. Petar Milanovic.)

in deep canyons were built for two main reasons: (1) due to its geomechanical characteristics, limestone is one of the best rocks for the construction of arch dams, and (2) the prevalence of karst did not leave much room for different choices.

In contrast to this practice of building dams in karst by fully engaging karst hydrogeologists and engineering geologists, there are numerous examples of dams built by civil engineers without any engagement of experienced karst specialists. As discussed by Schaefer (2009), at the time when many of the United States Army Corps of Engineers (USCOE) dams were constructed, it was thought that the solution features beneath the dams were of little dam safety significance. They were considered only an issue if the leakage volume impacted reservoir storage for water supply or hydropower. The following excerpt from the minutes of a conference with Dr. Arthur Casagrande in the Little Rock District Engineer's Office on January 25, 1941, concerning Clearwater Dam demonstrates the state of the art at this time.

> **Major T.F. Kern, District Engineer:** Dr. Casagrande, you spoke about seepage through the rock. Do you feel there is any danger of failure due to piping?
> **Dr. Casagrande:** No. In soil mechanics we understand by "piping" the rapid erosion of large channels in loose, sandy soils, resulting in undermining of the structure, usually causing its collapse, and always causing a very rapid escape of the impounded water. In the limestone bedrock the deep-seated solution channels are already in existence. Flow of water through

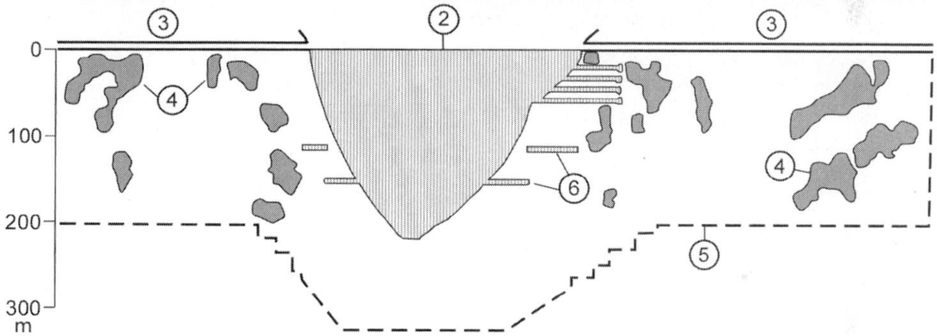

FIGURE 10.29 *Top*: Piva Dam in Montenegro built in the 1,500 m deep canyon of Piva River composed of highly karstified massive Triassic limestone. The arch dam is 220 m high and the reservoir volume is 790 million m^3. (Courtesy of Dr. Vladimir Beličević, Energoprojekt-Hidroinženjering, Belgrade.) *Bottom*: Grout curtain layout. (1) Dam body, (2) dam crest, (3) grouting gallery, (4) intensely karstified zones indicated by high grout mass consumption, (5) bottom of grout curtain, and (6) gallery. (Courtesy of Dr. Petar Milanovic; simplified from original engineering drawings by Energoprojekt-Hidroinženjering, Belgrade.)

> these channels will not cause rapid erosion, nor in other ways undermine the stability of the dam. The worst thing that could happen is that one cannot fill the reservoir. At Clearwater such extreme condition is, in my opinion, impossible within the range of human doubt; but the leakage may be large enough to make the use of the reservoir for power uneconomical.

Since that time, the dam engineering community has learned, as we normally do, from failures or incidents that solution features do create a significant risk to the safety of dams (Schaefer, 2009).

In 2005, the USCOE began a screening risk assessment process to evaluate its portfolio of dams. Six dams have been designated to be in Dam Safety Action Class I. This designation is for unsafe dams critically near failure or with extremely high risk. Three of the six dams are built over karst foundations and are experiencing significant distress. These dams are Wolf Creek Dam in southern Kentucky, Center Hill Dam in northern Tennessee, and Clearwater Dam in southeast Missouri. The risk evaluations also revealed numerous other dams in this region with serious issues related to karst foundations.

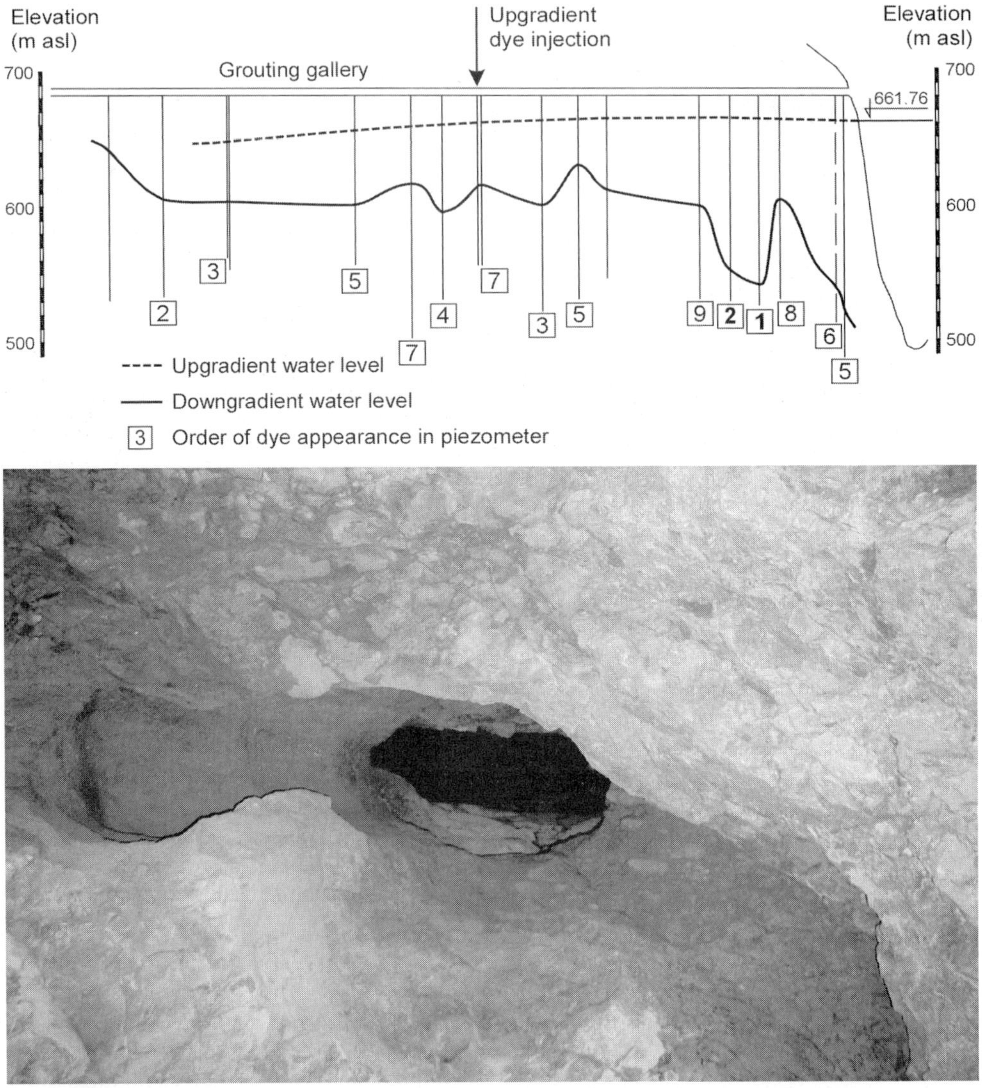

Figure 10.30 *Top*: Results of a dye tracing test showing the order of dye appearance at downgradeint piezometers, left abutment of Piva Dam, Montenegro. The test indicates a preferential flow path and likely breach of the grout curtain around piezometers with the lowest hydraulic head where the dye showed up first. The erratic karstification is clearly visible by the irregular pattern of dye appearance at various piezometers. *Bottom*: Typical example of cavities encountered by the grouting galleries. (Courtesy of Dr. Vladimir Beličević, Energoprojekt-Hidroinženjering, Belgrade.)

FIGURE 10.31 Grančarevo Dam built on the Trebišnjica River in the classic Dinaric karst of eastern Herzegovina. With the volume of 1.27×10^9 m^3 the Bileća Reservoir behind the dam is the largest one in the classic Dinaric karst; actual water loss from the reservoir is less than 150 L/s or significantly less than estimated in the design phase confirming the success of a double-row grout curtain. The curtain is 664-m long and 150-m deep. Grouting was performed after fractures and cavities were flushed out using batteries of three borings and injecting water and air under pressure into packer-sealed, 5-m long boring sections. The flushing was terminated after sustained clean water was coming out of a control boring. The grout was injected into the clean flushed borings and the process was repeated along the entire length of the grout curtain.

These dams have been designated in Dam Safety Action Class II. This classification is for dams characterized as unsafe or potentially unsafe with failure initiation foreseen (Schaefer, 2009).

The USCOE practice of building dam on karst foundations involved filling with clay any cavities that were uncovered during excavation of the dam footing, and not much more (Fig. 10.32). However, such fillings are easily washed away from the conduits and cavities because of the high pressure exerted by the reservoir water behind the dam. As a result, taxpayers in the United States are now paying hundreds of millions of dollars for extensive grouting and other remedy work at dams in karst, with little progress. The main reason for this losing battle at many dams in the United States and around the world is that it is very difficult to stop leaks under and around a dam in karst once the reservoir is filled; the steep hydraulic gradients maintain high groundwater velocity in the karst conduits so that successful grouting is rarely possible. More disturbing, however, is the fact that safety of earth-fill and rock-fill gravity dams can be seriously compromised by piping and fill collapse mechanisms caused by the presence of karst conduits beneath the dam (Fig. 10.33).

Unfortunately, both the practical experience and the results of various theoretical studies show that, once compromised, an improperly designed and constructed dam in karst may have to be abandoned for the safety and economical reasons. Of course, in some specific cases, investors with deep pockets may decide to implement what appears to be the ultimate solution as shown in Fig. 10.34.

Romanov et al. (2003) present results of a modeling study simulating processes of rock dissolution below a dam by coupling equations of dissolutional fracture widening to hydrodynamic flow. All model runs show similar characteristic behavior. Shortly after filling, the simulated reservoir exhibits a small leakage of about 0.01 m^3s^{-1}, which increases steadily until a breakthrough event occurs after several decades with an abrupt increase in leakage to about 1 m^3s^{-1} within the short time of a few years. Then, flow in the fractures becomes turbulent and the leakage increases to 10 m^3s^{-1} in a further time span of about 10 years. Small channels propagate downstream and leakage rises slowly until the first channel reaches the surface downgradient of the dam. Then, breakthrough occurs, the laminar flow changes to turbulent, and a dense net of fractures that carry flow is established.

A sensitivity analysis was performed on the dependence of breakthrough times on various parameters including the height of impounded water, the depth of grouting, the average aperture width of the fractures (a_0), and the chemical parameters, which are the equilibrium concentration of Ca with respect to calcite and the Ca concentration of the inflowing water. The results show that the most critical parameter is the average aperture width of the fractures. At fracture aperture widths of 0.01 cm, breakthrough times are above 500 years. For values of a_0>0.02 cm, breakthrough times are within the lifetime of the structure. Dam sites in gypsum exhibit similar breakthrough times. However, after breakthrough, owing to the much larger dissolution rates of gypsum, the time until excessive leakage is only a few years. The authors conclude that the results support the assumption that increasing leakage of dam sites in soluble carbonate rocks may be caused in part by dissolutional widening of fractures.

A very detailed discussion of various engineering problems and risks associated with dams and reservoirs in karst is provided by Milanovic (2004). This definitive guide, entitled *Water Resources Engineering in Karst* includes numerous examples from the classic Dinaric karst and around the world.

Figure 10.32 *Top*: Wolf Creek Dam construction photo, United States Army Corps of Engineers, Nashville District. Numerous cavities like these, encountered during excavation for the dam, were filled with soil and clay. *Bottom*: Wolf Creek Dam is on the Cumberland River in South Central Kentucky near Jamestown, Kentucky. It provides flood control, hydropower, recreation, and water supply. Construction began in 1941 and was interrupted by Second World War from 1943 to 1946. The reservoir was impounded in December 1950. The 5,736-ft-long dam is a combination earth-fill and concrete gravity section. (Courtesy of Don Dotson, AMEC Environment and Infrastructure, Inc.)

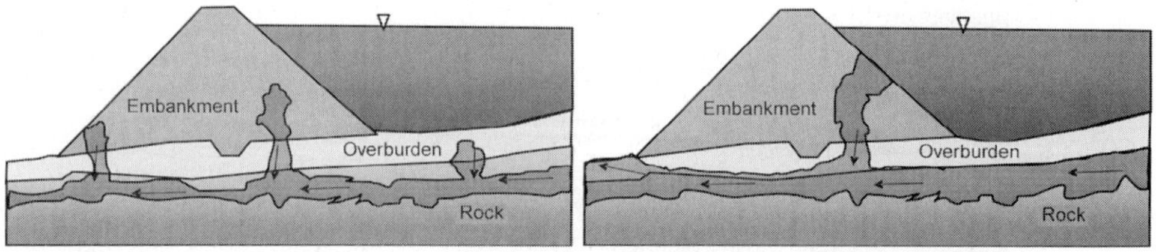

Figure 10.33 Some examples of conditions that may cause failure of dams improperly built on karst foundations. *Left*: Erosion into solution feature. *Right*: Solution feature with continuous soil exposure. (From Schaefer, 2009.)

10.4.3 Quarrying, Mining, and Tunneling

Water hazard common to quarrying, mining, and tunneling in karst is possibility of a sudden inflow of large quantity of water that can cause serious flooding to the point of endangering or losing human lives and the equipment. Catastrophic water inflow can happen both in the unsaturated zone after heavy rainfall and in the saturated zone after

Figure 10.34 Prevention of future leakage from the Akköprü Reservoir, Turkey, using thick reinforced concrete slab over the reservoir bottom and reservoir bank. Inset shows one of many alluvial ponors that were uncovered during field investigations for the feasibility study and final design. The reservoir will be formed behind the Akköprü Dam currently under construction. (Photographs courtesy of Dr. Petar Milanovic; see also Günay and Milanovic, 2005.)

FIGURE 10.35 Hydrotechnical tunnel in Dabarsko Polje, eastern Herzegovina, flooded during construction. (Courtesy of Dr. Petar Milanovic.)

FIGURE 10.36 Typical appearance of a karstic void partially filled with wet clay and silt; Dodoni Tunnel, northwestern Greece, 2000. Note the fault that facilitated development of the void. (From Marinos, 2005.)

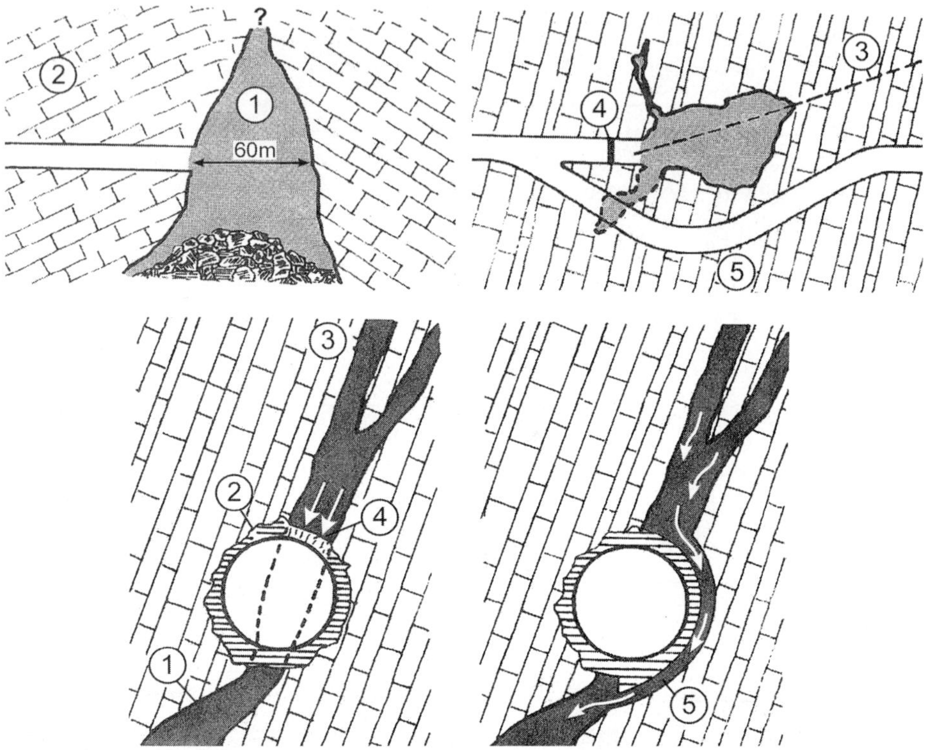

Figure 10.37 *Top*: Large caverns may require deviation of a tunnel. (1) Cavern, (2) karstified limestone, (3) fault, (4) concrete plug, (5) tunnel deviation. *Bottom*: Crossing a submerged karst channel with the tunnel requires that the channel remains drained to prevent possible lining damage due to episodes of excessive water pressure caused by heavy rainfall. (1) Karst channel, (2) tunnel, (3) water level in the channel, (4) part of tunnel lining under concentrated pressure, (5) drainage around the tunnel. (Courtesy of Dr. Petar Milanovic; see also Milanovic, 2004.)

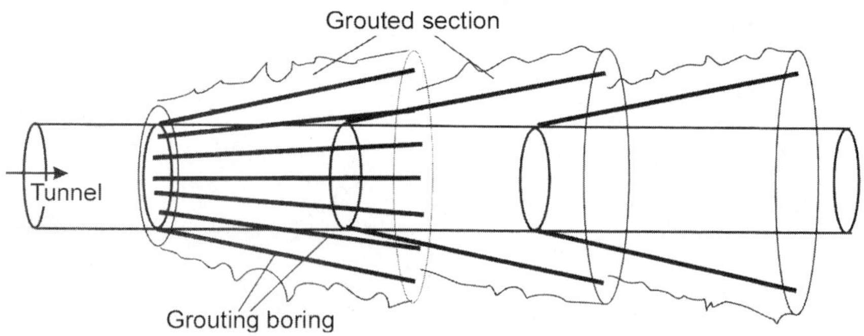

Figure 10.38 Concept of conical grouting ahead of tunnel when pilot borings encounter presence of karstified zone filled with water under high pressure. (Courtesy of Dr. Petar Milanovic; see also Milanovic, 2004.)

breaching a large cavity (cavern) filled with water. It is therefore mandatory to conduct comprehensive hydrogeologic and geotechnical investigations prior to any invasive activity in karst that creates open space in the subsurface. It is equally important to continue with targeted investigations and monitoring through completion of the specific project or activity. This includes continuous monitoring of rainfall and flows of any water (subsurface and subsurface) in the area, as well as geophysical surveying and drilling ahead of the advancing excavation. Only when sufficient field information and conceptual model of karst development in the wider area of the site are available, it would be possible to reduce the risks and design adequate drainage measures or adjust the existing drainage if needed. In any case, the project and construction teams should always be prepared for unexpected even when every aspect of the work is performed by the textbook.

As emphasized by Marinos (2005), groundwater is the main source of problems in tunnel construction and other underground works in karst associated with stability and safety issues. Groundwater control during construction and later during operation of the tunnel is one of the most challenging tasks faced by designers and contractors. Water can affect roof and face stability and, when in appreciable quantity, it will impede construction. The crossing of voids and caverns, whether empty, aquiferous, or filled with unconsolidated and erodible material, often requires site-specific solutions and real-time engineering.

Figure 10.39 Dr John Beck descending Bateman's Shaft that gives access to an eighteenth century lead mine drainage level "Lathkilldale Sough" (sough is an Derbyshire term for a mine drainage level). The level captures up to 220 L/s of flow from a surface river that, as a consequence, is now dry for several months of the year. As the river would otherwise be of high ecological value, efforts are being made to return flow to the surface. (Courtesy of Dr. John Gunn.)

Figures 10.35 through 10.38 illustrate some of the more common situations encountered during construction of tunnels, adits, and underground mining works in karst. There are many other aspects of safe design and implementation of such projects that directly or indirectly relate to water. It is therefore highly recommended that hydrogeologists working as part of the project team thoroughly familiarize themselves with key geotechnical and civil engineering principles of underground works in karst. Milanovic (2004) and Marinos (2005) provide very detail-related discussions illustrated with numerous examples from around the world.

Temporary or permanent drainage systems that facilitate tunneling, quarrying, and mining can have various local and regional impacts caused by lowering of the hydraulic head in the aquifer. This includes decreasing flows or drying of springs and surface water features, and formation of sinkholes. One potential benefit from dewatering works is that unwanted water may become a precious resource to various users in the affected area.

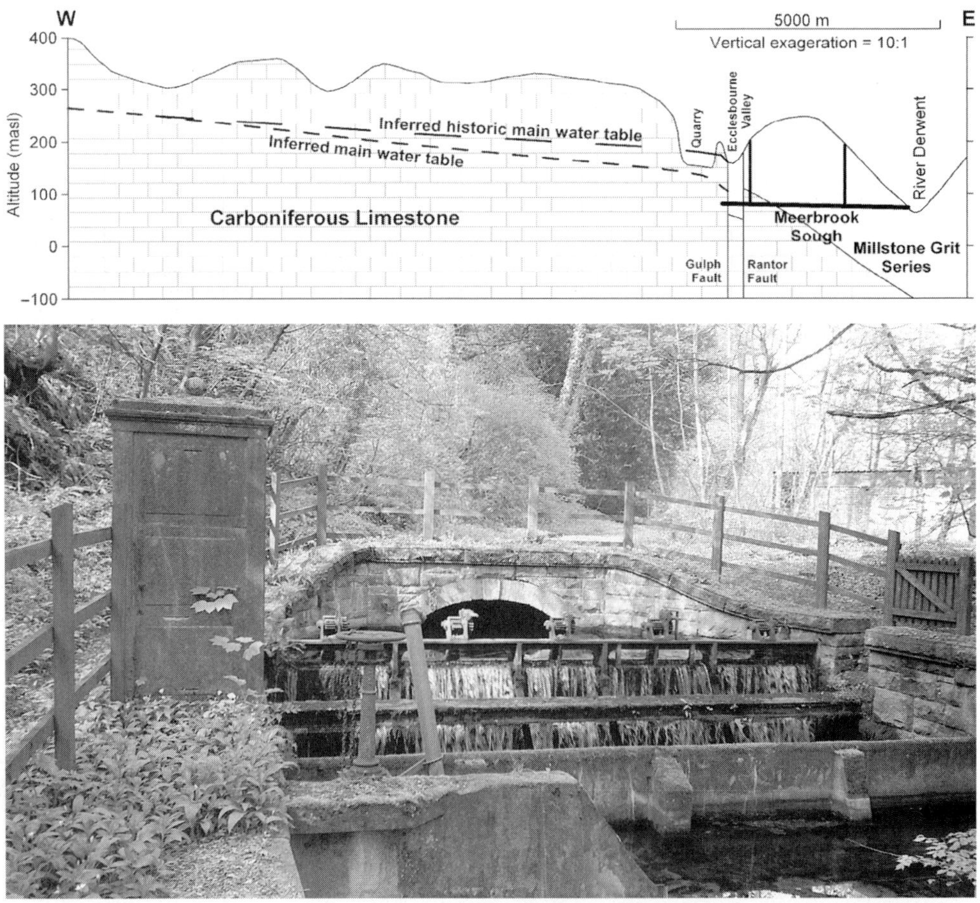

FIGURE 10.40 The "tail" (exit) of Meerbrook Sough, a lead mine drainage level that is now the largest public water supply groundwater source in the English Midlands with a licensed volume of 65 ML/day. (Courtesy of Martin Shepley, AMEC Environment and Infrastructure, Inc.)

This aspect of planned, ongoing, or completed projects should always be taken into consideration (see Figs. 10.39 and 10.40).

An illustrative review of various environmental impacts of quarrying in limestone including surface water and groundwater aspects is provided by Langer (2001).

10.5 References

Aller, L., Bennett, T., Lehr, J. H., and Petty, R. J., 1985. DRASTIC—A standardized system for evaluating ground water pollution potential using hydrogeologic settings. U.S. Environmental Protection Agency, Robert S. Kerr Environmental Research Laboratory, Ada, Oklahoma EPA/600/2–85/018, 163p.

Andersen, L. J., and Gosk, E., 1989. Applicability of vulnerability maps. *Environ Geol Water Sci*, 13(1): 39–43.

Arthur, J. D., Dabous, A. A., and Cowart, J. B., 2002. Mobilization of arsenic and other trace elements during aquifer storage and recovery, southwest Florida. In: Aiken, G. R., and Kuniansky, E. L. (eds.), *U.S. Geological Survey Artificial Recharge Workshop Proceedings, April 2–4, 2002*. Sacramento, CA. Geological Survey Open-File Report 02–89, pp. 47–50.

Banks, W. S. L., and Battigelli, D. A., 2002. Occurrence and distribution of microbiological contamination and enteric viruses in shallow ground water in Baltimore and Harford Counties, Maryland. U.S. Geological Survey Water-Resources Investigations Report 01–4216, Baltimore, MD, 39p.

Barr, T. C., 1985. Cave life of Kentucky. In: Dougherty, P.H. (ed.), *Caves and Karst of Kentucky, Kentucky Geological Survey Special Publication 12*. Series XI, Lexington, KY, pp. 146–175.

Beck, B. F., and Sinclair, W. C., 1986. Sinkholes in Florida: An Introduction. The Florida Sinkhole Research Institute, Report 85-86-4, Orlando, FL, 16p.

Böhlke, J-K, 2002. Groundwater recharge and agricultural contamination. *Hydrogeol J*, 10(1): 153–179.

Boyer, D. G., and, E Kuczynska, E., 2003. Storm and seasonal distributions of fecal coliforms and Cryptosporidium in a spring. *Journal of the American Water Resources Association* 39 (6), pp. 1449-1456.

Broyère, S., Jeannin, P.Y., and Dassargues, A., et al. 2001. Evaluation and validation of vulnerability concepts using a physically based approach. 7th Conference on Limestone Hydrology and Fissured Media, Besançon, Sep. 20–22, 2001. *Sci Tech Envir Mém. H. S.*, 13: 67–72.

Brune, G., 1975. *Major and Historical Springs of Texas*. Texas Water Development Board, Report 189, Austin, TX, 95p.

Buss, S. R., Rivett, M. O., Morgan, P., and Bemment, C. D., 2005. Attenuation of nitrate in the sub-surface environment. Science Report SC030155/SR2, Environment Agency. Bristol, England, 100p.

Chess, D. L. 1987. *Comparisons of Microbiology and Selected Anions for Surface and Subsurface Stream Waters for the Aqua Spring Watershed of Burnsville Cove*. Virginia. M.S. Thesis, The Pennsylvania State University.

Cohen, R. M., and Mercer, J. W., 1993. DNAPL site evaluation. C.K. Smoley, Boca Raton, FL, various paging.

Colborn, T., Dumanoski, D., and Meyers, J-P., 1997. Our stolen future: Are we threatening our fertility, intelligence and survival? A scientific detective story. Plume, Penguin Group, New York, 336p.

Craun, G. F., 1986. *Waterborne Diseases in the United States*. CRC Press, Boca Raton, FL, 192p.

Currens, J. C., 1999. Mass flux of agricultural nonpoint-source pollutants in a conduit-flow-dominated karst aquifer, *Logan County, Kentucky. Report of Investigations 1, Series XII, Kentucky Geological Survey*. University of Kentucky, Lexington, Kentucky, 151p + 2 plates.

Daughton, C. G., 2007. Pharmaceuticals and Personal Care Products (PPCPs) as environmental pollutants. National Exposure Research Laboratory, Office of Research and Development, Environmental Protection Agency, Las Vegas, Nevada. Presentation available at: http://www.epa.gov/nerlesd1/chemistry/pharma/index.htm. Accessed in 5th May 2007.

Doerfliger, N., and Zwahlen, F., 1998. Practical Guide, Groundwater Vulnerability Mapping in Karstic Regions (EPIK). Swiss Agency for the Environment, Forests and Landscape (SAEFL), Bern, 56p.

Doerfliger, N., Jeanin, P. Y., and Zwahlen, F., 1999. Water vulnerability assessment in karst environments: A new method of defining protection areas using a multi-attribute approach and GIS tools (EPIK method). *Environ Geol*, 39(2): 165–176.

Drew, D., and Hoetzl, H. (eds.), 1999a. Karst Hydrogeology and Human Activities, International Contributions to Hydrogeology Volume 20, International Association of Hydrogeologists, Balkema, Rotterdam, The Netherlands, 322p.

Drew, D., and Hötzl, H., 1999b. Conservation of karst terrains and karst waters: The future. In: Drew, D., and Hotzl, H. (eds.), *Karst Hydrogeology and Human Activities. Impacts, Consequences and Implications.*

International Contributions to Hydrogeology Vol. 20, International Association of Hydrogeologists, Balkema, Rotterdam, pp. 275–280.

Eckhardt, G., 2012. Does the Edwards act as a filter? Available at: http://www.edwardsaquifer.net. geology.html#quality. Accessed in February 2012.

Einsiedl, F., Radke, R. and Maloszewski, P., 2010. Occurrence and transport of pharmaceuticals in a karst groundwater system affected by domestic wastewater treatment plants. *J Contam Hydrol*, 117(1–4): 26–36.

Environment Agency, 2007. Underground, under threat. The state of groundwater in England and Wales. Environment Agency, Almondsburry, Bristol, 23p. Available at: http://www.environment-agency.gov.uk/

Focazio, M. J., Reilly, T. E., Rupert, M. G., and Helsel, D. R., 2002. Assessing ground-water vulnerability to contamination: providing scientifically defensible information for decision makers. U.S. Geological Survey Circular 1224, Reston, Virginia 33p.

Foster, S., Garduño, H., Kemper, K., Tiunhof, A., Nanni, M., and Dumars, C., 2002–2005. Groundwater quality protection; defining strategy and setting priorities. Sustainable Groundwater Management; Concepts & Tools, Briefing Note Series Note 8, The Global Water Partnership, The World Bank, Washington, D.C., 6p. Available at: www.worldbank.org/gwmate.

Franchi, A., 2009. Groundwater treatment. In: Kresic, N., *Groundwater Resources: Sustainability, Management, and Restoration*. McGraw-Hill, New York, pp. 437–482.

Fuller, M. L., 1910. Underground waters for farm use. U.S. Geological Survey Water-Supply Paper 255, Washington, D.C.

Gerba, C. P., 1988. Virus survival and transport in groundwater. *J Indust Microbiol Biotechnol*, 22(4): 247–251.

Gogu, R.C., 2000. *Advances in Groundwater Protection Strategy Using Vulnerability Mapping and Hydrogeological GIS Databases*. Ph.D. Thesis, University of Liège, Liège, France, 153p.

Goldscheider, N., 2002. *Hydrogeology and Vulnerability of Karst Systems—Examples from the Northern Alps and the Swabian Alb*. Ph.D Dissertation, University of Karlsruhe, Germany, 236p.

Goldscheider, N., Drew, D., and Worthington, S., 2007. Introduction. In: Goldscheider, N., and Drew, D. (eds.), *Methods in karst hydrogeology. International Contributions to Hydrogeology 26, International Association of Hydrogeologists*, Taylor & Francis, London, pp. 1–8.

Günay, G., and Milanovic, P., 2005. Karst engineering studies at the Akköprü Reservoir area, SW of Turkey. In: Stevanovic, Z., and Milanovic, P. (eds.), *Water Resources and Environmental Problems in Karst, Proceedings of the International Conference and Field Seminars*, Belgrade & Kotor, Serbia & Montenegro, 13–19 September 2005, Institute of Hydrogeology, University of Belgrade, Belgrade, pp. 651–658.

Hauwert, N. M., and Vickers, S., 1994. Barton Springs/Edwards Aquifer: Hydrogeology and groundwater quality. Barton Springs/Edwards Aquifer Conservation District, Austin, Texas, 91p. + Appendices.

Hinrichsen, D., Robey, B., and Upadhyay, U. D., 1997. Solutions for a water-short world. Population reports. Series M, No. 14., *Johns Hopkins School of Public Health, Population Information Program*. Baltimore, MD; Available at: http://www.infoforhealth.org/pr/m14edsum.shtml

Hötzl, H., 1996. Grundwasserschutz in Karstgebieten. *Grundwasser*, 1(1): 5–11.

Hötzl, H., Adams, B., Aldwell, R., et al. 1995. Regulations. In: COST 65: Hydrogeological aspects of groundwater protection in karstic areas, Final report (COST action 65). *European Commission, Directorate-General XII Science, Research and Development, Report EUR 16547 EN*. Brussels, Luxemburg, pp. 403–434.

Kraemer, S. R., Haitjema, H. M., and Kelson, V. A., 2005. Working with WhAEM2000; Capture zone delineation for a city wellfield in a valley fill glacial outwash aquifer supporting wellhead protection. EPA/600/R-00/022, United States Environmental Protection Agency, Office of Research and Development, Washington, D.C., 77p.

Kresic, N., 2007a. *Hydrogeology and Groundwater Modeling, Second edition*. CRC Press, Taylor & Francis Group, Boca Raton, FL, 807p.

Kresic, N., 2007b. Hydraulic methods. In: Goldscheider, N., and Drew, D. (eds.), *Methods in Karst Hydrogeology. International Contributions to Hydrogeology 26, International Association of Hydrogeologists*. Taylor & Francis, London, pp. 65–92.

Kresic, N., 2009. *Groundwater Resources: Sustainability, Management, and Restoration*. McGraw-Hill, New York, 852p.

Kresic, N., and Fotouhi, F., 2010. Springwater treatment. In: Kresic, N., and Stevanovic, Z. (eds.), *Groundwater Hydrology of Springs: Engineering, Theory, Management and Sustainability*. Elsevier, New York, pp. 269–304.

Kresic, N., Papic, P. and Golubovic, R., 1992. Elements of groundwater protection in karst environment. *J Environ Geol Water Sci*, 20(3): 157–164, Springer-Verlag, New York

Langer, W. H., 2001. Potential environmental impacts of quarrying stone in karst— A literature review. U.S. Geological Survey Open-File Report of–01–0484, 34p.

Laroche, E., Petit, F., Fournier, M., and Pawlak, B., 2010. Transport of antibiotic-resistant Escherichia coli in a public rural karst water supply. *J Hydrol*, 392(1–2): 12–21.

Lee, F. F., and Jones-Lee, A., 1993. Public health significance of waterborne pathogens. Report to California Environmental Protection Agency Comparative Risk Project, 22p.

Leibundgut, C., 1998. Vulnerability of karst aquifers (keynote paper). *IAHS Publ*, 247: 45–60.

Lindsey, B. D., Berndt, M. P., Katz, B. G., Ardis, A. F., and Skach, K. A., 2009. Factors affecting water quality in selected carbonate aquifers in the United States, 1993–2005. U.S. Geological Survey Scientific Investigations Report 2008–5240, Reston, VA, 117p.

Lopes, T.J., and Furlong, E.T., 2001. Occurrence and potential adverse effects of semivolatile organic compounds in streambed sediment, United States, 1992–1995. *Environ Toxicol Chem*, 20(4): 727–737.

Magiera, P., 2000. Methoden zur Abschätzung der Verschmutzungsempfindlichkeit des Grundwassers. *Grundwasser*, 3(2000): 103–114.

Marinos, P. G., 2005. Experiences in tunneling through karstic rock. In: Stevanovic, Z., and Milanovic, P. (eds.), *Water Resources and Environmental Problems in Karst, Proceedings of the International Conference and Field Seminars*. Belgrade & Kotor, Serbia & Montenegro, September 13–19, 2005, Institute of Hydrogeology, University of Belgrade, Belgrade, pp. 617–644.

Masters, R. W., Verstraeten, I. M., and Heberer, T., 2004. Fate and transport of pharmaceuticals and endocrine disrupting compounds during ground water recharge. *Ground Water Monitoring & Remediation*, Special Issue: Fate and transport of pharmaceuticals and endocrine disrupting compounds during ground water recharge, pp. 54–57.

Meinzer, O. E., 1927. Large springs in the United States. U.S. Geological Survey Water-Supply Paper 557, Washington, D.C., 94p.

Milanovic, P. T., 2004. *Water Resources Engineering in Karst*. CRC Press, Boca Raton, FL, 312p.

Nowell, L., and Capel, P., 2003. Semivolatile organic compounds (SVOC) in bed sediment from United States rivers and streams: Summary statistics; preliminary results from Cycle I of the National Water Quality Assessment Program (NAWQA), 1992–2001. Provisional data—subject to revision. Available at: http://ca.water.usgs.gov/pnsp/svoc/SVOC-SED_2001_Text.html

Panno, S. V., and Kelly, W. R., 2004. Nitrate and herbicide loading in two groundwater basins of Illinois' sinkhole plain. *J Hydrol*, 290(3–4): 229–242.

Papic, P., Kresic, N., and Golubovic, R., 1991. Acid rains and their influence on the quality of Petnica karstic spring water. Académie Serbe des sciences et des arts, Éditions speciales, Vol. DCXIV, Classe des sciences naturelles et mathématiques, Belgrade, pp. 95–105.

Peterson, E. W., Davis, R. K., Brahana, J. V., and Orndorff, H. A., 2002. Movement of nitrate through regolith covered karst terrane, northwest Arkansas. *J Hydrol*, 256(1–2): 35–47.

Porter, W. P., Jaeger, J. W., and Carlson, I. H., 1999. Endocrine, immune and behavioral effects of aldicarb (carbamate), atrazine (triazine) and nitrate (fertilizer) mixtures at groundwater concentrations. *Toxicol Ind Health*, 15: 133–150.

Romanov, D., Gabrovšek, F., and Dreybrodt, W., 2003. Dam sites in soluble rocks: A model of increasing leakage by dissolutional widening of fractures beneath a dam. *Eng Geol*, 70(1–2): 17–35.

Rupert, M. G., 1999. Improvements to the DRASTIC ground-water vulnerability mapping method. U.S. Geological Survey Fact Sheet FS-066–99, 6p.

Rupert, M. G., 2001. Calibration of the DRASTIC ground water vulnerability mapping method. *Ground Water*, 39(4): 625–630.

Schaefer, J. A., 2009. Risk evaluation of dams on karst foundations. *Proceedings of 29th Annual USSD Conference*, Nashville, TN, April 20–24, 2009, pp. 541–579.

Schindel, G., Johnson, S., Hoyt, J., Green, R. T., Alexander, E. C., and Krietler, C., 2009. *Hydrology of the Edwards Group: A Karst Aquifer Under Stress; A Field Trip Guide for the US EPA Groundwater Forum November 19*, 2009 San Antonio, TX, 57p.

Showcaves, 2012. Available at: http://www.showcaves.com/english/fr/karst/Pernod.html

Sowers, G. F., 1974. Foundation subsidence in soft limestones in tropical and subtropical environments. *Special Technical Publication G-7*. Law Engineering Testing Company, Atlanta, GA, various pagination.

Sowers, G. F., 1996. Building on sinkholes. *Design and Construction of Foundations on Karst Terrain*. American Society of Civil Engineers, ASCE Press, 202p.

Texas Administrative Code, 2006. Title 30 (Environmental Quality), Part 1 (Texas Commission on Environmental Quality), Chapter 330 (Municipal Solid Waste), Subhapter J (Groundwater Monitoring and Corrective Action), Rule 330.403 (Groundwater Monitoring Systems), effective March 27, 2006, 31 TexReg 2502.

The European Parliament and the Council of the European Union, 2000. Directive 2000/60/EC of the European Parliament and the Council of 23 October 2000 establishing a framework for community action in the field of water policy (EU Water Framework). *Official Journal of the European Union*, 22: L 327/1.

The European Parliament and the Council of the European Union, 2006. Directive 2006/118/EC on the protection of groundwater against pollution and deterioration. *Official Journal of the European Union*, 27: L 372/19–31.

Tihansky, A. B., 1999. Sinkholes, west-central Florida. In: Galloway, D., Jones, D. R., and Ingebritsen, S. E. (eds.), *Land Subsidence in the United States*. U.S. Geological Survey Circular 1182, Reston, VA pp. 121–140.

Trimmel, H. (ed.), 1998. *Die Karstlandschaften der Österreichischen Alpen und der Schutz ihres Lebensraumes und Ihrer Natürlichen Ressourcen*. Fachausschuß Karst, CIPRA-Österreich, Vienna, 119p.

Turkmen, S, Özgüler, E., Taga, H., and Karaogullarindan, T., 2002. Seepage problems in the karstic limestone foundation of the Kalecik Dam (south Turkey). *Eng Geol*, 63(3–4): 247–257.

USDA (United States Department of Agriculture), 2008. Sinkhole and sinkhole area treatment. Practice introduction. Natural Resources Conservation Service—Practice Code 527, 2p.

USEPA, 1993. A review of methods for assessing aquifer sensitivity and ground water vulnerability to pesticide contamination. U.S. Environmental Protection Agency, EPA/813/R–93/002, 147p.

USEPA, 2000. National Water Quality Inventory; 1998 Report to Congress; Ground water and drinking water chapters. EPA 816-R-00–013, Office of Water, Washington, D.C.

USEPA, 2003a. Atrazine interim reregistration eligibility decision (IRED), Q&A's—January 2003. Available at: http://www.epa.gov/pesticides/factsheets/atrazine.htm#q1. Accessed January 23, 2008.

USEPA, 2003b. Overview of the Clean Water Act and the Safe Drinking Water Act. Available at: http://www.epa.gov/OGWDW/dwa/electronic/ematerials/. Accessed in 18th September 2007.

USEPA, 2003c. An overview of the Safe Water Drinking Act. Available at: http://www.epa.gov/OGWDW/dwa/electronic/ematerials. Accessed in 18th September, 2007.

VA DCR (Virginia Department of Conservation and Recreation), 2008a. Natural Heritage Resources Fact Sheet. Karst Resources of the New River Basin. Virginia Karst Program, 10/08. Available at: http://www.dcr.virginia.gov/natural_heritage/karsthome.shtml

VA DCR (Virginia Department of Conservation and Recreation), 2012. Virginia Natural Heritage Kars Program. Available at: http://www.dcr.virginia.gov/natural_heritage/karsthome.shtml#

Vesper, D. J., Loop, C. M., and White, W. B., 2003. Contaminant transport in karst aquifers. *Speleogenesis and Evolution of Karst Aquifers*, 1(2); republished from *Theoretical and Applied Karstology*, 2001, 13–14: 101–111.

Vrba, J., and Zaporozec, A. (eds.), 1994. *Guidebook on mapping groundwater vulnerability*. International Contributions to Hydrogeology Vol. 16, International Association of Hydrogeologists (IAH), Swets & Zeitlinger Lisse, Munich, 156p.

Waltham, T., Bell, F., and Culshaw, M., 2004. *Sinkholes and Subsidence, Karst and Cavernous Rocks in Engineering and Construction*. Springer, Praxis Publishing, Chichester, UK, 300p.

WHG, 1996. Gesetz zur Ordnung des Wasserhaushalts (Wasserhaushaltsgesetz), BGBl, Bonn, Germany.

Wikipedia, 2007. Endocrine disruptor. Available at: http://en.wikipedia.org/wiki/Endocrine_disruptor. Accessed in 5th December 2007.

Worthington, S. R. H., Smart, C. C., and Ruland, W. W., 2003. Assessment of groundwater velocities to the municipal wells at Walkerton. Proceeding of the 2002 joint annual conference of the Canadian Geotechnical Society and the Canadian chapter of the IAH, Niagara Falls, Ontario, pp. 1081–1086.

Wyoming DEQ (Department of Environmental Quality), 2007. Wyoming's wellhead protection (WHP) program guidance document. Available at: http://www.wrds.uwyo.edu/wrds/deq/whp/whpcover.html. Accessed on November 21, 2007.

Zwahlen, F. (ed.), 2004. *COST Action 620. Vulnerability and Risk Mapping for the Protection of Carbonate (Karst) Aquifers, Final Report*. European Commission, Brussels, 297p.

CHAPTER 11

Restoration of Water in Karst

11.1 Introduction

The general debate on the definition and objectives of surface water and groundwater restoration is an ongoing one. Its nature is often, and somewhat misleadingly, simplified to emphasize the diverging interests of water "polluters" and "protectors," with the public caught somewhere in between. In karst, surface water and groundwater are inseparable and, in many cases, the complexity of their contamination and the feasibility of restoration do not support the above simplification. In addition, there are many different opinions at various regulatory levels (local, state, and federal) as to the objectives and goals of groundwater restoration. The following excerpts are from a memorandum written by a regulator in the United States, referencing the United States Environmental Protection Agency (USEPA) discussion paper entitled *Ground Water Use, Value, and Vulnerability as Factors in Setting Cleanup Goals* (Ground Water Task Force (GWTF), 2007):

> I think there is general agreement that it would be very difficult to set a national policy covering the many diverse issues raised in this discussion paper and by the commenters. There is abundant literature and discussion developed by the scientific community and commercial development interests that ground water has value; and its use, protection, and cleanup need to be addressed to varying degrees. Ground water Use, Value, and Vulnerability (UVV) has been, and most likely will remain, a state and local government issue because it is intrinsically bound to land use (e.g., zoning) and development. These have been traditionally viewed as being under the authority of state and local governments. Dealing with UVV issues is a site-specific process that depends on a large number of local factors, both scientific and political.
>
> The role of USEPA should be largely one of educator to the regulators, the regulated community, and the public. This education would be more technical than policy oriented. EPA could also act as discussion facilitator and, if asked, mediator on UVV issues affecting a site and, *as far as possible*, allow state and local regulators to continue to make decisions regarding setting cleanup goals protective of existing and potential drinking water supplies. These decisions would involve input from EPA and other Federal interests as well as from an active public participation program. Informed decisions by local regulators are the most effective way to address UVV issues.
>
> Addressing these issues needs to be a dynamic interactive process and not just guidance documents and policy statements. There are certainly plenty of those. A comprehensive national UVV guidance would be hard to formulate and would be ignored by many. Any national policy should be general in nature recognizing that each State has its own UVV issues most of which require unique solutions. (Pierce, 2004)

The main recommendation to the USEPA by this state regulator is to set an overall tone of good environmental stewardship and adopt a national policy statement regarding groundwater goals for the agency. This statement should reflect the view that (1) no aquifer will be degraded and (2) any degraded aquifer ultimately needs to be restored to its natural condition.

The above view of a state regulator is far from being a lonely one, and many stakeholders determined to protect the environment have commended this and similar views. However, what makes restoration of some already contaminated aquifers to their natural condition difficult are the nature of contamination, and the nature of certain complex hydrogeologic environments such as karst. Unfortunately, the nature of contamination can only, to a limited extent, be controlled or mitigated at a local "land use (e.g., zoning) and development" level, whereas hydrogeologic conditions cannot be controlled at all. In some cases, in order to restore contaminated karst and other aquifers to their natural condition, the local socioeconomic structure, as well as the laws of the entire society, would have to be radically changed. Even then, it may take tens or hundreds of years for aquifers to return to their natural condition, assuming that the definition of natural conditions means the absence of any anthropogenic substances in groundwater.

Two examples of nonpoint source groundwater contamination that require both the highest level regulatory and the local land-use changes in order to restore aquifers to their natural condition are the use of pesticides and the use of fertilizers described earlier. The following description of the main findings of a study performed in Slovenia, courtesy of Mihael Bricelj and Barbara Čenčur Curk, has far-fetching implications on the current practices of land application and household waste disposal in karst including use of septic tanks.

As discussed by Bricelj and Čenčur Curk (2005), pathogen pollution is the most important problem in groundwater protection, especially in karst areas where springs and wells are used for water supply. Solute tracers, particularly fluorescent dyes such as uranine, are commonly used to characterize the fate and transport of solute contaminants, while colloidal tracers are appropriate to simulate the transport of microorganisms and contaminants that are attached to suspended colloids. Thus, colloidal tracers such as bacteriophages and fluorescent polystyrene microspheres are used to evaluate the pathogen risk for karst groundwater resources.

The purpose of the research was to study the infiltration and migration of health-hazardous human viruses, such as enteroviruses, in the unsaturated zone of fractured and karstified rock, since these rocks present important aquifers in Slovenia. As a possible model for the behavior of health-hazardous viruses, salmonella bacteriophage P22H5 (400 to 800 nm) and fluorescent microspheres (5 μm) were used. Phages have served as useful models for the behavior of human enteric viruses in water treatment processes because of their similarity to enteric viruses in structure, size, and resistance to inactivation (Hedberg and Osterholm, 1993).

The bacteriophage P22H5 is a virulent mutant that propagates in mouse typhoid fever bacteria *Salmonella typhimurium* and rarely occurs in waters (Seeley and Primrose, 1982). From previous tracing experiments (Bricelj, 2003), it is well known that coliphages are a common constituent of waters contaminated with fecal matter and, for this reason, are not a suitable tracer. The phage tracer P22H5 was injected into the unsaturated zone in a karstic area where no background of phages for its host bacteria was present.

The experimental field site Sinji Vrh presents a 340-m-long artificial research tunnel, 5 to 25 m below the surface (Fig. 11.1). The unsaturated fractured and karstified

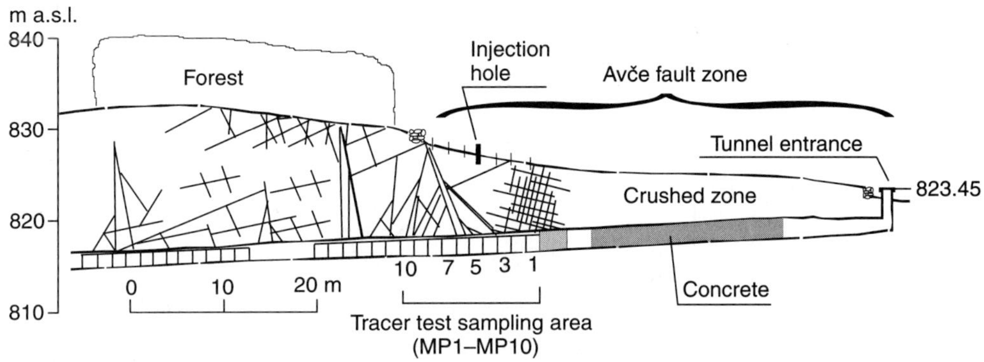

FIGURE 11.1 Longitudinal section of the experimental field site Sinji Vrh with tracer tests area and tracer test sampling points in the research tunnel. (Modified from Bricelj and Čenčur Curk, 2005.)

limestone has a negligible matrix porosity and very high fracture density with large conduits (Veselič and Čenčur Curk, 2001). For this study, the water seeping from the ceiling of the research tunnel was gathered in 1.5-m-long segments (MP1 to MP10) with a gathering surface of 2.2 m^2.

The phage tracer appeared first at sampling point 5 (MP5) after 4.1 days with the peak value of 3.1E + 9 plaque forming unit (pfu). One day later, a positive result was obtained at MP4 where the phage appeared with the peak value of 1.1E + 8 pfu. The tracer appeared at all sampling points within 22 days. Peak values occurred at the time of appearance only at MP2 and MP8. At MP3, the tracer appeared after 8 days, but the peak value was not reached until 50 days after the injection.

At the farthest sampling points from the injection hole (MP1, MP2, MP9, and MP10), the peak values were also within the lowest pfu values. The sampling of water was completed at MP4 after 324 days and at MP5 after 347 days. At the time of the last sample, there were still 4.2E + 2 pfu in MP4 and 9.8E + 2 pfu in MP5 in 1 mL of water sample. The recovery value was calculated for MP4 and MP5 and was less than 1% of the injected quantity at both sampling points (0.04% for MP4 and 0.91% for MP5). The results of appearance of the phage tracer at different sampling points are summarized in Table 11.1.

The results from the Sinji Vrh test site have shown that the unsaturated zone in the fractured and karstified rocks plays an important role in contaminant fate and transport including storage. The transfer of contaminants into deeper parts of the karst aquifer depends on the saturation rate of the soil and the unsaturated zone (precipitation events). In the test site area, fast conduits (large fractures or faults) exist in the unsaturated zone where water flows faster in the total conductive part of the rock, as in the case of MP4 and MP5. Tracer delay in microfracture system areas was also observed, especially at MP3, MP7, MP9, and MP10. At these points, the appearance of the peak value was delayed for 50, 31, 25, and 40 days, respectively. A very low tracer recovery rate is due to the dispersion of the tracer in directions where it could not be sampled, and the attenuation of the tracer dependent on mechanisms such as decay, filtration, sedimentation, and irreversible adsorption. After the injection of bacteriophages and microspheres, they remain in the fractures (channels) and microfracture systems of the unsaturated zone

Sampling Point	Appearance of Tracer (days)	Peak Value (pfu)	Appearance of Peak Value (days)
MP1	7	2.10E + 03	11.1
MP2	7	2.90E + 02	7
MP3	22.1	5.70E + 02	50
MP4	5	1.10E + 08	5
MP5	4.1	3.10E + 09	4.1
MP6	8	4.30E + 04	11.1
MP7	13.4	1.40E + 04	30.9
MP8	7	6.50E + 04	7
MP9	22.1	1.70E + 03	24.9
MP10	8	4.90E + 03	40

Note: The time is in days and quantity of phages in pfu mL^{-1}.
Modified from Bricelj and Čenčur Curk (2005).

Table 11.1 Appearance and Presence of the Phage Tracer at MP1 to MP10

and are rinsed by subsequent larger precipitation events several or more months after the injection (Fig. 11.2).

The GWTF, established in the fall of 2002 as part of the USEPA's *One Cleanup Program* to improve the planning and quality of USEPA cleanup programs dealing with brownfields, federal facilities, leaking underground storage tanks, and Resource Conservation and Recovery Act (RCRA) Corrective Action and Superfund, provides the following discussion on the importance of groundwater resources and their vulnerability (GWTF, 2007):

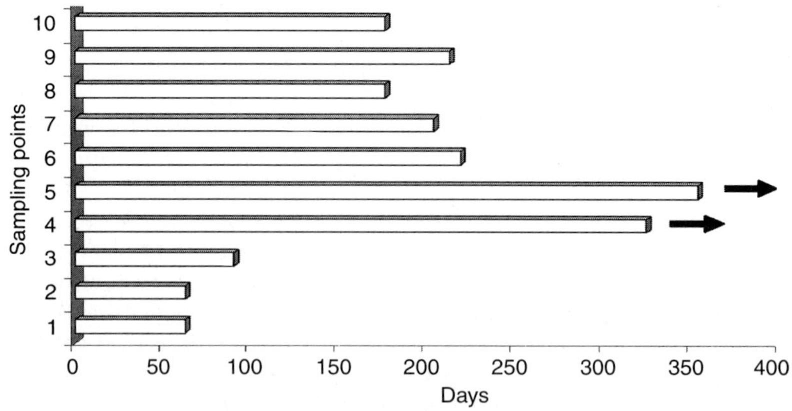

Figure 11.2 Presence of the phage tracer at sampling locations MP1 to MP10 after first appearance. *Arrows* denote continuing presence of phages at MP4 and MP5 on the last day of sampling. (Modified from Bricelj and Čenčur Curk, 2005.)

Ground water use typically refers to the current use(s) and functions of groundwater as well as future reasonably expected use(s). Groundwater use can generally be divided into drinking water, ecological, agricultural, industrial/commercial uses or functions, and recreational. Drinking water use includes both public supply and individual (household or domestic) water systems. Ecological use commonly refers to groundwater functions, such as providing base flow to surface water to support habitat; groundwater (most notably in karst settings) may also serve as an ecologic habitat in and of itself. Agricultural use generally refers to crop irrigation and live-stock watering. Industrial/commercial use refers to in any industrial process, such as for cooling water in manufacturing, or commercial uses, such as car wash facilities. Recreational use generally pertains to impacts on surface water caused by groundwater; however, groundwater in karst settings can be used for recreational purposes, such as cave diving. All of these uses and functions are considered "beneficial uses" of groundwater. Furthermore, within a range of reasonably expected uses and functions, the maximum (or highest) beneficial groundwater use refers to the use or function that warrants the most stringent groundwater cleanup levels.

Groundwater value is typically considered in three ways: for its current uses; for its future or reasonably expected uses; and for its intrinsic value. Current use value depends to a large part on need. Groundwater is more valuable where it is the only source of water, where it is less costly than treating and distributing surface water, or where it supports ecological habitat. Current use value can also consider the "costs" associated with impacts from contaminated groundwater on surrounding media (e.g., underlying drinking water aquifers, overlying air—particularly indoor air, and adjacent surface water). Future or reasonably expected values refer to the value people place on groundwater they expect to use in the future; the value will depend on the particular expected use or uses (e.g., drinking water, industrial). Society places an intrinsic value on groundwater, which is distinct from economic value. Intrinsic value refers to the value people place on just knowing clean groundwater exists and will be available for future generations, irrespective of current or expected uses. While the value of groundwater is often difficult to quantify, it will certainly increase as the expense of treating surface water increases, and as existing surface water and groundwater supplies reach capacity with continuing development.

Groundwater vulnerability refers to the relative ease with which a contaminant introduced into the environment can negatively impact groundwater quality and/or quantity. Vulnerability depends to a large extent upon local conditions including, for example, hydrogeology, contaminant properties, size or volume of a release, and location of the source of contamination. Shallow groundwater is generally more vulnerable than deep groundwater. Private (domestic) water supplies can be particularly vulnerable because (1) they are generally shallower than public water supplies, (2) regulatory agencies generally require little or no monitoring or testing for these wells, and (3) homeowners may be unaware of contamination unless there is a taste or odor problem.

As mentioned earlier in Chap. 10 in the example of atrazine, USEPA did not even consider private wells used for drinking water supply when allowing the continuing use of this pesticide (USEPA, 2003a). A similar thought-provoking topic discussed by Kresic and Mikszewski (2012) is the current regulatory policy related to arsenic (As) in private drinking water supplies in eastern New England. Arsenic occurs naturally in metasedimentary bedrock units in the regions that are extensively tapped by private water supply wells. In 2003, it was estimated that more than 100,000 people across eastern New England were using private water supplies with arsenic concentrations above the federal maximum contaminant level (MCL) of 10 μg/L (Ayotte et al., 2003). This represents a widespread exposure to a chemical universally acknowledged as poisonous.

In contrast to the policies of allowing atrazine and arsenic exposure, and not fully considering impacts of individual septic tank systems that may be quite serious as demonstrated by the study from Slovenia, environmental regulations require the expenditure of millions and hundreds of millions of dollars to remediate Superfund and state-led contaminated sites where the exposure often constitutes a very low risk (e.g., one in a million excess lifetime cancer risk), or is hypothetical in nature (e.g., potential future consumption of groundwater). For example, at the Visalia Pole Yard Superfund site, more than $30 million were spent to remediate groundwater contamination that was not posing an actual risk (a more detailed discussion is provided further in this chapter). This is a classic example of policy that permits self-inflicted risk while disproportionately targeting externally inflicted risk, ignoring the relative costs and benefits of the overall outcome. One potential declaration of this ideology is presented below (modified from Kresic and Mikszewski, 2012):

> When protecting human health and the environment, it is not our place to address risk related to naturally-occurring contamination, societal or individual lifestyle choices, but we will act aggressively to remedy any minimal level of risk caused by a third-party agent.

In other words, when point-source groundwater contamination is caused by known potentially responsible parties (PRPs) whose pockets are much deeper than those of individual households, the approaches to groundwater remediation would in many cases depend on the prevailing interpretation of the existing regulations at the local and state levels, and much less on the cost-benefit and risk analyses. As illustrated earlier, Georgia and quite a few other states in the United States, have adopted a zero tolerance for groundwater degradation by large polluters such as various industries and military installations; these PRPs are consequently required to restore their portions of contaminated aquifers to pristine natural conditions, often regardless of the underlying hydrogeologic characteristics, the risks, and the associated costs. Following the varying trends in groundwater restoration by individual states, different USEPA regions have also adopted varying strategies. For example, some regions have not yet approved a single technical impracticability (TI) waiver (see Sec. 11.3.2), whereas some regions have taken a more pragmatic approach to the feasibility of groundwater restoration in complex hydrogeologic environments such as karst by approving TI waivers at few groundwater contamination sites where the USEPA is the leading regulatory agency for cleanup.

It has been estimated that there are more than 20,000 megasites in the European countries, requiring extensive remediation (cleanup) of contaminated soils and groundwater, with the projected costs of tens to hundreds of billions U.S. dollars (Rügner and Bittens, 2006). At the vast majority of these sites, the cleanup efforts have not started yet as the regulatory agencies and various stakeholders are trying to develop restoration approaches that would balance risks to human health and the environment with socioeconomic realities. One reason for this strategy is that many of the megasites, legacy of the cold war and past industrial growth, are also owned by the governments, and the remediation costs would have to be paid by all citizens. In addition to the complex megasites, there are many tens of thousands of other sites with similar problems. For example, it is estimated that 14,000 sites in the small country of Denmark alone are contaminated from various chemical spills (Danish EPA, 1999) where typical contaminants include chlorinated solvents, gasoline/fuels, coal tar, and creosote (Gudbjerg, 2003). A more recent factor in developing pragmatic and realistic groundwater restoration strategies in the European countries is the complex question of carbon footprint, energy consumption, climate change, and the ongoing economic crisis.

In contrast to the European practice, because of the Superfund program (officially called Comprehensive Environmental Response, Compensation, and Liability Act (CERCLA) of 1980) and similar state-led initiatives, groundwater remediation has become a lucrative field in the United States with innumerable consultants and technology vendors offering solutions for a wide variety of contaminants and geologic settings. The cleanup of hazardous waste sites with groundwater contamination is a major element of professional hydrogeology in the United States. It is likely that every professional hydrogeologist will work on a groundwater remediation project at some point in his or her career. As discussed by Kresic and Mikszewski (2012), to the general public, invoking the term Superfund conjures up images of burning rivers, rusted 55-gallon drums leaking fluorescent fluids on the ground, contaminated wells, and mutated aquatic life. While in many instances these stereotyped attributes (mutations aside) are in fact accurate, often the perception of contamination is just as important as the actual data defining the nature and extent of contamination, and the risks associated with any existing or potential exposures to the contamination. Similarly, the perception of environmental cleanup is often more important than rigorous quantification of the costs and true environmental, social, and economic benefits of complicated remediation projects (see Sec. 11.4). For example, there are many installed pump-and-treat groundwater remediation systems in the United States that have very significant energy requirements, with the associated costs rising steadily. There are also examples where such systems are extracting hundreds of millions of gallons of water annually, with only few tens of pounds or less of the actual groundwater contaminant (see Fig. 11.3). In addition, these large quantities of water, after treatment, are usually being discharged to surface streams or sewers without being used or returned back to the aquifer. It is therefore likely that, even in the United States, the

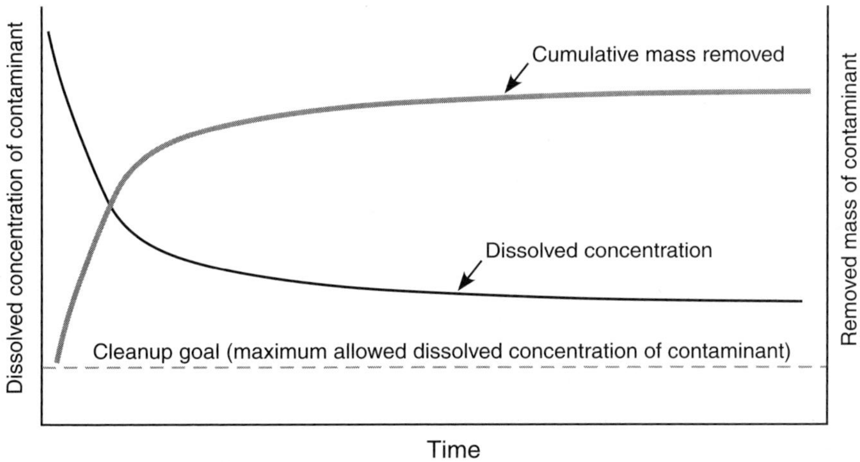

FIGURE 11.3 Typical graph illustrating long-term performance of pump and treat systems at contaminated sites. After substantial initial reduction of contaminant concentration and removal of contaminant mass, there is tailing of both metrics without achieving the cleanup goal (usually drinking water standard expressed as MCL or maximum contaminant level). The tailing is primarily result of slow release of contaminants from stagnant zones in porous media, back diffusion from the porous matrix, and slow dissolution of non-aqueous phase liquids (NAPLs) where present.

sustainability of certain groundwater remediation efforts will be increasingly questioned by various stakeholders, and not just by those that are paying the bill.

Water supply is usually a local responsibility and its costs are seldom fully covered. Most water utilities in both developed and developing countries rely heavily on subsidies and are far from being able to create their own capital. Since there is little political will to raise tariffs, and central governments are either unwilling or unable to provide financing, utilities are trapped in a vicious spiral of underspending on essential maintenance, resistance to fee increase, and inability to cope with groundwater contamination. In litigious societies such as the United States, this situation is fertile ground for the legal profession (attorneys). In societies that are more pragmatic or not so rich, central governments and water-sector agencies play key roles in the various disputes that arise from groundwater contamination. They also help both groundwater users and alleged polluters by creating a more flexible regulatory environment that emphasizes risk- and economy-based, sustainable approaches to the allocation of resources for surface water and groundwater restoration and management.

One possible roadmap to a more efficient and less confrontational approach to groundwater restoration is the establishment of a groundwater remediation trading system, similarly to the widely discussed carbon emissions trading. For example, large companies and parts of the government with seemingly deep pockets (such as departments of defense and energy), responsible for groundwater contamination at multiple sites, would be able to trade their existing and future remediation costs based on priorities developed by risk and cost-benefit analyses. This, of course, would include providing secure drinking water supplies to the affected groundwater users, thus eliminating risks to human health.

At the same time, some endemic groundwater contamination problems caused by small businesses unable to pay the full cost of groundwater cleanup, such as dry-cleaner shops, would be covered by common funds (this has already been practiced by some states in the United States). Such funds would also be used to support prevention of future groundwater contamination, including active protection of vulnerable groundwater recharge areas.

What this all may mean in an ideal world is that hundreds of millions of dollars would not be spent on a handful of sites where groundwater restoration to natural conditions may not be feasible within a reasonable timeframe or may be too costly for the given benefits. Rather, the funds and the resources would be spent at other sites where the cost-benefit ratio would be much more favorable. Common funds could also be used to fully protect valuable uncontaminated sources. Although this approach may seem farfetched given the complex regulatory and litigation framework in the United States, it may work elsewhere to the benefit of both the society and the environment.

11.2 Remediation Technologies

In general and regardless of the underlying hydrogeology, which may or may not limit their applicability, remediation technologies fall into one of the following categories (adapted from USEPA, 2010):

- *Source Treatment Remedies*—In situ or ex situ treatment of soil, sediment, sludge, and/or nonaqueous phase liquids (NAPLs) at or near the point of contaminant disposal. Example in situ technologies include soil vapor extraction (SVE), in situ thermal treatment, bioremediation, in situ chemical oxidation and flushing. Technologies are often combined; for example, in situ thermal treatment typically requires installation of SVE and multiphase extraction systems in addition to heating elements. Ex situ remediation primarily consists of soil excavation with subsequent treatment, incineration, and/or disposal.

- *Source Containment Remedies*—In situ containment of contaminated soil, sediment, sludge, or NAPL. Example technologies include landfill caps, barrier walls, and solidification/stabilization.
- *Groundwater Treatment Remedies*—In situ treatment of contaminants primarily existing in the dissolved aqueous phase at high concentrations within or immediately downgradient of the source area. Example technologies include air sparging, permeable reactive barriers (PRBs), bioremediation, and in situ chemical oxidation (the latter two technologies are also being used for source treatment).
- *Pump and Treat*—Extraction and ex situ treatment of contaminated groundwater to prevent further plume migration. Note that pump and treat was originally conceived to be an effective means of groundwater treatment; however, years of operational data demonstrate that it is better described as a plume containment remedy, with an ancillary benefit of enhanced groundwater flushing that results in a minor reduction in overall cleanup time.
- *Monitored Natural Attenuation (MNA)*—The reliance on natural attenuation factors, such as biodegradation, dissolution, dilution, sorption, and volatilization, to achieve remedial objectives within a reasonable time frame compared with that of other remedial technologies. MNA involves a detailed, controlled, and comprehensive monitoring system, and is often paired with source and/or groundwater treatment or containment to expedite site cleanup.

For a technical overview of each technology, the reader is referred to Appendix B of USEPA (2010) and Kresic (2009).

As explained later in Sec. 11.3.2, at the vast majority of contaminated karst sites in the United States, the involved regulatory agencies and consultants alike attempt their remediation in the same way as in any other, nonkarstic setting, by applying either a prescribed, cookie-cutter approach or the most popular current countrywide approach. Karst sites are therefore no different than all other sites in that, as we learn more and more over time about the behavior of contaminants in the subsurface and the efficacy of various technologies, the preferred technology from regulatory and commercial perspectives changes frequently. Trends in technology usage over time can be measured through public Record of Decision (ROD) documents produced by the USEPA for each Superfund site. As described in USEPA (2011a): "A ROD contains site history, site description, site characteristics, community participation, enforcement activities, past and present activities, contaminated media, the contaminants present, scope and role of response action and the remedy selected for cleanup."

As a result, RODs and other CERCLA decision documents [ROD amendments and Explanations of Significant Differences (ESDs)] provide informative snapshots at the time of the remediation marketplace, illustrating changes in the preferred remedial approaches

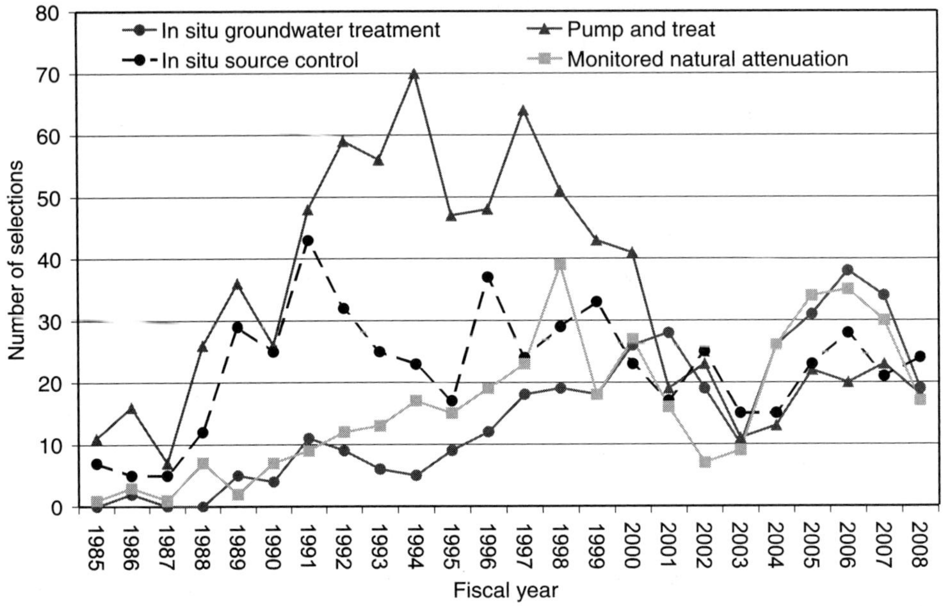

Figure 11.4 Graph illustrating total number of technology selections by fiscal year (FY) for the depicted remedy categories at Superfund sites. (From Kresic and Mikszewski, 2012; copyright Taylor & Francis, printed with permission.)

and individual technologies. Kresic and Mikszewski (2012) present a detailed analysis of the trends and reasons for changes in application of various remedial technologies. Several interesting observations are summarized below (see also Fig. 11.4):

- The use of pump and treat increased dramatically in the late 1980s and was the dominant remedial approach throughout the 1990s. Note that the early 1990s also represented the time frame with the greatest number of RODs published, peaking at 197 in 1991 (USEPA, 2010). The number of annual pump and treat selections decreased dramatically between 1997 and 2001, but the frequency of selection remained relatively constant throughout the 2000s. The late 1990s decline in pump and treat selections reflects an increased understanding of the technology's limited efficacy in terms of mass removal and cleanup time reduction, while the stabilization at approximately 20 selections per year in the 2000s reflects the ongoing need for the technology as a containment measure to prevent further migration of contaminant plumes.
- After peaking in 1991 and immediately declining, the annual number of in situ source control selections has remained relatively constant since 1993. The annual number of in situ source control remedies has closely paralleled that of pump and treat since 2001, and these two approaches are often used in tandem to reduce cleanup time frames.
- The use of in situ groundwater treatment and MNA steadily increased throughout the 1990s, and by the mid-2000s, these two technologies were selected more

often than in situ source control and pump and treat. Note that the reduced number of MNA selections between 2001 and 2003 was at least partially triggered by the 1999 publishing of USEPA guidance regarding the proper use of MNA (USEPA, 1999a). The rebound in 2004 was achieved after consultants and regulators alike adjusted to the new MNA framework and figured out how to best incorporate the technology.

In addition to the general increase in knowledge and understanding of physical, chemical, and biological processes, the increased selection of in situ groundwater treatment and MNA was driven by the following factors:

1. In situ groundwater treatment with technologies such as in situ chemical oxidation can remove significantly more contaminant mass than conventional pump and treat systems.
2. Cleanup time frames using MNA are often not remarkably different from those with pump and treat systems or those where source and/or groundwater treatment have reached a point of diminishing returns.
3. MNA is often perceived by PRP groups as a significantly less costly alternative than active remediation. Note that this is not always the case as MNA requires extensive monitoring over long periods of time.
4. PRP groups with experience operating pump and treat systems for years and years with no end in sight became more receptive to innovative technologies with higher capital costs to avoid perpetual operation of the existing or installation of additional pump and treat systems.

One complicating factor at many karst sites is that the karstified rock is covered by an uneven layer of residuum, which may be tens of feet thick. Since, in most cases, the contamination of the underlying karst aquifer had to come from the land surface and through the residuum, the unconsolidated residuum sediments are part of the problem as well. It is for this reason that various remedial technologies that do not work effectively in consolidated karsitified rocks are routinely applied at such sites to address the contaminated residuum. The following analysis by Kresic and Mikszewski (2012) is therefore generally applicable to all karst sites where the residuum is being aggressively remediated in hope that this will solve the contamination problem in the underlying karst aquifer.

SVE remains the dominant in situ source control option at all Superfund sites, but its use has decreased significantly since peaking in 1991. Aside from the reduced number of RODs, this decline was likely caused by an increased understanding of the limitations of ambient-temperature SVE, most notably the difficulty in treating contaminants with low volatility and those that reside deep below the water table, such as dense nonaqueous phase liquids (DNAPLs). SVE also has difficulty removing contaminants within low-permeability strata.

The annual number of bioremediation, multiphase extraction, and solidification/stabilization selections has remained relatively constant since the early 1990s; however, these technologies now comprise an increased percentage of in situ source control remedies. Over the course of the 2000s, thermal treatment has emerged as a more prevalent technology, rivaling the number of ambient temperature SVE selections in 2005. It is important to remember that thermal treatment almost always requires an SVE system, and the technology is therefore best thought of as "thermally-enhanced SVE." Therefore, even if the use of thermal treatment increases significantly in the coming years, the in situ source control marketplace can still be classified as SVE-centric.

Similar to SVE for source control, air sparging was the dominant groundwater technology during the 1990s. However, unlike SVE, air sparging appears to be somewhat obsolete, as its frequency of selection was greatly surpassed by both bioremediation and chemical treatment (most notably in situ chemical oxidation) in the 2000s. The limitations of air sparging are very similar to those of SVE; in fact, air sparging typically requires an SVE system to remove volatilized contaminants from the vadose zone, which is a limitation in its own right. It remains to be seen if the increasing trend toward in situ groundwater remedies continues into the new decade. As with air sparging and SVE, there may be a backlash against in situ chemical oxidation and bioremediation based on the increasing availability of performance data, which may not be as encouraging as originally anticipated.

Despite the continually evolving technological and political landscape, one thing that has remained constant in Superfund is the objective of restoring contaminant groundwater to its beneficial use. This beneficial use is often considered drinking water, as measured by MCLs, to ensure that a pristine groundwater resource is available to future generations. Unfortunately, as explained in the following chapter, the presence of NAPLs in the aquifer porous media poses great challenges for achieving this goal.

11.2.1 Nonaqueous Phase Liquids Problem

NAPLs are typically classified as either light nonaqueous phase liquids (LNAPLs), which have densities less than that of water, or dense nonaqueous phase liquids (DNAPLs), which have densities greater than that of water (see Table 11.2). As discussed in Sec. "Nonaqueous Phase Liquids" of Chap. 10, it is very important to make the distinction between the actual NAPL liquid in free phase, and the chemical of the same name dissolved in water. For example, most common organic contaminants, such as perchloroethylene (PCE), trichloroethylene (TCE), and benzene, can enter the subsurface as both pure NAPL as well as dissolved in percolating water. However, the fate and transport of pure NAPL and the same chemical dissolved in groundwater are quite different.

NAPLs may be present in the subsurface as continuous (contiguous) bodies of a relatively significant extent (volume), which occupy the pore space in the unconsolidated residuum material overlying a karst aquifer, or are present in fractures/conduits in the karst aquifer. In either case, they are referred to as free-phase NAPLs. As explained further, detecting free-phase DNAPLs in either the unconsolidated residuum or the karstified rock is in most cases impractical, and when it happens, it is often a matter of luck (Fig. 11.5). When this happens, DNAPLs are part of the contiguous invading bodies moving downward or have significant lateral extent in the form of a pool resting on low-permeable porous media (Fig. 11.6, *left*).

LNAPLs pools floating on the water table in saturated residuum are easier to detect; however, they too can surprise investigators assuming that all that is required to detect LNAPL is screening monitoring wells across the water table. Mobile LNAPL floating on the water table will move vertically as the groundwater elevation fluctuates. As the water table rises or falls, LNAPL will be retained in the soil pores, and it may leave behind a residual LNAPL "smear zone," or the entire pool may migrate vertically. A decline in groundwater elevation may trap the LNAPL pool below the water table when

IUPAC Name	Common or Alternative Name	Density
1,2,3-trichlorobenzene	1,2,6-trichlorobenzene	1.690
Tetrachloroethene	Perchloroethylene, tetrachloroethylene, PCE	1.623
Tetrachloromethane	Carbon tetrachloride	1.594
1,1,2-trichloroethene	1,1,2-trichloroethylene, TCE	1.464
1,2,4-trichlorobenzene	1,2,4-trichlorobenzol	1.450
1,1,2-trichloroethane	Methyl chloroform	1.440
1,1,1-trichloroethane	Methyl chloroform	1.339
1,2-dichlorobenzene	*o*-Dichlorobenzene	1.306
cis-1,2-dichloroethene	*cis*-1,2-dichloroethylene	1.284
trans-1,2-dichloroethene	*trans*-1,2-dichloroethylene	1.256
1,2-dichloroethane	1,2-ethylidene dichloride, glycol dichloride	1.235
1,1-dichloroethene	1,1-dichloroethylene, DCE	1.213
1,1-dichloroethane	1,1-ethylidene dichloride	1.176
Chlorobenzene	Monochlorobenzene	1.106
Pure water at 0°C		1.000
Naphthalene	Naphthene	0.997
Chloromethane	Methyl chloride	0.991
Chloroethane	Ethyl chloride	0.920
Chloroethene	Vinyl chloride, chloroethylene	0.910
Styrene	Vinyl benzene	0.906
1,2-dimethylbenzene	*o*-Xylene	0.880
Benzene	—	0.876
Ethylbenzene	—	0.867
1,2,4-trimethylbenzene	Pseudocumene	0.876
Methylbenzene	Toluene	0.867
1,3-dimethylbenzene	*m*-Xylene	0.864
1,4-dimethylbenzene	*p*-Xylene	0.861
2-methoxy-2-methylpropane	Methyl tert-butyl ether, MTBE	0.740

IUPAC, International Union of Pure and Applied Chemistry.
From Lawrence (2006).

TABLE 11.2 Density (in g/cm^3 at 20°C) of Selected Volatile Organic Compounds

groundwater elevations rise. For example, Oostrom et al. (2005) discuss the results of experiments and modeling, which show that viscous mobile LNAPL, subject to variable water table conditions, does not necessarily float on the water table and may not appear in an observation well. A similar situation may develop during product recovery efforts. LNAPL will flow toward a recovery well or trench in response to the gradient induced

FIGURE 11.5 *Top*: Typical highly weathered and karstified top portion of limestone beds covered with thin discontinuous residuum sediments, road cut near Knoxville, Tennessee. *Bottom*: Cavities at a deep dewatered construction site in Paleozoic limestone, Hartsville, Tennessee. (From the George Sowers collection, courtesy of AMEC Environment and Infrastructure, Inc.)

by water table depression. LNAPL may be retained below the water table, as the water table elevation returns to prepumping conditions (Newell et al., 1995).

At residual saturation, NAPLs in unconsolidated residuum sediments occur as disconnected single and multipore globules or ganglia (Fig. 11.6, *right*) within the larger pore spaces that have been cutoff and disconnected from the continuous NAPL body by the invading water (Cohen and Mercer, 1993; Pankow and Cherry, 1996). The actual saturation of pore space with NAPL can vary between zero and one, with the individual saturations of all fluids present (NAPLs and water) always summed to one in the saturated zone. Determining the percentage of NAPL saturation, even from the actual core samples, is difficult and requires application of complicated, laborious techniques and care in obtaining, preserving, and transporting the samples. In any case, numeric saturation data from discrete sampling points would have to be extrapolated and interpolated (contoured) in order to estimate the volume of DNAPL present in the corresponding

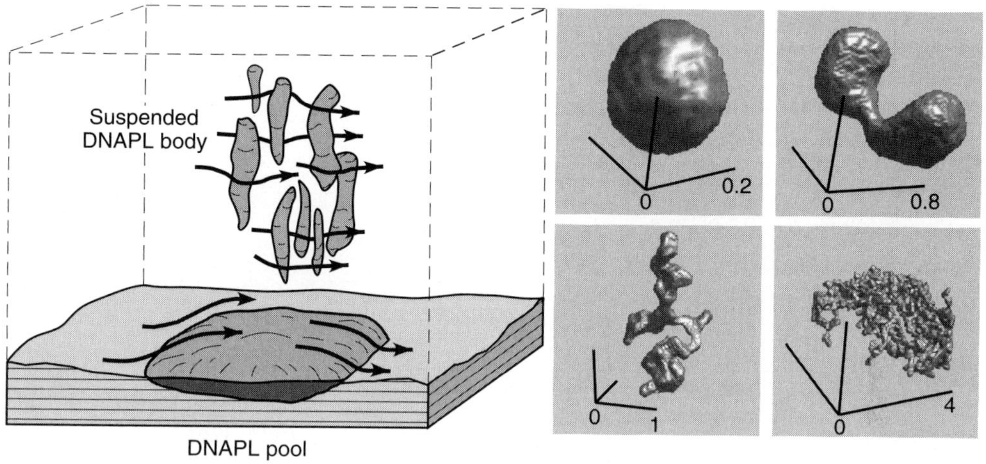

FIGURE 11.6 *Left*: DNAPL pool resting on an impermeable layer is exposed to the groundwater flux only at the top, while irregular, suspended, and disconnected DNAPL bodies have proportionally greater contact areas with the flowing groundwater and are dissolved (depleted) faster. (From Kresic, 2007; copyright Taylor & Francis, printed with permission.) *Right*: Electronic imaging of DNAPL blobs in porous media. Scale is in millimeters. (From Brusseau, 2005.)

volume of the aquifer. Cohen and Mercer (1993) present laboratory and field values of residual saturation for various NAPL fluids in intergranular porous media (clays, silts, sands, gravels, and their mixtures). They conclude that values of residual saturation generally range between 0.10 and 0.50 in saturated porous media, and tend to be higher in the preferential pathways of NAPL transport. Residual saturation also tends to increase with increasing pore aspect ratios and pore size heterogeneity, and with decreasing porosity, probably due to reduced pore connectivity and a decrease in mobile nonwetting fluid (NAPL) in smaller pore throats. Values of residual saturation in the vadose zone are generally smaller than in the saturated zone and range between 0.10 and 0.20. This is because NAPL drains more easily in the presence of air than in a water-saturated system. Residual saturation and retention capacity in the vadose zone increase with decreasing intrinsic permeability, effective porosity, and moisture content (Cohen and Mercer, 1993).

As discussed by Fountain (1998), below the water table, entry of DNAPL into water-filled pores requires overcoming an entry pressure resulting from interfacial tension between DNAPL and water. The required entry pressure increases with decreasing grain size (Table 11.3). The downward flow of DNAPL may therefore be interrupted each time a layer with a smaller grain size is encountered. DNAPL tends to flow laterally above the fine-grained layer, accumulating until there is sufficient thickness of DNAPL to overcome the entry pressure. Even very subtle variations in grain size distribution may produce significant deflection of DNAPL flow. This results in a series of horizontal lenses of DNAPL connected by narrow vertical pathways. As in the vadose zone, a small amount of DNAPL is retained as residual saturation in every pore through which it flows. If the DNAPL encounters a layer that has a sufficiently high entry pressure, the DNAPL will accumulate on the top of this layer, forming a "pool." Thus, DNAPL is typically found as multiple horizontal lenses, connected by sparse vertical pathways, with one

Porous Medium	Entry Pressure (cm)
Clean sand (K = 1×10^{-2} cm/s)	45
Silty sand (K = 1×10^{-4} cm/s)	286
Clay (K = 1×10^{-7} cm/s)	4,634
Fracture, 20 μ aperture	75
Fracture, 100 μ aperture	15
Fracture, 500 μ aperture	3

From Fountain (1998).

TABLE 11.3 Examples of DNAPL TCE Entry Pressure, in cm

or more pools above fine-grained layers (Fig. 11.7). Most of the horizontal lenses and vertical pathways will be at, or below, residual saturation; only pools will have higher saturations. The distinction between residual saturation and pools is important, since only the DNAPL in pools is expected to be mobile. However, changes in the interfacial tension between water and DNAPL, which are produced either by heat (all thermal methods) or by chemicals (surfactants or cosolvents), may mobilize DNAPL at residual saturation by reducing capillary forces.

One potential indication of the presence of NAPL in the saturated zone in a monitoring well screened in unconsolidated residuum sediments is the concentration of the dissolved contaminant greater than 1% to 10% of the compound's effective solubility (Feenstra and Cherry, 1988; Pankow and Cherry, 1996). One reasoning behind this widely accepted rule of thumb is that if NAPL is present, it will generally be present either as

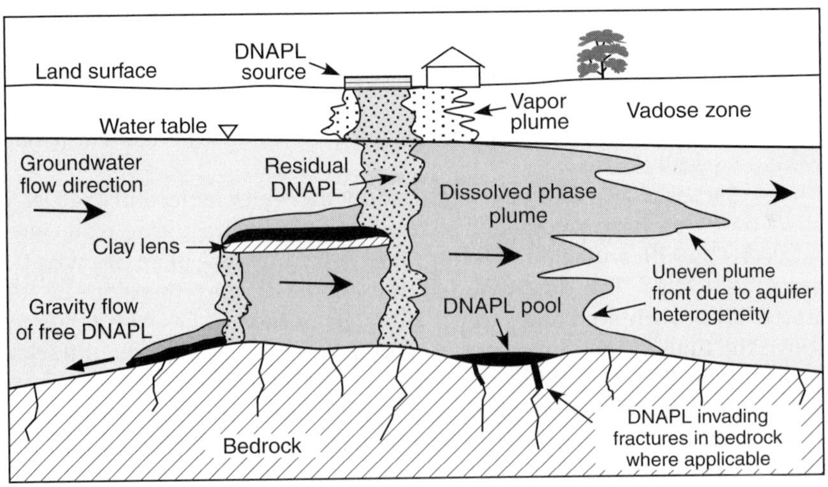

FIGURE 11.7 Schematic of possible migration pathways of free-phase DNAPL and the derived dissolved and vapor contaminant plumes. (From Kresic, 2007; copyright Taylor & Francis, printed with permission.)

a small lens in a small preferential pathway, as residual phase ganglia, or diffused from a preferential pathway into a fine-grained matrix. If a 10-ft well screen is close to or intersects one of these areas, the area where the NAPL is present will likely be thin when compared with the full length of the well screen. This is mainly because groundwater flow in unconsolidated sediments is generally laminar and will not mix quickly with the larger interval of the formation over short distances. As a consequence, the aqueous phase contamination dissolving from the NAPL into groundwater at a concentration close to its solubility limit will remain contained within a narrow (thin) interval some distance downgradient of the source zone. This contamination will be diluted in the monitoring well during sampling by the larger screened interval of the formation. Therefore, concentrations of a small percentage of solubility may indicate the presence of NAPL in the vicinity of the monitoring well. If well screens are short, there will be less dilution and the contaminant concentration will be a higher percentage of solubility before it indicates NAPL. This technique is subjective and must be applied very carefully because it may grossly overestimate or underestimate the presence and volumes of NAPL in the saturated zone if used alone. It should be considered only a part of the process used to determine if NAPL is present, not a method that by itself will indicate the presence or absence of NAPL (The Interstate Technology and Regulatory Council, ITRC, 2003). An informative discussion on the behavior and dissolved concentrations expected to be found in NAPL source zones in unconsolidated sediments is given by Anderson et al. (1987, 1992).

Another indirect method for detecting the potential presence of residual NAPL is to calculate the hypothetical pore-water concentration from the measured total soil concentration by assuming equilibrium chemical partitioning between the solid phase, the pore water, and the soil gas, and assuming that no NAPL is present in the collected sample. This pore-water concentration (C_w, in mg/L or μg/cm^3) can be expressed in terms of the total soil concentration (C_t, in μg/g dry weight) as (Pankow and Cherry, 1996)

$$C_w = \frac{C_t \rho_b}{K_d \rho_b + \theta_w + H_c \theta_a} \tag{11.1}$$

where ρ_b is the dry bulk density of the soil sample (g/cm^3); θ_w is the water-filled porosity (volume fraction); θ_a is the air-filled porosity (volume fraction); K_d is the partition coefficient between pore water and solids for the compound of interest (cm^3/g); H_c is the dimensionless Henry's gas law constant for the compound of interest (see Tables 11.4 and 11.5).

If no NAPL is present, there is a maximum amount of chemical, which can be contained in the soil sample at equilibrium with the soil pore water and air. In other words, for true aqueous solute (dissolved phase) in equilibrium, the calculated pore-water concentration (C_w) has to be equal to the solubility concentration (S_w) of the chemical: $C_w = S_w$. If the calculated pore-water concentration is higher than the solubility concentration, some NAPL phase of the chemical has to be present in the sample. Note that for an NAPL mixture, effective solubilities of individual compounds will be lower than their pure phase aqueous solubilities. Pankow and Cherry (1996) provide the following example of the application of Eq. (11.1):

- Measured TCE concentration (C_t) in the soil sample taken from the saturated zone (where θ_a equals zero) was 3,100 mg/kg, or 3,100 μg/g.

IUPAC Name	Common or Alternative Name	Water Solubility (mg/L) at 25°C	Henry's Law Constant (H)
2-methoxy-2-methylpropane	Methyl tert-butyl ether, MTBE	36,200	0.070
1,2-dichloroethane	1,2-ethylidene dichloride, glycol dichloride	8,600	0.140
Chloromethane	Methyl chloride	5,320	0.920
Chloroethane	Ethyl chloride, monochloroethane	6,710	1.11
cis-1,2-dichloroethene	*cis*-1,2-dichloroethylene, *cis*-1,2-DCE	6,400	0.460
1,1-dichloroethane	1,1-ethylidene dichloride	5,000	0.630
1,1,2-trichloroethane	Methyl chloroform	4,590	0.092
trans-1,2-dichloroethene	*trans*-1,2-dichloroethylene, *trans*-1,2-DCE	4,500	0.960
Chloroethene	Vinyl chloride, chloroethylene	2,700	2.68
1,1-dichloroethene	1,1-dichloroethylene, DCE	2,420	2.62
Benzene	—	1,780	0.557
1,1,1-trichloroethane	Methyl chloroform	1,290	1.76
1,1,2-trichloroethene	1,1,2-trichloroethylene, TCE	1,280	1.03
Tetrachloromethane	Carbon tetrachloride	1,200	2.99
Methylbenzene	Toluene	531	0.660
Chlorobenzene	Monochlorobenzene	495	0.320
Styrene	Vinyl benzene	321	0.286
Tetrachloroethene	Perchloroethylene, tetrachloroethylene, PCE	210	1.73
1,2-dimethylbenzene	*o*-Xylene	207	0.551
1,4-dimethylbenzene	*p*-Xylene	181	0.690
1,3-dimethylbenzene	*m*-Xylene	161	0.730
Ethylbenzene	—	161	0.843
1,2-dichlorobenzene	*o*-Dichlorobenzene	147	0.195
1,2,4-trimethylbenzene	Pseudocumene	57	0.524
1,2,4-trichlorobenzene	1,2,4-trichlorobenzol	37.9	0.277
Naphthalene	Naphthene	31.0	0.043
1,2,3-trichlorobenzene	1,2,6-trichlorobenzene	30.9	0.242

IUPAC, International Union of Pure and Applied Chemistry.
From Lawrence (2006).

Table 11.4 Aqueous Solubility (in mg/L at 25°C) and Henry's Law Constant (in kPa m^3/mol at 25°C) for Common Organic Contaminants

Chemical	Solubility[a] (mg/L)	Log K_{oc} (L/kg)	Vapor Pressure (mm Hg)	Henry's Constant (Dimensionless)
Benzene	1,780	1.5–2.2	76–95.2	0.22
Toluene	535	1.6–2.3	28.4	0.24
Ethylbenzene	161	2.0–3.0	9.5	0.35
m-Xylene	146	2.0–3.2	8.3	0.31
Ethanol	Miscible	0.20–1.21	49–56.5	0.00021–0.00026
Methanol	Miscible	0.44–0.92	121.6	0.00011
TBA	Miscible	1.57	40–42	0.00048–0.00059
MTBE	43,000–54,300	1.0–1.1	245–256	0.023–0.12
ETBE	26,000	1.0–2.2	152	0.11
TAME	20,000	1.3–2.2	68.3	0.052
DIPE	2,039–9,000	1.46–1.82	149–151	0.195–0.41

[a] Pure phase solubility; solubility in mixture is lower (Raoult's law).
From Nichols et al. (2000); copyright American Petroleum Institute, all rights reserved.

Table 11.5 Selected Properties of Some Common LNAPL Chemicals and Gasoline Additives

- The partition coefficient (K_d) for TCE was calculated from the fraction of organic carbon in the soil sample ($f_{oc} = 0.001$) and the so-called organic-carbon partition coefficient for TCE ($K_{oc} = 126$) as follows: $K_d = f_{oc} \times K_{oc} = 0.001 \times 126 = 0.126$.
- The bulk density (ρ_b) was estimated to be 1.86 g/cm^3.
- The total porosity, equal to water-filled porosity (θ_w), was estimated to be 0.3.

Inserting the above values into Eq. (11.1) gives the calculated value for the pore-water concentration of 10,790 mg/L, which is much higher than the TCE pure aqueous solubility of approximately 1,280 mg/L. The conclusion is that residual liquid TCE DNAPL is present in the sample.

Once LNAPLs or DNAPLs enter the karstified rock below the residuum, the whole narrative about their fate and transport, applied routinely at nonkarstic sites, is not valid any more. Informative and related discussions are presented by Wolfe et al. (1997), Loop and White (2001), and Vesper et al. (2003), with ample other literature from specialty groups and government agencies readily available (e.g., see ITRC, 2003 and 2004, for an overview).

As simply put by Vesper et al. (2003), LNAPLs will float on groundwater. This has led to some remarkable contamination problems such as a lake of gasoline nearly 2 m deep floating on the water table near Mechanicsburg, Pennsylvania (Rhindress, 1971). Conduits generally create a trough in the water table, so the LNAPL contamination in the conduit tributaries and in the fracture system tends to migrate toward the master conduit. During flood flow, the trough fills and conduits with free-face steams shift into a regime of pipe flow. Ponded LNAPL is lifted with the rising water and pressed against the ceiling. Any pockets in the ceiling will form traps for the LNAPL (Ewers et al., 1991). If the ceiling is tight, the LNAPL is forced through the obstructions as piston flow. A burst of LNAPL then continues down the conduit. Alternatively, if the ceiling is fractured, the piston flow drives the LNAPL upward where the vapors may rise into structures on the land surface. Fume problems in homes and other structures overlying conduit systems contaminated by LNAPLs may arise long after the initial spills (Stroud et al., 1986).

As emphasized by Vesper et al. (2003), many common light hydrocarbon compounds have substantial solubilities and vapor pressures (see Table 11.5). For accumulations of LNAPL ponded behind obstructions in conduits, the continuous sweep of fresh water beneath the pool will eventually dissolve and remove the pool. Likewise, the ponded LNAPL will gradually evaporate because of the vapor pressure of the compounds. In the process, the air-filled cave passages become contaminated with hydrocarbon fumes. The turnover in cave air due to barometric changes will eventually flush out the contaminants, but while they are present, they are a substantial hazard to cave explorers, especially those who explore by carbide lamp. One of the most serious incidents occurred in Howard's Cave, Georgia, in 1966 where a gasoline explosion set off by carbide lamps claimed three lives (Black, 1966).

The U.S. Geological Survey (USGS) has developed preliminary conceptual models emphasizing accumulation sites for chlorinated DNAPLs in karst aquifers (Wolfe et al., 1997). Although the models were developed for karst aquifers, which are hydrogeologically the most complex, one or more of them may be applicable for common situations found in other hydrogeologic settings. The mechanisms of DNAPL accumulation in karst settings illustrated in Fig. 11.8 include trapping in regolith, pooling at the top of bedrock,

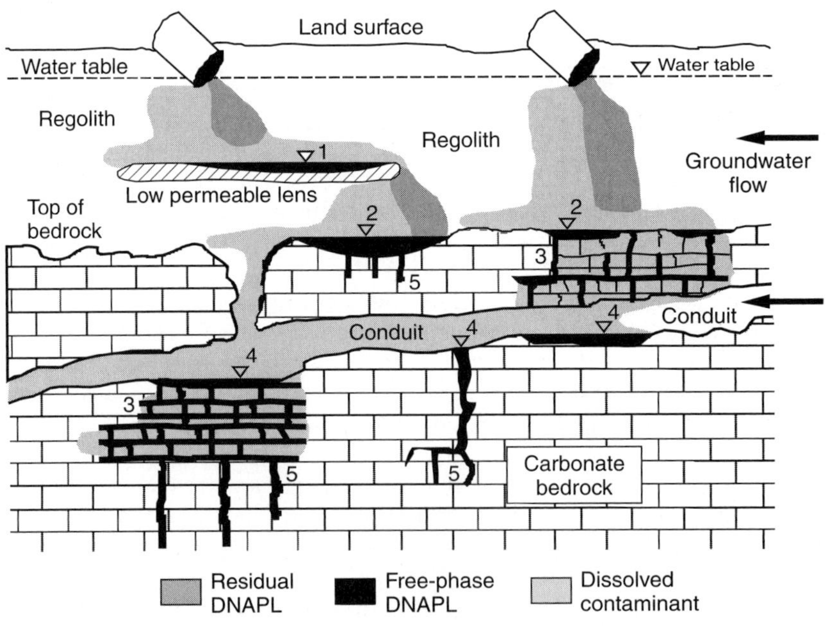

FIGURE 11.8 Distribution of potential DNAPL-accumulation sites in a hypothetical karst setting. (1) Pooling on low-permeability layer in regolith, (2) pooling on top of bedrock, (3) pooling in bedrock diffuse-flow zone, (4) pooling in conduit, (5) pooling in fractures isolated from flow. (From Wolfe et al., 1997.)

pooling in karst conduits, pooling in bedrock diffuse-flow zones, and pooling in isolation from active groundwater flow.

Trapping in regolith is most likely where the regolith is thick and relatively impermeable with few large cracks and fissures. Accumulation at the top of rock is favored by flat-lying strata with few fractures or karst features near the bedrock surface. Fractures or karst features near the bedrock surface encourage migration of chlorinated DNAPL into karst conduits or diffuse-flow zones in bedrock. DNAPL can migrate through one type of bedrock aquifer into an underlying aquifer of a different type or into openings that are isolated from significant groundwater flow (Wolfe et al., 1997). The most problematic, from the aquifer restoration perspective, in addition to an extremely low probability of actually finding DNAPL at any depth in karst aquifers, is its migration into and pooling in bedrock diffuse-flow zones and fractures isolated from major zones of groundwater flow. DNAPL accumulated in such regions will be dissolved very slowly into the slow flowing groundwater and will act as a long-term secondary source of groundwater contamination.

As emphasized by Wolfe et al. (1997), the concepts shown in Fig. 11.8 are intended to be starting points for the analysis of chlorinated-solvent contamination in karst settings and do not reduce the critical importance of careful characterization of the environmental settings and contaminant distributions at specific sites. These preliminary conceptual models are scale neutral. There is no minimum amount of DNAPL that could be stored in any of these environmental compartments, and the maximum amounts are a function of the size and nature of the release and the hydrogeologic character of specific sites. The models are mutually compatible in that more than one model may be applicable to a given site.

Unfortunately, "careful characterization of the contaminant distributions at specific sites" suggested by the USGS researchers is very complex and costly in karst as emphasized by ITRC (2003). At some sites, it is also fraught with the substantial risk of enlarging the extent of DNAPL contamination by investigation activities. Any fracture or solution channel may become a pathway for DNAPL migration, while "clean" fractures or solution channels can be found in very close proximity to the contaminated pathway if not interconnected. Once in a fracture or solution channel, the DNAPL can migrate significantly longer distances than might be expected in unconsolidated materials or the unfractured matrix of the bedrock, as the open spaces are usually larger and more continuous than those found in unconsolidated materials or unfractured rock. Even if an investigator finds one or more fractures containing DNAPL, it will be difficult to locate all fractures and conduits containing DNAPL and the extent of diffusion into the bedrock matrix. Limitations to drilling and sampling methods further hinder the ability to delineate the extent of chemical contamination in bedrock. For example, loss of volatiles as a result of sample handling or drilling, or DNAPL drag down as a result of drilling, may lead to misinterpretations regarding the extent and nature of bedrock contamination (ITRC, 2003).

If the releases or successive releases occur over time, the characteristics of the DNAPL itself could change (e.g., as manufacturing processes change). Changes in DNAPL composition and properties may result in altered flow potential and patterns. A phenomenon, observed at some creosote and manufactured gas plant sites, is that the specific gravity of the DNAPL can change over time due to weathering (i.e., preferential dissolution or other mechanisms). Some of the DNAPL components dissolve or are otherwise lost and the mixture becomes less dense than water. These neutrally buoyant and LNAPL components can migrate with the groundwater, although potentially slower than groundwater depending on the physical and chemical characteristics of the NAPL (ITRC, 2003).

In order to reasonably accurately characterize a DNAPL site, large amounts of data are usually needed. Even when using field analytical methods and dynamic work plans, large amounts of money can be spent on a DNAPL site investigation without ever determining the precise location of all of the DNAPL mass. Without a clear set of objectives, there can seem to be an almost endless need to keep collecting data and spending large amounts of money to collect these data. As reiterated by ITRC (2003), there is a need to decide when "enough is enough" and it is imperative that reasonable objectives be established for the project upfront so that the level of uncertainty remaining is acceptable. Depending on the project goals, it may be unnecessary (and wasteful) to attempt to locate all DNAPL at the site (ITRC, 2003).

The following paragraphs include paraphrased statements from an actual remedial investigation (RI) report for a Superfund site in karst that the author of this book peer-reviewed for a client. Similar statements are applicable to almost any karst settings in the United States subject to RI and can be included in any such report without worrying much at all about their accuracy or originality (except, understandably, for using site-specific numbers for quantities included in the statements).

The report integrates all available information and includes a *phenomenal volume of analytical data and subsurface characterization information* (emphasis added). Overall, the analytical data span a time frame extending from 1983 through 2012. Within the site, there are 568 groundwater monitoring wells.

Considering the complex architecture of the aquifer (in terms of karst development and structure), and the wide range of transmissivities observed for features dominating the flow regime, *it is unlikely that groundwater flow or contaminant transport at this site could ever be described using a numerical model* (emphasis added).

From the dye tracing results, it appears that there is considerable interconnection between distinct formations and hydrostratigraphic intervals within the site. This documents that mixing of groundwater does occur. However, dye results also document areas where shallow and deep flow components are isolated, indicating at least semi-confined conditions within the deeper flow component. It appears likely that the faults may serve to facilitate mixing locally, although the locations where this occurs along any single or group of faults is unknown and *possibly beyond resolution* (emphasis added).

The mapped faults (or any select fault) may serve to facilitate flow or to impede it. This may vary along any given fault as well. *Consequently, it is not possible to define the role each fault plays with respect to groundwater flow and contaminant nature and extent, or fate and transport* (emphasis added).

A well-developed epikarst consisting of residual pinnacles and incised cutters may provide a means for contaminants (DNAPL) to migrate into the bedrock or provide locations where DNAPL can accumulate. While it is recognized that the epikarst interval exists, the available data demonstrate that the top of rock surface is highly variable and unpredictable. The top of rock has been observed to vary as much as 60 ft vertically in boreholes less than 10 ft apart. The character of the overburden overlying pinnacles or filling cutters is equally variable, depending in part on the amount of residual chert left after dissolution of the host limestone. Further, the epikarst is underlain by (and connected to) an indefinable network of karst conduits of varying height and extent. Consequently, it is unlikely that the top of bedrock can be defined sufficiently to overcome these uncertainties either using surface geophysical methods or subsurface probing. *This may prove a limiting factor in attempting to locate and remediate DNAPL in the subsurface* (emphasis added).

Data from the site indicate that DNAPL has *migrated to great depth* and *also migrated laterally away from the source areas* (emphasis added). It appears that DNAPL released at the site has migrated as much as 1750 ft southeast at depths up to 240 ft, following bedding dip and/or faults. Although such lateral migration is documented in the literature, it was likely fortuitous placement of several wells that identified the degree of transport at the site. Other such pathways may likely exist, and it is also possible that the DNAPL is still migrating.

In addition to DNAPL migration at depth, DNAPL can also be transported through karst conduits for some distance as either emulsions or with entrained sediment through the karst flow system. *Consequently, DNAPL mass may be present at considerable distance from the source areas, providing for localized hot spots of dissolved-phase volatile organic compounds (VOCs)* (emphasis added).

Karst aquifers constitute the most complex settings for investigating and remediating contaminated groundwater according to studies by National Research Council (NRC, 1994 and 2005). The 2005 NRC study grouped hydrogeologic settings into five types based on media characteristics, degree of heterogeneity, permeability, and matrix porosity. According to the NRC, groundwater remediation in a Type V setting, which describes this site, is complicated by the potential for contaminant impacts over a large area because of the high aqueous flow rates that are possible within fractures or in conduits. Site characterization is typically difficult and costly, resulting in parameter estimates that may be highly uncertain. *It may not be possible to significantly reduce the uncertainties in the conceptual site model by performing additional site characterization beyond studies that have been completed to date* (emphasis added). Contaminant remediation is likely to be difficult, costly, and may be technically impracticable if DNAPL is present and contaminants have diffused into the rock matrix.

Unfortunately, regulatory agencies overseeing remediation efforts at many karst sites in the United States are unwilling to decide when "enough is enough" so that the endless data collection continues literary for decades (see Sec. 11.3.2), while decisions regarding the actual site remediation are put on hold. This is the main reason why some frustrated consultants create charts such as the one shown in Fig. 11.9. It is important to note that the colleague who volunteered this same chart for the work of Kresic and Mikszewski (2012) still wishes to remain anonymous as she or he is still afraid of alienating some of the regulators he or she must interact with on a regular basis.

In conclusion, delineation of the DNAPL source zone in fractured rock and karst aquifers is generally much more difficult than in other porous media. In relatively homogenous intergranular porous media, "clean" water in a sample provides reasonable evidence that there is no DNAPL further upgradient. In contrast, "clean" water in one well in a fractured unit provides information only on those fractures that are both upgradient and in hydraulic contact with the well (Fountain, 1998). The absence of an adequate conceptual model commonly leads to wasted effort and expense as data are collected which do little to illuminate the problem at hand. Standard techniques of site characterization developed for aqueous-phase contaminants or for porous granular media may provide irrelevant or erroneous results at DNAPL sites in fractured rock and karst settings (Cohen and Mercer, 1993; Barner and Uhlman, 1995; Wolfe et al., 1997; ITRC, 2003; Kresic and Mikszewski, 2012).

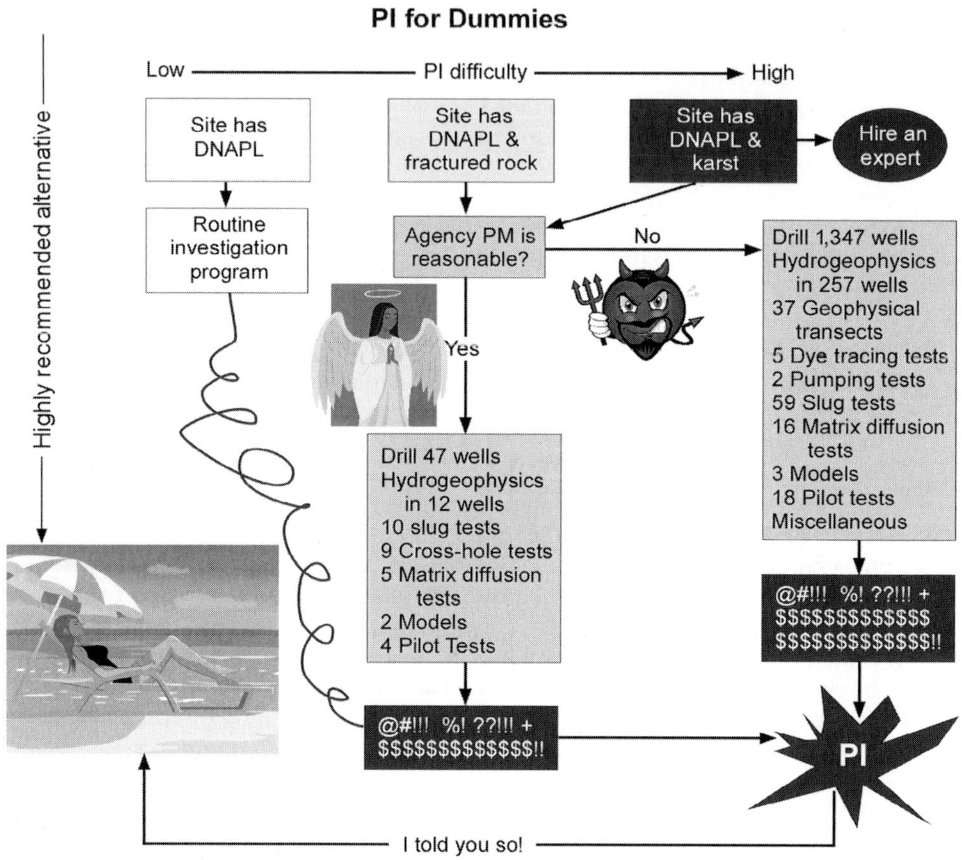

FIGURE 11.9 Political Impracticability (PI) of groundwater remediation in karst. Because of the lack of substantive guidance, regulations, or general policy, decisions on DNAPL-impacted groundwater in karst cannot be made without wasteful spending on excessive field investigations, remedial pilot tests, and data analysis. (Courtesy of Anonymous.)

There are risks associated with both the characterization and remediation of DNAPLs. Any invasive technique that penetrates a DNAPL pool (drilling, push-in tools, or similar) could provide a pathway for downward migration of the DNAPL (Pankow and Cherry, 1996). Any remediation technique that alters the water table, or decreases interfacial tension between DNAPL and water, may also enhance DNAPL mobility. The risks of mobilizing DNAPL are also dependent upon the hydrogeology of the site. Whether a competent confining layer is present or if there are water resources beneath the DNAPL source zone are two factors of particular importance. Thus, the risk must be evaluated for each site. Remediation may also increase the risk of DNAPL mobilization, and this risk should be carefully evaluated for each potential remediation technology (Fountain, 1998).

11.2.2 Source Zone Technologies

Although there are many different types of point sources of groundwater contamination, with varying spatial configurations and characteristics of contaminants, they can all be divided into two major groups from the practical remediation perspective: (1)

contamination with NAPLs and (2) contamination with chemicals readily or completely (infinitely) soluble in groundwater. Remediation strategies and technologies for these two groups of groundwater contamination sources will differ for the most part. As discussed earlier, widespread production, transportation, utilization, and disposal practices of NAPLs (both lighter and denser than water) have resulted in groundwater contamination at numerous sites throughout the world. The potential for continuing long-term contamination at many such sites is high due to toxicity, limited solubility (but much higher than drinking water limits), and significant migration potential of NAPLs in vadose zone, soil gas, groundwater, and as a separate phase. This migration can create secondary sources in the vadose and saturated zones, away from the locations of initial contaminant introduction into the subsurface. This is accentuated in karst environments as discussed throughout the preceding chapters. In addition, various NAPLs in source zones are often present in mixtures, which have different characteristics than pure chemicals, thus further complicating the analysis of their fate and transport mechanisms and the remediation options.

Many of the NAPL chemicals typically are proven or suspected carcinogens, and have low MCLs in drinking water. Considering that source concentrations can be four or five orders of magnitude greater than MCLs, restoration of source zones to pristine conditions represents a great challenge and may be unlikely in many cases. However, reduction of dissolved plumes emanating from the NAPL source zones may sometimes be a more realistic goal that could possibly be achieved through combinations of source and plume remediation. Various in situ methods (e.g., SVE, air sparging, thermal treatment, chemical oxidation, and surfactant and cosolvent flushing) have been used for removing or destroying NAPL mass contained in nonkarstic source zones with varying success. Source containment methods (slurry walls, sheet-pile walls, caps) are also used to remove or reduce the contaminant loading to the plume (Falta et al., 2007).

Source remediation at individual sites can cost anywhere from several hundred thousand dollars to tens of millions of dollars (e.g., see McDade et al., 2005), and it is rarely (if ever) possible to remove all of the contaminant. The benefit of source remediation efforts is that by removing source mass, they tend to reduce the mass discharge to the plume (Rao et al., 2001; Falta et al., 2005a; Jawitz et al., 2005). The reduced plume loading following source remediation may or may not be sufficient to allow natural attenuation processes to keep the plume concentrations within acceptable limits (Falta et al., 2005a, 2005b).

Costs for plume remediation are usually considered to be smaller than those for source remediation because of the lower capital costs. At sites where the source is nearly depleted by dissolution or other processes, plume remediation would tend to be the most cost-effective strategy for site management. However, if substantial source mass is present, in the absence of source remediation, the plume remediation systems must be operated for a long period. In this case, the operating costs (in terms of present worth) can be comparable to the costs of source remediation. As emphasized by the USEPA, a reasonable strategy for many sites would be some combination of source and plume remediation. Selection of the optimal remedy for a site, in terms of the degree of remediation, must consider the inherent coupling of the source remediation to the plume remediation (Falta et al., 2007).

As demonstrated by Lipson et al. (2005), back diffusion from the rock matrix to the open fractures can occur for very long periods of time following the removal of DNAPL sources in fractured bedrock. The back diffusion process takes a longer period of time to remove a certain amount of contaminant mass from the rock matrix than the forward diffusion process takes to initially place the mass into the rock matrix. This stems from the fact that the forward diffusion process is unidirectional, whereas the back diffusion is associated with local diffusion back into the open fractures while some forward diffusion is still taking place further in the rock matrix. The implication of this result is that the timescales associated with remediating fractured bedrock environments may not be associated with the time that DNAPL sources are present, but rather the time required for back diffusion of aqueous phase contamination from the rock matrix.

As discussed by Slough et al. (1999) and similar to many peer-reviewed papers in scientific journals, any measures taken to isolate or control the spread of contaminants, such as the installation of low-permeability walls around the site, steam flushing, flooding with solubility-enhancing chemicals, or even conventional pump and treat to control the migration of the dissolved contaminant, will increase in cost as the extent of the DNAPL contaminated zone increases. Moreover, if DNAPL is present in the rock matrix even in small quantities, the effectiveness of source-zone restoration technologies that rely on chemical or water flooding is likely to be poor because most of the flow induced will be channeled through the fracture network and contact with DNAPL in the matrix will be limited. As such, the authors hold little optimism that DNAPL-contaminated zones in porous fractured rock having nonnegligible matrix entry pressures can be restored with available technologies (Slough et al., 1999).

In situ thermal treatment and in situ chemical oxidation are increasingly used in the marketplace, are actively supported by the regulatory community, and are often touted as technologies capable of achieving closure at DNAPL sites. For example, in a 2009 status update report, USEPA declares that there are now five DNAPL sites in the entire United States where MCLs have been achieved through source zone remediation (USEPA, 2009a). However, it is important to note that none of these five sites involved contamination of bedrock groundwater, which remains the most complex remedial scenario. In addition, it is questionable whether two of these sites should even be classified as "complex" because of limited contaminant mass (Dry Clean USA No. 11502, Orlando, Florida, and Pasley Solvents and Chemicals, Inc., Hempstead, New York). The most complex project highlighted in the USEPA report is the Visalia Pole Yard remediation described in the next section.

Thermal Technologies

Thermal remediation technology has its origins in the petroleum industry, where for many years, subsurface heating and steam injection have been used to enhance oil recovery from high-gravity deposits and oil sands and oil shales. The application of thermal technology to groundwater remediation projects is a logical extension, as in many cases contamination is caused by NAPLs such as oils and chlorinated solvents. To reiterate an earlier point, in situ thermal treatment is an enhanced form of SVE, as volatilized contaminants must be captured and treated in the vapor phase. Thermal treatment also typically involves multiphase extraction wells. Thermal treatment has many advantages over ambient-temperature SVE, as thermal heating of contaminated zones greatly enhances mass removal by (from USACE, 2009)

- Increasing contaminant vapor pressures.
- Decreasing NAPL viscosities.
- Increasing contaminant solubilities and diffusivities.
- Increasing biological activities.

In addition, the thermal conductivity of soil is generally less variable than conventional remediation parameters such as soil permeability (hydraulic conductivity). Therefore, in situ thermal treatment can address low-permeability materials such as silt and clay where advective flow cannot be established. The three primary commercially available forms of in situ thermal treatment are

- Electrical resistivity heating (ERH).
- Steam-enhanced extraction (SEE).
- Conductive heating.

For detail description of these technologies, the reader is referred to USACE (2006, 2009), USEPA (1998a, 2004), Davis (1997, 1998), Powell et al. (2007), Kresic (2009), and Kresic and Mikszewski (2012).

The predominant design concerns for ERH systems are moisture loss and high groundwater flows (Kingston et al., 2009). To avoid excessive soil drying and the resultant loss of electrical conduction, ERH systems typically incorporate wetting systems around electrodes. High groundwater seepage velocities in excess of several feet per day can cause significant heat losses as warm water is continuously flushed out of the treatment area and replaced with cool water. This limitation is typical of karst aquifers with preferential flow paths. A management system consisting of upgradient pumping wells and/or downgradient injection wells can help reduce excessive groundwater flux and associated heat losses (Kingston et al., 2009).

SEE for hazardous waste remediation involves the use of steam injection wells to create a pressure gradient for recovery of NAPLs, and to heat the subsurface to volatilize and extract contaminants in the vapor phase (Fig. 11.10). Similar to ERH, SEE is capable of heating the subsurface to a maximum temperature equal to the boiling point of water or the steam distillation temperature at approximately 100°C. After NAPL has been displaced and recovered in the liquid phase, pressure cycling may be used to enhance mass removal in the vapor phase. Pressure cycling promotes contaminant volatilization by creating thermodynamically unstable conditions within soil pores (USACE, 2006).

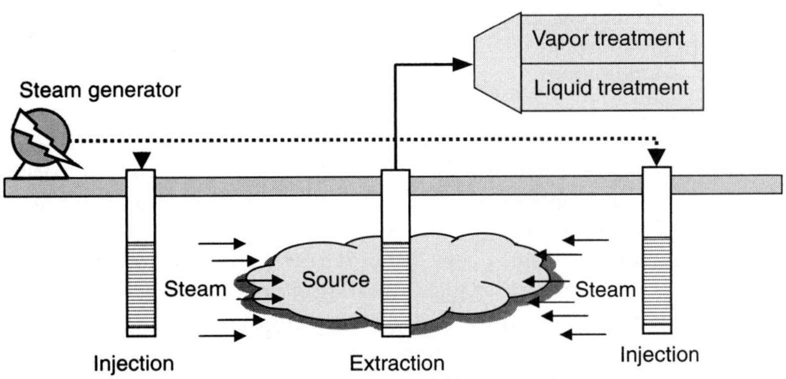

FIGURE 11.10 Schematic of steam-enhanced extraction (SEE) system. (From USEPA, 2004; Davis, 1998.)

As emphasized by Gudbjerg (2003) in case of nonkarstic granular porous media, SEE technology is expensive and difficult to implement and should only be used in highly contaminated source zones where rapid remediation is required. Steam injected in the saturated zone is strongly influenced by buoyancy, which causes the steam zone to rise to the unsaturated zone. Only in simple cases is it possible to predict steam zone development in the saturated zone without doing a full-scale complex numerical simulation. Downward migration of contaminant is a potential problem when remediating with steam. The capillary pressure decreases with temperature, thus facilitating downward migration of DNAPL in the saturated zone, but more importantly the accumulation of NAPL on the steam front can make the NAPL mobile. The latter is especially a problem in the unsaturated zone where the downward migrating NAPL only has to displace air. Mixing the injected steam with air may overcome this problem (Gudbjerg, 2003). Although there are no widely reported cases or analyses of the SEE applications in karst, all the above points are sure to be accentuated.

Conductive heating heats the subsurface primarily through conductive heat transfer. Heat is provided at point source vertical or horizontal heater units installed in the subsurface, termed wells and blankets, respectively. Heaters are typically operated at temperatures above 500°C, and heat spreads from these units through the subsurface by thermal conduction and fluid convection (i.e., hot water flow) (Kingston et al., 2009).

Unlike either ERH or SEE, thermal conductive heating is capable of heating the subsurface to temperatures significantly greater than the boiling point of water. Therefore, in addition to chlorinated solvents and light petroleum hydrocarbons, conductive heating can be used to remediate high boiling point contaminants with low volatility such as coal tar products and polychlorinated biphenyls (PCBs) (USACE, 2009). Similar to ERH, sites with high groundwater velocities, such as in karst, are challenging to remediate with conductive heating as heat losses may be excessive and prevent achievement of the target temperature. Installation of a groundwater management system for flux control may help mitigate this limitation (Kingston et al., 2009).

The performance of thermal remediation projects completed to date is comprehensively reviewed in a state of the practice overview presented in Kingston et al. (2009, overview report) and Kingston et al. (2010, detailed final report). With respect to remedial performance, Kingston et al. (2009) states:

> Despite the relatively large number of applications to date, there are limited data on post-treatment monitoring. Of the 182 sites, there was sufficient documentation to assess post-treatment groundwater quality improvements and source zone mass discharge reductions for only 14 applications.

This lack of available performance data makes it very difficult to define accurate performance expectations. This problem is not necessarily limited to thermal remediation technologies, as a majority of site remediation projects are either confidential or have involved parties that are very sensitive about the dissemination of site data and treatment results. Therefore, it becomes very difficult to practice evidence-based remediation, as the successes and failures of previous remediation projects are not well documented, and too much reliance is placed on the judgment of regulators and technology vendors. There is somewhat of a conflict of interest here, as both regulators and technology vendors have incentives to portray every remediation as a success regardless of its ability to efficiently meet the original objectives. While it is acknowledged that entities responsible for remediating sites do not want it widely known that they are involved (most likely because of fear of additional litigation), the practice of minimizing data distribution to the general public is very damaging to their own interests and additional effort should be expended to publish and consolidate data from remediation projects.

Based on the 14 applications with sufficient documentation, Kingston et al. (2009) concludes that in situ thermal treatment can achieve one-to-two orders of magnitude reductions in dissolved groundwater concentration and mass discharge when the source zone is appropriately defined. Better performance can be achieved by over-designing the system to extend beyond the delineated source zone, and by optimizing systems and allowing them to run for longer durations (Kingston et al., 2009). However, extending the treatment zone size and heating duration result in significant cost increases for a technology that already typically costs several or more million dollars to implement. For example, SEE heating at the Visalia Pole Yard site in Rosemead, CA, lasted for approximately 3 years and resulted in a thermal remediation cost of $21.5 million (energy cost excluded) for the 3.5-acre site (USACE, 2009).

As discussed by Kresic and Mikszewski (2012), the Visalia Pole Yard remediation project is often hailed by USEPA as an exemplary success story, a fact that has likely contributed to the rapid growth of in situ thermal technology in the 2000s. For example, in a publication prepared for the USEPA, Ryan (2010) states:

> The Visalia Pole Yard Superfund site attained all soil and groundwater remediation goals, becoming one of the best examples to date of a site with massive quantities of DNAPL in the saturated zone that has achieved and sustained drinking water standards following a source-mass depletion remedy.

To remediate groundwater contaminated by creosote and pentachlorophenol to applicable standards, the following activities were conducted over a 31-year period:

- Groundwater pump and treat with discharge to publicly owned treatment works (1975–1985).
- Construction and operation of an on-site groundwater treatment system to continue operating the pump and treat system (1985–1997).
- Steam-enhanced extraction (1997–2000).
- Installation and operation of an enhanced biodegradation system with continued pump and treat operation (2000–2004).
- Excavation of shallow soils from 0 to 10 ft below ground surface (bgs) (2006) (USEPA, 2009b).

The final closeout report for the site listed the total remedial cost between 1996 and 2006 as approximately $30 million (over two-thirds of which was the thermal remediation). Costs for the first 20 years of pump and treat operation and treatment building construction were not provided, nor was the total energy consumed by the thermal remediation (USEPA, 2009b). Without digging deeper, it may appear that the end result of clean groundwater justifies this significant expenditure, and that the remediation was necessary to protect human health and the environment and to restore beneficial use of the property and its underlying groundwater. Unfortunately, as detailed in the ROD (USEPA, 1994) and the final site closeout report (USEPA, 2009b), this was not the case.

First of all, no private or public drinking water was contaminated by the Visalia Pole Yard site. The following text is taken directly from the ROD (USEPA, 1994):

> The primary contribution to risk for each of these populations (on- and off-site occupational workers and off-site residents) is the estimated hypothetical future ingestion of groundwater from the intermediate aquifer. On-site wells are used for groundwater monitoring and treatment extraction purposes only. *Thus, groundwater exposures evaluated in this risk assessment are hypothetical* (emphasis added)

Kresic and Mikszewski (2012) acknowledge that contamination of potential future drinking water resources has long been justification for requiring remediation to MCLs, and the above quotation is only provided to illustrate that there were no actual exposures associated with the site that exceeded risk thresholds at the time of the ROD. In fact, the general public was not particularly interested in the cleanup of the Visalia Pole Yard site in the first place, as evidenced by the Community Relations Activities section of the site closeout report (USEPA, 2009b—emphasis added):

> Community involvement activities included the development of a Community Relations Plan (CRP), prior to initiation of the RI/FS activities. The CRP included development of a community profile and a list of key local contacts. *The community profile indicated the surrounding area was mainly businesses that had little interest in the site cleanup activities.* Copies of the Draft ROD were made available at the local public library, DTSC and USEPA Region IX Record Center. Notification of the issuance of the Draft ROD was made. A Public Notice was also placed in the local newspaper. A Public Meeting was held in Visalia, California on October 13, 1993, to provide information on the proposed cleanup. *There were no members of the public in attendance at the meeting.*

More disturbing is that the 5-year review specifically stated that "there are no specific redevelopment plans currently planned for the site." Site redevelopment opportunities were actually *severely limited* by a "Covenant to Restrict Use of Property, Environmental Restriction," which was required as part of the ROD. The following text related to this covenant is taken from USEPA (2009b):

> As remedial action objectives are based on industrial cleanup standards, prohibited Site Uses include: residences, human hospitals, schools, and day care centers for children. Prohibited Activities include: soil disturbance greater than ten feet below grade, *and the installation of water wells for any purpose* (emphasis added).

When considering all of the above information, it is unclear who or what exactly benefited from the remediation, other than the site environmental consultant, the thermal remediation vendor, and possibly the agency overseeing the remediation effort. The community was unaffected by the site and did not participate actively in site redevelopment plans (of which there were none), and future use of the groundwater that was remediated to site-specific standards (because of a hypothetical future exposure risk) is strictly prohibited by law. Additionally, the contaminated groundwater did not discharge to a surface water feature or result in any known adverse ecological impact. It therefore seems that upward of $30 million was spent simply to prove the point that remediation of DNAPL sites to some standard is feasible.

It is still puzzling, however, that the Visalia Pole Yard site closure report strictly prohibits any future use of groundwater underlying the site including installation of any wells. This means that no one can install a well to drink groundwater that was allegedly remediated to drinking water standards, and cannot even check by installing a monitoring well if the groundwater at the site currently or in some future meets the drinking water standard!

To put the above discussion in a karst perspective, it is important to understand that at the Visalia Pole Yard Superfund site, 3.5 acres were subject to thermal remediation at the cost upward of $21.5 million (energy costs excluded). In comparison, it is almost certain that DNAPL entering saturated zone of any karst aquifer will result in dissolved contaminant affecting more than just a few-acre area, with an unknown 3D extent of the residual and/or free DNAPL itself.

In Situ Chemical Oxidation

In situ chemical oxidation (ISCO) involves the introduction of a chemical oxidant into the subsurface for transforming groundwater or soil contaminants into less harmful chemical species. There are several different forms of oxidants that have been used for ISCO; however, the following four are the most commonly used: (1) permanganate (MnO_4^-), (2) hydrogen peroxide (H_2O_2) and iron (Fe) (Fenton-driven, or H_2O_2-derived oxidation), (3) persulfate ($S_2O_8^{2-}$), and (4) ozone (O_3). The type and physical form of the oxidant indicates the general materials handling and injection requirements. As discussed by Huling and Pivetz (2006), the persistence of the oxidant in the subsurface is important since this affects the contact time for advective and diffusive transport and ultimately the delivery of oxidant to targeted zones in the subsurface. For example, permanganate persists for long periods of time, and diffusion into low-permeability materials and greater transport distances through porous media are possible. H_2O_2 has been reported to persist in soil and aquifer material for minutes to hours, and the diffusive and advective transport distances will be relatively limited. Radical intermediates formed using hydrogen peroxide, persulfate, and ozone that are largely responsible for various contaminant transformations react very quickly and persist for very short periods of time (<1 second).

Table 11.6 lists the relative strengths of common oxidants. All the oxidants shown have enough oxidative power to remediate most organic contaminants. The standard

Chemical Species	Standard Oxidation Potential (Volts)	Relative Strength (Chlorine = 1)
Hydroxyl radical (OH^{-o})[a]	2.8	2.0
Sulfate radical (SO_4^{-o})	2.5	1.8
Ozone	2.1	1.5
Sodium persulfate	2.0	1.5
Hydrogen peroxide	1.8	1.3
Permanganate (Na/K)	1.7	1.2
Chlorine	1.4	1.0
Oxygen	1.2	0.9
Superoxide ion (O^{-o})[a]	−2.4	−1.8

[a] These radicals can be formed when ozone and H_2O_2 decompose.
From ITRC (2005).

Table 11.6 Oxidant Strengths

potentials are a useful general reference of the strength of an oxidant, but these values do not indicate how they will perform under field conditions. Four major factors play a role in determining whether an oxidant will react with a certain contaminant in the field: (1) kinetics, (2) thermodynamics, (3) stoichiometry, and (4) delivery of oxidants. On a microscale, kinetics or reaction rates are perhaps the most important. In fact, reactions that would be considered thermodynamically favorable based on E_0 values may be impractical under field conditions. The rates of oxidation reactions are dependent on many variables that must be considered simultaneously, including temperature, pH, concentration of the reactants, catalysts, reaction by-products, and system impurities such as natural organic matter and oxidant scavengers (ITRC, 2005). The oxidant chemicals react with the contaminants, producing innocuous substances such as carbon dioxide, water, and—in the case of chlorinated compounds—inorganic chloride. However, there may be many chemical reaction steps required to reach those endpoints, and some reaction intermediates, as in the case of polyaromatic hydrocarbons and organic pesticides, are not fully identified at this time.

The remediation of groundwater contamination using ISCO involves injecting oxidants and potentially amendments directly into the source zone and downgradient plume (Figs. 11.11 and 11.12). If the contamination is relatively shallow, in the residuum zone and immediately below in the karst aquifer, it may sometimes be possible to extensively (excessively) treat the residuum with the oxidant in an attempt to deliver it to the underlying karst aquifer over wider areas. Typical delivery methods include injection through temporary or permanent wells, gravity infiltration through trenches, and

FIGURE 11.11 Direct-push drilling for oxidant injection into shallow aquifer contaminated with chlorinated solvents. Inset: temporary injection point with the oxidant feed line. (Courtesy of ECC.)

FIGURE 11.12 Direct permanganate mixing with residuum soil contaminated with chlorinated solvents. (Courtesy of Rick Marotte, AMEC Environment and Infrastructure, Inc.)

direct mixing or tilling into soil. For deeper contamination, it will be necessary to install a dense array of bedrock wells at a much higher cost. In any case, the main challenge with ISCO in karst, similar to all other fluid-based remediation technologies, is the feasibility of delivering fluids to the target area, assuming such area is well defined, at the right dosage and residence time. In order to oxidize contaminants diffused into the rock matrix and/or moving with slow advective transport, the oxidant has to persist in such portions of the aquifer long enough. On the other hand, oxidant injected directly into or entering preferential flow paths (conduits) where groundwater and contaminant flow is rapid will obviously be quickly diluted and washed away from the target zone. For these reasons, ISCO applications in karst in many cases will not be feasible.

In general, the major benefits of ISCO are that minimal remedial waste material is generated, and that treatment can sometimes be completed in a relatively short time frame (i.e., weeks or months versus years). However, in case of NAPLs, oxidants that are in a water-based solution will only be able to react with the dissolved phase of the contaminant, since the two will not mix. As a result, treatment kinetics may become desorption limited, and the destruction of NAPLs, although possible, may require a cost-prohibitive oxidant dose and multiple applications (ITRC, 2005).

Contaminants amenable to treatment by ISCO include benzene, toluene, ethylbenzene, and xylenes (BTEX); methyl *tert*-butyl ether (MTBE); total petroleum hydrocarbons; chlorinated solvents (ethenes and ethanes); polyaromatic hydrocarbons; PCBs; chlorinated benzenes; phenols; organic pesticides (insecticides and herbicides); and munitions constituents (RDX, TNT, HMX, etc.)

As with any other fluid injection technology, ISCO application design should thoroughly address possible geochemical reactions with the aquifer (soil) porous media and all contaminants present. For example, naturally occurring or anthropogenic metals can be mobilized within the treatment zone due to a change in oxidation states and/or pH.

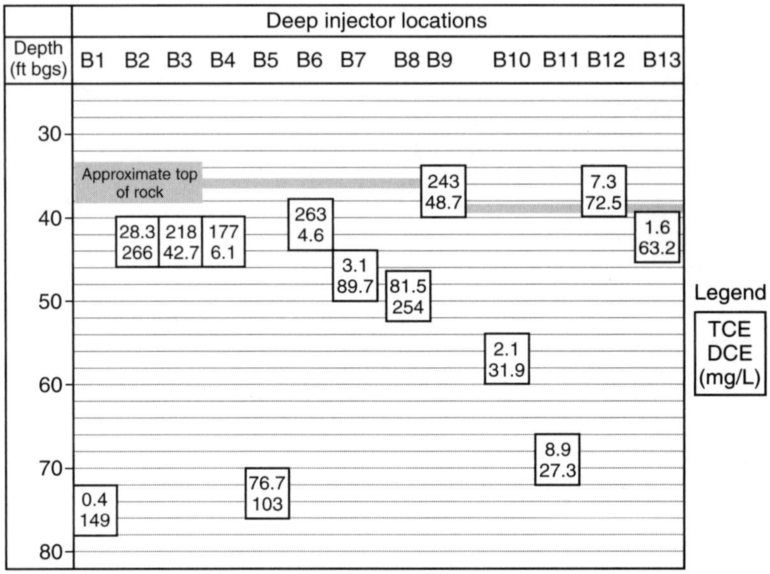

FIGURE 11.13 Concentrations of contaminants in deep injector wells, post-ISCO pilot test.

This is of particular concern when using persulfate as very low pH conditions (1.5 to 2.5) in water were observed due to persulfate decomposition. Natural soil buffering capacity can help alleviate this phenomenon but would have to be evaluated (tested) prior to ISCO application. Another observed problem at some sites is reduction of porous media permeability (hydraulic conductivity) due to chemical reactions that precipitate insoluble salts, such as MnO_2 in the case of ISCO with permanganate.

ISCO has other limitations that should also be recognized. It is possible that due to the total volume of oxidant required, it would not be cost-effective to use ISCO for site remediation. Site-specific information—including the applicability of ISCO to the specific contaminants, the concentration range, and hydrogeologic conditions—must be gathered and reviewed when evaluating the appropriateness of using ISCO for a remediation strategy. In any case, designing and implementing a full-scale ISCO remediation at a karst site (or any other site for that matter), without performing a pilot test, would in most cases be a waste of resources. Unfortunately, even when results of a pilot test are encouraging, there will be no guarantees that similar results will be achieved 50, 100, or 500 ft away due to the unpredictable nature of karst porous media.

Figure 11.13 shows the results of an extensive and expensive ISCO pilot test conducted at a karst site after which it was concluded that the technology was not feasible for full-scale implementation at this particular site. As can be seen from the reported dissolved contaminant concentrations after the test, it is virtually certain that significant mass of DNAPL remains in the karstified bedrock somewhere in the vicinity of all sampled wells.

An informative study of the effectiveness of source-zone remediation was performed by McGuire et al. (2006). Performance and rebound of intensive source depletion technologies were evaluated at 59 chlorinated solvent sites where remediation targeted DNAPL source zones. The four technologies included in the study were chemical oxidation, enhanced bioremediation, thermal treatment, and surfactant/cosolvent flushing. All results reported in this study were calculated from actual concentration versus time data at monitoring wells, and concentration reduction values reported in the literature were not used. Data were available for 147 wells at 26 enhanced bioremediation sites, 23 chemical oxidation sites, 6 thermal sites, and 4 surfactant/cosolvent sites. Approximately 75% of the source depletion projects were able to achieve a 70% reduction in parent compound concentrations. A median reduction in total chlorinated volatile organic compounds concentrations (parent plus daughter compounds) of 72% was observed at 12 chemical oxidation sites and 62% at 21 enhanced bioremediation sites. Note that enhanced bioremediation, which has primarily been used to treat dissolved phase contamination, is also being applied within and adjacent to DNAPL source zones to enhance dissolution rates (Parsons, 2004; U.S. DOE, 2002). A companion study by McDade et al. (2005) reports costs associated with the evaluated treatments at the 59 chlorinated solvent sites.

In the study by McGuire et al. (2006), rebound was assessed at sites having at least 1 year of posttreatment data and included 43 wells at 20 sites (10 enhanced bioremediation sites, 7 chemical oxidation sites, 2 surfactant/cosolvent sites, and 1 thermal site). On an individual well basis, rebound was observed in 20% of wells at enhanced bioremediation sites, in 81% of wells at chemical oxidation sites, and was not observed at surfactant/cosolvent and thermal sites. For example, concentrations in several wells at chemical oxidation sites rebounded by as much as 1 to 2 orders of magnitude throughout the posttreatment monitoring period. In fact, at 30% of the chemical oxidation rebound wells, rebound resulted in concentrations higher than pretreatment conditions. For rebound wells at enhanced bioremediation sites, the increased concentrations observed during the posttreatment period were still below pretreatment concentrations (McGuire et al., 2006).

Since the source depletion technologies evaluated in this study were applied in DNAPL source zones that had high initial dissolved concentrations, common regulatory standards, such as MCLs, were not achieved in any of the cases. Although several sites achieved MCLs at some wells, none of the sites attained and sustained MCLs for all chlorinated compounds at all wells. Given the inability of most source depletion technologies to achieve the primary remediation goal of returning groundwater to usable conditions, it is likely that some type of site management (e.g., institutional controls, long-term monitoring, MNA, or containment controls) will be necessary at many of these sites (McGuire et al., 2006).

11.2.3 Groundwater (Dissolved Phase) Technologies

The most serious conceptual error when applying any of the dissolved phase remedial technologies at karst sites is to treat the underlying karst aquifer as an equivalent porous medium. Unfortunately, this is also the most frequent error at most karst sites in the United States. Remedial systems are designed to deliver fluids to, or extract groundwater from, the porous media as if the flow of groundwater and transport of contaminants dissolved in groundwater were taking place relatively uniformly "everywhere" within the karst aquifer. This concept is usually cemented with plume maps similar to the one shown in Fig. 11.14. Such maps are created by default because most regulators insist, and everyone else expects, to see plume maps that look similar to maps created for countless other nonkarstic sites and are also common place in many textbooks. This crucial conceptual error then propagates to installation of pump and treat systems where the number of

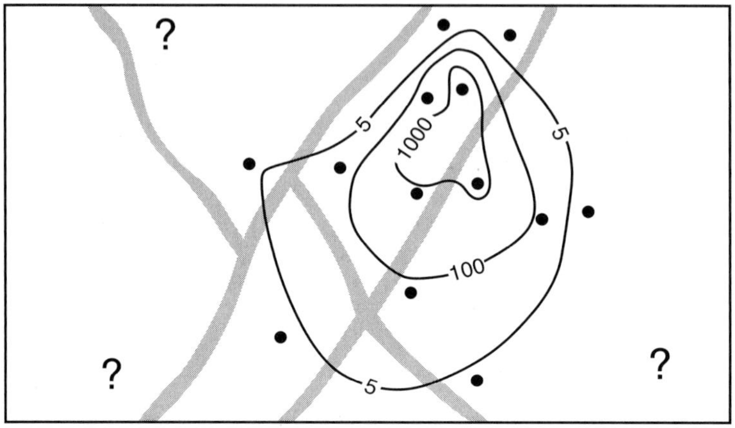

FIGURE 11.14 Plume map created by simple interpolation between sampling points (monitoring wells) without considering influence of possible preferential flow paths (depicted with gray shading).

wells and their locations are designed to capture the "nice-looking, normal" plume. The same is true with PRBs or injection of amendments for biostimulation, for example. If likely influence of any known or inferred preferential flow paths, in three dimensions and in transient conditions (baseflow and storm flow), is not evaluated appropriately or such analysis is disregarded altogether, there is high likelihood that any remediation system in karst will be ineffective. Unfortunately, in addition to various difficulties and high costs associated with defining preferential flow path in karst aquifers, the problem is compounded by the lack of applicable fate and transport numeric programs for coupled continuum conduit flow models. All these circumstances lead to the propagation of the same conceptual error again and again, with regulators blaming PRPs and their consultants for "hastily conceived and executed pilot tests" and the failure of full-scale remedial systems. As a result, before, during, and after agreeing to consider expenditure of more resources on additional investigations or installation of more recovery (and/or injection) wells, some frustrated consultants create charts such as the one shown earlier in Fig. 11.9.

Pump and Treat

Use of extraction wells to remediate groundwater in karst by removing dissolved contaminant mass is the last option any environmental professional should consider, and only after being requested to do so by the regulatory agency overseeing the remediation project. In a considerable number of cases, even regulatory agencies require this exercise only as a reference point for comparison with other remedial alternatives. This is because it has now been widely accepted by both the industry and the regulatory community that the use of pump and treat groundwater systems is ineffective in restoring aquifers to pristine conditions which by default means attaining MCL standards for all contaminants. Instead and unfortunately, pump and treat systems are still being installed en masse using other cited objectives such as plume containment or as "interim measures." Although the plume containment objective cannot be argued from the regulatory perspective, it still is important to address both its feasibility and the rationale in every site-specific karst settings. It is the experience of the author and many colleagues working

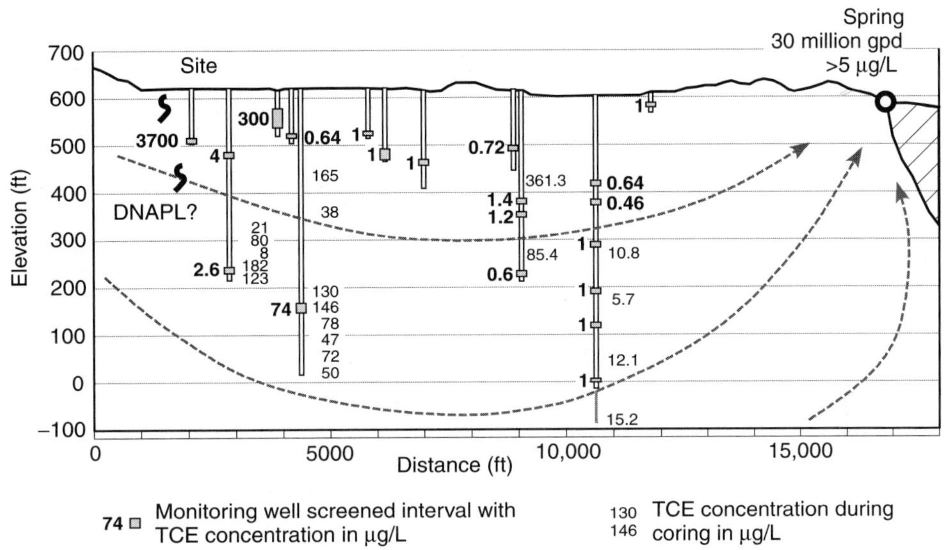

FIGURE 11.15 Conceptual schematic of groundwater flow toward a major ascending barrier karst spring showing contaminant concentration at borings and monitoring wells along a cross section between the source zone and the spring. Note that the deep bedrock well closest to the spring has all TCE concentrations less than or equal to 1 μg/L for the same sampling date, while the spring concentration is consistently >5 μg/L, emphasizing elusive nature of preferential flow paths in karst.

in karst that containment of contaminant plumes at any reasonably complex karst site is highly uncertain. The only true evidence that such containment is achieved would be contaminant concentrations less than MCL at all points of compliance monitoring and, most importantly, at all groundwater receptor locations because of which the pump and treat system has been installed in the first place. In other words, any other indirect "evidence" such as the hydraulic heads in the aquifer during pumping and the related potentiometric maps, however created, would be pointless as long as the contaminant concentrations at key locations, including springs and water supply wells, remain above MCLs.

As discussed throughout in Part I of the book, groundwater flow in karst aquifers is very complex and takes place through both the rock matrix and the preferential flow paths, which can be very deep and are often nondefinable at practical field (site) scales. Figures 11.15 and 11.16 depicting features from an actual site in the United States illustrate these points. DNAPL chemicals were disposed of at multiple locations, and their dissolved concentrations in excess of 1% to 10% solubility are found at many monitoring wells throughout the site in both the residuum and the karstified rock. Several very deep bedrock wells were drilled with great difficulty including loss of equipment and collapse of borehole walls. Several clusters of wells screened at different intervals in the bedrock were installed at the site together with many other single, shallower bedrock wells and few deeper multi-screen wells. Depth-discrete sampling during drilling revealed significant contaminant concentrations, well above MCL, down to boring bottoms exceeding 350 ft bgs, with the deepest such sample collected more than 700 ft bgs. The site has allegedly impacted a large karst spring miles away, where concentrations of COCs have

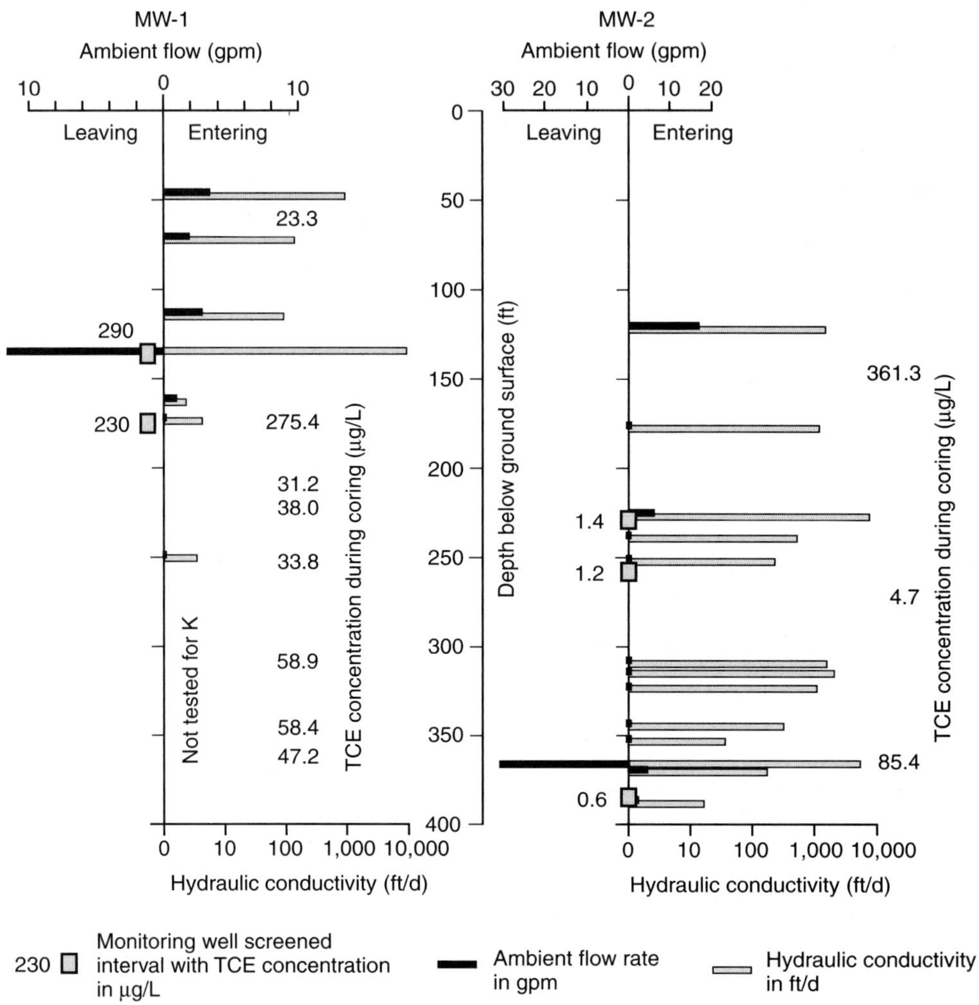

FIGURE 11.16 Results of hydrogeophysical logging and contaminant sampling at two adjacent monitoring wells located approximately along the general groundwater flow direction at the site presented in Fig. 11.15.

been consistently above MCL. Incidentally, the spring in question is used for water supply and the springwater is being treated to MCL standards by the treatment system paid for by the PRP.

Based on the cross section in Fig. 11.15 and the concept of groundwater flow in the thick karst aquifer, the spring acts as the ultimate, single-point containment of the contaminant plume(s) developed along very convoluted, indefinable 3D network of preferential flow paths. Notably, due to one reason or another, screen intervals of few deep bedrock monitoring wells do not monitor all contaminated discrete intervals sampled during investigative drilling. (It should be noted, however, that installation of monitoring wells and fully isolated multiple discrete screened intervals in a single well is extremely difficult and costly in karst.) In any case, the spring acts as one large pump and treat system for the contaminated aquifer as it continuously removes notable mass of the contaminant given its high discharge rate of >30 million gallons per day. In contrast, an uncertain number of shallow and deep extraction wells would have to be installed to first find all possible contaminant plumes at all possible depths and then intercept them before reaching the spring.

In conclusion, the main difficulties with achieving feasible plume containment in karst with pump and treat systems are (1) finding all preferential flow paths that facilitate transport of the dissolved contaminant to the points of compliance including receptors and (2) hydraulically stopping this transport by pumping both contaminated and clean water from the aquifer. Assuming that, after a sufficient expenditure of resources the plume containment can be achieved and can be fully verifiable, the following main question still remains: Is it all necessary from the risk perspective and who exactly will benefit from it? This includes comparison with logical alternatives such as groundwater treatment at the receptor locations (springs, water supply wells) and institutional controls of groundwater use. The overall cost of preventing contaminant migration to the receptors using the pump and treat system should be fully accounted for and compared with the cost of alternatives. This includes value of water that will be extracted from the aquifer prior to reaching downgradient receptors. Although in highly litigious societies with inflexible regulations and regulators this comparison may not achieve much, it is highly recommended in situations where sustainable approaches to groundwater restoration prevail.

Permeable Reactive Barriers

The concept of a PRB is relatively simple. Reactive material is placed in the subsurface where a plume of contaminated groundwater must flow through it, typically under its natural gradient (creating a passive treatment system) and treated water comes out the other side (Fig. 11.17). The PRB is not a barrier to the water, but it is a barrier to the contaminant. When properly designed and implemented, PRBs are capable of remediating a number of contaminants to regulatory concentration goals (USEPA, 1998b). PRBs can be

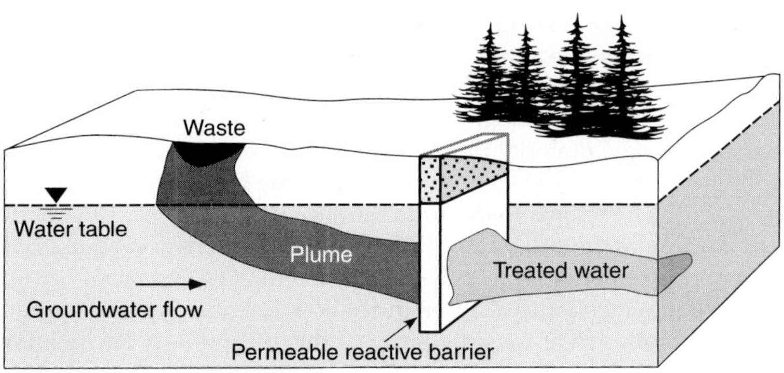

FIGURE 11.17 General schematic of a plume being treated by a permeable reactive barrier wall. (From USEPA, 1998b.)

installed in various configurations, using trenches or by injecting reactive materials into the subsurface via boreholes. If the injected materials serve to enhance biodegradation of the contaminant(s), such barriers are called bio-barriers.

The groundwater moves passively through the treatment zone, where the contaminants are degraded, precipitated, or absorbed by the treatment media. The treatment zone may contain metal-based catalysts for degrading volatile organics, chelators for immobilizing metals, nutrients and oxygen for microorganisms to enhance biodegradation, or other agents. Degradation reactions break down the contaminants in the groundwater into benign by-products. A precipitation wall reacts with the contaminants to form insoluble products that remain in the wall as the water passes. A sorption wall adsorbs or chelates contaminants to the wall surface (USEPA, 1998b).

To date, granular iron has been the most widely used reactive media in full-scale PRBs. The prevalent use of granular iron, or zero-valent iron (ZVI), stems mainly from its documented ability to abiotically degrade a variety of contaminant types, the most common of which are the chlorinated solvent compounds such as PCE and TCE (Wilkin and Puls, 2003). The abiotic process involves corrosion (oxidation) of ZVI and reduction in dissolved chlorinated hydrocarbons. The process induces highly reducing conditions that cause substitution of chlorine atoms by hydrogen in the structure of chlorinated hydrocarbons.

Alternative PRB designs using non-iron-based reactive materials to treat additional contaminants have been used increasingly. For example, reactive materials, such as compost, zeolites, activated carbon, apatite, and limestone, are now being used to control pH, metals, and radionuclides (ITRC, 2005). Table 11.7 is a partial list of materials that have been used as components within PRB systems. Each of the materials, including iron, which is shown for reference, conditions the aqueous system to either directly reduce the presence or mobility of the target chemical or promote its destruction or immobilization by other chemical or biological changes to the aqueous system. The observation that most of the materials listed are natural materials (e.g., not manufactured or enhanced by human intervention) is encouraging as PRBs can be promoted as remedies that take advantage of natural conditioning processes. The fact that most of the materials are well known to both the scientific community and the regulatory and public stakeholder community is also beneficial for receiving public approval for their use (ITRC, 2005).

The ability of iron to degrade a variety of contaminant types is being increasingly utilized in nontraditional technologies such as hydraulic injection of the nanoscale iron in an aqueous solution into the subsurface. When using granular iron, a biodegradable slurry containing iron and guar is injected through a borehole into the subsurface. Depending on the geologic material present, fractures are created or the material is intermixed with the more permeable soils. When using the biodegradable method, an enzyme is added after the injections to degrade the slurry over a short time period, leaving a lens of granular iron in the subsurface (ITRC, 2005). A discussion on use of nanoscale iron for environmental remediation is provided by Zhang (2003).

The economic benefits of PRBs in shallow unconsolidated aquifers with horizontal flow are the main driver for popularity of this technology. The passive functioning of a PRB means that relatively little energy or labor input (except for site monitoring) is necessary; therefore, the technology has a potential advantage over conventional groundwater treatment systems such as pump and treat. Regardless, a cost-benefit approach should be used to evaluate the economic feasibility of a PRB at a given site (ITRC, 2005).

Treatment Material Categories	Example Materials	Constituents Treated (Examples, Not Comprehensive)
Metal-enhanced reductive dechlorination for organic compounds	Zero-valent metals (Fe)	Chlorinated ethenes, ethanes, methanes, and propanes; chlorinated pesticides, Freons, nitrobenzene
Metal-enhanced reduction for metal contaminants	Zero-valent metals (Fe), basic oxygen furnace slag, ferric oxides	Cr, U, As, Tc, Pb, Cd, Mo, Hg, P, Se, Ni
Sorption and ion exchange	Zero-valent iron, granular-activated carbon, apatite (and related materials), bone char, zeolites, peat, humate	Chlorinated solvents (some), BTEX, Sr-90, Tc-99, U, Mo
pH control	Limestone, zero-valent iron	Cr, Mo, U, acidic water
In situ redox manipulation	Sodium dithionite, calcium polysulfide	Cr, chlorinated ethenes
Enhancements for bioremediation (including carbon, oxygen, and hydrogen sources)	(Includes solid, liquid, and gaseous sources) Oxygen-release compounds, hydrogen-release compounds, carbohydrates, lactate, zero-valent iron, compost, peat, sawdust, acetate, humate	Chlorinated ethenes and ethanes, nitrate, sulfate, perchlorate, Cr, MTBE, polyaromatic hydrocarbons

From ITRC (2005).

TABLE 11.7 Examples of Reactive Materials Used in PRBs

Most permeable barriers to date have been installed in trenches. However, trench emplacement of PRBs has a number of disadvantages. Trenches are limited to shallow treatment zones and require specialized trenching equipment, and the replacement and disposal costs of reactive material after contaminant breakthrough may be significant.

Most PRBs are younger than 15 years, and it is not known whether they will remain effective over the lifetime of the contaminant plume, which could be on the order of decades or more. Therefore, much research has focused on changes in PRB reaction rates over time. Additionally, some PRBs have had problems with permeability and hydraulics, most of which seem to be an artifact of the construction techniques for PRB installation or inadequate pre-design site characterization rather than chemical precipitation and clogging of the reactive media. As with any technology used to treat contaminants in the subsurface, successful implementation is contingent on effective site characterization, design, and construction. ITRC (2005) has produced a detail document highlighting many of the lessons learned over the 10-year history of iron-based PRB systems.

Research and deployment of bio-barrier systems are increasing in recent years, particularly for treatment of chlorinated solvents and petroleum hydrocarbon constituents such as BTEX and MTBE. Bio-barriers are often described as in situ bioremediation deployed with the PRB design concept (i.e., a continuous, linear, flow-through zone where treatment occurs). These systems may use solid, liquid, or gaseous amendments such as wood chips, compost, lactate, and molasses, to create an enhanced zone of biological activity where contaminant degradation occurs. In this way, the reactive treatment zone within a bio-barrier is created indirectly through the addition of amendments. This addition is usually achieved through vertical temporary borings or permanent wells, although other technologies including trenching can be applied depending on the site-specific conditions and depth of the contaminated zone.

Some bio-barrier designs, particularly those that require deep delivery and circulation of liquid amendments, can challenge the passive operation concept of PRBs. For example, although many bio-barriers are designed to deliver amendments into the subsurface using relatively passive techniques (i.e., slow injection or diffusion of oxygen or air), some bio-barriers require substantial energy input to deliver amendments to the proper aquifer depth and then to circulate and mix the amendments within the subsurface. Such designs function less passively than traditional PRBs and may incur greater operation and maintenance costs (ITRC, 2005).

The main limitation of using PRBs of any type in karst is barrier blowout due to high groundwater velocities and hydraulic pressures in karst conduits. Equally important concern is inability to control vertical component of groundwater flow and migration of contaminants that is always present in karst. Additionally, since most PRBs operate passively, site remediation may take several years or even decades, requiring the use of long-term institutional controls for site management. Therefore, a PRB should be considered within the context of overall and long-term site remediation goals (ITRC, 2005).

Bioremediation

In situ enhanced bioremediation of halogenated VOCs, and in particular of chlorinated aliphatic hydrocarbons (CAHs), is the most rapidly growing groundwater remediation technology. This is not surprising since VOCs are the most frequently occurring type of contaminant in soil and groundwater at Superfund and other hazardous waste sites in the United States. Innovative technologies, such as in situ bioremediation, are being developed and implemented in an effort to reduce the cost and time required to clean up those sites. In situ bioremediation is increasingly being selected to remediate hazardous waste sites because, when compared with other technologies, it is usually less expensive, does not require waste extraction or excavation, and is more publicly acceptable as it relies on natural processes to treat contaminants.

Engineered bioremediation is a technology of adjusting the concentration of various electron acceptors, electron donors (substrates), and nutrients in groundwater in order to stimulate biodegradation of contaminants by native (indigenous) microorganisms. This process is commonly referred to as biostimulation. Bioremediation also includes bioaugmentation, or the addition of microbes to the subsurface where organisms able to degrade specific contaminants are deficient. Such microbes may be "seeded" from populations already present at a site and grown in aboveground reactors, or they can be specially cultivated nonindigenous (exogenous) strains of bacteria having known capabilities to degrade specific contaminants.

An electron acceptor is a compound capable of accepting electrons during oxidation-reduction reactions. Microorganisms obtain energy by transferring electrons from electron donors, such as organic compounds (or sometimes reduced inorganic compounds, such as sulfide), to an electron acceptor. Electron acceptors are compounds that are reduced during the process and include oxygen, nitrate, iron (III), manganese (IV), sulfate, carbon dioxide, and, in some cases, CAHs, such as carbon tetrachloride (CT), PCE, TCE, DCE, and VC. Electron donors are compounds that are oxidized during the process and include fuel hydrocarbons and native organic carbon (USEPA, 2000). Hydrogen is the preferred electron donor for reductive dechlorination of chlorinated solvent PCE by the process of bacterial growth using PCE as the terminal electron acceptor (TEA). This process is termed dehalorespiration (Magnuson et al., 1998). The identification of *Dehalococcoides* (Dhc) was a watershed moment in the advancement of in situ bioremediation technologies, and led to the proliferation of bioremediation technologies in the 2000s.

Nutrients are elements required for microbial growth such as carbon, hydrogen, oxygen, nitrogen, and phosphorus. Substrate is a source of energy or molecular building block used by a microorganism to carry out biological processes and reproduce. Substrates include various forms of solid and liquid organic carbon such as carbohydrates.

The goal of engineered biodegradation is to promote growth, and stimulate the activity of those groups of microorganisms best capable of degrading certain contaminants. For example, biostimulation may be successful in degrading PCE and TCE, but it may result in a buildup of *cis*-DCE and VC, which cannot be as successfully degraded by the present bacteria. Another type of biostimulation would then have to be used downgradient from the first treatment zone to create conditions that promote biodegradation of *cis*-DCE and VC.

Biodegradation involves the production of energy in a reduction-oxidation (redox) reaction within a bacterial system. This includes respiration and other biological functions needed for cell maintenance and reproduction. Bacteria generally are categorized by (1) the means through which they derive energy, (2) the type of electron donors they require, or (3) the source of carbon that they require. For example, bacteria that are involved in the biodegradation of CAHs in the subsurface are chemotrophs (bacteria that derive their energy from chemical redox reactions) and use organic compounds as electron donors and sources of organic carbon (organoheterotrophs). However, lithotrophs (bacteria that use inorganic electron donors) and autotrophs (bacteria that use carbon dioxide as a carbon source) also may be involved in degradation of CAHs (USEPA, 2000).

Redox potentials measured in groundwater provide an indication of the relative dominance of the bacteria electron-acceptor classes (Table 11.8). These classes determine the

Dominance (as Determined by Relative Energy Generation)	Bacteria Electron-Acceptor Class	Predominant CAH Biodegradation Mechanism	Approximate Redox Potential (Volts)
Most dominant	Oxygen reducing (aerobes)	Aerobic oxidation	+ 0.82
↓	Nitrate reducing	Reductive dechlorination	+ 0.74
	Manganese (IV) reducing		+ 0.52
	Iron (III) reducing		−0.05
	Sulfate reducing		−0.22
Least dominant	Carbon dioxide reducing (methanotrophs)		−0.24

From USEPA (2000).

Table 11.8 Redox Potential as Indicator of Bacteria Electron-Acceptor Classes

Degradation Process	Compound[a]											
	Chloroethenes				Chloroethanes				Chloromethanes			
	PCE	TCE	DCE	VC	PCA	TCA	DCA	CA	CT	CF	MC	CM
Aerobic oxidation	N	N	P	Y	N	N	Y	Y	N	N	Y	P
Aerobic cometabolism	N	Y	Y	Y	P	Y	Y	Y	N	Y	Y	Y
Anaerobic oxidation	N	N	P	Y	N	N	Y	P	N	N	Y	P
Direct anaerobic reductive dechlorination	Y	Y	Y	Y	Y	Y	Y	Y	Y	Y	Y	Y
Cometabolic anaerobic reduction	Y	Y	Y	Y	P	Y	Y	P	Y	Y	Y	P
Abiotic transformation	Y	Y	Y	Y	Y	Y	Y	Y	Y	Y	Y	Y

[a] PCE, tetrachloroethene; TCE, trichloroethene; DCE, dichloroethene; VC, vinyl chloride; PCA, tetrachloroethane; TCA, trichloroethane; DCA, dichloroethane; CA, chloroethane; CT, carbon tetrachloride; CF, chloroform; MC, methylene chloride; CM, chloromethane.
N, not documented in the literature; Y, documented; P, potential for reaction to occur but not well documented in the literature.
From Parsons (2004).

TABLE 11.9 Potential Degradation Processes for Chlorinated Aliphatic Hydrocarbons

type of redox zone that will dominate in the subsurface (e.g., an aerobic zone will dominate when aerobes are present). The typical electron-acceptor classes of bacteria are listed in Table 11.8 in the order of those causing the largest energy generation to those causing the smallest energy generation during the redox reaction. A bacteria electron-acceptor class causing a redox reaction that generates relatively more energy will dominate over a bacteria electron-acceptor class that generates relatively less energy during the redox reaction (USEPA, 2000).

There are many potential reactions that may degrade CAHs in the subsurface, under both aerobic and anaerobic conditions (Table 11.9). Not all CAHs are amenable to degradation by each of these processes. However, anaerobic biodegradation processes may potentially degrade all of the common chloroethenes, chloroethanes, and chloromethanes. Enhanced in situ anaerobic bioremediation involves the delivery of an organic substrate into the subsurface for the purpose of stimulating microbial growth and development, creating an anaerobic groundwater treatment zone, and generating hydrogen through fermentation reactions. This creates conditions conducive to anaerobic biodegradation of chlorinated solvents dissolved in groundwater.

The most common chlorinated solvents released to the environment include tetrachloroethene (PCE, or perchloroethene), trichloroethene (TCE), trichloroethane (TCA), and carbon tetrachloride (CT). Because these chlorinated solvents exist in an oxidized state, they are generally not susceptible to aerobic oxidation processes (with the possible exception of cometabolism). However, oxidized compounds are susceptible to reduction under anaerobic conditions by either biotic (biological) or abiotic (chemical) processes. Enhanced anaerobic bioremediation is intended to exploit primarily biotic anaerobic processes to degrade CAHs in groundwater. Other common groundwater contaminants that are subject to reduction reactions are also susceptible to enhanced anaerobic bioremediation. They include chlorobenzenes, chlorinated pesticides (e.g., chlordane), PCBs and chlorinated cyclic hydrocarbons (e.g., pentachlorophenol), oxidizers such as perchlorate and chlorate, explosive and ordnance compounds, dissolved metals (e.g., hexavalent chromium), and nitrate and sulfate (Parsons, 2004).

Anaerobic reductive dechlorination is the degradation process targeted by enhanced anaerobic bioremediation. Through the addition of organic substrates to the subsurface, enhanced anaerobic bioremediation converts naturally aerobic or mildly anoxic aquifer zones to anaerobic and microbiologically diverse reactive zones, making them conducive to anaerobic degradation of CAHs. Examples of easily fermentable organic substrates typically used include alcohols, low-molecular-weight fatty acids (e.g., lactate), carbohydrates (e.g., sugars), vegetable oils, and plant debris (e.g., mulch). The substrates most commonly added for enhanced anaerobic bioremediation include lactate, molasses, Hydrogen Release Compound (HRC®), and vegetable oils. Substrates used less frequently include ethanol, methanol, benzoate, butyrate, high-fructose corn syrup, whey, bark mulch and compost, chitin, and gaseous hydrogen.

The most commonly used methods to deliver liquid substrates are via installed injection wells or direct-push well points, or by direct injection through temporary direct-push probes. Direct-push methods are commonly used for shallow groundwater applications in unconsolidated formations at depths less than approximately 50 ft. This technique is constrained by soil characteristics such as grain size or degree of cementation (i.e., rock fragments in residuum, or caliché inhibit use of direct-push technology). Direct injection of liquid substrates can be made through direct-push probes (e.g., Geoprobe®). This technique does not leave well points in place, and is only practical for long-lasting substrates such as HRC®, vegetable oil emulsions, or whey slurries. These substrates release carbon over periods of 6 months to several years, and typically require injection on 7.5- to 15-ft centers to treat the target zone (Parsons, 2004).

Permanent wells are typically used for continuous or multiple injections or recirculation of soluble substrate. Use of permanent injection wells is also necessary where direct-push technology is impractical, such as greater treatment depths and difficult lithology including karst. Existing monitoring or extraction wells from previous investigation or remediation activities may also be used where applicable.

Recirculation systems, consisting of closely spaced injection and extraction wells, are sometimes used to increase the retention time of contaminated groundwater in the treatment zone and to promote mixing of the substrate with the contaminant. The rate at which groundwater passes through the system depends on the rate of recirculation and the natural groundwater flux through the recirculation system. Therefore, design of recirculation systems must consider hydraulic conductivity, aquifer heterogeneity, and hydraulic gradient. Recirculation approaches may be the only effective method to achieve more uniform distribution of substrates and amendments at sites with difficult hydrogeological conditions (e.g., lack of a natural hydraulic gradient or pronounced heterogeneity). Recirculation may also be considered for shorter term applications that cannot be achieved through less aggressive, more passive methods. For example, recirculation may be useful to circulate groundwater from the greater contaminant plume through an established bioaugmented treatment zone (Parsons, 2004). The most critical design element of recirculation systems is prevention of biofouling and clogging of well screens.

Effective mixing of substrate with the contaminant plume is one of the most difficult design challenges for enhanced anaerobic bioremediation. Injection of large volumes of substrate may cause significant displacement of the contaminant plume, and sometimes a decrease in permeability of the porous media adjacent to injection points due to biofouling.

Depending on the frequency of substrate delivery to the subsurface, bioremediation systems can be grouped into passive, semi-passive, and active. Passive bio-barriers typically use slow-release, long-lasting substrates (e.g., HRC©, vegetable oils, or mulch) that can be either injected or otherwise placed in a trench, and that are designed to remain in place for long periods to maintain the reaction zone. Contaminant mass is delivered to the treatment zone via natural groundwater flow.

Semi-passive or active bio-barriers are similar to passive bio-barriers except that a soluble substrate is typically injected periodically (semi-passive) or via a recirculation system (active). Soluble substrates migrate with groundwater flow, are depleted more rapidly, and require frequent addition. However, these systems offer the advantage of being able to adjust the rate or type of substrate loading over time, and soluble substrates may be easier to distribute throughout larger volumes of the contaminant plume.

Biodegradation of an organic substrate depletes the aquifer of dissolved oxygen (DO) and other terminal electron acceptors (e.g., nitrate or sulfate), and lowers the oxidation-reduction potential (ORP) of groundwater, thereby stimulating conditions conducive to anaerobic degradation processes. After DO is consumed, anaerobic microorganisms typically use native electron acceptors (as available) in the following order of preference: nitrate, manganese and ferric iron oxyhydroxides, sulfate, and finally carbon dioxide. Anaerobic dechlorination has been demonstrated under nitrate, iron, and sulfate reducing conditions, but the most rapid biodegradation rates, affecting the widest range of CAHs, occur under methanogenic conditions (Bouwer, 1994).

Enhanced anaerobic bioremediation will not be effective in the following conditions:

- Sites with impacted receptors, or with short travel time or distance to potential receptors.
- The contaminant cannot be anaerobically degraded.
- Strongly reducing conditions cannot be generated.
- A microbial community capable of driving the process is not present or cannot be introduced to the subsurface.
- A fermentable carbon source cannot be successfully distributed throughout the subsurface treatment zone.
- There are unknown or inaccessible DNAPL sources.
- Difficult hydrogeologic characteristics that preclude cost-effective delivery of amendments, such as low permeability or a high degree of aquifer heterogeneity.
- Geochemical factors (e.g., unusually low or high pH) that inhibit the growth and development of dechlorinating bacteria.

Site Characteristics	Suitable for Enhanced Bioremediation	Suitability Uncertain	Suitability Unclear—Possible Red Flag—Requires Further Evaluation
DNAPL presence	Residual DNAPL or sorbed sources	Poorly defined sources may require additional characterization	May not be appropriate for aggressive treatment of pools of DNAPL
Plume size	Small, a few acres or less	Medium to large, a few acres plus. May require concurrent technology	Large plumes of many acres. May require concurrent technology
On-site or near site infrastructure	The risk of vapor intrusion from contaminants or biogenic gases is deemed acceptable	Target treatment zone in close proximity to sensitive infrastructure	Target treatment zone in an area where known vapor intrusion or high methane problem exists
Evidence of anaerobic dechlorination	Slow or stalled dechlorination	Limited evidence of anaerobic dechlorination	No evidence of any degradation
Depth	<50 ft to water	>100 ft to groundwater	Deep groundwater and deep contamination
Hydraulic conductivity	>1 ft/day ($>3 \times 10^{-4}$ cm/sec)	0.01–1 ft/day ($>3 \times 10^{-6}$ to 3×1^{-4} cm/sec)	<0.01 ft/day ($<3 \times 10^{-4}$ cm/sec)
Groundwater velocity	30 ft/yr to 5 ft/day	10–30 ft/yr, 5–10 ft/day	<10 ft/yr, >10 ft/day
pH	6.0–8.0	5.0–6.0, 8.0–9.0	<5.0, > 9.0
Solute concentration	<500 ppm	500–5,000 mg/L (with caution)	$>5,000$ mg/L or presence of mineral gypsum may not be suitable

From Parsons (2004).

Table 11.10 Suitability of Site Characteristics for Enhanced Anaerobic Bioremediation

Settings with the extremes of very high and very low rates of groundwater flow, which is typical of karst aquifers, impose significant limits to applying bioremediation. It may be impractical to maintain reducing conditions in high flow settings, due to the magnitude of groundwater and native electron-acceptor flux. On the other hand, it may be difficult to inject substrates into tight formations including low-porosity rock matrix. Under low flow settings, mixing of substrate with groundwater due to advection and dispersion may be limited (Parsons, 2004). Table 11.10 provides a summary of site characteristics related to suitability of enhanced bioremediation.

Engineered bioremediation is typically used to address the following remedial objectives: (1) remediation of source zones where good substrate/contaminant contact is possible; (2) reduction of mass flux from a source zone or across a specified boundary (e.g., plume containment); and (3) plume-wide treatment. Total treatment of an entire dissolved plume with active bioremediation may be feasible in some cases; ultimately, however, there will be an economic limit to the size of a plume that can be treated (Parsons, 2004).

While anaerobic dechlorination may be effective in degrading chlorinated solvents, secondary degradation of groundwater quality may occur. Degradation reactions or excessive changes in groundwater pH and redox conditions may lead to solubilization of metals (e.g., iron, manganese, and potentially arsenic), formation of undesirable fermentation products (e.g., aldehydes and ketones), and other potential impacts to secondary water quality (e.g., total dissolved solids). Many of these changes are not easily reversed, and in the case of a slow-release carbon source, it may take many years for the effects of the substrate addition to diminish. These issues should be considered during technology screening (Parsons, 2004).

In many cases, the sole use of an organic substrate (biostimulation) is sufficient to stimulate anaerobic reductive dechlorination. However, bioaugmentation may be considered at a site when an appropriate population of dechlorinating microorganisms is not present or sufficiently active to stimulate complete anaerobic reductive dechlorination of the CAH constituents present. Suthersan and Payne (2005) also show that at quite a few sites, biostimulation results in accumulation of *cis*-DCE and VC, which is not a desirable outcome and has to be addressed separately. To date, general experience with bioaugmentation is limited, and there is none widely publicized success in karst. There is also some disagreement among practitioners as to its benefits. Bioaugmentation involves the injection of a microbial amendment composed of nonnative organisms known to carry dechlorination of the targeted CAHs to completion. For example, the use of *Dehalococcoides*-related microorganisms has been linked to complete dechlorination of PCE to ethene in the field (Major et al., 2001; Hendrickson et al., 2002). Commercial bioaugmentation products that contain these microorganisms are available.

Monitored Natural Attenuation

MNA, sometimes referred to as passive bioremediation, is defined as use of natural fate and transport processes, such as contaminant dilution, volatilization, degradation, adsorption, diffusion, dispersion, immobilization, and chemical reactions with subsurface materials to reach site-specific remediation goals (Wiedemeier et al., 1998; USEPA, 1999a; Chapelle et al., 2007).

In current engineering practice, the effectiveness of MNA is evaluated on a site-by-site basis by considering three lines of evidence: (1) historical monitoring data showing decreasing concentrations and/or contaminant mass over time, (2) geochemical data showing that site conditions favor contaminant transformation or immobilization, or (3) site-specific laboratory studies documenting ongoing biodegradation processes. Various field and laboratory methods for assessing these three lines of evidence have been developed and are currently in use (Wiedemeier et al., 1998, 1999; Gilmore et al., 2006).

A major difficulty in confirming and measuring natural attenuation lies in its complexity and in the fact that most of the dominant and critical attenuation mechanisms cannot be determined directly. For example, in current practice, there is no direct measure that biodegradation is occurring. Instead, indicators that measure conditions suitable for biodegradation and breakdown products are collected to provide evidence that biodegradation is occurring. Therefore, multiple parameters are needed plus an understanding of their relationships to accurately assess MNA. The stakeholder or decision makers determine the amount of evidence needed in order to make a defensible decision. The question then arises, how much evidence is enough? Direct field measurement of key processes and degradation rates is therefore one of the research priorities because it would eliminate the question of where biodegradation is actually occurring along the groundwater flow path (Gilmore et al., 2006).

It has been emphasized by regulatory agencies and practitioners alike that MNA is not a "do nothing" approach because it involves the following (ITRC, 1999):

- Characterizing the fate and transport of the contaminants to evaluate the nature and extent of the natural attenuation processes.
- Ensuring that these processes reduce the mass, toxicity, and/or mobility of subsurface contamination in a way that reduces risk to human health and the environment to acceptable regulatory levels.
- Evaluating the factors that will affect the long-term performance of natural attenuation.
- Monitoring of the natural processes to ensure their continued effectiveness.

MNA has been the favorite first option for consideration at most groundwater contamination sites. Although it has been approved as the sole remedy at a handful of complex sites, it is always very attractive as a supplemental remedy for three main reasons: it is noninvasive, it does not require use of energy and working equipment, and has a much lower implementation cost compared with various engineered groundwater remediation systems. However, installation of monitoring wells, which is a necessary part of any MNA remedy, may involve a significant initial cost. One of the potentially most attractive aspects of MNA to general public is that it is "noninvasive": Unlike many elaborate engineered site cleanup facilities, it is "quietly" working below ground so that the land surface above ground may continue to be used. In its effort to educate the public on the benefits of natural attenuation and bioremediation in general, and alleviate concerns that MNA is not a "do-nothing" groundwater remedial alternative, USEPA has published various pamphlets (e.g., USEPA, 1996a), which include general explanations such as the following:

> Bioremediation, is a process in which naturally occurring microorganisms (yeast, fungi, or bacteria) break down, or degrade, hazardous substances into less toxic or nontoxic substances. Microorganisms, like humans, eat and digest organic substances for nutrition and energy. (In chemical terms, "organic" compounds are those that contain carbon and hydrogen atoms.) Certain microorganisms can digest organic substances such as fuels or solvents that are hazardous to humans. Biodegradation can occur in the presence of oxygen (aerobic conditions) or without oxygen (anaerobic conditions). In most subsurface environments, both aerobic and anaerobic biodegradation of contaminants occur. The microorganisms break down the organic contaminants into harmless products—mainly carbon dioxide and water in the case of aerobic biodegradation. Once the contaminants are degraded, the microorganism populations decline because they have used their food sources. Dead microorganisms or small populations in the absence of food pose no contamination risk.
>
> Many organic contaminants, like petroleum, can be biodegraded by microorganisms in the underground environment. For example, biodegradation processes can effectively cleanse soil and ground water of hydrocarbon fuels such as gasoline and the BTEX compounds—benzene, toluene, ethylbenzene, and xylenes. Biodegradation also can break down chlorinated solvents, like trichloroethylene (TCE), in ground water but the processes involved are harder to predict and are effective at a smaller percentage of sites compared to

> petroleum-contaminated sites. Chlorinated solvents, widely used for degreasing aircraft engines, automobile parts, and electronic components, are among the most often-found organic ground-water contaminants. When chlorinated compounds are biodegraded, it is important that the degradation be complete, because some products of the breakdown process can be more toxic than the original compounds.

A detailed explanation of various biodegradation processes and their characterization in the field and in the laboratory is beyond the scope of this book. As a result of the ever-increasing interest in the bioremediation of contaminated groundwater, there are many publications and resources available in the public domain and accessible via free download at various Web sites maintained by the U.S. government agencies. A good starting point are works by Wiedemeier et al. (1998, 1999), Azdapor-Keeley et al. (1999), Lawrence (2006), and Gilmore et al. (2006).

Byl and Williams (2000) provide very detail discussion of biodegradation of chlorinated ethenes in a karst aquifer in Tennessee and include an informative general overview of various related biodegradation mechanisms. Geochemical analysis at this site established that sulfate-reducing conditions, essential for reductive dechlorination of chlorinated solvents, existed in parts of the contaminated karst aquifer. Other areas of the aquifer fluctuated between anaerobic and aerobic conditions and contained compounds associated with cometabolism, such as ethane, methane, ammonia, and DO. A large, diverse bacteria population inhabits the contaminated aquifer. Bacteria known to biodegrade TCE and other chlorinated solvents, such as sulfate reducers, methanotrophs, and ammonia oxidizers, were identified from karst aquifer water using the RNA-hybridization technique. Results from microcosms using raw karst aquifer water found that aerobic cometabolism and anaerobic reductive-dechlorination degradation processes were possible when appropriate conditions were established in the microcosms.

These chemical and biological results provide circumstantial evidence that several biodegradation processes are active in the aquifer. Additional site hydrologic information was developed to determine if appropriate conditions persist long enough in the karst aquifer for these biodegradation processes to be significant. Continuous monitoring devices placed in four wells during the spring of 1998 indicated that pH, specific conductance, DO, and ORPs changed very little in areas isolated from active groundwater flow paths. These stable areas in the karst aquifer had geochemical conditions and bacteria conducive to reductive dechlorination of chlorinated ethenes. Other areas of the karst aquifer were associated with active groundwater flow paths and fluctuated between anaerobic and aerobic conditions in response to rain events. Associated with this dynamic environment were bacteria and geochemical conditions conducive to cometabolism. In summary, multiple lines of evidence developed from chemical, biological, and hydrologic data demonstrate that a variety of biodegradation processes are active in this karst aquifer (Byl and Williams, 2000).

One quantitative parameter needed to assess viability of sequential degradation or a complete destruction of an organic contaminant is the stoichiometric conversion ratio. Governed by the chemical structure of contaminants and their molecular weights, this ratio determines the relative masses of daughter products produced in the transformation processes. A compound degradation may be documented to occur at a particular site, by directly measuring the presence of its degradation products, or it may be assumed indirectly by measuring changes in concentrations of various groundwater constituents that

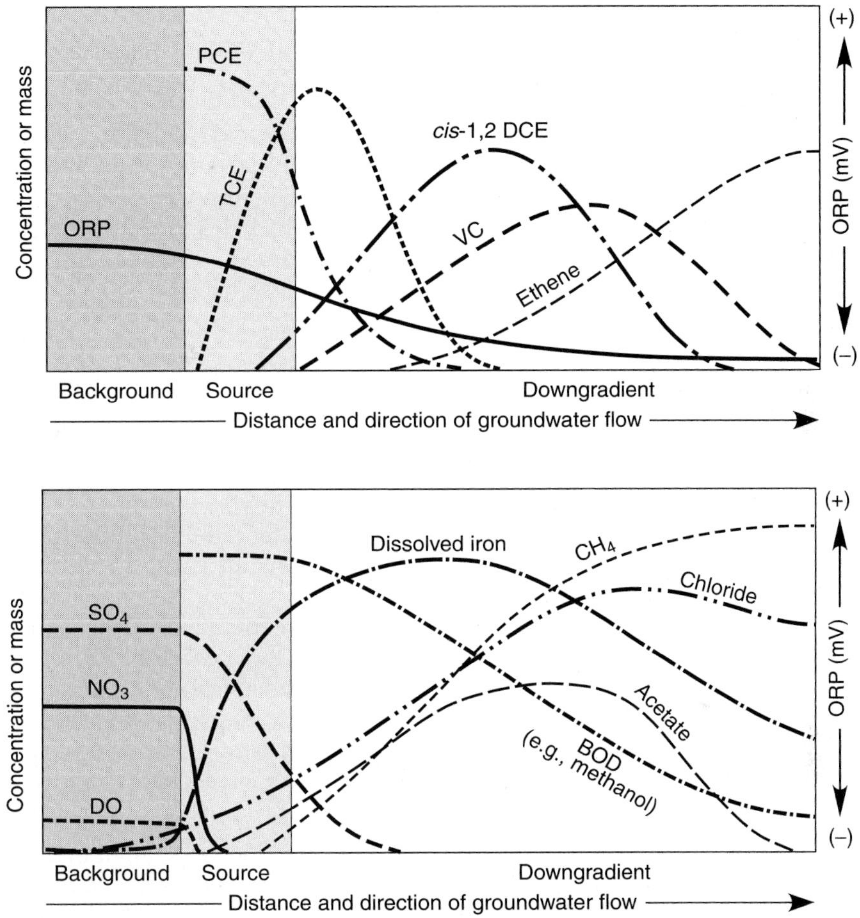

FIGURE 11.18 Common patterns of chlorinated solvent biodegradation in an anaerobic system. Interpretation: BOD (e.g., methanol) is supporting the growth of anaerobic bacteria as shown by the production of methane and acetate, and depletion of sulfate. Sulfate-reducing and possibly iron-reducing bacteria appear responsible for the initial dechlorination of PCE to DCE. As the sulfate concentration decreases, the activity of methanogenic bacteria increases. Under methanogenic/acetogenic conditions, 1,2-DCE and VC are dechlorinated to ethane. (ITRC, 1999; graphs by S. Jamal, Beak International, Inc., 1997.)

indicate such degradation is likely occurring (see Fig. 11.18). In either case, and regardless of the actual mechanism and reactions involved, the mass of the daughter product produced has to follow the exact chemical balance between the parent and the daughter products. Yield of the specific transformation, that is, parameter required by models simulating sequential degradation ("decay"), is the ratio between molecular weights of the daughter product and its parent chemical, multiplied by the mass conversion factor for the specific transformation. For example, yield values for the reductive dechlorination of chlorinated ethenes are TCE/PCE, 0.795; DCE/TCE, 0.737; VC/DCE, 0.645; ETH/VC, 0.450 (Aziz et al., 2000).

Many chemicals will undergo zero or first-order degradation in the dissolved, solid, and/or gaseous phase. A first-order degradation is described by the following equation:

$$C = C_0 e^{-kt} \tag{11.2}$$

where C is the concentration at time t; C_0 is the initial concentration at time $t = 0$; k is the first-order rate constant (units = time^{-1}); t is the time. First-order degradation can also be stated in terms of a chemical's half-life. The half-life is the time it takes for half the contaminant to degrade:

$$\frac{C}{C_0} = 0.5 = e^{-kt} \tag{11.3}$$

$$t_{1/2} = \frac{\ln 2}{k} = \frac{0.693}{k} \tag{11.4}$$

As can be seen, the contaminant half-life and the first-order degradation constant can both be used in quantitative analyses, but consistently, so that there is no confusion. For example, a 2-year half-life is equivalent to a first-order rate constant of 0.35 per year. Most practitioners, and most analytical equations, consistently use the first-order degradation constant in the same units of time as all other time-dependent parameters.

The key parameter in all calculations involving contaminant degradation is the degradation constant, k. Unfortunately, it is also one of the most difficult to accurately determine in the field, and it may change in time as geochemical conditions at the site change. It is also the least likely parameter to be accepted by regulatory agencies in case all other abiotic parameters, such as advection, recharge (very important when considering effects of potential dilution), sorption, and dispersion, are not reasonably accurately established for the particular site. In the case of sequential decay reactions, such as those involving chlorinated solvents and munitions constituents, degradation constants are certain to vary for different daughter products in both space and time. In karst aquifers, where water of different residence times and with different chemical signatures may mix more or less abruptly, some daughter products may be less biodegradable or not biodegradable at all in certain conditions. For example, Fig. 11.19 shows concentrations of TCE and *cis*-1,2 DCE collected at discrete depth intervals from two relatively close borings at the same karst site. As can be seen, while TCE concentrations are relatively comparable, the concentrations of its daughter product, *cis*-1,2 DCE, are quite different.

The USEPA published an informative work on different methods commonly used to determine the degradation rate constant k, including discussions on associated uncertainties and applicability of individual methods (Newell et al., 2002). The key point is that constant k may mean different things to different people and in different context. It is therefore very important to make a clear distinction between the general attenuation constant and the (bio)degradation constant. Although either constant can be used to quantitatively describe a general decrease in contaminant concentration in time (a first-order decay process), the degradation constant should be used only if the associated biological process is confirmed, and its rate quantified. The more general attenuation constant includes both abiotic and biotic processes, without making distinction between them. It is relatively easily determined from the measured contaminant concentrations at monitoring wells, at multiple times, by establishing a quantitative relationship between the concentration decrease and the time.

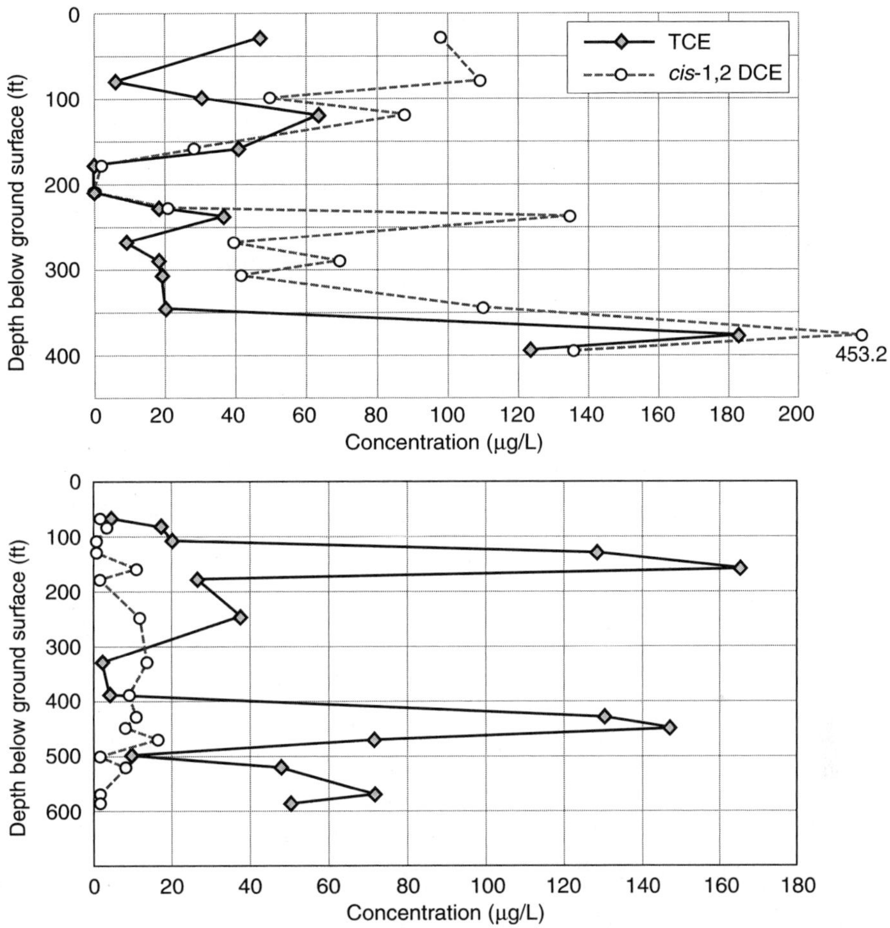

FIGURE 11.19 Changes in TCE and *cis*-1,2 DCE concentration with depth at two deep borings at the karst site presented in Fig. 11.15.

In summary, the biodegradation rate constant applies to both space and time, but only to one degradation mechanism. Quantification of this parameter is arguably the most critical part of contaminant fate and transport studies and remediation projects that consider biodegradation, in any form, as a potentially viable alternative. It can be performed in the laboratory, using controlled microcosm studies with soil and groundwater samples from the site, or in the field using extensive (and expensive) tracer studies.

For example, results of a microcosm study performed with crushed limestone and naphthalene-contaminated groundwater from a karst site in the southeastern United States shown in Fig. 11.20 demonstrate that naphthalene underwent extensive and relatively rapid biodegradation within 23 days of incubation in the live treatments. The lack of a comparable decrease in the killed controls confirmed that this activity was biotic. Since samples were not taken sooner than day 23, it is not possible to determine a rate. However, it is known that naphthalene reached the method detection limit (1 μg/L)

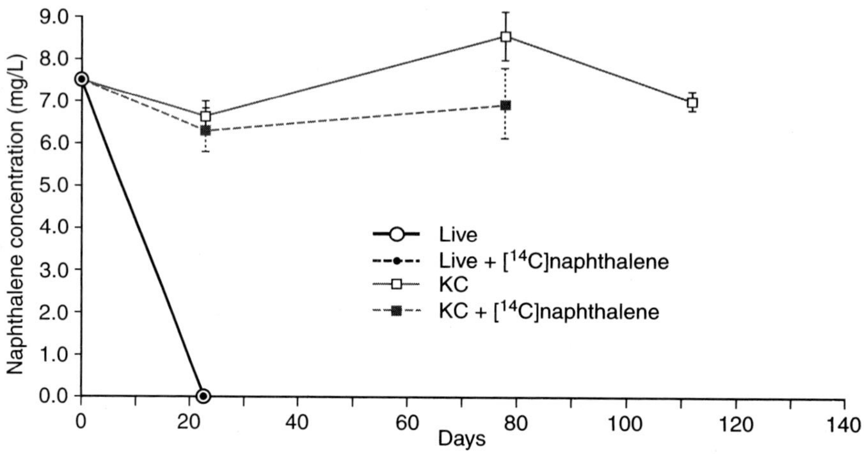

FIGURE 11.20 Naphthalene concentrations in the aerobic microcosms from location 1. (From DeLano et al., 2010; courtesy of Dr. David Freedman and Mihika Baruah, Clemson University, and AMEC Environment and Infrastructure, Inc.)

on or before day 23. Consequently, the first-order decay rate must be greater than or equal to 0.39 day^{-1}. The decrease in naphthalene coincided with a decrease in the percentage of oxygen in the headspace of the live treatments. Analysis of the distribution of ^{14}C from [^{14}C]naphthalene in the aerobic microcosm confirmed that mineralization was the predominant fate process and that CO_2 is the predominant product from aerobic biodegradation of naphthalene in this karst aquifer (Dr. David Freeman, 2009, written communication, Clemson University, and DeLano et al., 2010).

11.3 Alternative Remediation Metrics

Despite advances in technologies applicable to groundwater remediation (many termed innovative technologies), aquifer restoration at sites with complex geologic and contaminant characteristics has rarely been achieved. As discussed earlier, DNAPLs present unique challenges to site remediation because of their physical and chemical properties. To date, there are no documented, peer-reviewed case studies of DNAPL remediation below the water table in fractured rock and karst aquifers where concentrations were permanently reduced below MCLs. As a reminder of the problem, the following excerpt describing the general recalcitrance of DNAPLs to remediation is taken from USEPA (2009a):

> Due to their specific gravity, DNAPLs tend to sink in the subsurface. Their migration pathways tend to be complex and hard to predict due to the heterogeneous nature of the underlying soil and fractured bedrock. As a result, a complicated DNAPL architecture (shape and size) can develop that is made up of pools, ganglia, and globules in multiple soil layers and bedrock fracture zones. And because of their low solubility, tendency to displace water from larger soil pores, and tendency to diffuse into silt and clay [*and the bedrock matrix (added)*], DNAPLs can release dissolved constituents for long periods of time forming large groundwater plumes. Constituents in the migrating plume can diffuse into aquifer materials under certain conditions only to back diffuse out at a later time.

Taking these challenges into consideration, the Executive Summary of a 2003 national panel report on DNAPL remediation provides the following conclusions regarding "Appropriate Metrics for Performance Assessment" (USEPA, 2003b, p. xi):

> The Panel assessed the technical basis for using drinking water standards, such as Maximum Contaminant Levels (MCLs), as the single performance goal for successful DNAPL source-zone remediation and the use of chemical analyses in ground water samples from monitoring wells as the primary metric by which to judge performance of ground water remediation systems. Although an MCL goal may be consistent with prevailing state and federal laws for all ground water considered a potential source of drinking water and is a goal that is easily comprehended by the public, this goal is not likely to be achieved within a reasonable time frame in source zones at the vast majority of DNAPL sites. Thus, the exclusive reliance on this goal inhibits the application of source depletion technologies because achieving MCLs in the source zone is beyond the capabilities of currently available in-situ technologies in most geologic settings.

USEPA, 2000, estimates that site owners will spend billions of dollars over the next several decades remediating chlorinated solvents contamination. Similarly, the Department of Defense has an estimated 3,545 sites requiring further investigation and remediation, a number of which have high complexity (Deeb et al., 2011). Therefore, given the paucity of complex DNAPL sites that have been remediated to drinking water standards, there remains great demand for alternative remedial endpoints and metrics that can be used to set realistic performance expectations while also protecting human health and the environment.

Deeb et al. (2011) presents an overview of alternative remedial endpoints (i.e., other than MCLs) discussed in published literature, listing the following available options:

- Technical impracticability (TI) waivers.
- Other applicable or relevant and appropriate requirement (ARAR) waivers (greater risk, interim measures, equivalent standard of performance, inconsistent application of state standards, fund balancing).
- Alternative concentration limits.
- Groundwater management/containment zones.
- Groundwater reclassification/classification exemptions.
- MNA over long time frames.
- Adaptive site management.
- Remediation to the extent practicable.

The following sections highlight important concepts related to mass flux as an emerging remedial metric and to TI waivers.

11.3.1 Mass Flux

Mass flux is an emerging remedial metric that quantifies source or plume strength at a certain time and location. More often than not, *mass discharge* is actually the parameter of interest; however, mass flux is the more commonly used term. *Mass flux* is a solute flow rate measurement specific to an area, which is usually a subset of a plume cross section, and is expressed as mass per time per unit area (e.g., g/d/m^2). *Mass discharge* is calculated by integrating these mass flux measurements across the defined plane or cross section. Mass discharge therefore represents the total mass of the solute conveyed by groundwater through a defined plane, and is expressed in the useful units of mass per time (e.g., g/day) (ITRC, 2010).

There are three primary methods of mass flux determination, as adapted from ITRC, 2010:

1. Transect methods, in which individual monitoring points are used to integrate concentration and flow data along the plume cross section of interest.
2. Well capture/pump test methods, which uses flow and concentration data from extracted groundwater.
3. Passive flux meters, which are new in situ devices designed to directly estimate mass flux within monitoring wells.

Transect methods are the most commonly used method in unconsolidated aquifers because they are easier to implement in the field and data are less subject to significant temporal fluctuations. This, however, is not the case with karst aquifers where temporal and spatial variations in groundwater flow rates are significant and can be quite dramatic as discussed repeatedly in previous chapters. A conceptual diagram of the mass flux as it relates to karst aquifers is presented in Fig. 11.21.

Unfortunately, the task of defining contaminant mass flux in the most heterogeneous porous settings of them all, the karst, is extremely challenging to say the least. The largest

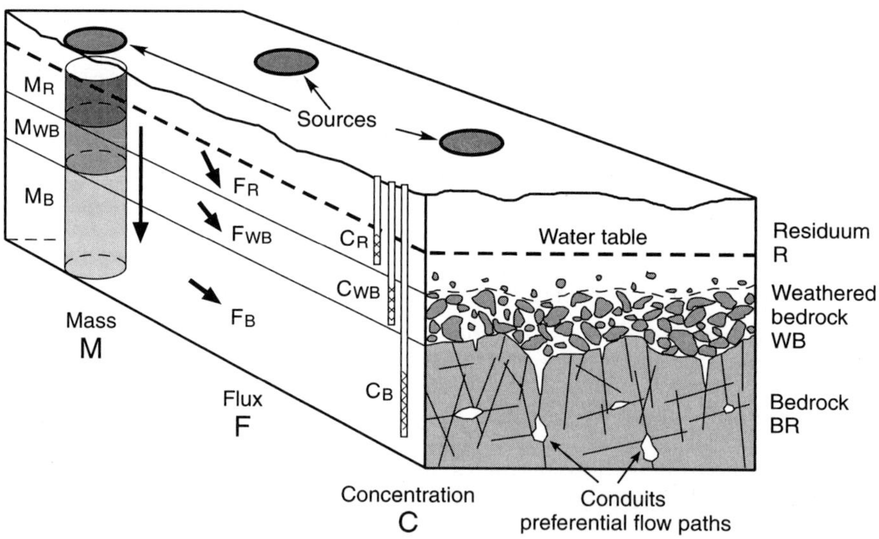

FIGURE 11.21 Schematic presentation of contaminant mass distribution in karst subsurface below DNAPL spill zones, including pathways of dissolved phase fate and transport. M_R, M_{WB}, and M_B—immobile contaminant mass in the residuum, weathered bedrock, and fractured (karst) bedrock, respectively; F_R, F_{WB}, and F_B—flux of dissolved contaminants through the residuum, weathered bedrock, and fractured (karst) bedrock, respectively; C_R, C_{WB}, and C_B—dissolved contaminant concentration measured at the point of compliance (e.g., monitoring well) in the residuum, weathered bedrock, and fractured (karst) bedrock, respectively. (From Kresic et al., 2005.)

source of error in traditional transect and passive flux meters methods of measuring flux is associated with aquifer heterogeneity and the subjective process of interpolating solute distribution. These flux measurement techniques rely on multiple point measurements from monitoring wells and subsequent interpolation, which potentially create error. For example, Fraser et al. (2005) evaluated mass flux versus sampling density for a naphthalene plume in the nonkarstic, unconsolidated sediments of the Borden research aquifer. When the sampling grid density was reduced from 1.7 points per meter squared (m^2) to 0.7 points per m^2, the range (as a standard deviation) in mass discharge increased to more than 50%. Guilbeault et al. (2005) showed that 75% of the mass flux occurred within 5% to 10% of the transect cross-section area for three nonkarstic plumes in Ontario, New Hampshire, and Florida, and that a spacing no larger than 15 to 30 cm was needed at some locations to identify high concentration zones. From these discussions, it is easily concluded that estimates of the mass flux in karst aquifers would in many cases be dubious at best even after spending large sums of money on the installation of monitoring wells.

Potential uses of mass flux and mass discharge estimates include the following:

- Evaluating source area strength, key to the overall conceptual site model.
- Evaluating plume stability, which is a critical step in MNA feasibility analysis (i.e., mass flux can help determine whether MNA will work or not).
- Assessing the relative contribution of different source areas to a contaminant plume. In this manner, sites may be prioritized in terms of the benefits of remediation to overall groundwater quality.
- Assessing the relative contribution of different aquifer intervals within a source area to a contaminant plume. In this manner, remediation can be performed more efficiently by targeting intervals of particular concern.
- Predicting and monitoring remedial performance, focusing on identifying the optimal transition points between remedial technologies (e.g., where in situ source remediation has reached a point of diminished return and transition to MNA is merited) (ITRC, 2010).

Mass discharge is now an official alternative remedial metric, as its use was specified in the ROD for Superfund Site 12A located in Tacoma, Washington, as quoted below from USEPA (2009c):

> The primary goals for the first tier of [Remedial Action Objective (RAO)] compliance are to address residual sources, minimize the risk to receptors due to contaminated surface soils and *achieve a contaminant discharge reduction of at least 90% from the high concentration source area* (emphasis added) near the Time Oil building to the dissolved-phase contaminant plume. Soil removal, ITR [in-situ thermal remediation] and EAB [in-situ enhanced anaerobic bioremediation] will be considered complete and the Remedy will be considered operational and functional when the tier 1 criteria have been met. Once the tier 1 criteria have been met, the operations and maintenance of OU1 will be turned over to the State of Washington.

This specific ROD (USEPA, 2009c) is also more progressive in that it provides a framework for discontinuing the existing pump and treats system and transitioning to MNA following the mass discharge reduction, and specifically mentions TI as an option if MNA cannot be shown to comply with ARARs in a reasonable time frame. However, consideration of TI would only occur *after* implementation of the $16,210,000 net present value source remediation remedy, and therefore is somewhat of a mute point (USEPA, 2009c). Regardless, this site strikes a positive chord by using mass discharge as an adaptive site management alternative endpoint, in that reaching the 90% mass discharge reduction triggers the transition from one remedial technology or approach to another (and in this case from one regulatory approach to another).

11.3.2 Technical Impracticability

Following is an example of an independent technical review of the remediation measures implemented for a carbonate aquifer that is karstified, and a small intermittent seepage spring that is being impacted with concentrations of TCE, DCE, and VC that are significantly above MCLs. A pump and treat system operated for approximately 10 years without achieving any of the remedial goals outlined in the ROD. The site owner instructed the environmental consultant to shorten the remedial time frame and eliminate significant long-term financial burden to the company by implementing other investigatory and remedial measures that would expedite the site closure. After performing two rounds of in situ injection of a biostimulation compound via 15 bedrock wells, the site consultant has contracted the independent reviewer for help in determining what the next remedial steps should be, specifically if another round of injections is merited. Key points from this review are provided below.

- Some monitoring wells responded more favorably, some seem not impacted at all by the injections.
- The small seep spring is responding only marginally to the remediation efforts. Changes of the concentration at the spring seem more related to the pumping than any injections—when pumping increases, the concentrations at the spring are going down and vice versa; this seems to indicate pumping is effective in containing a portion of the plume but not all of it.
- The concentrations at the spring are still well above MCL after the two rounds of expensive injections and regardless of pumping/no pumping.
- There is accumulation of DCE and VC at quite a few locations. This may indicate that less oxygenated portions of the aquifer are not favorable for degradation of the two (e.g., groundwater is not receiving enough oxygenated recharge from the land surface). This needs more analysis in terms of DO-ORP data. Well MW-X is a good example of accumulation of the two COCs, but the DO-ORP data shows reasonably high values, so the hypothesis does not hold in this case.
- Some wells like MW-XYZ are only seeing what is happening upgradient and are not being remediated by the injections (Fig. 11. 22). This well shows the same TCE concentration before and after the second injection, and a slug of DCE and VC that went through.

The reviewer then goes on to state that, if the measure of ultimate success of the remediation is the spring concentration below MCL, it cannot be achieved any time soon even with repeated injections. This is because there are portions of the aquifer not showing any

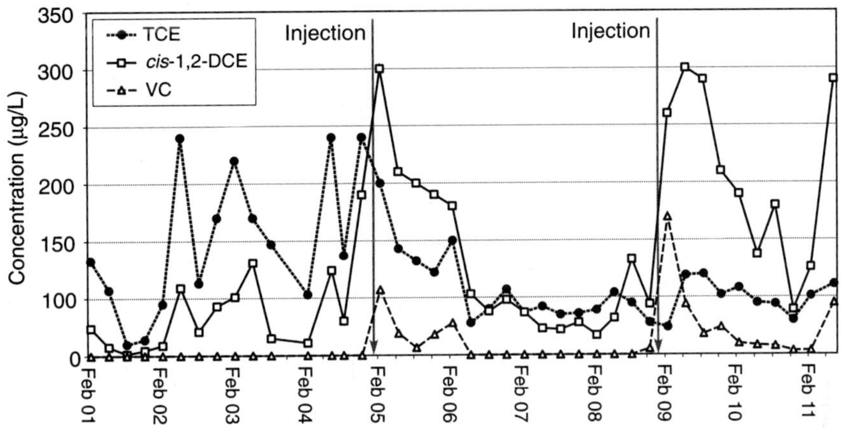

Figure 11.22 Graph of volatile organic compound concentrations at well MW-XYZ referenced in the text.

response to the injection; these portions are still contributing significant contaminant concentrations to the groundwater and ultimately to the spring. The reviewer also apologizes for quite a few additional questions she/he has regarding the site characterization and possible presence of DNAPL, but recommends that the same injection locations and protocol are not repeated without trying to answer four critical questions which have to be discussed with the client and the regulators:

(1) Is there possible measure of the success of remediation other than MCLs?

(2) Can this measure be guaranteed by the site consultant?

(3) How much money needs to be spent to achieve what is guaranteed?

(4) Is there a reasonable alternative still protective of human health and the environment?

Following is an excerpt from a 5-year review report for a Superfund site that illustrates a similar point:

> The TCE mass calculations in the 'Final Comprehensive Groundwater FS for OU 1, SAIC, April 2008' determined that there are millions of pounds of TCE in the SIA [Southeast Industrial Area] aquifer system. The calculations determined that in dissolved phase, sorbed phase, free phase and diffused mass of the residuum interval at the SIA, there could be 23,874,125 pounds of TCE. Even if these calculations are off by 50% there is a tremendous amount of contamination present in the SIA aquifer system. If only one million pounds of TCE is present, removing 625 pounds/year would take 1600 years to remove one million pounds if the removal and flow rate remain constant which is unlikely. If the removal rate could be increased to 5,000 pounds/year it would take 200 years to remove one million pounds, which is likely only a fraction of the contaminants present. Considering the contaminant mass, complexity and heterogeneity of the aquifer, and the limited success of previous remediation technologies, it appears that the SIA aquifer system is going to remain grossly impacted for hundreds of years.
>
> It would seem that a more effective course of action, rather than treatment actions that have a low expectation of real, positive results would be to focus on the protection of off-site receptors by increased off-site monitoring. (USACE, 2010.)

What the above 5-year review failed to emphasize is that this is a typical karst site together with a major impacted karst spring (flow rate >30 million gallons per day or 1.3 m^3/s), hydraulic conductivity of deep preferential flow paths (>300 ft bgs) greater than 1,000 and even 10,000 ft/day, and contaminated groundwater detected deep in the karst aquifer, with COC concentrations two orders of magnitude higher than MCLs. In spite of the overwhelming evidence of karst, USEPA is still characterizing the impacted groundwater system as (1) residuum, (2) weathered bedrock, and (3) competent bedrock (Wischkaemper, 2007). The failure to classify the system as karst is a fundamental error in the conceptual site model that obscures remedial decision-making processes.

This site has been investigated for more than two decades now, including multiple iterations and versions of the RI report, feasibility study, groundwater remediation pilot tests, and interim correction measures including pump and treat, with the estimated cost to this point upward of $100 million. Despite of all this, the USEPA has not yet issued a ROD for the site. In other words, the agency has not yet decided what the cleanup of the site should be. Additional investigations and pilot tests are being planned and are likely being performed as this is being written, partially because previous pilot tests were qualified as "hastily conceived and implemented" (Wischkaemper, 2007).

The above two examples are classic cases where remediation to MCLs is technically impracticable, meaning that remedial actions are either infeasible from an engineering perspective or unreliable in terms of meeting ARARs (the term used by USEPA at Superfund sites). The author of this book participated on a panel on TI held at a national groundwater conference co-organized by USEPA. Incidentally, the same $100 million plus site was a subject of discussion. After a remark by the author that $100 million plus spent on the site so far could have been used to better lives of the people living in the impacted community, including building libraries, schools, and possibly even a regional opera house (the community is drinking safe water provided by the PRP), one of the USEPA regulators replied that people in Europe cannot drink water from their faucets because it is contaminated. After this perplexing statement by the USEPA regulator, the panel discussion changed the subject.

The previous two examples are not an exception. There are numerous other sites in the United States contaminated with DNAPLs in difficult hydrogeologic settings such as fractured rock and karst aquifers with a very similar history of failed attempts to clean groundwater to MCLs (drinking water standards). USEPA recognized this early on and in 1993 issued a directive entitled *Guidance for Evaluating the Technical Impracticability of Ground-Water Restoration* (USEPA, 1993).

TI waivers are one of the six waivers of ARARs as defined in the *National Oil and Hazardous Substances Pollution Contingency Plan* (NCP) that are allowed under the federal Superfund statute to modify remedial action objectives. The limitations on groundwater restoration have also been addressed in the corrective action program under the RCRA and by various state initiatives.

These technical limitations have been widely recognized by the scientific and regulatory communities, and language in both federal and state statutes for groundwater cleanups considers these limitations. For example, 28 case studies presented by USEPA (1999c) showed that geologic complexity and TI were the factors responsible for controlling the cost and performance of remediation systems. Also, CERCLA legislation incorporated the concept of TI that states that USEPA may select a cleanup level that does not meet any ARAR, standard, or limitation if the agency makes a finding that compliance with such requirements is technically impracticable from an engineering perspective. USEPA incorporated this concept in the revised version (revised as of July 1, 1998) of the NCP calling it a remedial "alternative that does not meet any ARAR under federal environmental or state environmental or facility citing laws may be selected. . . [if] compliance with the requirement is technically impracticable from an engineering perspective" (40 CFR Sec 300.430(f)(1)(ii)C)(3)).

The following citation is from a memorandum by Elliott P. Laws, assistant administrator of USEPA dated July 31 (1995) and addressed to the regional administrators of all 10 EPA regions at the time, as well as the directors of various programs within the regions.

> During our meeting, we discussed the fundamental changes that have occurred in the program's approach to sites with contaminated groundwater where contamination may be 'technically impracticable' to restore to drinking water standards (e.g., where contaminants such as dense non-aqueous phase liquids (DNAPLs) warrant our use of a waiver of Federal and/or State cleanup standards (ARARs)). Based on the information now available on the special problems associated with DNAPL sites, OSWER expects that Technical Impracticability (TI) waivers will generally be appropriate for these sites. These situations demand a flexible, phased approach to groundwater remediation such as use of interim RODs, "no action" alternatives, natural attenuation, TI waivers, etc.
>
> To reiterate a major point of our discussion, I expect each Region to employ the TI waiver in appropriate remedy selection documents this fiscal year. I am concerned with preliminary data, indicating that about 30 out of 90 groundwater RODs planned for this fiscal year address sites with DNAPLs, but fewer than 10 TI waivers of ARARs have been planned for these RODs to date. I am concerned that these RODs may not fully reflect the current state of information about sites with DNAPLs present.
>
> Beginning immediately, RODs addressing DNAPL contamination that do not follow the policy in favor of TI waivers at such sites must include written justification for that departure from this policy. If you feel the data are incomplete on whether a TI waiver is justified, or that there is insufficient time this fiscal year to coordinate ROD changes, I am directing you to utilize an interim ROD or to postpone signing the ROD until the data become available and/or sufficient coordination among Federal/State/Tribal/community/PRP/other stakeholders can occur. I will adjust Regional Superfund accomplishment planning targets accordingly.
>
> Our Superfund policy guidances recognize that we can protect our groundwater resources and, at many sites, remediate large quantities of contaminated groundwater. However, they also identify situations, such as those described above, where technical, time, and cost limitations demand a more limited approach. I want to be sure you are taking command of these critical groundwater remedy selection decisions at both Federal facility and non-Federal facility Superfund sites. I have asked the Headquarters Superfund Regional Coordinators to follow up with Regional staff on this and other key remedy selection issues (land use designation, presumptive remedies, and adherence to lead policy) over the next few weeks. (Laws, 1995.)

In an extraordinary rejection of this policy statement, as of 2010, only 57 additional TI waivers have been issued since the first and only USEPA TI directive was published in 1993, equating to an annual average of 3.6 waivers per year (Deeb et al., 2011). For reference, 20 TI waivers were issued prior to directive's issuance in 1993, with an annual average of 3.3 waivers per year between 1988 and 1993. Thus, the annual number of TI waivers issued is approximately equal before and after the TI directive. As the number of ROD decision documents has decreased over time, it is also important to look at the TI statistics on a percent-selected basis. Figure 11.23 shows that the percent of CERCLA decision documents granting TI waivers increased only by approximately one percentage point in the late 1990s, which may or may not be attributable to the 1993

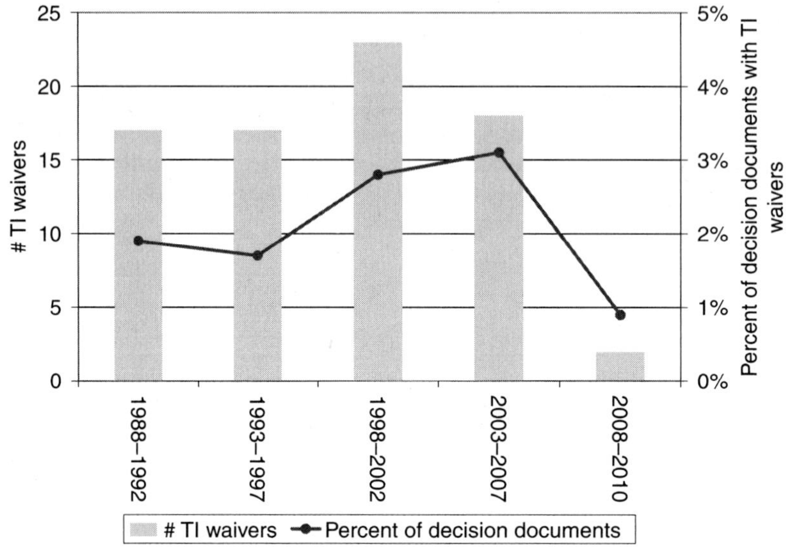

FIGURE 11.23 TI waivers granted between 1988 and 2010, using data published in the work of Deeb et al., 2011. (From Kresic and Mikszewski, 2012.)

guidance document. More alarmingly, in 2008 and 2009, the number and percent of TI waivers granted plummeted significantly. This recent trend proves that the USEPA has so far completely ignored the recommendations of its own Ground Water Task Force (2007):

> The decision-making process involved in determining cleanup goals appropriate for a DNAPL source zone, and whether remediation efforts should be undertaken to remove or treat the DNAPL source zone was a common theme in the comments received. Most task force members agreed that the current Superfund guidance on technical impracticability (TI) should be updated. Also, there was support for identifying mechanisms for acknowledging complex site conditions that would be useful in the decision-making process for cleanup programs other than Superfund.
>
> Recommendation. Develop guidance on how to acknowledge technical limitations posed by DNAPLs in EPA cleanup decisions, including updated guidance on the use of technical impracticability (TI) decisions in the Superfund program. The guidance should also discuss mechanisms for acknowledging technical limitations posed by site complexities other than DNAPLs.
>
> Long-term cleanup goals for Superfund sites and RCRA Corrective Action facilities do not always include attaining MCLs throughout the plume. For ground waters that are not designated by states as current or future sources of drinking water, drinking water standards are generally not used as cleanup levels and alternative cleanup goals are typically established, such as control of sources and containment of the plume. Also, where the remedy calls for on-site management of waste materials (such as a landfill), cleanup levels generally do not need to be attained in ground water beneath the waste management area. In such cases, attaining MCLs throughout the plume applies only to that portion of the plume outside the waste management area.
>
> Furthermore, both the Superfund and RCRA Corrective Action programs generally allow alternative cleanup goals to be established at sites where attaining MCLs throughout the plume is determined to be technically impracticable (TI). Both of these EPA cleanup programs also establish alternate cleanup limits (ACLs) in lieu of MCLs, under appropriate

> circumstances. However, ACLs defined under CERCLA are somewhat different from those in RCRA Corrective Action. Some state cleanup programs have provisions for establishing contaminated ground water containment or management zones. Within such a zone, active cleanup of contaminated ground water may be deferred or may not be required. The specifics of how containment or management zones are defined, and what alternative cleanup goals are applied, differ from state to state.
>
> For the reasons discussed above, sites where DNAPLs are present in the subsurface are very difficult to clean up to drinking water standards. Cleanup technologies applicable to these sites often include individual approaches or various combinations of approaches intended to control migration of contaminants (containment), remove contaminants from the subsurface (extraction), or treat contaminants in place (in-situ treatment). Each of these technology types have been used (with varying degrees of success) on DNAPLs in the source zone or on dissolved contaminants in the plume.

Unfortunately, not only did the agency fail to update its own TI waiver policy and thus selected to withhold related technical assistance to working professionals, but it appears that any references to TI, including related case studies, are completely absent from various USEPA Web sites. One cannot find any list of RODs where TI is a part of the remedy, or cannot use any search engine to learn more about the TI process, or somehow stumble upon a Superfund site with the issued TI waiver. The agency has mostly ignored TI in its annual reviews of applied technologies or any other technical publication (e.g., USEPA, 2010). It is also indicative that some USEPA regions have issued only a handful of TI waivers in almost 20 years, and Region 4 is yet to issue a single standing TI waiver. Interestingly, Region 4 covers several main karst states including Tennessee, Kentucky, Alabama, and Florida. It therefore appears that USEPA Region 4 is afraid of setting a precedent by acknowledging that groundwater remediation at a karst site may be technically impracticable. This policy is for the most part followed by the states in Region 4 including at their own sites. The only known document to provide a consolidated listing and description of sites with TI waivers is Deeb et al. (2011), which is a study funded by the Environmental Security Technology Certification Program through the Department of Defense, which as previously mentioned has a great number of complex sites for which cleanup would be funded by federal tax dollars.

The decreasing trend in annual TI waivers granted and the absence of substantive technical guidance regarding TI indicate that the USEPA adopted an informal policy against the use of TI waivers many years ago. This policy was formalized through two USEPA documents published in summer 2011. The first document, published in July 2011, is a Groundwater Road Map that summarizes the groundwater evaluation and remediation process at Superfund sites through a conceptual flowchart and related narrative (USEPA, 2011b). Surprisingly, TI documentation is an element of the road map. However, the road only leads to TI after the following steps have been completed:

- Remedial design.
- Remedial action.
- Remedy operation complete with performance monitoring, 5-year reviews, and potential system optimization.
- A feasibility evaluation for other technologies in the event that remedial goals are not being achieved.

In other words, the road map indicates that TI is a last resort option only to be considered after remedial actions have been conducted. Even if the best available technology has a very low probability of success in meeting MCLs (e.g., karst sites with DNAPL), some form of a costly remedial effort is recommended.

The second document, published in September 2011, is a memorandum from the director of the USEPA Office of Superfund Remediation and Technology Innovation clarifying the previously referenced July 31, 1995, memorandum from Elliott P. Laws, the former assistant administrator of USEPA. The following text is taken directly from the September 2011 memorandum (Woolford, 2011):

> This (memorandum) is to clarify that: 1) the 1995 memorandum was intended to apply only to remedy decisions made in Fiscal Year 1995 and 2) DNAPL contamination in and of itself should not be the sole basis for considering the use of a TI waiver at any given site. Improvements in science and technology have shown much progress in characterizing and successfully treating/extracting DNAPLs from subsurface areas. Recent remedy decisions have selected a variety of in situ technologies to address DNAPL contamination such as in situ chemical oxidation, in situ bioremediation and in situ thermal remediation. . . For the reasons stated above, *the 1995 memorandum entitled, Superfund Groundwater RODs: Implementing Change This Fiscal Year, July 31, 1995 (OSWER Directive 9335.3–03P) should no longer be considered when making current site decisions* (emphasis added).

This memorandum rescinds the 1995 mandate that USEPA regions utilize TI waivers at DNAPL sites or provide written justification otherwise. While it seems as though the 1995 directive was never truly followed, USEPA rejection of its policy prescription is now in writing. In addition, the memorandum formally endorses the use of aggressive *in situ* technologies such as thermal remediation and ISCO despite limited evidence that these technologies lead to cost-effective, better outcomes in the long term. While this memorandum is decidedly anti-TI, there is one important statement, quoted below, that conflicts with the road map publication:

> In situations where groundwater restoration is unattainable from an engineering perspective, considering a TI waiver may be an appropriate part of the remedy selection process (Woolford, 2011).

In other words, a TI waiver can theoretically be obtained at the remedy selection phase in lieu of aggressive remedial action. While DNAPL alone does not imply TI, "sufficient, science-based justification" can be used to invoke a TI waiver where appropriate (Woolford, 2011). The USEPA should therefore revise its road map to reflect that TI is a viable option at the initial remedy selection phase. Regardless of this caveat, it is likely that the 2011 memorandum will further diminish the usage of TI at complex Superfund sites (Kresic and Mikszewski, 2012).

As a consequence of inadequate technical and regulatory guidance from USEPA, including inconsistencies between various regions regarding TI waivers, working professionals developing conceptual site models for groundwater remediation at difficult sites are left between a "rock and a hard place." It is therefore not surprising that many of them are referring to TI as PI (political impracticability) and are producing for-internal-use-only flowcharts similar to the one shown earlier in Fig. 11.9.

One of two criteria needs to be met in order to apply for a TI waiver: (1) engineering infeasibility or (2) unreliability (USEPA, 1993). A remedial action can be considered infeasible from an engineering perspective if current engineering methods designed to meet the ARARs cannot be reasonably implemented. An action can be considered unreliable if it is shown that existing remedial alternatives are not likely to be protective in the future.

Together, these two criteria define the term *technical impracticability from an engineering perspective*. Furthermore, a TI waiver would only be granted based on demonstration that cleanup cannot be achieved within a reasonable time frame using the best available technology.

As discussed by the USEPA, a reasonable time frame for restoring groundwater to beneficial uses depends on the particular circumstances of the site and the restoration method employed. A comparison of restoration alternatives from the most aggressive to passive will provide information concerning the approximate range of time periods needed to attain groundwater cleanup levels. An excessively long restoration time frame, even with the most aggressive restoration methods, may indicate that groundwater restoration is technically impracticable from an engineering perspective (USEPA, 1996b). Notably, however, the various USEPA regions have differing views on the use of TI waivers, including definition of *reasonable*.

It should be noted that a reasonable time frame is sometimes generically applied to be 100 years, based on the order-of-magnitude number provided by USEPA (1993). However, there is no accepted definition of *reasonable* as applied to groundwater restoration because it is dependent on the applicable technologies and site-specific conditions such as hydrogeology. This leads to confusing and contradictory interpretations of what *reasonable* really means. In some instances, stakeholders have interpreted timescales on the order of 600 years to be *reasonable*, whereas others have viewed timescales greater than 30 years to be technically impracticable (Deeb et al., 2011).

Based on the author's experience, a remedial measure estimated to reduce the cleanup time to 250 years may be considered reasonable and practicable by a regulatory agency versus a 500-year cleanup time estimate under baseline conditions. It is not clear how a cleanup time of 250 years is reasonable. Consider, for example, that 250 years ago the year was 1761 and the United States of America did not even exist.

There are also definitive technical reasons why choosing between remedies with cleanup time frames in the hundreds of years is fundamentally wrong. Economically, the majority of the net present value of any remedial alternative will be captured in the first 30 years of costs indicating that long-term remedial cost differences 100 or more years in the future are irrelevant. The accuracy of numerical modeling at these timescales is also questionable, as quoted below from USEPA (1999b):

> The longer the time frame simulated, the greater the uncertainty associated with the modeling result. While the time to reach remedial objectives at all points in the Joint Site groundwater will likely be on the order of 100 years, simulations greater than the order of 50 years into the future are generally not reliable or useful. EPA has used simulations of 10–25 years for comparing remedial alternatives, even though the remedial action is not complete in that time frame under any of the alternatives. This provides a measure of each alternative's relative performance and progress at 25 years toward meeting the remedial objectives.

This USEPA statement directly contradicts other ROD decision documents where modeling is used to justify selection of one remedial approach over another when both alternatives have estimated cleanup times greater than 100 years. Inconsistent regulatory guidance regarding technical standards of practice and, more fundamentally, inconsistent interpretation of what constitutes a *reasonable* time frame will continue to hinder the advancement of TI as a viable remedial alternative until social and economic factors are considered in the cleanup process. This concept is discussed further in the next section.

While there is little that can be done if a cleanup time frame of several hundred years is considered *reasonable*, there are two primary methods of demonstrating TI at a site: field methods and modeling-based methods. Either approach can be applied *before* implementing a remedial measure as outlined in the only available guidance on TI waivers from USEPA (1993). Kresic and Mikszewski (2012) present two case studies illustrating the two approaches.

Interestingly, 2012 was the year of the two most important international conferences on the topic of hydrogeology and groundwater remediation. In Niagara Falls, Canada, the International Association of Hydrogeologists celebrated its 40th anniversary in September 2012 with more than 750 presentations at the congress, which featured some of the world's preeminent hydrogeologists as key note speakers. A session entitled "Remediation of Karst Aquifers" had to be cancelled because not a single abstract was submitted for it (the Karst Symposium within the conference alone featured more than 60 presentations). The world's most prestigious and most attended gathering of environmental remediation professionals, the Battelle's *International Conference on Remediation of Chlorinated and Recalcitrant Compounds*, was held in Monterey, California, in May 2012. Out of thousands of presentations on soil and groundwater remediation, only one was on karst aquifers, featuring a general overview which concluded that, paraphrased, groundwater contamination and remediation in karst should not be treated in the same manner as in other porous media. It is the opinion of this author that the main reason for this complete absence of karst remediation presentations at the two gatherings is quite simple: Environmental professionals much more prefer to present success stories (and even that only when allowed doing so by their clients paying the remediation bill).

11.4 The Way Forward: Sustainable Remediation

There are numerous noneconomic benefits to Superfund and state-led remediation in the United States, most notably related to public health and the reduction or reversal of damages to important natural resources (Kresic and Mikszewski, 2012). For example, Currie et al. (2011) found that Superfund cleanups can reduce the incidence of congenital anomalies by approximately 20% to 25% within the affected community. This was the first study to examine the impact of cleanups of hazardous waste sites on infant health, which has the important benefit of not requiring detailed knowledge of environmental factors that may affect adult health, including lifetime smoking behavior, lifetime exposure to ambient air pollution, and lifetime exposure to multiple hazardous waste sites (Currie et al., 2011). The reader is referred to USEPA (2011d) for a comprehensive listing of additional benefits of the Superfund program from USEPA's perspective.

The Currie et al. (2011) study demonstrates the need to eliminate actual (not hypothetical) contaminant exposures related to hazardous waste sites. However, how to most efficiently eliminate existing and prevent future exposures across the country is beyond the realm of Superfund as we know it. *Sustainable remediation* is an emerging field that has been specifically developed to counteract the irrational mind-set of the Visalia Pole Yard cleanup and other similar projects. The concept was originally introduced by the U.S. Sustainable Remediation Forum (SURF) in a 2008 white paper (SURF, 2009), and has since rapidly spread throughout the United States and European Union. In its Summer 2011 issue, the journal *Remediation* published the first guidance documents on the subject, which together present a framework for sustainable remediation, detail necessary steps in completing environmental footprint analysis and life-cycle assessments, and identify metrics for incorporating sustainable practices in remediation projects (see Simon, 2011, for an overview).

Sustainable remediation is defined as a remedy or combination of remedies whose net benefit on human health and the environment is maximized through the judicious use of limited resources (SURF, 2009). This is achieved by balancing economic growth, protection of the environment, and social responsibility to improve the quality of life for current and future generations (USEPA, 2011d). Thus, *the triple bottom line* of sustainable remediation is measured by environmental, social, and economic factors (Butler et al., 2011).

The Sustainable Remediation Framework, defined by Holland et al. (2011), is characterized by the following key elements:

- Process-based implementation that includes several decision points necessary to achieve the final outcome. This is an alternative to the more rigid and traditional goal-based implementation where all focus is placed on meeting a specific regulatory requirement (e.g., MCLs).
- Future-use planning that manages the entire site cleanup process with the preferred end use or future use in mind. This includes a transition strategy that efficiently moves the project from the remediation stage to the long-term use stage after achieving remedial objectives.
- A tiered sustainability evaluation that maximizes the positive sustainability impacts of the project, as measured through metrics such as carbon dioxide emissions, groundwater or energy use, local laborers trained, local suppliers utilized, or native species reintroduced to site habitat (see Butler et al., 2011, for additional metrics).
- Development of a sustainable conceptual site model that incorporates sustainable elements such as (1) resource input and output, (2) net environmental benefit, (3) stakeholder economic and social well-being, and (4) high-level land-use planning. The sustainable conceptual site model can be used to answer complex questions regarding the beneficial use of groundwater at the site, the feasibility of reaching site closure, and how achieving remedial goals will change site risk.
- Implementation of sustainable remedial measures, including but not limited to
  - Using in situ technologies that mimic natural processes and/or result in contaminant mineralization rather than phase transfer. For example, bioremediation, which can result in complete contaminant destruction in situ, is preferable to thermal remediation where phase change is often required to extract contaminants from the subsurface and to potentially dispose of remediation waste off site.
  - Minimizing or eliminating emissions and natural resource (e.g., energy, water) consumption; this includes minimizing transportation requirements wherever possible.
  - Using renewable energy to power remedial operations.
  - Recycling or reusing soil or demolition materials.

- Training and employing local workers.
- Conducting collaborative community events before, during, and after remedy selection and implementation.

Sustainability elements can also be incorporated into traditional site assessment activities, such as risk assessment. It is critical to remember the triple bottom line during sustainability evaluations, such that social and economic challenges faced by the community are given sufficient weight in decision-making processes. For example, very low excess cancer risks caused by site contamination (i.e., one in a million) are often used to justify site remediation while *zero* consideration is given to the American Cancer Society (ACS) estimate that nearly two-thirds of cancer deaths in the United States in 2011 will be attributable to tobacco use, overweight or obesity, physical inactivity, and poor nutrition—all of which are preventable causes (ACS, 2011). Sustainable remediation approaches would also consider and potentially address these health concerns in a manner that logically ties into the site cleanup. For example, site reuse plans could incorporate a community health center, community gardens for cultivation of fresh produce, or exercise facilities and/or nature trails. Community involvement activities could also include wellness and economic counseling. All of the above examples would represent a more judicious use of monies typically spent in a pyrrhic struggle to reach MCLs (Kresic and Mikszewski, 2012).

In the United States, there has been significant recent interest in "green remediation," a practice defined and encouraged by the USEPA. Green remediation considers all environmental effects of cleanup actions and incorporates options to minimize the environmental footprints of these actions (USEPA, 2011e, 2011f). Thus, green remediation shares several core elements with sustainable remediation, such as the minimization of energy use and emissions during remediation. However, there are critical fundamental differences between the two concepts that are often easy to disguise because of the way that the terms "green" and "sustainable" are often interchanged. For example, the following are key USEPA perspectives on green remediation, quoted directly from USEPA (2011e):

- Greener cleanups are not an alternative approach to setting cleanup levels and selecting remedies;
- Cleaning up sites for reuse supports sustainable development; and
- Reducing the environmental footprint of a cleanup does not justify changing the end point.

Simply put, green remediation does not incorporate sustainability considerations in selecting the remedial approach or the preferred end use. Green remediation is strictly process related, while sustainable remediation incorporates social and economic factors to address broader land management issues (USEPA, 2011e). This is unfortunate, as the greatest opportunity to realize sustainable outcomes are in the early stages of remedial implementation while setting the remedial specification and strategy (NICOLE, 2010).

As suggested by Kresic and Mikszewski (2012), despite being a prime example of LEED brain, green remediation accomplishes the critical objective of making consultants, regulators, and stakeholders consider the overall environmental footprint of a cleanup. Additionally, green remediation encourages the use of important technologies that

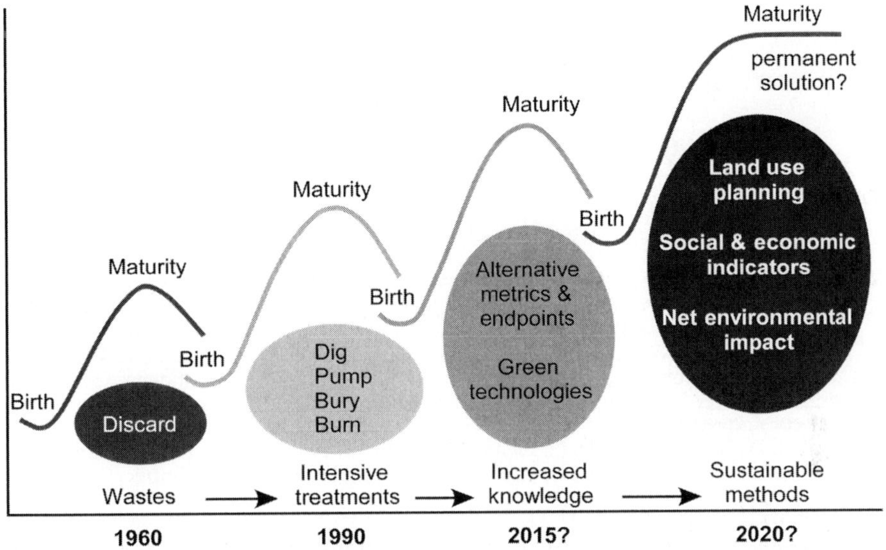

FIGURE 11.24 Conceptual diagram illustrating the evolution of societal thinking about waste and environmental cleanups. This version has been expanded from the original to include the phase we are currently entering (increased knowledge), in which green remediation practices are encouraged and alternative remedial endpoints and metrics are at least being actively considered. The next logical evolutionary step is the full embrace of sustainable remediation practices. (From Kresic and Mikszewski, 2012; adapted from SURF, 2009.)

undoubtedly play a role in sustainable development and is therefore another evolutionary step in the maturation of the remediation industry. Figure 11.24 is an expanded schematic of this evolutionary process originally presented in SURF (2009).

A transition toward sustainable remediation practices is critical in ensuring the long-term viability of environmental protection programs in the United States, European Union, and other countries. Ongoing budget and debt crises in many governments will continue to threaten financial resources dedicated to environmental cleanups, and the remediation industry has to do more with less in order to survive. One great danger is political embrace of the Forget About the Aquifer approach, in which all attempts at groundwater restoration will be abandoned in favor of cheaper water supply alternatives (Ronen et al., 2011). The shortcomings of remedial technologies and current cleanup ideologies highlighted by Kresic and Mikszewski (2012), as well as in this chapter, are not meant to discourage further cleanups or disparage the idea that pristine environmental resources have intrinsic value. Quite on the contrary, the hope of the author is to galvanize support for holistic and technically sound approaches that can achieve better remedial outcomes while also freeing up financial resources for assessment and proper management of as many contaminated sites as possible.

At the end, the author of this book would like to acknowledge that discussions and concepts presented throughout various chapters relate to the current state of the hydrologic, hydrogeologic, and remediation practices in karst, and focus on prevailing ideologies and recommendations for improvement. These topics are often controversial, and the author hopes that this book provokes thought and discussion on how we can evolve current policies and practices to achieve better outcomes at a lesser cost to society.

The author has no agenda or underlying motivation in these discussions, and it should be noted that this book was completed without financial support from any public or private entity.

11.5 References

ACS (American Cancer Society), 2011. *Cancer Facts and Figures 2011*. American Cancer Society, Atlanta, GA.

Anderson, M. R., Johnson, R. L., and Pankow, J. F., 1987. The dissolution of residual dense non-aqueous phase liquid (DNAPL) from a saturated porous medium. In: *Proceedings of the NWWA/API Conference on Petroleum Hydrocarbons and Organic Chemicals in Ground Water: Prevention, Detection, and Restoration*. National Water Well Association, American Petroleum Institute, Houston, TX, pp. 409–428.

Anderson, M. R., Johnson, R. L., and Pankow, J. F., 1992. Dissolution of dense chlorinated solvents into groundwater. 3. Modeling of contaminant plumes from fingers and pools of solvent. *Environ Sci Technol*, 26:901–908.

Ayotte, J. D., Montgomery, D. L., Flanagan, S. M., and Robinson, K. W., 2003. Arsenic in groundwater in eastern New England: Occurrence, controls, and human health implications. *Environ Sci Technol*, 37(10):2075–2083.

Azdapor-Keeley, A., Russell, H. H., and Sewell, W. W., 1999. Microbial processes affecting monitored natural attenuation of contaminants in the subsurface. Ground Water Issue, U.S. Environmental Protection Agency, Office of Research and Development, EPA/540/S-99/001, 18p.

Aziz, C. E., Newell, C. J., Gonzales, J. R., Haas, P., Clement, T. P., and Sun, Y., 2000. BIOCHLOR; Natural attenuation decision support system; User's manual, version 1.0. EPA/600/R-00/008, U.S. Environmental Protection Agency, Office of Research and Development, Washington, D.C., 46p.

Barner, W., and Uhlman, K., 1995. Contaminant transport mechanisms in karst terrains and implications on remediation. In: Beck, B. F. (ed.), *Karst Geohazards—Engineering and Environmental Problems in Karst Terrane*. A.A. Balkema, Rotterdam, pp. 207–212.

Black, D. F., 1966. Howard's Cave disaster. *National Speleological Society (NSS) News 24*, pp. 242–244.

Bouwer, E. J., 1994. Bioremediation of chlorinated solvents using alternate electron acceptors. In: Norris, R. D., et al. (eds.), *Handbook of Bioremediation*. Lewis Publishers, Boca Raton, FL, pp. 149–175.

Bricelj, M., 2003. Microbial tracers in groundwater research. *RMZ—Materials and Geoenvironment*, 50(1): 67–70.

Bricelj, M., and Čenčur Curk, B., 2005. Bacteriophage transport in the unsaturated zone of karstified limestone aquifers. In: Stevanovic, Z., and Milanovic, P. (eds.), Water Resources and Environmental Problems in Karst, Proceedings of the International Conference and Field Seminars, Belgrade & Kotor, Serbia & Montenegro, 13–19 September 2005, Institute of Hydrogeology, University of Belgrade, Belgrade, pp. 109–114.

Brusseau, M., 2005. DNAPL source zones. In: DNAPLs—Source zone behavior and mass flux measurement, Clue-In Seminar, August 10, 2005. U.S. Environmental Protection Agency, Technology Innovation Program. Available at: http://www.clu-in.org/conf/tio/dnapls1_081005/.

Butler, P. B., Larsen-Hallock, L., Lewis, R., Glenn, C., and Armstead, R., 2011. Metrics for integrating sustainability evaluations into remediation projects. *Remed J*, 21:81–87. doi: 10.1002/rem.20290

Byl, T. D., and Williams, S. D., 2000. Biodegradation of chlorinated ethenes at a karst site in middle Tennessee. *U.S. Geological Survey Water-Resources Investigations Report 99–4285*. Nashville, TN, 58p.

Chapelle, F. H., Novak, J., Parker, J., Campbell, B. G., and Widdowson, M. A., 2007. A framework for assessing the sustainability of monitored natural attenuation. *U.S. Geological Survey Circular 1303*, Reston, VA, 35p.

Cohen, R. M., and Mercer, J. W., 1993. DNAPL site evaluation. C.K. Smoley, Boca Raton, FL, various paging.

Currie, J., Greenstone, M., and Moretti, E., 2011. Superfund cleanups and infant health. Massachusetts Institute of Technology, Department of Economics Working Paper Series. Working Paper 11–02. Available at http://papers.ssrn.com/sol3/papers.cfm?abstract_id=1768233

Danish EPA, 1999. Natur og Miljøpolitisk Redegørelse 1999. Miljø- og Energiministeriet. (Nature and Environment Policy Statement 1999, in Danish). Environment and Energy, Copenhagen, Denmark. Available at: http://www.mem.dk/publikationer/nmpr99/

Davis, E., 1997. Ground water issue: How heat can enhance in-situ soil and aquifer remediation: Important chemical properties and guidance on choosing the appropriate technique, EPA 540/S-97/502. U.S. Environmental Protection Agency, Office of Research and Development, Ada, OK, 18p.

Davis, E. L., 1998. Steam injection for soil and aquifer remediation. EPA/540/S-97/505, U.S. Environmental Protection Agency, Office of Research and Development, Ada OK, 16p.

Deeb, R., Hawley, E., Kell, L., and O'Laskey, R., 2011. Assessing alternative endpoints for groundwater remediation at contaminated sites. ESTCP Project ER-200832, 232p. Available at: http://www.serdp.org/Program-Areas/Environmental-Restoration/Contaminated-Groundwater/Persistent-Contamination/ER-200832.

DeLano, J. M., Parker, J. T., Galloway, R., et al., 2010. MNA as an approved groundwater remedy for a carbonate bedrock aquifer contaminated with PAHs. *2010 GSA Annual Meeting & Exposition, Reaching New Peaks in Geoscience*, October 31–November 3, 2010, Denver, CO, Paper no. 58-6.

Ewers, R. O., Duda, A. J., Estes, E. K., Idstein, P. J., and Johnson, K. M, 1991. The transmission of light hydrocarbon contaminants in limestone karst aquifers. In: *Proceeding of the Third Conference on Hydrogeology, Ecology, Monitoring, and Management of Ground Water in Karst Terranes*. Association of Ground Water Scientists and Engineers, Nashville, TN, pp. 287–306.

Falta, R. W., Rao, P. S., and Basu, N., 2005a. Assessing the impacts of partial mass depletion in DNAPL source zones: I. Analytical modeling of source strength functions and plume response. *J Contam Hydrol*, 78(4):259–280.

Falta, R. W., Basu, N., and Rao, P. S., 2005b. Assessing the impacts of partial mass depletion in DNAPL source zones: II. Coupling source strength functions to plume evolution. *J Contam Hydrol*, 79(1): 45–66.

Falta, R. W., Stacy, M. B., Ahsanuzzaman, A. N. M., Wang, M., and Earle, R. C., 2007. REMChlor, remediation evaluation model for chlorinated solvents; User's manual, Version 1.0. U.S. Environmental Protection Agency, Center for Subsurface Modeling Support, National Risk Management Research Laboratory, Ada, OK, 79p.

Feenstra, S., and Cherry, J. A., 1988. Subsurface contamination by dense non-aqueous phase liquid (DNAPL) chemicals. In: *Proceedings of International Groundwater Symposium*, International Association of Hydrogeologists, May 1–4, 1988, Halifax, Nova Scotia, pp. 62–69.

Fountain, J. C., 1998. Technologies for dense nonaqueous phase liquid source zone remediation. *Technology Evaluation Report TE-98-02, Ground-Water Remediation Technologies Analysis Center (GWRTAC)*, Pittsburgh, PA, 62p.

Fraser, M., McLaren, R., and Barker, J., 2005. Multilevel monitoring wells to assess contaminant mass discharge: Magnitude of uncertainties based on Borden monitoring experience. *The Abstract Book of the 2005 Ground Water Summit Program*. National Ground Water Association, San Antonio, TX.

Gilmore, T., et al., 2006. Characterization and monitoring of natural attenuation of chlorinated solvents in ground water: A systems approach. WRSC-STI-2006–00084, Rev 1, Savannah River National Laboratory, Savannah River Site, Aiken, SC, 53p.

Gudbjerg, J., 2003. Remediation by steam injection. Ph.D. Thesis, Environment & Resources DTU, Technical University of Denmark, 137p.

Guilbeault, M. A., Parker, B. L., and Cherry, J. A., 2005. Mass and flux distributions from DNAPL zones in sandy aquifers. *Ground Water*, 43(1):70–86.

GWTF (Ground Water Task Force), 2007. Recommendations from the EPA Ground Water Task Force; Attachment B: Ground water use, value, and vulnerability as factors in setting cleanup goals. EPA 500-R-07–001, Office of Solid Waste and Emergency Response, pp. B1–B14.

Hedberg, C. W., and Osterholm M. T., 1993. Outbreaks of foodborne and waterborne viral gastroenteritis. *Clin Microbiol Rev*, 6:199–210.

Hendrickson, E. R., Payne, J. A., Young, R. M., et al., 2002. Molecular analysis of *Dehalococcoides* 16S ribosomal DNA from chloroethene-contaminated sites throughout North America and Europe. *Appl Environ Microbiol*, 68(2):485–495.

Holland, K. S., Lewis, R. E., Tipton, K., et al., 2011. Framework for integrating sustainability into remediation projects. *Remed J*, 21:7–38. doi: 10.1002/rem.20288

Huling, S. G., and Pivetz, B. E., 2006. In-situ chemical oxidation. Engineering Issue, EPA/600/R-06/072, U.S. Environmental Protection Agency, Office of Research and Development, National Risk Management Research Laboratory, Cincinnati, OH, 58p.

ITRC (The Interstate Technology & Regulatory Council), 1999. Natural attenuation of chlorinated solvents in groundwater: Principles and practices. Technical/Regulatory Guidelines, In Situ Bioremediation Work Team, Interstate Technology & Regulatory Council, Washington, D.C., 25p + appendices. Available at: http://www.itrcweb.org.

ITRC, 2003. An introduction to characterizing sites contaminated with DNAPLs. Technology Overview, Dense Nonaqueous Phase Liquids Team, Washington, D.C., 36p + appendices.

ITRC, 2004. Strategies for monitoring the performance of DNAPL source zone remedies. Technical and Regulatory Guidelines, Interstate Technology & Regulatory Council, Washington, D.C., 94p + appendices.

ITRC, 2005. *Technical and Regulatory Guidance for In Situ Chemical Oxidation of Contaminated Soil and Groundwater*, 2nd ed. In Situ Chemical Oxidation Team, Interstate Technology & Regulatory Council, Washington, D.C., 71p + appendices.

ITRC, 2010. Use and measurement of mass flux and mass discharge. Technology overview document. Interstate Technology & Regulatory Council, Washington, D.C., 89p + appendices. Available at: http://www.itrcweb.org/Documents/MASSFLUX1.pdf. Accessed July 20, 2011.

Jawitz, J. W., Fure, A. D., Demmy, G. G., Berglund, S., and Rao, P. S. C., 2005. Groundwater contaminant flux reduction resulting from nonaqueous phase liquid mass reduction. *Water Resources Res*, 41(10):W10408.

Kingston, J. T., Dahlen, P. R., Johnson, P. C., Foote, E., and Williams, S., 2009. State-of-the-practice overview: Critical evaluation of state-of-the-art in situ thermal treatment technologies for DNAPL source zone treatment. ESTCP Project ER-0314. Available at: http://cluin.org/techfocus/default.focus/sec/Thermal_Treatment%3A_In_Situ/cat/Guidance/

Kingston, J. T., Dahlen, P. R., Johnson, P. C., Foote, E., and Williams, S., 2010. Final report: Critical evaluation of state-of-the-art in situ thermal treatment technologies for DNAPL source zone treatment. ESTCP Project ER-0314, 1272pp. Available at: http://cluin.org/techfocus/default.focus/sec/Thermal_Treatment%3A_In_Situ/cat/Guidance/

Kresic, N., 2007. *Hydrogeology and Groundwater Modeling*, 2nd ed. CRC/Taylor & Francis, Boca Raton, FL, 807p.

Kresic, N., 2009. *Groundwater Resources. Sustainability, Management, and Restoration*. McGraw Hill, New York, 852p.

Kresic, N., and Mikszewski, A., 2012. *Hydrogeological Conceptual Site Models: Data Analysis and Visualization*. CRC/Taylor & Francis Group, Boca Raton, FL, 552p.

Kresic, N., O'Laskey, R., Deeb, R., et al., 2005. Technical impracticability (TI) of DNAPL remediation in karst. In: Stevanovic, Z., and Milanovic, P. (eds.), *Water Resources and Environmental Problems in Karst*, Proceedings of the International Conference and Field Seminars, Belgrade & Kotor, Serbia & Montenegro, September 13–19, 2005, Institute of Hydrogeology, University of Belgrade, Belgrade, pp. 63–65.

Lawrence, S. J., 2006. Description, properties, and degradation of selected volatile organic compounds detected in ground water—A review of selected literature. *U.S. Geological Survey, Open-File Report 2006–1338*, Reston, VA, 62p; a web-only publication at http://pubs.usgs.gov/ofr/2006/1338/.

Laws, E. P., 1995. Memorandum. Subject: Superfund Groundwater RODs: Implementing Change This Fiscal Year, July 31, 1995. EPA-540-F-99-005, OSWER-9335.5-03P, PB99-963220, Washington, D.C.

Lipson, D. S., Kueper, B. H., and Gefell, M. J., 2005. Matrix diffusion-derived plume attenuation in fractured bedrock. *Ground Water*, 43(1):30–39.

Loop, C. M., and White, W. B., 2001. A conceptual model for DNAPL transport in karst ground water basins. *Ground Water*, 39(1):119–127.

Magnuson, J. K., Stern, R. V., Gossett, J. M., Zinder, S. H., and Burris, D. R., 1998. Reductive dechlorination of tetrachloroethene to ethene by a two-component enzyme pathway. *Appl Environ Microbiol*, 64:1270–1275.

Major, D. W., McMaster, M. L., Cox, E. E., et al., 2001. Successful field demonstration of bioaugmentation to degrade PCE and TCE to ethene. *Proceedings of the Sixth International In Situ and On-Site Bioremediation Symposium*, San Diego, CA, Vol. 6, No. 8, pp. 27–34.

McDade, J. M., McGuire, T. M., and Newell, C. J., 2005. Analysis of DNAPL source-depletion costs at 36 field sites. *Remed J*, 15(2):9–18.

McGuire, T. M., McDade, J. M., and Newell, C. J., 2006. Performance of DNAPL source depletion at 59 chlorinated solvent-impacted sites. *Groundwater Monit Remediat*, 26(1):73–84.

Newell, C. J., et al., 1995. Light nonaqueous phase liquids. Ground Water Issue, EPA/540/S-95/500, Robert S. Kerr Environmental Research Laboratory, Ada, OK, 28p.

Newell, C. J., et al., 2002. Calculation and use of first-order rate constants for monitored natural attenuation studies. Ground Water Issue, EPA/540/S-02/500, U.S. Environmental Protection Agency, National Risk Management Research Laboratory, Cincinnati, OH, 27p.

Nichols, E. M., Beadle, S. C., and Einarson, M. D., 2000. Strategies for characterizing subsurface releases of gasoline containing MTBE. American Petroleum Institute (API), Regulatory and Scientific Affairs Publication Number 4699, Washington, D.C., various paging.

NICOLE (Network for Industrially Contaminated Land in Europe), 2010. NICOLE road map for sustainable remediation. Available at: http://www.nicole.org/documents/DocumentList.aspx?l= 2&w=n.

NRC (National Research Council), 1994. *Alternatives for Ground Water Cleanup*. National Academy Press, Washington, D.C., 315p.

NRC (National Research Council), 2005. Contaminants in the Subsurface: Source Zone Assessment and Remediation. National Academy Press. Washington, D.C., 358p.

Oostrom, M., White, M. D., Lenhard, R. J., van Geel, P. J., and Wietsma, T. W., 2005. A comparison of models describing residual NAPL formation in the vadose zone. *Vadose Zone J*, 4:163–174.

Pankow, J. F., and Cherry, J. A., 1996. *Dense Chlorinated Solvents and Other DNAPLs in Groundwater*. Waterloo Press, Guelph, Ont, Canada, 522p.

Parsons (Parsons Corporation), 2004. Principles and practices of enhanced anaerobic bioremediation of chlorinated solvents. Air Force Center for Environmental Excellence (AFCEE), Brooks City-Base, Texas; Naval Facilities Engineering Service Center Port Hueneme, California; Environmental Security Technology Certification Program, Arlington, VA, various paging.

Pierce, B., 2004. Comments & recommendations from the draft ground water use and vulnerability discussion paper. Memorandum to Guy Tomassoni, USEPA, Ken Lovelace, USEPA, September 30, 2004, Georgia Department of Natural Resources, Environmental Protection Division, Atlanta, GA, 4p. Available at: http://gwtf.clu-in.org/docs/options/comments/45.pdf.

Powell, T., Smith, G., Sturza, J., Lynch, K., and Truex, M., 2007. New advancements for in-situ treatment using electrical resistance heating. *Remed J*, 17:51–70. doi:10.1002/rem.20124.

Rao, P. S. C., Jawitz, J. W., Enfield, C. G., Falta, R. W., Annable, M. D., and Wood, A. L., 2001. Technology integration for contaminated site remediation: Cleanup goals and performance criteria. In: Groundwater quality: Natural and enhanced restoration of groundwater pollution. Publication No. 275, International Association of Hydrologic Sciences, Wallingford, UK, pp. 571–578.

Rhindress, R. C., 1971. Gasoline pollution of a karst aquifer. In: Parizek, R. R., White, W. B., and Langmuir, D. (eds.), *Hydrogeology and Geochemistry of Folded and Faulted Rocks of the Central Appalachian Type and Related Land Use Problems Earth and Mineral Sciences Experiment Station*, The Pennsylvania State University, Circular 82, pp. 171–175.

Ronen, D., Sorek, S., and Gilron, J., 2011. Rationales behind irrationality of decision making in groundwater quality management. *Ground Water*. doi:10.1111/j.1745–6584.2011.00823.x

Rügner, H., and Bittens, M., 2006. Revitalization of contaminated land and groundwater at megasites; SAFIRA II Research Program 2006–2012, UFZ Center for Environmental Research Leipzig-Halle GmbH, Germany.

Ryan, S., 2010. Dense nonaqueous phase liquid cleanup: Accomplishments at twelve NPL sites. National Network of Environmental Management Studies Fellow. Published by the US EPA on-line at http://cluin.org, 84p.

Seeley, N. D., and Primrose, S. B., 1982. The isolation of the bacteriophage from the environment. *J Appl Microbiol*, 53:1–17.

Simon, J. A., 2011. Editor's perspective—US sustainable remediation forum pushes forward with guidance on the state of the practice. *Remed J*, 21:1–5. doi:10.1002/rem.20287.

Slough, K. J., Sudicky, E. A., and Forsyth, P. A., 1999. Importance of rock matrix entry pressure on DNAPL migration in fractured geologic materials. *Ground Water*, 37(2):237–244.

Stroud, F. B., Gilbert, J., Powell, G. W., Crawford, N. C., Rigatti, M. J., and Johnson, P. C., 1986. U.S. Environmental Protection Agency emergency response to toxic fumes and contaminated ground water in karst topography: Bowling Green, Kentucky. In: Proceedings of the Environmental Problems in Karst Terranes and Their Solutions Conference, National Water Well Association, pp. 197–226.

SURF (U.S. Sustainable Remediation Forum), 2009. Sustainable remediation white paper—integrating sustainable principles, practices, and metrics into remediation projects. *Remed J*, 19:5–114. doi:10.1002/rem.20210

Suthersan, S. S., and Payne, F. C., 2005. In situ remediation engineering. CRC Press, Boca Raton, FL, 511p.

USACE (U.S. Army Corps of Engineers), 2006. Design: In situ thermal remediation. UFC 3–280-05. Unified Facilities Criteria (UFC). U.S. Army Corps of Engineers, Naval Facilities Engineering Command (NAVFAC), Air Force Civil Engineer Support Agency (AFCESA).

USACE, 2009. Design: In-Situ thermal remediation. Manual 1110–1-401536, 226p. Available at: http://www.usace.army.mil/inet/usace-docs/.

USACE, 2010. Five-Year Review Report for OU 1 SIA Groundwater Interim Remedial Action, OU 2 SIA Soils, OU 3 Ammunition Storage Area Soils and Groundwater, Anniston Army Depot, Calhoun County, Alabama, EPA ID: 321002027, Mobile, AL. Prepared for USEPA Region 4, Atlanta, GA.

U.S. DOE (U.S. Department of Energy), 2002. DNAPL bioremediation-RTDF. *Innovative Technology Summary Report, DOE/EM-0625. Office of Environmental Management*. Dover, DE, various paging.

USEPA (United States Environmental Protection Agency), 1993. Guidance for Evaluating the Technical Impracticability of Ground-Water Restoration, OWSER Directive 9234.2-5, EPA/540-R-93-080, September 1993.

USEPA (United States Environmental Protection Agency), 1994. EPA Superfund Record of Decision: Southern California Edison, Visalia Pole Yard Superfund Site, Visalia, CA. EPA ID: CAD980816466.

USEPA, 1995. Use of Risk-Based Decision Making in UST Corrective Action Programs, OWSER Directive 9610.17, Office of Solid Waste and Emergency Response, 20p.

USEPA, 1996a. A Citizen's guide to bioremediation. EPA 542-F-96–007, Office of Solid Waste and Emergency Response, 4p.

USEPA, 1996b. Presumptive response strategy and ex-situ treatment technologies for contaminated ground water at CERCLA sites, Final Guidance. OSWER Directive 9288.1-12, EPA 540/R-96/023.

USEPA, 1998a. Steam injection for soil and aquifer remediation. EPA/540/S-97/505, Office of Solid Waste and Emergency Response, U.S. Environmental Protection Agency, Washington, D.C., 16p.

USEPA, 1998b. Permeable reactive barrier technologies for contaminant remediation. EPA/600/R-98/125, Office of Solid Waste and Emergency Response, U.S. Environmental Protection Agency, Washington, D.C., 94p.

USEPA, 1999a. Use of monitored natural attenuation at superfund, RCRA corrective action, and underground storage tank sites. Directive 9200.4–17P. Office of Solid Waste and Emergency Response, 41p.

USEPA, 1999b. EPA Superfund Record of Decision: MONTROSE CHEMICAL CORP. and DEL AMO. EPA ID: CAD008242711 and CAD029544731 OU(s) 03 & 03, LOS ANGELES, CA 03/30/1999. Dual Site Groundwater Operable Unit. II: Decision Summary, pp. 11–7. EPA/ROD/R09–99/035

USEPA, 1999c. Groundwater cleanup: Overview of operating experience at 28 Sites. EPA/542/R-99/006, Office of Solid Waste and Emergency Response, Washington, D.C., various pagination.

USEPA, 2000. Engineered approaches to in situ bioremediation of chlorinated solvents: Fundamentals and field applications. EPA 542-R-00–008. Available at: http://cluin.org/download/remed/engappinsitbio.pdf. Accessed August 12, 2011.

USEPA, 2003a. Atrazine interim reregistration eligibility decision (IRED), Q&A's—January 2003. Available at: http://www.epa.gov/pesticides/factsheets/atrazine.htm#q1. Accessed January 23, 2008.

USEPA, 2003b. The DNAPL Remediation Challenge: Is There A Case For Source Depletion? Report Prepared by an Expert Panel to the Environmental Protection Agency, Office of Research and Development, Publication EPA/600/R-03/143. Available at: http://www.epa.gov/ada/download/reports/600R03143/600R03143.pdf.

USEPA, 2004. In situ thermal treatment of chlorinated solvents; Fundamentals and field applications. EPA 542/R-04/010. Office of Solid Waste and Emergency Response, U.S. Environmental Protection Agency, Washington, D.C., various paging.

USEPA, 2007. Treatment Technologies for Site Cleanup: Annual Status Report (Twelfth Edition). Office of Solid Waste and Emergency Response, EPA-542-R-07–012, 66p + appendices. Available at: http://www.clu-in.org/asr/.

USEPA, 2009a. DNAPL remediation: Selected projects where regulatory closure goals have been achieved. Office of Solid Waste and Emergency Response, EPA 542/R-09/008, 52p. Available at: http://www.clu-in.org/s.focus/c/pub/i/1719/.

USEPA, 2009b. Final Close Out Report: Southern California Edison Visalia Pole Yard Superfund Site, Visalia, Tulare County, CA. EPA Region 9.

USEPA, 2009c. Amendment #2 to the Record of Decision for the Commencement Bay–South Tacoma Channel Superfund Site, Operable Unit 1, Well 12A, EPA Region 10.

USEPA, 2010. Superfund Remedy Report, Thirteenth Edition. Office of Solid Waste and Emergency Response, EPA-542-R-10–004, 36p + appendices. Available at: http://www.clu-in.org/asr/.

USEPA, 2011a. Record of decision. Available at: http://www.epa.gov/superfund/cleanup/rod.htm.

USEPA, 2011b. Groundwater Road Map: Recommended Process for Restoring Contaminated Groundwater at Superfund Sites. OSWER 9283.1–34, 31p.

USEPA, 2011c. Green power equivalency calculator methodologies. Available at: http://www.epa.gov/greenpower/pubs/calcmeth.htm.

USEPA, 2011d. Beneficial effects of the superfund program. Office of Superfund Remediation and Technology Innovation. EPA Contract EP W-07–037. Available at: http://www.epa.gov/superfund/accomp/pdfs/SFBenefits-031011-Ver1.pdf.

USEPA, 2011e. US and EU perspectives on green and sustainable remediation part 2. CLU-IN Internet Seminar, Delivered: March 15, 2011, 10:00 AM – 12:00 PM, EDT (14:00–16:00 GMT). Available at: http://www.cluin.org/live/archive/#US_and_EU_Perspectives_on_Green_and_Sustainable_Remediation_Part_2.

USEPA, 2011f. Introduction to green remediation. Office of Superfund Remediation and Technology Innovation. Quick Reference Fact Sheet. Available at: http://cluin.org/greenremediation/.

Veselič, M., and Čenčur Curk, B., 2001. Test studies of flow and solute transport in the unsaturated fractured and karstified rock on the experimental field site Sinji Vrh, Slovenia. In: Seiler, K. P., and Wohnlich, S. (eds.), *New Approaches Characterizing Groundwater Flow*, Balkema, Lisse, The Netherlands, pp. 211–214.

Vesper, D. J., Loop, C. M., and White, W. B., 2003. Contaminant transport in karst aquifers. *Speleogenesis Evol Karst Aquifers* 1(2), 101–111 (republished from *Theoretical and Applied Karstology, 2001, 13–14*).

Wiedemeier, T. H., et al., 1998. Technical protocol for evaluating natural attenuation of chlorinated solvents in ground water. EPA/600/R-98/128, U.S. Environmental Protection Agency, Office of Research and Development, Washington, D.C., various pagination.

Wiedemeier, T. H., et al., 1999. Technical protocol for implementing intrinsic remediation with long-term monitoring for natural attenuation of fuel contamination dissolved in groundwater; Volume I (Revision 0), Air Force Center for Environmental Excellence (AFCEE), Technology Transfer Division, Brooks Air Force Base, San Antonio, TX, various pagination.

Wilkin, R. T., and Puls, R. W., 2003. Capstone report on the application, monitoring, and performance of permeable reactive barriers for groundwater remediation: Volume 1—Performance evaluation at two sites. EPA/600/R-03/045a. U.S. Environmental Protection Agency, Washington, D.C., various paging.

Wischkaemper, K., 2007. Technical Impracticability, So Far, Anniston Army Ammunition Depot in Anniston, AL. TSP Semi-Annual Meeting in Las Vegas, Wednesday November 7, 2007. Available at: www.epa.gov/tio/tsp/download/2007_fall_meeting/wed-wischkaemper.pdf

Wolfe, W. J., Haugh, C. J., Webbers, A., and Diehl, T. H., 1997. Preliminary conceptual models of the occurrence, fate, and transport of chlorinated solvents in karst regions of Tennessee. *U.S. Geological Survey Water-Resources Investigations Report 97–4097*, Nashville, TN, 80p.

Woolford, J. E., 2011. Memorandum. Subject: Clarification of OSWER's 1995 Technical Impracticability Waiver Policy. OSWER-#9335.5–32, Washington, D.C.

Zhang, W., 2003. Nanoscale iron particles for environmental remediation. *J Nanoparticle Res*, 5:323–332.

Index

Note: Page numbers referencing figures are italicized and followed by an "*f*"; page numbers referencing tables are italicized and followed by a "*t*".

D

E

H

I

J

N

Q

R

S

V

W

Z